RSMeans

Square Foot Costs

18th Annual Edition

1997

Senior Editor
John H. Ferguson, PE

Contributing Editors
Thomas J. Akins
Barbara Balboni
John H. Chiang, PE
Paul C. Crosscup
Mark H. Kaplan, Jr.
Robert W. Mewis
Melville J. Mossman, PE
John J. Moylan
Jeannene D. Murphy
Jesse R. Page
Stephen C. Plotner
Michael J. Regan
Kornelis Smit
William R. Tennyson, II
Phillip R. Waier, PE
James N. Wills
Rory Woolsey

Manager, Engineering Operations
Patricia L. Jackson, PE

Editorial Advisory Board

James E. Armstrong
Property and Utility Manager
Massachusetts Development
Finance Agency

William R. Barry
Chief Estimator
Stone & Webster Corporation

Charles Buddenhagen, PE
Executive Director
Division of Facilities
Georgetown University

Robert Cox
Assistant Professor
School of Construction
University of Florida

Roy F. Gilley III, AIA
Principal
Gilley Hinkel Architects

Kenneth K. Humphreys, PhD, PE, CCE
Executive Director-Retired
AACE International

Martin F. Joyce
Vice President, Utility Division
Bond Brothers, Inc.

Donald R. Wong
Project Manager Director
Whalen & Company, Inc.

President and CEO
Perry B. Sells

Vice President, Editorial Operations and Development
Roger J. Grant

Vice President, Sales and Marketing
John M. Shea

Vice President, Operations and CFO
Andrew J. Centauro

Production Manager
Karen L. O'Brien

Production Coordinator
Marion E. Schofield

Technical Support
Michele S. Able
Wayne D. Anderson
Thomas J. Dion
Michael H. Donelan
Gary L. Hoitt
Marla A. Marek
Paula Reale-Camelio
Kathryn S. Rodriguez
Ali Vaghar

Art Director
Helen A. Marcella

Book & Cover Design
Norman R. Forgit

R.S. Means Company, Inc. ("R.S. Means"), its authors, editors and engineers, apply diligence and judgment in locating and using reliable sources for the information published. However, R.S. Means makes no express or implied warranty or guarantee in connection with the content of the information contained herein, including the accuracy, correctness, value, sufficiency, or completeness of the data, methods and other information contained herein. R.S. Means makes no express or implied warranty of merchantability or fitness for a particular purpose. R.S. Means shall have no liability to any customer or third party for any loss, expense, or damage including consequential, incidental, special or punitive damages, including lost profits or lost revenue, caused directly or indirectly by any error or omission, or arising out of, or in connection with, the information contained herein.

No part of this publication may be reproduced, stored in a retrieval system, or transmitted in any form or by any means without prior written permission of R.S. Means Company, Inc.

Special thanks to the following firms that provided the inspirations for many illustrations:
  Home Planners, Inc.
  Design Basics, Inc.
  Donald A. Gardner Architects, Inc.
  Larry E. Belk Designs
  Larry W. Garnett & Associates, Inc.

First Printing

Foreword

Our Mission

Since 1942, R.S. Means Company, Inc. has been actively engaged in construction cost publishing and consulting throughout North America.

Today, over fifty years after the company began, our primary objective remains the same: to provide you, the construction industry professional, with the most current and comprehensive construction cost data possible.

Whether you are a contractor, an owner, an architect, an engineer, a facilities manager, or anyone else who needs a quick construction cost estimate, you'll find this publication to be a highly useful and necessary tool.

Today, with the constant flow of new construction methods and materials, it's difficult to find the time to look at and evaluate all the different construction cost possibilities. In addition, because labor and material costs keep changing, yesterday's cost information is not a reliable basis for today's estimate or budget.

That's why so many construction professionals turn to R.S. Means. We keep track of the costs for you along with a wide range of other key information, from city cost indexes . . . to productivity rates . . . to crew composition . . . to contractor's overhead and profit rates.

R.S. Means performs these functions by analyzing all facets of the industry. Data is collected and organized into a format that makes the cost information instantly accessible to you. From the preliminary budget to the detailed unit price estimate, you'll find that the data in this book is useful for all phases of construction cost determination.

The Staff, the Organization, and Our Services

When you purchase one of R.S. Means' publications, you are in effect hiring the services of a full-time staff of construction and engineering professionals.

A thoroughly experienced and highly qualified staff of professionals at R.S. Means works daily at collecting, analyzing, and disseminating comprehensive cost information for your needs. These staff members have years of practical construction experience and engineering training prior to joining the firm. As a result, you can count on them not only for the cost figures, but also for additional background reference information that will help you create a realistic estimate.

The Means organization is always prepared to assist you and help in the solution of construction problems through the services of its four major divisions: Construction and Cost Data Publishing, Electronic Products and Services, Consulting Services, and Educational Services.

Besides a full array of construction cost estimating books, Means also publishes a number of other reference works for the construction industry. Subjects include building automation, HVAC, roofing, plumbing, fire protection, hazardous materials and hazardous waste, business, project, and facilities management and many more.

In addition, you can access all of our construction cost data through your computer. For instance, MeansData™ for Spreadsheets is an electronic tool that lets you access over 40,000 lines of Means construction cost data right at your PC.

What's more, you can increase your knowledge and improve your construction estimating and management performance with a Means Construction Seminar or In-House Training Program. These two-day seminar programs offer unparalleled opportunities for everyone in your organization to get updated on a wide variety of construction related issues.

In short, R.S. Means can put the tools in your hands for constructing accurate and dependable construction estimates and budgets in a variety of ways.

Robert Snow Means Established a Tradition of Quality That Continues Today

Robert Snow Means took years to build up his company. He worked hard and always made certain that he delivered a quality product.

Today, at R.S. Means, we do more than talk about the quality of our data and the usefulness of our books. We stand behind all of our work, from historical cost indexes to construction techniques to current costs.

If you have any questions about our products or services, please call us toll-free at 1-800-448-8182. Our customer service representatives will be happy to assist you . . . as they would have over fifty years ago.

Table of Contents

Foreword	ii
How the Book Is Built: An Overview	iv
Residential Section	1
Introduction	3
Work Sheet	13
Economy	18
Average	26
Custom	38
Luxury	48
Adjustments	58
Commercial/Industrial/Institutional Section	65
Introduction	67
Examples	73
Building Types	78
Depreciation	218
Assemblies Section	219
Introduction	221
Foundations	233
Substructures	237
Superstructures	239
Exterior Closure	275
Roofing	313
Interior Construction	321
Conveying Systems	341
Mechanical	345
Electrical	399
Special Construction	413
Site Work	423
Reference Section	427
General Conditions	428
Location Factors	429
Historical Cost Indexes	434
Glossary	442
Abbreviations	446
Other Means Publications and Reference Works	Yellow Pages
Installing Contractor's Overhead & Profit	Inside Back Cover

RESIDENTIAL
- INTRODUCTION
- WORKSHEET
- ECONOMY
- AVERAGE
- CUSTOM
- LUXURY
- ADJUSTMENTS

COMM./INDUST./INSTIT.
- INTRODUCTION
- EXAMPLES
- BUILDING TYPES
- DEPRECIATION

ASSEMBLIES
- FOUNDATIONS — 1
- SUBSTRUCTURES — 2
- SUPERSTRUCTURES — 3
- EXTERIOR CLOSURE — 4
- ROOFING — 5
- INTERIOR CONSTRUCTION — 6
- CONVEYING SYSTEMS — 7
- MECHANICAL — 8
- ELECTRICAL — 9
- SPECIAL CONSTRUCTION — 11
- SITE WORK — 12

REFERENCE INFORMATION
- GENERAL CONDITIONS
- LOCATION FACTORS
- COST INDEXES
- GLOSSARY

How the Book Is Built: An Overview

A Powerful Construction Tool

You have in your hands one of the most powerful construction tools available today. A successful project is built on the foundation of an accurate and dependable estimate. This book will enable you to construct just such an estimate.

For the casual user the book is designed to be:

- quickly and easily understood so you can get right to your estimate
- filled with valuable information so you can understand the necessary factors that go into the cost estimate

For the regular user, the book is designed to be:

- a handy desk reference that can be quickly referred to for key costs
- a comprehensive, fully reliable source of current construction costs for typical building structures when only minimum dimensions, specifications and technical data are available. It is specifically useful in the conceptual or planning stage for preparing preliminary estimates which can be supported and developed into complete engineering estimates for construction.
- a source book for rapid estimating of replacement costs and preparing capital expenditure budgets

To meet all of these requirements we have organized the book into the following clearly defined sections.

How To Use the Book: The Details

This section contains an in-depth explanation of how the book is arranged . . . and how you can use it to determine a reliable construction cost estimate. It includes information about how we develop our cost figures and how to completely prepare your estimate.

Residential Section

Model buildings for four classes of construction—economy, average, custom and luxury—are developed and shown with complete costs per square foot.

Commercial, Industrial, Institutional Section

This section contains complete costs for 70 typical model buildings expressed as costs per square foot.

Assemblies Section

The cost data in this section has been organized in an "Assemblies" format. These assemblies are the functional elements of a building and are arranged according to the 12 divisions of the UniFormat classification system. The costs in this section include standard mark-ups for the overhead and profit of the installing contractor. The assemblies costs are those used to calculate the total building costs in the Commercial, Industrial, Institutional section.

Reference Section

This section includes information on General Conditions, Location Factors, Historical Cost Indexes, a Glossary, and a listing of Abbreviations used in this book.

Location Factors: Costs vary depending upon regional economy. You can adjust the "national average" costs in this book to 930 locations throughout the U.S. and Canada by using the data in this section.

Historical Cost Indexes: These indexes provide you with data to adjust construction costs over time. If you know costs for a project completed in the past, you can use these indexes to calculate a rough estimate of what it would cost to construct the same project today.

Glossary: A listing of common terms related to construction and their definitions is included.

Abbreviations: A listing of the abbreviations used throughout this book, along with the terms they represent, is included.

The Scope of This Book

This book is designed to be as comprehensive and as easy to use as possible. To that end we have made certain assumptions and limited its scope in two key ways:

1. We have established material prices based on a "national average."
2. We have computed labor costs based on a 30-city "national average" of union wage rates.

All costs and adjustments in this manual were developed from the *Means Building Construction Cost Data, Means Assemblies Cost Data, Means Residential Cost Data,* and *Means Light Commercial Cost Data.* Costs include labor, material, overhead, and profit, and general conditions and architect's fees where specified. No allowance is made for sales tax, financing, premiums for material and labor, or unusual business and weather conditions.

What's Behind the Numbers? The Development of Cost Data

The staff at R.S. Means continuously monitors developments in the construction industry in order to ensure reliable, thorough and up-to-date cost information.

While *overall* construction costs may vary relative to general economic conditions, price fluctuations within the industry are dependent upon many factors. Individual price variations may, in fact, be opposite to overall economic trends. Therefore, costs are continually monitored and complete updates are published yearly. Also, new items are frequently added in response to changes in materials and methods.

Costs—$ (U.S.)

All costs represent U.S. national averages and are given in U.S. dollars. The Means Location Factors can be used to adjust costs to a particular location. The Location Factors for Canada can be used to adjust U.S. national averages to local costs in Canadian dollars.

Factors Affecting Costs

Due to the almost limitless variation of building designs and combinations of construction methods and materials, costs in this book must be used with discretion. The editors have made every effort to develop costs for "typical" building types. By definition, a square foot estimate, due to the short time and limited amount of detail required, involves a relatively lower degree of accuracy than a detailed unit cost estimate, for example. The accuracy of an estimate will increase as the project is defined in more detailed components. The user should examine closely and compare the specific requirements of the project being estimated with the components included in the models in this book. Any differences should be accounted for in the estimate.

In many building projects, there may be factors that increase or decrease the cost beyond the range shown in this manual. Some of these factors are:

- Substitution of building materials or systems for those used in the model.
- Custom designed finishes, fixtures and/or equipment.
- Special structural qualities (allowance for earthquake, future expansion, high winds, long spans, unusual shape).
- Special foundations and substructures to compensate for unusual soil conditions.
- Inclusion of typical "owner supplied" equipment in the building cost (special testing and monitoring equipment for hospitals, kitchen equipment for hotels, offices that have cafeteria accommodations for employees, etc.).
- An abnormal shortage of labor or materials.
- Isolated building site or rough terrain that would affect the transportation of personnel, material or equipment.
- Unusual climatic conditions during the construction process.

We strongly urge that, for maximum accuracy, these factors be considered each time a structure is being evaluated.

If users require greater accuracy than this manual can provide, the editors recommend that the *Means Building Construction Cost Data* or the *Means Assemblies Cost Data* be consulted.

Final Checklist

Estimating can be a straightforward process provided you remember the basics. Here's a checklist of some of the items you should remember to do before completing your estimate.

Did you remember to . . .

- include the Location Factor for your city
- mark up the entire estimate sufficiently for your purposes
- include all components of your project in the final estimate
- double check your figures to be sure of your accuracy
- call R.S. Means if you have any questions about your estimate or the data you've found in our publications

Remember, R.S. Means stands behind its publications. If you have any questions about your estimate . . . about the costs you've used from our books . . . or even about the technical aspects of the job that may affect your estimate, feel free to call the R.S. Means editors at 1-617-585-7880 or 1-800-448-8182.

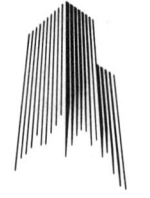

Residential Section

Table of Contents

Introduction
	Page
General	3
How to Use	4
Building Classes	6
Configurations	7
Building Types	8
Garage Types	10
Building Components	11
Exterior Wall Construction	12

Residential Cost Estimate Worksheets
	Page
Instructions	13
Model Residence Example	15

Economy
	Page
Illustrations	18
Specifications	19
1 Story	20
1-1/2 Story	21
2 Story	22
Bi-Level	23
Tri-Level	24
Wings & Ells	25

Average
	Page
Illustrations	26
Specifications	27
1 Story	28
1-1/2 Story	29
2 Story	30
2-1/2 Story	31
3 Story	32
Bi-Level	33
Tri-Level	34
Solid Wall (log home) 1 Story	35
Solid Wall (log home) 2 Story	36
Wings & Ells	37

Custom
	Page
Illustrations	38
Specifications	39
1 Story	40
1-1/2 Story	41
2 Story	42
2-1/2 Story	43
3 Story	44
Bi-Level	45
Tri-Level	46
Wings & Ells	47

Luxury
	Page
Illustrations	48
Specifications	49
1 Story	50
1-1/2 Story	51
2 Story	52
2-1/2 Story	53
3 Story	54
Bi-Level	55
Tri-Level	56
Wings & Ells	57

Residential Modifications
	Page
Kitchen Cabinets	58
Kitchen Counter Tops	58
Vanity Bases	59
Solid Surface Vanity Tops	59
Fireplaces & Chimneys	59
Windows & Skylights	59
Dormers	59
Appliances	60
Breezeway	60
Porches	60
Finished Attic	60
Alarm System	60
Sauna, Prefabricated	60
Garages	61
Swimming Pools	61
Wood & Coal Stoves	61
Sidewalks	61
Fencing	61
Carport	61
Insurance Exclusions	62
Architect/Designer Fees	62
Depreciation	63

Introduction to Residential Section

The residential section of this manual contains costs per square foot for four classes of construction in seven building types. Costs are listed for various exterior wall materials which are typical of the class and building type. There are cost tables for Wings and Ells with modification tables to adjust the base cost of each class of building. Non-standard items can easily be added to the standard structures.

Cost estimating for a residence is a three-step process:
   (1) Identification
   (2) Listing dimensions
   (3) Calculations

Guidelines and a sample cost estimating procedure are shown on the following pages.

Identification

To properly identify a residential building, the class of construction, type, and exterior wall material must be determined. Page 6 has drawings and guidelines for determining the class of construction. There are also detailed specifications and additional drawings at the beginning of each set of tables to further aid in proper building class and type identification.

Sketches for eight types of residential buildings and their configurations are shown on pages 8 and 9. Definitions of living area are next to each sketch. Sketches and definitions of garage types are on page 10.

Living Area

Base cost tables are prepared as costs per square foot of living area. The living area of a residence is that area which is suitable and normally designed for full time living. It does not include basement recreation rooms or finished attics, although these areas are often considered full time living areas by the owners.

Living area is calculated from the exterior dimensions without the need to adjust for exterior wall thickness. When calculating the living area of a 1-1/2 story, two story, three story or tri-level residence, overhangs and other differences in size and shape between floors must be considered.

Only the floor area with a ceiling height of six feet or more in a 1-1/2 story residence is considered living area. In bi-levels and tri-levels, the areas that are below grade are considered living area, even when these areas may not be completely finished.

Base Tables and Modifications

Base cost tables show the base cost per square foot without a basement, with one full bath and one full kitchen for economy and average homes and an additional half bath for custom and luxury models. Adjustments for finished and unfinished basements are part of the base cost tables. Adjustments for multi family residences, additional bathrooms, townhouses, alternate roofs, and air conditioning and heating systems are listed in modification tables below the base cost tables.

Costs for other modifications, including garages, breezeways and site improvements, are in the modification tables on pages 58 to 61.

Listing of Dimensions

To use this section of the manual only the dimensions used to calculate the horizontal area of the building and additions and modifications are needed. The dimensions, normally the length and width, can come from drawings or field measurements. For ease in calculation, consider measuring in tenths of feet, i.e., 9 ft. 6 in. = 9.5 ft., 9 ft. 4 in. = 9.3 ft.

In all cases, make a sketch of the building. Any protrusions or other variations in shape should be noted on the sketch with dimensions.

Calculations

The calculations portion of the estimate is a two-step activity:
   (1) The selection of appropriate costs from the tables
   (2) Computations

Selection of Appropriate Costs

To select the appropriate cost from the base tables, the following information is needed:
   (1) Class of construction
      (a) Economy
      (b) Average
      (c) Custom
      (d) Luxury
   (2) Type of residence
      (a) One story
      (b) 1-1/2 story
      (c) 2 story
      (d) 3 story
      (e) Bi-level
      (f) Tri-level
   (3) Occupancy
      (a) One family
      (b) Two family
      (c) Three family
   (4) Building configuration
      (a) Detached
      (b) Town/Row house
      (c) Semi-detached
   (5) Exterior wall construction
      (a) Wood frame
      (b) Brick veneer
      (c) Solid masonry
   (6) Living areas

Modifications are classified by class, type and size.

Computations

The computation process should take the following sequence:
   (1) Multiply the base cost by the area
   (2) Add or subtract the modifications
   (3) Apply the location modifier

When selecting costs, interpolate or use the cost that most nearly matches the structure under study. This applies to size, exterior wall construction and class.

How to Use the Residential Square Foot Cost Pages

The following is a detailed explanation of a sample entry in the Residential Square Foot Cost Section. Each bold number below corresponds to the item being described on the facing page with the appropriate component of the sample entry following in parenthesis.

Prices listed are costs that include overhead and profit of the installing contractor. Total model costs include an additional mark-up for General Contractors' overhead and profit, and fees specific to class of construction.

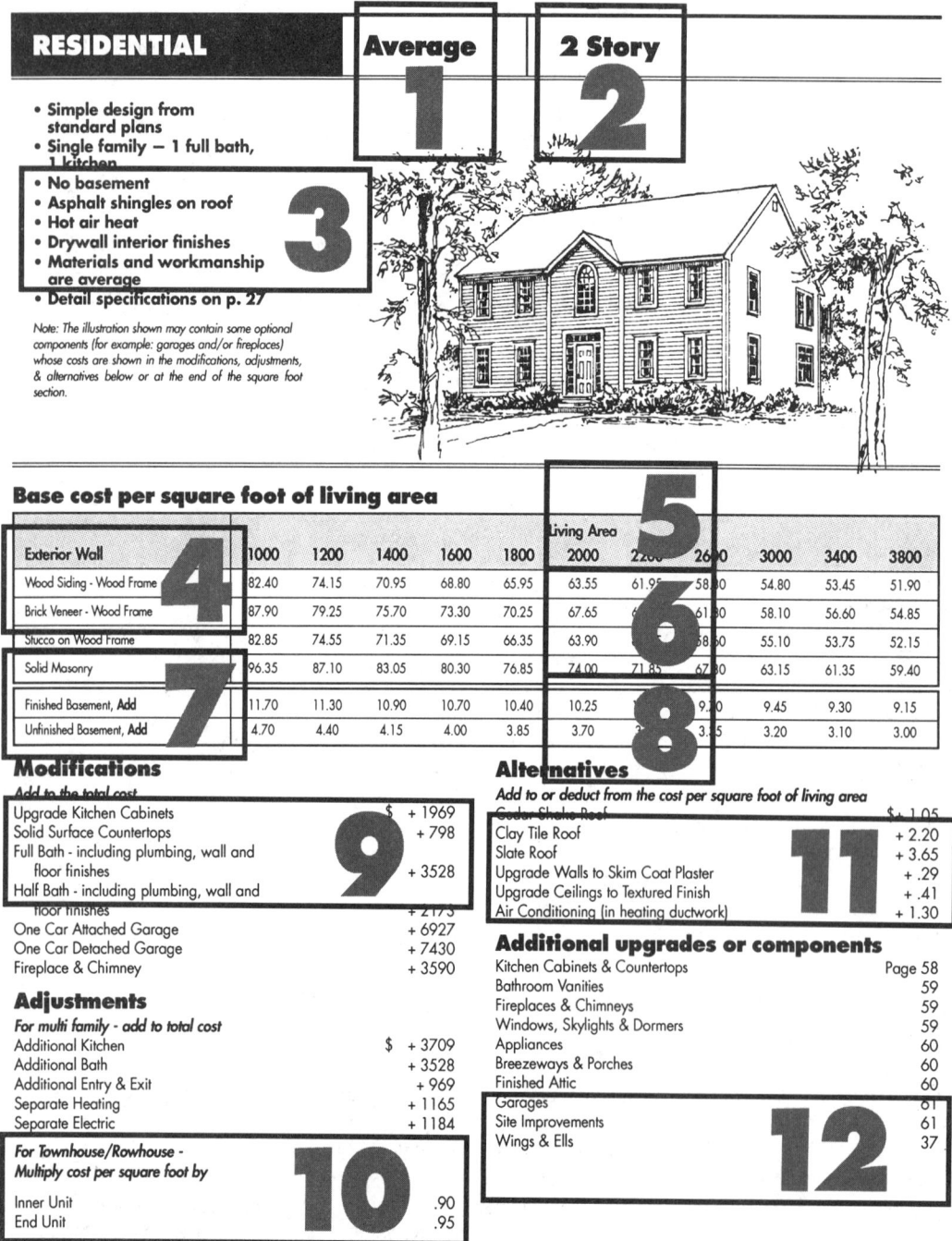

RESIDENTIAL — Average **1** — 2 Story **2**

- Simple design from standard plans
- Single family — 1 full bath, 1 kitchen
- No basement
- Asphalt shingles on roof
- Hot air heat
- Drywall interior finishes
- Materials and workmanship are average   **3**
- Detail specifications on p. 27

Note: The illustration shown may contain some optional components (for example: garages and/or fireplaces) whose costs are shown in the modifications, adjustments, & alternatives below or at the end of the square foot section.

Base cost per square foot of living area

						Living Area					
Exterior Wall **4**	1000	1200	1400	1600	1800	2000	2200	2600	3000	3400	3800
Wood Siding - Wood Frame	82.40	74.15	70.95	68.80	65.95	63.55	61.95	58.50	54.80	53.45	51.90
Brick Veneer - Wood Frame	87.90	79.25	75.70	73.30	70.25	67.65	65.95	61.30	58.10	56.60	54.85
Stucco on Wood Frame	82.85	74.55	71.35	69.15	66.35	63.90	62.20	58.50	55.10	53.75	52.15
Solid Masonry	96.35	87.10	83.05	80.30	76.85	74.00	71.85	67.30	63.15	61.35	59.40
Finished Basement, **Add**	11.70	11.30	10.90	10.70	10.40	10.25	10.00	9.70	9.45	9.30	9.15
Unfinished Basement, **Add**	4.70	4.40	4.15	4.00	3.85	3.70	3.60	3.35	3.20	3.10	3.00

5 **6** **7** **8**

Modifications
Add to the total cost

Upgrade Kitchen Cabinets	$ + 1969
Solid Surface Countertops	+ 798
Full Bath - including plumbing, wall and floor finishes	+ 3528
Half Bath - including plumbing, wall and floor finishes	+ 2173
One Car Attached Garage	+ 6927
One Car Detached Garage	+ 7430
Fireplace & Chimney	+ 3590

9

Adjustments
For multi family - add to total cost

Additional Kitchen	$ + 3709
Additional Bath	+ 3528
Additional Entry & Exit	+ 969
Separate Heating	+ 1165
Separate Electric	+ 1184

For Townhouse/Rowhouse - Multiply cost per square foot by

Inner Unit	.90
End Unit	.95

10

Alternatives
Add to or deduct from the cost per square foot of living area

Cedar Shake Roof	+ 1.05
Clay Tile Roof	+ 2.20
Slate Roof	+ 3.65
Upgrade Walls to Skim Coat Plaster	+ .29
Upgrade Ceilings to Textured Finish	+ .41
Air Conditioning (in heating ductwork)	+ 1.30

11

Additional upgrades or components

Kitchen Cabinets & Countertops	Page 58
Bathroom Vanities	59
Fireplaces & Chimneys	59
Windows, Skylights & Dormers	59
Appliances	60
Breezeways & Porches	60
Finished Attic	60
Garages	61
Site Improvements	61
Wings & Ells	37

12

Important: See the Reference Section for Location Factors (to adjust for your city) and Estimating Forms

1. Class of Construction (Average)

The class of construction depends upon the design and specifications of the plan. The four classes are economy, average, custom and luxury.

2. Type of Residence (2 Story)

The building type describes the number of stories or levels in the model. The seven building types are 1 story, 1-1/2 story, 2 story, 2-1/2 story, 3 story, bi-level and tri-level.

3. Specification Highlights (Hot Air Heat)

These specifications include information concerning the components of the model, including the number of baths, roofing types, HVAC systems, and materials and workmanship. If the components listed are not appropriate, modifications can be made by consulting the information shown lower on the page or the Assemblies Section.

4. Exterior Wall System (Wood Siding - Wood Frame)

This section includes the types of exterior wall systems and the structural frame used. The exterior wall systems shown are typical of the class of construction and the building type shown.

5. Living Areas (2000 SF)

The living area is that area of the residence which is suitable and normally designed for full time living. It does not include basement recreation rooms or finished attics. Living area is calculated from the exterior dimensions without the need to adjust for exterior wall thickness. When calculating the living area of a 1-1/2 story, 2 story, 3 story or tri-level residence, overhangs and other differences in size and shape between floors must be considered. Only the floor area with a ceiling height of six feet or more in a 1-1/2 story residence is considered living area. In bi-levels and tri-levels, the areas that are below grade are considered living area, even when these areas may not be completely finished. A range of various living areas for the residential model are shown to aid in selection of values from the matrix.

6. Base Costs per Square Foot of Living Area ($63.55)

Base cost tables show the cost per square foot of living area without a basement, with one full bath and one full kitchen for economy and average homes and an additional half bath for custom and luxury models. When selecting costs, interpolate or use the cost that most nearly matches the residence under consideration for size, exterior wall system and class of construction. Prices listed are costs that include overhead and profit of the installing contractor, a general contractor markup and an allowance for plans that vary by class of construction. For additional information on contractor overhead and architectural fees, see the Reference Section.

7. Basement Types (Finished)

The two types of basements are finished or unfinished. The specifications and components for both are shown on the Building Classes page in the Introduction to this section.

8. Additional Costs for Basements ($10.25)

These values indicate the additional cost per square foot of living area for either a finished or an unfinished basement.

9. Modifications and Adjustments (Solid Surface Countertops $798)

Modifications and Adjustments are costs added to or subtracted from the total cost of the residence. The total cost of the residence is equal to the cost per square foot of living area times the living area. Typical modifications and adjustments include kitchens, baths, garages and fireplaces.

10. Multiplier for Townhouse/Rowhouse (Inner Unit .90)

The multipliers shown adjust the base costs per square foot of living area for the common wall condition encountered in townhouses or rowhouses.

11. Alternatives (Skim Coat Plaster $.29)

Alternatives are costs added to or subtracted from the base cost per square foot of living area. Typical alternatives include variations in kitchens, baths, roofing, air conditioning and heating systems.

12. Additional Upgrades or Components (Wings & Ells page 37)

Costs for additional upgrades or components, including wings or ells, breezeways, porches, finished attics and site improvements, are shown in other locations in the Residential Square Foot Cost Section.

Building Classes

Economy Class

An economy class residence is usually built from stock plans. The materials and workmanship are sufficient to satisfy building codes. Low construction cost is more important than distinctive features. The overall shape of the foundation and structure is seldom other than square or rectangular.

An unfinished basement includes 7' high 8" thick foundation wall composed of either concrete block or cast-in-place concrete.

Included in the finished basement cost are inexpensive paneling or drywall as the interior finish on the foundation walls, a low cost sponge backed carpeting adhered to the concrete floor, a drywall ceiling, and overhead lighting.

Custom Class

A custom class residence is usually built from plans and specifications with enough features to give the building a distinction of design. Materials and workmanship are generally above average with obvious attention given to construction details. Construction normally exceeds building code requirements.

An unfinished basement includes a 7'-6" high 10" thick cast-in-place concrete foundation wall or a 7'-6" high 12" thick concrete block foundation wall.

A finished basement includes painted drywall on insulated 2" x 4" wood furring as the interior finish to the concrete walls, a suspended ceiling, carpeting adhered to the concrete floor, overhead lighting and heating.

Average Class

An average class residence is a simple design and built from standard plans. Materials and workmanship are average, but often exceed minimum building codes. There are frequently special features that give the residence some distinctive characteristics.

An unfinished basement includes 7'-6" high 8" thick foundation wall composed of either cast-in-place concrete or concrete block.

Included in the finished basement are plywood paneling or drywall on furring that is fastened to the foundation walls, sponge backed carpeting adhered to the concrete floor, a suspended ceiling, overhead lighting and heating.

Luxury Class

A luxury class residence is built from an architect's plan for a specific owner. It is unique both in design and workmanship. There are many special features, and construction usually exceeds all building codes. It is obvious that primary attention is placed on the owner's comfort and pleasure. Construction is supervised by an architect.

An unfinished basement includes 8' high 12" thick foundation wall that is composed of cast-in-place concrete or concrete block.

A finished basement includes painted drywall on 2" x 4" wood furring as the interior finish, suspended ceiling, tackless carpet on wood subfloor with sleepers, overhead lighting and heating.

Configurations

Detached House

This category of residence is a free-standing separate building with or without an attached garage. It has four complete walls.

Semi-Detached House

This category of residence has two living units side-by-side. The common wall is a fireproof wall. Semi-detached residences can be treated as a row house with two end units. Semi-detached residences can be any of the five types.

Town/Row House

This category of residence has a number of attached units made up of inner units and end units. The units are joined by common walls. The inner units have only two exterior walls. The common walls are fireproof. The end units have three walls and a common wall. Town houses/row houses can be any of the five types.

Building Types

One Story

This is an example of a one-story dwelling. The living area of this type of residence is confined to the ground floor. The headroom in the attic is usually too low for use as a living area.

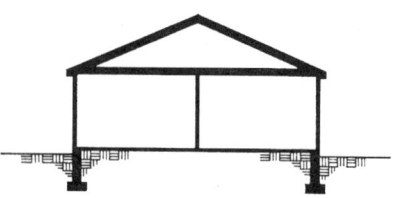

One-and-a-half Story

The living area in the upper level of this type of residence is 50% to 90% of the ground floor. This is made possible by a combination of this design's high-peaked roof and/or dormers. Only the upper level area with a ceiling height of 6 feet or more is considered living area. The living area of this residence is the sum of the ground floor area plus the area on the second level with a ceiling height of 6 feet or more.

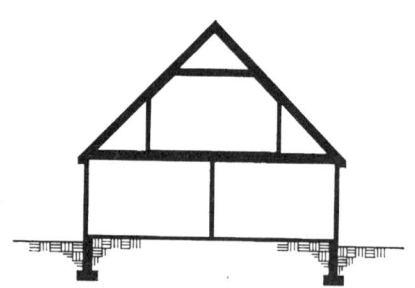

One Story with Finished Attic

The main living area in this type of residence is the ground floor. The upper level or attic area has sufficient headroom for comfortable use as a living area. This is made possible by a high peaked roof. The living area in the attic is less than 50% of the ground floor. The living area of this type of residence is the ground floor area only. The finished attic is considered an adjustment.

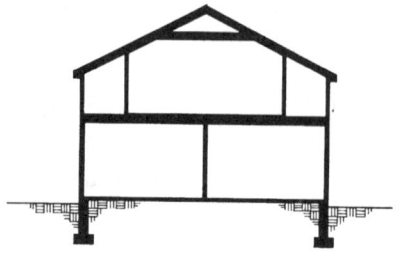

Two Story

This type of residence has a second floor or upper level area which is equal or nearly equal to the ground floor area. The upper level of this type of residence can range from 90% to 110% of the ground floor area, depending on setbacks or overhangs. The living area is the sum of the ground floor area and the upper level floor area.

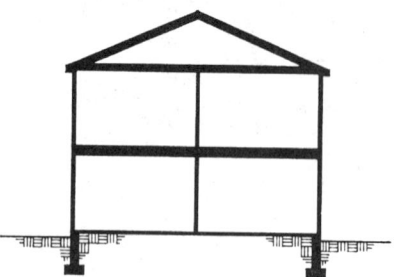

Two-and-one-half Story

This type of residence has two levels of equal or nearly equal area and a third level which has a living area that is 50% to 90% of the ground floor. This is made possible by a high peaked roof, extended wall heights and/or dormers. Only the upper level area with a ceiling height of 6 feet or more is considered living area. The living area of this residence is the sum of the ground floor area, the second floor area and the area on the third level with a ceiling height of 6 feet or more.

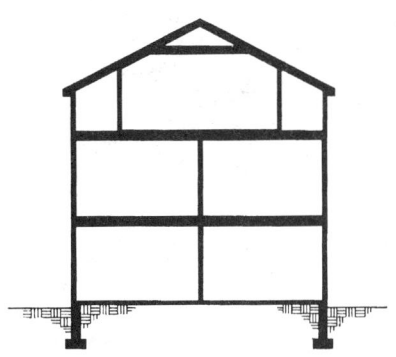

Bi-level

This type of residence has two living areas, one above the other. One area is about 4 feet below grade and the second is about 4 feet above grade. Both areas are equal in size. The lower level in this type of residence is originally designed and built to serve as a living area and not as a basement. Both levels have full ceiling heights. The living area is the sum of the lower level area and the upper level area.

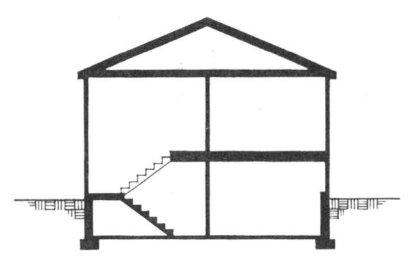

Three Story

This type of residence has three levels which are equal or nearly equal. As in the 2 story residence, the second and third floor areas may vary slightly depending on setbacks or overhangs. The living area is the sum of the ground floor area and the two upper level floor areas.

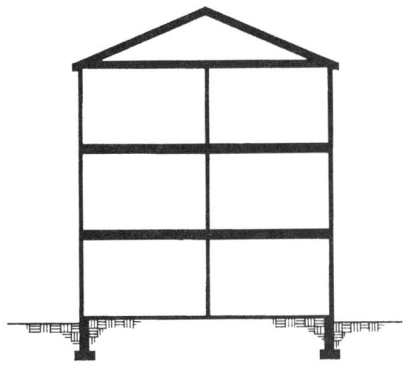

Tri-level

This type of residence has three levels of living area, one at grade level, one about 4 feet below grade and one about 4 feet above grade. All levels are originally designed to serve as living areas. All levels have full ceiling heights. The living area is a sum of the areas of each of the three levels.

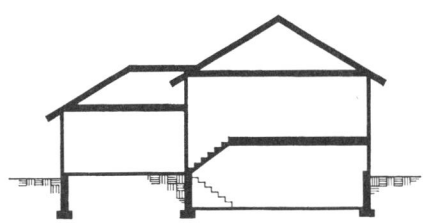

Garage Types

Attached Garage
Shares a common wall with the dwelling. Access is typically through a door between dwelling and garage.

Basement Garage
Constructed under the roof of the dwelling but below the living area.

Built-In Garage
Constructed under the second floor living space and above basement level of dwelling. Reduces gross square feet of living area.

Detached Garage
Constructed apart from the main dwelling. Shares no common area or wall with the dwelling.

Building Components

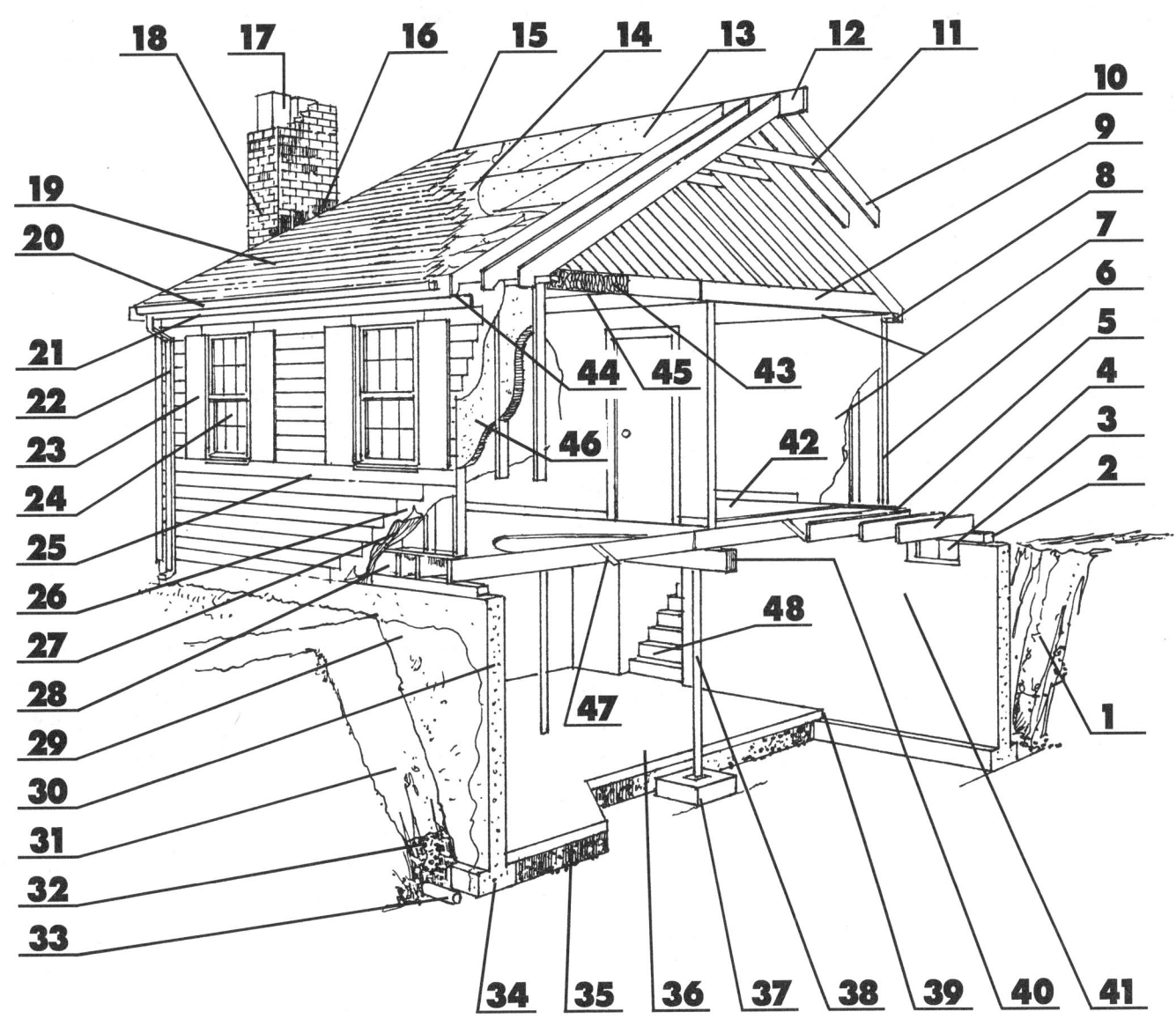

1. Excavation	11. Collar Ties	21. Fascia	31. Backfill	41. Sub-floor
2. Sill Plate	12. Ridge Rafter	22. Downspout	32. Drainage Stone	42. Finish Floor
3. Basement Window	13. Roof Sheathing	23. Shutter	33. Drainage Tile	43. Attic Insulation
4. Floor Joist	14. Roof Felt	24. Window	34. Wall Footing	44. Soffit
5. Shoe Plate	15. Roof Shingles	25. Wall Shingles	35. Gravel	45. Ceiling Strapping
6. Studs	16. Flashing	26. Building Paper	36. Concrete Slab	46. Wall Insulation
7. Drywall	17. Flue Lining	27. Wall Sheathing	37. Column Footing	47. Cross Bridging
8. Plate	18. Chimney	28. Fire Stop	38. Pipe Column	48. Bulkhead Stairs
9. Ceiling Joists	19. Roof Shingles	29. Dampproofing	39. Expansion Joint	
10. Rafters	20. Gutter	30. Foundation Wall	40. Girder	

Exterior Wall Construction

Typical Frame Construction

Typical wood frame construction consists of wood studs with insulation between them. A typical exterior surface is made up of sheathing, building paper and exterior siding consisting of wood, vinyl, aluminum or stucco over the wood sheathing.

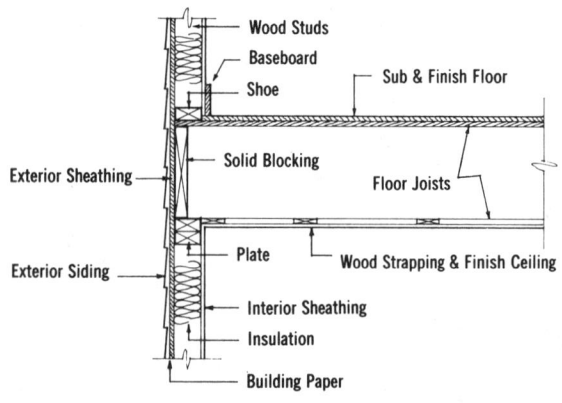

Brick Veneer

Typical brick veneer construction consists of wood studs with insulation between them. A typical exterior surface is sheathing, building paper and an exterior of brick tied to the sheathing, with metal strips.

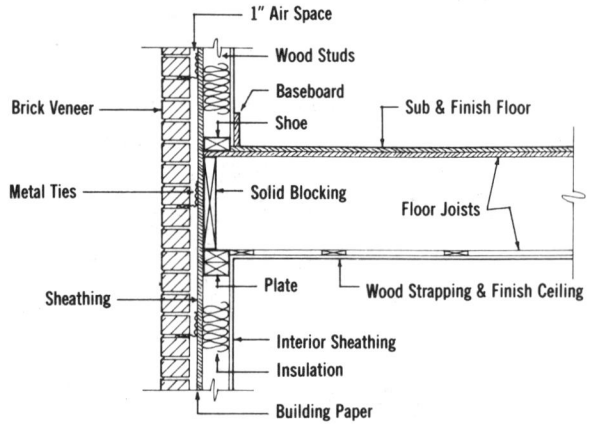

Stone

Typical solid masonry construction consists of a stone or block wall covered on the exterior with brick, stone or other masonry.

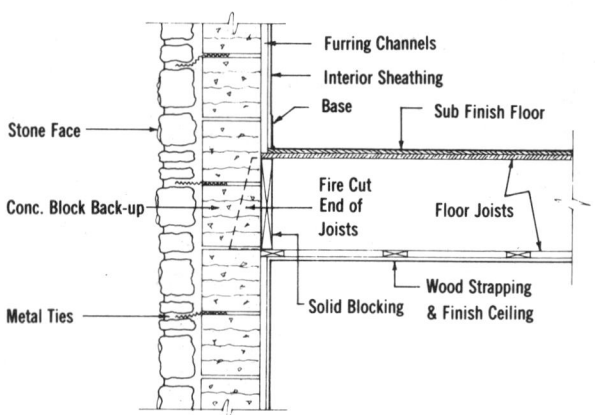

Residential Cost Estimate Worksheets

Worksheet Instructions

The residential cost estimate worksheet can be used as an outline for developing a residential construction or replacement cost. It is also useful for insurance appraisals. The design of the worksheet helps eliminate errors and omissions. To use the worksheet, follow the example below.

1. Fill out the owner's name, residence address, the estimator or appraiser's name, some type of project identifying number or code, and the date.
2. Determine from the plans, specifications, owner's description, photographs or any other means possible the class of construction. The models in this book use economy, average, custom and luxury as classes. Fill in the appropriate box.
3. Fill in the appropriate box for the residence type, configuration, occupancy, and exterior wall. If you require clarification, the pages preceding this worksheet describe each of these.
4. Next, the living area of the residence must be established. The heated or air conditioned space of the residence, not including the basement, should be measured. It is easiest to break the structure up into separate components as shown in the example. The main house (A), a one-and-one-half story wing (B), and a one story wing (C). The breezeway (D), garage (E) and open covered porch (F) will be treated differently. Data entry blocks for the living area are included on the worksheet for your use. Keep each level of each component separate, and fill out the blocks as shown.
5. By using the information on the worksheet, find the model, wing or ell in the following square foot cost pages that best matches the class, type, exterior finish and size of the residence being estimated. Use the *modifications, adjustments, and alternatives* to determine the adjusted cost per square foot of living area for each component.
6. For each component, multiply the cost per square foot by the living area square footage. If the residence is a townhouse/rowhouse, a multiplier should be applied based upon the configuration.
7. The second page of the residential cost estimate worksheet has space for the additional components of a house. The cost for additional bathrooms, finished attic space, breezeways, porches, fireplaces, appliance or cabinet upgrades, and garages should be added on this page. The information for each of these components is found with the model being used, or in the *modifications, adjustments, and alternatives* pages.
8. Add the total from page one of the estimate worksheet and the items listed on page two. The sum is the adjusted total building cost.
9. Depending on the use of the final estimated cost, one of the remaining two boxes should be filled out. Any additional items or exclusions should be added or subtracted at this time. The data contained in this book is a national average. Construction costs are different throughout the country. To allow for this difference, a location factor based upon the first three digits of the residence's zip code must be applied. The location factor is a multiplier that increases or decreases the adjusted total building cost. Find the appropriate location factor and calculate the local cost. If depreciation is a concern, a dollar figure should be subtracted at this point.
10. No residence will match a model exactly. Many differences will be found. At this level of estimating, a variation of plus or minus 10% should be expected.

Adjustments Instructions

No residence matches a model exactly in shape, material, or specifications.

The common differences are:
1. Two or more exterior wall systems
   a. Partial basement
   b. Partly finished basement
2. Specifications or features that are between two classes
3. Crawl space instead of a basement

EXAMPLES

Below are quick examples. See pages 14-16 for complete examples of cost adjustments for these differences:

1. Residence "A" is an average one-story structure with 1,600 S.F. of living area and no basement. Three walls are wood siding on wood frame, and the fourth wall is brick veneer on wood frame. The brick veneer wall is 35% of the exterior wall area.

   Use page 28 to calculate the Base Cost per S.F. of Living Area.
   Wood Siding for 1,600 S.F. = $64.10 per S.F.
   Brick Veneer for 1,600 S.F. = $72.70 per S.F.
   .65 ($64.10) + .35 ($72.70) = $67.12 per S.F. of Living Area.

2. a. Residence "B" is the same as Residence "A"; However, it has an unfinished basement under 50% of the building. To adjust the $67.12 per S.F. of living area for this partial basement, use page 28.
   $67.12 + .50 ($6.10) = $70.17 per S.F. of Living Area.

   b. Residence "C" is the same as Residence "A"; However, it has a full basement under the entire building. 640 S.F. or 40% of the basement area is finished.

   Using Page 30:
   $67.12 + .40 ($16.30) + .60 ($6.10) = $77.30 per S.F. of Living Area.

3. When specifications or features of a building are between classes, estimate the percent deviation, and use two tables to calculate the cost per S.F.

   A two-story residence with wood siding and 1,800 S.F. of livin area has features 30% better than Average, but 70% less than Custom.

   From pages 30 and 42:
   Custom 1,800 S.F. Base Cost   = $86.15 per S.F.
   Average 1,800 S.F. Base Cost = $65.95 per S.F.
   DIFFERENCE   = $20.20 per S.F.

   Cost is $65.95 + .30 ($20.20) = $72.01 per S.F. of Living Area.

4. To add the cost of a crawl space, use the cost of an unfinished basement as a maximum. For specific costs of components to be added or deducted, such as vapor barrier, underdrain, and floor, see the "Assemblies" section, pages 233 to 425.

Model Residence Example

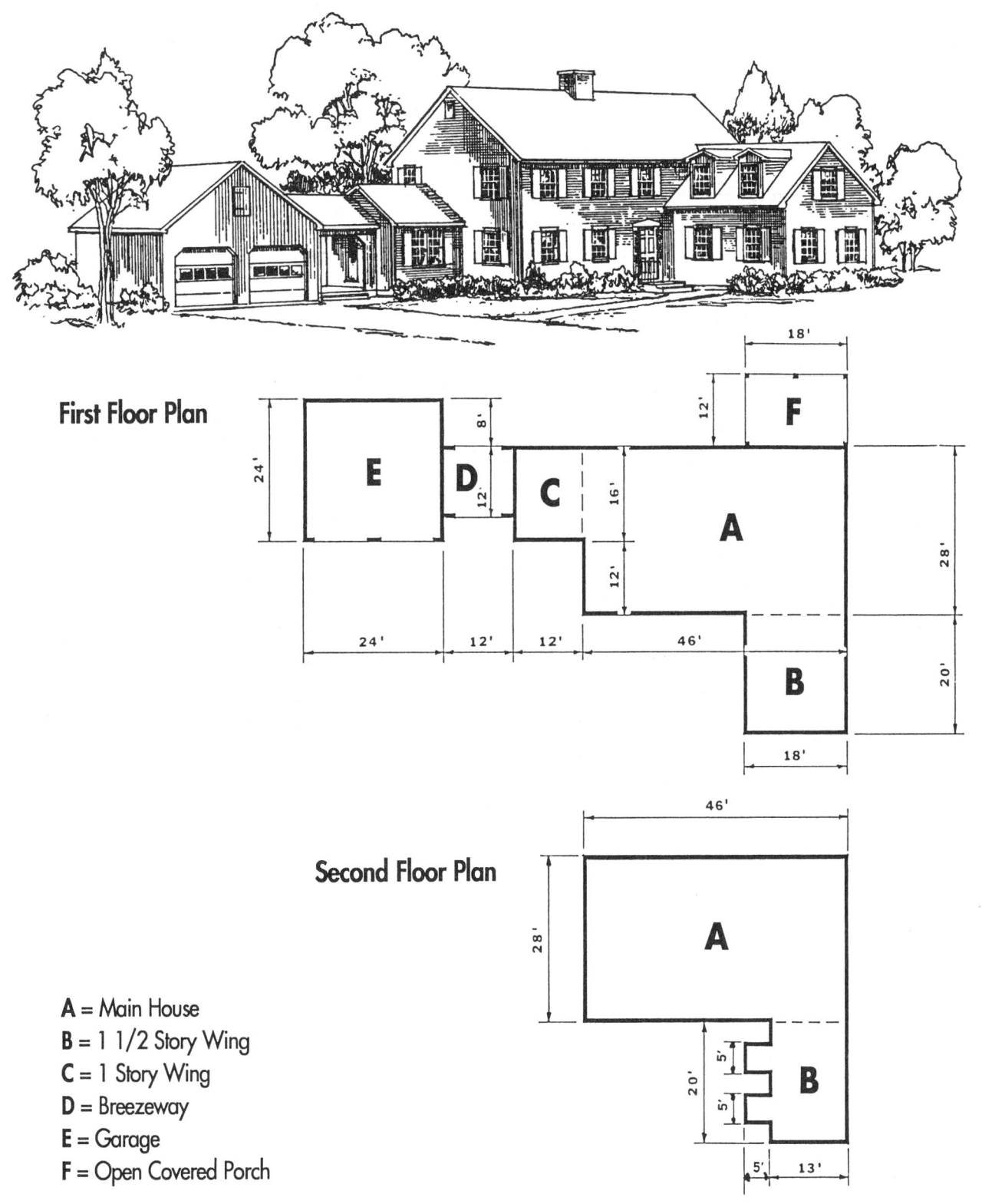

First Floor Plan

Second Floor Plan

A = Main House
B = 1 1/2 Story Wing
C = 1 Story Wing
D = Breezeway
E = Garage
F = Open Covered Porch

Means Forms
RESIDENTIAL COST ESTIMATE

OWNER'S NAME: **Albert Westenberg**	APPRAISER: **Nicole Wojtowicz**
RESIDENCE ADDRESS: **300 Sygiel Road**	PROJECT: **#55**
CITY, STATE, ZIP CODE: **Three Rivers, MA 01080**	DATE: **January 2, 1997**

CLASS OF CONSTRUCTION
- ☐ ECONOMY
- ☑ AVERAGE
- ☐ CUSTOM
- ☐ LUXURY

RESIDENCE TYPE
- ☐ 1 STORY
- ☐ 1-1/2 STORY
- ☑ 2 STORY
- ☐ 2-1/2 STORY
- ☐ 3 STORY
- ☐ BI-LEVEL
- ☐ TRI-LEVEL

CONFIGURATION
- ☑ DETACHED
- ☐ TOWN/ROW HOUSE
- ☐ SEMI-DETACHED

OCCUPANCY
- ☑ ONE FAMILY
- ☐ TWO FAMILY
- ☐ THREE FAMILY
- ☐ OTHER

EXTERIOR WALL SYSTEM
- ☑ WOOD SIDING - WOOD FRAME
- ☐ BRICK VENEER - WOOD FRAME
- ☐ STUCCO ON WOOD FRAME
- ☐ PAINTED CONCRETE BLOCK
- ☐ SOLID MASONRY (AVERAGE & CUSTOM)
- ☐ STONE VENEER - WOOD FRAME
- ☐ SOLID BRICK (LUXURY)
- ☐ SOLID STONE (LUXURY)

* LIVING AREA (Main Building)
First Level	1288	S.F.
Second level	1288	S.F.
Third Level		S.F.
Total	2576	S.F.

* LIVING AREA (Wing or Ell) (B)
First Level	360	S.F.
Second level	310	S.F.
Third Level		S.F.
Total	670	S.F.

* LIVING AREA (WING or ELL) (C)
First Level	192	S.F.
Second level		S.F.
Third Level		S.F.
Total	192	S.F.

* Basement Area is not part of living area.

MAIN BUILDING — COSTS PER S.F. LIVING AREA

Cost per Square Foot of Living Area, from Page **30**	$	**58.30**
Basement Addition: _____ % Finished, **100** % Unfinished	+	**3.35**
Roof Cover Adjustment: **Cedar Shake** Type, Page **30** (Add or Deduct)	(+)	**1.05**
Central Air Conditioning: ☐ Separate Ducts ☑ Heating Ducts, Page **30**	+	**1.30**
Heating System Adjustment: _____ Type, Page _____ (Add or Deduct)	( )	—
Main Building: Adjusted Cost per S.F. of Living Area	$	**64.00**

MAIN BUILDING TOTAL COST: $ **64.00** /S.F. x **2,576** S.F. x **1** = $ **164,864**
(Cost per S.F. Living Area) (Living Area) (Town/Row House Multiplier — Use 1 for Detached) (TOTAL COST)

WING OR ELL (B) 1-1/2 STORY — COSTS PER S.F. LIVING AREA

Cost per Square Foot of Living Area, from Page **37** (Wood Siding)	$	**51.65**
Basement Addition: **100** % Finished, _____ % Unfinished	+	**14.45**
Roof Cover Adjustment: _____ Type, Page _____ (Add or Deduct)	( )	—
Central Air Conditioning: ☐ Separate Ducts ☑ Heating Ducts, Page **29**	+	**1.65**
Heating System Adjustment: _____ Type, Page _____ (Add or Deduct)	( )	—
Wing or Ell (B): Adjusted Cost per S.F. of Living Area	$	**67.75**

WING OR ELL (B) TOTAL COST: $ **67.75** /S.F. x **670** S.F. x = $ **45,393**

WING OR ELL (C) 1 STORY — COSTS PER S.F. LIVING AREA

Cost per Square Foot of Living Area, from Page **37** (Wood Siding)	$	**77.30**
Basement Addition: _____ % Finished, _____ % Unfinished	+	—
Roof Cover Adjustment: _____ Type, Page _____ (Add or Deduct)	( )	—
Central Air Conditioning: ☐ Separate Ducts ☐ Heating Ducts, Page _____	+	—
Heating System Adjustment: _____ Type, Page _____ (Add or Deduct)	( )	—
Wing or Ell ( ): Adjusted Cost per S.F. of Living Area	$	**77.30**

WING OR ELL (C) TOTAL COST: $ **77.30** /S.F. x **192** S.F. x = $ **14,842**

TOTAL THIS PAGE: **225,099**

Page 1 of 2

Means Forms
RESIDENTIAL COST ESTIMATE

Total Page 1				$	225,099
		QUANTITY	UNIT COST		
Additional Bathrooms: 2 Full 1 Half 2 @ 3,528 1 @ 2,173					9,229
Finished Attic: N/A Ft. x Ft.		S.F.		+	
Breezeway: ☑ Open ☐ Enclosed 12 Ft. x 12 Ft.		144 S.F.	13.85	+	1,994
Covered Porch: ☑ Open ☐ Enclosed 18 Ft. x 12 Ft.		216 S.F.	20.80	+	4,493
Fireplace: ☑ Interior Chimney ☐ Exterior Chimney ☑ No. of Flues (2) ☑ Additional Fireplaces 1 - 2nd Story				+	6,050
Appliances:				+	—
Kitchen Cabinets Adjustments: (±)					—
☑ Garage ☐ Carport: 2 Car(s) Description Wood, Attached (±)					9,831
Miscellaneous:				+	

ADJUSTED TOTAL BUILDING COST $ 256,696

REPLACEMENT COST		
ADJUSTED TOTAL BUILDING COST	$	256,696
Site Improvements		
(A) Paving & Sidewalks	$	
(B) Landscaping	$	
(C) Fences	$	
(D) Swimming Pools	$	
(E) Miscellaneous	$	
TOTAL	$	256,696
Location Factor	x	1.07
Location Replacement Cost	$	274,665
Depreciation -10 %	- $	27,466
LOCAL DEPRECIATED COST	$	247,199

INSURANCE COST		
ADJUSTED TOTAL BUILDING COST	$	
Insurance Exclusions		
(A) Footings, Site work, Underground Piping	- $	
(B) Architects Fees	- $	
Total Building Cost Less Exclusion	$	
Location Factor	x	
LOCAL INSURABLE REPLACEMENT COST	$	

SKETCH AND ADDITIONAL CALCULATIONS

| RESIDENTIAL | Economy | Building Types |

1 Story

1-1/2 Story

2 Story

Bi-Level

Tri-Level

RESIDENTIAL — Economy — Specifications

	Components
1 Site Work	Site preparation for slab or excavation for lower level; 4' deep trench excavation for foundation wall.
2 Foundations	Continuous reinforced concrete footing, 8" deep x 18" wide; dampproofed and insulated 8" thick reinforced concrete block foundation wall, 4' deep; 4" concrete slab on 4" crushed stone base and polyethylene vapor barrier, trowel finish.
3 Framing	Exterior walls—2" x 4" wood studs, 16" O.C.; 1/2" insulation board sheathing; wood truss roof 24" O.C. or 2" x 6" rafters 16" O.C. with 1/2" plywood sheathing, 4 in 12 or 8 in 12 roof pitch.
4 Exterior Walls	Beveled wood siding and #15 felt building paper on insulated wood frame walls; sliding sash or double hung wood windows; 2 flush solid core wood exterior doors. **Alternates:** • Brick veneer on wood frame, has 4" veneer of common brick. • Stucco on wood frame has 1" thick colored stucco finish. • Painted concrete block has 8" concrete block sealed and painted on exterior and furring on the interior for the drywall.
5 Roofing	20 year asphalt shingles; #15 felt building paper; aluminum gutters, downspouts, drip edge and flashings.
6 Interiors	Walls and ceilings—1/2" taped and finished drywall, primed and painted with 2 coats; painted baseboard and trim; rubber backed carpeting 80%, asphalt tile 20%; hollow core wood interior doors.
7 Specialties	Economy grade kitchen cabinets—6 L.F. wall and base with plastic laminate counter top and kitchen sink; 30 gallon electric water heater.
8 Mechanical	1 lavatory, white, wall hung; 1 water closet, white, 1 bathtub, enameled steel, white; gas fired warm air heat.
9 Electrical	100 Amp. service; romex wiring; incandescent lighting fixtures, switches, receptacles.
10 Overhead and Profit	General Contractor overhead and profit.

Adjustments

Unfinished Basement:
7' high 8" concrete block or cast-in-place concrete.

Finished Basement:
Includes inexpensive paneling or drywall on foundation walls. Inexpensive sponge backed carpeting on concrete floor, drywall ceiling, and lighting.

RESIDENTIAL — Economy — 1 Story

- Mass produced from stock plans
- Single family — 1 full bath, 1 kitchen
- No basement
- Asphalt shingles on roof
- Hot air heat
- Drywall interior finishes
- Materials and workmanship are sufficient to meet codes
- Detail specifications on page 19

Note: The illustration shown may contain some optional components (for example: garages and/or fireplaces) whose costs are shown in the modifications, adjustments, & alternatives below or at the end of the square foot section.

© Home Planners, Inc.

Base cost per square foot of living area

Exterior Wall	600	800	1000	1200	1400	1600	1800	2000	2400	2800	3200
Wood Siding - Wood Frame	71.45	65.15	59.95	55.65	52.15	49.65	48.40	46.90	43.50	41.10	39.45
Brick Veneer - Wood Frame	76.30	69.45	63.85	59.15	55.35	52.70	51.30	49.60	46.00	43.40	41.60
Stucco on Wood Frame	70.00	63.85	58.75	54.55	51.20	48.75	47.50	46.05	42.75	40.45	38.80
Painted Concrete Block	72.40	66.00	60.70	56.35	52.80	50.25	49.00	47.45	44.00	41.60	39.90
Finished Basement, Add	15.75	14.85	14.20	13.65	13.15	12.85	12.65	12.40	12.05	11.80	11.55
Unfinished Basement, Add	7.80	7.05	6.50	5.95	5.60	5.30	5.15	4.95	4.65	4.40	4.20

Modifications

Add to the total cost

Upgrade Kitchen Cabinets	$ + 228
Solid Surface Countertops	+ 452
Full Bath - including plumbing, wall and floor finishes	+ 2811
Half Bath - including plumbing, wall and floor finishes	+ 1731
One Car Attached Garage	+ 6219
One Car Detached Garage	+ 6733
Fireplace & Chimney	+ 3205

Adjustments

For multi family - add to total cost

Additional Kitchen	$ + 1987
Additional Bath	+ 2811
Additional Entry & Exit	+ 969
Separate Heating	+ 1135
Separate Electric	+ 611

For Townhouse/Rowhouse - Multiply cost per square foot by

Inner Unit	.95
End Unit	.97

Alternatives

Add to or deduct from the cost per square foot of living area

Composition Roll Roofing	$ − .45
Cedar Shake Roof	+ 2.30
Upgrade Walls and Ceilings to Skim Coat Plaster	+ .37
Upgrade Ceilings to Textured Finish	+ .41
Air Conditioning (in heating ductwork)	+ 2.15

Additional upgrades or components

Kitchen Cabinets & Countertops	Page 58
Bathroom Vanities	59
Fireplaces & Chimneys	59
Windows, Skylights & Dormers	59
Appliances	60
Breezeways & Porches	60
Finished Attic	60
Garages	61
Site Improvements	61
Wings & Ells	25

Important: See the Reference Section for Location Factors (to adjust for your city) and Estimating Forms

RESIDENTIAL | Economy | 1-1/2 Story

- Mass produced from stock plans
- Single family — 1 full bath, 1 kitchen
- No basement
- Asphalt shingles on roof
- Hot air heat
- Drywall interior finishes
- Materials and workmanship are sufficient to meet codes
- Detail specifications on page 19

Note: The illustration shown may contain some optional components (for example: garages and/or fireplaces) whose costs are shown in the modifications, adjustments, & alternatives below or at the end of the square foot section.

Base cost per square foot of living area

Exterior Wall	600	800	1000	1200	1400	1600	1800	2000	2400	2800	3200
Wood Siding - Wood Frame	82.20	68.70	61.00	57.90	55.65	51.80	50.10	47.95	43.90	42.50	40.75
Brick Veneer - Wood Frame	88.95	73.60	65.55	62.10	59.70	55.45	53.60	51.25	46.80	45.25	43.30
Stucco on Wood Frame	80.15	67.25	59.65	56.60	54.45	50.70	49.05	46.95	43.00	41.65	40.00
Painted Concrete Block	83.50	69.70	61.90	58.70	56.45	52.55	50.80	48.60	44.45	43.05	41.25
Finished Basement, **Add**	12.10	10.30	9.85	9.50	9.25	8.90	8.70	8.50	8.15	7.95	7.80
Unfinished Basement, **Add**	6.85	5.25	4.85	4.60	4.40	4.05	3.90	3.75	3.40	3.25	3.10

Modifications

Add to the total cost

Upgrade Kitchen Cabinets	$ + 228
Solid Surface Countertops	+ 452
Full Bath - including plumbing, wall and floor finishes	+ 2811
Half Bath - including plumbing, wall and floor finishes	+ 1731
One Car Attached Garage	+ 6219
One Car Detached Garage	+ 6733
Fireplace & Chimney	+ 3205

Adjustments

For multi family - add to total cost

Additional Kitchen	$ + 1987
Additional Bath	+ 2811
Additional Entry & Exit	+ 969
Separate Heating	+ 1135
Separate Electric	+ 611

For Townhouse/Rowhouse -
Multiply cost per square foot by

Inner Unit	.95
End Unit	.97

Alternatives

Add to or deduct from the cost per square foot of living area

Composition Roll Roofing	$ − .30
Cedar Shake Roof	+ 1.65
Upgrade Walls and Ceilings to Skim Coat Plaster	+ .38
Upgrade Ceilings to Textured Finish	+ .41
Air Conditioning (in heating ductwork)	+ 1.60

Additional upgrades or components

Kitchen Cabinets & Countertops	Page 58
Bathroom Vanities	59
Fireplaces & Chimneys	59
Windows, Skylights & Dormers	59
Appliances	60
Breezeways & Porches	60
Finished Attic	60
Garages	61
Site Improvements	61
Wings & Ells	25

Important: See the Reference Section for Location Factors (to adjust for your city) and Estimating Forms

RESIDENTIAL Economy 2 Story

- Mass produced from stock plans
- Single family — 1 full bath, 1 kitchen
- No basement
- Asphalt shingles on roof
- Hot air heat
- Drywall interior finishes
- Materials and workmanship are sufficient to meet codes
- Detail specifications on page 19

Note: The illustration shown may contain some optional components (for example: garages and/or fireplaces) whose costs are shown in the modifications, adjustments, & alternatives below or at the end of the square foot section.

Base cost per square foot of living area

Exterior Wall	\multicolumn{11}{c}{Living Area}
	1000	1200	1400	1600	1800	2000	2200	2600	3000	3400	3800
Wood Siding - Wood Frame	66.50	59.65	56.95	55.10	52.90	50.65	49.25	46.20	43.25	42.05	40.80
Brick Veneer - Wood Frame	71.75	64.50	61.45	59.40	56.95	54.60	53.00	49.60	46.40	45.05	43.60
Stucco on Wood Frame	64.95	58.15	55.55	53.80	51.65	49.50	48.15	45.20	42.30	41.15	39.95
Painted Concrete Block	67.50	60.60	57.85	55.95	53.70	51.45	50.00	46.90	43.90	42.65	41.35
Finished Basement, **Add**	8.20	7.85	7.65	7.40	7.25	7.10	6.95	6.70	6.50	6.40	6.25
Unfinished Basement, **Add**	4.20	3.90	3.70	3.50	3.35	3.25	3.10	2.90	2.70	2.60	2.50

Modifications

Add to the total cost

Upgrade Kitchen Cabinets	$ + 228
Solid Surface Countertops	+ 452
Full Bath - including plumbing, wall and floor finishes	+ 2811
Half Bath - including plumbing, wall and floor finishes	+ 1731
One Car Attached Garage	+ 6219
One Car Detached Garage	+ 6733
Fireplace & Chimney	+ 3205

Adjustments

For multi family - add to total cost

Additional Kitchen	$ + 1987
Additional Bath	+ 2811
Additional Entry & Exit	+ 969
Separate Heating	+ 1135
Separate Electric	+ 611

For Townhouse/Rowhouse -
Multiply cost per square foot by

Inner Unit	.93
End Unit	.96

Alternatives

Add to or deduct from the cost per square foot of living area

Composition Roll Roofing	$ - .25
Cedar Shake Roof	+ 1.15
Upgrade Walls and Ceilings to Skim Coat Plaster	+ .38
Upgrade Ceilings to Textured Finish	+ .41
Air Conditioning (in heating ductwork)	+ 1.30

Additional upgrades or components

Kitchen Cabinets & Countertops	Page 58
Bathroom Vanities	59
Fireplaces & Chimneys	59
Windows, Skylights & Dormers	59
Appliances	60
Breezeways & Porches	60
Finished Attic	60
Garages	61
Site Improvements	61
Wings & Ells	25

RESIDENTIAL Economy Bi-Level

- Mass produced from stock plans
- Single family — 1 full bath, 1 kitchen
- No basement
- Asphalt shingles on roof
- Hot air heat
- Drywall interior finishes
- Materials and workmanship are sufficient to meet codes
- Detail specifications on page 19

Note: The illustration shown may contain some optional components (for example: garages and/or fireplaces) whose costs are shown in the modifications, adjustments, & alternatives below or at the end of the square foot section.

Base cost per square foot of living area

Exterior Wall	1000	1200	1400	1600	1800	2000	2200	2600	3000	3400	3800
Wood Siding - Wood Frame	61.40	54.95	52.55	50.90	48.90	46.85	45.65	42.95	40.25	39.20	38.10
Brick Veneer - Wood Frame	65.30	58.55	55.95	54.10	51.95	49.80	48.45	45.45	42.60	41.40	40.20
Stucco on Wood Frame	60.25	53.85	51.50	49.90	48.00	46.00	44.80	42.15	39.50	38.50	37.45
Painted Concrete Block	62.20	55.65	53.20	51.55	49.50	47.45	46.20	43.45	40.70	39.65	38.50
Finished Basement, **Add**	8.20	7.85	7.65	7.40	7.25	7.10	6.95	6.70	6.50	6.40	6.25
Unfinished Basement, **Add**	4.20	3.90	3.70	3.50	3.35	3.25	3.10	2.90	2.70	2.60	2.50

Modifications

Add to the total cost

Upgrade Kitchen Cabinets	$ + 228
Solid Surface Countertops	+ 452
Full Bath - including plumbing, wall and floor finishes	+ 2811
Half Bath - including plumbing, wall and floor finishes	+ 1731
One Car Attached Garage	+ 6219
One Car Detached Garage	+ 6733
Fireplace & Chimney	+ 3540

Adjustments

For multi family - add to total cost

Additional Kitchen	$ + 1987
Additional Bath	+ 2811
Additional Entry & Exit	+ 969
Separate Heating	+ 1135
Separate Electric	+ 611

For Townhouse/Rowhouse - Multiply cost per square foot by

Inner Unit	.94
End Unit	.97

Alternatives

Add to or deduct from the cost per square foot of living area

Composition Roll Roofing	$ − .25
Cedar Shake Roof	+ 1.15
Upgrade Walls and Ceilings to Skim Coat Plaster	+ .36
Upgrade Ceilings to Textured Finish	+ .41
Air Conditioning (in heating ductwork)	+ 1.30

Additional upgrades or components

Kitchen Cabinets & Countertops	Page 58
Bathroom Vanities	59
Fireplaces & Chimneys	59
Windows, Skylights & Dormers	59
Appliances	60
Breezeways & Porches	60
Finished Attic	60
Garages	61
Site Improvements	61
Wings & Ells	25

Important: See the Reference Section for Location Factors (to adjust for your city) and Estimating Forms

RESIDENTIAL | Economy | Tri-Level

- Mass produced from stock plans
- Single family — 1 full bath, 1 kitchen
- No basement
- Asphalt shingles on roof
- Hot air heat
- Drywall interior finishes
- Materials and workmanship are sufficient to meet codes
- Detail specifications on page 19

Note: The illustration shown may contain some optional components (for example: garages and/or fireplaces) whose costs are shown in the modifications, adjustments, & alternatives below or at the end of the square foot section.

Base cost per square foot of living area

Exterior Wall	1200	1500	1800	2000	2200	2400	2800	3200	3600	4000	4400
Wood Siding - Wood Frame	56.00	51.50	48.00	46.65	44.80	42.95	41.65	39.80	37.70	37.05	35.45
Brick Veneer - Wood Frame	59.60	54.75	50.90	49.45	47.45	45.50	44.05	42.05	39.80	39.10	37.35
Stucco on Wood Frame	54.90	50.50	47.10	45.80	43.95	42.20	40.90	39.15	37.10	36.45	34.85
Solid Masonry	56.70	52.15	48.55	47.20	45.30	43.45	42.10	40.25	38.15	37.45	35.80
Finished Basement, **Add***	9.90	9.50	9.10	8.90	8.75	8.60	8.40	8.25	8.05	7.95	7.80
Unfinished Basement, **Add***	4.70	4.35	4.00	3.85	3.70	3.55	3.40	3.25	3.10	3.05	2.90

*Basement under middle level only.

Modifications

Add to the total cost

Upgrade Kitchen Cabinets	$ + 228
Solid Surface Countertops	+ 452
Full Bath - including plumbing, wall and floor finishes	+ 2811
Half Bath - including plumbing, wall and floor finishes	+ 1731
One Car Attached Garage	+ 6219
One Car Detached Garage	+ 6733
Fireplace & Chimney	+ 3540

Adjustments

For multi family - add to total cost

Additional Kitchen	$ + 1987
Additional Bath	+ 2811
Additional Entry & Exit	+ 969
Separate Heating	+ 1135
Separate Electric	+ 611

For Townhouse/Rowhouse - Multiply cost per square foot by

Inner Unit	.93
End Unit	.96

Alternatives

Add to or deduct from the cost per square foot of living area

Composition Roll Roofing	$ - .30
Cedar Shake Roof	+ 1.65
Upgrade Walls and Ceilings to Skim Coat Plaster	+ .33
Upgrade Ceilings to Textured Finish	+ .41
Air Conditioning (in heating ductwork)	+ 1.05

Additional upgrades or components

Kitchen Cabinets & Countertops	Page 58
Bathroom Vanities	59
Fireplaces & Chimneys	59
Windows, Skylights & Dormers	59
Appliances	60
Breezeways & Porches	60
Finished Attic	60
Garages	61
Site Improvements	61
Wings & Ells	25

RESIDENTIAL | Economy | Wings & Ells

1 Story — Base cost per square foot of living area

Exterior Wall	50	100	200	300	400	500	600	700
Wood Siding - Wood Frame	98.10	74.35	64.05	52.40	49.05	47.05	45.70	45.90
Brick Veneer - Wood Frame	110.30	83.10	71.30	57.25	53.40	51.10	49.55	49.65
Stucco on Wood Frame	94.40	71.75	61.85	50.95	47.70	45.80	44.50	44.80
Painted Concrete Block	100.55	76.10	65.50	53.35	49.95	47.85	46.45	46.65
Finished Basement, Add	24.30	19.85	17.95	14.85	14.25	13.85	13.60	13.40
Unfinished Basement, Add	14.90	11.15	9.55	6.95	6.45	6.10	5.90	5.75

1-1/2 Story — Base cost per square foot of living area

Exterior Wall	100	200	300	400	500	600	700	800
Wood Siding - Wood Frame	78.00	62.60	53.10	46.90	44.05	42.60	40.80	40.35
Brick Veneer - Wood Frame	88.90	71.35	60.35	52.55	49.30	47.55	45.50	45.00
Stucco on Wood Frame	74.70	59.95	50.90	45.15	42.45	41.10	39.40	38.95
Painted Concrete Block	80.15	64.35	54.55	48.00	45.05	43.55	41.75	41.30
Finished Basement, Add	16.25	14.35	13.10	11.75	11.40	11.15	10.90	10.85
Unfinished Basement, Add	9.25	7.65	6.60	5.45	5.15	4.95	4.75	4.70

2 Story — Base cost per square foot of living area

Exterior Wall	100	200	400	600	800	1000	1200	1400
Wood Siding - Wood Frame	80.35	59.45	50.35	40.65	37.65	35.90	34.70	35.05
Brick Veneer - Wood Frame	92.55	68.15	57.60	45.50	42.05	39.95	38.60	38.80
Stucco on Wood Frame	76.65	56.80	48.10	39.15	36.35	34.65	33.55	33.90
Painted Concrete Block	82.75	61.15	51.75	41.55	38.55	36.70	35.45	35.80
Finished Basement, Add	12.15	9.90	9.00	7.45	7.10	6.95	6.80	6.70
Unfinished Basement, Add	7.45	5.55	4.80	3.50	3.20	3.05	2.95	2.90

Base costs do not include bathroom or kitchen facilities. Use Modifications/Adjustments/Alternatives on pages 58-61 where appropriate.

Important: See the Reference Section for Location Factors (to adjust for your city) and Estimating Forms.

RESIDENTIAL | Average | Building Types

1 Story

1-1/2 Story

2 Story

2-1/2 Story

Bi-Level

Tri-Level

RESIDENTIAL — Average Specifications

	Components
1 Site Work	Site preparation for slab or excavation for lower level; 4' deep trench excavation for foundation wall.
2 Foundations	Continuous reinforced concrete footing, 8" deep x 18" wide; dampproofed and insulated 8" thick reinforced concrete foundation wall, 4' deep; 4" concrete slab on 4" crushed stone base and polyethylene vapor barrier, trowel finish.
3 Framing	Exterior walls—2" x 4" wood studs, 16" O.C.; 1/2" plywood sheathing; 2" x 6" rafters 16" O.C. with 1/2" plywood sheathing, 4 in 12 or 8 in 12 roof pitch, 2" x 6" ceiling joists; 1/2" plywood subfloor on 1" x 2" wood sleepers 16" O.C.; 2" x 8" floor joists with 5/8" plywood subfloor on models with more than one level.
4 Exterior Walls	Beveled wood siding and #15 felt building paper on insulated wood frame walls; double hung windows; 3 flush solid core wood exterior doors with storms. **Alternates:** • Brick veneer on wood frame, has 4" veneer of common brick. • Stucco on wood frame has 1" thick colored stucco finish. • Solid masonry has a 6" concrete block load bearing wall with insulation and a brick or stone exterior. Also solid brick or stone walls.
5 Roofing	25 year asphalt shingles; #15 felt building paper; aluminum gutters, downspouts, drip edge and flashings.
6 Interiors	Walls and ceilings—1/2" taped and finished drywall, primed and painted with 2 coats; painted baseboard and trim, finished hardwood floor 40%, carpet with 1/2" underlayment 40%, vinyl tile with 1/2" underlayment 15%, ceramic tile with 1/2" underlayment 5%; hollow core and louvered interior doors.
7 Specialties	Average grade kitchen cabinets—14 L.F. wall and base with plastic laminate counter top and kitchen sink; 40 gallon electric water heater.
8 Mechanical	1 lavatory, white, wall hung; 1 water closet, white; 1 bathtub, enameled steel, white; gas fired warm air heat.
9 Electrical	100 Amp. service; romex wiring; incandescent lighting fixtures, switches, receptacles.
10 Overhead and Profit	General Contractor overhead and profit.

Adjustments

Unfinished Basement:
7'-6" high cast-in-place concrete 8" thick or 8" concrete block.

Finished Basement:
Includes plywood paneling or drywall on furring on foundation walls, sponge backed carpeting on concrete floor, suspended ceiling, lighting and heating.

RESIDENTIAL — Average — 1 Story

- Simple design from standard plans
- Single family — 1 full bath, 1 kitchen
- No basement
- Asphalt shingles on roof
- Hot air heat
- Drywall interior finishes
- Materials and workmanship are average
- Detail specifications on p. 27

Note: The illustration shown may contain some optional components (for example: garages and/or fireplaces) whose costs are shown in the modifications, adjustments, & alternatives below or at the end of the square foot section.

Base cost per square foot of living area

Exterior Wall	600	800	1000	1200	1400	1600	1800	2000	2400	2800	3200
Wood Siding - Wood Frame	90.10	82.50	76.25	71.10	67.10	64.10	62.45	60.70	56.65	53.90	51.95
Brick Veneer - Wood Frame	100.60	92.45	85.75	80.20	75.85	72.70	70.90	68.95	64.65	61.75	59.60
Stucco on Wood Frame	95.90	88.30	81.95	76.80	72.75	69.80	68.10	66.35	62.25	59.50	57.55
Solid Masonry	107.70	98.80	91.50	85.40	80.60	77.10	75.15	73.00	68.35	65.10	62.75
Finished Basement, **Add**	20.15	18.95	18.10	17.30	16.70	16.30	16.05	15.70	15.20	14.85	14.55
Unfinished Basement, **Add**	8.60	7.85	7.30	6.75	6.40	6.10	5.95	5.75	5.45	5.20	5.00

Modifications

Add to the total cost

Upgrade Kitchen Cabinets	$ + 1969
Solid Surface Countertops	+ 798
Full Bath - including plumbing, wall and floor finishes	+ 3528
Half Bath - including plumbing, wall and floor finishes	+ 2173
One Car Attached Garage	+ 6927
One Car Detached Garage	+ 7430
Fireplace & Chimney	+ 3215

Adjustments

For multi family - add to total cost

Additional Kitchen	$ + 3709
Additional Bath	+ 3528
Additional Entry & Exit	+ 969
Separate Heating	+ 1165
Separate Electric	+ 1184

For Townhouse/Rowhouse - Multiply cost per square foot by

Inner Unit	.92
End Unit	.96

Alternatives

Add to or deduct from the cost per square foot of living area

Cedar Shake Roof	$+ 2.05
Clay Tile Roof	+ 4.35
Slate Roof	+ 7.30
Upgrade Walls to Skim Coat Plaster	+ .25
Upgrade Ceilings to Textured Finish	+ .41
Air Conditioning (in heating ductwork)	+ 2.20

Additional upgrades or components

Kitchen Cabinets & Countertops	Page 58
Bathroom Vanities	59
Fireplaces & Chimneys	59
Windows, Skylights & Dormers	59
Appliances	60
Breezeways & Porches	60
Finished Attic	60
Garages	61
Site Improvements	61
Wings & Ells	37

RESIDENTIAL | Average | 1-1/2 Story

- Simple design from standard plans
- Single family — 1 full bath, 1 kitchen
- No basement
- Asphalt shingles on roof
- Hot air heat
- Drywall interior finishes
- Materials and workmanship are average
- Detail specifications on p. 27

Note: The illustration shown may contain some optional components (for example: garages and/or fireplaces) whose costs are shown in the modifications, adjustments, & alternatives below or at the end of the square foot section.

Base cost per square foot of living area

Exterior Wall	600	800	1000	1200	1400	1600	1800	2000	2400	2800	3200
Wood Siding - Wood Frame	101.55	86.10	76.75	73.05	70.45	65.85	63.85	61.25	56.55	54.90	52.80
Brick Veneer - Wood Frame	108.65	91.20	81.50	77.55	74.65	69.70	67.55	64.75	59.60	57.80	55.50
Stucco on Wood Frame	102.10	86.50	77.10	73.45	70.80	66.15	64.15	61.55	56.80	55.15	53.05
Solid Masonry	118.60	98.45	88.20	83.80	80.65	75.10	72.65	69.60	63.95	61.90	59.30
Finished Basement, **Add**	16.50	14.05	13.45	13.05	12.75	12.25	11.95	11.70	11.25	11.00	10.75
Unfinished Basement, **Add**	7.40	5.85	5.45	5.15	4.95	4.60	4.45	4.30	4.00	3.80	3.65

Modifications

Add to the total cost

Upgrade Kitchen Cabinets	$ + 1969
Solid Surface Countertops	+ 798
Full Bath - including plumbing, wall and floor finishes	+ 3528
Half Bath - including plumbing, wall and floor finishes	+ 2173
One Car Attached Garage	+ 6927
One Car Detached Garage	+ 7430
Fireplace & Chimney	+ 3215

Adjustments

For multi family - add to total cost

Additional Kitchen	$ + 3709
Additional Bath	+ 3528
Additional Entry & Exit	+ 969
Separate Heating	+ 1165
Separate Electric	+ 1184

For Townhouse/Rowhouse - Multiply cost per square foot by

Inner Unit	.92
End Unit	.96

Alternatives

Add to or deduct from the cost per square foot of living area

Cedar Shake Roof	$+ 1.50
Clay Tile Roof	+ 3.15
Slate Roof	+ 5.25
Upgrade Walls to Skim Coat Plaster	+ .28
Upgrade Ceilings to Textured Finish	+ .41
Air Conditioning (in heating ductwork)	+ 1.65

Additional upgrades or components

Kitchen Cabinets & Countertops	Page 58
Bathroom Vanities	59
Fireplaces & Chimneys	59
Windows, Skylights & Dormers	59
Appliances	60
Breezeways & Porches	60
Finished Attic	60
Garages	61
Site Improvements	61
Wings & Ells	37

Important: See the Reference Section for Location Factors (to adjust for your city) and Estimating Forms

RESIDENTIAL — Average — 2 Story

- Simple design from standard plans
- Single family — 1 full bath, 1 kitchen
- No basement
- Asphalt shingles on roof
- Hot air heat
- Drywall interior finishes
- Materials and workmanship are average
- Detail specifications on p. 27

Note: The illustration shown may contain some optional components (for example: garages and/or fireplaces) whose costs are shown in the modifications, adjustments, & alternatives below or at the end of the square foot section.

Base cost per square foot of living area

Exterior Wall	1000	1200	1400	1600	1800	2000	2200	2600	3000	3400	3800
Wood Siding - Wood Frame	82.40	74.15	70.95	68.80	65.95	63.55	61.95	58.30	54.80	53.45	51.90
Brick Veneer - Wood Frame	87.90	79.25	75.70	73.30	70.25	67.65	65.85	61.80	58.10	56.60	54.85
Stucco on Wood Frame	82.85	74.55	71.35	69.15	66.35	63.90	62.25	58.60	55.10	53.75	52.15
Solid Masonry	96.35	87.10	83.05	80.30	76.85	74.00	71.85	67.30	63.15	61.35	59.40
Finished Basement, Add	11.70	11.30	10.90	10.70	10.40	10.25	10.05	9.70	9.45	9.30	9.15
Unfinished Basement, Add	4.70	4.40	4.15	4.00	3.85	3.70	3.60	3.35	3.20	3.10	3.00

Modifications

Add to the total cost

Upgrade Kitchen Cabinets	$ + 1969
Solid Surface Countertops	+ 798
Full Bath - including plumbing, wall and floor finishes	+ 3528
Half Bath - including plumbing, wall and floor finishes	+ 2173
One Car Attached Garage	+ 6927
One Car Detached Garage	+ 7430
Fireplace & Chimney	+ 3590

Adjustments

For multi family - add to total cost

Additional Kitchen	$ + 3709
Additional Bath	+ 3528
Additional Entry & Exit	+ 969
Separate Heating	+ 1165
Separate Electric	+ 1184

For Townhouse/Rowhouse - Multiply cost per square foot by

Inner Unit	.90
End Unit	.95

Alternatives

Add to or deduct from the cost per square foot of living area

Cedar Shake Roof	$+ 1.05
Clay Tile Roof	+ 2.20
Slate Roof	+ 3.65
Upgrade Walls to Skim Coat Plaster	+ .29
Upgrade Ceilings to Textured Finish	+ .41
Air Conditioning (in heating ductwork)	+ 1.30

Additional upgrades or components

Kitchen Cabinets & Countertops	Page 58
Bathroom Vanities	59
Fireplaces & Chimneys	59
Windows, Skylights & Dormers	59
Appliances	60
Breezeways & Porches	60
Finished Attic	60
Garages	61
Site Improvements	61
Wings & Ells	37

RESIDENTIAL — Average — 2-1/2 Story

- Simple design from standard plans
- Single family — 1 full bath, 1 kitchen
- No basement
- Asphalt shingles on roof
- Hot air heat
- Drywall interior finishes
- Materials and workmanship are average
- Detail specifications on p. 27

Note: The illustration shown may contain some optional components (for example: garages and/or fireplaces) whose costs are shown in the modifications, adjustments, & alternatives below or at the end of the square foot section.

Base cost per square foot of living area

Exterior Wall	1200	1400	1600	1800	2000	2400	2800	3200	3600	4000	4400
Wood Siding - Wood Frame	81.90	77.45	69.95	69.00	66.75	62.65	59.95	56.65	55.20	52.20	51.40
Brick Veneer - Wood Frame	87.75	82.75	74.80	73.90	71.35	66.80	63.90	60.25	58.60	55.35	54.45
Stucco on Wood Frame	82.40	77.90	70.35	69.40	67.15	63.00	60.30	56.95	55.50	52.45	51.65
Solid Masonry	96.00	90.15	81.65	80.75	77.75	72.60	69.45	65.25	63.35	59.80	58.70
Finished Basement, Add	10.00	9.55	9.20	9.15	8.95	8.55	8.45	8.15	8.00	7.80	7.75
Unfinished Basement, Add	3.90	3.65	3.40	3.35	3.20	3.00	2.90	2.70	2.60	2.50	2.45

Modifications
Add to the total cost

Upgrade Kitchen Cabinets	$ + 1969
Solid Surface Countertops	+ 798
Full Bath - including plumbing, wall and floor finishes	+ 3528
Half Bath - including plumbing, wall and floor finishes	+ 2173
One Car Attached Garage	+ 6927
One Car Detached Garage	+ 7430
Fireplace & Chimney	+ 4085

Adjustments
For multi family - add to total cost

Additional Kitchen	$ + 3709
Additional Bath	+ 3528
Additional Entry & Exit	+ 969
Separate Heating	+ 1165
Separate Electric	+ 1184

For Townhouse/Rowhouse - Multiply cost per square foot by

Inner Unit	.90
End Unit	.95

Alternatives
Add to or deduct from the cost per square foot of living area

Cedar Shake Roof	$+ .90
Clay Tile Roof	+ 1.90
Slate Roof	+ 3.15
Upgrade Walls to Skim Coat Plaster	+ .28
Upgrade Ceilings to Textured Finish	+ .41
Air Conditioning (in heating ductwork)	+ 1.20

Additional upgrades or components

Kitchen Cabinets & Countertops	Page 58
Bathroom Vanities	59
Fireplaces & Chimneys	59
Windows, Skylights & Dormers	59
Appliances	60
Breezeways & Porches	60
Finished Attic	60
Garages	61
Site Improvements	61
Wings & Ells	37

Important: See the Reference Section for Location Factors (to adjust for your city) and Estimating Forms

RESIDENTIAL — Average — 3 Story

- Simple design from standard plans
- Single family — 1 full bath, 1 kitchen
- No basement
- Asphalt shingles on roof
- Hot air heat
- Drywall interior finishes
- Materials and workmanship are average
- Detail specifications on p. 27

Note: The illustration shown may contain some optional components (for example: garages and/or fireplaces) whose costs are shown in the modifications, adjustments, & alternatives below or at the end of the square foot section.

Base cost per square foot of living area

Exterior Wall	1500	1800	2100	2500	3000	3500	4000	4500	5000	5500	6000
Wood Siding - Wood Frame	76.80	69.15	66.35	64.30	59.65	57.90	54.80	51.70	50.85	49.50	48.40
Brick Veneer - Wood Frame	82.30	74.20	71.10	68.85	63.75	61.80	58.35	54.95	54.05	52.55	51.30
Stucco on Wood Frame	70.10	62.90	60.55	58.75	54.60	53.15	50.50	47.70	47.00	45.80	44.90
Solid Masonry	90.00	81.35	77.80	75.20	69.55	67.30	63.30	59.55	58.50	56.80	55.30
Finished Basement, Add	8.60	8.30	8.10	7.90	7.65	7.45	7.25	7.10	7.00	6.95	6.80
Unfinished Basement, Add	3.20	3.00	2.80	2.70	2.50	2.40	2.30	2.20	2.10	2.10	2.00

Modifications

Add to the total cost

Upgrade Kitchen Cabinets	$ + 1969
Solid Surface Countertops	+ 798
Full Bath - including plumbing, wall and floor finishes	+ 3528
Half Bath - including plumbing, wall and floor finishes	+ 2173
One Car Attached Garage	+ 6927
One Car Detached Garage	+ 7430
Fireplace & Chimney	+ 4085

Adjustments

For multi family - add to total cost

Additional Kitchen	$ + 3709
Additional Bath	+ 3528
Additional Entry & Exit	+ 969
Separate Heating	+ 1165
Separate Electric	+ 1184

For Townhouse/Rowhouse - Multiply cost per square foot by

Inner Unit	.88
End Unit	.94

Alternatives

Add to or deduct from the cost per square foot of living area

Cedar Shake Roof	$ + .70
Clay Tile Roof	+ 1.45
Slate Roof	+ 2.45
Upgrade Walls to Skim Coat Plaster	+ .28
Upgrade Ceilings to Textured Finish	+ .41
Air Conditioning (in heating ductwork)	+ 1.20

Additional upgrades or components

Kitchen Cabinets & Countertops	Page 58
Bathroom Vanities	59
Fireplaces & Chimneys	59
Windows, Skylights & Dormers	59
Appliances	60
Breezeways & Porches	60
Finished Attic	60
Garages	61
Site Improvements	61
Wings & Ells	37

Important: See the Reference Section for Location Factors (to adjust for your city) and Estimating Forms

RESIDENTIAL — Average — Bi-Level

- Simple design from standard plans
- Single family — 1 full bath, 1 kitchen
- No basement
- Asphalt shingles on roof
- Hot air heat
- Drywall interior finishes
- Materials and workmanship are average
- Detail specifications on p. 27

Note: The illustration shown may contain some optional components (for example: garages and/or fireplaces) whose costs are shown in the modifications, adjustments, & alternatives below or at the end of the square foot section.

Base cost per square foot of living area

Exterior Wall	1000	1200	1400	1600	1800	2000	2200	2600	3000	3400	3800
Wood Siding - Wood Frame	76.65	68.80	65.95	64.00	61.50	59.25	57.85	54.55	51.40	50.25	48.80
Brick Veneer - Wood Frame	80.75	72.65	69.55	67.40	64.70	62.35	60.75	57.20	53.85	52.55	51.05
Stucco on Wood Frame	77.00	69.15	66.30	64.30	61.75	59.55	58.10	54.80	51.60	50.40	49.00
Solid Masonry	86.55	78.00	74.55	72.20	69.20	66.70	64.85	60.95	57.30	55.80	54.15
Finished Basement, Add	11.70	11.30	10.90	10.70	10.40	10.25	10.05	9.70	9.45	9.30	9.15
Unfinished Basement, Add	4.70	4.40	4.15	4.00	3.85	3.70	3.60	3.35	3.20	3.10	3.00

Modifications

Add to the total cost

Upgrade Kitchen Cabinets	$ + 1969
Solid Surface Countertops	+ 798
Full Bath - including plumbing, wall and floor finishes	+ 3528
Half Bath - including plumbing, wall and floor finishes	+ 2173
One Car Attached Garage	+ 6927
One Car Detached Garage	+ 7430
Fireplace & Chimney	+ 3215

Adjustments

For multi family - add to total cost

Additional Kitchen	$ + 3709
Additional Bath	+ 3528
Additional Entry & Exit	+ 969
Separate Heating	+ 1165
Separate Electric	+ 1184

For Townhouse/Rowhouse - Multiply cost per square foot by

Inner Unit	.91
End Unit	.96

Alternatives

Add to or deduct from the cost per square foot of living area

Cedar Shake Roof	$+ 1.05
Clay Tile Roof	+ 2.20
Slate Roof	+ 3.65
Upgrade Walls to Skim Coat Plaster	+ .27
Upgrade Ceilings to Textured Finish	+ .41
Air Conditioning (in heating ductwork)	+ 1.30

Additional upgrades or components

Kitchen Cabinets & Countertops	Page 58
Bathroom Vanities	59
Fireplaces & Chimneys	59
Windows, Skylights & Dormers	59
Appliances	60
Breezeways & Porches	60
Finished Attic	60
Garages	61
Site Improvements	61
Wings & Ells	37

Important: See the Reference Section for Location Factors (to adjust for your city) and Estimating Forms

RESIDENTIAL — Average — Tri-Level

- Simple design from standard plans
- Single family — 1 full bath, 1 kitchen
- No basement
- Asphalt shingles on roof
- Hot air heat
- Drywall interior finishes
- Materials and workmanship are average
- Detail specifications on p. 27

Note: The illustration shown may contain some optional components (for example: garages and/or fireplaces) whose costs are shown in the modifications, adjustments, & alternatives below or at the end of the square foot section.

© Design Basics, Inc.

Base cost per square foot of living area

Exterior Wall	1200	1500	1800	2100	2400	2700	3000	3400	3800	4200	4600
Wood Siding - Wood Frame	71.75	66.35	62.10	58.85	56.35	54.90	53.50	52.05	49.70	47.70	46.70
Brick Veneer - Wood Frame	75.55	69.75	65.20	61.65	59.00	57.45	55.90	54.35	51.85	49.70	48.65
Stucco on Wood Frame	72.05	66.65	62.40	59.10	56.55	55.10	53.70	52.25	49.85	47.85	46.90
Solid Masonry	80.85	74.60	69.55	65.60	62.70	61.00	59.25	57.60	54.85	52.55	51.40
Finished Basement, **Add***	13.45	12.90	12.35	11.95	11.65	11.50	11.30	11.15	10.90	10.75	10.65
Unfinished Basement, **Add***	5.30	4.90	4.55	4.30	4.15	4.00	3.90	3.80	3.65	3.55	3.45

*Basement under middle level only.

Modifications

Add to the total cost

Upgrade Kitchen Cabinets	$ + 1969
Solid Surface Countertops	+ 798
Full Bath - including plumbing, wall and floor finishes	+ 3528
Half Bath - including plumbing, wall and floor finishes	+ 2173
One Car Attached Garage	+ 6927
One Car Detached Garage	+ 7430
Fireplace & Chimney	+ 3590

Adjustments

For multi family - add to total cost

Additional Kitchen	$ + 3709
Additional Bath	+ 3528
Additional Entry & Exit	+ 969
Separate Heating	+ 1165
Separate Electric	+ 1184

For Townhouse/Rowhouse - Multiply cost per square foot by

Inner Unit	.90
End Unit	.95

Alternatives

Add to or deduct from the cost per square foot of living area

Cedar Shake Roof	$+ 1.50
Clay Tile Roof	+ 3.15
Slate Roof	+ 5.25
Upgrade Walls to Skim Coat Plaster	+ .24
Upgrade Ceilings to Textured Finish	+ .41
Air Conditioning (in heating ductwork)	+ 1.10

Additional upgrades or components

Kitchen Cabinets & Countertops	Page 58
Bathroom Vanities	59
Fireplaces & Chimneys	59
Windows, Skylights & Dormers	59
Appliances	60
Breezeways & Porches	60
Finished Attic	60
Garages	61
Site Improvements	61
Wings & Ells	37

Important: See the Reference Section for Location Factors (to adjust for your city) and Estimating Forms

RESIDENTIAL — Solid Wall — 1 Story

- Post and beam frame
- Log exterior walls
- Simple design from standard plans
- Single family — 1 full bath, 1 kitchen
- No basement
- Asphalt shingles on roof
- Hot air heat
- Drywall interior finishes
- Materials and workmanship are average
- Detail specification on page 27

Note: The illustration shown may contain some optional components (for example: garages and/or fireplaces) whose costs are shown in the modifications, adjustments, & alternatives below or at the end of the square foot section.

Base cost per square foot of living area

Exterior Wall	600	800	1000	1200	1400	1600	1800	2000	2400	2800	3200
6" Log - Solid Wall	87.80	79.40	72.80	67.60	63.40	59.80	58.10	56.90	53.00	49.95	48.30
8" Log - Solid Wall	87.20	78.75	72.20	67.10	62.95	59.35	57.70	56.50	52.70	49.45	47.90
Finished Basement, **Add**	20.15	18.95	18.10	17.30	16.70	16.30	16.05	15.70	15.20	14.85	14.55
Unfinished Basement, **Add**	8.60	7.85	7.30	6.75	6.40	6.10	5.95	5.75	5.45	5.20	5.00

Modifications

Add to the total cost

Upgrade Kitchen Cabinets	$ + 1969
Solid Surface Countertops	+ 798
Full Bath - including plumbing, wall and floor finishes	+ 3528
Half Bath - including plumbing, wall and floor finishes	+ 2173
One Car Attached Garage	+ 6927
One Car Detached Garage	+ 7430
Fireplace & Chimney	+ 3215

Adjustments

For multi family - add to total cost

Additional Kitchen	$ + 3709
Additional Bath	+ 3528
Additional Entry & Exit	+ 969
Separate Heating	+ 1165
Separate Electric	+ 1184

For Townhouse/Rowhouse - Multiply cost per square foot by

Inner Unit	.92
End Unit	.96

Alternatives

Add to or deduct from the cost per square foot of living area

Cedar Shake Roof	$ + 2.05
Air Conditioning (in heating ductwork)	+ 2.15

Additional upgrades or components

Kitchen Cabinets & Countertops	Page 58
Bathroom Vanities	59
Fireplaces & Chimneys	59
Windows, Skylights & Dormers	59
Appliances	60
Breezeways & Porches	60
Finished Attic	60
Garages	61
Site Improvements	61
Wings & Ells	37

Important: See the Reference Section for Location Factors (to adjust for your city) and Estimating Forms

RESIDENTIAL | Solid Wall | 2 Story

- Post and beam frame
- Log exterior walls
- Simple design from standard plans
- Single family — 1 full bath, 1 kitchen
- No basement
- Asphalt shingles on roof
- Hot air heat
- Drywall interior finishes
- Materials and workmanship are average
- Detail specification on page 27

Note: The illustration shown may contain some optional components (for example: garages and/or fireplaces) whose costs are shown in the modifications, adjustments, & alternatives below or at the end of the square foot section.

Base cost per square foot of living area

Exterior Wall	1000	1200	1400	1600	1800	2000	2200	2600	3000	3400	3800
6" Log-Solid	87.90	72.45	69.15	66.70	64.00	61.40	59.65	56.10	52.00	50.65	49.30
8" Log-Solid	84.40	75.70	72.00	69.70	66.90	63.95	62.20	58.40	54.45	53.00	51.35
Finished Basement, **Add**	11.70	11.30	10.90	10.70	10.40	10.25	10.05	9.70	9.45	9.30	9.15
Unfinished Basement, **Add**	4.70	4.40	4.15	4.00	3.85	3.60	3.60	3.35	3.20	3.10	3.00

Modifications

Add to the total cost

Upgrade Kitchen Cabinets	$ + 1969
Solid Surface Countertops	+ 798
Full Bath - including plumbing, wall and floor finishes	+ 3528
Half Bath - including plumbing, wall and floor finishes	+ 2173
One Car Attached Garage	+ 6927
One Car Detached Garage	+ 7430
Fireplace & Chimney	+ 3590

Adjustments

For multi family - add to total cost

Additional Kitchen	$ + 3709
Additional Bath	+ 3528
Additional Entry & Exit	+ 969
Separate Heating	+ 1165
Separate Electric	+ 1184

For Townhouse/Rowhouse - Multiply cost per square foot by

Inner Unit	.92
End Unit	.96

Alternatives

Add to or deduct from the cost per square foot of living area

Cedar Shake Roof	$ + 1.05
Air Conditioning (in heating ductwork)	+ 1.30

Additional upgrades or components

Kitchen Cabinets & Countertops	Page 58
Bathroom Vanities	59
Fireplaces & Chimneys	59
Windows, Skylights & Dormers	59
Appliances	60
Breezeways & Porches	60
Finished Attic	60
Garages	61
Site Improvements	61
Wings & Ells	37

RESIDENTIAL — Average — Wings & Ells

1 Story — Base cost per square foot of living area

Exterior Wall	\multicolumn{8}{c}{Living Area}
	50	100	200	300	400	500	600	700
Wood Siding - Wood Frame	116.55	89.25	77.30	64.55	60.65	58.30	56.70	57.25
Brick Veneer - Wood Frame	124.00	93.00	79.55	64.25	59.80	57.20	55.40	55.80
Stucco on Wood Frame	117.50	89.85	77.80	64.85	60.90	58.55	56.95	57.45
Solid Masonry	147.40	111.25	95.65	76.75	71.65	68.55	67.40	67.45
Finished Basement, Add	32.30	26.10	23.50	19.15	18.30	17.80	17.45	17.20
Unfinished Basement, Add	15.70	11.95	10.35	7.75	7.25	6.90	6.70	6.55

1-1/2 Story — Base cost per square foot of living area

Exterior Wall	100	200	300	400	500	600	700	800
Wood Siding - Wood Frame	95.90	77.05	65.75	58.85	55.40	53.75	51.65	51.00
Brick Veneer - Wood Frame	124.95	95.05	79.25	69.20	64.40	61.90	59.05	58.10
Stucco on Wood Frame	114.50	86.65	72.30	63.75	59.40	57.15	54.55	53.60
Solid Masonry	141.05	107.85	89.95	77.55	72.10	69.15	65.95	64.90
Finished Basement, Add	21.90	19.30	17.55	15.65	15.15	14.80	14.45	14.40
Unfinished Basement, Add	9.90	8.35	7.30	6.15	5.80	5.60	5.40	5.40

2 Story — Base cost per square foot of living area

Exterior Wall	100	200	400	600	800	1000	1200	1400
Wood Siding - Wood Frame	96.30	71.90	61.25	50.55	47.05	44.95	43.55	44.20
Brick Veneer - Wood Frame	128.05	90.55	73.60	58.80	54.00	51.15	49.20	49.45
Stucco on Wood Frame	116.30	82.15	66.60	54.10	49.75	47.20	45.45	45.85
Solid Masonry	146.05	103.40	84.30	65.90	60.40	57.15	54.90	54.95
Finished Basement, Add	17.35	14.25	12.95	10.80	10.35	10.10	9.95	9.80
Unfinished Basement, Add	7.95	6.05	5.25	3.95	3.70	3.55	3.45	3.35

Base costs do not include bathroom or kitchen facilities. Use Modifications/Adjustments/Alternatives on pages 58-61 where appropriate.

Important: See the Reference Section for Location Factors (to adjust for your city) and Estimating Forms.

RESIDENTIAL | Custom | Building Types

1 Story

1-1/2 Story

2 Story

2-1/2 Story

Bi-Level

Tri-Level

RESIDENTIAL   Custom   Specifications

	Components
1 Site Work	Site preparation for slab or excavation for lower level; 4' deep trench excavation for foundation wall.
2 Foundations	Continuous reinforced concrete footing, 8" deep x 18" wide; dampproofed and insulated 8" thick reinforced concrete foundation wall, 4' deep; 4" concrete slab on 4" crushed stone base and polyethylene vapor barrier, trowel finish.
3 Framing	Exterior walls—2" x 6" wood studs, 16" O.C.; 1/2" plywood sheathing; 2" x 8" rafters 16" O.C. with 1/2" plywood sheathing, 4 in 12, 6 in 12 or 8 in 12 roof pitch, 2" x 6" or 2" x 8" ceiling joists; 1/2" plywood subfloor on 1" x 3" wood sleepers 16" O.C.; 2" x 10" floor joists with 5/8" plywood subfloor on models with more than one level.
4 Exterior Walls	Horizontal beveled wood siding and #15 felt building paper on insulated wood frame walls; double hung windows; 3 flush solid core wood exterior doors with storms. **Alternates:** • Brick veneer on wood frame, has 4" veneer high quality face brick or select common brick. • Stone veneer on wood frame has exterior veneer of field stone or 2" thick limestone. • Solid masonry has a 8" concrete block wall with insulation and a brick or stone facing. It may be a solid brick or stone structure.
5 Roofing	30 year asphalt shingles; #15 felt building paper; aluminum gutters, downspouts and drip edge; copper flashings.
6 Interiors	Walls and ceilings—5/8" drywall, skim coat plaster, primed and painted with 2 coats; hardwood baseboard and trim; sanded and finished, hardwood floor 70%, ceramic tile with 1/2" underlayment 20%, vinyl tile with 1/2" underlayment 10%; wood panel interior doors, primed and painted with 2 coats.
7 Specialties	Custom grade kitchen cabinets—20 L.F. wall and base with plastic laminate counter top and kitchen sink; 4 L.F. bathroom vanity; 75 gallon electric water heater; medicine cabinet.
8 Mechanical	Gas fired warm air heat/air conditioning; one full bath including bathtub, corner shower, built in lavatory and water closet; one 1/2 bath including built in lavatory and water closet.
9 Electrical	100 Amp. service; romex wiring; incandescent lighting fixtures, switches, receptacles.
10 Overhead and Profit	General Contractor overhead and profit.

Adjustments

Unfinished Basement:
7'-6" high cast-in-place concrete 10" thick or 12" concrete block.

Finished Basement:
Includes painted drywall on 2" x 4" wood furring with insulation, suspended ceiling, carpeting on concrete floor, heating and lighting.

RESIDENTIAL — Custom — 1 Story

- A distinct residence from designer's plans
- Single family — 1 full bath, 1 half bath, 1 kitchen
- No basement
- Asphalt shingles on roof
- Forced hot air heat/air conditioning
- Drywall interior finishes
- Materials and workmanship are above average
- Detail specifications on page 39

Note: The illustration shown may contain some optional components (for example: garages and/or fireplaces) whose costs are shown in the modifications, adjustments, & alternatives below or at the end of the square foot section.

© Design Basics, Inc.

Base cost per square foot of living area

Exterior Wall	800	1000	1200	1400	1600	1800	2000	2400	2800	3200	3600
Wood Siding - Wood Frame	112.45	102.55	94.35	88.25	83.60	80.95	78.20	72.30	68.30	65.35	62.15
Brick Veneer - Wood Frame	121.70	111.45	102.95	96.60	91.75	89.00	86.10	80.05	75.90	72.80	69.45
Stone Veneer - Wood Frame	126.05	115.40	106.55	99.80	94.80	91.95	88.85	82.55	78.20	74.90	71.45
Solid Masonry	126.35	115.65	106.75	100.00	95.00	92.15	89.05	82.75	78.35	75.05	71.55
Finished Basement, Add	28.10	26.75	25.50	24.55	23.90	23.55	23.05	22.30	21.75	21.25	20.85
Unfinished Basement, Add	11.80	11.15	10.50	10.05	9.70	9.50	9.25	8.85	8.60	8.35	8.10

Modifications

Add to the total cost

Upgrade Kitchen Cabinets	$ + 563
Solid Surface Countertops	+ 1140
Full Bath - including plumbing, wall and floor finishes	+ 4180
Half Bath - including plumbing, wall and floor finishes	+ 2575
Two Car Attached Garage	+ 11,236
Two Car Detached Garage	+ 11,684
Fireplace & Chimney	+ 3380

Adjustments

For multi family - add to total cost

Additional Kitchen	$ + 7694
Additional Full Bath & Half Bath	+ 6755
Additional Entry & Exit	+ 969
Separate Heating & Air Conditioning	+ 3695
Separate Electric	+ 1184

For Townhouse/Rowhouse - Multiply cost per square foot by

Inner Unit	.90
End Unit	.95

Alternatives

Add to or deduct from the cost per square foot of living area

Cedar Shake Roof	$+ 1.75
Clay Tile Roof	+ 4.00
Slate Roof	+ 6.95
Upgrade Ceilings to Textured Finish	+ .41

Additional upgrades or components

Kitchen Cabinets & Countertops	Page 58
Bathroom Vanities	59
Fireplaces & Chimneys	59
Windows, Skylights & Dormers	59
Appliances	60
Breezeways & Porches	60
Finished Attic	60
Garages	61
Site Improvements	61
Wings & Ells	47

Important: See the Reference Section for Location Factors (to adjust for your city) and Estimating Forms

RESIDENTIAL | Custom | 1-1/2 Story

- A distinct residence from designer's plans
- Single family — 1 full bath, 1 half bath, 1 kitchen
- No basement
- Asphalt shingles on roof
- Forced hot air heat/air conditioning
- Drywall interior finishes
- Materials and workmanship are above average
- Detail specifications on page 39

Note: The illustration shown may contain some optional components (for example: garages and/or fireplaces) whose costs are shown in the modifications, adjustments, & alternatives below or at the end of the square foot section.

Base cost per square foot of living area

Exterior Wall	\multicolumn{11}{c}{Living Area}
	1000	1200	1400	1600	1800	2000	2400	2800	3200	3600	4000
Wood Siding - Wood Frame	103.30	97.25	92.85	86.50	83.35	79.60	73.00	70.35	67.35	65.15	61.95
Brick Veneer - Wood Frame	107.15	100.85	96.30	89.65	86.30	82.45	75.50	72.70	69.55	67.20	63.90
Stone Veneer - Wood Frame	111.70	105.10	100.35	93.35	89.80	85.80	78.45	75.50	72.10	69.70	66.25
Solid Masonry	111.95	105.40	100.65	93.60	90.05	86.00	78.65	75.70	72.30	69.85	66.40
Finished Basement, **Add**	18.70	18.00	17.45	16.70	16.30	15.95	15.20	14.80	14.40	14.20	13.90
Unfinished Basement, **Add**	7.95	7.60	7.35	6.95	6.75	6.55	6.20	6.00	5.80	5.70	5.50

Modifications

Add to the total cost

Upgrade Kitchen Cabinets	+ 563
Solid Surface Countertops	+ 1140
Full Bath - including plumbing, wall and floor finishes	+ 4180
Half Bath - including plumbing, wall and floor finishes	+ 2575
Two Car Attached Garage	+ 11,236
Two Car Detached Garage	+ 11,684
Fireplace & Chimney	+ 3380

Adjustments

For multi family - add to total cost

Additional Kitchen	$ + 7694
Additional Full Bath & Half Bath	+ 6755
Additional Entry & Exit	+ 969
Separate Heating & Air Conditioning	+ 3695
Separate Electric	+ 1184

*For Townhouse/Rowhouse -
Multiply cost per square foot by*

Inner Unit	.90
End Unit	.95

Alternatives

Add to or deduct from the cost per square foot of living area

Cedar Shake Roof	$+ 1.25
Clay Tile Roof	+ 2.90
Slate Roof	+ 5.00
Upgrade Ceilings to Textured Finish	+ .41

Additional upgrades or components

Kitchen Cabinets & Countertops	Page 58
Bathroom Vanities	59
Fireplaces & Chimneys	59
Windows, Skylights & Dormers	59
Appliances	60
Breezeways & Porches	60
Finished Attic	60
Garages	61
Site Improvements	61
Wings & Ells	47

Important: See the Reference Section for Location Factors (to adjust for your city) and Estimating Forms

RESIDENTIAL | Custom | 2 Story

- A distinct residence from designer's plans
- Single family — 1 full bath, 1 half bath, 1 kitchen
- No basement
- Asphalt shingles on roof
- Forced hot air heat/air conditioning
- Drywall interior finishes
- Materials and workmanship are above average
- Detail specifications on page 39

Note: The illustration shown may contain some optional components (for example: garages and/or fireplaces) whose costs are shown in the modifications, adjustments, & alternatives below or at the end of the square foot section.

Base cost per square foot of living area

Exterior Wall	1200	1400	1600	1800	2000	2400	2800	3200	3600	4000	4400
Wood Siding - Wood Frame	98.95	93.80	90.25	86.15	82.80	76.80	72.40	69.00	66.85	64.95	63.00
Brick Veneer - Wood Frame	103.05	97.65	93.90	89.60	86.15	79.80	75.15	71.55	69.35	67.25	65.25
Stone Veneer - Wood Frame	107.90	102.20	98.25	93.70	90.05	83.40	78.35	74.60	72.25	70.00	67.90
Solid Masonry	108.20	102.50	98.55	93.95	90.30	83.60	78.60	74.80	72.45	70.20	68.10
Finished Basement, Add	15.00	14.45	14.05	13.65	13.40	12.80	12.30	11.95	11.80	11.50	11.35
Unfinished Basement, Add	6.40	6.10	5.90	5.70	5.60	5.30	5.00	4.85	4.75	4.65	4.55

Modifications

Add to the total cost

Upgrade Kitchen Cabinets	+ 563
Solid Surface Countertops	+ 1140
Full Bath - including plumbing, wall and floor finishes	+ 4180
Half Bath - including plumbing, wall and floor finishes	+ 2575
Two Car Attached Garage	+ 11,236
Two Car Detached Garage	+ 11,684
Fireplace & Chimney	+ 3815

Adjustments

For multi family - add to total cost

Additional Kitchen	$ + 7694
Additional Full Bath & Half Bath	+ 6755
Additional Entry & Exit	+ 969
Separate Heating & Air Conditioning	+ 3695
Separate Electric	+ 1184

For Townhouse/Rowhouse - Multiply cost per square foot by

Inner Unit	.87
End Unit	.93

Alternatives

Add to or deduct from the cost per square foot of living area

Cedar Shake Roof	$+.85
Clay Tile Roof	+ 2.00
Slate Roof	+ 3.50
Upgrade Ceilings to Textured Finish	+ .41

Additional upgrades or components

Kitchen Cabinets & Countertops	Page 58
Bathroom Vanities	59
Fireplaces & Chimneys	59
Windows, Skylights & Dormers	59
Appliances	60
Breezeways & Porches	60
Finished Attic	60
Garages	61
Site Improvements	61
Wings & Ells	47

RESIDENTIAL — Custom — 2-1/2 Story

- A distinct residence from designer's plans
- Single family — 1 full bath, 1 half bath, 1 kitchen
- No basement
- Asphalt shingles on roof
- Forced hot air heat/air conditioning
- Drywall interior finishes
- Materials and workmanship are above average
- Detail specifications on page 39

Note: The illustration shown may contain some optional components (for example: garages and/or fireplaces) whose costs are shown in the modifications, adjustments, & alternatives below or at the end of the square foot section.

Base cost per square foot of living area

Exterior Wall	1500	1800	2100	2400	2800	3200	3600	4000	4500	5000	5500
Wood Siding - Wood Frame	100.25	89.70	84.35	80.55	76.75	72.10	69.95	66.00	64.15	62.10	60.05
Brick Veneer - Wood Frame	104.55	93.65	87.90	83.90	79.95	75.00	72.70	68.55	66.55	64.40	62.25
Stone Veneer - Wood Frame	109.60	98.35	92.05	87.85	83.75	78.40	75.95	71.55	69.45	67.10	64.80
Solid Masonry	109.95	98.65	92.35	88.10	84.00	78.65	76.20	71.75	69.60	67.30	65.00
Finished Basement, Add	11.95	11.40	10.80	10.45	10.20	9.75	9.55	9.25	9.10	8.90	8.75
Unfinished Basement, Add	5.15	4.85	4.55	4.40	4.25	4.05	3.90	3.80	3.70	3.60	3.50

Modifications

Add to the total cost

Upgrade Kitchen Cabinets	+ 563
Solid Surface Countertops	+ 1140
Full Bath - including plumbing, wall and floor finishes	+ 4180
Half Bath - including plumbing, wall and floor finishes	+ 2575
Two Car Attached Garage	+ 11,236
Two Car Detached Garage	+ 11,684
Fireplace & Chimney	+ 4310

Adjustments

For multi family - add to total cost

Additional Kitchen	$ + 7694
Additional Full Bath & Half Bath	+ 6755
Additional Entry & Exit	+ 969
Separate Heating & Air Conditioning	+ 3695
Separate Electric	+ 1184

For Townhouse/Rowhouse - Multiply cost per square foot by

Inner Unit	.87
End Unit	.94

Alternatives

Add to or deduct from the cost per square foot of living area

Cedar Shake Roof	$+.75
Clay Tile Roof	+ 1.75
Slate Roof	+ 3.00
Upgrade Ceilings to Textured Finish	+ .41

Additional upgrades or components

Kitchen Cabinets & Countertops	Page 58
Bathroom Vanities	59
Fireplaces & Chimneys	59
Windows, Skylights & Dormers	59
Appliances	60
Breezeways & Porches	60
Finished Attic	60
Garages	61
Site Improvements	61
Wings & Ells	47

Important: See the Reference Section for Location Factors (to adjust for your city) and Estimating Forms

RESIDENTIAL | Custom | 3 Story

- A distinct residence from designer's plans
- Single family — 1 full bath, 1 half bath, 1 kitchen
- No basement
- Asphalt shingles on roof
- Forced hot air heat/air conditioning
- Drywall interior finishes
- Materials and workmanship are above average
- Detail specifications on page 39

Note: The illustration shown may contain some optional components (for example: garages and/or fireplaces) whose costs are shown in the modifications, adjustments, & alternatives below or at the end of the square foot section.

Base cost per square foot of living area

Exterior Wall	1500	1800	2100	2500	3000	3500	4000	4500	5000	5500	6000
Wood Siding - Wood Frame	100.30	89.90	85.70	82.40	76.20	73.45	69.25	65.10	63.85	62.00	60.50
Brick Veneer - Wood Frame	104.70	94.00	89.55	86.10	79.50	76.60	72.10	67.75	66.40	64.45	62.80
Stone Veneer - Wood Frame	109.95	98.90	94.10	90.45	83.45	80.35	75.50	70.85	69.45	67.40	65.55
Solid Masonry	110.30	99.20	94.40	90.70	83.70	80.60	75.75	71.10	69.65	67.55	65.75
Finished Basement, Add	10.50	10.00	9.60	9.35	8.90	8.65	8.35	8.05	7.95	7.85	7.70
Unfinished Basement, Add	4.50	4.25	4.05	3.95	3.70	3.55	3.40	3.30	3.20	3.15	3.05

Modifications

Add to the total cost

Upgrade Kitchen Cabinets	+ 563
Solid Surface Countertops	+ 1140
Full Bath - including plumbing, wall and floor finishes	+ 4180
Half Bath - including plumbing, wall and floor finishes	+ 2575
Two Car Attached Garage	+ 11,236
Two Car Detached Garage	+ 11,684
Fireplace & Chimney	+ 4310

Adjustments

For multi family - add to total cost

Additional Kitchen	$ + 7694
Additional Full Bath & Half Bath	+ 6755
Additional Entry & Exit	+ 969
Separate Heating & Air Conditioning	+ 3695
Separate Electric	+ 1184

For Townhouse/Rowhouse -
Multiply cost per square foot by

Inner Unit	.85
End Unit	.93

Alternatives

Add to or deduct from the cost per square foot of living area

Cedar Shake Roof	$+ .60
Clay Tile Roof	+ 1.35
Slate Roof	+ 2.30
Upgrade Ceilings to Textured Finish	+ .41

Additional upgrades or components

Kitchen Cabinets & Countertops	Page 58
Bathroom Vanities	59
Fireplaces & Chimneys	59
Windows, Skylights & Dormers	59
Appliances	60
Breezeways & Porches	60
Finished Attic	60
Garages	61
Site Improvements	61
Wings & Ells	47

Important: See the Reference Section for Location Factors (to adjust for your city) and Estimating Forms

RESIDENTIAL — Custom — Bi-Level

- A distinct residence from designer's plans
- Single family — 1 full bath, 1 half bath, 1 kitchen
- No basement
- Asphalt shingles on roof
- Forced hot air heat/air conditioning
- Drywall interior finishes
- Materials and workmanship are above average
- Detail specifications on page 39

Note: The illustration shown may contain some optional components (for example: garages and/or fireplaces) whose costs are shown in the modifications, adjustments, & alternatives below or at the end of the square foot section.

CUSTOM

Base cost per square foot of living area

Exterior Wall	1200	1400	1600	1800	2000	2400	2800	3200	3600	4000	4400
Wood Siding - Wood Frame	91.95	87.30	84.00	80.30	77.10	71.80	67.75	64.60	62.70	61.00	59.15
Brick Veneer - Wood Frame	95.05	90.20	86.75	82.90	79.60	74.05	69.80	66.55	64.60	62.75	60.85
Stone Veneer - Wood Frame	98.70	93.60	90.05	85.95	82.55	76.70	72.25	68.80	66.75	64.80	62.85
Solid Masonry	98.90	93.85	90.25	86.15	82.80	76.85	72.40	68.95	66.90	64.90	62.95
Finished Basement, **Add**	15.00	14.45	14.05	13.65	13.40	12.80	12.30	11.95	11.80	11.50	11.35
Unfinished Basement, **Add**	6.40	6.10	5.90	5.70	5.60	5.30	5.00	4.85	4.75	4.65	4.55

Modifications

Add to the total cost

Upgrade Kitchen Cabinets	+ 563
Solid Surface Countertops	+ 1140
Full Bath - including plumbing, wall and floor finishes	+ 4180
Half Bath - including plumbing, wall and floor finishes	+ 2575
Two Car Attached Garage	+ 11,236
Two Car Detached Garage	+ 11,684
Fireplace & Chimney	+ 3380

Adjustments

For multi family - add to total cost

Additional Kitchen	$ + 7694
Additional Full Bath & Half Bath	+ 6755
Additional Entry & Exit	+ 969
Separate Heating & Air Conditioning	+ 3695
Separate Electric	+ 1184

For Townhouse/Rowhouse - Multiply cost per square foot by

Inner Unit	.89
End Unit	.95

Alternatives

Add to or deduct from the cost per square foot of living area

Cedar Shake Roof	$+ .85
Clay Tile Roof	+ 2.00
Slate Roof	+ 3.50
Upgrade Ceilings to Textured Finish	+ .41

Additional upgrades or components

Kitchen Cabinets & Countertops	Page 58
Bathroom Vanities	59
Fireplaces & Chimneys	59
Windows, Skylights & Dormers	59
Appliances	60
Breezeways & Porches	60
Finished Attic	60
Garages	61
Site Improvements	61
Wings & Ells	47

Important: See the Reference Section for Location Factors (to adjust for your city) and Estimating Forms

RESIDENTIAL | Custom | Tri-Level

- A distinct residence from designer's plans
- Single family — 1 full bath, 1 half bath, 1 kitchen
- No basement
- Asphalt shingles on roof
- Forced hot air heat/air conditioning
- Drywall interior finishes
- Materials and workmanship are above average
- Detail specifications on page 39

Note: The illustration shown may contain some optional components (for example: garages and/or fireplaces) whose costs are shown in the modifications, adjustments, & alternatives below or at the end of the square foot section.

Base cost per square foot of living area

Exterior Wall	1200	1500	1800	2100	2400	2800	3200	3600	4000	4500	5000
Wood Siding - Wood Frame	94.30	86.15	79.80	75.00	71.30	68.60	65.40	62.15	60.95	57.45	55.65
Brick Veneer - Wood Frame	97.40	88.90	82.30	77.30	73.45	70.65	67.30	63.90	62.65	59.05	57.15
Stone Veneer - Wood Frame	101.00	92.20	85.25	80.00	75.95	73.10	69.60	66.00	64.65	60.90	58.90
Solid Masonry	101.25	92.45	85.45	80.15	76.10	73.25	69.75	66.15	64.75	61.05	59.00
Finished Basement, **Add***	18.70	17.80	17.00	16.35	15.95	15.65	15.20	14.85	14.70	14.35	14.10
Unfinished Basement, **Add***	7.90	7.45	7.00	6.70	6.45	6.30	6.10	5.90	5.85	5.65	5.50

*Basement under middle level only.

Modifications
Add to the total cost

Upgrade Kitchen Cabinets	+ 563
Solid Surface Countertops	+ 1140
Full Bath - including plumbing, wall and floor finishes	+ 4180
Half Bath - including plumbing, wall and floor finishes	+ 2575
Two Car Attached Garage	+ 11,236
Two Car Detached Garage	+ 11,684
Fireplace & Chimney	+ 3815

Adjustments
For multi family - add to total cost

Additional Kitchen	$ + 7694
Additional Full Bath & Half Bath	+ 6755
Additional Entry & Exit	+ 969
Separate Heating & Air Conditioning	+ 3695
Separate Electric	+ 1184

For Townhouse/Rowhouse -
Multiply cost per square foot by

Inner Unit	.87
End Unit	.94

Alternatives
Add to or deduct from the cost per square foot of living area

Cedar Shake Roof	$ + 1.25
Clay Tile Roof	+ 2.90
Slate Roof	+ 5.00
Upgrade Ceilings to Textured Finish	+ .41

Additional upgrades or components

Kitchen Cabinets & Countertops	Page 58
Bathroom Vanities	59
Fireplaces & Chimneys	59
Windows, Skylights & Dormers	59
Appliances	60
Breezeways & Porches	60
Finished Attic	60
Garages	61
Site Improvements	61
Wings & Ells	47

RESIDENTIAL | Custom | Wings & Ells

1 Story — Base cost per square foot of living area

Exterior Wall	Living Area							
	50	100	200	300	400	500	600	700
Wood Siding - Wood Frame	146.50	110.75	95.25	78.00	72.90	69.85	67.80	68.35
Brick Veneer - Wood Frame	156.90	118.15	101.40	82.10	76.60	73.30	71.05	71.55
Stone Veneer - Wood Frame	169.15	126.95	108.70	87.00	81.00	77.40	74.95	75.30
Solid Masonry	169.95	127.50	109.20	87.30	81.30	77.65	75.25	75.55
Finished Basement, Add	45.75	37.10	33.50	27.50	26.30	25.60	25.10	24.75
Unfinished Basement, Add	32.40	22.40	17.85	13.60	12.50	11.85	11.45	11.10

1-1/2 Story — Base cost per square foot of living area

Exterior Wall	Living Area							
	100	200	300	400	500	600	700	800
Wood Siding - Wood Frame	120.05	96.55	82.15	72.85	68.45	66.30	63.65	62.90
Brick Veneer - Wood Frame	129.30	103.95	88.35	77.65	72.90	70.55	67.65	66.85
Stone Veneer - Wood Frame	140.25	112.75	95.65	83.35	78.15	75.50	72.35	71.50
Solid Masonry	141.00	113.30	96.15	83.75	78.50	75.80	72.65	71.80
Finished Basement, Add	30.40	26.80	24.40	21.75	21.05	20.55	20.10	20.05
Unfinished Basement, Add	19.40	14.80	12.55	10.70	10.00	9.60	9.20	9.10

2 Story — Base cost per square foot of living area

Exterior Wall	Living Area							
	100	200	400	600	800	1000	1200	1400
Wood Siding - Wood Frame	122.25	90.60	76.75	62.35	57.85	55.10	53.30	54.00
Brick Veneer - Wood Frame	132.60	98.00	82.95	66.45	61.55	58.55	56.55	57.20
Stone Veneer - Wood Frame	144.90	106.75	90.25	71.35	65.90	62.65	60.45	60.95
Solid Masonry	145.70	107.35	90.75	71.65	66.20	62.90	60.75	61.20
Finished Basement, Add	22.90	18.55	16.75	13.75	13.15	12.80	12.60	12.40
Unfinished Basement, Add	16.20	11.20	8.90	6.80	6.25	5.95	5.70	5.55

Base costs do not include bathroom or kitchen facilities. Use Modifications/Adjustments/Alternatives on pages 58-61 where appropriate.

Important: See the Reference Section for Location Factors (to adjust for your city) and Estimating Forms.

RESIDENTIAL | Luxury | Building Types

1 Story

1-1/2 Story

2 Story

2-1/2 Story

Bi-Level

Tri-Level

RESIDENTIAL Luxury Specifications

	Components
1 Site Work	Site preparation for slab or excavation for lower level; 4' deep trench excavation for foundation wall.
2 Foundations	Continuous reinforced concrete footing, 8" deep x 18" wide; dampproofed and insulated 12" thick reinforced concrete foundation wall, 4' deep; 4" concrete slab on 4" crushed stone base and polyethylene vapor barrier, trowel finish.
3 Framing	Exterior walls—2" x 6" wood studs, 16" O.C.; 1/2" plywood sheathing; 2" x 8" rafters 16" O.C. with 1/2" plywood sheathing, 4 in 12, 6 in 12 or 8 in 12 roof pitch, 2" x 6" or 2" x 8" ceiling joists; 1/2" plywood subfloor on 1" x 3" wood sleepers 16" O.C.; 2" x 10" floor joists with 5/8" plywood subfloor on models with more than one level.
4 Exterior Walls	Face brick veneer and #15 felt building paper on insulated wood frame walls; double hung windows; 3 flush solid core wood exterior doors with storms. **Alternates:** • Wood siding on wood frame, has top quality cedar or redwood siding or hand split cedar shingles or shakes. • Solid brick may have solid brick exterior wall or brick on concrete block. • Solid stone concrete block with selected fieldstone or limestone exterior.
5 Roofing	Red cedar shingles; #15 felt building paper; aluminum gutters, downspouts and drip edge; copper flashings.
6 Interiors	Walls and ceilings—5/8" drywall, skim coat plaster, primed and painted with 2 coats; hardwood baseboard and trim; sanded and finished, hardwood floor 70%, ceramic tile with 1/2" underlayment 20%, vinyl tile with 1/2" underlayment 10%; wood panel interior doors, primed and painted with 2 coats.
7 Specialties	Luxury grade kitchen cabinets—25 L.F. wall and base with plastic laminate counter top and kitchen sink; 6 L.F. bathroom vanity; 75 gallon electric water heater; medicine cabinet.
8 Mechanical	Gas fired warm air heat/air conditioning; one full bath including bathtub, corner shower, built in lavatory and water closet; one 1/2 bath including built in lavatory and water closet.
9 Electrical	100 Amp. service; romex wiring; incandescent lighting fixtures, switches, receptacles.
10 Overhead and Profit	General Contractor overhead and profit.

Adjustments

Unfinished Basement:
8" high cast-in-place concrete walls 12" thick or 12" concrete block.

Finished Basement:
Includes painted drywall on 2" x 4" wood furring with insulation, suspended ceiling, carpeting on subfloor with sleepers, heating and lighting.

RESIDENTIAL | Luxury | 1 Story

- Unique residence built from an architect's plan
- Single family — 1 full bath, 1 half bath, 1 kitchen
- No basement
- Cedar shakes on roof
- Forced hot air heat/air conditioning
- Double drywall interior
- Many special features
- Extraordinary materials and workmanship
- Detail specifications on page 49

Note: The illustration shown may contain some optional components (for example: garages and/or fireplaces) whose costs are shown in the modifications, adjustments, & alternatives below or at the end of the square foot section.

Base cost per square foot of living area

Exterior Wall	1000	1200	1400	1600	1800	2000	2400	2800	3200	3600	4000
Wood Siding - Wood Frame	126.65	117.15	110.05	104.65	101.50	98.30	91.55	86.90	83.55	79.90	76.90
Brick Veneer - Wood Frame	130.00	120.15	112.80	107.25	104.00	100.60	93.65	88.90	85.35	81.60	78.50
Solid Brick	137.25	126.70	118.75	112.80	109.35	105.65	98.30	93.15	89.30	85.30	81.95
Solid Stone	138.85	128.15	120.05	114.05	110.60	106.80	99.35	94.10	90.20	86.05	82.70
Finished Basement, **Add**	32.75	31.20	29.95	29.20	28.75	28.10	27.15	26.50	25.80	25.30	24.85
Unfinished Basement, **Add**	12.55	11.85	11.30	10.95	10.75	10.40	10.00	9.65	9.40	9.10	8.90

Modifications

Add to the total cost

Upgrade Kitchen Cabinets	$+ 744
Solid Surface Countertops	+ 1282
Full Bath - including plumbing, wall and floor finishes	+ 4849
Half Bath - including plumbing, wall and floor finishes	+ 2987
Two Car Attached Garage	+ 16,806
Two Car Detached Garage	+ 16,685
Fireplace & Chimney	+ 4800

Adjustments

For multi family - add to total cost

Additional Kitchen	$ + 9837
Additional Full Bath & Half Bath	+ 7836
Additional Entry & Exit	+ 1499
Separate Heating & Air Conditioning	+ 3995
Separate Electric	+ 1184

For Townhouse/Rowhouse -
Multiply cost per square foot by

Inner Unit	.90
End Unit	.95

Alternatives

Add to or deduct from the cost per square foot of living area

Heavyweight Asphalt Shingles	$− 1.75
Clay Tile Roof	+ 2.30
Slate Roof	+ 5.25
Upgrade Ceilings to Textured Finish	+ .41

Additional upgrades or components

Kitchen Cabinets & Countertops	Page 58
Bathroom Vanities	59
Fireplaces & Chimneys	59
Windows, Skylights & Dormers	59
Appliances	60
Breezeways & Porches	60
Finished Attic	60
Garages	61
Site Improvements	61
Wings & Ells	57

RESIDENTIAL | Luxury | 1-1/2 Story

- Unique residence built from an architect's plan
- Single family — 1 full bath, 1 half bath, 1 kitchen
- No basement
- Cedar shakes on roof
- Forced hot air heat/air conditioning
- Double drywall interior
- Many special features
- Extraordinary materials and workmanship
- Detail specifications on page 49

Note: The illustration shown may contain some optional components (for example: garages and/or fireplaces) whose costs are shown in the modifications, adjustments, & alternatives below or at the end of the square foot section.

© Larry E. Belk Designs

Base cost per square foot of living area

Exterior Wall	1000	1200	1400	1600	1800	2000	2400	2800	3200	3600	4000
Wood Siding - Wood Frame	119.20	111.95	106.70	99.45	95.75	91.45	83.85	80.75	77.30	74.70	71.10
Brick Veneer - Wood Frame	123.05	115.60	110.20	102.60	98.70	94.25	86.35	83.10	79.50	76.80	73.05
Solid Brick	131.45	123.45	117.65	109.40	105.15	100.35	91.75	88.20	84.25	81.35	77.30
Solid Stone	133.35	125.20	119.35	110.95	106.60	101.80	92.95	89.40	85.30	82.40	78.30
Finished Basement, Add	22.95	22.05	21.40	20.50	19.95	19.50	18.55	18.10	17.55	17.30	16.85
Unfinished Basement, Add	8.95	8.55	8.30	7.85	7.60	7.40	6.95	6.75	6.55	6.40	6.20

Modifications

Add to the total cost

Upgrade Kitchen Cabinets	$+ 744
Solid Surface Countertops	+ 1282
Full Bath - including plumbing, wall and floor finishes	+ 4849
Half Bath - including plumbing, wall and floor finishes	+ 2987
Two Car Attached Garage	+ 16,806
Two Car Detached Garage	+ 16,685
Fireplace & Chimney	+ 4800

Adjustments

For multi family - add to total cost

Additional Kitchen	$ + 9837
Additional Full Bath & Half Bath	+ 7836
Additional Entry & Exit	+ 1499
Separate Heating & Air Conditioning	+ 3995
Separate Electric	+ 1184

*For Townhouse/Rowhouse -
Multiply cost per square foot by*

Inner Unit	.90
End Unit	.95

Alternatives

Add to or deduct from the cost per square foot of living area

Heavyweight Asphalt Shingles	$- 1.25
Clay Tile Roof	+ 1.65
Slate Roof	+ 3.75
Upgrade Ceilings to Textured Finish	+ .41

Additional upgrades or components

Kitchen Cabinets & Countertops	Page 58
Bathroom Vanities	59
Fireplaces & Chimneys	59
Windows, Skylights & Dormers	59
Appliances	60
Breezeways & Porches	60
Finished Attic	60
Garages	61
Site Improvements	61
Wings & Ells	57

Important: See the Reference Section for Location Factors (to adjust for your city) and Estimating Forms

RESIDENTIAL — Luxury — 2 Story

- Unique residence built from an architect's plan
- Single family — 1 full bath, 1 half bath, 1 kitchen
- No basement
- Cedar shakes on roof
- Forced hot air heat/air conditioning
- Double drywall interior
- Many special features
- Extraordinary materials and workmanship
- Detail specifications on page 49

Note: The illustration shown may contain some optional components (for example: garages and/or fireplaces) whose costs are shown in the modifications, adjustments, & alternatives below or at the end of the square foot section.

Base cost per square foot of living area

Exterior Wall	1200	1400	1600	1800	2000	2400	2800	3200	3600	4000	4400
Wood Siding - Wood Frame	112.85	106.85	102.70	98.05	94.10	87.30	82.30	78.40	75.95	73.75	71.55
Brick Veneer - Wood Frame	117.00	110.75	106.40	101.50	97.45	90.30	85.05	80.95	78.45	76.10	73.85
Solid Brick	125.95	119.10	114.40	109.00	104.70	96.85	91.00	86.55	83.80	81.15	78.70
Solid Stone	127.95	120.95	116.20	110.70	106.35	98.35	92.30	87.80	85.05	82.30	79.80
Finished Basement, **Add**	18.45	17.70	17.25	16.70	16.40	15.65	15.05	14.60	14.40	14.05	13.85
Unfinished Basement, **Add**	7.15	6.85	6.65	6.40	6.30	5.95	5.65	5.45	5.35	5.20	5.15

Modifications

Add to the total cost

Upgrade Kitchen Cabinets	$+ 744
Solid Surface Countertops	+ 1282
Full Bath - including plumbing, wall and floor finishes	+ 4849
Half Bath - including plumbing, wall and floor finishes	+ 2987
Two Car Attached Garage	+ 16,806
Two Car Detached Garage	+ 16,685
Fireplace & Chimney	+ 5255

Adjustments

For multi family - add to total cost

Additional Kitchen	$ + 9837
Additional Full Bath & Half Bath	+ 7836
Additional Entry & Exit	+ 1499
Separate Heating & Air Conditioning	+ 3995
Separate Electric	+ 1184

For Townhouse/Rowhouse -
Multiply cost per square foot by

Inner Unit	.86
End Unit	.93

Alternatives

Add to or deduct from the cost per square foot of living area

Heavyweight Asphalt Shingles	$− .85
Clay Tile Roof	+ 1.15
Slate Roof	+ 2.60
Upgrade Ceilings to Textured Finish	+ .41

Additional upgrades or components

Kitchen Cabinets & Countertops	Page 58
Bathroom Vanities	59
Fireplaces & Chimneys	59
Windows, Skylights & Dormers	59
Appliances	60
Breezeways & Porches	60
Finished Attic	60
Garages	61
Site Improvements	61
Wings & Ells	57

Important: See the Reference Section for Location Factors (to adjust for your city) and Estimating Forms

RESIDENTIAL — Luxury — 2-1/2 Story

- Unique residence built from an architect's plan
- Single family — 1 full bath, 1 half bath, 1 kitchen
- No basement
- Cedar shakes on roof
- Forced hot air heat/air conditioning
- Double drywall interior
- Many special features
- Extraordinary materials and workmanship
- Detail specifications on page 49

Note: The illustration shown may contain some optional components (for example: garages and/or fireplaces) whose costs are shown in the modifications, adjustments, & alternatives below or at the end of the square foot section.

© Larry W. Garnett & Associates, Inc.

LUXURY

Base cost per square foot of living area

Exterior Wall	1500	1800	2100	2500	3000	3500	4000	4500	5000	5500	6000
Wood Siding - Wood Frame	113.45	101.40	95.35	90.30	84.40	79.25	74.45	72.35	70.00	67.70	65.75
Brick Veneer - Wood Frame	117.70	105.35	98.85	93.65	87.45	82.00	77.00	74.80	72.35	69.90	67.85
Solid Brick	127.00	113.95	106.50	100.85	94.00	88.00	82.55	80.05	77.35	74.65	72.40
Solid Stone	129.15	115.85	108.25	102.50	95.50	89.35	83.80	81.20	78.45	75.70	73.45
Finished Basement, **Add**	14.65	14.00	13.25	12.80	12.25	11.75	11.35	11.10	10.90	10.70	10.55
Unfinished Basement, **Add**	5.75	5.45	5.10	4.90	4.65	4.45	4.25	4.15	4.05	3.95	3.90

Modifications

Add to the total cost

Upgrade Kitchen Cabinets	$+ 744
Solid Surface Countertops	+ 1282
Full Bath - including plumbing, wall and floor finishes	+ 4849
Half Bath - including plumbing, wall and floor finishes	+ 2987
Two Car Attached Garage	+ 16,806
Two Car Detached Garage	+ 16,685
Fireplace & Chimney	+ 5755

Adjustments

For multi family - add to total cost

Additional Kitchen	$ + 9837
Additional Full Bath & Half Bath	+ 7836
Additional Entry & Exit	+ 1499
Separate Heating & Air Conditioning	+ 3995
Separate Electric	+ 1184

For Townhouse/Rowhouse - Multiply cost per square foot by

Inner Unit	.86
End Unit	.93

Alternatives

Add to or deduct from the cost per square foot of living area

Heavyweight Asphalt Shingles	$- .75
Clay Tile Roof	+ 1.00
Slate Roof	+ 2.25
Upgrade Ceilings to Textured Finish	+ .41

Additional upgrades or components

Kitchen Cabinets & Countertops	Page 58
Bathroom Vanities	59
Fireplaces & Chimneys	59
Windows, Skylights & Dormers	59
Appliances	60
Breezeways & Porches	60
Finished Attic	60
Garages	61
Site Improvements	61
Wings & Ells	57

Important: See the Reference Section for Location Factors (to adjust for your city) and Estimating Forms

RESIDENTIAL — Luxury — 3 Story

- Unique residence built from an architect's plan
- Single family — 1 full bath, 1 half bath, 1 kitchen
- No basement
- Cedar shakes on roof
- Forced hot air heat/air conditioning
- Double drywall interior
- Many special features
- Extraordinary materials and workmanship
- Detail specifications on page 49

Note: The illustration shown may contain some optional components (for example: garages and/or fireplaces) whose costs are shown in the modifications, adjustments, & alternatives below or at the end of the square foot section.

Base cost per square foot of living area

Exterior Wall	1500	1800	2100	2500	3000	3500	4000	4500	5000	5500	6000
Wood Siding - Wood Frame	113.20	101.30	96.50	92.65	85.60	82.50	77.80	73.15	71.70	69.60	67.90
Brick Veneer - Wood Frame	117.65	105.45	100.40	96.35	88.95	85.70	80.65	75.80	74.25	72.05	70.25
Solid Brick	127.30	114.40	108.75	104.35	96.20	92.55	86.90	81.55	79.85	77.45	75.30
Solid Stone	129.45	116.40	110.60	106.15	97.85	94.10	88.30	82.85	81.10	78.60	76.45
Finished Basement, **Add**	12.90	12.30	11.80	11.50	10.95	10.60	10.20	9.90	9.75	9.55	9.40
Unfinished Basement, **Add**	5.05	4.80	4.55	4.45	4.20	4.05	3.85	3.70	3.65	3.55	3.50

Modifications

Add to the total cost

Upgrade Kitchen Cabinets	$+ 744
Solid Surface Countertops	+ 1282
Full Bath - including plumbing, wall and floor finishes	+ 4849
Half Bath - including plumbing, wall and floor finishes	+ 2987
Two Car Attached Garage	+ 16,806
Two Car Detached Garage	+ 16,685
Fireplace & Chimney	+ 5755

Adjustments

For multi family - add to total cost

Additional Kitchen	$ + 9837
Additional Full Bath & Half Bath	+ 7836
Additional Entry & Exit	+ 1499
Separate Heating & Air Conditioning	+ 3995
Separate Electric	+ 1184

For Townhouse/Rowhouse -
Multiply cost per square foot by

Inner Unit	.84
End Unit	.92

Alternatives

Add to or deduct from the cost per square foot of living area

Heavyweight Asphalt Shingles	$ − .60
Clay Tile Roof	+ .75
Slate Roof	+ 1.75
Upgrade Ceilings to Textured Finish	+ .41

Additional upgrades or components

Kitchen Cabinets & Countertops	Page 58
Bathroom Vanities	59
Fireplaces & Chimneys	59
Windows, Skylights & Dormers	59
Appliances	60
Breezeways & Porches	60
Finished Attic	60
Garages	61
Site Improvements	61
Wings & Ells	57

RESIDENTIAL | Luxury | Bi-Level

- Unique residence built from an architect's plan
- Single family — 1 full bath, 1 half bath, 1 kitchen
- No basement
- Cedar shakes on roof
- Forced hot air heat/air conditioning
- Double drywall interior
- Many special features
- Extraordinary materials and workmanship
- Detail specifications on page 49

Note: The illustration shown may contain some optional components (for example: garages and/or fireplaces) whose costs are shown in the modifications, adjustments, & alternatives below or at the end of the square foot section.

Base cost per square foot of living area

Exterior Wall	1200	1400	1600	1800	2000	2400	2800	3200	3600	4000	4400
Wood Siding - Wood Frame	105.50	100.05	96.15	91.80	88.10	82.00	77.40	73.80	71.60	69.60	67.50
Brick Veneer - Wood Frame	108.60	102.95	98.90	94.40	90.65	84.25	79.45	75.70	73.45	71.35	69.20
Solid Brick	115.30	109.25	104.90	100.10	96.10	89.15	83.90	79.90	77.50	75.15	72.90
Solid Stone	116.80	110.60	106.25	101.35	97.30	90.25	84.90	80.85	78.40	76.00	73.70
Finished Basement, Add	18.45	17.70	17.25	16.70	16.40	15.65	15.05	14.60	14.40	14.05	13.85
Unfinished Basement, Add	7.15	6.85	6.65	6.40	6.30	5.95	5.65	5.45	5.35	5.20	5.15

Modifications

Add to the total cost

Upgrade Kitchen Cabinets	$+ 744
Solid Surface Countertops	+ 1282
Full Bath - including plumbing, wall and floor finishes	+ 4849
Half Bath - including plumbing, wall and floor finishes	+ 2987
Two Car Attached Garage	+ 16,806
Two Car Detached Garage	+ 16,685
Fireplace & Chimney	+ 4800

Adjustments

For multi family - add to total cost

Additional Kitchen	$ + 9837
Additional Full Bath & Half Bath	+ 7836
Additional Entry & Exit	+ 1499
Separate Heating & Air Conditioning	+ 3995
Separate Electric	+ 1184

For Townhouse/Rowhouse -
Multiply cost per square foot by

Inner Unit	.89
End Unit	.94

Alternatives

Add to or deduct from the cost per square foot of living area

Heavyweight Asphalt Shingles	$− .85
Clay Tile Roof	+ 1.15
Slate Roof	+ 2.60
Upgrade Ceilings to Textured Finish	+ .41

Additional upgrades or components

Kitchen Cabinets & Countertops	Page 58
Bathroom Vanities	59
Fireplaces & Chimneys	59
Windows, Skylights & Dormers	59
Appliances	60
Breezeways & Porches	60
Finished Attic	60
Garages	61
Site Improvements	61
Wings & Ells	57

Important: See the Reference Section for Location Factors (to adjust for your city) and Estimating Forms

RESIDENTIAL | Luxury | Tri-Level

- Unique residence built from an architect's plan
- Single family — 1 full bath, 1 half bath, 1 kitchen
- No basement
- Cedar shakes on roof
- Forced hot air heat/air conditioning
- Double drywall interior
- Many special features
- Extraordinary materials and workmanship
- Detail specifications on page 49

Note: The illustration shown may contain some optional components (for example: garages and/or fireplaces) whose costs are shown in the modifications, adjustments, & alternatives below or at the end of the square foot section.

Base cost per square foot of living area

Exterior Wall	1500	1800	2100	2400	2800	3200	3600	4000	4500	5000	5500
Wood Siding - Wood Frame	99.20	91.90	86.40	82.05	78.90	75.30	71.55	70.10	66.20	64.15	61.90
Brick Veneer - Wood Frame	102.00	94.40	88.65	84.20	80.95	77.20	73.30	71.85	67.80	65.65	63.35
Solid Brick	108.00	99.85	93.60	88.85	85.40	81.30	77.15	75.55	71.25	68.85	66.45
Solid Stone	109.40	101.10	94.70	89.85	86.40	82.30	78.05	76.40	72.00	69.55	67.10
Finished Basement, **Add***	21.80	20.80	20.00	19.45	19.10	18.55	18.10	17.90	17.40	17.10	16.85
Unfinished Basement, **Add***	8.35	7.90	7.55	7.30	7.10	6.90	6.65	6.55	6.35	6.20	6.05

*Basement under middle level only.

Modifications

Add to the total cost

Upgrade Kitchen Cabinets	$+ 744
Solid Surface Countertops	+ 1282
Full Bath - including plumbing, wall and floor finishes	+ 4849
Half Bath - including plumbing, wall and floor finishes	+ 2987
Two Car Attached Garage	+ 16,806
Two Car Detached Garage	+ 16,685
Fireplace & Chimney	+ 5255

Adjustments

For multi family - add to total cost

Additional Kitchen	$ + 9837
Additional Full Bath & Half Bath	+ 7836
Additional Entry & Exit	+ 1499
Separate Heating & Air Conditioning	+ 3995
Separate Electric	+ 1184

For Townhouse/Rowhouse -
Multiply cost per square foot by

Inner Unit	.86
End Unit	.93

Alternatives

Add to or deduct from the cost per square foot of living area

Heavyweight Asphalt Shingles	$– 1.25
Clay Tile Roof	+ 1.65
Slate Roof	+ 3.75
Upgrade Ceilings to Textured Finish	+ .41

Additional upgrades or components

Kitchen Cabinets & Countertops	Page 58
Bathroom Vanities	59
Fireplaces & Chimneys	59
Windows, Skylights & Dormers	59
Appliances	60
Breezeways & Porches	60
Finished Attic	60
Garages	61
Site Improvements	61
Wings & Ells	57

RESIDENTIAL — Luxury — Wings & Ells

1 Story — Base cost per square foot of living area

Exterior Wall	50	100	200	300	400	500	600	700
Wood Siding - Wood Frame	164.85	125.95	109.00	90.55	85.05	81.70	79.45	80.15
Brick Veneer - Wood Frame	175.25	133.35	115.20	94.70	88.75	85.15	82.75	83.35
Solid Brick	197.80	149.50	128.65	103.65	96.80	92.70	89.90	90.30
Solid Stone	202.90	153.10	131.65	105.65	98.60	94.35	91.50	91.85
Finished Basement, **Add**	57.55	46.40	41.75	34.00	32.45	31.50	30.85	30.45
Unfinished Basement, **Add**	25.05	19.60	17.30	13.50	12.70	12.25	11.95	11.75

1-1/2 Story — Base cost per square foot of living area

Exterior Wall	100	200	300	400	500	600	700	800
Wood Siding - Wood Frame	133.95	108.35	92.70	82.85	78.10	75.80	72.90	72.05
Brick Veneer - Wood Frame	143.30	115.80	98.90	87.65	82.60	80.00	76.90	76.00
Solid Brick	163.40	131.90	112.35	98.15	92.25	89.15	85.55	84.55
Solid Stone	167.95	135.55	115.35	100.50	94.40	91.20	87.50	86.50
Finished Basement, **Add**	38.05	33.40	30.30	26.90	25.95	25.35	24.75	24.70
Unfinished Basement, **Add**	16.15	13.85	12.30	10.65	10.20	9.90	9.60	9.55

2 Story — Base cost per square foot of living area

Exterior Wall	100	200	400	600	800	1000	1200	1400
Wood Siding - Wood Frame	134.85	100.60	85.60	70.35	65.45	62.50	60.55	61.40
Brick Veneer - Wood Frame	145.30	108.00	91.80	74.50	69.20	65.95	63.85	64.60
Solid Brick	167.85	124.15	105.25	83.45	77.25	73.50	71.00	71.50
Solid Stone	172.95	127.75	108.25	85.45	79.05	75.15	72.60	73.10
Finished Basement, **Add**	28.80	23.20	20.90	17.00	16.25	15.75	15.45	15.25
Unfinished Basement, **Add**	12.55	9.80	8.65	6.75	6.35	6.15	6.00	5.85

Base costs do not include bathroom or kitchen facilities. Use Modifications/Adjustments/Alternatives on pages 58-61 where appropriate.

Important: See the Reference Section for Location Factors (to adjust for your city) and Estimating Forms.

RESIDENTIAL — Modifications/Adjustments/Alternatives

Kitchen cabinets - Base units, hardwood *(Cost per Unit)*

	Economy	Average	Custom	Luxury
24" deep, 35" high,				
One top drawer, one door below				
12" wide	$ 97	$129	$170	$225
15" wide	102	136	180	240
18" wide	107	142	190	250
21" wide	115	153	205	270
24" wide	125	166	220	290
four drawers				
12" wide	105	140	185	245
15" wide	128	171	225	300
18" wide	136	181	240	315
24" wide	151	201	265	350
Two top drawers, two doors below				
27" wide	157	209	280	365
30" wide	168	226	297	390
33" wide	170	227	300	395
36" wide	174	232	310	405
42" wide	188	250	335	440
48" wide	197	263	350	460
Range or sink base				
Two doors below				
30" wide	108	144	190	250
33" wide	125	166	220	290
36" wide	134	178	235	310
42" wide	142	189	250	330
48" wide	156	208	275	365
Corner Base Cabinet				
36" wide	140	187	250	325
Lazy Susan				
With revolving door	240	320	425	560

Kitchen cabinets - Wall cabinets, hardwood *(Cost per Unit)*

	Economy	Average	Custom	Luxury
12" deep, 2 doors				
12" high				
30" wide	$ 89	$119	$160	$210
36" wide	95	127	170	220
15" high				
30" wide	92	123	165	215
33" wide	98	131	175	230
36" wide	99	132	175	230
24" high				
30" wide	112	149	200	260
36" wide	119	159	210	280
42" wide	125	167	220	290
30" high, 1 door				
12" wide	91	121	160	210
15" wide	91	121	160	210
18" wide	95	127	170	220
24" wide	108	144	190	250
30" high, 2 doors				
27" wide	122	162	215	285
30" wide	130	173	230	305
36" wide	143	191	255	335
42" wide	153	204	270	355
48" wide	164	219	290	385
Corner wall, 30" high				
24" wide	128	170	225	300
30" wide	145	193	255	340
36" wide	158	211	280	370
Broom closet				
84" high, 24" deep				
18" wide	158	210	280	370
Oven Cabinet				
84" high, 24" deep				
27" wide	216	288	385	505

Kitchen countertops *(Cost per L.F.)*

	Economy	Average	Custom	Luxury
Solid surface				
24" wide, no backsplash	$55	$61	$ 83	$103
with backsplash	63	70	93	112
Stock plastic laminate, 24" wide				
with backsplash	10	13	20	25
Custom plastic laminate, no splash				
7/8" thick, alum. molding	19	25	35	45
1-1/4" thick, no splash	21	29	40	50
Marble				
1/2"—3/4" thick w/splash	34	46	60	100
Maple, laminated				
1-1/2" thick w/splash	34	46	60	80
Stainless steel				
(per S.F.)	66	88	115	155
Cutting blocks, recessed				
16" x 20" x 1" (each)	56	74	100	130

ADJUSTMENTS

RESIDENTIAL — Modifications/Adjustments/Alternatives

Vanity bases (Cost per Unit)

	Economy	Average	Custom	Luxury
2 door, 30" high, 21" deep				
24" wide	$155	$206	$275	$360
30" wide	179	239	320	420
36" wide	233	310	410	545
48" wide	311	415	550	725

Solid surface vanity tops (Cost Each)

	Economy	Average	Custom	Luxury
Center bowl				
22" x 25"	$254	$343	$481	$521
22" x 31"	293	396	527	577
22" x 37"	340	459	573	633
22" x 49"	420	567	666	745

Fireplaces & Chimneys (Cost per Unit)

	1-1/2 Story	2 Story	3 Story
Economy (prefab metal)			
Exterior chimney & 1 fireplace	$3205	$3540	$3875
Interior chimney & 1 fireplace	3070	3410	3575
Average (masonry)			
Exterior chimney & 1 fireplace	3215	3590	4085
Interior chimney & 1 fireplace	3015	3380	3680
For more than 1 flue, add	240	390	660
For more than 1 fireplace, add	2280	2280	2280
Custom (masonry)			
Exterior chimney & 1 fireplace	3380	3815	4310
Interior chimney & 1 fireplace	3170	3590	3870
For more than 1 flue, add	265	455	625
For more than 1 fireplace, add	2425	2425	2425
Luxury (masonry)			
Exterior chimney & 1 fireplace	4800	5255	5755
Interior chimney & 1 fireplace	4585	5010	5300
For more than 1 flue, add	395	660	925
For more than 1 fireplace, add	3780	3780	3780

Windows & Skylights (Cost Each)

	Economy	Average	Custom	Luxury
Fixed Picture Windows				
3'-6" x 4'-0"	$ 177	$ 221	$ 256	$ 391
4'-0" x 6'-0"	314	393	455	645
5'-0" x 6'-0"	390	488	565	830
6'-0" x 6'-0"	470	587	680	1150
Bay/Bow Windows				
8'-0" x 5'-0"	696	870	995	1118
10'-0" x 5'-0"	1206	1508	1725	1900
10'-0" x 6'-0"	1417	1771	2025	2175
12'-0" x 6'-0"	1609	2011	2300	2750
Palladian Windows				
3'-2" x 6'-4"		2205	2450	2854
4'-0" x 6'-0"		2228	2475	2883
5'-5" x 6'-10"		2813	3125	3641
8'-0" x 6'-0"		3443	3825	4456
Skylights				
46" x 21-1/2"	336	373	525	780
46" x 28"	367	408	563	818
57" x 44"	486	540	725	985

Dormers (Cost/S.F. of plan area)

	Economy	Average	Custom	Luxury
Framing and Roofing Only				
Gable dormer, 2" x 6" roof frame	$17	$19	$22	$34
2" x 8" roof frame	18	20	23	36
Shed dormer, 2" x 6" roof frame	11	12	14	22
2" x 8" roof frame	12	13	15	23
2" x 10" roof frame	13	14	16	24

ADJUSTMENTS

RESIDENTIAL — Modifications/Adjustments/Alternatives

Appliances (Cost per Unit)

	Economy	Average	Custom	Luxury
Range				
30" free standing, 1 oven	$ 365	$ 724	$1134	$1400
2 oven	770	1060	1465	1725
30" built-in, 1 oven	615	858	1116	1525
2 oven	1000	1230	1485	1900
21" free standing				
1 oven	375	432	500	535
Counter Top Ranges				
4 burner standard	284	390	529	635
As above with griddle	390	490	600	670
Microwave Oven	173	519	1185	1800
Combination Range, Refrigerator, Sink				
30" wide	820	1158	1473	1650
60" wide	2375	2607	2927	3010
72" wide	2700	2851	3311	3430
Comb. Range, Refrig., Sink, Microwave Oven & Ice Maker	3975	4654	5030	6129
Compactor				
4 to 1 compaction	380	462	544	630
Deep Freeze				
15 to 23 C.F.	510	543	718	895
30 C.F.	825	912	1005	1025
Dehumidifier, portable, auto.				
15 pint	198	209	235	246
30 pint	270	284	299	313
Washing Machine, automatic	470	772	1079	1250
Water Heater				
Electric, glass lined				
30 gal.	213	281	353	450
80 gal.	410	615	799	910
Water Heater, Gas, glass lined				
30 gal.	274	348	522	680
50 gal.	430	632	764	855
Water Softener, automatic				
30 grains/gal.	480	503	533	538
100 grains/gal.	955	1021	1051	1056
Dishwasher, built-in				
2 cycles	300	411	489	635
4 or more cycles	500	713	919	930
Dryer, automatic	470	589	827	1100
Garage Door Opener	239	289	324	335
Garbage Disposal	75	122	173	258
Heater, Electric, built-in				
1250 watt ceiling type	113	154	188	195
1250 watt wall type	107	119	135	184
Wall type w/blower				
1500 watt	162	174	181	199
3000 watt	228	251	263	289
Hood For Range, 2 speed				
30" wide	105	204	321	385
42" wide	232	301	359	430
Humidifier, portable				
7 gal. per day	99	103	109	120
15 gal. per day	160	170	179	197
Ice Maker, automatic				
13 lb. per day	575	610	645	685
51 lb. per day	1075	1182	1223	1271
Refrigerator, no frost				
10–12 C.F.	425	557	706	895
14–16 C.F.	645	752	975	1125
18–20 C.F.	760	1043	1344	1375
21–29 C.F.	885	1611	2429	3150
Sump Pump, 1/3 H.P.	161	289	394	460

Breezeway (Cost per S.F.)

Class	Type	Area (S.F.) 50	100	150	200
Economy	Open	$ 14.10	$12.05	$10.10	$ 9.90
	Enclosed	67.85	52.45	43.55	38.15
Average	Open	18.00	15.85	13.85	12.60
	Enclosed	76.45	56.90	46.55	41.00
Custom	Open	25.50	22.40	19.55	17.90
	Enclosed	104.80	77.95	63.75	56.05
Luxury	Open	27.60	24.20	21.90	21.10
	Enclosed	111.10	82.40	66.65	59.70

Porches (Cost per S.F.)

Class	Type	Area (S.F.) 25	50	100	200	300
Economy	Open	$ 41.75	$28.00	$21.85	$18.50	$15.80
	Enclosed	83.55	58.20	44.00	34.30	29.40
Average	Open	51.10	32.60	24.95	20.80	20.80
	Enclosed	101.00	68.40	51.80	40.10	33.95
Custom	Open	66.85	44.35	33.40	29.15	26.15
	Enclosed	132.90	90.80	69.10	53.75	46.40
Luxury	Open	72.45	47.35	35.00	31.45	27.90
	Enclosed	141.75	99.60	73.90	57.35	49.45

Finished attic (Cost per S.F.)

Class	Area (S.F.) 400	500	600	800	1000
Economy	$11.20	$10.75	$10.30	$10.15	$ 9.85
Average	17.25	16.75	16.40	16.15	15.75
Custom	20.95	20.50	20.10	19.75	19.35
Luxury	27.25	26.65	26.05	25.45	24.95

Alarm system (Cost per System)

	Burglar Alarm	Smoke Detector
Economy	$ 335	$ 49
Average	390	61
Custom	654	124
Luxury	1025	153

Sauna, prefabricated
(Cost per unit, including heater and controls — 7' high)

Size	Cost
6' x 4'	$3525
6' x 5'	3725
6' x 6'	4125
6' x 9'	5125
8' x 10'	6425
8' x 12'	6925
10' x 12'	7575

RESIDENTIAL — Modifications/Adjustments/Alternatives

Garages*
(Costs include exterior wall systems comparable with the quality of the residence. Included in the cost is an allowance for one door, manual overhead garage door(s) and electrical fixture.)

Class	Detached One Car	Detached Two Car	Detached Three Car	Attached One Car	Attached Two Car	Attached Three Car	Built-in One Car	Built-in Two Car	Basement One Car	Basement Two Car
Economy										
Wood	$ 6733	$ 9177	$12,595	$ 6219	$ 8993	$13,048	$ -1219	$ -2075	$ 598	$ 995
Masonry	8865	11,786	15,624	6884	10,106	13,921	-1406	-2722		
Average										
Wood	7430	10,029	14,159	6927	9831	13,900	-1260	-2362	650	1232
Masonry	9375	12,438	15,540	8113	10,743	14,487	-1441	-3187		
Custom										
Wood	8629	11,684	15,890	7910	11,236	15,430	-1392	-2578	796	1522
Masonry	11,133	13,896	18,609	8997	11,922	16,787	-1753	-3454		
Luxury										
Wood	13,703	16,806	20,957	13,078	16,685	20,830	-1586	-2941	3601	5309
Masonry	20,045	25,272	30,281	17,734	23,128	28,413	-2086	-3982		

*See the Introduction to this section for definitions of garage types.

Swimming pools *(Cost per S.F.)*

Residential (includes equipment)	
In-ground	$ 17-34
Deck equipment	1.30
Paint pool, preparation & 3 coats (epoxy)	2.40
Rubber base paint	2.22
Pool Cover	.31
Swimming Pool Heaters *(Cost per unit)*	
(not including wiring, external piping, base or pad)	
Gas	
155 MBH	2350
190 MBH	3150
500 MBH	7675
Electric	
15 KW 7200 gallon pool	1825
24 KW 9600 gallon pool	2425
54 KW 24,000 gallon pool	3625

Wood & coal stoves

Wood Only	
Free Standing (minimum)	$ 1050
Fireplace Insert (minimum)	1206
Coal Only	
Free Standing	1363
Fireplace Insert	1492
Wood and Coal	
Free Standing	2803
Fireplace Insert	2872

Sidewalks *(Cost per S.F.)*

Concrete, 3000 psi with wire mesh	4" thick	$ 2.14
	5" thick	2.47
	6" thick	2.75
Precast concrete patio blocks (natural)	2" thick	6.40
Precast concrete patio blocks (colors)	2" thick	6.55
Flagstone, bluestone	1" thick	7.23
Flagstone, bluestone	1-1/2" thick	8.60
Slate (natural, irregular)	3/4" thick	6.49
Slate (random rectangular)	1/2" thick	8.10
Seeding		
Fine grading & seeding includes lime, fertilizer & seed	per S.Y.	1.53
Lawn Sprinkler System	per S.F.	.46

Fencing *(Cost per L.F.)*

Chain Link, 4' high, galvanized	$ 9.30
Gate, 4' high (each)	105.00
Cedar Picket, 3' high, 2 rail	8.05
Gate (each)	101.00
3 Rail, 4' high	8.95
Gate (each)	109.00
Cedar Stockade, 3 Rail, 6' high	9.20
Gate (each)	109.00
Board & Battens, 2 sides 6' high, pine	11.95
6' high, cedar	20.00
No. 1 Cedar, basketweave, 6' high	24.00
Gate, 6' high (each)	118.00

Carport *(Cost per S.F.)*

Economy	$ 5.26
Average	8.03
Custom	12.09
Luxury	13.90

ADJUSTMENTS

RESIDENTIAL — Adjustments: Fees/Exclusions

Insurance Exclusions

Insurance exclusions are a matter of policy coverage and are not standard or universal. When making an appraisal for insurance purposes, it is recommended that some time be taken in studying the insurance policy to determine the specific items to be excluded. Most homeowners insurance policies have, as part of their provisions, statements that read "the policy permits the insured to exclude from value, in determining whether the amount of the insurance equals or exceeds 80% of its replacement cost, such items as building excavation, basement footings, underground piping and wiring, piers and other supports which are below the undersurface of the lowest floor or where there is no basement below the ground." Loss to any of these items, however, is covered.

Costs for Excavation, Spread and Strip Footings and Underground Piping

This chart shows excluded items expressed as a percentage of total building cost.

Class	\multicolumn{4}{c}{Number of Stories}
	1	1-1/2	2	3
Luxury	3.4%	2.7%	2.4%	1.9%
Custom	3.5%	2.7%	2.4%	1.9%
Average	5.0%	4.3%	3.8%	3.3%
Economy	5.0%	4.4%	4.1%	3.5%

Architect/Designer Fees

Architect/designer fees as presented in the following chart are typical ranges for 4 classes of residences. Factors affecting these ranges include economic conditions, size and scope of project, site selection, and standardization. Where superior quality and detail is required or where closer supervision is required, fees may run higher than listed. Lower quality or simplicity may dictate lower fees.

The editors have included architect/designers fees from the "mean" column in calculating costs.

Listed below are average costs expressed as a range and mean of total building cost for 4 classes of residences.

Class		Range	Mean
Luxury	— Architecturally designed and supervised	5%–15%	10%
Custom	— Architecturally modified designer plans	$1500–$2000	$1800
Average	— Designer plans	$800–$1200	$1000
Economy	— Stock plans	$25–$400	$250

RESIDENTIAL Adjustments: Depreciation

Depreciation as generally defined is "the loss of value due to any cause." Specifically, depreciation can be broken down into three categories: **Physical Depreciation, Functional Obsolescence,** and **Economic Obsolescence.**

Physical Depreciation is the loss of value from the wearing out of the building's components. Such causes may be decay, dry rot, cracks, or structural defects.

Functional Obsolescence is the loss of value from features which render the residence less useful or desirable. Such features may be higher than normal ceilings, design and/or style changes, or outdated mechanical systems.

Economic Obsolescence is the loss of value from external features which render the residence less useful or desirable. These features may be changes in neighborhood socio-economic grouping, zoning changes, legislation, etc.

Depreciation as it applies to the residential portion of this manual deals with the observed physical condition of the residence being appraised. It is in essence *"the cost to cure."*

The ultimate method to arrive at *"the cost to cure"* is to analyze each component of the residence.

For example:

Component, Roof Covering — Cost = $1800
Average Life = 15 years
Actual Life = 5 years
Depreciation = 5 ÷ 15 = .33 or 33%
$1800 × 33% = $594 = Amount of Depreciation

The following table, however, can be used as a guide in estimating the % depreciation using the actual age and general condition of the residence.

Depreciation Table — Residential

Age in Years	Good	Average	Poor
2	2%	3%	10%
5	4	6	20
10	7	10	25
15	10	15	30
20	15	20	35
25	18	25	40
30	24	30	45
35	28	35	50
40	32	40	55
45	36	45	60
50	40	50	65

Commercial/Industrial/Institutional Section

Table of Contents

Table No.	Page
Introduction	
General	67
How To Use	68
Examples	
Example 1	73
Example 2	75
Building Types	
M.010 Apartment, 1-3 Story	78
M.020 Apartment, 4-7 Story	80
M.030 Apartment, 8-24 Story	82
M.040 Auditorium	84
M.050 Bank	86
M.060 Bowling Alley	88
M.070 Bus Terminal	90
M.080 Car Wash	92
M.090 Church	94
M.100 Club, Country	96
M.110 Club, Social	98
M.120 College, Classroom, 2-3 Story	100
M.130 College, Dorm., 2-3 Story	102
M.140 College, Dorm., 4-8 Story	104
M.150 College, Laboratory	106
M.160 College, Student Union	108
M.170 Community Center	110
M.180 Courthouse, 1 Story	112
M.190 Courthouse, 2-3 Story	114
M.200 Factory, 1 Story	116

Table No.	Page
M.210 Factory, 3 story	118
M.220 Fire Station, 1 Story	120
M.230 Fire Station, 2 Story	122
M.240 Fraternity/Sorority House	124
M.250 Funeral Home	126
M.260 Garage, Auto Sales	128
M.270 Garage, Parking	130
M.280 Garage, Underground Parking	132
M.290 Garage, Repair	134
M.300 Garage, Service Station	136
M.310 Gymnasium	138
M.320 Hangar, Aircraft	140
M.330 Hospital, 2-3 Story	142
M.340 Hospital, 4-8 Story	144
M.350 Hotel, 4-7 Story	146
M.360 Hotel, 8-24 Story	148
M.370 Jail	150
M.380 Laundromat	152
M.390 Library	154
M.400 Medical Office, 1 Story	156
M.410 Medical Office, 2 Story	158
M.420 Motel, 1 Story	160
M.430 Motel, 2-3 Story	162
M.440 Movie Theatre	164
M.450 Nursing Home	166
M.460 Office, 2-4 Story	168

Table No.	Page
M.470 Office, 5-10 Story	170
M.480 Office, 11-20 Story	172
M.490 Police Station	174
M.500 Post Office	176
M.510 Racquetball Court	178
M.520 Religious Education	180
M.530 Restaurant	182
M.540 Restaurant, Fast Food	184
M.550 Rink, Hockey/Indoor Soccer	186
M.560 School, Elementary	188
M.570 School, High, 2-3 Story	190
M.580 School, Jr. High, 2-3 Story	192
M.590 School, Vocational	194
M.600 Store, Convenience	196
M.610 Store, Department, 1 Story	198
M.620 Store, Department, 3 Story	200
M.630 Store, Retail	202
M.640 Supermarket	204
M.650 Swimming Pool, Enclosed	206
M.660 Telephone Exchange	208
M.670 Town Hall, 1 Story	210
M.680 Town Hall, 2-3 Story	212
M.690 Warehouse	214
M.700 Warehouse, Mini	216
Depreciation	218

Introduction to the Commercial/Industrial/Institutional Section

General

The Commercial/Industrial/Institutional section of this manual contains base building costs per square foot of floor area for 70 model buildings. Each model has a table of square foot costs for combinations of exterior wall and framing systems. This table is supplemented by a list of common additives and their unit costs. A breakdown of the component costs used to develop the base cost for the model is on the opposite page. The total cost derived from the base cost and additives must be modified using the appropriate location factor from the Reference Section.

This section may be used directly to estimate the construction cost of most types of buildings knowing only the floor area, exterior wall construction and framing systems. To adjust the base cost for components which are different than the model, use the tables from the Assemblies Section.

Building Identification & Model Selection

The building models in this section represent structures by use. Occupancy, however, does not necessarily identify the building, i.e., a restaurant could be a converted warehouse. In all instances, the building should be described and identified by its own physical characteristics. The model selection should also be guided by comparing specifications with the model. In the case of converted use, data from one model may be used to supplement data from another.

Adjustments

The base cost tables represent the base cost per square foot of floor area for buildings without a basement and without unusual special features. Basement costs and other common additives are listed below the base cost table. Cost adjustments can also be made to the model by using the tables from the Assemblies Section. This table is for example only.

Dimensions

All base cost tables are developed so that measurement can be readily made during the inspection process. Areas are calculated from exterior dimensions and story heights are measured from the top surface of one floor to the top surface of the floor above. Roof areas are measured by horizontal area covered and costs related to inclines are converted with appropriate factors. The precision of measurement is a matter of the user's choice and discretion. For ease in calculation, consideration should be given to measuring in tenths of a foot, i.e., 9 ft. 6 in. = 9.5 ft., 9 ft. 4 in. = 9.3 ft.

Floor Area

The expression "Floor Area" as used in this section includes the sum of floor areas at grade level and above. Basement costs are calculated separately. The user must exercise his own judgment, where the lowest level floor is slightly below grade, whether to consider it at grade level or make the basement adjustment.

How to Use the Commercial/Industrial/Institutional Section

The following is a detailed explanation of a sample entry in the Commercial/Industrial/Institutional Square Foot Cost Section. Each bold number below corresponds to the item being described on the following page with the appropriate component or cost of the sample entry following in parenthesis.

Prices listed are costs that include overhead and profit of the installing contractor and additional mark-ups for General Conditions and Architects' Fees.

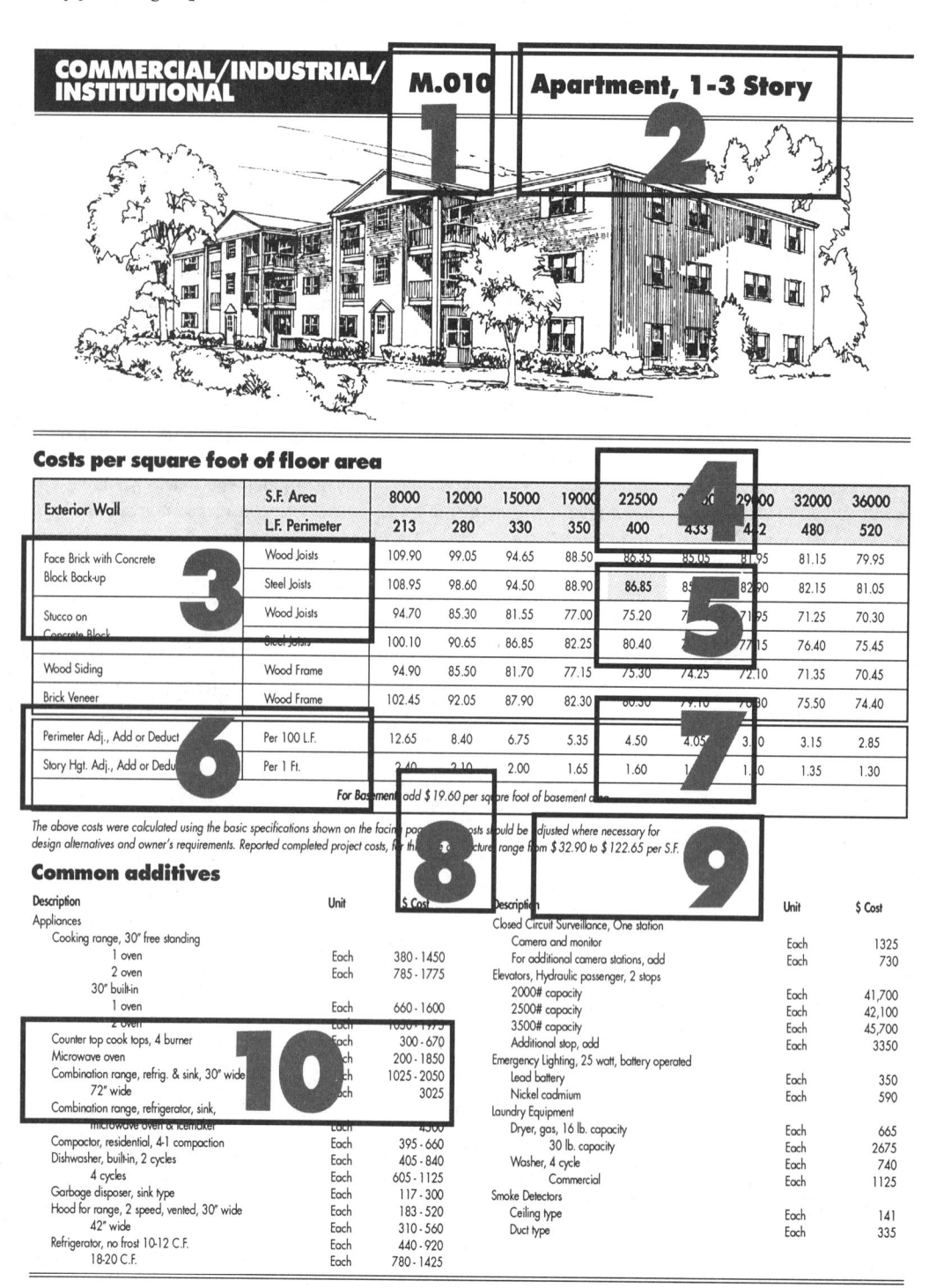

COMMERCIAL/INDUSTRIAL/INSTITUTIONAL

① M.010 **② Apartment, 1-3 Story**

Costs per square foot of floor area

Exterior Wall		S.F. Area	8000	12000	15000	19000	22500	26000	29000	32000	36000
		L.F. Perimeter	213	280	330	350	400	433	452	480	520
Face Brick with Concrete Block Back-up	③	Wood Joists	109.90	99.05	94.65	88.50	86.35	85.05	81.95	81.15	79.95
		Steel Joists	108.95	98.60	94.50	88.90	**86.85**	85.	82.90	82.15	81.05
Stucco on Concrete Block		Wood Joists	94.70	85.30	81.55	77.00	75.20	7	⑤ 71.95	71.25	70.30
		Steel Joists	100.10	90.65	86.85	82.25	80.40		77.15	76.40	75.45
Wood Siding		Wood Frame	94.90	85.50	81.70	77.15	75.30	74.25	72.10	71.35	70.45
Brick Veneer		Wood Frame	102.45	92.05	87.90	82.30	80.30	79.10	76.80	75.50	74.40
Perimeter Adj., Add or Deduct	⑥	Per 100 L.F.	12.65	8.40	6.75	5.35	4.50	4.05	⑦ 3.30	3.15	2.85
Story Hgt. Adj., Add or Deduct		Per 1 Ft.	2.40	2.10	2.00	1.65	1.60	1.	1.40	1.35	1.30

For Basement, add $19.60 per square foot of basement area

The above costs were calculated using the basic specifications shown on the facing page. ⑧ Costs should be adjusted where necessary for design alternatives and owner's requirements. Reported completed project costs, for this ⑨ structure range from $32.90 to $122.65 per S.F.

Common additives

Description	Unit	$ Cost	Description	Unit	$ Cost
Appliances			Closed Circuit Surveillance, One station		
Cooking range, 30" free standing			Camera and monitor	Each	1325
1 oven	Each	380 - 1450	For additional camera stations, add	Each	730
2 oven	Each	785 - 1775	Elevators, Hydraulic passenger, 2 stops		
30" built-in			2000# capacity	Each	41,700
1 oven	Each	660 - 1600	2500# capacity	Each	42,100
2 oven	Each	1050-1975	3500# capacity	Each	45,700
Counter top cook tops, 4 burner	Each	300 - 670	Additional stop, add	Each	3350
Microwave oven	⑩ Each	200 - 1850	Emergency Lighting, 25 watt, battery operated		
Combination range, refrig. & sink, 30" wide	Each	1025 - 2050	Lead battery	Each	350
72" wide	Each	3025	Nickel cadmium	Each	590
Combination range, refrigerator, sink,			Laundry Equipment		
microwave oven & icemaker	Each	4500	Dryer, gas, 16 lb. capacity	Each	665
Compactor, residential, 4-1 compaction	Each	395 - 660	30 lb. capacity	Each	2675
Dishwasher, built-in, 2 cycles	Each	405 - 840	Washer, 4 cycle	Each	740
4 cycles	Each	605 - 1125	Commercial	Each	1125
Garbage disposer, sink type	Each	117 - 300	Smoke Detectors		
Hood for range, 2 speed, vented, 30" wide	Each	183 - 520	Ceiling type	Each	141
42" wide	Each	310 - 560	Duct type	Each	335
Refrigerator, no frost 10-12 C.F.	Each	440 - 920			
18-20 C.F.	Each	780 - 1425			

Location Factors appear in the Reference Section

68

1 Model Number (M.010)

"M" distinguishes this section of the book and stands for model. The number designation is a sequential number.

2 Type of Building (Apartment, 1-3 Story)

There are 70 different types of commercial/industrial/institutional buildings highlighted in this section.

3 Exterior Wall Construction and Building Framing Options (Face Brick with Concrete Block Back-up and Open Web Steel Bar Joists)

Three or more commonly used exterior walls, and in most cases, two typical building framing systems are presented for each type of building. The model selected should be based on the actual characteristics of the building being estimated.

4 Total Square Foot of Floor Area and Base Perimeter Used to Compute Base Costs (22,500 Square Feet and 400 Linear Feet)

Square foot of floor area is the total gross area of all floors at grade, and above, and does not include a basement. The perimeter in linear feet used for the base cost is for a generally rectangular economical building shape.

5 Cost per Square Foot of Floor Area ($86.85)

The highlighted cost is for a building of the selected exterior wall and framing system and floor area. Costs for buildings with floor areas other than those calculated may be interpolated between the costs shown.

6 Building Perimeter and Story Height Adjustments

Square foot costs for a building with a perimeter or floor to floor story height significantly different from the model used to calculate the base cost may be adjusted, add or deduct, to reflect the actual building geometry.

7 Cost per Square Foot of Floor Area for the Perimeter and/or Height Adjustment ($4.50 for Perimeter Difference and $1.60 for Story Height Difference)

Add (or deduct) $4.50 to the base square foot cost for each 100 feet of perimeter difference between the model and the actual building. Add (or deduct) $1.60 to the base square foot cost for each 1 foot of story height difference between the model and the actual building.

8 Optional Cost per Square Foot of Basement Floor Area ($19.60)

The cost of an unfinished basement for the building being estimated is $19.60 times the gross floor area of the basement.

9 Range of Cost per Square Foot of Floor Area for Similar Buildings ($32.90 to $122.65)

Many different buildings of the same type have been built using similar materials and systems. Means historical cost data of actual construction projects indicates a range of $32.90 to $122.65 for this type of building.

10 Common Additives

Common components and/or systems used in this type of building are listed. These costs should be added to the total building cost. Additional selections may be found in the Assemblies Section.

How to Use the Commercial/Industrial/Institutional Section *(Continued)*

The following is a detailed explanation of a specification and costs for a model building in the Commercial/Industrial/Institutional Square Foot Cost Section. Each bold number below corresponds to the item being described on the following page with the appropriate component of the sample entry following in parenthesis.

Prices listed are costs that include overhead and profit of the installing contractor.

1 Model costs calculated for a 3 story building with 10' story height and 22,500 square feet of floor area

2 Apartment, 1-3 Story

			Unit	Unit Cost	Cost Per S.F.	% Of Sub-Total
1.0 Foundations						
.1	Footings & Foundations	Poured concrete; strip and spread footings and 4' foundation wall	S.F. Ground	6.63	2.21	
.4	Piles & Caissons	N/A	—	—	—	3.6%
.9	Excavation & Backfill	Site preparation for slab and trench for foundation wall and footing	S.F. Ground	.89	.30	
2.0 Substructure						
.1	Slab on Grade	4" reinforced concrete with vapor barrier and granular base	S.F. Slab	3.05	1.02	1.5%
.2	Special Substructures	N/A	—	—	—	
3.0 Superstructure						
.1	Columns & Beams	Gypsum board fireproofing on columns; steel columns in 3.5 and 3.7	S.F. Floor	.80	.80	
.4	Structural Walls	N/A	—	—	—	
.5	Elevated Floors	Open web steel joists, slab form, concrete, interior steel columns	S.F. Floor	8.81	5.87	13.1%
.7	Roof	Open web steel joists with rib metal deck, interior steel columns	S.F. Roof	4.26	1.42	
.9	Stairs	Concrete filled metal pan	Flight	4925	1.09	
4.0 Exterior Closure						
.1	Walls	Face brick with concrete block backup 88% of wall	S.F. Wall	16.90	7.93	
.5	Exterior Wall Finishes	N/A	—	—	—	13.3%
.6	Doors	Aluminum and glass	Each	1085	.19	
.7	Windows & Glazed Walls	Aluminum horizontal sliding 12% of wall	Each	282	1.20	
5.0 Roofing						
.1	Roof Coverings	Built-up tar and gravel with flashing	S.F. Roof	2.49		
.7	Insulation	Perlite/EPS composite		1.18		.7%
6.0 Interior Construction						
.1	Partitions	Gypsum board on metal studs 10 S.F. Floor/L.F. Partition	S.F. Partition	2.10	2.83	
.4	Interior Doors	15% solid core wood, 85% hollow core wood 80 S.F. Floor/Door	Each	353	4.42	
.5	Wall Finishes	70% paint, 25% vinyl wall covering, 5% ceramic tile	S.F. Surface	.98		23.9%
.6	Floor Finishes	60% carpet, 30% vinyl composition tile, 10% ceramic tile	S.F. Floor		3.81	
.7	Ceiling Finishes	Painted gypsum board on resilient channels	S.F. Ceiling		2.68	
.9	Interior Surface/Exterior Wall	Painted gypsum board on furring 80% of wall	S.F. Wall		1.20	
7.0 Conveying						
.1	Elevators	One hydraulic passenger elevator	Each	61,050	2.74	
.2	Special Conveyors	N/A	—	—	—	
8.0 Mechanical						
.1	Plumbing	Kitchen, bathroom and service fixtures, supply and drainage 1 Fixture/200 S.F. Floor	Each	1746	8.73	
.2	Fire Protection	Wet pipe sprinkler system	S.F. Floor	1.61	1.61	
.3	Heating	Oil fired hot water, baseboard radiation	S.F. Floor	4.51	4.51	29.6%
.4	Cooling	Chilled water, air cooled condenser system	S.F. Floor	5.81	5.81	
.5	Special Systems	N/A	—	—	—	
9.0 Electrical						
.1	Service & Distribution	400 ampere service, panel board and feeders	S.F. Floor	.77	.77	
.2	Lighting & Power	Incandescent fixtures, receptacles, switches, A.C. and misc. power	S.F. Floor	4.02	4.02	7.7%
.4	Special Electrical	Alarm systems and emergency lighting	S.F. Floor	.59	.59	
11.0 Special Construction						
.1	Specialties	Kitchen cabinets	S.F. Floor	1.22	1.22	1.7%
12.0 Site Work						
.1	Earthwork	N/A				
.3	Utilities	N/A				0.0%
.5	Roads & Parking	N/A				
.7	Site Improvements	N/A				
				Sub-Total	69.93	100%
	GENERAL CONDITIONS (Overhead & Profit)			15%	10.49	
	ARCHITECT FEES			8%	6.43	
				Total Building Cost	**86.85**	

70

1 Building Description
(Model costs are calculated for a three-story apartment building with 10' story height and 22,500 square feet of floor area)

The model highlighted is described in terms of building type, number of stories, typical story height and square footage.

2 Type of Building
(Apartment, 1-3 Story)

3 Division 6.0 Interior Construction
(6.4 Interior Doors)

System costs are presented in divisions according to the 12-division UniFormat classifications. Each of the component systems are listed.

4 Specification Highlights
(15% solid core wood; 85% hollow core wood)

All systems in each subdivision are described with the material and proportions used.

5 Quality Criteria
(80 S.F. Floor/Door)

The criteria used in determining quantities for the calculations are shown.

6 Unit (Each)

The unit of measure shown in this column is the unit of measure of the particular system shown that corresponds to the unit cost.

7 Unit Cost ($353)

The cost per unit of measure of each system subdivision.

8 Cost per Square Foot ($4.42)

The cost per square foot for each system is the unit cost of the system times the total number of units divided by the total square feet of building area.

9 % of Sub-Total (23.9%)

The percent of sub-total is the total cost per square foot of all systems in the division divided by the sub-total cost per square foot of the building.

10 Sub-Total ($69.93)

The sub-total is the total of all the system costs per square foot.

11 General Conditions (Overhead & Profit) (15%) (Architects' Fees) (8%)

General Conditions to cover the overhead and profit of the General Contractor are added as a percentage of the sub-total. An Architect's Fee, also as a percentage of the sub-total, is also added. These values vary with the building type.

12 Total Building Cost ($86.85)

The total building cost per square foot of building area is the sum of the square foot costs of all the systems plus the General Contractor's overhead and profit and the Architect's fee. The total building cost is the amount which appears shaded in the Cost per Square Foot of Floor Area table shown previously.

Examples

Example 1

This example illustrates the use of the base cost tables. The base cost is adjusted for different exterior wall systems, different story height and a partial basement.

CII APPRAISAL FIELD DATA FORM

COMMERCIAL
INDUSTRIAL
INSTITUTIONAL

Use sheets A & D for App. Form 2
Use sheets A, B & C for App. Form 3

CII APPRAISAL

1. SUBJECT PROPERTY: Westernberg Holdings, Inc.
2. BUILDING: #623
3. ADDRESS: 55 Pine wood Rd., Bondsville, N.J. 07410
4. BUILDING USE: Apartments
5. DATE: Jan. 2 1997
6. APPRAISER: SJS
7. YEAR BUILT: 1972

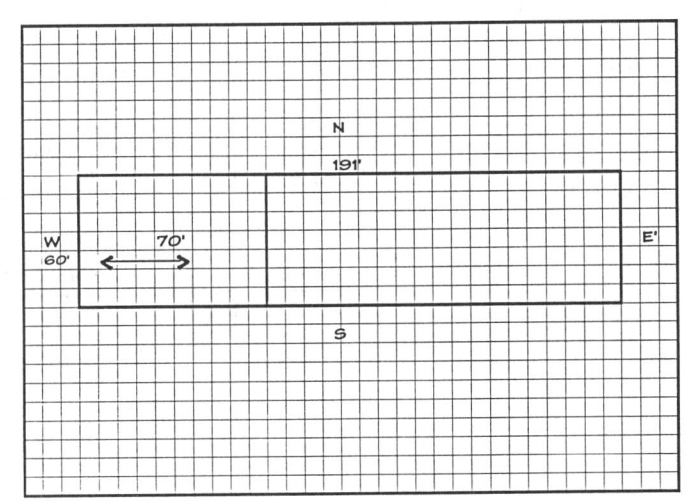

(Building footprint diagram: 60' × 191', oriented with N, S, E, W labels; 70' interior dimension shown)

8. EXTERIOR WALL CONSTRUCTION: North wall - Face brick w/concrete block back up
   East, West & South wall - Decorative concrete block
9. FRAME: Steel
10. GROUND FLOOR AREA: 60 × 191   11,460 S.F.
11. GROSS FLOOR AREA (EXCL. BASEMENT): 80,220 (7 Floors) S.F.
12. NUMBER OF STORIES: 7
13. STORY HEIGHT: 11'-4"
14. PERIMETER: 502 L.F.
15. BASEMENT AREA: 4,200 S.F.
16. GENERAL COMMENTS:

Example 1 (continued)

Base Cost per square area: *(from square foot table)*

Specify source: Page: **80**, Model # **M.020**, Area **80,220** S.F., Exterior wall **See Below**, Frame **Steel** — Identify the source of base costs

Adjustments for exterior wall variation:
North wall is face brick with concrete block back up.
$\frac{191}{502}$ = 38 % ~~East, West,~~ South = 62 % dec. concrete block.

$38\% = \frac{\text{Area of north wall}}{\text{Total wall area}}$

Size Adjustment: (.38 × 85.60) + (.62 × 80.75) = 32.53 + 50.06 = 82.59
(Interpolate)

$80.75 from page 80; 80,000 S.F. Decorative Concrete Block

Height Adjustment: **82.59** + **1.00** = **83.59**

Adjusted Base Cost per square foot: **83.59**

$85.60 from page 80; 80,000 S.F. Face Brick and Block

Building Cost $ **83.59** × **80,220** = **6,705,590**
Adjusted Base Cost per square foot / Floor Area

$1.00 from page 80; 80,000 S.F. Story Height Adjustment

Basement Cost **19.85** × **4,200** = **83,370**
Basement Cost / Basement Area

$19.85 from page 80; Basement Addition

Lump Sum Additions
70 Smoke detectors @ $141.00 each **9,870**

$141.00 from page 80; Common Additives

TOTAL BUILDING COST *(Sum of above costs)* **6,798,830**
Modifications: *(complexity, workmanship, size)* +/- ____ % ____
Location Modifier: City **Bondsville, NJ 07410** Date **January 1997** × **1.13**

Page 431

Local cost of replacement **7,682,678**
Less depreciation: **20%** **1,536,536**

Page 218

Local cost of replacement less depreciation $ **6,146,142**

Example 2
This example shows how to modify a model building

Model #M.020, 4–7 Story Apartment, page 80, matches the example quite closely. The model specifies a 6-story building with a 10' story height.

The following adjustments must be made:
- add stone ashlar wall
- adjust model to a 5-story building
- change partitions
- change floor finish
- change plumbing

CII APPRAISAL

1. SUBJECT PROPERTY: **Westernberg Holdings, Inc.**
2. BUILDING: **#475**
3. ADDRESS: **55 Pine wood Rd., Bondsville, N.J. 07410**
4. BUILDING USE: **Apartments**
5. DATE: **Jan. 2 1997**
6. APPRAISER: **SJS**
7. YEAR BUILT: **1972**

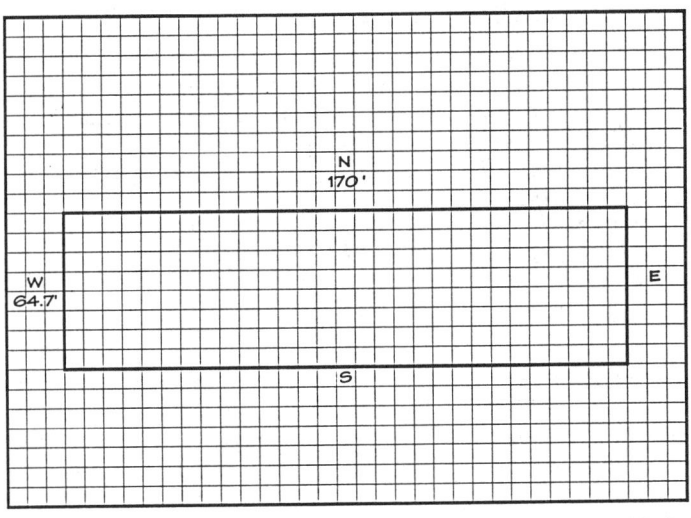

N 170'
W 64.7'
E
S

Two exterior wall types

8. EXTERIOR WALL CONSTRUCTION: **North wall - 4" stone Ashlar w/8" block back up (Split Face)**
   East, West & South walls - 12" Decorative concrete block
9. FRAME: **Steel**
10. GROUND FLOOR AREA: **11,000** S.F.
11. GROSS FLOOR AREA (EXCL. BASEMENT): **55,000** S.F.
12. NUMBER OF STORIES: **5**
13. STORY HEIGHT: **11 Ft.**
14. PERIMETER: **469.4** L.F.
15. BASEMENT AREA: **None** S.F.
16. GENERAL COMMENTS:

EXAMPLES

Example 2 (continued)

Square foot area (excluding basement) from item 11 __55,000__ S.F.
Perimeter from item 14 __469.4__ Lin. Ft.
Item 17—Model square foot costs (from model sub-total) $ __72.55__

FIELD DESCRIPTION & CALCULATION SECTION

NO.	SYSTEM/COMPONENT	DESCRIPTION	UNIT	UNIT COST	NEW SF COST	MODEL SF COST	+/- CHANGE
1.0	**FOUNDATION**						
.1	Footings and Foundations: L.F. Length Bay Size Same as model		S.F. Gnd.	6.72	1.34	1.12	+.22
.4	Piles & Caissons:						
.9	Excavation & Backfill: Depth Area 11,000 S.F.		S.F. Gnd.	.85	.17	.14	+.03
2.0	**SUBSTRUCTURE**						
.1	Slab on grade: Material Thickness Same as model		S.F. Slab	3.05	.61	.51	+.10
.2	Special Substructures: Type						
3.0	**SUPERSTRUCTURE**						
.1	Columns, Beams, & Joists: Size Type Length Fireproof only		S.F. Floor	1.50	1.50	1.50	—
	Size Type Length		L.F.				
.4	Structure Walls:		S.F.				
.5	Elevated Floors: Same as model		S.F. Floor	11.50	9.20	9.58	-.38
.7	Roof:		S.F. Roof	4.82	.96	.80	+.16
.9	Stairs: 13 Flights		Flight	4080	.96	1.09	-.13
4.0	**EXTERIOR CLOSURE**						
.1	Walls: North wall - 4" stone ashlar Material on 8" conc. block Thickness 36 % of Wall		S.F. Wall	15.36	6.20	7.27	-1.07
	East, West & South walls - 12" Material split ribbed block Thickness 64 % of Wall		S.F. Wall				
	Material Thickness % of Wall		S.F. Wall				
.5	Exterior Wall Finishes: Type		S.F. Surf				
.6	Doors: Type Same as model Number 4		Each	3122	.23	.21	+.02
	Type Number		Each				
.7	Windows & Glazed Walls: Type % of Wall		S.F. Wind.				
	Type 14 % of Wall Each		S.F. Wind.	282	1.24	1.32	-.08
5.0	**ROOFING**						
.1	Roof Coverings: Material Same as model		S.F. Roof	2.46	.49	.41	+.08
.7	Insulation: Material Thickness		S.F. Insul.	1.18	.24	.20	+.04
.8	Openings & Specialities:		S.F. Opng.				
6.0	**INTERIOR CONSTRUCTION**						
.1	Partitions: 9' High Material Gypsum board on metal stud Density 8 S.F. Floor/L.F. Part.		S.F. Part.	3.80	4.28	3.60	+.68
	Material Density		S.F. Part.				
.4	Interior Doors: Type Number		Each	424	5.31	5.31	—
.5	Wall Finishes: Material Same as model but % of Wall 9'-0" High		S.F. Wall	1.07	2.41	2.13	+.28
	Material		S.F. Wall				
.6	Floor Finishes: Material Hardwood 60 % of Floor Carpet 30 % of Floor		S.F. Floor	3.81	5.62	3.81	+1.81
	Material Tile 10 % of Floor		S.F. Floor				
.7	Ceiling Finishes: Material Same as model % of Ceiling		S.F. Cell.	2.64	2.64	2.64	—

From page 81, Model #M.020 Sub-total

$$\frac{\$.85/S.F. \text{ ground area}}{5 \text{ stories}}$$

$$\frac{\$.84/S.F. \text{ ground area}}{6 \text{ stories}}$$

(Page 81, Line 3.1)

$$\frac{\text{Area of Elevated Floors}}{\$11.50/S.F. \times 4 \times 11,000 \text{ S.F.}}$$
$$\frac{55,000}{\text{Building Floor Area}}$$

$$\frac{\$4.82/S.F. \text{ roof}}{5 \text{ stories}}$$

Windows are 14% of wall area, 15 S.F./window

$$\frac{282}{15} \times .14 \times 469.4 \times 11 \times 5 \div 55,000 = 1.24$$

From Table 6.1-510 page 325 and 6.1-580 page 327

$$\frac{\$3.80/S.F. \text{ partition} \times 9 \text{ ft. high}}{8 \text{ S.F. floor/L.F. partition}}$$

From Table 6.6-100 pages 336 & 337

Oak Strip, Fin. (max. price) @ $6.56/S.F.
Carpet (26 oz.) @ $3.34/S.F.
Ceramic Tile @ $6.76/S.F.

.60 x $7.05 = $3.94
.30 x $6.76 = 1.00
.10 x $6.59 = .68
          $5.62

North wall is 36% of wall area, e.g., $\frac{170 \times 11}{469.4 \times 11}$
From Table 4.1-242-2350 page 289
Stone ashlar @ $21.25/S.F. wall
East, west and south walls are 64% of wall area
From Table 4.1-212-1530 page 284
Split ribbed block @ $12.05/S.F. wall

New wall cost
.36 × $21.25/S.F. + .64 × $12.05/S.F. = $15.36/S.F. wall

Convert cost per S.F. wall to cost per S.F. of floor
Windows are 14% of wall area

$$\$6.20 \text{ S.F. floor} = \frac{\$15.36/S.F. \text{ wall} \times .86 \times 469.4 \text{ L.F.} \times 11 \text{ ft. high} \times 5 \text{ stories}}{55,000 \text{ S.F. of floor}}$$

Example 2 (Continued)

NO.	SYSTEM/COMPONENET					Unit	Unit Cost	New S.F. Cost	Model S.F. Cost	+/- Change
6.0	**INTERIOR CONSTRUCTION**									
.9	Interior Surface/Exterior Wall:	Material		86 % of Wall	Board on furring	S.F. Surf.	2.81	1.13	1.13	—
		Material		% of Wall		S.F. Surf.				
7.0	**CONVEYING**									
.1	Elevators: 2	Type Same as model	Capacity		Stops	Each	120,300	4.37	4.01	+.36
.2	Special Conveyors:	Type				Each				
8.0	**MECHANICAL**									
.1	Plumbing: Kitchen, bath + service fixtures 1 Fixture /200					S.F. Floor	1713	8.57	7.97	+.60
.2	Fire Protection:	Same as model				S.F.	1.59	1.59	1.59	—
.3	Heating:	Type Same as model				S.F.	3.99	3.99	3.99	—
.4	Cooling:	Type Same as model				S.F.	5.60	5.60	5.60	—
.5	Special Systems:					Each				
9.0	**ELECTRICAL**									
.1	Service Distribution:	Same as model				S.F.	.85	.85	.85	—
.2	Lighting & Power:					S.F.	3.99	3.99	3.99	—
.4	Special Electrical:					S.F.	.33	.33	.33	—
11.0	**SPECIAL CONSTRUCTION**									
.1	Specialties:					S.F.	1.46	1.46	1.46	—
12.0	**SITE WORK**									
.1	Earthwork:									
.3	Utilities:									
.5	Roads & Parking:									
.7	Site Improvement:									
								Total Change		+2.72

ITEM		
18	Adjusted S.F. cost $ 75.27 item 17 +/- changes	
19	Building area - from item 11 55,000 S.F. x adjusted S.F. cost	$ 4,139,850
20	Basement area - from item 15 — S.F. x S.F. cost $	$ —
21	Base building sub-total - item 20 + item 19	$ 4,139,850
22	Miscellaneous addition (quality, etc.)	$ —
23	Sub-total - item 22 + item 21	$ 4,139,850
24	General conditions - 15 % of item 23	$ 620,978
25	Sub-total - item 24 + item 23	$ 4,760,828
26	Architects fees 5.5 % of item 25	$ 261,846
27	Sub-total - item 26 + item 27	$ 5,022,674
28	Location modifier zip code 07410	x 1.13
29	Local replacement cost - item 28 x item 27	$ 5,675,622
30	Depreciation 20 % of item 29	$ 1,135,124
31	Depreciated local replacement cost - item 29 less item 30	$ 4,540,498
32	Exclusions	$ —
33	Net depreciated replacement cost - item 31 less item 32	$ 4,540,498

469.4 L.F. × 11 ft. high × 5 stories

$$\frac{\text{Unit Cost} \qquad 86\%}{\$2.81/\text{S.F. wall} \times \text{total S.F. wall} \times .86}{55,000}$$

From page 431

Use table on page 218

COMMERCIAL/INDUSTRIAL/INSTITUTIONAL

M.010 Apartment, 1-3 Story

Costs per square foot of floor area

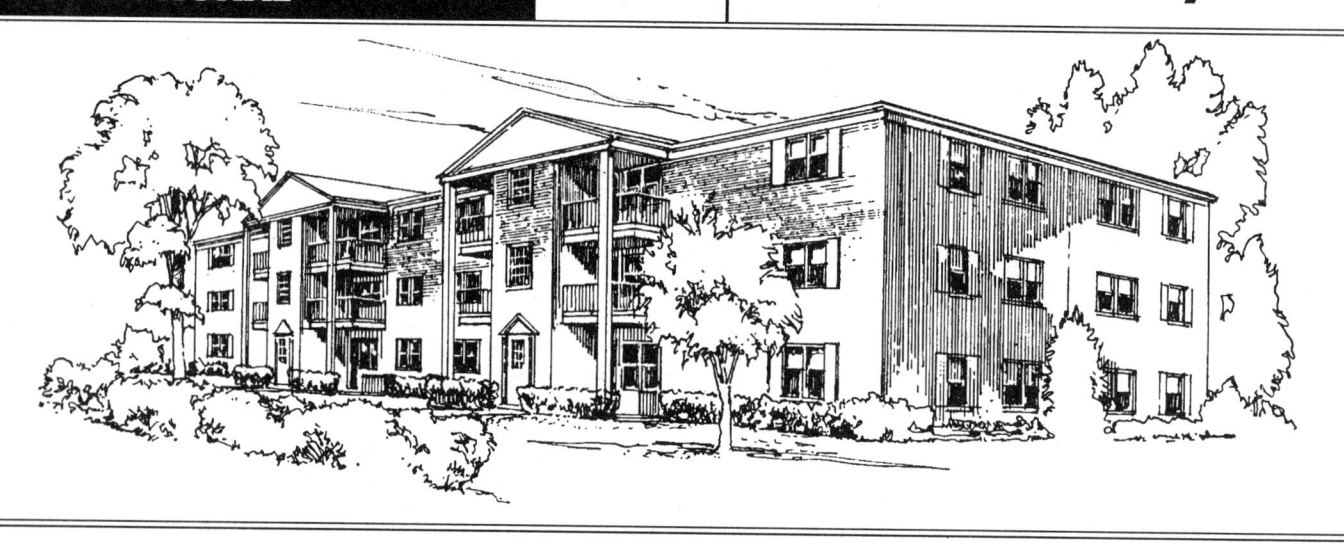

Exterior Wall	S.F. Area	8000	12000	15000	19000	22500	25000	29000	32000	36000
	L.F. Perimeter	213	280	330	350	400	433	442	480	520
Face Brick with Concrete Block Back-up	Wood Joists	109.90	99.05	94.65	88.50	86.35	85.05	81.95	81.15	79.95
	Steel Joists	108.95	98.60	94.50	88.90	86.85	85.65	82.90	82.15	81.05
Stucco on Concrete Block	Wood Joists	94.70	85.30	81.55	77.00	75.20	74.10	71.95	71.25	70.30
	Steel Joists	100.10	90.65	86.85	82.25	80.40	79.30	77.15	76.40	75.45
Wood Siding	Wood Frame	94.90	85.50	81.70	77.15	75.30	74.25	72.10	71.35	70.45
Brick Veneer	Wood Frame	102.45	92.05	87.90	82.30	80.30	79.10	76.30	75.50	74.40
Perimeter Adj., Add or Deduct	Per 100 L.F.	12.65	8.40	6.75	5.35	4.50	4.05	3.50	3.15	2.85
Story Hgt. Adj., Add or Deduct	Per 1 Ft.	2.40	2.10	2.00	1.65	1.60	1.60	1.40	1.35	1.30

For Basement, add $19.60 per square foot of basement area

The above costs were calculated using the basic specifications shown on the facing page. These costs should be adjusted where necessary for design alternatives and owner's requirements. Reported completed project costs, for this type of structure, range from $32.90 to $122.65 per S.F.

Common additives

Description	Unit	$ Cost
Appliances		
Cooking range, 30" free standing		
1 oven	Each	380 - 1450
2 oven	Each	785 - 1775
30" built-in		
1 oven	Each	660 - 1600
2 oven	Each	1050 - 1975
Counter top cook tops, 4 burner	Each	300 - 670
Microwave oven	Each	200 - 1850
Combination range, refrig. & sink, 30" wide	Each	1025 - 2050
72" wide	Each	3025
Combination range, refrigerator, sink, microwave oven & icemaker	Each	4500
Compactor, residential, 4-1 compaction	Each	395 - 660
Dishwasher, built-in, 2 cycles	Each	405 - 840
4 cycles	Each	605 - 1125
Garbage disposer, sink type	Each	117 - 300
Hood for range, 2 speed, vented, 30" wide	Each	183 - 520
42" wide	Each	310 - 560
Refrigerator, no frost 10-12 C.F.	Each	440 - 920
18-20 C.F.	Each	780 - 1425

Description	Unit	$ Cost
Closed Circuit Surveillance, One station		
Camera and monitor	Each	1325
For additional camera stations, add	Each	730
Elevators, Hydraulic passenger, 2 stops		
2000# capacity	Each	41,700
2500# capacity	Each	42,100
3500# capacity	Each	45,700
Additional stop, add	Each	3350
Emergency Lighting, 25 watt, battery operated		
Lead battery	Each	350
Nickel cadmium	Each	590
Laundry Equipment		
Dryer, gas, 16 lb. capacity	Each	665
30 lb. capacity	Each	2675
Washer, 4 cycle	Each	740
Commercial	Each	1125
Smoke Detectors		
Ceiling type	Each	141
Duct type	Each	335

Model costs calculated for a 3 story building with 10′ story height and 22,500 square feet of floor area

Apartment, 1-3 Story

				Unit	Unit Cost	Cost Per S.F.	% Of Sub-Total
1.0 Foundations							
.1	Footings & Foundations	Poured concrete; strip and spread footings and 4′ foundation wall		S.F. Ground	6.63	2.21	
.4	Piles & Caissons	N/A		—	—	—	3.6%
.9	Excavation & Backfill	Site preparation for slab and trench for foundation wall and footing		S.F. Ground	.89	.30	
2.0 Substructure							
.1	Slab on Grade	4″ reinforced concrete with vapor barrier and granular base		S.F. Slab	3.05	1.02	1.5%
.2	Special Substructures	N/A		—	—	—	
3.0 Superstructure							
.1	Columns & Beams	Gypsum board fireproofing on columns; steel columns in 3.5 and 3.7		S.F. Floor	.80	.80	
.4	Structural Walls	N/A		—	—	—	
.5	Elevated Floors	Open web steel joists, slab form, concrete, interior steel columns		S.F. Floor	8.81	5.87	13.1%
.7	Roof	Open web steel joists with rib metal deck, interior steel columns		S.F. Roof	4.26	1.42	
.9	Stairs	Concrete filled metal pan		Flight	4925	1.09	
4.0 Exterior Closure							
.1	Walls	Face brick with concrete block backup	88% of wall	S.F. Wall	16.90	7.93	
.5	Exterior Wall Finishes	N/A		—	—	—	13.3%
.6	Doors	Aluminum and glass		Each	1085	.19	
.7	Windows & Glazed Walls	Aluminum horizontal sliding	12% of wall	Each	282	1.20	
5.0 Roofing							
.1	Roof Coverings	Built-up tar and gravel with flashing		S.F. Roof	2.49	.83	
.7	Insulation	Perlite/EPS composite		S.F. Roof	1.18	.39	1.7%
.8	Openings & Specialties	N/A		—	—	—	
6.0 Interior Construction							
.1	Partitions	Gypsum board and sound deadening board on metal studs	10 S.F. of Floor/L.F. Partition	S.F. Partition	3.18	2.83	
.4	Interior Doors	15% solid core wood, 85% hollow core wood	80 S.F. Floor/Door	Each	353	4.42	
.5	Wall Finishes	70% paint, 25% vinyl wall covering, 5% ceramic tile		S.F. Surface	.98	1.74	23.9%
.6	Floor Finishes	60% carpet, 30% vinyl composition tile, 10% ceramic tile		S.F. Floor	3.81	3.81	
.7	Ceiling Finishes	Painted gypsum board on resilient channels		S.F. Ceiling	2.68	2.68	
.9	Interior Surface/Exterior Wall	Painted gypsum board on furring	80% of wall	S.F. Wall	2.81	1.20	
7.0 Conveying							
.1	Elevators	One hydraulic passenger elevator		Each	61,650	2.74	3.9%
.2	Special Conveyors	N/A		—	—	—	
8.0 Mechanical							
.1	Plumbing	Kitchen, bathroom and service fixtures, supply and drainage	1 Fixture/200 S.F. Floor	Each	1746	8.73	
.2	Fire Protection	Wet pipe sprinkler system		S.F. Floor	1.61	1.61	
.3	Heating	Oil fired hot water, baseboard radiation		S.F. Floor	4.51	4.51	29.6%
.4	Cooling	Chilled water, air cooled condenser system		S.F. Floor	5.81	5.81	
.5	Special Systems	N/A		—	—	—	
9.0 Electrical							
.1	Service & Distribution	400 ampere service, panel board and feeders		S.F. Floor	.77	.77	
.2	Lighting & Power	Incandescent fixtures, receptacles, switches, A.C. and misc. power		S.F. Floor	4.02	4.02	7.7%
.4	Special Electrical	Alarm systems and emergency lighting		S.F. Floor	.59	.59	
11.0 Special Construction							
.1	Specialties	Kitchen cabinets		S.F. Floor	1.22	1.22	1.7%
12.0 Site Work							
.1	Earthwork	N/A		—	—	—	
.3	Utilities	N/A		—	—	—	0.0%
.5	Roads & Parking	N/A		—	—	—	
.7	Site Improvements	N/A		—	—	—	
				Sub-Total		69.93	100%
	GENERAL CONDITIONS (Overhead & Profit)				15%	10.49	
	ARCHITECT FEES				8%	6.43	
				Total Building Cost		86.85	

BUILDING TYPES

COMMERCIAL/INDUSTRIAL/INSTITUTIONAL   M.020   Apartment, 4-7 Story

Costs per square foot of floor area

Exterior Wall	S.F. Area	40000	45000	50000	55000	60000	70000	80000	90000	100000
	L.F. Perimeter	366	400	433	466	500	566	505	550	594
Face Brick with Concrete Block Back-up	Steel Frame	91.40	90.75	90.15	89.65	89.25	88.65	85.60	85.15	84.70
	R/Conc. Frame	96.15	95.40	94.75	94.25	93.80	93.10	89.65	89.10	88.65
Decorative Concrete Block	Steel Frame	85.25	84.70	84.20	83.80	83.50	83.00	80.75	80.40	80.05
	R/Conc. Frame	87.45	86.85	86.35	86.00	85.65	85.15	82.85	82.50	82.15
Precast Concrete Panels	Steel Frame	87.50	86.85	86.30	85.90	85.55	84.95	82.30	81.90	81.50
	R/Conc. Frame	89.50	88.85	88.30	87.90	87.55	86.95	84.25	83.85	83.50
Perimeter Adj., Add or Deduct	Per 100 L.F.	4.10	3.60	3.25	3.00	2.70	2.30	2.05	1.80	1.65
Story Hgt. Adj., Add or Deduct	Per 1 Ft.	1.40	1.35	1.30	1.30	1.25	1.25	1.00	.90	.90

For Basement, add $19.85 per square foot of basement area

The above costs were calculated using the basic specifications shown on the facing page. These costs should be adjusted where necessary for design alternatives and owner's requirements. Reported completed project costs, for this type of structure, range from $37.30 to $101.45 per S.F.

Common additives

Description	Unit	$ Cost
Appliances		
Cooking range, 30" free standing		
1 oven	Each	380 - 1450
2 oven	Each	785 - 1775
30" built-in		
1 oven	Each	660 - 1600
2 oven	Each	1050 - 1975
Counter top cook tops, 4 burner	Each	300 - 670
Microwave oven	Each	200 - 1850
Combination range, refrig. & sink, 30" wide	Each	1025 - 2050
72" wide	Each	3025
Combination range, refrigerator, sink, microwave oven & icemaker	Each	4500
Compactor, residential, 4-1 compaction	Each	395 - 660
Dishwasher, built-in, 2 cycles	Each	405 - 840
4 cycles	Each	605 - 1125
Garbage disposer, sink type	Each	117 - 300
Hood for range, 2 speed, vented, 30" wide	Each	183 - 520
42" wide	Each	310 - 560
Refrigerator, no frost 10-12 C.F.	Each	440 - 920
18-20 C.F.	Each	780 - 1425

Description	Unit	$ Cost
Closed Circuit Surveillance, One station		
Camera and monitor	Each	1325
For additional camera stations, add	Each	730
Elevators, Electric passenger, 5 stops		
2000# capacity	Each	88,800
3500# capacity	Each	94,800
5000# capacity	Each	98,300
Additional stop, add	Each	5225
Emergency Lighting, 25 watt, battery operated		
Lead battery	Each	350
Nickel cadmium	Each	590
Laundry Equipment		
Dryer, gas, 16 lb. capacity	Each	665
30 lb. capacity	Each	2675
Washer, 4 cycle	Each	740
Commercial	Each	1125
Smoke Detectors		
Ceiling type	Each	141
Duct type	Each	335

Model costs calculated for a 6 story building with 10'-4" story height and 60,000 square feet of floor area

Apartment, 4-7 Story

			Unit	Unit Cost	Cost Per S.F.	% Of Sub-Total
1.0 Foundations						
.1	Footings & Foundations	Poured concrete; strip and spread footings and 4' foundation wall	S.F. Ground	6.72	1.12	1.7%
.4	Piles & Caissons	N/A	—	—	—	
.9	Excavation & Backfill	Site preparation for slab and trench for foundation wall and footing	S.F. Ground	.85	.14	
2.0 Substructure						
.1	Slab on Grade	4" reinforced concrete with vapor barrier and granular base	S.F. Slab	3.05	.51	0.7%
.2	Special Substructures	N/A	—	—	—	
3.0 Superstructure						
.1	Columns & Beams	Gypsum board fireproofing on columns, steel columns in 3.5 and 3.7	S.F. Floor	1.50	1.50	
.4	Structural Walls	N/A	—	—	—	
.5	Elevated Floors	Open web steel joists, slab form, concrete, steel columns	S.F. Floor	11.50	9.58	17.9%
.7	Roof	Open web steel joists with rib metal deck, steel columns	S.F. Roof	4.82	.80	
.9	Stairs	Concrete filled metal pan	Flight	4080	1.09	
4.0 Exterior Closure						
.1	Walls	Face brick with concrete block backup 86% of wall	S.F. Wall	16.36	7.27	
.5	Exterior Wall Finishes	N/A	—	—	—	12.1%
.6	Doors	Aluminum and glass	Each	3122	.21	
.7	Windows & Glazed Walls	Aluminum horizontal sliding 14% of wall	Each	282	1.32	
5.0 Roofing						
.1	Roof Coverings	Built-up tar and gravel with flashing	S.F. Roof	2.46	.41	
.7	Insulation	Perlite/EPS composite	S.F. Roof	1.18	.20	0.8%
.8	Openings & Specialties	N/A	—	—	—	
6.0 Interior Construction						
.1	Partitions	Gypsum board and sound deadening board on metal studs 8 S.F. Floor/L.F. Partitions	S.F. Partition	3.60	3.60	
.4	Interior Doors	15% solid core wood, 85% hollow core wood 80 S.F Floor/Door	Each	424	5.31	
.5	Wall Finishes	70% paint, 25% vinyl wall covering, 5% ceramic tile	S.F. Surface	1.07	2.13	25.7%
.6	Floor Finishes	60% carpet, 30% vinyl composition tile, 10% ceramic tile	S.F. Floor	3.81	3.81	
.7	Ceiling Finishes	Painted gypsum board on resilient channels	S.F. Ceiling	2.64	2.64	
.9	Interior Surface/Exterior Wall	Painted gypsum board on furring 80% of wall	S.F. Wall	2.81	1.12	
7.0 Conveying						
.1	Elevators	Two geared passenger elevators	Each	120,300	4.01	5.5%
.2	Special Conveyors	N/A	—	—	—	
8.0 Mechanical						
.1	Plumbing	Kitchen, bathroom and service fixtures, supply and drainage 1 Fixture/215 S.F. Floor	Each	1713	7.97	
.2	Fire Protection	Standpipe and wet pipe sprinkler system	S.F. Floor	1.59	1.59	
.3	Heating	Oil fired hot water, baseboard radiation	S.F. Floor	3.99	3.99	26.5%
.4	Cooling	Chilled water, air cooled condenser system	S.F. Floor	5.60	5.60	
.5	Special Systems	N/A	—	—	—	
9.0 Electrical						
.1	Service & Distribution	1000 ampere service, panel board and feeders	S.F. Floor	.85	.85	
.2	Lighting & Power	Incandescent fixtures, receptacles, switches, A.C. and misc. power	S.F. Floor	3.99	3.99	7.1%
.4	Special Electrical	Alarm systems, emergency lighting, and intercom	S.F. Floor	.33	.33	
11.0 Special Construction						
.1	Specialties	Kitchen cabinets	S.F. Floor	1.46	1.46	2.0%
12.0 Site Work						
.1	Earthwork	N/A	—	—	—	
.3	Utilities	N/A	—	—	—	0.0%
.5	Roads & Parking	N/A	—	—	—	
.7	Site Improvements	N/A	—	—	—	
			Sub-Total		72.55	100%
	GENERAL CONDITIONS (Overhead & Profit)			15%	10.88	
	ARCHITECT FEES			7%	5.82	
			Total Building Cost		89.25	

BUILDING TYPES

COMMERCIAL/INDUSTRIAL/INSTITUTIONAL — M.030 — Apartment, 8-24 Story

Costs per square foot of floor area

Exterior Wall	S.F. Area	95000	112000	129000	145000	162000	181000	210000	220000	240000
	L.F. Perimeter	345	386	406	442	480	522	510	526	560
Ribbed Precast Concrete Panel	Steel Frame	98.05	96.35	94.50	93.60	92.80	92.10	89.90	89.60	89.15
	R/Conc. Frame	94.30	92.70	90.95	90.10	89.35	88.65	86.60	86.30	85.90
Face Brick with Concrete Block Back-up	Steel Frame	93.65	91.90	89.90	88.95	88.15	87.40	84.95	84.65	84.20
	R/Conc. Frame	97.00	95.25	93.25	92.35	91.50	90.80	88.40	88.10	87.60
Stucco on Concrete Block	Steel Frame	86.30	84.90	83.50	82.80	82.15	81.55	80.05	79.80	79.45
	R/Conc. Frame	89.60	88.25	86.85	86.15	85.50	84.95	83.45	83.20	82.90
Perimeter Adj., Add or Deduct	Per 100 L.F.	4.20	3.55	3.10	2.75	2.45	2.20	1.85	1.80	1.65
Story Hgt. Adj., Add or Deduct	Per 1 Ft.	1.40	1.35	1.20	1.20	1.15	1.10	.90	.90	.90

For Basement, add $19.65 per square foot of basement area

The above costs were calculated using the basic specifications shown on the facing page. These costs should be adjusted where necessary for design alternatives and owner's requirements. Reported completed project costs, for this type of structure, range from $50.30 to $115.95 per S.F.

Common additives

Description	Unit	$ Cost
Appliances		
Cooking range, 30" free standing		
1 oven	Each	380 - 1450
2 oven	Each	785 - 1775
30" built-in		
1 oven	Each	660 - 1600
2 oven	Each	1050 - 1975
Counter top cook tops, 4 burner	Each	300 - 670
Microwave oven	Each	200 - 1850
Combination range, refrig. & sink, 30" wide	Each	1025 - 2050
72" wide	Each	3025
Combination range, refrigerator, sink, microwave oven & icemaker	Each	4500
Compactor, residential, 4-1 compaction	Each	395 - 660
Dishwasher, built-in, 2 cycles	Each	405 - 840
4 cycles	Each	605 - 1125
Garbage disposer, sink type	Each	117 - 300
Hood for range, 2 speed, vented, 30" wide	Each	183 - 520
42" wide	Each	310 - 560
Refrigerator, no frost 10-12 C.F.	Each	440 - 920
18-20 C.F.	Each	780 - 1425

Description	Unit	$ Cost
Closed Circuit Surveillance, One station		
Camera and monitor	Each	1325
For additional camera stations, add	Each	730
Elevators, Electric passenger, 10 stops		
3000# capacity	Each	196,000
4000# capacity	Each	197,500
5000# capacity	Each	201,500
Additional stop, add	Each	5225
Emergency Lighting, 25 watt, battery operated		
Lead battery	Each	350
Nickel cadmium	Each	590
Laundry Equipment		
Dryer, gas, 16 lb. capacity	Each	665
30 lb. capacity	Each	2675
Washer, 4 cycle	Each	740
Commercial	Each	1125
Smoke Detectors		
Ceiling type	Each	141
Duct type	Each	335

Model costs calculated for a 15 story building with 10'-6" story height and 162,000 square feet of floor area

Apartment, 8-24 Story

			Unit	Unit Cost	Cost Per S.F.	% Of Sub-Total
1.0 Foundations						
.1	Footings & Foundations	Poured concrete; strip and spread footings and 4' foundation wall	S.F. Ground	10.65	.71	1.0%
.4	Piles & Caissons	N/A	—	—	—	
.9	Excavation & Backfill	Site preparation for slab and trench for foundation wall and footing	S.F. Ground	.85	.06	
2.0 Substructure						
.1	Slab on Grade	4" reinforced concrete with vapor barrier and granular base	S.F. Slab	3.05	.20	0.3%
.2	Special Substructures	N/A	—	—	—	
3.0 Superstructure						
.1	Columns & Beams	Gypsum board fireproofing on columns; steel columns in 3.5 and 3.7	S.F. Floor	2.53	2.53	
.4	Structural Walls	N/A	—	—	—	
.5	Elevated Floors	Open web steel joists, slab form, concrete, interior steel columns	S.F. Floor	8.52	7.95	15.5%
.7	Roof	Open web steel joists with rib metal deck, interior steel columns	S.F. Roof	3.64	.24	
.9	Stairs	Concrete filled metal pan	Flight	4080	1.08	
4.0 Exterior Closure						
.1	Walls	Ribbed precast concrete panel 87% of wall	S.F. Wall	15.47	6.28	
.5	Exterior Wall Finishes	N/A	—	—	—	11.0%
.6	Doors	Aluminum and glass	Each	1330	1.01	
.7	Windows & Glazed Walls	Aluminum horizontal sliding 13% of wall	Each	282	1.09	
5.0 Roofing						
.1	Roof Coverings	Built-up tar and gravel with flashing	S.F. Roof	2.40	.16	
.7	Insulation	Perlite/EPS composite	S.F. Roof	1.18	.08	0.3%
.8	Openings & Specialties	N/A	—	—	—	
6.0 Interior Construction						
.1	Partitions	Gypsum board on concrete block and metal studs 10 S.F. of Floor/L.F. Partition	S.F. Partition	7.53	7.53	
.4	Interior Doors	15% solid core wood, 85% hollow core wood 80 S.F. Floor/Door	Each	424	5.31	
.5	Wall Finishes	70% paint, 25% vinyl wall covering, 5% ceramic tile	S.F. Surface	1.05	2.11	29.4%
.6	Floor Finishes	60% carpet, 30% vinyl composition tile, 10% ceramic tile	S.F. Floor	3.81	3.81	
.7	Ceiling Finishes	Painted gypsum board on resilient channels	S.F. Ceiling	2.64	2.64	
.9	Interior Surface/Exterior Wall	Painted gypsum board on furring 80% of wall	S.F. Wall	2.81	1.00	
7.0 Conveying						
.1	Elevators	Four geared passenger elevators	Each	213,030	5.26	6.9%
.2	Special Conveyors	N/A	—	—	—	
8.0 Mechanical						
.1	Plumbing	Kitchen, bathroom and service fixtures, supply and drainage 1 Fixture/210 S.F. Floor	Each	1709	8.14	
.2	Fire Protection	Standpipe and wet pipe sprinkler systems	S.F. Floor	1.91	1.91	
.3	Heating	Oil fired hot water, baseboard radiation	S.F. Floor	3.99	3.99	25.8%
.4	Cooling	Chilled water, air cooled condenser system	S.F. Floor	5.60	5.60	
.5	Special Systems	N/A	—	—	—	
9.0 Electrical						
.1	Service & Distribution	1200 ampere service, panel board and feeders	S.F. Floor	.44	.44	
.2	Lighting & Power	Incandescent fixtures, receptacles, switches, A.C. and misc. power	S.F. Floor	4.30	4.30	8.0%
.4	Special Electrical	Alarm systems, emergency lighting, antenna, intercom and security television	S.F. Floor	1.36	1.36	
11.0 Special Construction						
.1	Specialties	Kitchen cabinets	S.F. Floor	1.35	1.35	1.8%
12.0 Site Work						
.1	Earthwork	N/A	—	—	—	
.3	Utilities	N/A	—	—	—	0.0%
.5	Roads & Parking	N/A	—	—	—	
.7	Site Improvements	N/A	—	—	—	
				Sub-Total	76.14	100%
	GENERAL CONDITIONS (Overhead & Profit)			15%	11.42	
	ARCHITECT FEES			6%	5.24	
				Total Building Cost	92.80	

BUILDING TYPES

COMMERCIAL/INDUSTRIAL/INSTITUTIONAL M.040 Auditorium

Costs per square foot of floor area

Exterior Wall	S.F. Area	12000	15000	18000	21000	24000	27000	30000	33000	36000
	L.F. Perimeter	440	500	540	590	640	665	700	732	770
Face Brick with Concrete Block Back-up	Steel Frame	115.70	112.20	108.85	106.90	105.45	103.40	102.15	101.00	100.20
	Bearing Wall	111.80	108.65	105.65	103.90	102.60	100.75	99.65	98.60	97.90
Precast Concrete	Steel Frame	107.45	104.75	102.15	100.65	99.50	97.95	97.00	96.15	95.50
Decorative Concrete Block	Bearing Wall	95.60	93.45	91.45	90.25	89.30	88.10	87.35	86.65	86.20
Concrete Block	Steel Frame	103.55	101.80	100.25	99.30	98.55	97.65	97.05	96.50	96.15
	Bearing Wall	92.85	90.95	89.15	88.10	87.30	86.25	85.60	85.00	84.60
Perimeter Adj., Add or Deduct	Per 100 L.F.	7.75	6.20	5.20	4.45	3.90	3.50	3.10	2.80	2.60
Story Hgt. Adj., Add or Deduct	Per 1 Ft.	1.25	1.10	1.00	.95	.85	.85	.80	.75	.75

For Basement, add $16.00 per square foot of basement area

The above costs were calculated using the basic specifications shown on the facing page. These costs should be adjusted where necessary for design alternatives and owner's requirements. Reported completed project costs, for this type of structure, range from $48.65 to $128.95 per S.F.

Common additives

Description	Unit	$ Cost
Closed Circuit Surveillance, One station		
Camera and monitor	Each	1325
For additional camera stations, add	Each	730
Emergency Lighting, 25 watt, battery operated		
Lead battery	Each	350
Nickel cadmium	Each	590
Seating		
Auditorium chair, all veneer	Each	127
Veneer back, padded seat	Each	152
Upholstered, spring seat	Each	178
Classroom, movable chair & desk	Set	65 - 120
Lecture hall, pedestal type	Each	127 - 370
Smoke Detectors		
Ceiling type	Each	141
Duct type	Each	335
Sound System		
Amplifier, 250 watts	Each	1500
Speaker, ceiling or wall	Each	130
Trumpet	Each	244

Location Factors appear in the Reference Section

Auditorium

Model costs calculated for a 1 story building with 24' story height and 24,000 square feet of floor area

			Unit	Unit Cost	Cost Per S.F.	% Of Sub-Total
1.0 Foundations						
.1	Footings & Foundations	Poured concrete; strip and spread footings and 4' foundation wall	S.F. Ground	2.27	2.27	3.9%
.4	Piles & Caissons	N/A	—	—	—	
.9	Excavation & Backfill	Site preparation for slab and trench for foundation wall and footing	S.F. Ground	.89	.89	
2.0 Substructure						
.1	Slab on Grade	6" reinforced concrete with vapor barrier and granular base	S.F. Slab	3.61	3.61	4.5%
.2	Special Substructures	N/A	—	—	—	
3.0 Superstructure						
.1	Columns & Beams	Steel columns	S.F. Floor	.37	.37	
.4	Structural Walls	N/A				
.5	Elevated Floors	Open web steel joists, slab form, concrete (balcony)	S.F. Floor	10.27	1.28	14.4%
.7	Roof	Metal deck on steel truss	S.F. Roof	9.24	9.24	
.9	Stairs	Concrete filled metal pan	Flight	5750	.72	
4.0 Exterior Closure						
.1	Walls	Precast concrete panel 80% of wall (adjusted for end walls)	S.F. Wall	15.88	8.13	
.5	Exterior Wall Finishes	N/A	—	—	—	15.5%
.6	Doors	Double aluminum and glass and hollow metal	Each	2350	1.18	
.7	Windows & Glazed Walls	Glass curtain wall 20% of wall	S.F. Window	25	3.26	
5.0 Roofing						
.1	Roof Coverings	Built-up tar and gravel with flashing	S.F. Roof	2.01	2.01	
.7	Insulation	Perlite/EPS composite	S.F. Roof	1.18	1.18	4.2%
.8	Openings & Specialties	Gravel stop and roof hatches	S.F. Roof	.24	.24	
6.0 Interior Construction						
.1	Partitions	Concrete Block and toilet partitions 40 S.F. Floor/L.F. Partition	S.F. Partition	5.05	2.02	
.4	Interior Doors	Single leaf hollow metal 400 S.F. Floor/Door	Each	530	1.33	
.5	Wall Finishes	70% paint, 30% epoxy coating	S.F. Surface	2.15	1.72	19.3%
.6	Floor Finishes	70% hardwood, 30% carpet	S.F. Floor	7.71	7.71	
.7	Ceiling Finishes	Fiberglass board, suspended	S.F. Floor	1.87	1.87	
.9	Interior Surface/Exterior Wall	Paint 80% of wall	S.F. Wall	1.93	.99	
7.0 Conveying						
.1	Elevators	One hydraulic passenger elevator	Each	55,200	2.30	2.8%
.2	Special Conveyors	N/A	—	—	—	
8.0 Mechanical						
.1	Plumbing	Toilet and service fixtures, supply and drainage 1 Fixture/800 S.F. Floor	Each	2848	3.56	
.2	Fire Protection	Wet pipe sprinkler system	S.F. Floor	1.57	1.57	
.3	Heating	Included in 8.4	—	—	—	20.7%
.4	Cooling	Single zone rooftop unit, gas heating, electric cooling	S.F. Floor	11.53	11.53	
.5	Special Systems	N/A	—	—	—	
9.0 Electrical						
.1	Service & Distribution	800 ampere service, panel board and feeders	S.F. Floor	1.37	1.37	
.2	Lighting & Power	Fluorescent fixtures, receptacles, switches, A.C. and misc. power	S.F. Floor	7.20	7.20	14.7%
.4	Special Electrical	Alarm systems and emergency lighting, and public address system	S.F. Floor	3.33	3.33	
11.0 Special Construction						
.1	Specialties	N/A	—	—	—	0.0%
12.0 Site Work						
.1	Earthwork	N/A	—	—	—	
.3	Utilities	N/A	—	—	—	0.0%
.5	Roads & Parking	N/A	—	—	—	
.7	Site Improvements	N/A	—	—	—	
			Sub-Total		80.88	100%
	GENERAL CONDITIONS (Overhead & Profit)			15%	12.13	
	ARCHITECT FEES			7%	6.49	
			Total Building Cost		99.50	

BUILDING TYPES

COMMERCIAL/INDUSTRIAL/INSTITUTIONAL M.050 Bank

Costs per square foot of floor area

Exterior Wall	S.F. Area	2000	2700	3400	4100	4800	5500	6200	6900	7600
	L.F. Perimeter	180	208	236	256	280	303	317	337	357
Face Brick with Concrete Block Back-up	Steel Frame	141.05	131.90	126.55	121.90	119.10	116.90	114.35	112.85	111.60
	R/Conc. Frame	152.85	143.70	138.35	133.75	130.90	128.70	126.15	124.65	123.40
Precast Concrete Panel	Steel Frame	130.45	122.85	118.40	114.55	112.25	110.45	108.35	107.10	106.10
	R/Conc. Frame	142.30	134.70	130.20	126.40	124.05	122.25	120.15	118.90	117.90
Limestone with Concrete Block Back-up	Steel Frame	147.25	137.25	131.35	126.25	123.15	120.70	117.90	116.20	114.85
	R/Conc. Frame	159.10	149.05	143.15	138.05	134.95	132.55	129.70	128.05	126.65
Perimeter Adj., Add or Deduct	Per 100 L.F.	28.40	21.05	16.70	13.85	11.85	10.35	9.20	8.25	7.50
Story Hgt. Adj., Add or Deduct	Per 1 Ft.	2.75	2.35	2.10	1.90	1.80	1.70	1.55	1.50	1.45
	For Basement, add $18.10 per square foot of basement area									

The above costs were calculated using the basic specifications shown on the facing page. These costs should be adjusted where necessary for design alternatives and owner's requirements. Reported completed project costs, for this type of structure, range from $75.10 to $184.90 per S.F.

Common additives

Description	Unit	$ Cost
Bulletproof Teller Window, 44" x 60"	Each	5150
60" x 48"	Each	5475
Closed Circuit Surveillance, One station		
Camera and monitor	Each	1325
For additional camera stations, add	Each	730
Counters, Complete	Station	4125
Door & Frame, 3' x 6'-8", bullet resistant steel with vision panel	Each	3075 - 4175
Drive-up Window, Drawer & micr., not incl. glass	Each	5950 - 9100
Emergency Lighting, 25 watt, battery operated		
Lead battery	Each	350
Nickel cadmium	Each	590
Night Depository	Each	7450 - 12,200
Package Receiver, painted	Each	1375
stainless steel	Each	2150
Partitions, Bullet resistant to 8' high	L.F.	226 - 370
Pneumatic Tube Systems, 2 station	Each	23,600
With TV viewer	Each	43,800

Description	Unit	$ Cost
Service Windows, Pass thru, steel		
24" x 36"	Each	1950
48" x 48"	Each	2875
72" x 40"	Each	3825
For stainless steel frames, add	Each	
Smoke Detectors		
Ceiling type	Each	141
Duct type	Each	335
Twenty-four Hour Teller		
Automatic deposit cash & memo	Each	40,900
Vault Front, Door & frame		
1 hour test, 32"x 78"	Opening	3225
2 hour test, 32" door	Opening	3625
40" door	Opening	4150
4 hour test, 32" door	Opening	3800
40" door	Opening	4475
Time lock, two movement, add	Each	1575

Bank

Model costs calculated for a 1 story building with 14' story height and 4,100 square feet of floor area

				Unit	Unit Cost	Cost Per S.F.	% Of Sub-Total
1.0 Foundations							
.1	Footings & Foundations	Poured concrete; strip and spread footings and 4' foundation wall		S.F. Ground	6.08	6.08	
.4	Piles & Caissons	N/A		—	—	—	7.5%
.9	Excavation & Backfill	Site preparation for slab and trench for foundation wall and footing		S.F. Ground	1.75	1.75	
2.0 Substructure							
.1	Slab on Grade	4" reinforced concrete with vapor barrier and granular base		S.F. Slab	3.05	3.05	2.9%
.2	Special Substructures	N/A		—	—	—	
3.0 Superstructure							
.1	Columns & Beams	Concrete columns		L.F. Columns	80	3.93	
.4	Structural Walls	N/A		—	—	—	
.5	Elevated Floors	N/A		—	—	—	12.9%
.7	Roof	Cast-in-place concrete slab		S.F. Roof	9.59	9.59	
.9	Stairs	N/A		—	—	—	
4.0 Exterior Closure							
.1	Walls	Face brick with concrete block backup	80% of wall	S.F. Wall	21	14.90	
.5	Exterior Wall Finishes	N/A		—	—	—	19.1%
.6	Doors	Double aluminum and glass and hollow metal		Each	1883	.92	
.7	Windows & Glazed Walls	Horizontal aluminum sliding	20% of wall	Each	357	4.16	
5.0 Roofing							
.1	Roof Coverings	Built-up tar and gravel with flashing		S.F. Roof	2.68	2.68	
.7	Insulation	Perlite/EPS composite		S.F. Roof	1.18	1.18	4.0%
.8	Openings & Specialties	Gravel stop		L.F. Perimeter	5.55	.35	
6.0 Interior Construction							
.1	Partitions	Gypsum board on metal studs	20 S.F. of Floor/L.F. Partition	S.F. Partition	2.92	1.46	
.4	Interior Doors	Single leaf hollow metal	200 S.F. Floor/Door	Each	530	2.65	
.5	Wall Finishes	50% vinyl wall covering, 50% paint		S.F. Surface	1.00	1.00	13.1%
.6	Floor Finishes	50% carpet, 50% vinyl composition tile		S.F. Floor	3.55	3.55	
.7	Ceiling Finishes	Mineral fiber tile on concealed zee bars		S.F. Ceiling	3.06	3.06	
.9	Interior Surface/Exterior Wall	Painted gypsum board on furring	80% of wall	S.F. Wall	2.80	1.96	
7.0 Conveying							
.1	Elevators	N/A		—	—	—	0.0%
.2	Special Conveyors	N/A		—	—	—	
8.0 Mechanical							
.1	Plumbing	Toilet and service fixtures, supply and drainage	1 Fixture/580 S.F. Floor	Each	3253	5.61	
.2	Fire Protection	Wet pipe sprinkler system		S.F. Floor	2.04	2.04	
.3	Heating	Included in 8.4		—	—	—	16.9%
.4	Cooling	Single zone rooftop unit, gas heating, electric cooling		S.F. Floor	10.08	10.08	
.5	Special Systems	N/A		—	—	—	
9.0 Electrical							
.1	Service & Distribution	200 ampere service, panel board and feeders		S.F. Floor	1.47	1.47	
.2	Lighting & Power	Fluorescent fixtures, receptacles, switches, A.C. and misc. power		S.F. Floor	6.38	6.38	9.1%
.4	Special Electrical	Alarm systems, emergency lighting, and security television		S.F. Floor	1.68	1.68	
11.0 Special Construction							
.1	Specialties	Vault door, automatic teller, drive up window, closed circuit T.V., night depository		S.F. Floor	15.24	15.24	14.5%
12.0 Site Work							
.1	Earthwork	N/A		—	—	—	
.3	Utilities	N/A		—	—	—	0.0%
.5	Roads & Parking	N/A		—	—	—	
.7	Site Improvements	N/A		—	—	—	
				Sub-Total		104.77	100%
	GENERAL CONDITIONS (Overhead & Profit)				15%	15.72	
	ARCHITECT FEES				11%	13.26	
				Total Building Cost		**133.75**	

BUILDING TYPES

COMMERCIAL/INDUSTRIAL/INSTITUTIONAL

M.060 Bowling Alley

Costs per square foot of floor area

Exterior Wall	S.F. Area	12000	14000	16000	18000	20000	22000	24000	26000	28000
	L.F. Perimeter	460	491	520	540	566	593	620	645	670
Concrete Block	Steel Roof Deck	63.35	61.65	60.30	59.15	58.30	57.60	57.05	56.55	56.10
Decorative Concrete Block	Steel Roof Deck	66.20	64.30	62.75	61.45	60.45	59.65	59.05	58.40	57.95
Face Brick on Concrete Block	Steel Roof Deck	71.65	69.25	67.35	65.65	64.40	63.45	62.65	61.95	61.30
Jumbo Brick on Concrete Block	Steel Roof Deck	68.45	66.35	64.65	63.15	62.10	61.20	60.55	59.90	59.30
Stucco on Concrete Block	Steel Roof Deck	64.70	62.90	61.45	60.25	59.30	58.55	58.00	57.45	57.00
Precast Concrete Panel	Steel Roof Deck	64.50	62.70	61.30	60.05	59.15	58.40	57.85	57.30	56.85
Perimeter Adj., Add or Deduct	Per 100 L.F.	2.65	2.30	2.00	1.75	1.60	1.50	1.35	1.25	1.15
Story Hgt. Adj., Add or Deduct	Per 1 Ft.	.60	.55	.50	.45	.45	.40	.40	.40	.35

For Basement, add $18.15 per square foot of basement area

The above costs were calculated using the basic specifications shown on the facing page. These costs should be adjusted where necessary for design alternatives and owner's requirements. Reported completed project costs, for this type of structure, range from $34.30 to $84.25 per S.F.

Common additives

Description	Unit	$ Cost
Bowling Alleys, Incl. alley, pinsetter,		
scorer, counter & misc. supplies, average	Lane	44,900
For automatic scorer, add (maximum)	Lane	7475
Emergency Lighting, 25 watt, battery operated		
Lead battery	Each	350
Nickel cadmium	Each	590
Lockers, Steel, Single tier, 60" or 72"	Opening	134 - 213
2 tier, 60" or 72" total	Opening	83 - 104
5 tier, box lockers	Opening	44 - 56
Locker bench, lam., maple top only	L.F.	18.40
Pedestals, steel pipe	Each	43
Seating		
Auditorium chair, all veneer	Each	127
Veneer back, padded seat	Each	152
Upholstered, spring seat	Each	178
Sound System		
Amplifier, 250 watts	Each	1500
Speaker, ceiling or wall	Each	130
Trumpet	Each	244

Model costs calculated for a 1 story building with 14' story height and 20,000 square feet of floor area

Bowling Alley

			Unit	Unit Cost	Cost Per S.F.	% Of Sub-Total
1.0 Foundations						
.1	Footings & Foundations	Poured concrete; strip and spread footings and 4' foundation wall	S.F. Ground	2.07	2.07	
.4	Piles & Caissons	N/A	—	—	—	6.7%
.9	Excavation & Backfill	Site preparation for slab and trench for foundation wall and footing	S.F. Ground	1.09	1.09	
2.0 Substructure						
.1	Slab on Grade	4" reinforced concrete with vapor barrier and granular base	S.F. Slab	3.05	3.05	6.4%
.2	Special Substructures	N/A	—	—	—	
3.0 Superstructure						
.1	Columns & Beams	Columns included in 3.7	—	—	—	
.4	Structural Walls	N/A	—	—	—	
.5	Elevated Floors	N/A	—	—	—	10.9%
.7	Roof	Metal deck on open web steel joists with columns and beams	S.F. Roof	5.16	5.16	
.9	Stairs	N/A	—	—	—	
4.0 Exterior Closure						
.1	Walls	Concrete block 90% of wall	S.F. Wall	7.32	2.61	
.5	Exterior Wall Finishes	N/A	—	—	—	9.6%
.6	Doors	Double aluminum and glass and hollow metal	Each	1362	.34	
.7	Windows & Glazed Walls	Horizontal pivoted 10% of wall	Each	961	1.59	
5.0 Roofing						
.1	Roof Coverings	Built-up tar and gravel with flashing	S.F. Roof	2.04	2.04	
.7	Insulation	Perlite/EPS composite	S.F. Roof	1.18	1.18	7.0%
.8	Openings & Specialties	Gravel stop and hatches	S.F. Roof	.08	.08	
6.0 Interior Construction						
.1	Partitions	Concrete block 50 S.F. Floor/L.F. Partition	S.F. Partition	5.60	1.12	
.4	Interior Doors	Hollow metal single leaf 1000 S.F. Floor/Door	Each	530	.53	
.5	Wall Finishes	Paint and block filler	S.F. Surface	1.92	.77	8.5%
.6	Floor Finishes	Vinyl tile 25% of floor	S.F. Floor	1.88	.47	
.7	Ceiling Finishes	Suspended fiberglass board 25% of area	S.F. Ceiling	1.87	.47	
.9	Interior Surface/Exterior Wall	Paint and block filler 90% of wall	S.F. Wall	1.93	.69	
7.0 Conveying						
.1	Elevators	N/A	—	—	—	0.0%
.2	Special Conveyors	N/A	—	—	—	
8.0 Mechanical						
.1	Plumbing	Toilet and service fixtures, supply and drainage 1 Fixture/2200 S.F. Floor	Each	2706	1.23	
.2	Fire Protection	Sprinklers, light hazard	S.F. Floor	1.41	1.41	
.3	Heating	Included in 8.4	—	—	—	33.1%
.4	Cooling	Single zone rooftop unit, gas heating, electric cooling	S.F. Floor	13.05	13.05	
.5	Special Systems	N/A	—	—	—	
9.0 Electrical						
.1	Service & Distribution	800 ampere service, panel board and feeders	S.F. Floor	1.64	1.64	
.2	Lighting & Power	Fluorescent fixtures, receptacles, switches, A.C. and misc. power	S.F. Floor	6.52	6.52	17.8%
.4	Special Electrical	Alarm systems and emergency lighting	S.F. Floor	.26	.26	
11.0 Special Construction						
.1	Specialties	N/A	—	—	—	0.0%
12.0 Site Work						
.1	Earthwork	N/A	—	—	—	
.3	Utilities	N/A	—	—	—	0.0%
.5	Roads & Parking	N/A	—	—	—	
.7	Site Improvements	N/A	—	—	—	
				Sub-Total	47.37	100%
	GENERAL CONDITIONS (Overhead & Profit)			15%	7.11	
	ARCHITECT FEES			7%	3.82	
				Total Building Cost	58.30	

BUILDING TYPES

COMMERCIAL/INDUSTRIAL/INSTITUTIONAL — M.070 Bus Terminal

Costs per square foot of floor area

Exterior Wall	S.F. Area	6000	8000	10000	12000	14000	16000	18000	20000	22000
	L.F. Perimeter	320	386	453	520	540	597	610	660	710
Face Brick with Concrete Block Back-up	Bearing Walls	87.85	84.25	82.20	80.75	78.10	77.30	75.45	74.90	74.40
	Steel Frame	88.55	85.00	82.90	81.50	78.90	78.10	76.30	75.70	75.25
Decorative Concrete Block	Bearing Walls	82.05	79.00	77.20	76.00	73.90	73.20	71.75	71.25	70.85
	Steel Frame	82.70	79.70	77.95	76.75	74.65	74.00	72.60	72.10	71.70
Precast Concrete Panels	Bearing Walls	81.30	78.30	76.55	75.40	73.35	72.70	71.25	70.80	70.40
	Steel Frame	81.95	79.00	77.30	76.15	74.15	73.50	72.10	71.60	71.25
Perimeter Adj., Add or Deduct	Per 100 L.F.	8.00	6.05	4.85	4.00	3.45	3.00	2.65	2.40	2.20
Story Hgt. Adj., Add or Deduct	Per 1 Ft.	1.35	1.25	1.10	1.10	1.00	.95	.85	.85	.80
Basement—Not Applicable										

The above costs were calculated using the basic specifications shown on the facing page. These costs should be adjusted where necessary for design alternatives and owner's requirements. Reported completed project costs, for this type of structure, range from $42.15 to $101.00 per S.F.

Common additives

Description	Unit	$ Cost
Directory Boards, Plastic, glass covered		
30" x 20"	Each	500
36" x 48"	Each	910
Aluminum, 24" x 18"	Each	445
36" x 24"	Each	535
48" x 32"	Each	630
48" x 60"	Each	1425
Emergency Lighting, 25 watt, battery operated		
Lead battery	Each	350
Nickel cadmium	Each	590
Benches, Hardwood	L.F.	69 - 108
Ticket Printer	Each	5325
Turnstiles, One way		
4 arm, 46" dia., manual	Each	395
Electric	Each	1475
High security, 3 arm		
65" dia., manual	Each	3075
Electric	Each	4100
Gate with horizontal bars		
65" dia., 7' high transit type	Each	3575

Model costs calculated for a 1 story building with 14' story height and 12,000 square feet of floor area

Bus Terminal

				Unit	Unit Cost	Cost Per S.F.	% Of Sub-Total
1.0 Foundations							
.1	Footings & Foundations	Poured concrete; strip and spread footings and 4' foundation wall		S.F. Ground	3.72	3.72	
.4	Piles & Caissons	N/A		—	—	—	7.1%
.9	Excavation & Backfill	Site preparation for slab and trench for foundation wall and footing		S.F. Ground	.96	.96	
2.0 Substructure							
.1	Slab on Grade	6" reinforced concrete with vapor barrier and granular base		S.F. Slab	3.61	3.61	5.5%
.2	Special Substructures	N/A		—	—	—	
3.0 Superstructure							
.1	Columns & Beams	Columns included in 3.7		—	—	—	
.4	Structural Walls	N/A		—	—	—	
.5	Elevated Floors	N/A		—	—	—	4.9%
.7	Roof	Metal deck on open web steel joists with columns and beams		S.F. Roof	3.24	3.24	
.9	Stairs	N/A		—	—	—	
4.0 Exterior Closure							
.1	Walls	Face brick with concrete block backup	70% of wall	S.F. Wall	21	9.05	
.5	Exterior Wall Finishes	N/A		—	—	—	20.0%
.6	Doors	Double aluminum and glass		Each	3525	1.18	
.7	Windows & Glazed Walls	Store front	30% of wall	S.F. Window	16.05	2.92	
5.0 Roofing							
.1	Roof Coverings	Built-up tar and gravel with flashing		S.F. Roof	2.32	2.32	
.7	Insulation	Perlite/EPS composite		S.F. Roof	1.18	1.18	5.7%
.8	Openings & Specialties	Gravel stop		L.F. Perimeter	5.55	.24	
6.0 Interior Construction							
.1	Partitions	Lightweight concrete block	15 S.F. Floor/L.F. Partition	S.F. Partition	5.58	3.72	
.4	Interior Doors	Hollow metal	150 S.F. Floor/Door	Each	530	3.53	
.5	Wall Finishes	Glazed coating		S.F. Surface	1.13	1.51	26.5%
.6	Floor Finishes	Quarry tile and vinyl composition tile		S.F. Floor	5.11	5.11	
.7	Ceiling Finishes	Mineral fiber tile on concealed zee bars		S.F. Ceiling	3.06	3.06	
.9	Interior Surface/Exterior Wall	Glazed coating	70% of wall	S.F. Wall	1.14	.48	
7.0 Conveying							
.1	Elevators	N/A		—	—	—	0.0%
.2	Special Conveyors	N/A		—	—	—	
8.0 Mechanical							
.1	Plumbing	Toilet and service fixtures, supply and drainage	1 Fixture/850 S.F. Floor	Each	2278	2.68	
.2	Fire Protection	Wet pipe sprinkler system		S.F. Floor	1.57	1.57	
.3	Heating	Included in 8.4		—	—	—	21.4%
.4	Cooling	Single zone rooftop unit, gas heating, electric cooling		S.F. Floor	9.73	9.73	
.5	Special Systems	N/A		—	—	—	
9.0 Electrical							
.1	Service & Distribution	400 ampere service, panel board and feeders		S.F. Floor	1.01	1.01	
.2	Lighting & Power	Fluorescent fixtures, receptacles, switches, A.C. and misc. power		S.F. Floor	4.36	4.36	8.9%
.4	Special Electrical	Alarm systems and emergency lighting		S.F. Floor	.46	.46	
11.0 Special Construction							
.1	Specialties	N/A		—	—	—	0.0%
12.0 Site Work							
.1	Earthwork	N/A		—	—	—	
.3	Utilities	N/A		—	—	—	0.0%
.5	Roads & Parking	N/A		—	—	—	
.7	Site Improvements	N/A		—	—	—	
				Sub-Total		65.64	100%
	GENERAL CONDITIONS (Overhead & Profit)				15%	9.85	
	ARCHITECT FEES				7%	5.26	
				Total Building Cost		80.75	

BUILDING TYPES

COMMERCIAL/INDUSTRIAL/INSTITUTIONAL M.080 Car Wash

Costs per square foot of floor area

Exterior Wall	S.F. Area	600	800	1000	1200	1600	2000	2400	3000	4000
	L.F. Perimeter	100	114	128	139	164	189	214	250	314
Brick with Concrete Block Back-up	Steel Frame	190.30	182.15	177.25	173.15	168.65	165.95	164.15	162.20	160.55
	Bearing Walls	189.30	181.15	176.25	172.15	167.65	164.95	163.15	161.20	159.55
Concrete Block	Steel Frame	163.40	159.15	156.60	154.45	152.10	150.70	149.80	148.75	147.90
	Bearing Walls	162.45	158.15	155.60	153.45	151.15	149.70	148.80	147.75	146.90
Galvanized Steel Siding	Steel Frame	161.85	158.15	155.95	154.10	152.10	150.85	150.05	149.15	148.40
Metal Sandwich Panel	Steel Frame	162.35	157.75	155.00	152.65	150.15	148.60	147.60	146.45	145.55
Perimeter Adj., Add or Deduct	Per 100 L.F.	56.25	42.20	33.75	28.15	21.10	16.85	14.05	11.25	8.45
Story Hgt. Adj., Add or Deduct	Per 1 Ft.	3.50	3.00	2.70	2.45	2.15	2.00	1.85	1.75	1.65
	Basement—Not Applicable									

The above costs were calculated using the basic specifications shown on the facing page. These costs should be adjusted where necessary for design alternatives and owner's requirements. Reported completed project costs, for this type of structure, range from $66.80 to $196.25 per S.F.

Common additives

Description	Unit	$ Cost
Air Compressors, Electric		
1-1/2 H.P., standard controls	Each	3325
Dual controls	Each	3550
5 H.P., 115/230 volt, standard control	Each	4500
Dual controls	Each	4725
Emergency Lighting, 25 watt, battery operated		
Lead battery	Each	350
Nickel cadmium	Each	590
Fence, Chain link, 6' high		
9 ga. wire, galvanized	L.F.	13.30
6 ga. wire	L.F.	17.20
Gate	Each	241
Product Dispenser with vapor recovery		
for 6 nozzles	Each	16,500
Laundry Equipment		
Dryer, gas, 16 lb. capacity	Each	665
30 lb. capacity	Each	2675
Washer, 4 cycle	Each	740
Commercial	Each	1125

Description	Unit	$ Cost
Lockers, Steel, single tier, 60" or 72"	Opening	134 - 213
2 tier, 60" or 72" total	Opening	83 - 104
5 tier, box lockers	Opening	44 - 56
Locker bench, lam. maple top only	L.F.	18.40
Pedestals, steel pipe	Each	43
Paving, Bituminous		
Wearing course plus base course	Sq. Yard	6.95
Safe, Office type, 4 hour rating		
30" x 18" x 18"	Each	3475
62" x 33" x 20"	Each	7575
Sidewalks, Concrete 4" thick	S.F.	2.56
Yard Lighting,		
20' aluminum pole		
with 400 watt		
high pressure sodium		
fixture	Each	2165

Model costs calculated for a 1 story building with 12' story height and 800 square feet of floor area

Car Wash

				Unit	Unit Cost	Cost Per S.F.	% Of Sub-Total
1.0 Foundations							
.1	Footings & Foundations	Poured concrete; strip and spread footings and 4' foundation wall		S.F. Ground	7.73	7.73	
.4	Piles & Caissons	N/A		–	–	–	6.2%
.9	Excavation & Backfill	Site preparation for slab and trench for foundation wall and footing		S.F. Ground	1.43	1.43	
2.0 Substructure							
.1	Slab on Grade	5" reinforced concrete with vapor barrier and granular base		S.F. Slab	3.27	3.27	2.2%
.2	Special Substructures	N/A		–	–	–	
3.0 Superstructure							
.1	Columns & Beams	Steel columns included in 3.7		–	–	–	
.4	Structural Walls	N/A		–	–	–	
.5	Elevated Floors	N/A		–	–	–	2.1%
.7	Roof	Metal deck, open web steel joists, beams, columns		S.F. Roof	3.18	3.18	
.9	Stairs	N/A		–	–	–	
4.0 Exterior Closure							
.1	Walls	Face brick with concrete block backup	70% of wall	S.F. Wall	21	25.50	
.5	Exterior Wall Finishes	N/A		–	–	–	25.4%
.6	Doors	Steel overhead hollow metal		Each	1695	8.47	
.7	Windows & Glazed Walls	Horizontal pivoted steel	5% of wall	Each	386	3.67	
5.0 Roofing							
.1	Roof Coverings	Built-up tar and gravel with flashing		S.F. Roof	3.69	3.69	
.7	Insulation	Perlite/EPS composite		S.F. Roof	1.18	1.18	3.3%
.8	Openings & Specialties	N/A		–	–	–	
6.0 Interior Construction							
.1	Partitions	Concrete block	20 S.F. Floor/S.F. Partition	S.F. Partition	5.70	2.85	
.4	Interior Doors	Hollow metal	600 S.F. Floor/Door	Each	530	.88	
.5	Wall Finishes	Paint		S.F. Surface	.76	.76	
.6	Floor Finishes	N/A		–	–	–	3.0%
.7	Ceiling Finishes	N/A		–	–	–	
.9	Interior Surface/Exterior Wall	N/A		–	–	–	
7.0 Conveying							
.1	Elevators	N/A		–	–	–	0.0%
.2	Special Conveyors	N/A		–	–	–	
8.0 Mechanical							
.1	Plumbing	Toilet and service fixtures, supply and drainage	1 Fixture/160 S.F. Floor	Each	5057	31.61	
.2	Fire Protection	N/A		–	–	–	
.3	Heating	Unit heaters		S.F. Floor	12.20	12.20	29.7%
.4	Cooling	N/A		–	–	–	
.5	Special Systems	N/A		–	–	–	
9.0 Electrical							
.1	Service & Distribution	400 ampere service, panel board and feeders		S.F. Floor	13.61	13.61	
.2	Lighting & Power	Fluorescent fixtures, receptacles, switches and misc. power		S.F. Floor	26	26.74	28.1%
.4	Special Electrical	Emergency power		S.F. Floor	1.26	1.26	
11.0 Special Construction							
.1	Specialties	N/A		–	–	–	0.0%
12.0 Site Work							
.1	Earthwork	N/A		–	–	–	
.3	Utilities	N/A		–	–	–	0.0%
.5	Roads & Parking	N/A		–	–	–	
.7	Site Improvements	N/A		–	–	–	
				Sub-Total		148.03	100%
	GENERAL CONDITIONS (Overhead & Profit)				15%	22.20	
	ARCHITECT FEES				7%	11.92	
				Total Building Cost		**182.15**	

BUILDING TYPES

COMMERCIAL/INDUSTRIAL/INSTITUTIONAL — M.090 Church

Costs per square foot of floor area

Exterior Wall	S.F. Area	2000	7000	12000	17000	22000	27000	32000	37000	42000
	L.F. Perimeter	180	340	470	540	640	740	762	793	860
Decorative Concrete Brick	Wood Arch	142.00	108.30	100.85	95.25	93.15	91.80	89.15	87.35	86.65
	Steel Truss	138.90	105.25	97.80	92.15	90.10	88.75	86.10	84.25	83.60
Stone with Concrete Block Back-up	Wood Arch	162.95	119.60	110.00	102.60	99.90	98.20	94.70	92.35	91.40
	Steel Truss	156.00	112.65	103.05	95.65	92.95	91.25	87.75	85.35	84.45
Face Brick with Concrete Block Back-up	Wood Arch	163.05	119.70	110.05	102.65	99.95	98.25	94.75	92.35	91.45
	Steel Truss	156.10	112.75	103.05	95.70	93.00	91.30	87.75	85.40	84.50
Perimeter Adj., Add or Deduct	Per 100 L.F.	36.60	10.45	6.10	4.30	3.30	2.70	2.30	2.00	1.75
Story Hgt. Adj., Add or Deduct	Per 1 Ft.	2.20	1.15	.95	.75	.70	.65	.55	.55	.50

For Basement, add $20.00 per square foot of basement area

The above costs were calculated using the basic specifications shown on the facing page. These costs should be adjusted where necessary for design alternatives and owner's requirements. Reported completed project costs, for this type of structure, range from $46.70 to $172.05 per S.F.

Common additives

Description	Unit	$ Cost
Altar, Wood, custom design, plain	Each	1600
Deluxe	Each	9175
Granite or marble, average	Each	6425
Deluxe	Each	14,700
Ark, Prefabricated, plain	Each	3100
Deluxe	Each	107,000
Baptistry, Fiberglass, incl. plumbing	Each	2400 - 4725
Bells & Carillons, 48 bells	Each	357,500
24 bells	Each	137,500
Confessional, Prefabricated wood		
Single, plain	Each	3825
Deluxe	Each	7975
Double, plain	Each	7600
Deluxe	Each	18,100
Emergency Lighting, 25 watt, battery operated		
Lead battery	Each	350
Nickel cadmium	Each	590
Lecterns, Wood, plain	Each	495
Deluxe	Each	1550

Description	Unit	$ Cost
Pews/Benches, Hardwood	L.F.	69 - 108
Pulpits, Prefabricated, hardwood	Each	1275 - 7450
Railing, Hardwood	L.F.	84
Steeples, translucent fiberglass		
30" square, 15' high	Each	2950
25' high	Each	4350
Painted fiberglass, 24" square, 14' high	Each	3000
28' high	Each	4150
Aluminum		
20' high, 3'- 6" base	Each	4275
35' high, 8'- 0" base	Each	16,200
60' high, 14'- 0" base	Each	36,400

Church

Model costs calculated for a 1 story building with 24' story height and 17,000 square feet of floor area

			Unit	Unit Cost	Cost Per S.F.	% Of Sub-Total
1.0 Foundations						
.1	Footings & Foundations	Poured concrete; strip and spread footings and 4' foundation wall	S.F. Ground	4.70	4.70	
.4	Piles & Caissons	N/A	–	–	–	7.0%
.9	Excavation & Backfill	Site preparation for slab and trench for foundation wall and footing	S.F. Ground	.89	.89	
2.0 Substructure						
.1	Slab on Grade	4" reinforced concrete with vapor barrier and granular base	S.F. Slab	3.05	3.05	3.8%
.2	Special Substructures	N/A	–	–	–	
3.0 Superstructure						
.1	Columns & Beams	N/A	–	–	–	
.4	Structural Walls	N/A	–	–	–	
.5	Elevated Floors	N/A	–	–	–	16.6%
.7	Roof	Wood deck on laminated wood arches	S.F. Roof	12.49	13.34	
.9	Stairs	N/A	–	–	–	
4.0 Exterior Closure						
.1	Walls	Face brick with concrete block backup 80% of wall (adjusted for end walls)	S.F. Wall	25	15.43	
.5	Exterior Wall Finishes	N/A	–	–	–	23.1%
.6	Doors	Double hollow metal swinging, single hollow metal	Each	906	.32	
.7	Windows & Glazed Walls	Aluminum, top hinged, in-swinging and curtain wall panels 20% of wall	S.F. Window	18.43	2.81	
5.0 Roofing						
.1	Roof Coverings	Asphalt shingles with flashing	S.F. Roof	1.50	1.50	
.7	Insulation	Polystyrene	S.F. Ground	1.12	1.34	3.8%
.8	Openings & Specialties	Gutters and downspouts	S.F. Ground	.21	.21	
6.0 Interior Construction						
.1	Partitions	Plaster on metal studs 40 S.F. Floor/L.F. Partitions	S.F. Partition	6.63	3.98	
.4	Interior Doors	Hollow metal 400 S.F. Floor/Door	Each	530	1.33	
.5	Wall Finishes	Paint	S.F. Surface	.72	.86	15.9%
.6	Floor Finishes	Carpet	S.F. Floor	4.73	4.73	
.7	Ceiling Finishes	N/A	–	–	–	
.9	Interior Surface/Exterior Wall	Painted plaster 80% of wall	S.F. Wall	3.12	1.90	
7.0 Conveying						
.1	Elevators	N/A	–	–	–	0.0%
.2	Special Conveyors	N/A	–	–	–	
8.0 Mechanical						
.1	Plumbing	Kitchen, toilet and service fixtures, supply and drainage 1 Fixture/2430 S.F. Floor	Each	2551	1.05	
.2	Fire Protection	Wet pipe sprinkler system	S.F. Floor	1.57	1.57	
.3	Heating	Oil fired hot water, wall fin radiation	S.F. Floor	6.05	6.05	20.7%
.4	Cooling	Split systems with air cooled condensing units	S.F. Floor	8.02	8.02	
.5	Special Systems	N/A	–	–	–	
9.0 Electrical						
.1	Service & Distribution	200 ampere service, panel board and feeders	S.F. Floor	.31	.31	
.2	Lighting & Power	Fluorescent fixtures, receptacles, switches, A.C. and misc. power	S.F. Floor	5.23	5.23	9.1%
.4	Special Electrical	Alarm systems, sound system and emergency lighting	S.F. Floor	1.81	1.81	
11.0 Special Construction						
.1	Specialties	N/A	–	–	–	0.0%
12.0 Site Work						
.1	Earthwork	N/A	–	–	–	
.3	Utilities	N/A	–	–	–	0.0%
.5	Roads & Parking	N/A	–	–	–	
.7	Site Improvements	N/A	–	–	–	
			Sub-Total		80.43	100%
	GENERAL CONDITIONS (Overhead & Profit)			15%	12.06	
	ARCHITECT FEES			11%	10.16	
			Total Building Cost		102.65	

BUILDING TYPES

COMMERCIAL/INDUSTRIAL/INSTITUTIONAL M.100 Club, Country

Costs per square foot of floor area

Exterior Wall	S.F. Area	2000	4000	6000	8000	12000	15000	18000	20000	22000
	L.F. Perimeter	180	280	340	386	460	535	560	600	640
Stone Ashlar Veneer On Concrete Block	Wood Truss	140.95	124.45	115.80	110.70	104.85	103.15	100.75	100.00	99.50
	Steel Joists	138.80	122.30	113.70	108.55	102.75	101.05	98.60	97.90	97.35
Stucco on Concrete Block	Wood Truss	128.55	114.80	108.00	104.05	99.60	98.25	96.45	95.90	95.45
	Steel Joists	126.65	112.90	106.15	102.20	97.70	96.40	94.55	94.05	93.60
Brick Veneer	Wood Frame	137.55	121.75	113.60	108.75	103.30	101.65	99.35	98.70	98.20
Wood Shingles	Wood Frame	129.05	115.10	108.25	104.15	99.65	98.25	96.40	95.85	95.45
Perimeter Adj., Add or Deduct	Per 100 L.F.	23.30	11.65	7.75	5.80	3.90	3.10	2.60	2.35	2.10
Story Hgt. Adj., Add or Deduct	Per 1 Ft.	2.55	2.00	1.60	1.35	1.10	1.00	.90	.85	.80

For Basement, add $16.85 per square foot of basement area

The above costs were calculated using the basic specifications shown on the facing page. These costs should be adjusted where necessary for design alternatives and owner's requirements. Reported completed project costs, for this type of structure, range from $46.95 to $153.65 per S.F.

Common additives

Description	Unit	$ Cost
Bar, Front bar	L.F.	278
Back bar	L.F.	220
Booth, Upholstered, custom, straight	L.F.	127 - 207
"L" or "U" shaped	L.F.	143 - 225
Fireplaces, Brick not incl. chimney		
or foundation, 30" x 24" opening	Each	1850
Chimney, standard brick		
Single flue 16" x 20"	V.L.F.	55
20" x 20"	V.L.F.	63
2 flue, 20" x 32"	V.L.F.	92
Lockers, Steel, single tier, 60" or 72"	Opening	134 - 213
2 tier, 60" or 72" total	Opening	83 - 104
5 tier, box lockers	Opening	44 - 56
Locker bench, lam. maple top only	L.F.	18.40
Pedestals, steel pipe	Each	43
Refrigerators, Prefabricated, walk-in		
7'-6" High, 6' x 6'	S.F.	117
10' x 10'	S.F.	91
12' x 14'	S.F.	81
12' x 20'	S.F.	72

Description	Unit	$ Cost
Sauna, Prefabricated, complete, 6' x 4'	Each	3725
6' x 9'	Each	5400
8' x 8'	Each	5725
10' x 12'	Each	8025
Smoke Detectors		
Ceiling type	Each	141
Duct type	Each	335
Sound System		
Amplifier, 250 watts	Each	1500
Speaker, ceiling or wall	Each	130
Trumpet	Each	244
Steam Bath, Complete, to 140 C.F.	Each	1175
To 300 C.F.	Each	1325
To 800 C.F.	Each	3375
To 2500 C.F.	Each	5525
Swimming Pool Complete, gunite	S.F.	45 - 54
Tennis Court, Complete with fence		
Bituminous	Each	19,700 - 35,800
Clay	Each	20,300 - 29,100

Model costs calculated for a 1 story building with 12' story height and 6,000 square feet of floor area

Club, Country

			Unit	Unit Cost	Cost Per S.F.	% Of Sub-Total
1.0 Foundations						
.1	Footings & Foundations	Poured concrete; strip and spread footings and 4' foundation wall	S.F. Ground	5.19	5.19	
.4	Piles & Caissons	N/A	—	—	—	6.8%
.9	Excavation & Backfill	Site preparation for slab and trench for foundation wall and footing	S.F. Ground	.96	.96	
2.0 Substructure						
.1	Slab on Grade	4" reinforced concrete with vapor barrier and granular base	S.F. Slab	3.05	3.05	3.4%
.2	Special Substructures	N/A	—	—	—	
3.0 Superstructure						
.1	Columns & Beams	N/A	—	—	—	
.4	Structural Walls	Load bearing partition walls, see item 6.1	—	—	—	
.5	Elevated Floors	N/A	—	—	—	4.8%
.7	Roof	Wood truss with plywood sheathing	S.F. Ground	4.35	4.35	
.9	Stairs	N/A	—	—	—	
4.0 Exterior Closure						
.1	Walls	Stone ashlar veneer on concrete block 65% of wall	S.F. Wall	21	9.39	
.5	Exterior Wall Finishes	N/A	—	—	—	17.0%
.6	Doors	Double aluminum and glass, hollow metal	Each	1594	1.60	
.7	Windows & Glazed Walls	Aluminum horizontal sliding 35% of wall	Each	282	4.47	
5.0 Roofing						
.1	Roof Coverings	Asphalt shingles	S.F. Roof	.98	1.21	
.7	Insulation	N/A	—	—	—	1.8%
.8	Openings & Specialties	Gutters and downspouts	S.F. Roof	.39	.39	
6.0 Interior Construction						
.1	Partitions	Gypsum board on metal studs, load bearing 14 S.F. Floor/L.F. Partition	S.F. Partition	2.91	2.08	
.4	Interior Doors	Single leaf wood 140 S.F. Floor/Door	Each	399	2.85	
.5	Wall Finishes	40% vinyl wall covering, 40% paint, 20% ceramic tile	S.F. Surface	1.82	2.60	18.7%
.6	Floor Finishes	50% carpet, 30% hardwood tile, 20% ceramic tile	S.F. Floor	5.33	5.33	
.7	Ceiling Finishes	Gypsum plaster on wood furring	S.F. Ceiling	2.83	2.83	
.9	Interior Surface/Exterior Wall	Painted gypsum board on furring 65% of wall	S.F. Wall	2.80	1.24	
7.0 Conveying						
.1	Elevators	N/A	—	—	—	0.0%
.2	Special Conveyors	N/A	—	—	—	
8.0 Mechanical						
.1	Plumbing	Kitchen, toilet and service fixtures, supply and drainage 1 Fixture/125 S.F. Floor	Each	1878	15.03	
.2	Fire Protection	Wet pipe sprinkler system	S.F. Floor	2.04	2.04	
.3	Heating	Included in 8.4	—	—	—	42.1%
.4	Cooling	Multizone rooftop unit, gas heating, electric cooling	S.F. Floor	21	21.25	
.5	Special Systems	N/A	—	—	—	
9.0 Electrical						
.1	Service & Distribution	200 ampere service, panel board and feeders	S.F. Floor	1.21	1.21	
.2	Lighting & Power	Fluorescent fixtures, receptacles, switches, A.C. and misc. power	S.F. Floor	3.24	3.24	5.4%
.4	Special Electrical	Alarm systems and emergency lighting	S.F. Floor	.42	.42	
11.0 Special Construction						
.1	Specialties	N/A	—	—	—	0.0%
12.0 Site Work						
.1	Earthwork	N/A	—	—	—	
.3	Utilities	N/A	—	—	—	0.0%
.5	Roads & Parking	N/A	—	—	—	
.7	Site Improvements	N/A	—	—	—	
			Sub-Total		90.73	100%
	GENERAL CONDITIONS (Overhead & Profit)			15%	13.61	
	ARCHITECT FEES			11%	11.46	
			Total Building Cost		115.80	

BUILDING TYPES

COMMERCIAL/INDUSTRIAL/INSTITUTIONAL M.110 Club, Social

Costs per square foot of floor area

Exterior Wall	S.F. Area	4000	8000	12000	17000	22000	27000	32000	37000	42000
	L.F. Perimeter	280	386	520	585	640	740	840	940	940
Stone Ashlar on Concrete Block	Steel Joists	109.90	94.10	90.15	84.45	81.15	79.95	79.10	78.55	76.75
	Wood Joists	107.85	92.55	88.70	83.20	80.00	78.85	78.05	77.45	75.75
Face Brick on Concrete Block	Steel Joists	109.95	94.15	90.15	84.50	81.15	79.95	79.15	78.55	76.75
	Wood Joists	108.25	92.85	89.00	83.45	80.20	79.05	78.25	77.70	75.95
Decorative Concrete Block	Steel Joists	103.80	89.90	86.40	81.45	78.60	77.60	76.85	76.30	74.80
	Wood Joists	101.70	88.20	84.75	80.00	77.25	76.25	75.55	75.05	73.55
Perimeter Adj., Add or Deduct	Per 100 L.F.	13.85	6.95	4.65	3.25	2.50	2.10	1.75	1.50	1.30
Story Hgt. Adj., Add or Deduct	Per 1 Ft.	2.25	1.55	1.40	1.10	.95	.85	.85	.80	.75

For Basement, add $16.50 per square foot of basement area

The above costs were calculated using the basic specifications shown on the facing page. These costs should be adjusted where necessary for design alternatives and owner's requirements. Reported completed project costs, for this type of structure, range from $45.75 to $119.05 per S.F.

Common additives

Description	Unit	$ Cost
Bar, Front bar	L.F.	278
Back bar	L.F.	220
Booth, Upholstered, custom, straight	L.F.	127 - 207
"L" or "U" shaped	L.F.	143 - 225
Emergency Lighting, 25 watt, battery operated		
Lead battery	Each	350
Nickel cadmium	Each	590
Flagpoles, Complete		
Aluminum, 20' high	Each	1050
40' High	Each	2350
70' High	Each	6675
Fiberglass, 23' High	Each	1375
39'-5" High	Each	2400
59' High	Each	6575
Kitchen Equipment		
Broiler	Each	3300
Coffee urn, twin 6 gallon	Each	5500
Cooler, 6 ft. long, reach-in	Each	2600
Dishwasher, 10-12 racks per hr.	Each	2500
Food warmer, counter 1.2 kW	Each	715

Description	Unit	$ Cost
Kitchen Equipment, cont.		
Freezer, 44 C.F., reach-in	Each	6375
Ice cube maker, 50 lb. per day	Each	1675
Lockers, Steel, single tier, 60" or 72"	Opening	134 - 213
2 tier, 60" or 72" total	Opening	83 - 104
5 tier, box lockers	Opening	44 - 56
Locker bench, lam. maple top only	L.F.	18.40
Pedestals, steel pipe	Each	43
Refrigerators, Prefabricated, walk-in		
7'-6" High, 6' x 6'	S.F.	117
10' x 10'	S.F.	91
12' x 14'	S.F.	81
12' x 20'	S.F.	72
Smoke Detectors		
Ceiling type	Each	141
Duct type	Each	335
Sound System		
Amplifier, 250 watts	Each	1500
Speaker, ceiling or wall	Each	130
Trumpet	Each	244

Club, Social

Model costs calculated for a 1 story building with 12' story height and 22,000 square feet of floor area

				Unit	Unit Cost	Cost Per S.F.	% Of Sub-Total
1.0 Foundations							
.1	Footings & Foundations	Poured concrete; strip and spread footings and 4' foundation wall		S.F. Ground	3.29	3.29	6.3%
.4	Piles & Caissons	N/A		—	—	—	
.9	Excavation & Backfill	Site preparation for slab and trench for foundation wall and footing		S.F. Ground	.89	.89	
2.0 Substructure							
.1	Slab on Grade	4" reinforced concrete with vapor barrier and granular base		S.F. Slab	3.05	3.05	4.6%
.2	Special Substructures	N/A		—	—	—	
3.0 Superstructure							
.1	Columns & Beams	N/A		—	—	—	
.4	Structural Walls	Load bearing partition walls, see item 6.1		—	—	—	5.7%
.5	Elevated Floors	N/A		—	—	—	
.7	Roof	Metal deck on open web steel joists		S.F. Roof	3.76	3.76	
.9	Stairs	N/A		—	—	—	
4.0 Exterior Closure							
.1	Walls	Stone, ashlar veneer on concrete block	65% of wall	S.F. Wall	21	4.82	
.5	Exterior Wall Finishes	N/A		—	—	—	13.4%
.6	Doors	Double aluminum and glass doors		Each	1594	.43	
.7	Windows & Glazed Walls	Window wall	35% of wall	S.F. Window	29	3.60	
5.0 Roofing							
.1	Roof Coverings	Built-up tar and gravel with flashing		S.F. Roof	2.06	2.06	
.7	Insulation	Perlite/EPS composite		S.F. Roof	1.18	1.18	5.2%
.8	Openings & Specialties	Gravel stop		L.F. Perimeter	5.55	.16	
6.0 Interior Construction							
.1	Partitions	Lightweight concrete block	14 S.F. Floor/L.F. Partition	S.F. Partition	5.59	3.99	
.4	Interior Doors	Single leaf wood	140 S.F. Floor/Door	Each	399	2.85	
.5	Wall Finishes	65% paint, 25% vinyl wall covering, 10% ceramic tile		S.F. Surface	1.20	1.72	26.9%
.6	Floor Finishes	60% carpet, 35% hardwood, 15% ceramic tile		S.F. Floor	5.40	5.40	
.7	Ceiling Finishes	Mineral fiber tile on concealed zee bars		S.F. Ceiling	3.06	3.06	
.9	Interior Surface/Exterior Wall	Painted gypsum board on furring	65% of wall	S.F. Wall	3.12	.71	
7.0 Conveying							
.1	Elevators	N/A		—	—	—	0.0%
.2	Special Conveyors	N/A		—	—	—	
8.0 Mechanical							
.1	Plumbing	Kitchen, toilet and service fixtures, supply and drainage	1 Fixture/1050 S.F. Floor	Each	2289	2.18	
.2	Fire Protection	Wet pipe sprinkler system		S.F. Floor	1.57	1.57	
.3	Heating	Included in 8.4		—	—	—	31.5%
.4	Cooling	Multizone rooftop unit, gas heating, electric cooling		S.F. Floor	17.00	17.00	
.5	Special Systems	N/A		—	—	—	
9.0 Electrical							
.1	Service & Distribution	400 ampere service, panel board and feeders		S.F. Floor	.67	.67	
.2	Lighting & Power	Fluorescent fixtures, receptacles, switches, A.C. and misc. power		S.F. Floor	3.31	3.31	6.4%
.4	Special Electrical	Alarm systems and emergency lighting		S.F. Floor	.23	.23	
11.0 Special Construction							
.1	Specialties	N/A		—	—	—	0.0%
12.0 Site Work							
.1	Earthwork	N/A		—	—	—	
.3	Utilities	N/A		—	—	—	0.0%
.5	Roads & Parking	N/A		—	—	—	
.7	Site Improvements	N/A		—	—	—	
				Sub-Total		65.93	100%
	GENERAL CONDITIONS (Overhead & Profit)				15%	9.89	
	ARCHITECT FEES				7%	5.33	
				Total Building Cost		**81.15**	

BUILDING TYPES

COMMERCIAL/INDUSTRIAL/INSTITUTIONAL — M.120 — College, Classroom, 2-3 Story

Costs per square foot of floor area

Exterior Wall	S.F. Area	30000	45000	60000	75000	90000	105000	120000	135000	150000
	L.F. Perimeter	550	650	800	950	1100	1250	1210	1330	1450
Face Brick with Concrete Block Back-up	Steel Frame	94.10	88.55	86.50	85.20	84.40	83.85	82.05	81.70	81.35
	Bearing Walls	94.60	88.25	85.95	84.55	83.65	83.05	80.85	80.45	80.10
Decorative Concrete Block	Steel Frame	90.05	85.40	83.55	82.40	81.75	81.20	79.85	79.50	79.25
	Bearing Walls	90.55	85.05	83.00	81.75	80.95	80.40	78.65	78.25	77.95
Stucco on Concrete Block	Steel Frame	88.50	83.95	82.20	81.10	80.40	79.90	78.65	78.30	78.05
	Bearing Walls	89.75	84.40	82.40	81.20	80.40	79.85	78.20	77.80	77.50
Perimeter Adj., Add or Deduct	Per 100 L.F.	2.70	1.85	1.35	1.10	.90	.75	.70	.65	.55
Story Hgt. Adj., Add or Deduct	Per 1 Ft.	1.10	.85	.80	.75	.75	.70	.60	.60	.60

For Basement, add $19.70 per square foot of basement area

The above costs were calculated using the basic specifications shown on the facing page. These costs should be adjusted where necessary for design alternatives and owner's requirements. Reported completed project costs, for this type of structure, range from $64.40 to $159.35 per S.F.

Common additives

Description	Unit	$ Cost
Carrels Hardwood	Each	655 - 850
Clock System		
20 Room	Each	11,700
50 Room	Each	28,300
Elevators, Hydraulic passenger, 2 stops		
1500# capacity	Each	40,800
2500# capacity	Each	42,100
3500# capacity	Each	45,700
Additional stop, add	Each	3350
Emergency Lighting, 25 watt, battery operated		
Lead battery	Each	350
Nickel cadmium	Each	590
Flagpoles, Complete		
Aluminum, 20' high	Each	1050
40' High	Each	2350
70' High	Each	6675
Fiberglass, 23' High		1375
39'-5" High	Each	2400
59' High	Each	6575

Description	Unit	$ Cost
Lockers, Steel, single tier, 60" or 72"	Opening	134 - 213
2 tier, 60" or 72" total	Opening	83 - 104
5 tier, box lockers	Opening	44 - 56
Locker bench, lam. maple top only	L.F.	18.40
Pedestals, steel pipe	Each	43
Seating		
Auditorium chair, all veneer	Each	127
Veneer back, padded seat	Each	152
Upholstered, spring seat	Each	178
Classroom, movable chair & desk	Set	65 - 120
Lecture hall, pedestal type	Each	127 - 370
Smoke Detectors		
Ceiling type	Each	141
Duct type	Each	335
Sound System		
Amplifier, 250 watts	Each	1500
Speaker, ceiling or wall	Each	130
Trumpet	Each	244
TV Antenna, Master system, 12 outlet	Outlet	227
30 outlet	Outlet	146
100 outlet	Outlet	140

College, Classroom, 2-3 Story

Model costs calculated for a 2 story building with 12' story height and 90,000 square feet of floor area

				Unit	Unit Cost	Cost Per S.F.	% Of Sub-Total
1.0	**Foundations**						
.1	Footings & Foundations	Poured concrete; strip and spread footings and 4' foundation wall		S.F. Ground	2.50	1.25	
.4	Piles & Caissons	N/A		—	—	—	2.6%
.9	Excavation & Backfill	Site preparation for slab and trench for foundation wall and footing		S.F. Ground	.85	.43	
2.0	**Substructure**						
.1	Slab on Grade	4" reinforced concrete with vapor barrier and granular base		S.F. Slab	3.05	1.52	2.3%
.2	Special Substructures	N/A		—	—	—	
3.0	**Superstructure**						
.1	Columns & Beams	Interior columns included in 3.7		—	—	—	
.4	Structural Walls	Concrete block bearing walls		S.F. Wall	5.55	1.63	
.5	Elevated Floors	Open web steel joists, slab form, concrete		S.F. Floor	9.18	4.59	14.1%
.7	Roof	Metal deck on open web steel joists, columns		S.F. Roof	4.85	2.42	
.9	Stairs	Concrete filled metal pan		Flight	5750	.64	
4.0	**Exterior Closure**						
.1	Walls	Decorative concrete block	65% of wall	S.F. Wall	9.76	1.86	
.5	Exterior Wall Finishes	N/A		—	—	—	7.0%
.6	Doors	Double glass and aluminum with transom		Each	2750	.18	
.7	Windows & Glazed Walls	Window wall	35% of wall	S.F. Wall	25	2.57	
5.0	**Roofing**						
.1	Roof Coverings	Built-up tar and gravel with flashing		S.F. Roof	1.96	.98	
.7	Insulation	Perlite/EPS composite		S.F. Roof	1.18	.59	2.5%
.8	Openings & Specialties	Gravel stop		L.F. Perimeter	5.55	.07	
6.0	**Interior Construction**						
.1	Partitions	Concrete block	20 S.F. Floor/L.F. Partition	S.F. Partition	5.06	2.53	
.4	Interior Doors	Single leaf hollow metal	200 S.F. Floor/Door	Each	530	2.65	
.5	Wall Finishes	95% paint, 5% ceramic tile		S.F. Surface	2.09	2.09	20.9%
.6	Floor Finishes	70% vinyl composition tile, 25% carpet, 5% ceramic tile		S.F. Floor	3.05	3.05	
.7	Ceiling Finishes	Mineral fiber tile on concealed zee bars		S.F. Ceiling	3.06	3.06	
.9	Interior Surface/Exterior Wall	Paint and block filler	65% of wall	S.F. Wall	1.93	.37	
7.0	**Conveying**						
.1	Elevators	Two hydraulic passenger elevators		Each	49,950	1.11	1.7%
.2	Special Conveyors	N/A		—	—	—	
8.0	**Mechanical**						
.1	Plumbing	Toilet and service fixtures, supply and drainage	1 Fixture/455 S.F. Floor	Each	2479	5.45	
.2	Fire Protection	Sprinklers, light hazard		S.F. Floor	1.24	1.24	
.3	Heating	Included in 8.4		—	—	—	30.0%
.4	Cooling	Multizone unit, gas heating, electric cooling		S.F. Floor	13.05	13.05	
.5	Special Systems	N/A		—	—	—	
9.0	**Electrical**						
.1	Service & Distribution	1600 ampere service, panel board and feeders		S.F. Floor	1.19	1.19	
.2	Lighting & Power	Fluorescent fixtures, receptacles, switches, A.C. and misc. power		S.F. Floor	7.51	7.51	16.3%
.4	Special Electrical	Alarm systems, communications systems and emergency lighting		S.F. Floor	2.04	2.04	
11.0	**Special Construction**						
.1	Specialties	Chalkboards, counters, cabinets		S.F. Floor	1.73	1.73	2.6%
12.0	**Site Work**						
.1	Earthwork	N/A		—	—	—	
.3	Utilities	N/A		—	—	—	0.0%
.5	Roads & Parking	N/A		—	—	—	
.7	Site Improvements	N/A		—	—	—	
				Sub-Total		65.80	100%
	GENERAL CONDITIONS (Overhead & Profit)				15%	9.87	
	ARCHITECT FEES				7%	5.28	
				Total Building Cost		80.95	

BUILDING TYPES

COMMERCIAL/INDUSTRIAL/INSTITUTIONAL

M.130 College, Dormitory, 2-3 Story

Costs per square foot of floor area

Exterior Wall	S.F. Area	20000	30000	40000	50000	60000	70000	80000	90000	100000
	L.F. Perimeter	341	454	476	550	600	628	684	721	772
Face Brick with Concrete Block Back-up	R/Conc. Frame	93.55	90.00	85.65	84.25	82.80	81.45	80.90	80.10	79.70
	Steel Frame	96.80	93.25	88.95	87.50	86.05	84.75	84.15	83.35	82.95
Decorative Concrete Block	R/Conc. Frame	86.85	84.10	81.10	80.05	79.00	78.10	77.70	77.15	76.85
	Steel Frame	89.85	87.05	84.05	83.00	81.95	81.05	80.65	80.10	79.80
Precast Concrete Panels	R/Conc. Frame	89.35	86.30	82.80	81.55	80.40	79.30	78.85	78.20	77.85
	Steel Frame	92.55	89.50	86.00	84.75	83.60	82.50	82.05	81.40	81.05
Perimeter Adj., Add or Deduct	Per 100 L.F.	5.60	3.75	2.80	2.20	1.85	1.60	1.35	1.25	1.10
Story Hgt. Adj., Add or Deduct	Per 1 Ft.	1.45	1.25	1.00	.90	.85	.75	.70	.65	.65

For Basement, add $19.85 per square foot of basement area

The above costs were calculated using the basic specifications shown on the facing page. These costs should be adjusted where necessary for design alternatives and owner's requirements. Reported completed project costs, for this type of structure, range from $44.05 to $123.05 per S.F.

Common additives

Description	Unit	$ Cost
Carrels Hardwood	Each	655 - 850
Closed Circuit Surveillance, One station		
Camera and monitor	Each	1325
For additional camera stations, add	Each	730
Elevators, Hydraulic passenger, 2 stops		
2000# capacity	Each	41,700
2500# capacity	Each	42,100
3500# capacity	Each	45,700
Additional stop, add	Each	3350
Emergency Lighting, 25 watt, battery operated		
Lead battery	Each	350
Nickel cadmium	Each	590
Furniture	Student	1900 - 3600
Intercom System, 25 station capacity		
Master station	Each	1800
Intercom outlets	Each	113
Handset	Each	299

Description	Unit	$ Cost
Kitchen Equipment		
Broiler	Each	3300
Coffee urn, twin 6 gallon	Each	5500
Cooler, 6 ft. long	Each	2600
Dishwasher, 10-12 racks per hr.	Each	2500
Food warmer	Each	715
Freezer, 44 C.F., reach-in	Each	6375
Ice cube maker, 50 lb. per day	Each	1675
Range with 1 oven	Each	2250
Laundry Equipment		
Dryer, gas, 16 lb. capacity	Each	665
30 lb. capacity	Each	2675
Washer, 4 cycle	Each	740
Commercial	Each	1125
Smoke Detectors		
Ceiling type	Each	141
Duct type	Each	335
TV Antenna, Master system, 12 outlet	Outlet	227
30 outlet	Outlet	146
100 outlet	Outlet	140

Model costs calculated for a 3 story building with 12' story height and 40,000 square feet of floor area

College, Dormitory, 2-3 Story

				Unit	Unit Cost	Cost Per S.F.	% Of Sub-Total
1.0 Foundations							
.1	Footings & Foundations	Poured concrete; strip and spread footings and 4' foundation wall		S.F. Ground	4.23	1.41	2.5%
.4	Piles & Caissons	N/A		—	—	—	
.9	Excavation & Backfill	Site preparation for slab and trench for foundation wall and footing		S.F. Ground	.89	.30	
2.0 Substructure							
.1	Slab on Grade	4" reinforced concrete with vapor barrier and granular base		S.F. Slab	3.05	1.02	1.5%
.2	Special Substructures	N/A		—	—	—	
3.0 Superstructure							
.1	Columns & Beams	Concrete columns		L.F. Column	70	3.05	
.4	Structural Walls	N/A		—	—	—	
.5	Elevated Floors	Concrete flat plate		S.F. Floor	8.66	5.77	18.0%
.7	Roof	Concrete flat plate		S.F. Roof	8.31	2.77	
.9	Stairs	Concrete		Flight	2950	.96	
4.0 Exterior Closure							
.1	Walls	Face brick with concrete block backup	80% of wall	S.F. Wall	21	7.30	
.5	Exterior Wall Finishes	N/A		—	—	—	13.7%
.6	Doors	Double glass & aluminum doors		Each	3500	.53	
.7	Windows & Glazed Walls	Aluminum horizontal sliding	20% of wall	Each	458	1.71	
5.0 Roofing							
.1	Roof Coverings	Built-up tar and gravel with flashing		S.F. Roof	2.16	.72	
.7	Insulation	Perlite/EPS composite		S.F. Roof	1.18	.39	1.7%
.8	Openings & Specialties	Gravel stop		L.F. Perimeter	5.55	.07	
6.0 Interior Construction							
.1	Partitions	Gypsum board on metal studs, concrete block	9 S.F. Floor/L.F. Partition	S.F. Partition	4.00	4.44	
.4	Interior Doors	Single leaf wood	90 S.F. Floor/Door	Each	399	4.43	
.5	Wall Finishes	95% paint, 5% ceramic tile		S.F. Surface	.76	1.68	24.2%
.6	Floor Finishes	80% carpet, 10% vinyl composition tile, 10% ceramic tile		S.F. Floor	5.14	5.14	
.7	Ceiling Finishes	90% paint, 10% suspended fiberglass board		S.F. Ceiling	.53	.53	
.9	Interior Surface/Exterior Wall	Paint	80% of wall	S.F. Wall	1.93	.66	
7.0 Conveying							
.1	Elevators	One hydraulic passenger elevator		Each	64,000	1.60	2.3%
.2	Special Conveyors	N/A		—	—	—	
8.0 Mechanical							
.1	Plumbing	Toilet and service fixtures, supply and drainage	1 Fixture/455 S.F. Floor	Each	1965	4.32	
.2	Fire Protection	Wet pipe sprinkler system		S.F. Floor	1.37	1.37	
.3	Heating	Included in 8.4		—	—	—	19.6%
.4	Cooling	Rooftop multizone unit system		S.F. Floor	7.95	7.95	
.5	Special Systems	N/A		—	—	—	
9.0 Electrical							
.1	Service & Distribution	600 ampere service, panel board and feeders		S.F. Floor	.69	.69	
.2	Lighting & Power	Fluorescent fixtures, receptacles, switches, A.C. and misc. power		S.F. Floor	5.67	5.67	11.1%
.4	Special Electrical	Alarm systems, communications systems and emergency lighting		S.F. Floor	1.40	1.40	
11.0 Special Construction							
.1	Specialties	Closet shelving, mirrors		S.F. Floor	3.75	3.75	5.4%
12.0 Site Work							
.1	Earthwork	N/A		—	—	—	
.3	Utilities	N/A		—	—	—	0.0%
.5	Roads & Parking	N/A		—	—	—	
.7	Site Improvements	N/A		—	—	—	
				Sub-Total		69.63	100%
	GENERAL CONDITIONS (Overhead & Profit)				15%	10.44	
	ARCHITECT FEES				7%	5.58	
				Total Building Cost		85.65	

BUILDING TYPES

COMMERCIAL/INDUSTRIAL/INSTITUTIONAL — M.140 — College, Dormitory, 4-8 Story

Costs per square foot of floor area

Exterior Wall	S.F. Area	50000	70000	90000	100000	110000	130000	150000	170000	190000
	L.F. Perimeter	372	461	550	533	566	633	650	703	756
Face Brick with Concrete Block Back-up	R/Conc. Frame	92.00	89.35	87.85	86.05	85.50	84.70	83.40	82.90	82.50
	Steel Frame	97.70	95.05	93.55	91.75	91.20	90.45	89.10	88.60	88.20
Decorative Concrete Block	R/Conc. Frame	86.35	84.35	83.20	81.95	81.60	81.00	80.10	79.75	79.45
	Steel Frame	92.15	90.15	88.95	87.75	87.35	86.80	85.90	85.55	85.20
Precast Concrete Panels With Exposed Aggregate	R/Conc. Frame	90.70	88.20	86.80	85.10	84.60	83.90	82.65	82.20	81.80
	Steel Frame	96.40	93.85	92.45	90.80	90.30	89.55	88.35	87.85	87.45
Perimeter Adj., Add or Deduct	Per 100 L.F.	4.25	3.00	2.35	2.15	1.95	1.65	1.40	1.30	1.15
Story Hgt. Adj., Add or Deduct	Per 1 Ft.	1.20	1.10	1.00	.85	.85	.80	.70	.70	.65

For Basement, add $19.60 per square foot of basement area

The above costs were calculated using the basic specifications shown on the facing page. These costs should be adjusted where necessary for design alternatives and owner's requirements. Reported completed project costs, for this type of structure, range from $61.35 to $134.40 per S.F.

Common additives

Description	Unit	$ Cost
Carrels Hardwood	Each	655 - 850
Closed Circuit Surveillance, One station		
Camera and monitor	Each	1325
For additional camera stations, add	Each	730
Elevators, Electric passenger, 5 stops		
2000# capacity	Each	88,800
2500# capacity	Each	91,800
3500# capacity	Each	92,800
Additional stop, add	Each	5225
Emergency Lighting, 25 watt, battery operated		
Lead battery	Each	350
Nickel cadmium	Each	590
Furniture	Student	1900 - 3600
Intercom System, 25 station capacity		
Master station	Each	1800
Intercom outlets	Each	113
Handset	Each	299

Description	Unit	$ Cost
Kitchen Equipment		
Broiler	Each	3300
Coffee urn, twin, 6 gallon	Each	5500
Cooler, 6 ft. long	Each	2600
Dishwasher, 10-12 racks per hr.	Each	2500
Food warmer	Each	715
Freezer, 44 C.F., reach-in	Each	6375
Ice cube maker, 50 lb. per day	Each	1675
Range with 1 oven	Each	2250
Laundry Equipment		
Dryer, gas, 16 lb. capacity	Each	665
30 lb. capacity	Each	2675
Washer, 4 cycle	Each	740
Commercial	Each	1125
Smoke Detectors		
Ceiling type	Each	141
Duct type	Each	335
TV Antenna, Master system, 12 outlet	Outlet	227
30 outlet	Outlet	146
100 outlet	Outlet	140

Location Factors appear in the Reference Section

Model costs calculated for a 6 story building with 12' story height and 110,000 square feet of floor area

College, Dormitory, 4-8 Story

				Unit	Unit Cost	Cost Per S.F.	% Of Sub-Total
1.0 Foundations							
.1	Footings & Foundations	Poured concrete; strip and spread footings and 4' foundation wall		S.F. Ground	6.48	1.08	1.7%
.4	Piles & Caissons	N/A		—	—	—	
.9	Excavation & Backfill	Site preparation for slab and trench for foundation wall and footing		S.F. Ground	.85	.14	
2.0 Substructure							
.1	Slab on Grade	4" reinforced concrete with vapor barrier and granular base		S.F. Slab	3.05	.51	0.7%
.2	Special Substructures	N/A		—	—	—	
3.0 Superstructure							
.1	Columns & Beams	Steel columns with fireproofing		L.F. Column	121	1.76	
.4	Structural Walls	N/A		—	—	—	
.5	Elevated Floors	Concrete slab with metal deck and beams		S.F. Floor	12.01	10.01	20.6%
.7	Roof	Concrete slab with metal deck and beams		S.F. Roof	11.28	1.88	
.9	Stairs	Concrete filled metal pan		Flight	5750	.94	
4.0 Exterior Closure							
.1	Walls	Decorative concrete block	80% of wall	S.F. Wall	10.73	3.18	
.5	Exterior Wall Finishes	N/A		—	—	—	6.7%
.6	Doors	Double glass & aluminum doors		Each	2296	.17	
.7	Windows & Glazed Walls	Aluminum horizontal sliding	20% of wall	Each	282	1.39	
5.0 Roofing							
.1	Roof Coverings	Built-up tar and gravel with flashing		S.F. Roof	2.10	.35	
.7	Insulation	Perlite/EPS composite		S.F. Roof	1.18	.20	0.8%
.8	Openings & Specialties	Gravel stop		L.F. Perimeter	5.55	.03	
6.0 Interior Construction							
.1	Partitions	Concrete block	9 S.F. Floor/L.F. Partition	S.F. Partition	4.69	5.21	
.4	Interior Doors	Single leaf wood	90 S.F. Floor/Door	Each	399	4.43	
.5	Wall Finishes	95% paint, 5% ceramic tile		S.F. Surface	1.28	2.84	26.4%
.6	Floor Finishes	80% carpet, 10% vinyl composition tile, 10% ceramic tile		S.F. Floor	5.14	5.14	
.7	Ceiling Finishes	Mineral fiber tile on concealed zee bars, paint		S.F. Ceiling	.53	.53	
.9	Interior Surface/Exterior Wall	Paint and block filler	80% of wall	S.F. Wall	1.93	.57	
7.0 Conveying							
.1	Elevators	Four geared passenger elevator		Each	123,200	4.48	6.3%
.2	Special Conveyors	N/A		—	—	—	
8.0 Mechanical							
.1	Plumbing	Toilet and service fixtures, supply and drainage	1 Fixture/390 S.F. Floor	Each	1903	4.88	
.2	Fire Protection	Sprinklers, light hazard		S.F. Floor	1.32	1.32	
.3	Heating	Oil fired hot water, wall fin radiation		S.F. Floor	2.69	2.69	20.4%
.4	Cooling	Chilled water, air cooled condenser system		S.F. Floor	5.60	5.60	
.5	Special Systems	N/A		—	—	—	
9.0 Electrical							
.1	Service & Distribution	1000 ampere service, panel board and feeders		S.F. Floor	.47	.47	
.2	Lighting & Power	Fluorescent fixtures, receptacles, switches, A.C. and misc. power		S.F. Floor	5.59	5.59	11.1%
.4	Special Electrical	Alarm systems, communications systems and emergency lighting		S.F. Floor	1.81	1.81	
11.0 Special Construction							
.1	Specialties	Closet shelving, mirrors		S.F. Floor	3.79	3.79	5.3%
12.0 Site Work							
.1	Earthwork	N/A		—	—	—	
.3	Utilities	N/A		—	—	—	0.0%
.5	Roads & Parking	N/A		—	—	—	
.7	Site Improvements	N/A		—	—	—	
				Sub-Total		70.99	100%
	GENERAL CONDITIONS (Overhead & Profit)				15%	10.65	
	ARCHITECT FEES				7%	5.71	
				Total Building Cost		87.35	

BUILDING TYPES

COMMERCIAL/INDUSTRIAL/INSTITUTIONAL

M.150 College, Laboratory

Costs per square foot of floor area

Exterior Wall	S.F. Area	12000	20000	28000	37000	45000	57000	68000	80000	92000
	L.F. Perimeter	470	600	698	793	900	1060	1127	1200	1320
Face Brick with Concrete Brick Back-up	Steel Frame	154.20	127.00	114.50	106.60	102.65	98.70	95.50	93.00	91.50
	Bearing Walls	151.55	124.35	111.85	103.95	100.00	96.05	92.85	90.35	88.85
Decorative Concrete Block	Steel Frame	149.45	123.40	111.50	104.00	100.20	96.45	93.45	91.15	89.80
	Bearing Walls	146.65	120.55	108.70	101.20	97.40	93.65	90.65	88.35	86.95
Stucco on Concrete Block	Steel Frame	148.00	122.25	110.55	103.25	99.50	95.75	92.85	90.60	89.25
	Bearing Walls	145.20	119.45	107.75	100.45	96.70	92.95	90.05	87.80	86.45
Perimeter Adj., Add or Deduct	Per 100 L.F.	6.15	3.70	2.65	2.00	1.65	1.30	1.10	.90	.85
Story Hgt. Adj., Add or Deduct	Per 1 Ft.	1.20	.90	.75	.65	.60	.55	.50	.45	.45

For Basement, add $18.15 per square foot of basement area

The above costs were calculated using the basic specifications shown on the facing page. These costs should be adjusted where necessary for design alternatives and owner's requirements. Reported completed project costs, for this type of structure, range from $89.85 to $168.95 per S.F.

Common additives

Description	Unit	$ Cost
Cabinets, Base, door units, metal	L.F.	154
Drawer units	L.F.	267
Tall storage cabinets, open	L.F.	275
With doors	L.F.	310
Wall, metal 12-1/2" deep, open	L.F.	104
With doors	L.F.	179
Carrels Hardwood	Each	655 - 850
Countertops, not incl. base cabinets, acid proof	S.F.	23 - 33
Stainless steel	S.F.	73
Fume Hood, Not incl. ductwork	L.F.	825 - 1700
Ductwork	Hood	1650 - 6050
Glassware Washer, Distilled water rinse	Each	5500 - 19,500
Seating		
Auditorium chair, all veneer	Each	127
Veneer back, padded seat	Each	152
Upholstered, spring seat	Each	178
Classroom, movable chair & desk	Set	65 - 120
Lecture hall, pedestal type	Each	127 - 370

Description	Unit	$ Cost
Safety Equipment, Eye wash, hand held	Each	360
Deluge shower	Each	193
Sink, One piece plastic		
Flask wash, freestanding	Each	1575
Tables, acid resist. top, drawers	L.F.	132
Titration Unit, Four 2000 ml reservoirs	Each	5875
Alternate Pricing Method: As % of lab furniture		
Plumbing, final connections, simple		10%
Moderately complex		15%
Complex		20%

Model costs calculated for a 1 story building with 12' story height and 45,000 square feet of floor area

College, Laboratory

			Unit	Unit Cost	Cost Per S.F.	% Of Sub-Total
1.0 Foundations						
.1	Footings & Foundations	Poured concrete; strip and spread footings and 4' foundation wall	S.F. Ground	5.54	5.54	
.4	Piles & Caissons	N/A	—	—	—	8.1%
.9	Excavation & Backfill	Site preparation for slab and trench for foundation wall and footing	S.F. Ground	.89	.89	
2.0 Substructure						
.1	Slab on Grade	4" reinforced concrete with vapor barrier and granular base	S.F. Slab	3.05	3.05	3.9%
.2	Special Substructures	N/A	—	—	—	
3.0 Superstructure						
.1	Columns & Beams	N/A	—	—	—	
.4	Structural Walls	Grouted concrete block wall	S.F. Wall	2.34	2.34	
.5	Elevated Floors	Metal deck and concrete trench cover (5680 S.F.)	S.F. Cover	3.22	.41	6.9%
.7	Roof	Metal deck on open web steel joists	S.F. Roof	2.72	2.72	
.9	Stairs	N/A	—	—	—	
4.0 Exterior Closure						
.1	Walls	Face brick with concrete block backup 75% of wall	S.F. Wall	21	3.83	
.5	Exterior Wall Finishes	N/A	—	—	—	8.2%
.6	Doors	Glass and metal doors and entrances with transom	Each	2538	1.12	
.7	Windows & Glazed Walls	Window wall 25% of wall	S.F. Window	25	1.51	
5.0 Roofing						
.1	Roof Coverings	Built-up tar and gravel with flashing	S.F. Roof	1.89	1.89	
.7	Insulation	Perlite/EPS composite	S.F. Roof	1.18	1.18	4.3%
.8	Openings & Specialties	Gravel stop and skylight	S.F. Roof	.31	.31	
6.0 Interior Construction						
.1	Partitions	Concrete block 10 S.F. Floor/L.F. Partition	S.F. Partition	5.32	5.32	
.4	Interior Doors	Single leaf-kalamein fire doors 820 S.F. Floor/Door	Each	608	.74	
.5	Wall Finishes	60% paint, 40% epoxy coating	S.F. Surface	1.72	3.44	21.4%
.6	Floor Finishes	60% epoxy, 20% carpet, 20% vinyl composition tile	S.F. Floor	3.99	3.99	
.7	Ceiling Finishes	Mineral fiber tile on concealed zee runners	S.F. Ceiling	3.06	3.06	
.9	Interior Surface/Exterior Wall	Paint and block filler 75% of wall	S.F. Wall	1.93	.35	
7.0 Conveying						
.1	Elevators	N/A	—	—	—	0.0%
.2	Special Conveyors	N/A	—	—	—	
8.0 Mechanical						
.1	Plumbing	Toilet and service fixtures, supply and drainage 1 Fixture/260 S.F. Floor	Each	3593	13.82	
.2	Fire Protection	Sprinklers, light hazard	S.F. Floor	1.41	1.41	
.3	Heating	Included in 8.4	—	—	—	35.7%
.4	Cooling	Multizone unit, gas heating, electric cooling	S.F. Floor	13.05	13.05	
.5	Special Systems	N/A	—	—	—	
9.0 Electrical						
.1	Service & Distribution	600 ampere service, panel board and feeders	S.F. Floor	.71	.71	
.2	Lighting & Power	Fluorescent fixtures, receptacles, switches, A.C. and misc. power	S.F. Floor	6.45	6.45	9.7%
.4	Special Electrical	Alarm systems and emergency lighting	S.F. Floor	.52	.52	
11.0 Special Construction						
.1	Specialties	Cabinets, fume hoods, lockers	S.F. Floor	1.39	1.39	1.8%
12.0 Site Work						
.1	Earthwork	N/A	—	—	—	
.3	Utilities	N/A	—	—	—	0.0%
.5	Roads & Parking	N/A	—	—	—	
.7	Site Improvements	N/A	—	—	—	
			Sub-Total		79.04	100%
	GENERAL CONDITIONS (Overhead & Profit)			15%	11.86	
	ARCHITECT FEES			10%	9.10	
			Total Building Cost		100.00	

BUILDING TYPES

COMMERCIAL/INDUSTRIAL/INSTITUTIONAL

M.160 College, Student Union

Costs per square foot of floor area

Exterior Wall	S.F. Area	15000	20000	25000	30000	35000	40000	45000	50000	55000
	L.F. Perimeter	354	425	457	513	568	583	629	644	683
Brick Face with Concrete Block Back-up	Steel Frame	97.50	94.50	91.35	89.90	88.90	87.35	86.65	85.55	85.10
	R/Conc. Frame	94.55	91.55	88.40	87.00	85.95	84.40	83.70	82.65	82.15
Precast Concrete Panel	Steel Frame	94.35	91.65	88.95	87.70	86.80	85.45	84.85	83.90	83.50
	R/Conc. Frame	91.70	88.95	86.15	84.90	84.00	82.60	82.00	81.05	80.65
Limestone Face Concrete Block Back-up	Steel Frame	100.05	96.75	93.35	91.75	90.65	88.90	88.15	86.95	86.45
	R/Conc. Frame	97.10	93.80	90.35	88.80	87.70	85.95	85.20	84.00	83.50
Perimeter Adj., Add or Deduct	Per 100 L.F.	5.50	4.15	3.30	2.75	2.35	2.05	1.85	1.65	1.55
Story Hgt. Adj., Add or Deduct	Per 1 Ft.	1.40	1.25	1.05	1.05	.95	.85	.85	.75	.75

For Basement, add $20.85 per square foot of basement area

The above costs were calculated using the basic specifications shown on the facing page. These costs should be adjusted where necessary for design alternatives and owner's requirements. Reported completed project costs, for this type of structure, range from $72.85 to $149.40 per S.F.

Common additives

Description	Unit	$ Cost
Carrels Hardwood	Each	655 - 850
Elevators, Hydraulic passenger, 2 stops		
2000# capacity	Each	41,700
2500# capacity	Each	42,100
3500# capacity	Each	45,700
Emergency Lighting, 25 watt, battery operated		
Lead battery	Each	350
Nickel cadmium	Each	590
Escalators, Metal		
32" wide, 10' story height	Each	92,200
20' story height	Each	105,300
48" wide, 10' story height	Each	97,300
20' story height	Each	110,800
Glass		
32" wide, 10' story height	Each	86,200
20' story height	Each	98,800
48" wide, 10' story height	Each	91,800
20' story height	Each	103,800

Description	Unit	$ Cost
Lockers, Steel, Single tier, 60" or 72"	Opening	134 - 213
2 tier, 60" or 72" total	Opening	83 - 104
5 tier, box lockers	Opening	44 - 56
Locker bench, lam. maple top only	L.F.	18.40
Pedestals, steel pipe	Each	43
Sound System		
Amplifier, 250 watts	Each	1500
Speaker, ceiling or wall	Each	130
Trumpet	Each	244

Model costs calculated for a 2 story building with 12' story height and 25,000 square feet of floor area

College, Student Union

				Unit	Unit Cost	Cost Per S.F.	% Of Sub-Total
1.0 Foundations							
.1	Footings & Foundations	Poured concrete; strip and spread footings and 4' foundation wall		S.F. Ground	4.54	2.27	
.4	Piles & Caissons	N/A		—	—	—	3.8%
.9	Excavation & Backfill	Site preparation for slab and trench for foundation wall and footing		S.F. Ground	.89	.44	
2.0 Substructure							
.1	Slab on Grade	4" reinforced concrete with vapor barrier and granular base		S.F. Slab	3.05	1.52	2.1%
.2	Special Substructures	N/A		—	—	—	
3.0 Superstructure							
.1	Columns & Beams	Concrete columns		L.F. Column	113	2.72	
.4	Structural Walls	N/A		—	—	—	
.5	Elevated Floors	Concrete flat plate		S.F. Floor	9.92	4.96	18.1%
.7	Roof	Concrete flat plate		S.F. Roof	9.18	4.59	
.9	Stairs	Concrete		Flight	4550	.73	
4.0 Exterior Closure							
.1	Walls	Face brick with concrete block backup	75% of wall	S.F. Wall	21	7.01	
.5	Exterior Wall Finishes	N/A		—	—	—	14.1%
.6	Doors	Double aluminum and glass		Each	2170	.35	
.7	Windows & Glazed Walls	Window wall	25% of wall	S.F. Window	25	2.76	
5.0 Roofing							
.1	Roof Coverings	Built-up tar and gravel with flashing		S.F. Roof	2.18	1.09	
.7	Insulation	Perlite/EPS composite		S.F. Roof	1.18	.59	2.5%
.8	Openings & Specialties	Gravel stop, hatches, gutters and downspouts		S.F. Roof	.30	.15	
6.0 Interior Construction							
.1	Partitions	Gypsum board on metal studs	14 S.F. Floor/L.F. Partition	S.F. Partition	3.84	2.74	
.4	Interior Doors	Single leaf hollow metal	140 S.F. Floor/Door	Each	530	3.79	
.5	Wall Finishes	50% paint, 50% vinyl wall covering		S.F. Surface	1.00	1.43	
.6	Floor Finishes	50% carpet, 50% vinyl composition tile		S.F. Floor	3.88	3.88	20.3%
.7	Ceiling Finishes	Suspended fiberglass board		S.F. Ceiling	1.87	1.87	
.9	Interior Surface/Exterior Wall	Paint	75% of wall	S.F. Wall	1.93	.85	
7.0 Conveying							
.1	Elevators	One hydraulic passenger elevator		Each	49,750	1.99	2.8%
.2	Special Conveyors	N/A		—	—	—	
8.0 Mechanical							
.1	Plumbing	Toilet and service fixtures, supply and drainage	1 Fixture/1040 S.F. Floor	Each	1788	1.72	
.2	Fire Protection	Wet pipe sprinkler system		S.F. Floor	1.43	1.43	
.3	Heating	Included in 8.4		—	—	—	22.5%
.4	Cooling	Multizone unit, gas heating, electric cooling		S.F. Floor	13.05	13.05	
.5	Special Systems	N/A		—	—	—	
9.0 Electrical							
.1	Service & Distribution	600 ampere service, panel board and feeders		S.F. Floor	1.27	1.27	
.2	Lighting & Power	Fluorescent fixtures, receptacles, switches, A.C. and misc. power		S.F. Floor	7.49	7.49	13.8%
.4	Special Electrical	Alarm systems, communications systems and emergency lighting		S.F. Floor	1.16	1.16	
11.0 Special Construction							
.1	Specialties	N/A		—	—	—	0.0%
12.0 Site Work							
.1	Earthwork	N/A		—	—	—	
.3	Utilities	N/A		—	—	—	
.5	Roads & Parking	N/A		—	—	—	0.0%
.7	Site Improvements	N/A		—	—	—	
				Sub-Total		71.85	100%
	GENERAL CONDITIONS (Overhead & Profit)				15%	10.78	
	ARCHITECT FEES				7%	5.77	
				Total Building Cost		88.40	

BUILDING TYPES

COMMERCIAL/INDUSTRIAL/INSTITUTIONAL   M.170   Community Center

Costs per square foot of floor area

Exterior Wall	S.F. Area	4000	6000	8000	10000	12000	14000	16000	18000	20000
	L.F. Perimeter	260	340	420	453	460	510	560	610	600
Face Brick with Concrete Block Back-up	Bearing Walls	94.40	89.40	86.90	82.90	79.15	78.05	77.25	76.60	74.50
	Steel Frame	91.50	87.05	84.85	81.35	78.10	77.15	76.45	75.90	74.05
Decorative Concrete Block	Bearing Walls	81.65	77.90	76.00	73.05	70.30	69.45	68.85	68.40	66.85
	Steel Frame	82.45	79.15	77.55	75.05	72.75	72.05	71.55	71.20	69.90
Tilt Up Concrete Wall Panels	Bearing Walls	82.05	78.60	76.90	74.30	71.85	71.10	70.60	70.15	68.80
	Steel Frame	79.15	76.30	74.85	72.75	70.85	70.20	69.75	69.45	68.35
Perimeter Adj., Add or Deduct	Per 100 L.F.	13.10	8.70	6.50	5.25	4.35	3.75	3.25	2.85	2.65
Story Hgt. Adj., Add or Deduct	Per 1 Ft.	1.75	1.50	1.40	1.20	1.00	.95	.90	.90	.80

For Basement, add $18.25 per square foot of basement area

The above costs were calculated using the basic specifications shown on the facing page. These costs should be adjusted where necessary for design alternatives and owner's requirements. Reported completed project costs, for this type of structure, range from $45.60 to $144.10 per S.F.

Common additives

Description	Unit	$ Cost
Bar, Front bar	L.F.	278
Back bar	L.F.	220
Booth, Upholstered, custom straight	L.F.	127 - 207
"L" or "U" shaped	L.F.	143 - 225
Bowling Alleys, incl. alley, pinsetter		
Scorer, counter & misc. supplies, average	Lane	44,900
For automatic scorer, add	Lane	7475
Emergency Lighting, 25 watt, battery operated		
Lead battery	Each	350
Nickel cadmium	Each	590
Kitchen Equipment		
Broiler	Each	3300
Coffee urn, twin 6 gallon	Each	5500
Cooler, 6 ft. long	Each	2600
Dishwasher, 10-12 racks per hr.	Each	2500
Food warmer	Each	715
Freezer, 44 C.F., reach-in	Each	6375
Ice cube maker, 50 lb. per day	Each	1675
Range with 1 oven	Each	2250

Description	Unit	$ Cost
Movie Equipment		
Projector, 35mm	Each	10,700 - 14,200
Screen, wall or ceiling hung	S.F.	6.85 - 10.80
Partitions, Folding leaf, wood		
Acoustic type	S.F.	47 - 78
Seating		
Auditorium chair, all veneer	Each	127
Veneer back, padded seat	Each	152
Upholstered, spring seat	Each	178
Classroom, movable chair & desk	Set	65 - 120
Lecture hall, pedestal type	Each	127 - 370
Sound System		
Amplifier, 250 watts	Each	1500
Speaker, ceiling or wall	Each	130
Trumpet	Each	244
Stage Curtains, Medium weight	S.F.	13.20 - 400
Curtain Track, Light duty	L.F.	45
Swimming Pools, Complete, gunite	S.F.	45 - 54

Model costs calculated for a 1 story building with 12' story height and 10,000 square feet of floor area

Community Center

			Unit	Unit Cost	Cost Per S.F.	% Of Sub-Total
1.0 Foundations						
.1	Footings & Foundations	Poured concrete; strip and spread footings and 4' foundation wall	S.F. Ground	5.97	5.97	
.4	Piles & Caissons	N/A	—	—	—	10.5%
.9	Excavation & Backfill	Site preparation for slab and trench for foundation wall and footing	S.F. Ground	.96	.96	
2.0 Substructure						
.1	Slab on Grade	4" reinforced concrete with vapor barrier and granular base	S.F. Slab	3.05	3.05	4.6%
.2	Special Substructures	N/A	—	—	—	
3.0 Superstructure						
.1	Columns & Beams	N/A	—	—	—	
.4	Structural Walls	Concrete block	S.F. Wall	5.55	.91	
.5	Elevated Floors	N/A	—	—	—	7.1%
.7	Roof	Metal deck on open web steel joists	S.F. Roof	3.76	3.76	
.9	Stairs	N/A	—	—	—	
4.0 Exterior Closure						
.1	Walls	Face brick with concrete block backup 80% of wall	S.F. Wall	21	9.26	
.5	Exterior Wall Finishes	N/A	—	—	—	17.0%
.6	Doors	Double aluminum and glass and hollow metal	Each	1449	.58	
.7	Windows & Glazed Walls	Aluminum sliding 20% of wall	Each	413	1.40	
5.0 Roofing						
.1	Roof Coverings	Built-up tar and gravel with flashing	S.F. Roof	2.35	2.35	
.7	Insulation	Perlite/EPS composite	S.F. Roof	1.18	1.18	6.3%
.8	Openings & Specialties	Gravel stop and hatches	S.F. Roof	.61	.61	
6.0 Interior Construction						
.1	Partitions	Gypsum board on metal studs, toilet partitions 14 S.F. Floor/L.F. Partition	S.F. Partition	5.46	3.90	
.4	Interior Doors	Single leaf hollow metal 140 S.F. Floor/Door	Each	530	3.79	
.5	Wall Finishes	Paint	S.F. Surface	.50	.71	24.5%
.6	Floor Finishes	50% carpet, 50% vinyl tile	S.F. Floor	3.88	3.88	
.7	Ceiling Finishes	Mineral fiber tile on concealed zee bars	S.F. Ceiling	3.06	3.06	
.9	Interior Surface/Exterior Wall	Paint 80% of wall	S.F. Wall	1.93	.84	
7.0 Conveying						
.1	Elevators	N/A	—	—	—	0.0%
.2	Special Conveyors	N/A	—	—	—	
8.0 Mechanical						
.1	Plumbing	Kitchen, toilet and service fixtures, supply and drainage 1 Fixture/910 S.F. Floor	Each	3512	3.86	
.2	Fire Protection	Wet pipe sprinkler system	S.F. Floor	1.57	1.57	
.3	Heating	Included in 8.4	—	—	—	21.7%
.4	Cooling	Single zone rooftop unit, gas heating, electric cooling	S.F. Floor	8.97	8.97	
.5	Special Systems	N/A	—	—	—	
9.0 Electrical						
.1	Service & Distribution	200 ampere service, panel board and feeders	S.F. Floor	.60	.60	
.2	Lighting & Power	Incandescent fixtures, receptacles, switches, A.C. and misc. power	S.F. Floor	2.80	2.80	5.7%
.4	Special Electrical	Alarm systems and emergency lighting	S.F. Floor	.40	.40	
11.0 Special Construction						
.1	Specialties	Built-in coat racks, fume hoods, freezer, kitchen equipment	S.F. Floor	1.72	1.72	2.6%
12.0 Site Work						
.1	Earthwork	N/A	—	—	—	
.3	Utilities	N/A	—	—	—	0.0%
.5	Roads & Parking	N/A	—	—	—	
.7	Site Improvements	N/A	—	—	—	
			Sub-Total		66.13	100%
	GENERAL CONDITIONS (Overhead & Profit)			15%	9.92	
	ARCHITECT FEES			9%	6.85	
			Total Building Cost		82.90	

BUILDING TYPES

COMMERCIAL/INDUSTRIAL/INSTITUTIONAL — M.180 Courthouse, 1 Story

Costs per square foot of floor area

Exterior Wall	S.F. Area	16000	23000	30000	37000	44000	51000	58000	65000	72000
	L.F. Perimeter	597	763	821	968	954	1066	1090	1132	1220
Limestone with Concrete Block Back-up	R/Conc. Frame	119.30	115.55	111.25	110.05	106.90	106.25	104.75	103.75	103.35
	Steel Frame	118.20	114.45	110.15	108.95	105.80	105.15	103.70	102.65	102.30
Face Brick with Concrete Block Back-up	R/Conc. Frame	114.75	111.50	107.90	106.90	104.30	103.70	102.50	101.65	101.35
	Steel Frame	113.70	110.40	106.80	105.80	103.20	102.60	101.40	100.55	100.25
Stone with Concrete Block Back-up	R/Conc. Frame	114.70	111.45	107.85	106.85	104.25	103.65	102.45	101.60	101.25
	Steel Frame	113.65	110.40	106.80	105.80	103.20	102.60	101.40	100.55	100.25
Perimeter Adj., Add or Deduct	Per 100 L.F.	4.05	2.80	2.15	1.70	1.45	1.25	1.10	1.00	.90
Story Hgt. Adj., Add or Deduct	Per 1 Ft.	1.40	1.25	1.00	.95	.80	.80	.70	.65	.65

For Basement, add $17.80 per square foot of basement area

The above costs were calculated using the basic specifications shown on the facing page. These costs should be adjusted where necessary for design alternatives and owner's requirements. Reported completed project costs, for this type of structure, range from $78.65 to $146.60 per S.F.

Common additives

Description	Unit	$ Cost
Benches, Hardwood	L.F.	69 - 108
Clock System		
20 room	Each	11,700
50 room	Each	28,300
Closed Circuit Surveillance, One station		
Camera and monitor	Each	1325
For additional camera stations, add	Each	730
Directory Boards, Plastic, glass covered		
30" x 20"	Each	500
36" x 48"	Each	910
Aluminum, 24" x 18"	Each	445
36" x 24"	Each	535
48" x 32"	Each	630
48" x 60"	Each	1425
Emergency Lighting, 25 watt, battery operated		
Lead battery	Each	350
Nickel cadmium	Each	590

Description	Unit	$ Cost
Flagpoles, Complete		
Aluminum, 20' high	Each	1050
40' high	Each	2350
70' high	Each	6675
Fiberglass, 23' high	Each	1375
39'-5" high	Each	2400
59' high	Each	6575
Intercom System, 25 station capacity		
Master station	Each	1800
Intercom outlets	Each	113
Handset	Each	299
Safe, Office type, 4 hour rating		
30" x 18" x 18"	Each	3475
62" x 33" x 20"	Each	7575
Smoke Detectors		
Ceiling type	Each	141
Duct type	Each	335

Model costs calculated for a 1 story building with 14' story height and 30,000 square feet of floor area

Courthouse, 1 Story

				Unit	Unit Cost	Cost Per S.F.	% Of Sub-Total
1.0 Foundations							
.1	Footings & Foundations	Poured concrete; strip and spread footings and 4' foundation wall		S.F. Ground	2.44	2.44	
.4	Piles & Caissons	N/A		—	—	—	3.7%
.9	Excavation & Backfill	Site preparation for slab and trench for foundation wall and footing		S.F. Ground	.89	.89	
2.0 Substructure							
.1	Slab on Grade	4" reinforced concrete with vapor barrier and granular base		S.F. Slab	3.05	3.05	3.4%
.2	Special Substructures	N/A		—	—	—	
3.0 Superstructure							
.1	Columns & Beams	Concrete		L.F. Column	45	1.15	
.4	Structural Walls	N/A		—	—	—	
.5	Elevated Floors	N/A		—	—	—	14.1%
.7	Roof	Cast-in-place concrete waffle slab		S.F. Roof	11.57	11.57	
.9	Stairs	N/A		—	—	—	
4.0 Exterior Closure							
.1	Walls	Limestone panels with concrete block backup	75% of wall	S.F. Wall	30	8.81	
.5	Exterior Wall Finishes	N/A		—	—	—	12.2%
.6	Doors	Double wood		Each	1242	.29	
.7	Windows & Glazed Walls	Aluminum with insulated glass	25% of wall	Each	458	1.91	
5.0 Roofing							
.1	Roof Coverings	Built-up tar and gravel with flashing		S.F. Roof	2.11	2.11	
.7	Insulation	Perlite/EPS composite		S.F. Roof	1.18	1.18	3.7%
.8	Openings & Specialties	Gravel stop, hatches		S.F. Roof	.05	.05	
6.0 Interior Construction							
.1	Partitions	Plaster on metal studs, toilet partitions	10 S.F. Floor/L.F. Partition	S.F. Partition	7.88	9.46	
.4	Interior Doors	Single leaf wood	100 S.F. Floor/Door	Each	399	3.99	
.5	Wall Finishes	70% paint, 20% wood paneling, 10% vinyl wall covering		S.F. Surface	1.74	4.17	
.6	Floor Finishes	60% hardwood, 20% carpet, 20% terrazzo		S.F. Floor	8.35	8.35	36.6%
.7	Ceiling Finishes	Gypsum plaster on metal lath, suspended		S.F. Ceiling	6.23	6.23	
.9	Interior Surface/Exterior Wall	Painted plaster	75% of wall	S.F. Wall	3.12	.90	
7.0 Conveying							
.1	Elevators	N/A		—	—	—	0.0%
.2	Special Conveyors	N/A		—	—	—	
8.0 Mechanical							
.1	Plumbing	Toilet and service fixtures, supply and drainage	1 Fixture/1110 S.F. Floor	Each	2897	2.61	
.2	Fire Protection	Wet pipe sprinkler system		S.F. Floor	1.41	1.41	
.3	Heating	Included in 8.4		—	—	—	18.8%
.4	Cooling	Multizone unit, gas heating, electric cooling		S.F. Floor	13.05	13.05	
.5	Special Systems	N/A		—	—	—	
9.0 Electrical							
.1	Service & Distribution	400 ampere service, panel board and feeders		S.F. Floor	.57	.57	
.2	Lighting & Power	Fluorescent fixtures, receptacles, switches, A.C. and misc. power		S.F. Floor	5.83	5.83	7.5%
.4	Special Electrical	Alarm systems and emergency lighting		S.F. Floor	.37	.37	
11.0 Special Construction							
.1	Specialties	N/A		—	—	—	0.0%
12.0 Site Work							
.1	Earthwork	N/A		—	—	—	
.3	Utilities	N/A		—	—	—	0.0%
.5	Roads & Parking	N/A		—	—	—	
.7	Site Improvements	N/A		—	—	—	
				Sub-Total		90.39	100%
	GENERAL CONDITIONS (Overhead & Profit)				15%	13.56	
	ARCHITECT FEES				7%	7.30	
				Total Building Cost		**111.25**	

BUILDING TYPES

COMMERCIAL/INDUSTRIAL/INSTITUTIONAL — M.190 — Courthouse, 2-3 Story

Costs per square foot of floor area

Exterior Wall	S.F. Area	30000	40000	45000	50000	60000	70000	80000	90000	100000
	L.F. Perimeter	410	493	535	533	600	666	733	800	795
Limestone with Concrete Block Back-up	R/Conc. Frame	122.95	119.00	117.70	115.15	113.35	112.10	111.10	110.40	108.60
	Steel Frame	122.40	118.45	117.10	114.60	112.80	111.55	110.55	109.80	108.00
Face Brick with Concrete Block Back-up	R/Conc. Frame	118.70	115.15	114.00	111.85	110.25	109.15	108.25	107.60	106.10
	Steel Frame	118.15	114.60	113.45	111.30	109.70	108.60	107.70	107.05	105.55
Stone with Concrete Block Back-up	R/Conc. Frame	118.65	115.15	113.95	111.85	110.25	109.15	108.25	107.60	106.10
	Steel Frame	118.10	114.60	113.40	111.30	109.70	108.55	107.70	107.05	105.50
Perimeter Adj., Add or Deduct	Per 100 L.F.	5.60	4.20	3.75	3.35	2.80	2.40	2.10	1.85	1.70
Story Hgt. Adj., Add or Deduct	Per 1 Ft.	1.80	1.60	1.55	1.40	1.30	1.25	1.20	1.15	1.05

For Basement, add $18.05 per square foot of basement area

The above costs were calculated using the basic specifications shown on the facing page. These costs should be adjusted where necessary for design alternatives and owner's requirements. Reported completed project costs, for this type of structure, range from $78.65 to $146.60 per S.F.

Common additives

Description	Unit	$ Cost
Benches, Hardwood	L.F.	69 - 108
Clock System		
20 room	Each	11,700
50 room	Each	28,300
Closed Circuit Surveillance, One station		
Camera and monitor	Each	1325
For additional camera stations, add	Each	730
Directory Boards, Plastic, glass covered		
30" x 20"	Each	500
36" x 48"	Each	910
Aluminum, 24" x 18"	Each	445
36" x 24"	Each	535
48" x 32"	Each	630
48" x 60"	Each	1425
Elevators, Hydraulic passenger, 2 stops		
1500# capacity	Each	40,800
2500# capacity	Each	42,100
3500# capacity	Each	45,700
Additional stop, add	Each	3350

Description	Unit	$ Cost
Emergency Lighting, 25 watt, battery operated		
Lead battery	Each	350
Nickel cadmium	Each	590
Flagpoles, Complete		
Aluminum, 20' high	Each	1050
40' high	Each	2350
70' high	Each	6675
Fiberglass, 23' high	Each	1375
39'-5" high	Each	2400
59' high	Each	6575
Intercom System, 25 station capacity		
Master station	Each	1800
Intercom outlets	Each	113
Handset	Each	299
Safe, Office type, 4 hour rating		
30" x 18" x 18"	Each	3475
62" x 33" x 20"	Each	7575
Smoke Detectors		
Ceiling type	Each	141
Duct type	Each	335

Model costs calculated for a 3 story building with 12' story height and 60,000 square feet of floor area

Courthouse, 2-3 Story

				Unit	Unit Cost	Cost Per S.F.	% Of Sub-Total
1.0 Foundations							
.1	Footings & Foundations	Poured concrete; strip and spread footings and 4' foundation wall		S.F. Ground	4.02	1.34	
.4	Piles & Caissons	N/A		—	—	—	1.8%
.9	Excavation & Backfill	Site preparation for slab and trench for foundation wall and footing		S.F. Ground	.89	.30	
2.0 Substructure							
.1	Slab on Grade	4" reinforced concrete with vapor barrier and granular base		S.F. Slab	3.05	1.02	1.1%
.2	Special Substructures	N/A		—	—	—	
3.0 Superstructure							
.1	Columns & Beams	Steel columns		L.F. Column	43	1.31	
.4	Structural Walls	N/A		—	—	—	
.5	Elevated Floors	Concrete slab with metal deck and beams		S.F. Floor	13.35	8.90	16.6%
.7	Roof	Concrete slab with metal deck and beams		S.F. Roof	10.35	3.45	
.9	Stairs	Concrete filled metal pan		Flight	6625	1.10	
4.0 Exterior Closure							
.1	Walls	Face brick with concrete block backup	75% of wall	S.F. Wall	22	6.13	
.5	Exterior Wall Finishes	N/A		—	—	—	11.2%
.6	Doors	Double aluminum and glass and hollow metal		Each	2152	.22	
.7	Windows & Glazed Walls	Horizontal pivoted steel	25% of wall	Each	360	3.60	
5.0 Roofing							
.1	Roof Coverings	Built-up tar and gravel with flashing		S.F. Roof	2.04	.68	
.7	Insulation	Perlite/EPS composite		S.F. Roof	1.18	.39	1.3%
.8	Openings & Specialties	Gravel stop		L.F. Perimeter	5.55	.06	
6.0 Interior Construction							
.1	Partitions	Plaster on metal studs, toilet partitions	10 S.F. Floor/L.F. Partition	S.F. Partition	7.75	7.75	
.4	Interior Doors	Single leaf wood	100 S.F. Floor/Door	Each	399	3.99	
.5	Wall Finishes	70% paint, 20% wood paneling, 10% vinyl wall covering		S.F. Surface	1.74	3.48	34.4%
.6	Floor Finishes	60% hardwood, 20% terrazzo, 20% carpet		S.F. Floor	8.35	8.35	
.7	Ceiling Finishes	Gypsum plaster on metal lath, suspended		S.F. Ceiling	6.23	6.23	
.9	Interior Surface/Exterior Wall	Painted plaster	75% of wall	S.F. Wall	3.12	.84	
7.0 Conveying							
.1	Elevators	Five hydraulic passenger elevator		Each	73,080	6.09	6.8%
.2	Special Conveyors	N/A		—	—	—	
8.0 Mechanical							
.1	Plumbing	Toilet and service fixtures, supply and drainage	1 Fixture/665 S.F. Floor	Each	1615	2.43	
.2	Fire Protection	Wet pipe sprinkler system		S.F. Floor	1.37	1.37	
.3	Heating	Included in 8.4		—	—	—	18.9%
.4	Cooling	Multizone unit, gas heating, electric cooling		S.F. Floor	13.05	13.05	
.5	Special Systems	N/A		—	—	—	
9.0 Electrical							
.1	Service & Distribution	800 ampere service, panel board and feeders		S.F. Floor	.68	.68	
.2	Lighting & Power	Fluorescent fixtures, receptacles, switches, A.C. and misc. power		S.F. Floor	5.91	5.91	7.9%
.4	Special Electrical	Alarm systems and emergency lighting		S.F. Floor	.47	.47	
11.0 Special Construction							
.1	Specialties	N/A		—	—	—	0.0%
12.0 Site Work							
.1	Earthwork	N/A		—	—	—	
.3	Utilities	N/A		—	—	—	
.5	Roads & Parking	N/A		—	—	—	0.0%
.7	Site Improvements	N/A		—	—	—	
				Sub-Total		89.14	100%
	GENERAL CONDITIONS (Overhead & Profit)				15%	13.37	
	ARCHITECT FEES				7%	7.19	
				Total Building Cost		109.70	

BUILDING TYPES

COMMERCIAL/INDUSTRIAL/INSTITUTIONAL — M.200 Factory, 1 Story

Costs per square foot of floor area

Exterior Wall	S.F. Area	12000	18000	24000	30000	36000	42000	48000	54000	60000
	L.F. Perimeter	460	580	713	730	826	880	965	1006	1045
Concrete Block	Steel Frame	66.65	63.00	61.40	59.05	58.25	57.30	56.85	56.20	55.70
	Bearing Walls	65.70	62.10	60.50	58.10	57.35	56.40	55.95	55.30	54.75
Precast Concrete Panels	Steel Frame	70.15	65.95	64.10	61.25	60.35	59.20	58.65	57.90	57.25
Insulated Metal Panels	Steel Frame	67.80	64.00	62.30	59.75	58.95	57.95	57.45	56.75	56.20
Face Brick on Common Brick	Steel Frame	79.95	74.20	71.70	67.50	66.20	64.60	63.80	62.65	61.70
Tilt-up Concrete Panel	Steel Frame	67.20	63.50	61.85	59.40	58.60	57.60	57.15	56.45	55.95
Perimeter Adj., Add or Deduct	Per 100 L.F.	3.05	2.05	1.55	1.20	1.00	.90	.75	.65	.60
Story Hgt. Adj., Add or Deduct	Per 1 Ft.	.45	.40	.35	.30	.30	.25	.25	.20	.20

For Basement, add $17.90 per square foot of basement area

The above costs were calculated using the basic specifications shown on the facing page. These costs should be adjusted where necessary for design alternatives and owner's requirements. Reported completed project costs, for this type of structure, range from $24.90 to $96.20 per S.F.

Common additives

Description	Unit	$ Cost
Clock System		
20 room	Each	11,700
50 room	Each	28,300
Dock Bumpers, Rubber blocks		
4-1/2" thick, 10" high, 14" long	Each	58
24" long	Each	74
36" long	Each	93
12" high, 14" long	Each	75
24" long	Each	86
36" long	Each	100
6" thick, 10" high, 14" long	Each	72
24" long	Each	94
36" long	Each	121
20" high, 11" long	Each	123
Dock Boards, Heavy		
60" x 60" Aluminum, 5,000# cap.	Each	1075
9000# cap.	Each	1325
15,000# cap.	Each	1475

Description	Unit	$ Cost
Dock Levelers, Hinged 10 ton cap.		
6' x 8'	Each	3875
7' x 8'	Each	4100
Partitions, Woven wire, 10 ga., 1-1/2" mesh		
4' wide x 7' high	Each	115
8' high	Each	122
10' High	Each	148
Platform Lifter, Portable, 6'x 6'		
3000# cap.	Each	5725
4000# cap.	Each	8025
Fixed, 6' x 8', 5000# cap.	Each	8325

Model costs calculated for a 1 story building with 20' story height and 30,000 square feet of floor area

Factory, 1 Story

				Unit	Unit Cost	Cost Per S.F.	% Of Sub-Total
1.0 Foundations							
.1	Footings & Foundations	Poured concrete; strip and spread footings and 4' foundation wall		S.F. Ground	2.24	2.24	6.5%
.4	Piles & Caissons	N/A		—	—	—	
.9	Excavation & Backfill	Site preparation for slab and trench for foundation wall and footing		S.F. Ground	.89	.89	
2.0 Substructure							
.1	Slab on Grade	4" reinforced concrete with vapor barrier and granular base		S.F. Slab	3.05	3.05	6.4%
.2	Special Substructures	N/A		—	—	—	
3.0 Superstructure							
.1	Columns & Beams	Steel columns included in 3.7		—	—	—	
.4	Structural Walls	N/A		—	—	—	
.5	Elevated Floors	N/A		—	—	—	10.0%
.7	Roof	Metal deck, open web steel joists, beams and columns		S.F. Roof	4.82	4.82	
.9	Stairs	N/A		—	—	—	
4.0 Exterior Closure							
.1	Walls	Concrete block	75% of wall	S.F. Wall	4.71	1.72	
.5	Exterior Wall Finishes	N/A		—	—	—	9.5%
.6	Doors	Double aluminum and glass, hollow metal, steel overhead		Each	1289	.65	
.7	Windows & Glazed Walls	Industrial horizontal pivoted steel	25% of wall	Each	578	2.20	
5.0 Roofing							
.1	Roof Coverings	Built-up tar and gravel with flashing		S.F. Roof	1.96	1.96	
.7	Insulation	Perlite/EPS composite		S.F. Roof	1.18	1.18	7.2%
.8	Openings & Specialties	Gravel stop, hatches, gutters and downspouts		S.F. Roof	.33	.33	
6.0 Interior Construction							
.1	Partitions	Concrete block, toilet partitions	60 S.F. Floor/L.F. Partition	S.F. Partition	8.60	1.72	
.4	Interior Doors	Single leaf hollow metal and fire doors	600 S.F. Floor/Door	Each	530	.88	
.5	Wall Finishes	Paint		S.F. Surface	1.07	.43	8.9%
.6	Floor Finishes	Vinyl composition tile	10% of floor	S.F. Floor	2.40	.24	
.7	Ceiling Finishes	Fiberglass board on exposed grid system	10% of area	S.F. Ceiling	3.06	.31	
.9	Interior Surface/Exterior Wall	Paint and block filler	75% of wall	S.F. Wall	1.93	.70	
7.0 Conveying							
.1	Elevators	N/A		—	—	—	0.0%
.2	Special Conveyors	N/A		—	—	—	
8.0 Mechanical							
.1	Plumbing	Toilet and service fixtures, supply and drainage	1 Fixture/1000 S.F. Floor	Each	2820	2.82	
.2	Fire Protection	Sprinklers, ordinary hazard		S.F. Floor	2.03	2.03	
.3	Heating	Oil fired hot water, unit heaters		S.F. Floor	5.63	5.63	37.4%
.4	Cooling	Chilled water, air cooled condenser system		S.F. Floor	7.39	7.39	
.5	Special Systems	N/A		—	—	—	
9.0 Electrical							
.1	Service & Distribution	600 ampere service, panel board and feeders		S.F. Floor	.78	.78	
.2	Lighting & Power	High intensity discharge fixtures, receptacles, switches, A.C. and misc. power		S.F. Floor	5.73	5.73	14.1%
.4	Special Electrical	Alarm systems and emergency lighting		S.F. Floor	.27	.27	
11.0 Special Construction							
.1	Specialties	N/A		—	—	—	0.0%
12.0 Site Work							
.1	Earthwork	N/A		—	—	—	
.3	Utilities	N/A		—	—	—	0.0%
.5	Roads & Parking	N/A		—	—	—	
.7	Site Improvements	N/A		—	—	—	
				Sub-Total		47.97	100%
	GENERAL CONDITIONS (Overhead & Profit)				15%	7.20	
	ARCHITECT FEES				7%	3.88	
				Total Building Cost		59.05	

BUILDING TYPES

COMMERCIAL/INDUSTRIAL/INSTITUTIONAL — M.210 Factory, 3 Story

Costs per square foot of floor area

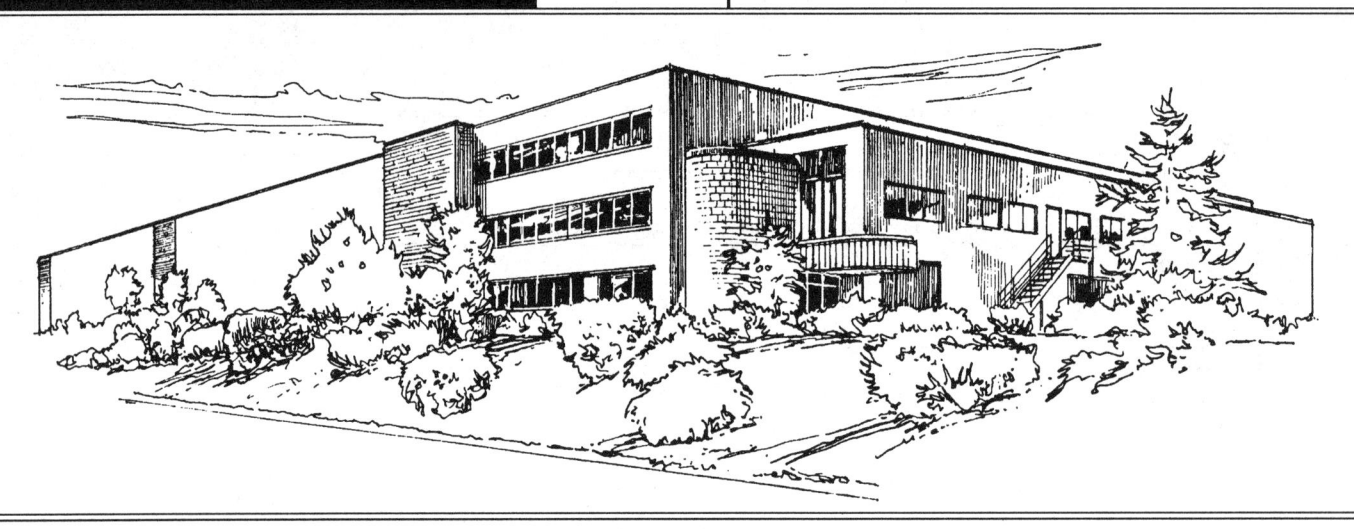

Exterior Wall	S.F. Area	20000	30000	40000	50000	60000	70000	80000	90000	100000
	L.F. Perimeter	362	410	493	576	600	628	660	700	744
Face Brick Common Brick Back-up	Steel Frame	83.55	75.90	73.30	71.75	69.30	67.70	66.55	65.75	65.15
	Concrete Frame	83.20	75.60	73.00	71.40	69.00	67.40	66.20	65.45	64.85
Face Brick Concrete Block Back-up	Steel Frame	81.85	74.70	72.25	70.70	68.45	66.95	65.85	65.10	64.60
	Concrete Frame	81.40	74.25	71.75	70.25	68.00	66.50	65.40	64.65	64.10
Stucco on Concrete Block	Steel Frame	74.15	68.85	66.95	65.80	64.20	63.10	62.35	61.80	61.40
	Concrete Frame	73.65	68.35	66.45	65.30	63.70	62.65	61.85	61.30	60.90
Perimeter Adj., Add or Deduct	Per 100 L.F.	6.95	4.65	3.45	2.80	2.30	2.00	1.75	1.55	1.40
Story Hgt. Adj., Add or Deduct	Per 1 Ft.	1.90	1.45	1.35	1.25	1.05	.95	.90	.85	.80

For Basement, add $19.15 per square foot of basement area

The above costs were calculated using the basic specifications shown on the facing page. These costs should be adjusted where necessary for design alternatives and owner's requirements. Reported completed project costs, for this type of structure, range from $24.90 to $96.20 per S.F.

Common additives

Description	Unit	$ Cost
Clock System		
20 room	Each	11,700
50 room	Each	28,300
Dock Bumpers, Rubber blocks		
4-1/2" thick, 10" high, 14" long	Each	58
24" long	Each	74
36" long	Each	93
12" high, 14" long	Each	75
24" long	Each	86
36" long	Each	100
6" thick, 10" high, 14" long	Each	72
24" long	Each	94
36" long	Each	121
20" high, 11" long	Each	123
Dock Boards, Heavy		
60" x 60" Aluminum, 5,000# cap.	Each	1075
9000# cap.	Each	1325
15,000# cap.	Each	1475

Description	Unit	$ Cost
Dock Levelers, Hinged 10 ton cap.		
6'x 8'	Each	3875
7'x 8'	Each	4100
Elevator, Hydraulic freight, 2 stops		
3500# capacity	Each	52,400
4000# capacity	Each	55,300
Additional stop, add	Each	3325
Partitions, Woven wire, 10 ga., 1-1/2" mesh		
4' Wide x 7" high	Each	115
8' High	Each	122
10' High	Each	148
Platform Lifter, Portable, 6'x 6'		
3000# cap.	Each	5725
4000# cap.	Each	8025
Fixed, 6'x 8', 5000# cap.	Each	8325

Model costs calculated for a 3 story building with 12' story height and 90,000 square feet of floor area

Factory, 3 Story

				Unit	Unit Cost	Cost Per S.F.	% Of Sub-Total
1.0 Foundations							
.1	Footings & Foundations	Poured concrete; strip and spread footings and 4' foundation wall		S.F. Ground	5.13	1.71	3.7%
.4	Piles & Caissons	N/A		—	—	—	
.9	Excavation & Backfill	Site preparation for slab and trench for foundation wall and footing		S.F. Ground	.89	.30	
2.0 Substructure							
.1	Slab on Grade	4" reinforced concrete with vapor barrier and granular base		S.F. Slab	5.30	1.77	3.3%
.2	Special Substructures	N/A		—	—	—	
3.0 Superstructure							
.1	Columns & Beams	Concrete columns		L.F. Column	101	2.55	
.4	Structural Walls	N/A		—	—	—	
.5	Elevated Floors	Concrete flat slab		S.F. Floor	10.72	7.15	24.7%
.7	Roof	Concrete flat slab		S.F. Roof	9.46	3.15	
.9	Stairs	Concrete		Flight	4550	.40	
4.0 Exterior Closure							
.1	Walls	Face brick with common brick backup	70% of wall	S.F. Wall	24	4.81	
.5	Exterior Wall Finishes	N/A		—	—	—	15.7%
.6	Doors	Double aluminum & glass, hollow metal, overhead doors		Each	1211	.23	
.7	Windows & Glazed Walls	Industrial, horizontal pivoted steel	30% of wall	Each	641	3.37	
5.0 Roofing							
.1	Roof Coverings	Built-up tar and gravel with flashing		S.F. Roof	1.86	.62	
.7	Insulation	Perlite/EPS composite		S.F. Roof	1.18	.39	2.1%
.8	Openings & Specialties	Gravel stop, hatches, gutters and downspouts		S.F. Roof	.30	.10	
6.0 Interior Construction							
.1	Partitions	Gypsum board on metal studs, toilet partition	50 S.F. Floor/L.F. Partition	S.F. Partition	4.85	.97	
.4	Interior Doors	Single leaf fire doors	500 S.F. Floor/Door	S.F. Door	608	1.22	
.5	Wall Finishes	Paint		S.F. Surface	.50	.20	8.2%
.6	Floor Finishes	90% metallic hardener, 10% vinyl composition tile		S.F. Floor	1.83	1.83	
.7	Ceiling Finishes	Fiberglass board on exposed grid systems	10% of area	S.F. Ceiling	1.87	.19	
.9	Interior Surface/Exterior Wall	N/A		—	—	—	
7.0 Conveying							
.1	Elevators	Two hydraulic freight elevators		Each	77,400	1.72	3.2%
.2	Special Conveyors	N/A		—	—	—	
8.0 Mechanical							
.1	Plumbing	Toilet and service fixtures, supply and drainage	1 Fixture/1345 S.F. Floor	Each	2716	2.02	
.2	Fire Protection	Wet pipe sprinkler system		S.F. Floor	1.80	1.80	
.3	Heating	Oil fired hot water, unit heaters		S.F. Floor	2.83	2.83	26.1%
.4	Cooling	Chilled water, air cooled condenser system		S.F. Floor	7.39	7.39	
.5	Special Systems	N/A		—	—	—	
9.0 Electrical							
.1	Service & Distribution	800 ampere service, panel board and feeders		S.F. Floor	.46	.46	
.2	Lighting & Power	High intensity discharge fixtures, switches, A.C. and misc. power		S.F. Floor	6.06	6.06	13.0%
.4	Special Electrical	Alarm systems and emergency lighting		S.F. Floor	.45	.45	
11.0 Special Construction							
.1	Specialties	N/A		—	—	—	0.0%
12.0 Site Work							
.1	Earthwork	N/A		—	—	—	
.3	Utilities	N/A		—	—	—	0.0%
.5	Roads & Parking	N/A		—	—	—	
.7	Site Improvements	N/A		—	—	—	
				Sub-Total		53.69	**100%**
	GENERAL CONDITIONS (Overhead & Profit)				15%	8.05	
	ARCHITECT FEES				6%	3.71	
				Total Building Cost		**65.45**	

BUILDING TYPES

COMMERCIAL/INDUSTRIAL/INSTITUTIONAL M.220 Fire Station, 1 Story

Costs per square foot of floor area

Exterior Wall	S.F. Area	4000	4500	5000	5500	6000	6500	7000	7500	8000
	L.F. Perimeter	260	280	300	320	320	336	353	370	386
Face Brick Concrete Block Back-up	Steel Joists	94.00	92.10	90.65	89.45	86.80	85.80	84.95	84.25	83.55
	Precast Conc.	94.95	93.10	91.65	90.40	87.80	86.80	85.95	85.25	84.55
Decorative Concrete Block	Steel Joists	84.20	82.75	81.60	80.65	78.75	78.00	77.35	76.85	76.30
	Precast Conc.	85.30	83.85	82.75	81.80	79.90	79.15	78.50	78.00	77.45
Limestone with Concrete Block Back-up	Steel Joists	98.10	96.05	94.45	93.15	90.15	89.05	88.15	87.40	86.65
	Precast Conc.	99.25	97.20	95.60	94.25	91.30	90.20	89.30	88.55	87.75
Perimeter Adj., Add or Deduct	Per 100 L.F.	12.50	11.10	9.95	9.10	8.30	7.65	7.10	6.65	6.25
Story Hgt. Adj., Add or Deduct	Per 1 Ft.	1.65	1.60	1.50	1.50	1.35	1.30	1.25	1.25	1.25

For Basement, add $21.40 per square foot of basement area

The above costs were calculated using the basic specifications shown on the facing page. These costs should be adjusted where necessary for design alternatives and owner's requirements. Reported completed project costs, for this type of structure, range from $41.40 to $122.40 per S.F.

Common additives

Description	Unit	$ Cost
Appliances		
Cooking range, 30" free standing		
1 oven	Each	380 - 1450
2 oven	Each	785 - 1775
30" built-in		
1 oven	Each	660 - 1600
2 oven	Each	1050 - 1975
Counter top cook tops, 4 burner	Each	300 - 670
Microwave oven	Each	200 - 1850
Combination range, refrig. & sink, 30" wide	Each	1025 - 2050
60" wide	Each	2675
72" wide	Each	3025
Combination range refrigerator, sink microwave oven & icemaker	Each	4500
Compactor, residential, 4-1 compaction	Each	395 - 660
Dishwasher, built-in, 2 cycles	Each	405 - 840
4 cycles	Each	605 - 1125
Garbage disposer, sink type	Each	117 - 300
Hood for range, 2 speed, vented, 30" wide	Each	183 - 520
42" wide	Each	310 - 560

Description	Unit	$ Cost
Appliances, cont.		
Refrigerator, no frost 10-12 C.F.	Each	440 - 920
14-16 C.F.	Each	660 - 1150
18-20 C.F.	Each	780 - 1425
Lockers, Steel, single tier, 60" or 72"	Opening	134 - 213
2 tier, 60" or 72" total	Opening	83 - 104
5 tier, box lockers	Opening	44 - 56
Locker bench, lam. maple top only	L.F.	18.40
Pedestals, steel pipe	Each	43
Sound System		
Amplifier, 250 watts	Each	1500
Speaker, ceiling or wall	Each	130
Trumpet	Each	244

Location Factors appear in the Reference Section

Fire Station, 1 Story

Model costs calculated for a 1 story building with 14' story height and 6,000 square feet of floor area

				Unit	Unit Cost	Cost Per S.F.	% Of Sub-Total
1.0	**Foundations**						
.1	Footings & Foundations	Poured concrete; strip and spread footings and 4' foundation wall		S.F. Ground	4.56	4.56	
.4	Piles & Caissons	N/A		—	—	—	8.1%
.9	Excavation & Backfill	Site preparation for slab and trench for foundation wall and footing		S.F. Ground	1.09	1.09	
2.0	**Substructure**						
.1	Slab on Grade	4" reinforced concrete with vapor barrier and granular base		S.F. Slab	3.61	3.61	5.2%
.2	Special Substructures	N/A		—	—	—	
3.0	**Superstructure**						
.1	Columns & Beams	N/A		—	—	—	
.4	Structural Walls	N/A		—	—	—	
.5	Elevated Floors	N/A		—	—	—	3.7%
.7	Roof	Metal deck, open web steel joists, beams		S.F. Roof	2.56	2.56	
.9	Stairs	N/A		—	—	—	
4.0	**Exterior Closure**						
.1	Walls	Face brick with concrete block backup	75% of wall	S.F. Wall	21	11.93	
.5	Exterior Wall Finishes	N/A		—	—	—	22.2%
.6	Doors	Single aluminum and glass, overhead, hollow metal	15% of wall	S.F. Door	20	2.26	
.7	Windows & Glazed Walls	Aluminum insulated glass	10% of wall	Each	578	1.35	
5.0	**Roofing**						
.1	Roof Coverings	Built-up tar and gravel with flashing		S.F. Roof	2.51	2.51	
.7	Insulation	Perlite/EPS composite		S.F. Roof	1.18	1.18	6.3%
.8	Openings & Specialties	Gravel stop		S.F. Roof	.74	.74	
6.0	**Interior Construction**						
.1	Partitions	Concrete block, toilet partitions	17 S.F. Floor/L.F. Partition	S.F. Partition	5.71	3.36	
.4	Interior Doors	Single leaf hollow metal	500 S.F. Floor/Door	Each	530	1.06	
.5	Wall Finishes	Paint		S.F. Surface	1.08	1.27	13.7%
.6	Floor Finishes	50% vinyl tile, 50% paint		S.F. Floor	1.74	1.74	
.7	Ceiling Finishes	Fiberglass board on exposed grid, suspended	50% of area	S.F. Ceiling	3.06	1.53	
.9	Interior Surface/Exterior Wall	Acrylic glazed coating	75% of wall	S.F. Wall	1.14	.64	
7.0	**Conveying**						
.1	Elevators	N/A		—	—	—	0.0%
.2	Special Conveyors	N/A		—	—	—	
8.0	**Mechanical**						
.1	Plumbing	Kitchen, toilet and service fixtures, supply and drainage	1 Fixture/375 S.F. Floor	Each	2336	6.23	
.2	Fire Protection	Wet pipe sprinkler system		S.F. Floor	2.04	2.04	
.3	Heating	Included in 8.4		—	—	—	33.5%
.4	Cooling	Rooftop multizone unit system		S.F. Floor	15.13	15.13	
.5	Special Systems	N/A		—	—	—	
9.0	**Electrical**						
.1	Service & Distribution	200 ampere service, panel board and feeders		S.F. Floor	.87	.87	
.2	Lighting & Power	Fluorescent fixtures, receptacles, switches, A.C. and misc. power		S.F. Floor	3.96	3.96	7.3%
.4	Special Electrical	Alarm systems		S.F. Floor	.27	.27	
11.0	**Special Construction**						
.1	Specialties	N/A		—	—	—	0.0%
12.0	**Site Work**						
.1	Earthwork	N/A		—	—	—	
.3	Utilities	N/A		—	—	—	0.0%
.5	Roads & Parking	N/A		—	—	—	
.7	Site Improvements	N/A		—	—	—	
				Sub-Total		69.89	**100%**
	GENERAL CONDITIONS (Overhead & Profit)				15%	10.48	
	ARCHITECT FEES				8%	6.43	
				Total Building Cost		86.80	

BUILDING TYPES

COMMERCIAL/INDUSTRIAL/INSTITUTIONAL — M.230 Fire Station, 2 Story

Costs per square foot of floor area

Exterior Wall	S.F. Area	6000	7000	8000	9000	10000	11000	12000	13000	14000
	L.F. Perimeter	220	240	260	280	286	303	320	336	353
Face Brick with Concrete Block Back-up	Steel Joists	96.85	93.95	91.80	90.10	87.55	86.30	85.35	84.40	83.70
	Precast Conc.	100.65	97.80	95.65	93.95	91.45	90.25	89.25	88.35	87.65
Decorative Concrete Block	Steel Joists	86.90	84.55	82.85	81.45	79.50	78.55	77.75	77.05	76.45
	Precast Conc.	91.95	89.65	87.95	86.60	84.70	83.70	82.95	82.20	81.65
Limestone with Concrete Block Back-up	Steel Joists	101.50	98.30	95.90	94.00	91.15	89.80	88.70	87.65	86.90
	Precast Conc.	105.25	102.10	99.75	97.85	95.05	93.70	92.60	91.60	90.80
Perimeter Adj., Add or Deduct	Per 100 L.F.	14.20	12.15	10.65	9.45	8.50	7.75	7.10	6.55	6.05
Perimeter Adj., Add or Deduct	Per 100 L.F.	1.85	1.75	1.65	1.60	1.50	1.40	1.35	1.30	1.30

For Basement, add $20.10 per square foot of basement area

The above costs were calculated using the basic specifications shown on the facing page. These costs should be adjusted where necessary for design alternatives and owner's requirements. Reported completed project costs, for this type of structure, range from $41.40 to $122.40 per S.F.

Common additives

Description	Unit	$ Cost
Appliances		
Cooking range, 30" free standing		
1 oven	Each	380 - 1450
2 oven	Each	785 - 1775
30" built-in		
1 oven	Each	660 - 1600
2 oven	Each	1050 - 1975
Counter top cook tops, 4 burner	Each	300 - 670
Microwave oven	Each	200 - 1850
Combination range, refrig. & sink, 30" wide	Each	1025 - 2050
60" wide	Each	2675
72" wide	Each	3025
Combination range, refrigerator, sink, microwave oven & icemaker	Each	4500
Compactor, residential, 4-1 compaction	Each	395 - 660
Dishwasher, built-in, 2 cycles	Each	405 - 840
4 cycles	Each	605 - 1125
Garbage disposer, sink type	Each	117 - 300
Hood for range, 2 speed, vented, 30" wide	Each	183 - 520
42" wide	Each	310 - 560

Description	Unit	$ Cost
Appliances, cont.		
Refrigerator, no frost 10-12 C.F.	Each	440 - 920
14-16 C.F.	Each	660 - 1150
18-20 C.F.	Each	780 - 1425
Elevators, Hydraulic passenger, 2 stops		
1500# capacity	Each	40,800
2500# capacity	Each	42,100
3500# capacity	Each	45,700
Lockers, Steel, single tier, 60" or 72"	Opening	134 - 213
2 tier, 60" or 72" total	Opening	83 - 104
5 tier, box lockers	Opening	44 - 56
Locker bench, lam. maple top only	L.F.	18.40
Pedestals, steel pipe	Each	43
Sound System		
Amplifier, 250 watts	Each	1500
Speaker, ceiling or wall	Each	130
Trumpet	Each	244

Model costs calculated for a 2 story building with 14' story height and 10,000 square feet of floor area

Fire Station, 2 Story

				Unit	Unit Cost	Cost Per S.F.	% Of Sub-Total
1.0 Foundations							
.1	Footings & Foundations	Poured concrete; strip and spread footings and 4' foundation wall		S.F. Ground	4.90	2.45	4.7%
.4	Piles & Caissons	N/A		—	—	—	
.9	Excavation & Backfill	Site preparation for slab and trench for foundation wall and footing		S.F. Ground	1.09	.55	
2.0 Substructure							
.1	Slab on Grade	4" reinforced concrete with vapor barrier and granular base		S.F. Slab	3.05	1.52	2.4%
.2	Special Substructures	N/A		—	—	—	
3.0 Superstructure							
.1	Columns & Beams	N/A		—	—	—	
.4	Structural Walls	Included in 6.1		—	—	—	
.5	Elevated Floors	Open web steel joists, slab form, concrete		S.F. Floor	7.76	3.88	10.3%
.7	Roof	Metal deck on open web steel joists		S.F. Roof	2.72	1.36	
.9	Stairs	Concrete filled metal pan		Flight	6625	1.33	
4.0 Exterior Closure							
.1	Walls	Decorative concrete block	75% of wall	S.F. Wall	12.20	7.33	16.7%
.5	Exterior Wall Finishes	N/A		—	—	—	
.6	Doors	Single aluminum and glass, steel overhead, hollow metal	15% of wall	S.F. Door	14.90	1.79	
.7	Windows & Glazed Walls	Aluminum insulated glass	10% of wall	Each	458	1.59	
5.0 Roofing							
.1	Roof Coverings	Tar and gravel with flashing		S.F. Roof	2.58	1.29	3.2%
.7	Insulation	Perlite/EPS composite		S.F. Roof	1.18	.59	
.8	Openings & Specialties	Gravel stop		L.F. Perimeter	5.55	.16	
6.0 Interior Construction							
.1	Partitions	Concrete block, toilet partitions	10 S.F. Floor/L.F. Partition	S.F. Partition	5.64	3.32	13.3%
.4	Interior Doors	Single leaf hollow metal	500 S.F. Floor/Door	Each	608	1.22	
.5	Wall Finishes	Paint		S.F. Surface	.50	.59	
.6	Floor Finishes	50% vinyl tile, 50% paint		S.F. Floor	1.74	1.74	
.7	Ceiling Finishes	Fiberglass board on exposed grid, suspended	50% of area	S.F. Ceiling	1.87	.94	
.9	Interior Surface/Exterior Wall	Acrylic glazed coating	75% of wall	S.F. Wall	1.14	.68	
7.0 Conveying							
.1	Elevators	One hydraulic passenger elevator		Each	49,800	4.98	7.8%
.2	Special Conveyors	N/A		—	—	—	
8.0 Mechanical							
.1	Plumbing	Kitchen toilet and service fixtures, supply and drainage	1 Fixture/400 S.F. Floor	Each	1864	4.66	
.2	Fire Protection	Wet pipe sprinkler system		S.F. Floor	1.73	1.73	33.5%
.3	Heating	Included in 8.4		—	—	—	
.4	Cooling	Rooftop multizone unit system		S.F. Floor	15.13	15.13	
.5	Special Systems	N/A		—	—	—	
9.0 Electrical							
.1	Service & Distribution	100 ampere service, panel board and feeders		S.F. Floor	.60	.60	8.1%
.2	Lighting & Power	Fluorescent fixtures, receptacles, switches, A.C. and misc. power		S.F. Floor	4.11	4.11	
.4	Special Electrical	Alarm systems and emergency lighting		S.F. Floor	.49	.49	
11.0 Special Construction							
.1	Specialties	N/A		—	—	—	0.0%
12.0 Site Work							
.1	Earthwork	N/A		—	—	—	
.3	Utilities	N/A		—	—	—	0.0%
.5	Roads & Parking	N/A		—	—	—	
.7	Site Improvements	N/A		—	—	—	
				Sub-Total		64.03	100%
	GENERAL CONDITIONS (Overhead & Profit)				15%	9.60	
	ARCHITECT FEES				8%	5.87	
				Total Building Cost		79.50	

BUILDING TYPES

COMMERCIAL/INDUSTRIAL/INSTITUTIONAL — M.240 — Fraternity/Sorority House

Costs per square foot of floor area

Exterior Wall	S.F. Area	4000	5000	6000	8000	10000	12000	14000	16000	18000
	L.F. Perimeter	180	205	230	260	300	340	353	386	420
Cedar Beveled Siding	Wood Frame	95.75	90.75	87.40	82.40	79.65	77.90	75.95	74.90	74.15
Aluminum Siding	Wood Frame	93.70	88.85	85.60	80.75	78.15	76.45	74.55	73.55	72.80
Board and Batten	Wood Frame	93.20	89.00	86.20	81.90	79.55	78.10	76.30	75.45	74.80
Face Brick on Block	Wood Joists	108.50	102.35	98.25	91.60	88.20	85.95	83.10	81.75	80.75
Stucco on Block	Wood Joists	95.40	90.40	87.05	82.00	79.30	77.55	75.55	74.50	73.75
Decorative Block	Wood Joists	98.80	93.55	90.00	84.60	81.75	79.85	77.70	76.55	75.75
Perimeter Adj., Add or Deduct	Per 100 L.F.	8.70	6.95	5.80	4.35	3.50	2.90	2.50	2.20	1.95
Story Hgt. Adj., Add or Deduct	Per 1 Ft.	1.10	1.00	.90	.80	.75	.70	.65	.60	.60

For Basement, add $13.10 per square foot of basement area

The above costs were calculated using the basic specifications shown on the facing page. These costs should be adjusted where necessary for design alternatives and owner's requirements. Reported completed project costs, for this type of structure, range from $58.55 to $112.85 per S.F.

Common additives

Description	Unit	$ Cost
Appliances		
Cooking range, 30" free standing		
1 oven	Each	380 - 1450
2 oven	Each	785 - 1775
30" built-in		
1 oven	Each	660 - 1600
2 oven	Each	1050 - 1975
Counter top cook tops, 4 burner	Each	300 - 670
Microwave oven	Each	200 - 1850
Combination range, refrig. & sink, 30" wide	Each	1025 - 2050
60" wide	Each	2675
72" wide	Each	3025
Combination range, refrigerator, sink, microwave oven & icemaker	Each	4500
Compactor, residential, 4-1 compaction	Each	395 - 660
Dishwasher, built-in, 2 cycles	Each	405 - 840
4 cycles	Each	605 - 1125
Garbage disposer, sink type	Each	117 - 300
Hood for range, 2 speed, vented, 30" wide	Each	183 - 520
42" wide	Each	310 - 560

Description	Unit	$ Cost
Appliances, cont.		
Refrigerator, no frost 10-12 C.F.	Each	440 - 920
14-16 C.F.	Each	660 - 1150
18-20 C.F.	Each	780 - 1425
Elevators, Hydraulic passenger, 2 stops		
1500# capacity	Each	40,800
2500# capacity	Each	42,100
3500# capacity	Each	45,700
Laundry Equipment		
Dryer, gas, 16 lb. capacity	Each	665
30 lb. capacity	Each	2675
Washer, 4 cycle	Each	740
Commercial	Each	1125
Sound System		
Amplifier, 250 watts	Each	1500
Speaker, ceiling or wall	Each	130
Trumpet	Each	244

Fraternity/Sorority House

Model costs calculated for a 2 story building with 10' story height and 10,000 square feet of floor area

				Unit	Unit Cost	Cost Per S.F.	% Of Sub-Total
1.0	**Foundations**						
.1	Footings & Foundations	Poured concrete; strip and spread footings and 4' foundation wall		S.F. Ground	4.36	2.18	
.4	Piles & Caissons	N/A		—	—	—	4.3%
.9	Excavation & Backfill	Site preparation for slab and trench for foundation wall and footing		S.F. Ground	1.09	.55	
2.0	**Substructure**						
.1	Slab on Grade	4" reinforced concrete with vapor barrier and granular base		S.F. Slab	3.05	1.52	2.4%
.2	Special Substructures	N/A		—	—	—	
3.0	**Superstructure**						
.1	Columns & Beams	N/A		—	—	—	
.4	Structural Walls	Included in 6.1		—	—	—	
.5	Elevated Floors	Plywood on wood joists		S.F. Floor	2.98	1.49	5.2%
.7	Roof	Plywood on wood rafters (pitched)		S.F. Roof	2.39	1.34	
.9	Stairs	Wood		Flight	1264	.50	
4.0	**Exterior Closure**						
.1	Walls	Cedar bevel siding on wood studs, insulated	80% of wall	S.F. Wall	7.31	3.51	
.5	Exterior Wall Finishes	N/A		—	—	—	9.6%
.6	Doors	Solid core wood		Each	1357	.95	
.7	Windows & Glazed Walls	Double hung wood	20% of wall	Each	340	1.63	
5.0	**Roofing**						
.1	Roof Coverings	Asphalt shingles with flashing (pitched)		S.F. Ground	1.30	.65	
.7	Insulation	Fiberglass sheets		S.F. Ground	.96	.48	2.1%
.8	Openings & Specialties	Gutters and downspouts		S.F. Ground	.40	.20	
6.0	**Interior Construction**						
.1	Partitions	Gypsum board on wood studs	25 S.F. Floor/L.F. Partition	S.F. Partition	3.03	.97	
.4	Interior Doors	Single leaf wood	200 S.F. Floor/Door	Each	399	2.00	
.5	Wall Finishes	Paint		S.F. Surface	.50	.32	20.4%
.6	Floor Finishes	50% hardwood, 50% carpet		S.F. Floor	6.14	6.14	
.7	Ceiling Finishes	Gypsum board on wood furring		S.F. Ceiling	2.83	2.83	
.9	Interior Surface/Exterior Wall	Painted gypsum board	80% of wall	S.F. Wall	1.18	.71	
7.0	**Conveying**						
.1	Elevators	One hydraulic passenger elevator		Each	49,800	4.98	7.8%
.2	Special Conveyors	N/A		—	—	—	
8.0	**Mechanical**						
.1	Plumbing	Kitchen toilet and service fixtures, supply and drainage	1 Fixture/150 S.F. Floor	Each	757	5.05	
.2	Fire Protection	Wet pipe sprinkler system		S.F. Floor	1.73	1.73	
.3	Heating	Oil fired hot water, baseboard radiation		S.F. Floor	4.08	4.08	25.8%
.4	Cooling	Split system with air cooled condensing unit		S.F. Floor	5.49	5.49	
.5	Special Systems	N/A		—	—	—	
9.0	**Electrical**						
.1	Service & Distribution	600 ampere service, panel board and feeders		S.F. Floor	2.80	2.80	
.2	Lighting & Power	Fluorescent fixtures, receptacles, switches, A.C. and misc. power		S.F. Floor	5.89	5.89	22.4%
.4	Special Electrical	Alarm, communication system and generator set		S.F. Floor	5.57	5.57	
11.0	**Special Construction**						
.1	Specialties	N/A		—	—	—	0.0%
12.0	**Site Work**						
.1	Earthwork	N/A		—	—	—	
.3	Utilities	N/A		—	—	—	0.0%
.5	Roads & Parking	N/A		—	—	—	
.7	Site Improvements	N/A		—	—	—	
				Sub-Total		63.56	100%
	GENERAL CONDITIONS (Overhead & Profit)				15%	9.53	
	ARCHITECT FEES				9%	6.56	
				Total Building Cost		**79.65**	

BUILDING TYPES

COMMERCIAL/INDUSTRIAL/INSTITUTIONAL — M.250 Funeral Home

Costs per square foot of floor area

Exterior Wall	S.F. Area	4000	5000	6000	7000	8000	9000	10000	11000	12000
	L.F. Perimeter	260	300	340	353	386	420	425	435	460
Vertical Redwood Siding	Wood Frame	81.40	78.50	76.60	74.35	73.20	72.40	71.00	70.00	69.50
Brick Veneer	Wood Frame	88.55	85.15	82.90	80.00	78.65	77.65	75.85	74.55	73.90
Aluminum Siding	Wood Frame	80.00	77.30	75.45	73.35	72.30	71.55	70.30	69.35	68.85
Brick on Block	Wood Truss	92.50	88.90	86.50	83.30	81.85	80.80	78.80	77.35	76.65
Limestone on Block	Wood Truss	98.60	94.50	91.75	87.95	86.30	85.10	82.70	81.00	80.15
Stucco on Block	Wood Truss	82.25	79.40	77.55	75.30	74.20	73.40	72.10	71.10	70.60
Perimeter Adj., Add or Deduct	Per 100 L.F.	5.95	4.70	3.95	3.40	2.95	2.60	2.40	2.15	2.00
Story Hgt. Adj., Add or Deduct	Per 1 Ft.	.85	.80	.75	.65	.65	.65	.55	.55	.50

For Basement, add $16.90 per square foot of basement area

The above costs were calculated using the basic specifications shown on the facing page. These costs should be adjusted where necessary for design alternatives and owner's requirements. Reported completed project costs, for this type of structure, range from $57.10 to $161.20 per S.F.

Common additives

Description	Unit	$ Cost
Autopsy Table, Standard	Each	6425
Deluxe	Each	8875
Directory Boards, Plastic, glass covered		
30" x 20"	Each	500
36" x 48"	Each	910
Aluminum, 24" x 18"	Each	445
36" x 24"	Each	535
48" x 32"	Each	630
48" x 60"	Each	1425
Emergency Lighting, 25 watt, battery operated		
Lead battery	Each	350
Nickel cadmium	Each	590
Mortuary Refrigerator, End operated		
Two capacity	Each	11,000
Six capacity	Each	17,900

Description	Unit	$ Cost
Planters, Precast concrete		
48" diam., 24" high	Each	505
7" diam., 36" high	Each	1950
Fiberglass, 36" diam., 24" high	Each	540
60" diam., 24" high	Each	945
Smoke Detectors		
Ceiling type	Each	141
Duct type	Each	335

Funeral Home

Model costs calculated for a 1 story building with 10' story height and 10,000 square feet of floor area

				Unit	Unit Cost	Cost Per S.F.	% Of Sub-Total
1.0 Foundations							
.1	Footings & Foundations	Poured concrete; strip and spread footings and 4' foundation wall		S.F. Ground	3.05	3.05	7.1%
.4	Piles & Caissons	N/A		—	—	—	
.9	Excavation & Backfill	Site preparation for slab and trench for foundation wall and footing		S.F. Ground	.96	.96	
2.0 Substructure							
.1	Slab on Grade	4" reinforced concrete with vapor barrier and granular base		S.F. Slab	3.05	3.05	5.4%
.2	Special Substructures	N/A		—	—	—	
3.0 Superstructure							
.1	Columns & Beams	N/A		—	—	—	
.4	Structural Walls	N/A		—	—	—	
.5	Elevated Floors	N/A		—	—	—	6.0%
.7	Roof	Plywood on wood rafters (pitched)		S.F. Ground	3.38	3.38	
.9	Stairs	N/A		—	—	—	
4.0 Exterior Closure							
.1	Walls	1" x 4" vertical T & G redwood siding on wood studs	90% of wall	S.F. Wall	7.97	3.05	
.5	Exterior Wall Finishes	N/A		—	—	—	8.6%
.6	Doors	Wood swinging double doors, single leaf hollow metal		Each	1298	.78	
.7	Windows & Glazed Walls	Double hung wood	10% of wall	Each	294	1.04	
5.0 Roofing							
.1	Roof Coverings	Asphalt shingles with flashing		S.F. Roof	1.53	1.53	
.7	Insulation	Fiberglass sheets		S.F. Ground	.86	.86	4.7%
.8	Openings & Specialties	Gutters and downspouts		S.F. Ground	.29	.29	
6.0 Interior Construction							
.1	Partitions	Gypsum board on wood studs with sound deadening board	15 S.F. Floor/L.F. Partition	S.F. Partition	4.76	2.54	
.4	Interior Doors	Single leaf wood	150 S.F. Floor/Door	Each	399	2.66	
.5	Wall Finishes	50% wallpaper, 25% wood paneling, 25% paint		S.F. Surface	1.88	2.00	32.3%
.6	Floor Finishes	70% carpet, 30% terrazzo		S.F. Floor	8.53	8.53	
.7	Ceiling Finishes	Mineral fiberboard on wood furring		S.F. Ceiling	2.12	2.12	
.9	Interior Surface/Exterior Wall	Painted gypsum board on wood furring	90% of wall	S.F. Wall	1.47	.44	
7.0 Conveying							
.1	Elevators	N/A		—	—	—	0.0%
.2	Special Conveyors	N/A		—	—	—	
8.0 Mechanical							
.1	Plumbing	Toilet and service fixtures, supply and drainage	1 Fixture/770 S.F. Floor	Each	2926	3.80	
.2	Fire Protection	Wet pipe sprinkler system		S.F. Floor	1.57	1.57	
.3	Heating	Included in 8.4		—	—	—	28.9%
.4	Cooling	Multizone rooftop unit, gas heating, electric cooling		S.F. Floor	11.02	11.02	
.5	Special Systems	N/A		—	—	—	
9.0 Electrical							
.1	Service & Distribution	200 ampere service, panel board and feeders		S.F. Floor	.52	.52	
.2	Lighting & Power	Fluorescent fixtures, receptacles, switches, A.C. and misc. power		S.F. Floor	3.25	3.25	7.0%
.4	Special Electrical	Alarm systems and emergency lighting		S.F. Floor	.22	.22	
11.0 Special Construction							
.1	Specialties	N/A		—	—	—	0.0%
12.0 Site Work							
.1	Earthwork	N/A		—	—	—	
.3	Utilities	N/A		—	—	—	0.0%
.5	Roads & Parking	N/A		—	—	—	
.7	Site Improvements	N/A		—	—	—	
				Sub-Total		56.66	100%
	GENERAL CONDITIONS (Overhead & Profit)				15%	8.50	
	ARCHITECT FEES				9%	5.84	
				Total Building Cost		71.00	

BUILDING TYPES

127

COMMERCIAL/INDUSTRIAL/INSTITUTIONAL   M.260   Garage, Auto Sales

Costs per square foot of floor area

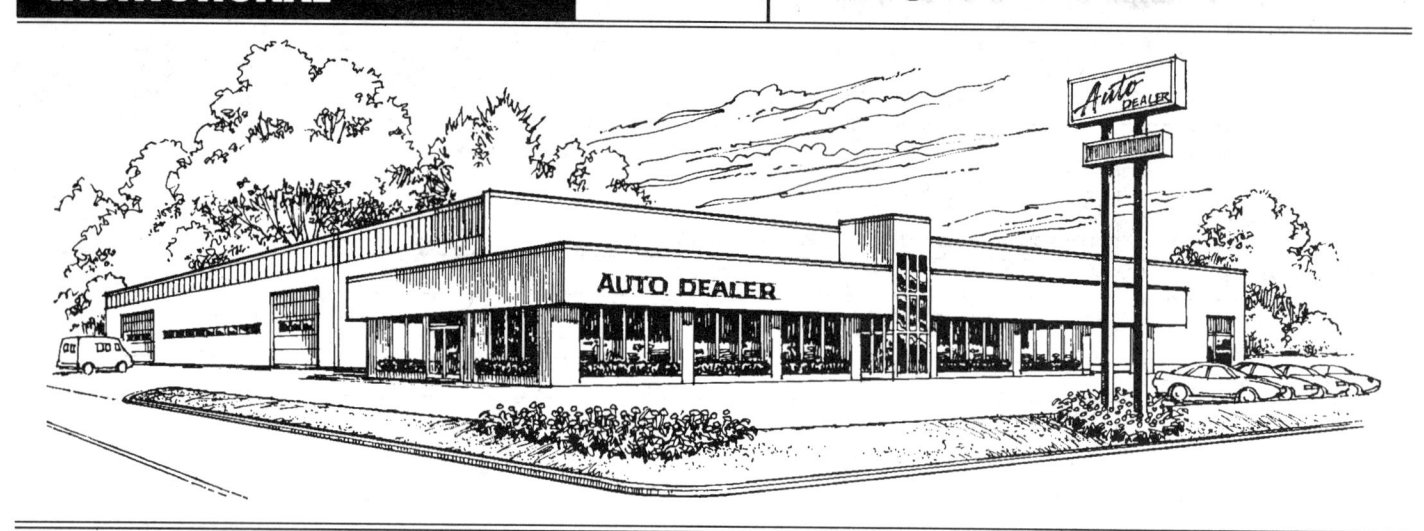

Exterior Wall	S.F. Area	12000	14000	16000	19000	21000	23000	26000	28000	30000
	L.F. Perimeter	440	474	510	556	583	607	648	670	695
Metal Panel Curtain Walls	Steel Frame	62.50	60.85	59.70	58.20	57.35	56.60	55.85	55.30	54.85
Tilt-up Concrete Wall	Steel Frame	60.95	59.40	58.30	56.95	56.20	55.50	54.80	54.25	53.85
Face Brick with Concrete Block Back-up	Bearing Walls	65.70	63.65	62.20	60.30	59.20	58.25	57.25	56.55	56.05
	Steel Frame	67.90	65.85	64.40	62.50	61.45	60.45	59.45	58.75	58.25
Stucco on Concrete Block	Bearing Walls	59.40	57.80	56.65	55.20	54.40	53.65	52.90	52.40	52.00
	Steel Frame	61.90	60.30	59.15	57.70	56.90	56.20	55.45	54.90	54.45
Perimeter Adj., Add or Deduct	Per 100 L.F.	3.55	3.05	2.60	2.25	2.05	1.85	1.60	1.50	1.45
Story Hgt. Adj., Add or Deduct	Per 1 Ft.	.80	.75	.70	.65	.65	.60	.55	.55	.50

For Basement, add $19.55 per square foot of basement area

The above costs were calculated using the basic specifications shown on the facing page. These costs should be adjusted where necessary for design alternatives and owner's requirements. Reported completed project costs, for this type of structure, range from $30.70 to $88.35 per S.F.

Common additives

Description	Unit	$ Cost
Emergency Lighting, 25 watt, battery operated		
Lead battery	Each	350
Nickel cadmium	Each	590
Smoke Detectors		
Ceiling type	Each	141
Duct type	Each	335
Sound System		
Amplifier, 250 watts	Each	1500
Speaker, ceiling or wall	Each	130
Trumpet	Each	244

BUILDING TYPES

Location Factors appear in the Reference Section

Model costs calculated for a 1 story building with 14' story height and 21,000 square feet of floor area

Garage, Auto Sales

				Unit	Unit Cost	Cost Per S.F.	% Of Sub-Total
1.0 Foundations							
.1	Footings & Foundations	Poured concrete; strip and spread footings and 4' foundation wall		S.F. Ground	2.19	2.19	
.4	Piles & Caissons	N/A		—	—	—	6.8%
.9	Excavation & Backfill	Site preparation for slab and trench for foundation wall and footing		S.F. Ground	.96	.96	
2.0 Substructure							
.1	Slab on Grade	4" reinforced concrete with vapor barrier and granular base		S.F. Slab	3.61	3.61	7.7%
.2	Special Substructures	N/A		—	—	—	
3.0 Superstructure							
.1	Columns & Beams	Steel columns included in 3.7		—	—	—	
.4	Structural Walls	Metal siding support		S.F. Wall	3.06	.83	
.5	Elevated Floors	N/A		—	—	—	13.1%
.7	Roof	Metal deck, open web steel joists, beams, columns		S.F. Floor	5.28	5.28	
.9	Stairs	N/A		—	—	—	
4.0 Exterior Closure							
.1	Walls	Metal panel	70% of wall	S.F. Wall	8.78	2.39	
.5	Exterior Wall Finishes	N/A		—	—	—	17.6%
.6	Doors	Double aluminum and glass, hollow metal, steel overhead		Each	2435	2.09	
.7	Windows & Glazed Walls	Window wall	30% of wall	S.F. Window	31	3.72	
5.0 Roofing							
.1	Roof Coverings	Built-up tar and gravel with flashing		S.F. Roof	1.93	1.93	
.7	Insulation	Perlite/EPS composite		S.F. Roof	1.18	1.18	7.0%
.8	Openings & Specialties	Gravel stop and skylight		L.F. Perimeter	5.55	.15	
6.0 Interior Construction							
.1	Partitions	Gypsum board on metal studs	28 S.F. Floor/L.F. Partition	S.F. Partition	2.92	1.25	
.4	Interior Doors	Hollow metal	280 S.F. Floor/Door	Each	530	1.89	
.5	Wall Finishes	Paint		S.F. Surface	.50	.43	13.4%
.6	Floor Finishes	50% vinyl tile, 50% paint		S.F. Floor	1.74	1.74	
.7	Ceiling Finishes	Fiberglass board on exposed grid, suspended	50% of area	S.F. Ceiling	1.87	.94	
.9	Interior Surface/Exterior Wall	N/A		—	—	—	
7.0 Conveying							
.1	Elevators	N/A		—	—	—	0.0%
.2	Special Conveyors	N/A		—	—	—	
8.0 Mechanical							
.1	Plumbing	Toilet and service fixtures, supply and drainage	1 Fixture/1500 S.F. Floor	Each	3075	2.05	
.2	Fire Protection	Wet pipe sprinkler system		S.F. Floor	2.09	2.09	
.3	Heating	Gas fired hot water, unit heaters (service area)		S.F. Floor	2.56	2.56	22.9%
.4	Cooling	Single zone rooftop unit, gas heating, electric (office and showroom)		S.F. Floor	3.73	3.73	
.5	Special Systems	Underfloor garage exhaust system		S.F. Floor	.21	.20	
9.0 Electrical							
.1	Service & Distribution	200 ampere service, panel board and feeders		S.F. Floor	.26	.26	
.2	Lighting & Power	Fluorescent fixtures, receptacles, switches, A.C. and misc. power		S.F. Floor	4.00	4.00	9.5%
.4	Special Electrical	Alarm systems and emergency lighting		S.F. Floor	.19	.19	
11.0 Special Construction							
.1	Specialties	Hoists, compressor, fuel pump		S.F. Floor	.95	.95	2.0%
12.0 Site Work							
.1	Earthwork	N/A		—	—	—	
.3	Utilities	N/A		—	—	—	
.5	Roads & Parking	N/A		—	—	—	0.0%
.7	Site Improvements	N/A		—	—	—	
				Sub-Total		46.61	100%
	GENERAL CONDITIONS (Overhead & Profit)				15%	6.99	
	ARCHITECT FEES				7%	3.75	
				Total Building Cost		57.35	

BUILDING TYPES

129

COMMERCIAL/INDUSTRIAL/INSTITUTIONAL

M.270 Garage, Parking

Costs per square foot of floor area

Exterior Wall	S.F. Area	85000	115000	145000	175000	205000	235000	265000	295000	325000
	L.F. Perimeter	529	638	723	823	923	951	1037	1057	1132
Face Brick with Concrete Block Back-up	Steel Frame	32.15	31.50	31.00	30.75	30.55	30.15	30.10	29.80	29.70
	R/Conc. Frame	26.60	25.95	25.45	25.20	25.00	24.60	24.55	24.25	24.15
Precast Concrete	Steel Frame	32.40	31.75	31.30	31.05	30.85	30.50	30.40	30.10	30.00
	R/Conc. Frame	26.45	25.80	25.35	25.10	24.90	24.55	24.45	24.15	24.05
Reinforced Concrete	Steel Frame	31.55	31.05	30.65	30.45	30.30	30.00	29.90	29.70	29.65
	R/Conc. Frame	25.40	24.90	24.50	24.30	24.15	23.85	23.80	23.55	23.50
Perimeter Adj., Add or Deduct	Per 100 L.F.	.90	.65	.55	.45	.40	.35	.30	.25	.25
Story Hgt. Adj., Add or Deduct	Per 1 Ft.	.35	.30	.25	.25	.25	.20	.20	.20	.20
	Basement—Not Applicable									

The above costs were calculated using the basic specifications shown on the facing page. These costs should be adjusted where necessary for design alternatives and owner's requirements. Reported completed project costs, for this type of structure, range from $17.45 to $72.20 per S.F.

Common additives

Description	Unit	$ Cost
Automatic Gates, 8' arm, 1 way	Each	3225
2 way	Each	3525
Booth for attendant, average	Each	4400 - 7325
Elevators, Electric passenger, 5 stops		
2000# capacity	Each	88,800
3500# capacity	Each	94,800
5000# capacity	Each	98,300
Fee Indicator, 1" display	Each	1750
Ticket Printer & Dispenser, Standard	Each	5325
Rate computing	Each	6950
Card control station, single period	Each	855
4 period	Each	945
Key station on pedestal	Each	535
Coin station, multiple coins	Each	3975
Painting, Parking stalls	Stall	4.39
Parking Barriers		
Timber with saddles, 4" x 4"	L.F.	5.40
Precast concrete, 6" x 10" x 6'	Each	31
Traffic Signs, directional, 12" x 18", high densit	Each	39

Location Factors appear in the Reference Section

Model costs calculated for a 5 story building with 10' story height and 145,000 square feet of floor area

Garage, Parking

			Unit	Unit Cost	Cost Per S.F.	% Of Sub-Total
1.0 Foundations						
.1	Footings & Foundations	Poured concrete; strip and spread footings and 4' foundation wall	S.F. Ground	8.35	1.67	8.9%
.4	Piles & Caissons	N/A	—	—	—	
.9	Excavation & Backfill	Site preparation for slab and trench for foundation wall and footing	S.F. Ground	.89	.18	
2.0 Substructure						
.1	Slab on Grade	6" reinforced concrete with vapor barrier and granular base	S.F. Slab	3.82	.76	3.6%
.2	Special Substructures	N/A	—	—	—	
3.0 Superstructure						
.1	Columns & Beams	Precast concrete columns	S.F. Floor	4.56	4.56	
.4	Structural Walls	Concrete block elevator shaft	S.F. Wall	9.26	.48	
.5	Elevated Floors	Double tee precast concrete slab	S.F. Floor	7.55	6.04	53.9%
.7	Roof	N/A	—	—	—	
.9	Stairs	Concrete	Flight	2425	.17	
4.0 Exterior Closure						
.1	Walls	Face brick with concrete block backup 40% of story height	S.F. Wall	21	2.12	
.5	Exterior Wall Finishes	N/A	—	—	—	10.2%
.6	Doors	N/A	—	—	—	
.7	Windows & Glazed Walls	N/A	—	—	—	
5.0 Roofing						
.1	Roof Coverings	N/A	—	—	—	
.7	Insulation	N/A	—	—	—	0.0%
.8	Openings & Specialties	N/A	—	—	—	
6.0 Interior Construction						
.1	Partitions	N/A	—	—	—	
.4	Interior Doors	N/A	—	—	—	
.5	Wall Finishes	N/A	—	—	—	0.0%
.6	Floor Finishes	N/A	—	—	—	
.7	Ceiling Finishes	N/A	—	—	—	
.9	Interior Surface/Exterior Wall	N/A	—	—	—	
7.0 Conveying						
.1	Elevators	Two hydraulic passenger elevators	Each	82,650	1.14	5.5%
.2	Special Conveyors	N/A	—	—	—	
8.0 Mechanical						
.1	Plumbing	Toilet and service fixtures, supply and drainage 1 Fixture/18,125 S.F. Floor	Each	14,681	.81	
.2	Fire Protection	Standpipes and hose systems	S.F. Floor	.04	.04	
.3	Heating	N/A	—	—	—	3.9%
.4	Cooling	N/A	—	—	—	
.5	Special Systems	N/A	—	—	—	
9.0 Electrical						
.1	Service & Distribution	400 ampere service, panel board and feeders	S.F. Floor	.14	.14	
.2	Lighting & Power	Fluorescent fixtures, receptacles, switches and misc. power	S.F. Floor	1.76	1.76	9.7%
.4	Special Electrical	Alarm systems and emergency lighting	S.F. Floor	.12	.12	
11.0 Special Construction						
.1	Specialties	Ticket dispensers, booths automatic gates	S.F. Floor	.89	.89	4.3%
12.0 Site Work						
.1	Earthwork	N/A	—	—	—	
.3	Utilities	N/A	—	—	—	0.0%
.5	Roads & Parking	N/A	—	—	—	
.7	Site Improvements	N/A	—	—	—	
			Sub-Total		20.88	100%
	GENERAL CONDITIONS (Overhead & Profit)			15%	3.13	
	ARCHITECT FEES			6%	1.44	
			Total Building Cost		**25.45**	

BUILDING TYPES

COMMERCIAL/INDUSTRIAL/INSTITUTIONAL

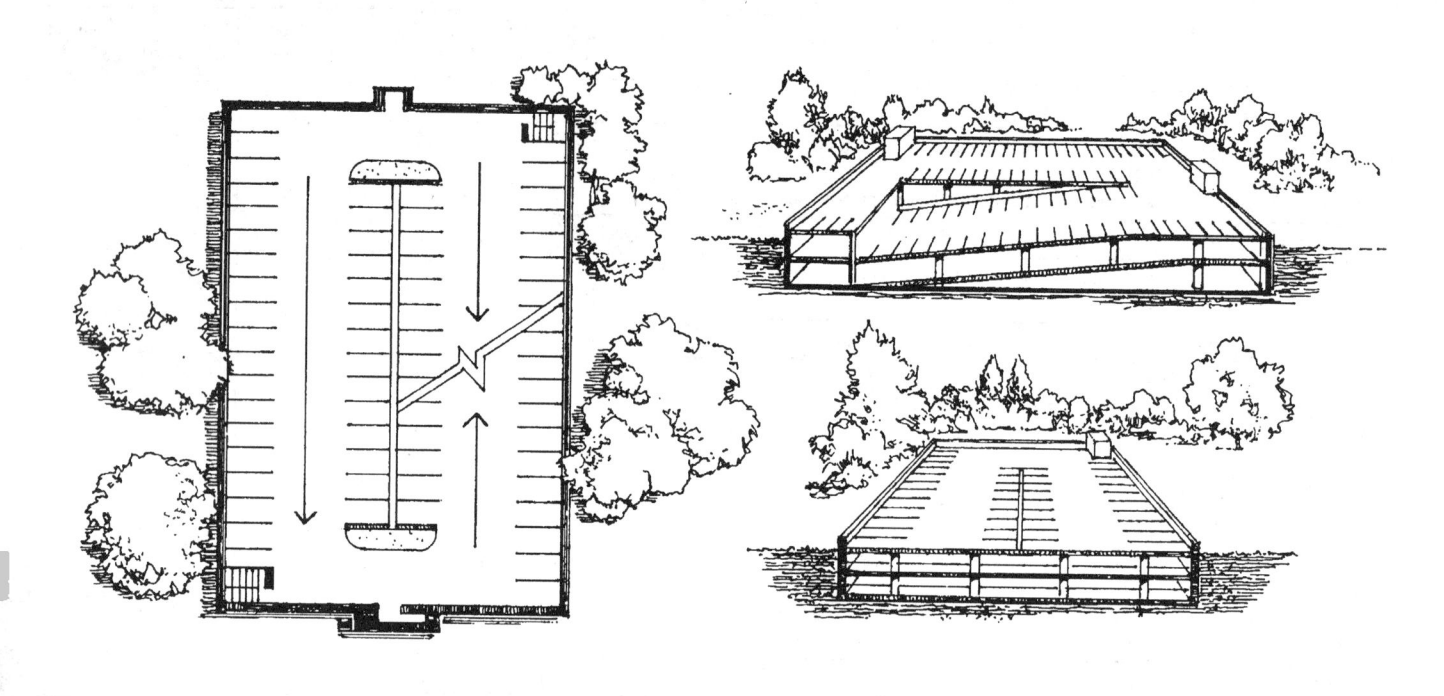

M.280 Garage, Underground Parking

Costs per square foot of floor area

Exterior Wall	S.F. Area	20000	30000	40000	50000	75000	100000	125000	150000	175000
	L.F. Perimeter	400	500	600	650	775	900	1000	1100	1185
Reinforced Concrete	R/Conc. Frame	46.75	44.30	43.05	41.65	39.85	38.95	38.30	37.80	37.45
Perimeter Adj., Add or Deduct	Per 100 L.F.	3.15	2.10	1.55	1.25	.80	.65	.50	.40	.35
Story Hgt. Adj., Add or Deduct	Per 1 Ft.	1.20	1.00	.90	.75	.60	.55	.45	.40	.40
Basement—Not Applicable										

The above costs were calculated using the basic specifications shown on the facing page. These costs should be adjusted where necessary for design alternatives and owner's requirements. Reported completed project costs, for this type of structure, range from $27.75 to $65.85 per S.F.

Common additives

Description	Unit	$ Cost
Automatic Gates, 8' arm, 1 way	Each	3225
2 way	Each	3525
Booth for attendant	Each	4400 - 7325
Elevators, Hydraulic passenger, 2 stops		
1500# capacity	Each	40,800
2500# capacity	Each	42,100
3500# capacity	Each	45,700
Fee Indicator, 1" display	Each	1750
Ticket Printer & Dispenser, Standard	Each	5325
Rate computing	Each	6950
Card control station, single period	Each	855
4 period	Each	945
Key station on pedestal	Each	535
Coin station, multiple coins	Each	3975
Painting, Parking stalls	Stall	4.39
Parking Barriers		
Timber with saddles, 4" x 4"	L.F.	5.40
Precast concrete, 6" x 10" x 6'	Each	31
Traffic Signs, directional, 12" x 18"	Each	39

Model costs calculated for a 2 story building with 10' story height and 100,000 square feet of floor area

Garage, Underground Parking

			Unit	Unit Cost	Cost Per S.F.	% Of Sub-Total
1.0 Foundations						
.1	Footings & Foundations	Poured concrete; strip and spread footings and waterproofing	S.F. Ground	4.74	2.37	16.9%
.4	Piles & Caissons	N/A	—	—	—	
.9	Excavation & Backfill	Excavation 24' deep	S.F. Ground	5.85	2.92	
2.0 Substructure						
.1	Slab on Grade	5" reinforced concrete with vapor barrier and granular base	S.F. Slab	3.27	1.63	5.2%
.2	Special Substructures	N/A	—	—	—	
3.0 Superstructure						
.1	Columns & Beams	Concrete columns	L.F. Column	121	.61	
.4	Structural Walls	N/A	—	—	—	
.5	Elevated Floors	Cast-in-place concrete beam and slab	S.F. Floor	18.98	7.73	51.8%
.7	Roof	Cast-in-place concrete beam and slab	S.F. Roof	15.40	7.70	
.9	Stairs	Concrete	Flight	3750	.19	
4.0 Exterior Closure						
.1	Walls	Cast-in place concrete	S.F. Wall	13.56	2.44	
.5	Exterior Wall Finishes	N/A	—	—	—	8.1%
.6	Doors	Steel overhead, hollow metal	Each	2795	.11	
.7	Windows & Glazed Walls	N/A	—	—	—	
5.0 Roofing						
.1	Roof Coverings	Neoprene membrane	S.F. Roof	3.10	1.55	4.9%
.7	Insulation	N/A	—	—	—	
.8	Openings & Specialties	N/A	—	—	—	
6.0 Interior Construction						
.1	Partitions	N/A	—	—	—	
.4	Interior Doors	N/A	—	—	—	
.5	Wall Finishes	N/A	—	—	—	
.6	Floor Finishes	N/A	—	—	—	0.0%
.7	Ceiling Finishes	N/A	—	—	—	
.9	Interior Surface/Exterior Wall	N/A	—	—	—	
7.0 Conveying						
.1	Elevators	Two hydraulic passenger elevators	Each	50,000	1.00	3.2%
.2	Special Conveyors	N/A	—	—	—	
8.0 Mechanical						
.1	Plumbing	Drainage in parking areas, toilets, & service fixtures 1 Fixture/5000 S.F. Floor	Each	.79	.79	
.2	Fire Protection	Dry standpipe system, class 1	S.F. Floor	.08	.08	
.3	Heating	N/A	—	—	—	3.1%
.4	Cooling	Exhaust fans	S.F. Floor	.12	.12	
.5	Special Systems	N/A	—	—	—	
9.0 Electrical						
.1	Service & Distribution	200 ampere service, panel board and feeders	S.F. Floor	.08	.08	
.2	Lighting & Power	Fluorescent fixtures, receptacles, switches and misc. power	S.F. Floor	1.48	1.48	6.0%
.4	Special Electrical	Alarm systems and emergency lighting	S.F. Floor	.32	.32	
11.0 Special Construction						
.1	Specialties	Ticket dispensers, booths, automatic gates	S.F. Floor	.24	.24	0.8%
12.0 Site Work						
.1	Earthwork	N/A	—	—	—	
.3	Utilities	N/A	—	—	—	0.0%
.5	Roads & Parking	N/A	—	—	—	
.7	Site Improvements	N/A	—	—	—	
			Sub-Total		31.36	100%
	GENERAL CONDITIONS (Overhead & Profit)			15%	4.70	
	ARCHITECT FEES			8%	2.89	
			Total Building Cost		**38.95**	

BUILDING TYPES

COMMERCIAL/INDUSTRIAL/INSTITUTIONAL   M.290   Garage, Repair

Costs per square foot of floor area

Exterior Wall	S.F. Area	2000	4000	6000	8000	10000	12000	14000	16000	18000
	L.F. Perimeter	180	260	340	420	500	580	586	600	610
Concrete Block	Wood Joists	89.60	77.50	73.40	71.35	70.20	69.40	67.35	65.95	64.80
	Steel Joists	89.60	77.90	73.95	71.95	70.80	70.05	68.05	66.65	65.50
Poured Concrete	Wood Joists	96.55	82.70	78.00	75.70	74.35	73.40	70.85	69.05	67.60
	Steel Joists	97.25	83.40	78.70	76.40	75.05	74.10	71.55	69.80	68.30
Insulated Metal Panels	Wood Frame	96.65	82.75	78.10	75.75	74.40	73.45	70.90	69.10	67.65
	Steel Frame	99.05	85.15	80.45	78.15	76.75	75.85	73.25	71.50	70.05
Perimeter Adj., Add or Deduct	Per 100 L.F.	13.85	6.90	4.60	3.50	2.75	2.30	1.95	1.75	1.55
Story Hgt. Adj., Add or Deduct	Per 1 Ft.	1.05	.75	.65	.65	.60	.55	.50	.45	.40

For Basement, add $18.85 per square foot of basement area

The above costs were calculated using the basic specifications shown on the facing page. These costs should be adjusted where necessary for design alternatives and owner's requirements. Reported completed project costs, for this type of structure, range from $38.85 to $116.70 per S.F.

Common additives

Description	Unit	$ Cost
Air Compressors		
Electric 1-1/2 H.P., standard controls	Each	3325
Dual controls	Each	3550
5 H.P. 115/230 Volt, standard controls	Each	4500
Dual controls	Each	4725
Product Dispenser		
with vapor recovery for 6 nozzles	Each	16,500
Hoists, Single post		
8000# cap., swivel arm	Each	6150
Two post, adjustable frames, 11,000# cap.	Each	8650
24,000# cap.	Each	12,800
7500# Frame support	Each	7350
Four post, roll on ramp	Each	6800
Lockers, Steel, single tier, 60" or 72"	Opening	134 - 213
2 tier, 60" or 72" total	Opening	83 - 104
5 tier, box lockers	Opening	44 - 56
Locker bench, lam. maple top only	L.F.	18.40
Pedestals, steel pipe	Each	43
Lube Equipment		
3 reel type, with pumps, no piping	Each	8450
Spray Painting Booth, 26' long, complete	Each	15,500

Model costs calculated for a 1 story building with 14' story height and 4,000 square feet of floor area

Garage, Repair

					Unit	Unit Cost	Cost Per S.F.	% Of Sub-Total
1.0 Foundations								
.1	Footings & Foundations	Poured concrete; strip and spread footings and 4' foundation wall			S.F. Ground	4.50	4.50	8.7%
.4	Piles & Caissons	N/A			—	—	—	
.9	Excavation & Backfill	Site preparation for slab and trench for foundation wall and footing			S.F. Ground	.96	.96	
2.0 Substructure								
.1	Slab on Grade	4" reinforced concrete with vapor barrier and granular base			S.F. Slab	3.05	3.05	4.9%
.2	Special Substructures	N/A			—	—	—	
3.0 Superstructure								
.1	Columns & Beams	N/A			—	—	—	
.4	Structural Walls	N/A			—	—	—	4.8%
.5	Elevated Floors	N/A			—	—	—	
.7	Roof	Metal deck on open web steel joists			S.F. Roof	3.03	3.03	
.9	Stairs	N/A			—	—	—	
4.0 Exterior Closure								
.1	Walls	Concrete block		80% of wall	L.F. Wall	7.32	5.33	
.5	Exterior Wall Finishes	N/A			—	—	—	12.7%
.6	Doors	Steel overhead and hollow metal		15% of wall	S.F. Door	13.11	1.79	
.7	Windows & Glazed Walls	Hopper type commercial steel		5% of wall	Each	282	.86	
5.0 Roofing								
.1	Roof Coverings	Built-up tar and gravel			S.F. Roof	2.50	2.50	
.7	Insulation	Perlite/EPS composite			S.F. Roof	1.18	1.18	6.5%
.8	Openings & Specialties	Gravel stop and skylight			S.F. Roof	.42	.42	
6.0 Interior Construction								
.1	Partitions	Concrete block, toilet partitions	50 S.F. Floor/L.F. Partition		S.F. Partition	5.70	1.14	
.4	Interior Doors	Single leaf hollow metal	3000 S.F. Floor/Door		Each	530	.18	
.5	Wall Finishes	Paint			S.F. Surface	.90	.36	6.5%
.6	Floor Finishes	90% metallic floor hardener, 10% vinyl composition tile			S.F. Floor	.71	.71	
.7	Ceiling Finishes	Gypsum board on wood joists in office and washrooms		10% of area	S.F. Ceiling	2.83	.28	
.9	Interior Surface/Exterior Wall	Paint		80% of wall	S.F. Wall	1.93	1.41	
7.0 Conveying								
.1	Elevators	N/A			—	—	—	0.0%
.2	Special Conveyors	N/A			—	—	—	
8.0 Mechanical								
.1	Plumbing	Toilet and service fixtures, supply and drainage	1 Fixture/500 S.F. Floor		Each	2795	5.59	
.2	Fire Protection	Sprinklers, ordinary hazard			S.F. Floor	2.19	2.19	
.3	Heating	Oil fired hot water, unit heaters			S.F. Floor	5.12	5.12	33.3%
.4	Cooling	Split systems with air cooled condensing units			S.F. Floor	7.12	7.12	
.5	Special Systems	Garage exhaust system			S.F. Floor	.79	.80	
9.0 Electrical								
.1	Service & Distribution	100 ampere service, panel board and feeders			S.F. Floor	.47	.47	
.2	Lighting & Power	Fluorescent fixtures, receptacles, switches, A.C. and misc. power			S.F. Floor	4.04	4.04	7.9%
.4	Special Electrical	Alarm systems and emergency lighting			S.F. Floor	.45	.45	
11.0 Special Construction								
.1	Specialties	Hoists			S.F. Floor	9.23	9.23	14.7%
12.0 Site Work								
.1	Earthwork	N/A			—	—	—	
.3	Utilities	N/A			—	—	—	0.0%
.5	Roads & Parking	N/A			—	—	—	
.7	Site Improvements	N/A			—	—	—	
					Sub-Total		62.71	100%
	GENERAL CONDITIONS (Overhead & Profit)					15%	9.41	
	ARCHITECT FEES					8%	5.78	
					Total Building Cost		77.90	

BUILDING TYPES

COMMERCIAL/INDUSTRIAL/INSTITUTIONAL M.300 | Garage, Service Station

Costs per square foot of floor area

Exterior Wall	S.F. Area	600	800	1000	1200	1400	1600	1800	2000	2200
	L.F. Perimeter	100	120	126	140	153	160	170	180	190
Face Brick with Concrete Block Back-up	Wood Truss	124.70	115.20	104.65	99.90	96.30	92.30	89.70	87.65	86.00
	Steel Joists	120.05	110.55	99.95	95.25	91.65	87.60	85.05	83.00	81.35
Enameled Sandwich Panel Tile on Concrete Block	Steel Frame	107.70	99.50	90.80	86.80	83.75	80.50	78.35	76.65	75.30
	Steel Joists	132.75	122.05	109.80	104.45	100.30	95.65	92.70	90.35	88.40
Aluminum Siding Wood Siding	Wood Frame	105.40	97.85	90.05	86.40	83.60	80.70	78.75	77.25	76.00
	Wood Frame	106.15	98.50	90.60	86.95	84.10	81.15	79.20	77.65	76.40
Perimeter Adj., Add or Deduct	Per 100 L.F.	57.55	43.15	34.50	28.75	24.65	21.55	19.20	17.25	15.70
Story Hgt. Adj., Add or Deduct	Per 1 Ft.	4.40	3.95	3.30	3.10	2.85	2.60	2.50	2.40	2.25
	Basement—Not Applicable									

The above costs were calculated using the basic specifications shown on the facing page. These costs should be adjusted where necessary for design alternatives and owner's requirements. Reported completed project costs, for this type of structure, range from $25.20 to $116.80 per S.F.

Common additives

Description	Unit	$ Cost
Air Compressors		
Electric 1-1/2 H.P., standard controls	Each	3325
Dual controls	Each	3550
5 H.P. 115/230 volt, standard controls	Each	4500
Dual controls	Each	4725
Product Dispenser		
with vapor recovery for 6 nozzles	Each	16,500
Hoists, Single post		
8000# cap. swivel arm	Each	6150
Two post, adjustable frames, 11,000# cap.	Each	8650
24,000# cap.	Each	12,800
7500# cap.	Each	7350
Four post, roll on ramp	Each	6800
Lockers, Steel, single tier, 60" or 72"	Opening	134 - 213
2 tier, 60" or 72" total	Opening	83 - 104
5 tier, box lockers	Each	44 - 56
Locker bench, lam. maple top only	L.F.	18.40
Pedestals, steel pipe	Each	43
Lube Equipment		
3 reel type, with pumps, no piping	Each	8450

Model costs calculated for a 1 story building with 10' story height and 1,400 square feet of floor area

Garage, Service Station

				Unit	Unit Cost	Cost Per S.F.	% Of Sub-Total
1.0 Foundations							
.1	Footings & Foundations	Poured concrete; strip and spread footings and 4' foundation wall		S.F. Ground	7.14	7.14	
.4	Piles & Caissons	N/A		—	—	—	11.1%
.9	Excavation & Backfill	Site preparation for slab and trench for foundation wall and footing		S.F. Ground	1.43	1.43	
2.0 Substructure							
.1	Slab on Grade	4" reinforced concrete with vapor barrier and granular base		S.F. Slab	3.05	3.05	3.9%
.2	Special Substructures	N/A		—	—	—	
3.0 Superstructure							
.1	Columns & Beams	N/A		—	—	—	
.4	Structural Walls	N/A		—	—	—	
.5	Elevated Floors	N/A		—	—	—	5.6%
.7	Roof	Plywood on wood trusses		S.F. Ground	4.35	4.35	
.9	Stairs	N/A		—	—	—	
4.0 Exterior Closure							
.1	Walls	Face brick with concrete block backup	60% of wall	S.F. Wall	21	13.97	
.5	Exterior Wall Finishes	N/A		—	—	—	32.1%
.6	Doors	Steel overhead, aluminum & glass and hollow metal	20% of wall	S.F. Door	24	5.41	
.7	Windows & Glazed Walls	Store front and metal top hinged outswinging	20% of wall	S.F. Window	25	5.49	
5.0 Roofing							
.1	Roof Coverings	Asphalt shingles with flashing		S.F. Ground	1.15	1.15	
.7	Insulation	Perlite/EPS composite		S.F. Ground	.86	.86	2.6%
.8	Openings & Specialties	N/A		—	—	—	
6.0 Interior Construction							
.1	Partitions	Concrete block, toilet partitions	25 S.F. Floor/L.F. Partition	S.F. Partition	9.47	3.03	
.4	Interior Doors	Single leaf hollow metal	700 S.F. Floor/Door	Each	530	.76	
.5	Wall Finishes	Paint		S.F. Surface	1.08	.69	9.8%
.6	Floor Finishes	Vinyl composition tile	35% of floor area	S.F. Floor	2.40	.84	
.7	Ceiling Finishes	Painted gypsum board on wood joists in sales area & washrooms	35% of floor area	S.F. Ceiling	2.83	.99	
.9	Interior Surface/Exterior Wall	Paint	60% of wall	S.F. Wall	1.93	1.27	
7.0 Conveying							
.1	Elevators	N/A		—	—	—	0.0%
.2	Special Conveyors	N/A		—	—	—	
8.0 Mechanical							
.1	Plumbing	Toilet and service fixtures, supply and drainage	1 Fixture/235 S.F. Floor	Each	1727	7.35	
.2	Fire Protection	N/A		—	—	—	
.3	Heating	Oil fired hot water, wall fin radiation		S.F. Floor	5.12	5.12	25.4%
.4	Cooling	Split systems with air cooled condensing units		S.F. Floor	7.28	7.28	
.5	Special Systems	N/A		—	—	—	
9.0 Electrical							
.1	Service & Distribution	100 ampere service, panel board and feeders		S.F. Floor	2.12	2.12	
.2	Lighting & Power	Fluorescent fixtures, receptacles, switches, A.C. and misc. power		S.F. Floor	4.42	4.42	9.5%
.4	Special Electrical	Alarm systems and emergency lighting		S.F. Floor	.81	.81	
11.0 Special Construction							
.1	Specialties	N/A		—	—	—	0.0%
12.0 Site Work							
.1	Earthwork	N/A		—	—	—	
.3	Utilities	N/A		—	—	—	0.0%
.5	Roads & Parking	N/A		—	—	—	
.7	Site Improvements	N/A		—	—	—	
				Sub-Total		77.53	100%
	GENERAL CONDITIONS (Overhead & Profit)				15%	11.63	
	ARCHITECT FEES				8%	7.14	
				Total Building Cost		96.30	

BUILDING TYPES

COMMERCIAL/INDUSTRIAL/INSTITUTIONAL — M.310 Gymnasium

Costs per square foot of floor area

Exterior Wall	S.F. Area	12000	16000	20000	25000	30000	35000	40000	45000	50000
	L.F. Perimeter	440	520	600	700	708	780	841	910	979
Reinforced Concrete Block	Lam. Wood Arches	89.90	86.95	85.20	83.80	81.50	80.70	79.95	79.45	79.05
	Rigid Steel Frame	87.15	84.25	82.50	81.10	78.80	78.00	77.25	76.75	76.35
Face Brick with Concrete Block Back-up	Lam. Wood Arches	107.00	102.15	99.20	96.90	92.50	91.10	89.75	88.90	88.20
	Rigid Steel Frame	104.30	99.40	96.50	94.20	89.80	88.35	87.05	86.20	85.45
Metal Sandwich Panels	Lam. Wood Arches	97.00	93.30	91.05	89.30	86.10	85.05	84.05	83.40	82.85
	Rigid Steel Frame	94.30	90.60	88.35	86.55	83.40	82.35	81.35	80.70	80.15
Perimeter Adj., Add or Deduct	Per 100 L.F.	3.75	2.80	2.25	1.80	1.50	1.25	1.10	.95	.90
Story Hgt. Adj., Add or Deduct	Per 1 Ft.	.50	.45	.45	.40	.35	.30	.30	.25	.25
Basement—Not Applicable

The above costs were calculated using the basic specifications shown on the facing page. These costs should be adjusted where necessary for design alternatives and owner's requirements. Reported completed project costs, for this type of structure, range from $42.55 to $127.65 per S.F.

Common additives

Description	Unit	$ Cost
Bleachers, Telescoping, manual		
To 15 tier	Seat	65 - 76
16-20 tier	Seat	72 - 83
21-30 tier	Seat	88 - 104
For power operation, add	Seat	13.40 - 19.40
Gym Divider Curtain, Mesh top		
Manual roll-up	S.F.	8.45
Gym Mats		
2" naugahyde covered	S.F.	4.19
2" nylon	S.F.	7.90
1-1/2" wall pads	S.F.	7.70
1" wrestling mats	S.F.	4.76
Scoreboard		
Basketball, one side	Each	3025 - 21,000
Basketball Backstop		
Wall mtd., 6' extended, fixed	Each	1125 - 1900
Swing up, wall mtd.	Each	1700 - 2500

Description	Unit	$ Cost
Lockers, Steel, single tier, 60" or 72"	Opening	134 - 213
2 tier, 60" or 72" total	Opening	83 - 104
5 tier, box lockers	Opening	44 - 56
Locker bench, lam. maple top only	L.F.	18.40
Pedestals, steel pipe	Each	43
Sound System		
Amplifier, 250 watts	Each	1500
Speaker, ceiling or wall	Each	130
Trumpet	Each	244
Emergency Lighting, 25 watt, battery operated		
Lead battery	Each	350
Nickel cadmium	Each	590

Model costs calculated for a 1 story building with 25' story height and 20,000 square feet of floor area

Gymnasium

			Unit	Unit Cost	Cost Per S.F.	% Of Sub-Total
1.0 Foundations						
.1	Footings & Foundations	Poured concrete; strip and spread footings and 4' foundation wall	S.F. Ground	2.54	2.54	5.0%
.4	Piles & Caissons	N/A	—	—	—	
.9	Excavation & Backfill	Site preparation for slab and trench for foundation wall and footing	S.F. Ground	.89	.89	
2.0 Substructure						
.1	Slab on Grade	4" reinforced concrete with vapor barrier and granular base	S.F. Slab	3.05	3.05	4.4%
.2	Special Substructures	N/A	—	—	—	
3.0 Superstructure						
.1	Columns & Beams	N/A	—	—	—	
.4	Structural Walls	N/A	—	—	—	
.5	Elevated Floors	N/A	—	—	—	17.8%
.7	Roof	Wood deck on laminated wood arches	S.F. Ground	12.34	12.34	
.9	Stairs	N/A	—	—	—	
4.0 Exterior Closure						
.1	Walls	Reinforced concrete block (end walls included) 90% of wall	S.F. Wall	7.75	5.23	
.5	Exterior Wall Finishes	N/A	—	—	—	11.0%
.6	Doors	Aluminum and glass, hollow metal, steel overhead	Each	896	.27	
.7	Windows & Glazed Walls	Metal horizontal pivoted 10% of wall	Each	280	2.10	
5.0 Roofing						
.1	Roof Coverings	EPDM, 60 mils, fully adhered	S.F. Ground	1.71	1.71	
.7	Insulation	Polyisocyanurate	—	1.20	1.24	4.3%
.8	Openings & Specialties	N/A	—	—	—	
6.0 Interior Construction						
.1	Partitions	Concrete block, toilet partitions 50 S.F. Floor/L.F. Partition	S.F. Partition	6.20	1.24	
.4	Interior Doors	Single leaf hollow metal 500 S.F. Floor/Door	Each	530	1.06	
.5	Wall Finishes	50% paint, 50% ceramic tile	S.F. Surface	3.08	1.23	21.2%
.6	Floor Finishes	90% hardwood, 10% ceramic tile	S.F. Floor	9.41	9.41	
.7	Ceiling Finishes	Mineral fiber tile on concealed zee bars 15% of area	S.F. Ceiling	3.06	.46	
.9	Interior Surface/Exterior Wall	Paint 90% of wall	S.F. Wall	1.93	1.30	
7.0 Conveying						
.1	Elevators	N/A	—	—	—	0.0%
.2	Special Conveyors	N/A	—	—	—	
8.0 Mechanical						
.1	Plumbing	Toilet and service fixtures, supply and drainage 1 Fixture/515 S.F. Floor	Each	2904	5.64	
.2	Fire Protection	Wet pipe sprinkler system	S.F. Floor	1.57	1.57	
.3	Heating	Included in 8.4	—	—	—	24.4%
.4	Cooling	Single zone rooftop unit, gas heating, electric cooling	S.F. Floor	9.73	9.73	
.5	Special Systems	N/A	—	—	—	
9.0 Electrical						
.1	Service & Distribution	400 ampere service, panel board and feeders	S.F. Floor	.61	.61	
.2	Lighting & Power	Fluorescent fixtures, receptacles, switches, A.C. and misc. power	S.F. Floor	5.29	5.29	10.4%
.4	Special Electrical	Alarm systems, sound system and emergency lighting	S.F. Floor	1.31	1.31	
11.0 Special Construction						
.1	Specialties	Bleachers, sauna, weight room	S.F. Floor	1.03	1.03	1.5%
12.0 Site Work						
.1	Earthwork	N/A	—	—	—	
.3	Utilities	N/A	—	—	—	0.0%
.5	Roads & Parking	N/A	—	—	—	
.7	Site Improvements	N/A	—	—	—	
			Sub-Total		69.25	100%
	GENERAL CONDITIONS (Overhead & Profit)			15%	10.39	
	ARCHITECT FEES			7%	5.56	
			Total Building Cost		85.20	

BUILDING TYPES

COMMERCIAL/INDUSTRIAL/INSTITUTIONAL — M.320 Hangar, Aircraft

Costs per square foot of floor area

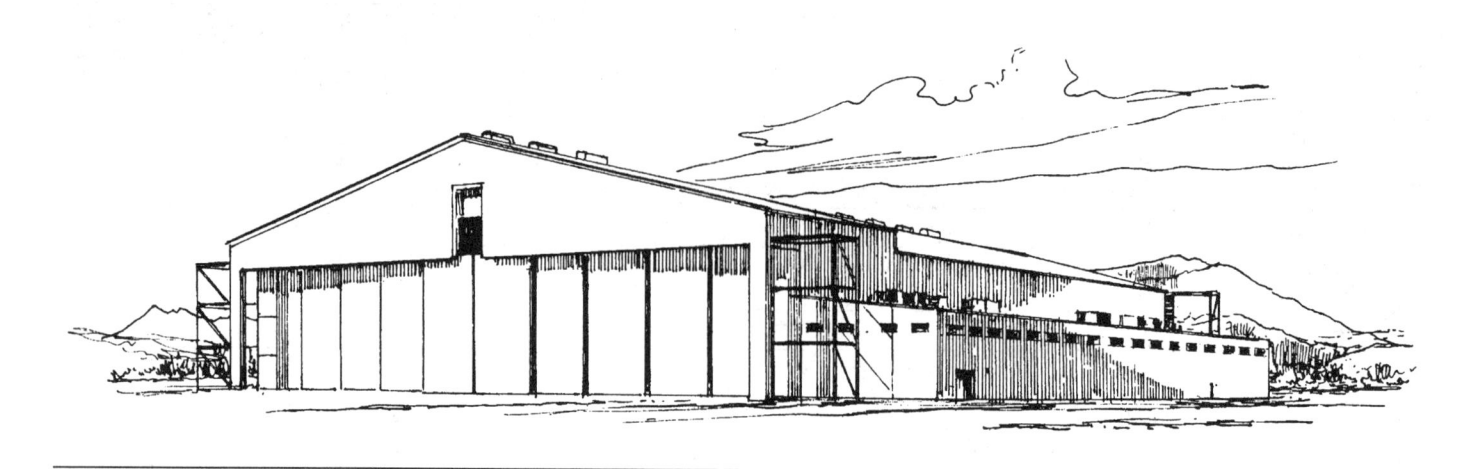

Exterior Wall	S.F. Area	5000	10000	15000	20000	30000	40000	50000	75000	100000
	L.F. Perimeter	300	410	500	580	710	830	930	1150	1300
Concrete Block Reinforced	Steel Frame	71.95	63.90	60.65	58.80	56.55	55.30	54.40	53.00	52.05
	Bearing Walls	75.25	67.10	63.80	61.95	59.70	58.40	57.50	56.10	55.10
Precast Concrete	Steel Frame	76.95	67.30	63.45	61.20	58.50	57.05	55.95	54.30	53.10
	Bearing Walls	80.00	70.35	66.50	64.25	61.60	60.10	59.00	57.35	56.15
Galv. Steel Siding	Steel Frame	71.50	63.60	60.40	58.60	56.40	55.15	54.25	52.90	51.90
Metal Sandwich Panel	Steel Frame	81.90	72.85	69.20	67.10	64.55	63.15	62.10	60.60	59.50
Perimeter Adj., Add or Deduct	Per 100 L.F.	8.45	4.25	2.80	2.10	1.40	1.05	.85	.55	.40
Story Hgt. Adj., Add or Deduct	Per 1 Ft.	.80	.55	.45	.40	.35	.30	.25	.20	.20

Basement—Not Applicable

The above costs were calculated using the basic specifications shown on the facing page. These costs should be adjusted where necessary for design alternatives and owner's requirements. Reported completed project costs, for this type of structure, range from $24.75 to $112.55 per S.F.

Common additives

Description	Unit	$ Cost
Closed Circuit Surveillance, One station		
Camera and monitor	Each	1325
For additional camera stations, add	Each	730
Emergency Lighting, 25 watt, battery operated		
Lead battery	Each	350
Nickel cadmium	Each	590
Lockers, Steel, single tier, 60" or 72"	Opening	134 - 213
2 tier, 60" or 72" total	Opening	83 - 104
5 tier, box lockers	Opening	44 - 56
Locker bench, lam. maple top only	L.F.	18.40
Pedestals, steel pipe	Each	43
Safe, Office type, 4 hour rating		
30" x 18" x 18"	Each	3475
62" x 33" x 20"	Each	7575
Sound System		
Amplifier, 250 watts	Each	1500
Speaker, ceiling or wall	Each	130
Trumpet	Each	244

Model costs calculated for a 1 story building with 24' story height and 20,000 square feet of floor area

Hangar, Aircraft

				Unit	Unit Cost	Cost Per S.F.	% Of Sub-Total
1.0 Foundations							
.1	Footings & Foundations	Poured concrete; strip and spread footings and 4' foundation wall		S.F. Ground	2.54	2.54	7.3%
.4	Piles & Caissons	N/A		—	—	—	
.9	Excavation & Backfill	Site preparation for slab and trench for foundation wall and footing		S.F. Ground	.96	.96	
2.0 Substructure							
.1	Slab on Grade	6" reinforced concrete with vapor barrier and granular base		S.F. Slab	4.46	4.46	9.4%
.2	Special Substructures	N/A		—	—	—	
3.0 Superstructure							
.1	Columns & Beams	Steel columns included in 3.7		—	—	—	
.4	Structural Walls	Metal siding support		S.F. Floor	.28	.28	
.5	Elevated Floors	N/A		—	—	—	11.1%
.7	Roof	Metal deck, open web steel joists, beams, columns		S.F. Roof	5.02	5.02	
.9	Stairs	N/A		—	—	—	
4.0 Exterior Closure							
.1	Walls	Concrete block	50% of wall	S.F. Wall	5.09	1.77	
.5	Exterior Wall Finishes	N/A		—	—	—	27.7%
.6	Doors	Steel overhead and sliding	30% of wall	S.F. Door	28	5.85	
.7	Windows & Glazed Walls	Industrial horizontal pivoted steel	20% of wall	Each	961	5.57	
5.0 Roofing							
.1	Roof Coverings	Elastomeric membrane		S.F. Roof	1.80	1.80	
.7	Insulation	Fiberboard		S.F. Roof	.93	.93	6.2%
.8	Openings & Specialties	Gravel stop and hatches		S.F. Roof	.21	.21	
6.0 Interior Construction							
.1	Partitions	Concrete block, toilet partitions	200 S.F. Floor/L.F. Partition	S.F. Partition	5.69	.46	
.4	Interior Doors	Single leaf hollow metal	5000 S.F. Floor/Door	Each	530	.11	
.5	Wall Finishes	Paint		S.F. Surface	.80	.08	
.6	Floor Finishes	N/A		—	—	—	1.4%
.7	Ceiling Finishes	N/A		—	—	—	
.9	Interior Surface/Exterior Wall	N/A		—	—	—	
7.0 Conveying							
.1	Elevators	N/A		—	—	—	0.0%
.2	Special Conveyors	N/A		—	—	—	
8.0 Mechanical							
.1	Plumbing	Toilet and service fixtures, supply and drainage	1 Fixture/1000 S.F. Floor	Each	2070	2.07	
.2	Fire Protection	Sprinklers, extra hazard		S.F. Floor	3.23	3.23	
.3	Heating	Unit heaters		S.F. Floor	4.67	4.67	21.3%
.4	Cooling	Exhaust fan		S.F. Floor	.20	.20	
.5	Special Systems	N/A		—	—	—	
9.0 Electrical							
.1	Service & Distribution	200 ampere service, panel board and feeders		S.F. Floor	.53	.53	
.2	Lighting & Power	High intensity discharge fixtures, receptacles, switches and misc. power		S.F. Floor	6.53	6.53	15.6%
.4	Special Electrical	Alarm systems and emergency lighting		S.F. Floor	.35	.35	
11.0 Special Construction							
.1	Specialties	N/A		—	—	—	0.0%
12.0 Site Work							
.1	Earthwork	N/A		—	—	—	
.3	Utilities	N/A		—	—	—	0.0%
.5	Roads & Parking	N/A		—	—	—	
.7	Site Improvements	N/A		—	—	—	
				Sub-Total		47.62	100%
	GENERAL CONDITIONS (Overhead & Profit)				15%	7.14	
	ARCHITECT FEES				7%	3.84	
				Total Building Cost		58.60	

BUILDING TYPES

COMMERCIAL/INDUSTRIAL/INSTITUTIONAL — M.330 — Hospital, 2-3 Story

Costs per square foot of floor area

Exterior Wall	S.F. Area	25000	40000	55000	70000	85000	100000	115000	130000	145000
	L.F. Perimeter	388	520	566	666	766	866	878	962	1045
Face Brick with Structural Facing Tile	Steel Frame	133.50	126.50	121.10	119.10	117.80	116.90	115.20	114.60	114.15
	R/Conc. Frame	139.40	132.40	127.00	125.00	123.70	122.80	121.05	120.50	120.00
Face Brick with Concrete Block Back-up	Steel Frame	128.70	122.50	117.95	116.15	115.00	114.25	112.85	112.30	111.90
	R/Conc. Frame	134.60	128.35	123.85	122.05	120.90	120.10	118.70	118.20	117.80
Precast Concrete Panels	Steel Frame	127.65	121.60	117.25	115.55	114.45	113.65	112.30	111.80	111.45
	R/Conc. Frame	133.55	127.50	123.15	121.40	120.30	119.55	118.20	117.70	117.30
Perimeter Adj., Add or Deduct	Per 100 L.F.	5.70	3.55	2.60	2.05	1.65	1.45	1.20	1.10	1.00
Story Hgt. Adj., Add or Deduct	Per 1 Ft.	1.70	1.40	1.10	1.05	.95	.95	.85	.80	.80

For Basement, add $19.70 per square foot of basement area

The above costs were calculated using the basic specifications shown on the facing page. These costs should be adjusted where necessary for design alternatives and owner's requirements. Reported completed project costs, for this type of structure, range from $96.00 to $212.20 per S.F.

Common additives

Description	Unit	$ Cost
Cabinet, Base, door units, metal	L.F.	154
Drawer units	L.F.	267
Tall storage cabinets, 7' high, open	L.F.	275
With doors	L.F.	310
Wall, metal 12-1/2" deep, open	L.F.	104
With doors	L.F.	179
Closed Circuit TV (Patient monitoring)		
One station camera & monitor	Each	1325
For additional camera, add	Each	730
For automatic iris for low light, add	Each	1900
Doctors In-Out Register, 200 names	Each	12,700
Comb. control & recall, 200 names	Each	15,900
Recording register	Each	6025
Transformers	Each	268
Pocket pages	Each	920
Hubbard Tank, with accessories		
Stainless steel, 125 GPM 45 psi	Each	18,200
For electric hoist, add	Each	1975
Mortuary Refrigerator, End operated		
2 capacity	Each	11,000
6 capacity	Each	17,900

Description	Unit	$ Cost
Nurses Call Station		
Single bedside call station	Each	199
Ceiling speaker station	Each	96
Emergency call station	Each	132
Pillow speaker	Each	209
Double bedside call station	Each	360
Duty station	Each	219
Standard call button	Each	107
Master control station for 20 stations	Each	3800
Sound System		
Amplifier, 250 watts	Each	1500
Speaker, ceiling or wall	Each	130
Trumpet	Each	244
Station, Dietary with ice	Each	11,200
Sterilizers		
Single door, steam	Each	120,000
Double door, steam	Each	154,000
Portable, countertop, steam	Each	3025 - 4725
Gas	Each	29,700
Automatic washer/sterilizer	Each	40,900

Model costs calculated for a 3 story building with 12' story height and 55,000 square feet of floor area

Hospital, 2-3 Story

				Unit	Unit Cost	Cost Per S.F.	% Of Sub-Total
1.0 Foundations							
.1	Footings & Foundations	Poured concrete; strip and spread footings and 4' foundation wall		S.F. Ground	4.86	1.62	
.4	Piles & Caissons	N/A		—	—	—	1.9%
.9	Excavation & Backfill	Site preparation for slab and trench for foundation wall and footing		S.F. Ground	.89	.30	
2.0 Substructure							
.1	Slab on Grade	4" reinforced concrete with vapor barrier and granular base		S.F. Slab	3.05	1.02	1.0%
.2	Special Substructures	N/A		—	—	—	
3.0 Superstructure							
.1	Columns & Beams	Concrete columns		L.F. Column	77	1.40	
.4	Structural Walls	N/A		—	—	—	
.5	Elevated Floors	Cast-in-place concrete slab		S.F. Floor	12.17	8.11	14.1%
.7	Roof	Cast-in-place concrete slab		S.F. Roof	11.72	3.91	
.9	Stairs	Concrete filled metal pan		Flight	3675	.87	
4.0 Exterior Closure							
.1	Walls	Face brick and structural facing tile	85% of wall	S.F. Wall	29	9.24	
.5	Exterior Wall Finishes	N/A		—	—	—	10.4%
.6	Doors	Double aluminum and glass and sliding doors		Each	1898	.21	
.7	Windows & Glazed Walls	Aluminum sliding	15% of wall	Each	458	1.11	
5.0 Roofing							
.1	Roof Coverings	Built-up tar and gravel with flashing		S.F. Roof	2.10	.70	
.7	Insulation	Perlite/EPS composite		S.F. Roof	1.17	.39	1.1%
.8	Openings & Specialties	Gravel stop and hatches		S.F. Roof	.21	.07	
6.0 Interior Construction							
.1	Partitions	Concrete block, gypsum board on metal studs	9 S.F. Floor/L.F. Partition	S.F. Partition	3.54	3.93	
.4	Interior Doors	Single leaf hollow metal	90 S.F. Floor/Door	Each	530	5.89	
.5	Wall Finishes	40% vinyl wall covering, 35% ceramic tile, 25% epoxy coating		S.F. Surface	2.50	5.55	24.2%
.6	Floor Finishes	60% vinyl tile, 20% ceramic, 20% terrazzo		S.F. Floor	5.86	5.86	
.7	Ceiling Finishes	Plaster on suspended metal lath		S.F. Ceiling	3.06	3.06	
.9	Interior Surface/Exterior Wall	Glazed coating	85% of wall	S.F. Wall	1.14	.27	
7.0 Conveying							
.1	Elevators	Two hydraulic hospital elevators		Each	68,200	2.48	2.4%
.2	Special Conveyors	N/A		—	—	—	
8.0 Mechanical							
.1	Plumbing	Kitchen, toilet and service fixtures, supply and drainage	1 Fixture/265 S.F. Floor	Each	4102	15.48	
.2	Fire Protection	Wet pipe sprinkler system		S.F. Floor	1.37	1.37	
.3	Heating	Oil fired hot water, wall fin radiation		S.F. Floor	2.96	2.96	28.5%
.4	Cooling	Chilled water, fan coil units		S.F. Floor	8.95	8.95	
.5	Special Systems	N/A		—	—	—	
9.0 Electrical							
.1	Service & Distribution	1200 ampere service, panel board and feeders		S.F. Floor	1.02	1.02	
.2	Lighting & Power	Fluorescent fixtures, receptacles, switches, A.C. and misc. power		S.F. Floor	7.21	7.21	10.6%
.4	Special Electrical	Alarm systems, communications systems and emergency lighting		S.F. Floor	2.48	2.48	
11.0 Special Construction							
.1	Specialties	Conductive flooring, oxygen piping, curtain partitions		S.F. Floor	5.86	5.86	5.8%
12.0 Site Work							
.1	Earthwork	N/A		—	—	—	
.3	Utilities	N/A		—	—	—	0.0%
.5	Roads & Parking	N/A		—	—	—	
.7	Site Improvements	N/A		—	—	—	
				Sub-Total		101.32	100%
	GENERAL CONDITIONS (Overhead & Profit)				15%	15.20	
	ARCHITECT FEES				9%	10.48	
				Total Building Cost		127.00	

BUILDING TYPES

COMMERCIAL/INDUSTRIAL/INSTITUTIONAL — M.340 — Hospital, 4-8 Story

Costs per square foot of floor area

Exterior Wall	S.F. Area	100000	125000	150000	175000	200000	225000	250000	275000	300000
	L.F. Perimeter	594	705	816	783	866	950	1033	1116	1200
Face Brick with Structural Facing Tile	Steel Frame	115.80	113.70	112.30	109.10	108.25	107.55	107.05	106.60	106.30
	R/Conc. Frame	118.45	116.35	114.95	111.75	110.90	110.20	109.65	109.25	108.95
Face Brick with Concrete Block Back-up	Steel Frame	112.80	110.85	109.55	106.80	106.05	105.40	104.95	104.55	104.25
	R/Conc. Frame	115.45	113.50	112.15	109.45	108.70	108.05	107.55	107.20	106.90
Precast Concrete Panels With Exposed Aggregate	Steel Frame	111.15	109.35	108.10	105.75	105.05	104.45	104.00	103.65	103.35
	R/Conc. Frame	113.20	111.40	110.15	107.80	107.05	106.50	106.05	105.70	105.40
Perimeter Adj., Add or Deduct	Per 100 L.F.	2.70	2.15	1.75	1.50	1.30	1.20	1.05	1.00	.90
Story Hgt. Adj., Add or Deduct	Per 1 Ft.	1.30	1.20	1.15	.95	.90	.90	.90	.85	.85

For Basement, add $20.80 per square foot of basement area

The above costs were calculated using the basic specifications shown on the facing page. These costs should be adjusted where necessary for design alternatives and owner's requirements. Reported completed project costs, for this type of structure, range from $96.40 to $216.55 per S.F.

Common additives

Description	Unit	$ Cost
Cabinets, Base, door units, metal	L.F.	154
Drawer units	L.F.	267
Tall storage cabinets, 7' high, open	L.F.	275
With doors	L.F.	310
Wall, metal 12-1/2" deep, open	L.F.	104
With doors	L.F.	179
Closed Circuit TV (Patient monitoring)		
One station camera & monitor	Each	1325
For additional camera add	Each	730
For automatic iris for low light add	Each	1900
Doctors In-Out Register, 200 names	Each	12,700
Comb. control & recall, 200 names	Each	15,900
Recording register	Each	6025
Transformers	Each	268
Pocket pages	Each	920
Hubbard Tank, with accessories		
Stainless steel, 125 GPM 45 psi	Each	18,200
For electric hoist, add	Each	1975
Mortuary Refrigerator, End operated		
2 capacity	Each	11,000
6 capacity	Each	17,900

Description	Unit	$ Cost
Nurses Call Station		
Single bedside call station	Each	199
Ceiling speaker station	Each	96
Emergency call station	Each	132
Pillow speaker	Each	209
Double bedside call station	Each	360
Duty station	Each	219
Standard call button	Each	107
Master control station for 20 stations	Each	3800
Sound System		
Amplifier, 250 watts	Each	1500
Speaker, ceiling or wall	Each	130
Trumpet	Each	244
Station, Dietary with ice	Each	11,200
Sterilizers		
Single door, steam	Each	120,000
Double door, steam	Each	154,000
Portable, counter top, steam	Each	3025 - 4725
Gas	Each	29,700
Automatic washer/sterilizer	Each	40,900

Model costs calculated for a 6 story building with 12' story height and 200,000 square feet of floor area

Hospital, 4-8 Story

				Unit	Unit Cost	Cost Per S.F.	% Of Sub-Total
1.0 Foundations							
.1	Footings & Foundations	Poured concrete; strip and spread footings and 4' foundation wall		S.F. Ground	4.38	.73	1.0%
.4	Piles & Caissons	N/A		—	—	—	
.9	Excavation & Backfill	Site preparation for slab and trench for foundation wall and footing		S.F. Ground	.89	.15	
2.0 Substructure							
.1	Slab on Grade	4" reinforced concrete with vapor barrier and granular base		S.F. Slab	3.05	.51	0.6%
.2	Special Substructures	N/A		—	—	—	
3.0 Superstructure							
.1	Columns & Beams	Fireproofed steel columns		L.F. Column	54	.95	
.4	Structural Walls	N/A		—	—	—	
.5	Elevated Floors	Concrete slab with metal deck and beams		S.F. Floor	9.97	8.31	12.2%
.7	Roof	Metal deck, open web steel joists, beams, interior columns		S.F. Roof	4.63	.77	
.9	Stairs	Concrete filled metal pan		Flight	4105	.53	
4.0 Exterior Closure							
.1	Walls	Face brick and structural facing tile	70% of wall	S.F. Wall	29	6.41	
.5	Exterior Wall Finishes	N/A		—	—	—	10.6%
.6	Doors	Double aluminum and glass and sliding doors		Each	3705	.52	
.7	Windows & Glazed Walls	Aluminum sliding	30% of wall	Each	357	2.23	
5.0 Roofing							
.1	Roof Coverings	Built-up tar and gravel with flashing		S.F. Roof	2.10	.35	
.7	Insulation	Perlite/EPS composite		S.F. Roof	1.18	.20	0.6%
.8	Openings & Specialties	Gravel stop and hatches		S.F. Roof	.06	.01	
6.0 Interior Construction							
.1	Partitions	Gypsum board on metal studs with sound deadening board	9 S.F. Floor/L.F. Partition	S.F. Partition	4.76	5.29	
.4	Interior Doors	Single leaf hollow metal	90 S.F. Floor/Door	Each	530	5.89	
.5	Wall Finishes	40% vinyl wall covering, 35% ceramic tile, 25% epoxy coating		S.F. Surface	2.50	5.55	30.0%
.6	Floor Finishes	60% vinyl tile, 20% ceramic, 20% terrazzo		S.F. Floor	5.86	5.86	
.7	Ceiling Finishes	Plaster on suspended metal lath		S.F. Ceiling	3.06	3.06	
.9	Interior Surface/Exterior Wall	Glazed coating	70% of wall	S.F. Wall	1.14	.25	
7.0 Conveying							
.1	Elevators	Six geared hospital elevators		Each	131,666	3.95	4.6%
.2	Special Conveyors	N/A		—	—	—	
8.0 Mechanical							
.1	Plumbing	Kitchen, toilet and service fixtures, supply and drainage	1 Fixture/275 S.F. Floor	Each	4149	15.09	
.2	Fire Protection	Standpipe and wet pipe sprinkler systems		S.F. Floor	1.22	1.22	
.3	Heating	Oil fired hot water, wall fin radiation		S.F. Floor	2.96	2.96	24.8%
.4	Cooling	Chilled water, fan coil units		S.F. Floor	2.04	2.04	
.5	Special Systems	N/A		—	—	—	
9.0 Electrical							
.1	Service & Distribution	3600 ampere service, panel board and feeders		S.F. Floor	.97	.97	
.2	Lighting & Power	Fluorescent fixtures, receptacles, switches, A.C. and misc. power		S.F. Floor	6.62	6.62	11.4%
.4	Special Electrical	Alarm systems, communications systems and emergency lighting		S.F. Floor	2.27	2.27	
11.0 Special Construction							
.1	Specialties	Conductive flooring, oxygen piping, curtain partitions		S.F. Floor	3.67	3.67	4.2%
12.0 Site Work							
.1	Earthwork	N/A		—	—	—	
.3	Utilities	N/A		—	—	—	0.0%
.5	Roads & Parking	N/A		—	—	—	
.7	Site Improvements	N/A		—	—	—	
				Sub-Total		86.36	100%
	GENERAL CONDITIONS (Overhead & Profit)				15%	12.95	
	ARCHITECT FEES				9%	8.94	
				Total Building Cost		108.25	

BUILDING TYPES

COMMERCIAL/INDUSTRIAL/INSTITUTIONAL — M.350 Hotel, 4-7 Story

Costs per square foot of floor area

Exterior Wall	S.F. Area	35000	55000	75000	95000	115000	135000	155000	175000	195000
	L.F. Perimeter	314	401	497	555	639	722	716	783	850
Face Brick with Concrete Block Back-up	Steel Frame	101.65	94.90	92.00	89.45	88.30	87.45	85.65	85.10	84.70
	R/Conc. Frame	105.00	98.20	95.30	92.80	91.60	90.80	89.00	88.45	88.00
Glass and Metal Curtain Walls	Steel Frame	95.60	89.95	87.50	85.50	84.55	83.85	82.55	82.10	81.75
	R/Conc. Frame	99.05	93.40	90.95	89.00	88.00	87.30	86.00	85.55	85.20
Precast Concrete Panels	Steel Frame	103.75	96.60	93.55	90.85	89.60	88.70	86.75	86.15	85.70
	R/Conc. Frame	107.85	100.55	97.45	94.70	93.45	92.55	90.50	89.95	89.45
Perimeter Adj., Add or Deduct	Per 100 L.F.	5.90	3.75	2.75	2.15	1.80	1.55	1.35	1.20	1.05
Story Hgt. Adj., Add or Deduct	Per 1 Ft.	1.65	1.35	1.25	1.10	1.05	1.00	.85	.85	.80

For Basement, add $19.65 per square foot of basement area

The above costs were calculated using the basic specifications shown on the facing page. These costs should be adjusted where necessary for design alternatives and owner's requirements. Reported completed project costs, for this type of structure, range from $64.20 to $123.80 per S.F.

Common additives

Description	Unit	$ Cost
Bar, Front bar	L.F.	278
Back bar	L.F.	220
Booth, Upholstered, custom, straight	L.F.	127 - 207
"L" or "U" shaped	L.F.	143 - 225
Closed Circuit Surveillance, One station		
Camera and monitor	Each	1325
For additional camera stations, add	Each	730
Directory Boards, Plastic, glass covered		
30" x 20"	Each	500
36" x 48"	Each	910
Aluminum, 24" x 18"	Each	445
48" x 32"	Each	630
48" x 60"	Each	1425
Elevators, Electric passenger, 5 stops		
3500# capacity	Each	94,800
5000# capacity	Each	98,300
Additional stop, add	Each	5225
Emergency Lighting, 25 watt, battery operated		
Lead battery	Each	350
Nickel cadmium	Each	590

Description	Unit	$ Cost
Laundry Equipment		
Folders, blankets & sheets, king size	Each	49,800
Ironers, 110" single roll	Each	27,100
Combination washer extractor 50#	Each	9725
125#	Each	24,400
Sauna, Prefabricated, complete		
6' x 4'	Each	3725
6' x 6'	Each	4350
6' x 9'	Each	5400
8' x 8'	Each	5725
10' x 12'	Each	8025
Smoke Detectors		
Ceiling type	Each	141
Duct type	Each	335
Sound System		
Amplifier, 250 watts	Each	1500
Speaker, ceiling or wall	Each	130
Trumpet	Each	244
TV Antenna, Master system, 12 outlet	Outlet	227
30 outlet	Outlet	146
100 outlet	Outlet	140

Model costs calculated for a 6 story building with 10' story height and 135,000 square feet of floor area

Hotel, 4-7 Story

				Unit	Unit Cost	Cost Per S.F.	% Of Sub-Total
1.0 Foundations							
.1	Footings & Foundations	Poured concrete; strip and spread footings and 4' foundation wall		S.F. Ground	6.78	1.13	
.4	Piles & Caissons	N/A		–	–	–	1.8%
.9	Excavation & Backfill	Site preparation for slab and trench for foundation wall and footing		S.F. Ground	.89	.15	
2.0 Substructure							
.1	Slab on Grade	4" reinforced concrete with vapor barrier and granular base		S.F. Slab	3.05	.51	0.7%
.2	Special Substructures	N/A					
3.0 Superstructure							
.1	Columns & Beams	Included in 3.5 and 3.7		–	–	–	
.4	Structural Walls	N/A		–	–	–	
.5	Elevated Floors	Concrete slab with metal deck, beams, columns		S.F. Floor	10.08	8.40	13.2%
.7	Roof	Metal deck, open web steel joists, beams, columns		S.F. Roof	4.63	.77	
.9	Stairs	Concrete		Flight	2950	.31	
4.0 Exterior Closure							
.1	Walls	Face brick with concrete block backup	80% of wall	S.F. Wall	21	5.47	
.5	Exterior Wall Finishes	N/A		–	–	–	10.6%
.6	Doors	Glass and metal doors and entrances		Each	2056	.15	
.7	Windows & Glazed Walls	Window wall	20% of wall	S.F. Window	31	1.99	
5.0 Roofing							
.1	Roof Coverings	Built-up tar and gravel with flashing		S.F. Roof	2.10	.35	
.7	Insulation	Perlite/EPS composite		S.F. Roof	1.18	.20	0.8%
.8	Openings & Specialties	Gravel stop and hatches		S.F. Roof	.24	.04	
6.0 Interior Construction							
.1	Partitions	Gypsum board and sound deadening board, steel studs	9 S.F. Floor/L.F. Partition	S.F. Partition	4.76	4.23	
.4	Interior Doors	Single leaf hollow metal	90 S.F. Floor/Door	Each	530	5.89	
.5	Wall Finishes	50% paint, 45% vinyl cover, 5% ceramic tile		S.F. Surface	1.18	2.09	25.2%
.6	Floor Finishes	80% carpet, 10% vinyl composition tile, 10% ceramic tile		S.F. Floor	3.97	3.97	
.7	Ceiling Finishes	Mineral fiber tile applied with adhesive		S.F. Ceiling	1.18	1.18	
.9	Interior Surface/Exterior Wall	Painted gypsum board on furring	80% of wall	S.F. Wall	2.80	.72	
7.0 Conveying							
.1	Elevators	Four geared passenger elevators		Each	120,150	3.56	5.0%
.2	Special Conveyors	N/A		–	–	–	
8.0 Mechanical							
.1	Plumbing	Kitchen, toilet and service fixtures, supply and drainage	1 Fixture/155 S.F. Floor	Each	1647	10.63	
.2	Fire Protection	Standpipes and hose systems and sprinklers, light hazard		S.F. Floor	1.39	1.39	
.3	Heating	Oil fired hot water, wall fin radiation		S.F. Floor	2.96	2.96	32.2%
.4	Cooling	Chilled water, fan coil units		S.F. Floor	8.14	8.14	
.5	Special Systems	N/A		–	–	–	
9.0 Electrical							
.1	Service & Distribution	2000 ampere service, panel board and feeders		S.F. Floor	.75	.75	
.2	Lighting & Power	Fluorescent fixtures, receptacles, switches, A.C. and misc. power		S.F. Floor	5.02	5.02	10.5%
.4	Special Electrical	Alarm systems, communications systems and emergency lighting		S.F. Floor	1.76	1.76	
11.0 Special Construction							
.1	Specialties	N/A		–	–	–	0.0%
12.0 Site Work							
.1	Earthwork	N/A		–	–	–	
.3	Utilities	N/A		–	–	–	
.5	Roads & Parking	N/A		–	–	–	0.0%
.7	Site Improvements	N/A		–	–	–	
				Sub-Total		71.76	100%
	GENERAL CONDITIONS (Overhead & Profit)				15%	10.76	
	ARCHITECT FEES				6%	4.93	
				Total Building Cost		87.45	

BUILDING TYPES

COMMERCIAL/INDUSTRIAL/INSTITUTIONAL — M.360 — Hotel, 8-24 Story

Costs per square foot of floor area

Exterior Wall	S.F. Area	140000	243000	346000	450000	552000	655000	760000	860000	965000
	L.F. Perimeter	403	587	672	800	936	1073	1213	1195	1312
Face Brick with Concrete Block Back-up	Steel Frame	86.05	81.85	78.75	77.50	76.85	76.40	76.05	74.95	74.75
	R/Conc. Frame	88.80	84.55	81.45	80.25	79.60	79.15	78.75	77.70	77.45
Face Brick Veneer On Steel Studs	Steel Frame	84.95	80.90	78.00	76.85	76.20	75.75	75.40	74.45	74.20
	R/Conc. Frame	88.00	83.95	81.05	79.90	79.30	78.85	78.50	77.50	77.25
Glass and Metal Curtain Walls	Steel Frame	92.10	87.00	84.05	82.70	82.00	81.45	81.05	80.20	80.00
	R/Conc. Frame	95.15	90.05	87.10	85.80	85.05	84.50	84.10	83.25	83.05
Perimeter Adj., Add or Deduct	Per 100 L.F.	3.40	2.00	1.40	1.05	.90	.75	.65	.60	.50
Story Hgt. Adj., Add or Deduct	Per 1 Ft.	1.35	1.10	.90	.85	.80	.75	.75	.65	.60

For Basement, add $19.85 per square foot of basement area

The above costs were calculated using the basic specifications shown on the facing page. These costs should be adjusted where necessary for design alternatives and owner's requirements. Reported completed project costs, for this type of structure, range from $70.55 to $123.30 per S.F.

Common additives

Description	Unit	$ Cost
Bar, Front bar	L.F.	278
Back bar	L.F.	220
Booth, Upholstered, custom, straight	L.F.	127 - 207
"L" or "U" shaped	L.F.	143 - 225
Closed Circuit Surveillance, One station		
Camera and monitor	Each	1325
For additional camera stations, add	Each	730
Directory Boards, Plastic, glass covered		
30" x 20"	Each	500
36" x 48"	Each	910
Aluminum, 24" x 18"	Each	445
48" x 32"	Each	630
48" x 60"	Each	1425
Elevators, Electric passenger, 10 stops		
3500# capacity	Each	196,000
5000# capacity	Each	201,500
Additional stop, add	Each	5225
Emergency Lighting, 25 watt, battery operated		
Lead battery	Each	350
Nickel cadmium	Each	590

Description	Unit	$ Cost
Laundry Equipment		
Folders, blankets & sheets, king size	Each	49,800
Ironers, 110" single roll	Each	27,100
Combination washer & extractor 50#	Each	9725
125#	Each	24,400
Sauna, Prefabricated, complete		
6' x 4'	Each	3725
6' x 6'	Each	4350
6' x 9'	Each	5400
8' x 8'	Each	5725
10' x 12'	Each	8025
Smoke Detectors		
Ceiling type	Each	141
Duct type	Each	335
Sound System		
Amplifier, 250 watts	Each	1500
Speaker, ceiling or wall	Each	130
Trumpet	Each	244
TV Antenna, Master system, 12 outlet	Outlet	227
30 outlet	Outlet	146
100 outlet	Outlet	140

Model costs calculated for a 15 story building with 10' story height and 450,000 square feet of floor area

Hotel, 8-24 Story

				Unit	Unit Cost	Cost Per S.F.	% Of Sub-Total
1.0 Foundations							
.1	Footings & Foundations	Poured concrete; strip and spread footings and 4' foundation wall		S.F. Ground	12.45	.83	
.4	Piles & Caissons	N/A		—	—	—	1.3%
.9	Excavation & Backfill	Site preparation for slab and trench for foundation wall and footing		S.F. Ground	.89	.06	
2.0 Substructure							
.1	Slab on Grade	4" reinforced concrete with vapor barrier and granular base		S.F. Slab	3.05	.20	0.3%
.2	Special Substructures	N/A		—	—	—	
3.0 Superstructure							
.1	Columns & Beams	Steel columns included in 3.5 and 3.7		—	—	—	
.4	Structural Walls	N/A		—	—	—	
.5	Elevated Floors	Open web steel joists, slab form, concrete, columns		S.F. Floor	9.77	9.12	15.3%
.7	Roof	Metal deck, open web steel joists, beams, columns		S.F. Roof	3.64	.24	
.9	Stairs	Concrete filled metal pan		Flight	4925	1.00	
4.0 Exterior Closure							
.1	Walls	N/A		—	—	—	
.5	Exterior Wall Finishes	N/A		—	—	—	5.7%
.6	Doors	Glass and metal doors and entrances		Each	1842	.13	
.7	Windows & Glazed Walls	Glass and metal curtain walls	100% of wall	S.F. Wall	14.00	3.73	
5.0 Roofing							
.1	Roof Coverings	Built-up tar and gravel with flashing		S.F. Roof	2.10	.14	
.7	Insulation	Perlite/EPS composite		S.F. Roof	1.18	.08	0.3%
.8	Openings & Specialties	Gravel stop and hatches		S.F. Roof	.15	.01	
6.0 Interior Construction							
.1	Partitions	Gypsum board and sound deadening board, steel studs	9 S.F. Floor/L.F. Partition	S.F. Partition	4.76	4.23	
.4	Interior Doors	Single leaf hollow metal	90 S.F. Floor/Door	Each	530	5.89	
.5	Wall Finishes	50% paint, 45% vinyl cover, 5% ceramic tile		S.F. Surface	1.18	2.09	30.4%
.6	Floor Finishes	80% carpet, 10% vinyl composition tile, 10% ceramic tile		S.F. Floor	3.97	3.97	
.7	Ceiling Finishes	Mineral fiber tile applied with adhesive to suspended gypsum board		S.F. Ceiling	3.82	3.82	
.9	Interior Surface/Exterior Wall	Painted gypsum board on furring	80% of wall	S.F. Wall	2.80	.60	
7.0 Conveying							
.1	Elevators	One geared freight, six geared passenger elevators		Each	213,750	2.85	4.2%
.2	Special Conveyors	N/A		—	—	—	
8.0 Mechanical							
.1	Plumbing	Kitchen, toilet and service fixtures, supply and drainage	1 Fixture/165 S.F. Floor	Each	1631	9.89	
.2	Fire Protection	Standpipes and hose systems and sprinklers, light hazard		S.F. Floor	2.20	2.20	
.3	Heating	Oil fired hot water, wall fin radiation		S.F. Floor	1.23	1.23	31.6%
.4	Cooling	Chilled water, fan coil units		S.F. Floor	8.14	8.14	
.5	Special Systems	N/A		—	—	—	
9.0 Electrical							
.1	Service & Distribution	2400 ampere service, panel board and feeders		S.F. Floor	.80	.80	
.2	Lighting & Power	Fluorescent fixtures, receptacles, switches, A.C. and misc. power		S.F. Floor	4.95	4.95	10.9%
.4	Special Electrical	Alarm systems, communications systems and emergency lighting		S.F. Floor	1.66	1.66	
11.0 Special Construction							
.1	Specialties	N/A		—	—	—	0.0%
12.0 Site Work							
.1	Earthwork	N/A		—	—	—	
.3	Utilities	N/A		—	—	—	
.5	Roads & Parking	N/A		—	—	—	0.0%
.7	Site Improvements	N/A		—	—	—	

	Sub-Total	67.86	100%
GENERAL CONDITIONS (Overhead & Profit)	15%	10.18	
ARCHITECT FEES	6%	4.66	
Total Building Cost		**82.70**	

BUILDING TYPES

COMMERCIAL/INDUSTRIAL/INSTITUTIONAL — M.370 Jail

Costs per square foot of floor area

Exterior Wall	S.F. Area	7000	13000	20000	26000	32000	45000	58000	72000	85000
	L.F. Perimeter	196	273	366	408	475	620	609	713	766
Face Brick with Concrete Block Back-up	Steel Frame	197.05	179.00	171.85	166.65	164.55	161.90	156.40	155.15	153.55
	R/Conc. Frame	193.70	175.70	168.55	163.35	161.20	158.60	153.10	151.80	150.20
Stucco on Concrete Block	Steel Frame	185.05	170.05	164.10	160.00	158.25	156.10	152.05	151.00	149.80
	R/Conc. Frame	181.55	166.55	160.60	156.50	154.80	152.60	148.55	147.50	146.30
Reinforced Concrete	Steel Frame	189.00	173.00	166.70	162.25	160.35	158.05	153.50	152.40	151.05
	R/Conc. Frame	185.50	169.50	163.20	158.75	156.85	154.55	150.00	148.90	147.55
Perimeter Adj., Add or Deduct	Per 100 L.F.	21.50	11.55	7.55	5.75	4.70	3.35	2.60	2.10	1.75
Story Hgt. Adj., Add or Deduct	Per 1 Ft.	3.25	2.40	2.10	1.80	1.70	1.60	1.20	1.15	1.05

For Basement, add $26.45 per square foot of basement area

The above costs were calculated using the basic specifications shown on the facing page. These costs should be adjusted where necessary for design alternatives and owner's requirements. Reported completed project costs, for this type of structure, range from $99.60 to $228.30 per S.F.

Common additives

Description	Unit	$ Cost
Clock System		
20 room	Each	11,700
50 room	Each	28,300
Closed Circuit Surveillance, One station		
Camera and monitor	Each	1325
For additional camera stations, add	Each	730
Elevators, Hydraulic passenger, 2 stops		
1500# capacity	Each	40,800
2500# capacity	Each	42,100
3500# capacity	Each	45,700
Additional stop, add	Each	3350
Emergency Generators, Complete system, gas		
15 KW	Each	11,800
70 KW	Each	26,800
85 KW	Each	31,400
115 KW	Each	55,500
170 KW	Each	93,500
Diesel, 50 KW	Each	24,000
100 KW	Each	34,700
150 KW	Each	42,400
350 KW	Each	64,000

Description	Unit	$ Cost
Emergency Lighting, 25 watt, battery operated		
Nickel cadmium	Each	590
Flagpoles, Complete		
Aluminum, 20' high	Each	1050
40' High	Each	2350
70' High	Each	6675
Fiberglass, 23' High	Each	1375
39'-5" High	Each	2400
59' High	Each	6575
Laundry Equipment	Each	
Folders, blankets, & sheets	Each	49,800
Ironers, 110" single roll	Each	27,100
Combination washer & extractor, 50#	Each	9725
125#	Each	24,400
Safe, Office type, 4 hour rating		
30" x 18" x 18"	Each	3475
62" x 33" x 20"	Each	7575
Sound System		
Amplifier, 250 watts	Each	1500
Speaker, ceiling or wall	Each	130
Trumpet	Each	244

Jail

Model costs calculated for a 3 story building with 12' story height and 32,000 square feet of floor area

				Unit	Unit Cost	Cost Per S.F.	% Of Sub-Total
1.0 Foundations							
.1	Footings & Foundations	Poured concrete; strip and spread footings and 4' foundation wall		S.F. Ground	3.36	1.12	
.4	Piles & Caissons	N/A		—	—	—	1.1%
.9	Excavation & Backfill	Site preparation for slab and trench for foundation wall and footing		S.F. Ground	.96	.32	
2.0 Substructure							
.1	Slab on Grade	4" reinforced concrete with vapor barrier and granular base		S.F. Slab	3.05	1.02	0.8%
.2	Special Substructures	N/A		—	—	—	
3.0 Superstructure							
.1	Columns & Beams	Steel columns		L.F. Column	41	1.39	
.4	Structural Walls	N/A		—	—	—	
.5	Elevated Floors	Concrete slab with metal deck and beams		S.F. Floor	13.70	9.13	11.6%
.7	Roof	Concrete slab with metal deck and beams		S.F. Roof	12.30	4.10	
.9	Stairs	Concrete filled metal pan		Flight	5750	.90	
4.0 Exterior Closure							
.1	Walls	Face brick with reinforced concrete block backup	85% of wall	S.F. Wall	21	9.67	
.5	Exterior Wall Finishes	N/A		—	—	—	12.0%
.6	Doors	Metal		Each	3525	.11	
.7	Windows & Glazed Walls	Bullet resisting and metal horizontal pivoted	15% of wall	S.F. Window	77	6.22	
5.0 Roofing							
.1	Roof Coverings	Built-up tar and gravel with flashing		S.F. Roof	2.19	.73	
.7	Insulation	Perlite/EPS composite		S.F. Roof	1.18	.39	0.9%
.8	Openings & Specialties	Gravel stop and hatches		S.F. Roof	.30	.10	
6.0 Interior Construction							
.1	Partitions	Concrete block	45 S.F. of Floor/L.F. Partition	S.F. Partition	5.59	1.49	
.4	Interior Doors	Hollow metal fire doors	930 S.F. Floor/Door	Each	530	.57	
.5	Wall Finishes	Paint		S.F. Surface	1.09	.58	5.1%
.6	Floor Finishes	70% vinyl composition tile, 20% carpet, 10% ceramic tile		S.F. Floor	2.41	2.41	
.7	Ceiling Finishes	Mineral tile on zee runners	30% of area	S.F. Ceiling	3.06	.92	
.9	Interior Surface/Exterior Wall	Paint	85% of wall	S.F. Wall	1.93	.88	
7.0 Conveying							
.1	Elevators	One hydraulic passenger elevator		Each	95,680	2.99	2.2%
.2	Special Conveyors	N/A		—	—	—	
8.0 Mechanical							
.1	Plumbing	Toilet and service fixtures, supply and drainage	1 Fixture/78 S.F. Floor	Each	1904	24.42	
.2	Fire Protection	Standpipes and hose systems		S.F. Floor	.36	.36	
.3	Heating	Included in 8.4		—	—	—	25.9%
.4	Cooling	Rooftop multizone unit systems		S.F. Floor	9.94	9.94	
.5	Special Systems	N/A		—	—	—	
9.0 Electrical							
.1	Service & Distribution	600 ampere service, panel board and feeders		S.F. Floor	.87	.87	
.2	Lighting & Power	Fluorescent fixtures, receptacles, switches, A.C. and misc. power		S.F. Floor	5.98	5.98	6.1%
.4	Special Electrical	Alarm systems and emergency lighting		S.F. Floor	1.28	1.28	
11.0 Special Construction							
.1	Specialties	Prefabricated cells, visitor cubicles		S.F. Floor	45	45.84	34.3%
12.0 Site Work							
.1	Earthwork	N/A		—	—	—	
.3	Utilities	N/A		—	—	—	0.0%
.5	Roads & Parking	N/A		—	—	—	
.7	Site Improvements	N/A		—	—	—	
				Sub-Total		133.73	100%
	GENERAL CONDITIONS (Overhead & Profit)				15%	20.06	
	ARCHITECT FEES				7%	10.76	
				Total Building Cost		**164.55**	

BUILDING TYPES

COMMERCIAL/INDUSTRIAL/INSTITUTIONAL M.380 Laundromat

Costs per square foot of floor area

Exterior Wall	S.F. Area	1000	2000	3000	4000	5000	10000	15000	20000	25000
	L.F. Perimeter	126	179	219	253	283	400	490	568	632
Decorative Concrete Block	Steel Frame	110.20	99.20	94.25	91.30	89.30	84.30	82.10	80.80	79.90
	Bearing Walls	115.30	103.75	98.45	95.40	93.30	88.00	85.65	84.35	83.35
Face Brick with Concrete Block Back-up	Steel Frame	134.40	117.45	109.75	105.25	102.15	94.40	91.00	89.05	87.60
	Bearing Walls	134.65	117.50	109.70	105.10	102.00	94.15	90.70	88.70	87.25
Metal Sandwich Panel Precast Concrete Panel	Steel Frame	113.85	102.95	98.00	95.10	93.10	88.15	85.95	84.70	83.75
	Bearing Walls	110.85	100.55	95.90	93.15	91.30	86.60	84.55	83.35	82.45
Perimeter Adj., Add or Deduct	Per 100 L.F.	30.10	15.05	10.00	7.50	6.00	3.05	1.95	1.50	1.20
Story Hgt. Adj., Add or Deduct	Per 1 Ft.	2.00	1.45	1.15	1.00	.90	.65	.50	.45	.40

For Basement, add $17.80 per square foot of basement area

The above costs were calculated using the basic specifications shown on the facing page. These costs should be adjusted where necessary for design alternatives and owner's requirements. Reported completed project costs, for this type of structure, range from $57.95 to $141.15 per S.F.

Common additives

Description	Unit	$ Cost
Closed Circuit Surveillance, One station		
Camera and monitor	Each	1325
For additional camera stations, add	Each	730
Emergency Lighting, 25 watt, battery operated		
Lead battery	Each	350
Nickel cadmium	Each	590
Laundry Equipment		
Dryers, coin operated 30 lb.	Each	2675
Double stacked	Each	5450
50 lb.	Each	2800
Dry cleaner 20 lb.	Each	33,400
30 lb.	Each	45,600
Washers, coin operated	Each	1125
Washer/extractor 20 lb.	Each	3900
30 lb.	Each	9175
50 lb.	Each	9725
75 lb.	Each	18,300
Smoke Detectors		
Ceiling type	Each	141
Duct type	Each	335

Location Factors appear in the Reference Section

Laundromat

Model costs calculated for a 1 story building with 12' story height and 3,000 square feet of floor area

				Unit	Unit Cost	Cost Per S.F.	% Of Sub-Total
1.0 Foundations							
.1	Footings & Foundations	Poured concrete; strip and spread footings and 4' foundation wall		S.F. Ground	5.08	5.08	8.1%
.4	Piles & Caissons	N/A		—	—	—	
.9	Excavation & Backfill	Site preparation for slab and trench for foundation wall and footing		S.F. Ground	1.09	1.09	
2.0 Substructure							
.1	Slab on Grade	5" reinforced concrete with vapor barrier and granular base		S.F. Slab	3.27	3.27	4.3%
.2	Special Substructures	N/A		—	—	—	
3.0 Superstructure							
.1	Columns & Beams	Fireproofing; steel columns included in 3.7		—	—	—	
.4	Structural Walls	N/A		—	—	—	
.5	Elevated Floors	N/A		—	—	—	4.0%
.7	Roof	Metal deck, open web steel joists, beams, columns		S.F. Roof	3.03	3.03	
.9	Stairs	N/A		—	—	—	
4.0 Exterior Closure							
.1	Walls	Decorative concrete block	90% of wall	S.F. Wall	8.71	6.87	
.5	Exterior Wall Finishes	N/A		—	—	—	13.2%
.6	Doors	Double aluminum and glass		Each	2750	.92	
.7	Windows & Glazed Walls	Store front	10% of wall	S.F. Window	26	2.30	
5.0 Roofing							
.1	Roof Coverings	Built-up tar and gravel with flashing		S.F. Roof	3.12	3.12	
.7	Insulation	Perlite/EPS composite		S.F. Roof	1.18	1.18	5.8%
.8	Openings & Specialties	Gravel stop, gutters and downspouts		S.F. Roof	.11	.11	
6.0 Interior Construction							
.1	Partitions	Gypsum board on metal studs	60 S.F. Floor/L.F. Partition	S.F. Partition	2.94	.49	
.4	Interior Doors	Single leaf wood	750 S.F. Floor/Door	Each	399	.53	
.5	Wall Finishes	Paint		S.F. Surface	.51	.17	8.5%
.6	Floor Finishes	Vinyl composition tile		S.F. Floor	1.35	1.35	
.7	Ceiling Finishes	Fiberglass board on exposed grid system		S.F. Ceiling	1.76	1.76	
.9	Interior Surface/Exterior Wall	Painted gypsum board on furring	90% of wall	S.F. Wall	2.81	2.22	
7.0 Conveying							
.1	Elevators	N/A		—	—	—	0.0%
.2	Special Conveyors	N/A		—	—	—	
8.0 Mechanical							
.1	Plumbing	Toilet and service fixtures, supply and drainage	1 Fixture/600 S.F. Floor	Each	10,218	17.03	
.2	Fire Protection	Sprinkler, ordinary hazard		S.F. Floor	2.19	2.19	
.3	Heating	Included in 8.4		—	—	—	35.1%
.4	Cooling	Rooftop single zone unit systems		S.F. Floor	7.82	7.82	
.5	Special Systems	N/A		—	—	—	
9.0 Electrical							
.1	Service & Distribution	200 ampere service, panel board and feeders		S.F. Floor	2.19	2.19	
.2	Lighting & Power	Fluorescent fixtures, receptacles, switches, A.C. and misc. power		S.F. Floor	13.53	13.53	21.0%
.4	Special Electrical	Alarm systems and emergency lighting		S.F. Floor	.34	.34	
11.0 Special Construction							
.1	Specialties	N/A		—	—	—	0.0%
12.0 Site Work							
.1	Earthwork	N/A		—	—	—	
.3	Utilities	N/A		—	—	—	0.0%
.5	Roads & Parking	N/A		—	—	—	
.7	Site Improvements	N/A		—	—	—	
				Sub-Total		76.59	100%
	GENERAL CONDITIONS (Overhead & Profit)				15%	11.49	
	ARCHITECT FEES				7%	6.17	
				Total Building Cost		94.25	

BUILDING TYPES

COMMERCIAL/INDUSTRIAL/INSTITUTIONAL — M.390 Library

Costs per square foot of floor area

Exterior Wall	S.F. Area	7000	10000	13000	16000	19000	22000	25000	28000	31000
	L.F. Perimeter	240	300	336	386	411	435	472	510	524
Face Brick with Concrete Block Back-up	R/Conc. Frame	106.70	100.85	95.80	93.55	90.75	88.60	87.50	86.70	85.30
	Steel Frame	103.40	97.50	92.50	90.25	87.45	85.30	84.20	83.40	81.95
Limestone with Concrete Block	R/Conc. Frame	115.95	108.90	102.75	100.05	96.55	93.95	92.60	91.60	89.80
	Steel Frame	112.60	105.60	99.45	96.75	93.25	90.65	89.30	88.30	86.50
Precast Concrete Panels	R/Conc. Frame	99.00	94.05	90.00	88.10	85.85	84.20	83.25	82.60	81.45
	Steel Frame	95.70	90.75	86.70	84.80	82.55	80.85	79.95	79.30	78.15
Perimeter Adj., Add or Deduct	Per 100 L.F.	14.05	9.80	7.60	6.15	5.20	4.45	3.95	3.50	3.20
Story Hgt. Adj., Add or Deduct	Per 1 Ft.	2.10	1.85	1.60	1.50	1.35	1.20	1.15	1.10	1.05

For Basement, add $26.65 per square foot of basement area

The above costs were calculated using the basic specifications shown on the facing page. These costs should be adjusted where necessary for design alternatives and owner's requirements. Reported completed project costs, for this type of structure, range from $54.55 to $142.45 per S.F.

Common additives

Description	Unit	$ Cost
Carrels Hardwood	Each	655 - 850
Closed Circuit Surveillance, One station		
Camera and monitor	Each	1325
For additional camera stations, add	Each	730
Elevators, Hydraulic passenger, 2 stops		
1500# capacity	Each	40,800
2500# capacity	Each	42,100
3500# capacity	Each	45,700
Emergency Lighting, 25 watt, battery operated		
Lead battery	Each	350
Nickel cadmium	Each	590
Flagpoles, Complete		
Aluminum, 20' high	Each	1050
40' high	Each	2350
70' high	Each	6675
Fiberglass, 23' high	Each	1375
39'-5" high	Each	2400
59' high	Each	6575

Description	Unit	$ Cost
Library Furnishings		
Bookshelf, 90" high, 10" shelf double face	L.F.	122
single face	L.F.	95
Charging desk, built-in with counter		
Plastic laminated top	L.F.	460
Reading table, laminated		
top 60" x 36"	Each	625

Library

Model costs calculated for a 2 story building with 14' story height and 22,000 square feet of floor area

				Unit	Unit Cost	Cost Per S.F.	% Of Sub-Total
1.0 Foundations							
.1	Footings & Foundations	Poured concrete; strip and spread footings and 4' foundation wall		S.F. Ground	4.22	2.11	
.4	Piles & Caissons	N/A		–	–	–	3.6%
.9	Excavation & Backfill	Site preparation for slab and trench for foundation wall and footing		S.F. Ground	.96	.48	
2.0 Substructure							
.1	Slab on Grade	4" reinforced concrete with vapor barrier and granular base		S.F. Slab	3.05	1.52	2.1%
.2	Special Substructures	N/A		–	–	–	
3.0 Superstructure							
.1	Columns & Beams	Concrete columns		L.F. Column	67	2.07	
.4	Structural Walls	N/A		–	–	–	
.5	Elevated Floors	Concrete waffle slab		S.F. Floor	12.15	6.07	20.0%
.7	Roof	Concrete waffle slab		S.F. Roof	11.32	5.66	
.9	Stairs	Concrete filled metal pan		Flight	5325	.48	
4.0 Exterior Closure							
.1	Walls	Face brick with concrete block backup	90% of wall	S.F. Wall	21	10.69	
.5	Exterior Wall Finishes	N/A		–	–	–	17.9%
.6	Doors	Double aluminum and glass, single leaf hollow metal		Each	3525	.32	
.7	Windows & Glazed Walls	Window wall	10% of wall	S.F. Wall	32	1.78	
5.0 Roofing							
.1	Roof Coverings	Built-up tar and gravel with flashing		S.F. Roof	2.10	1.05	
.7	Insulation	Perlite/EPS composite		S.F. Roof	1.18	.59	2.5%
.8	Openings & Specialties	Gravel stop and hatches		L.F. Roof	.26	.13	
6.0 Interior Construction							
.1	Partitions	Gypsum board on metal studs	30 S.F. Floor/L.F. Partition	S.F. Partition	3.85	1.54	
.4	Interior Doors	Single leaf wood	300 S.F. Floor/Door	Each	399	1.33	
.5	Wall Finishes	Paint		S.F. Surface	.50	.40	15.6%
.6	Floor Finishes	50% carpet, 50% vinyl tile		S.F. Floor	3.42	3.42	
.7	Ceiling Finishes	Mineral fiber on concealed zee bars		S.F. Ceiling	3.06	3.06	
.9	Interior Surface/Exterior Wall	Painted gypsum board on furring	90% of wall	S.F. Wall	2.80	1.40	
7.0 Conveying							
.1	Elevators	One hydraulic passenger elevator		Each	51,920	2.36	3.3%
.2	Special Conveyors	N/A		–	–	–	
8.0 Mechanical							
.1	Plumbing	Toilet and service fixtures, supply and drainage	1 Fixture/1835 S.F. Floor	Each	2697	1.47	
.2	Fire Protection	Wet pipe sprinkler system		S.F. Floor	1.43	1.43	
.3	Heating	Included in 8.4		–	–	–	24.0%
.4	Cooling	Multizone unit, gas heating, electric cooling		S.F. Floor	14.15	14.15	
.5	Special Systems	N/A		–	–	–	
9.0 Electrical							
.1	Service & Distribution	400 ampere service, panel board and feeders		S.F. Floor	.65	.65	
.2	Lighting & Power	Fluorescent fixtures, receptacles, switches, A.C. and misc. power		S.F. Floor	6.68	6.68	11.0%
.4	Special Electrical	Alarm systems and emergency lighting		S.F. Floor	.51	.51	
11.0 Special Construction							
.1	Specialties	N/A		–	–	–	0.0%
12.0 Site Work							
.1	Earthwork	N/A		–	–	–	
.3	Utilities	N/A		–	–	–	
.5	Roads & Parking	N/A		–	–	–	0.0%
.7	Site Improvements	N/A		–	–	–	
				Sub-Total		71.35	100%
	GENERAL CONDITIONS (Overhead & Profit)				15%	10.70	
	ARCHITECT FEES				8%	6.55	
				Total Building Cost		**88.60**	

BUILDING TYPES

COMMERCIAL/INDUSTRIAL/INSTITUTIONAL

M.400 Medical Office, 1 Story

Costs per square foot of floor area

Exterior Wall	S.F. Area	4000	5500	7000	8500	10000	11500	13000	14500	16000
	L.F. Perimeter	280	320	380	440	453	503	510	522	560
Face Brick with Concrete Block Back-up	Steel Joists	101.30	95.55	93.45	92.10	89.25	88.45	86.50	85.10	84.65
	Wood Joists	105.35	99.60	97.50	96.15	93.25	92.50	90.50	89.15	88.65
Stucco on Concrete Block	Steel Joists	92.65	88.40	86.75	85.70	83.65	83.05	81.65	80.65	80.30
	Wood Joists	96.70	92.45	90.80	89.75	87.70	87.10	85.70	84.70	84.35
Brick Veneer	Wood Frame	102.85	97.55	95.55	94.30	91.65	90.95	89.15	87.85	87.40
Wood Siding	Wood Frame	95.85	91.70	90.15	89.10	87.10	86.55	85.20	84.25	83.90
Perimeter Adj., Add or Deduct	Per 100 L.F.	9.85	7.20	5.60	4.60	3.95	3.45	3.05	2.70	2.45
Story Hgt. Adj., Add or Deduct	Per 1 Ft.	2.10	1.75	1.60	1.55	1.35	1.30	1.20	1.10	1.05

For Basement, add $17.10 per square foot of basement area

The above costs were calculated using the basic specifications shown on the facing page. These costs should be adjusted where necessary for design alternatives and owner's requirements. Reported completed project costs, for this type of structure, range from $46.00 to $118.25 per S.F.

Common additives

Description	Unit	$ Cost
Cabinets, Hospital, base		
Laminated plastic	L.F.	244
Stainless steel	L.F.	289
Counter top, laminated plastic	L.F.	47
Stainless steel	L.F.	101
For drop-in sink, add	Each	435
Nurses station, door type		
Laminated plastic	L.F.	272
Enameled steel	L.F.	245
Stainless steel	L.F.	305
Wall cabinets, laminated plastic	L.F.	177
Enameled steel	L.F.	193
Stainless steel	L.F.	275

Description	Unit	$ Cost
Directory Boards, Plastic, glass covered		
30" x 20"	Each	500
36" x 48"	Each	910
Aluminum, 24" x 18"	Each	445
36" x 24"	Each	535
48" x 32"	Each	630
48" x 60"	Each	1425
Heat Therapy Unit		
Humidified, 26" x 78" x 28"	Each	2525
Smoke Detectors		
Ceiling type	Each	141
Duct type	Each	335
Tables, Examining, vinyl top		
with base cabinets	Each	1200 - 3250
Utensil Washer, Sanitizer	Each	8425
X-Ray, Mobile	Each	8800 - 49,500

BUILDING TYPES

Location Factors appear in the Reference Section

Medical Office, 1 Story

Model costs calculated for a 1 story building with 10' story height and 7,000 square feet of floor area

				Unit	Unit Cost	Cost Per S.F.	% Of Sub-Total
1.0 Foundations							
.1	Footings & Foundations	Poured concrete; strip and spread footings and 4' foundation wall		S.F. Ground	3.55	3.55	
.4	Piles & Caissons	N/A		—	—	—	5.8%
.9	Excavation & Backfill	Site preparation for slab and trench for foundation wall and footing		S.F. Ground	.96	.96	
2.0 Substructure							
.1	Slab on Grade	4" reinforced concrete with vapor barrier and granular base		S.F. Slab	3.05	3.05	3.9%
.2	Special Substructures	N/A		—	—	—	
3.0 Superstructure							
.1	Columns & Beams	N/A		—	—	—	
.4	Structural Walls	N/A		—	—	—	
.5	Elevated Floors	N/A		—	—	—	5.6%
.7	Roof	Plywood on wood trusses		S.F. Ground	4.35	4.35	
.9	Stairs	N/A		—	—	—	
4.0 Exterior Closure							
.1	Walls	Face brick with concrete block backup	70% of wall	S.F. Wall	21	8.15	
.5	Exterior Wall Finishes	N/A		—	—	—	17.3%
.6	Doors	Aluminum and glass doors and entrance with transoms		Each	906	1.55	
.7	Windows & Glazed Walls	Wood double hung	30% of wall	Each	392	3.76	
5.0 Roofing							
.1	Roof Coverings	Asphalt shingles with flashing (Pitched)		S.F. Ground	1.22	1.22	
.7	Insulation	Fiber glass sheeets		S.F. Ground	.86	.86	3.1%
.8	Openings & Specialties	Gutters and downspouts		S.F. Ground	.37	.37	
6.0 Interior Construction							
.1	Partitions	Gypsum bd. & sound deadening bd. on wood studs w/insul.	6 S.F. Floor/L.F. Partition	S.F. Partition	4.76	6.35	
.4	Interior Doors	Single leaf wood	60 S.F. Floor/Door	Each	399	6.65	
.5	Wall Finishes	50% paint, 50% vinyl wall covering		S.F. Surface	1.00	2.67	29.4%
.6	Floor Finishes	50% carpet, 50% vinyl composition tile		S.F. Floor	4.01	4.01	
.7	Ceiling Finishes	Mineral fiber tile on concealed zee bars		S.F. Ceiling	2.12	2.12	
.9	Interior Surface/Exterior Wall	Painted gypsum board on furring	70% of wall	S.F. Wall	2.81	1.07	
7.0 Conveying							
.1	Elevators	N/A		—	—	—	0.0%
.2	Special Conveyors	N/A		—	—	—	
8.0 Mechanical							
.1	Plumbing	Toilet and service fixtures, supply and drainage	1 Fixture/195 S.F. Floor	Each	1673	8.58	
.2	Fire Protection	Wet pipe sprinkler system		S.F. Floor	2.04	2.04	
.3	Heating	Included in 8.4		—	—	—	26.9%
.4	Cooling	Multizone unit, gas heating, electric cooling		S.F. Floor	10.28	10.28	
.5	Special Systems	N/A		—	—	—	
9.0 Electrical							
.1	Service & Distribution	200 ampere service, panel board and feeders		S.F. Floor	.87	.87	
.2	Lighting & Power	Fluorescent fixtures, receptacles, switches, A.C. and misc. power		S.F. Floor	4.44	4.44	8.0%
.4	Special Electrical	Alarm systems and emergency lighting		S.F. Floor	.88	.88	
11.0 Special Construction							
.1	Specialties	N/A		—	—	—	0.0%
12.0 Site Work							
.1	Earthwork	N/A		—	—	—	
.3	Utilities	N/A		—	—	—	
.5	Roads & Parking	N/A		—	—	—	0.0%
.7	Site Improvements	N/A		—	—	—	
				Sub-Total		77.78	100%
	GENERAL CONDITIONS (Overhead & Profit)				15%	11.67	
	ARCHITECT FEES				9%	8.05	
				Total Building Cost		97.50	

BUILDING TYPES

157

COMMERCIAL/INDUSTRIAL/INSTITUTIONAL

M.410 Medical Office, 2 Story

Costs per square foot of floor area

Exterior Wall	S.F. Area	4000	5500	7000	8500	10000	11500	13000	14500	16000
	L.F. Perimeter	180	210	240	270	286	311	336	361	386
Face Brick with Concrete Block Back-up	Steel Joists	116.80	110.95	107.60	105.50	103.00	101.70	100.75	99.95	99.30
	Wood Joists	119.25	113.40	110.05	107.95	105.45	104.15	103.20	102.35	101.75
Stucco on Concrete Block	Steel Joists	112.25	107.05	104.10	102.20	100.05	98.95	98.10	97.35	96.80
	Wood Joists	110.15	105.70	103.10	101.50	99.65	98.70	97.95	97.35	96.85
Brick Veneer	Wood Frame	116.15	110.80	107.70	105.75	103.50	102.30	101.45	100.65	100.10
Wood Siding	Wood Frame	109.35	105.00	102.55	100.95	99.20	98.20	97.50	96.90	96.45
Perimeter Adj., Add or Deduct	Per 100 L.F.	17.20	12.50	9.80	8.10	6.85	6.00	5.25	4.75	4.30
Story Hgt. Adj., Add or Deduct	Per 1 Ft.	2.55	2.15	1.90	1.80	1.60	1.55	1.45	1.40	1.35

For Basement, add $18.90 per square foot of basement area

The above costs were calculated using the basic specifications shown on the facing page. These costs should be adjusted where necessary for design alternatives and owner's requirements. Reported completed project costs, for this type of structure, range from $46.00 to $118.25 per S.F.

Common additives

Description	Unit	$ Cost
Cabinets, Hospital, base		
Laminated plastic	L.F.	244
Stainless steel	L.F.	289
Counter top, laminated plastic	L.F.	47
Stainless steel	L.F.	101
For drop-in sink, add	Each	435
Nurses station, door type		
Laminated plastic	L.F.	272
Enameled steel	L.F.	245
Stainless steel	L.F.	305
Wall cabinets, laminated plastic	L.F.	177
Enameled steel	L.F.	193
Stainless steel	L.F.	275
Elevators, Hydraulic passenger, 2 stops		
1500# capacity	Each	40,800
2500# capacity	Each	42,100
3500# capacity	Each	45,700

Description	Unit	$ Cost
Directory Boards, Plastic, glass covered		
30" x 20"	Each	500
36" x 48"	Each	910
Aluminum, 24" x 18"	Each	445
36" x 24"	Each	535
48" x 32"	Each	630
48" x 60"	Each	1425
Emergency Lighting, 25 watt, battery operated		
Lead battery	Each	350
Nickel cadmium	Each	590
Heat Therapy Unit		
Humidified, 26" x 78" x 28"	Each	2525
Smoke Detectors		
Ceiling type	Each	141
Duct type	Each	335
Tables, Examining, vinyl top		
with base cabinets	Each	1200 - 3250
Utensil Washer, Sanitizer	Each	8425
X-Ray, Mobile	Each	8800 - 49,500

Location Factors appear in the Reference Section

Model costs calculated for a 2 story building with 10' story height and 7,000 square feet of floor area

Medical Office, 2 Story

				Unit	Unit Cost	Cost Per S.F.	% Of Sub-Total
1.0 Foundations							
.1	Footings & Foundations	Poured concrete; strip and spread footings and 4' foundation wall		S.F. Ground	5.46	2.73	
.4	Piles & Caissons	N/A		—	—	—	3.8%
.9	Excavation & Backfill	Site preparation for slab and trench for foundation wall and footing		S.F. Ground	.89	.44	
2.0 Substructure							
.1	Slab on Grade	4" reinforced concrete with vapor barrier and granular base		S.F. Slab	3.05	1.52	1.8%
.2	Special Substructures	N/A		—	—	—	
3.0 Superstructure							
.1	Columns & Beams	Included in 3.5 and 3.7		—	—	—	
.4	Structural Walls	N/A		—	—	—	
.5	Elevated Floors	Open web steel joists, slab form, concrete, columns		S.F. Floor	7.15	3.58	7.7%
.7	Roof	Metal deck, open web steel joists, beams, columns		S.F. Roof	2.87	1.43	
.9	Stairs	Concrete filled metal pan		Flight	4925	1.41	
4.0 Exterior Closure							
.1	Walls	Concrete block, insulated	70% of wall	S.F. Floor	4.46	4.46	
.5	Exterior Wall Finishes	Stucco on concrete block	70% of wall	S.F. Wall	6.40	3.07	14.9%
.6	Doors	Aluminum and glass doors with transoms		Each	3525	1.01	
.7	Windows & Glazed Walls	Outward projecting metal	30% of wall	Each	282	3.87	
5.0 Roofing							
.1	Roof Coverings	Built-up tar and gravel with flashing		S.F. Roof	2.54	1.27	
.7	Insulation	Perlite/EPS composite		S.F. Roof	1.18	.59	2.6%
.8	Openings & Specialties	Gravel stop and hatches		S.F. Roof	.52	.26	
6.0 Interior Construction							
.1	Partitions	Gypsum bd. & sound deadening bd. on wood studs w/insul.	6 S.F. Floor/L.F. Partition	S.F. Partition	4.76	6.35	
.4	Interior Doors	Single leaf wood	60 S.F. Floor/Door	Each	399	6.65	
.5	Wall Finishes	50% paint, 50% vinyl wall coating, 5% ceramic tile		S.F. Surface	1.00	2.67	27.9%
.6	Floor Finishes	50% carpet, 50% vinyl asbestos tile		S.F. Floor	4.01	4.01	
.7	Ceiling Finishes	Mineral fiber tile on concealed zee bars		S.F. Ceiling	2.12	2.12	
.9	Interior Surface/Exterior Wall	Painted gypsum board on furring	70% of wall	S.F. Wall	2.81	1.35	
7.0 Conveying							
.1	Elevators	One hydraulic hospital elevator		Each	58,800	8.40	10.1%
.2	Special Conveyors	N/A		—	—	—	
8.0 Mechanical							
.1	Plumbing	Toilet and service fixtures, supply and drainage	1 Fixture/160 S.F. Floor	Each	1225	7.66	
.2	Fire Protection	Wet pipe sprinkler system		S.F. Floor	1.73	1.73	
.3	Heating	Included in 8.4		—	—	—	23.7%
.4	Cooling	Multizone unit, gas heating, electric cooling		S.F. Floor	10.28	10.28	
.5	Special Systems	N/A		—	—	—	
9.0 Electrical							
.1	Service & Distribution	200 ampere service, panel board and feeders		S.F. Floor	.87	.87	
.2	Lighting & Power	Fluorescent fixtures, receptacles, switches, A.C. and misc. power		S.F. Floor	4.44	4.44	7.5%
.4	Special Electrical	Alarm systems and emergency lighting		S.F. Floor	.88	.88	
11.0 Special Construction							
.1	Specialties	N/A		—	—	—	0.0%
12.0 Site Work							
.1	Earthwork	N/A		—	—	—	
.3	Utilities	N/A		—	—	—	0.0%
.5	Roads & Parking	N/A		—	—	—	
.7	Site Improvements	N/A		—	—	—	
				Sub-Total		83.05	100%
	GENERAL CONDITIONS (Overhead & Profit)				15%	12.46	
	ARCHITECT FEES				9%	8.59	
				Total Building Cost		104.10	

BUILDING TYPES

COMMERCIAL/INDUSTRIAL/INSTITUTIONAL — M.420 Motel, 1 Story

Costs per square foot of floor area

Exterior Wall	S.F. Area	2000	3000	4000	6000	8000	10000	12000	14000	16000
	L.F. Perimeter	240	260	280	380	480	560	580	660	740
Brick Veneer	Wood Frame	105.50	92.20	85.50	82.35	80.80	79.10	76.30	75.70	75.30
Aluminum Siding	Wood Frame	93.10	83.25	78.25	75.80	74.60	73.35	71.30	70.85	70.55
Wood Siding	Wood Frame	93.35	83.40	78.40	75.95	74.70	73.45	71.40	70.95	70.65
Wood Shingles	Wood Frame	95.40	84.85	79.60	77.00	75.70	74.40	72.20	71.70	71.40
Precast Concrete Block	Wood Truss	92.05	82.45	77.65	75.25	74.05	72.80	70.85	70.40	70.15
Brick on Concrete Block	Wood Truss	109.65	95.15	87.90	84.55	82.85	81.05	77.95	77.35	76.90
Perimeter Adj., Add or Deduct	Per 100 L.F.	17.50	11.60	8.75	5.80	4.35	3.50	2.90	2.50	2.20
Story Hgt. Adj., Add or Deduct	Per 1 Ft.	3.10	2.20	1.80	1.65	1.55	1.45	1.25	1.20	1.20

For Basement, add $13.05 per square foot of basement area

The above costs were calculated using the basic specifications shown on the facing page. These costs should be adjusted where necessary for design alternatives and owner's requirements. Reported completed project costs, for this type of structure, range from $37.75 to $91.15 per S.F.

Common additives

Description	Unit	$ Cost
Closed Circuit Surveillance, One station		
Camera and monitor	Each	1325
For additional camera stations, add	Each	730
Emergency Lighting, 25 watt, battery operated		
Lead battery	Each	350
Nickel cadmium	Each	590
Laundry Equipment		
Dryer, gas, 16 lb. capacity	Each	665
30 lb. capacity	Each	2675
Washer, 4 cycle	Each	740
Commercial	Each	1125
Sauna, Prefabricated, complete		
6' x 4'	Each	3725
6' x 6'	Each	4350
6' x 9'	Each	5400
8' x 8'	Each	5725
8' x 10'	Each	6800
10' x 12'	Each	8025
Smoke Detectors		
Ceiling type	Each	141
Duct type	Each	335

Description	Unit	$ Cost
Swimming Pools, Complete, gunite	S.F.	45 - 54
TV Antenna, Master system, 12 outlet	Outlet	227
30 outlet	Outlet	146
100 outlet	Outlet	140

Location Factors appear in the Reference Section

Model costs calculated for a 1 story building with 9' story height and 8,000 square feet of floor area

Motel, 1 Story

				Unit	Unit Cost	Cost Per S.F.	% Of Sub-Total
1.0 Foundations							
.1	Footings & Foundations	Poured concrete; strip and spread footings and 4' foundation wall		S.F. Ground	5.38	5.38	
.4	Piles & Caissons	N/A		—	—	—	9.7%
.9	Excavation & Backfill	Site preparation for slab and trench for foundation wall and footing		S.F. Ground	.96	.96	
2.0 Substructure							
.1	Slab on Grade	4" reinforced concrete with vapor barrier and granular base		S.F. Slab	3.05	3.05	4.6%
.2	Special Substructures	N/A		—	—	—	
3.0 Superstructure							
.1	Columns & Beams	N/A		—	—	—	
.4	Structural Walls	N/A		—	—	—	
.5	Elevated Floors	N/A		—	—	—	6.6%
.7	Roof	Plywood on wood trusses		S.F. Ground	4.35	4.35	
.9	Stairs	N/A		—	—	—	
4.0 Exterior Closure							
.1	Walls	Face brick on wood studs with sheathing, insulation and paper	80% of wall	S.F. Wall	17.41	7.52	
.5	Exterior Wall Finishes	N/A		—	—	—	19.3%
.6	Doors	Wood solid core		Each	906	2.49	
.7	Windows & Glazed Walls	Wood double hung	20% of wall	Each	346	2.67	
5.0 Roofing							
.1	Roof Coverings	Asphalt shingles with flashing (pitched)		S.F. Ground	1.03	1.03	
.7	Insulation	Fiberglass sheets		S.F. Ground	.86	.86	3.5%
.8	Openings & Specialties	Gutters and downspouts		S.F. Ground	.41	.41	
6.0 Interior Construction							
.1	Partitions	Gypsum bd. and sound deadening bd. on wood studs	9 S.F. Floor/L.F. Partition	S.F. Partition	4.76	4.23	
.4	Interior Doors	Single leaf hollow metal	300 S.F. Floor/Door	Each	345	1.15	
.5	Wall Finishes	90% paint, 10% ceramic tile		S.F. Surface	.96	1.70	25.2%
.6	Floor Finishes	85% carpet, 15% vinyl composition tile		S.F. Floor	5.55	5.55	
.7	Ceiling Finishes	Painted gypsum board on furring		S.F. Ceiling	2.83	2.83	
.9	Interior Surface/Exterior Wall	Painted gypsum board on furring	80% of wall	S.F. Wall	2.81	1.06	
7.0 Conveying							
.1	Elevators	N/A		—	—	—	0.0%
.2	Special Conveyors	N/A		—	—	—	
8.0 Mechanical							
.1	Plumbing	Toilet and service fixtures, supply and drainage	1 Fixture/90 S.F. Floor	Each	937	10.42	
.2	Fire Protection	Wet pipe sprinkler system		S.F. Floor	2.04	2.04	
.3	Heating	Included in 8.4		—	—	—	23.6%
.4	Cooling	Through the wall electric heating and cooling units		S.F. Floor	3.03	3.03	
.5	Special Systems	N/A		—	—	—	
9.0 Electrical							
.1	Service & Distribution	100 ampere service, panel board and feeders		S.F. Floor	.52	.52	
.2	Lighting & Power	Fluorescent fixtures, receptacles, switches and misc. power		S.F. Floor	4.13	4.13	7.5%
.4	Special Electrical	Alarm systems		S.F. Floor	.27	.27	
11.0 Special Construction							
.1	Specialties	N/A		—	—	—	0.0%
12.0 Site Work							
.1	Earthwork	N/A		—	—	—	
.3	Utilities	N/A		—	—	—	0.0%
.5	Roads & Parking	N/A		—	—	—	
.7	Site Improvements	N/A		—	—	—	
				Sub-Total		65.65	100%
	GENERAL CONDITIONS (Overhead & Profit)				15%	9.85	
	ARCHITECT FEES				7%	5.30	
				Total Building Cost		**80.80**	

BUILDING TYPES

COMMERCIAL/INDUSTRIAL/INSTITUTIONAL — M.430 Motel, 2-3 Story

Costs per square foot of floor area

Exterior Wall	S.F. Area	25000	37000	49000	61000	73000	81000	88000	96000	104000
	L.F. Perimeter	433	593	606	720	835	911	978	1054	1074
Decorative Concrete Block	Wood Joists	75.30	73.45	70.50	69.85	69.40	69.10	68.95	68.85	68.35
	Precast Conc.	79.65	77.75	74.85	74.15	73.70	73.40	73.30	73.15	72.65
Stucco on Concrete Block	Wood Joists	74.35	72.50	69.65	69.00	68.55	68.30	68.15	68.00	67.55
	Precast Conc.	79.15	77.30	74.50	73.80	73.35	73.10	72.95	72.85	72.40
Wood Siding	Wood Frame	73.85	72.05	69.30	68.70	68.25	68.00	67.85	67.70	67.25
Brick Veneer	Wood Frame	78.05	75.90	72.30	71.50	71.00	70.70	70.55	70.35	69.75
Perimeter Adj., Add or Deduct	Per 100 L.F.	2.65	1.80	1.35	1.10	.90	.80	.75	.65	.60
Story Hgt. Adj., Add or Deduct	Per 1 Ft.	.95	.90	.70	.65	.65	.60	.60	.60	.55

For Basement, add $16.60 per square foot of basement area

The above costs were calculated using the basic specifications shown on the facing page. These costs should be adjusted where necessary for design alternatives and owner's requirements. Reported completed project costs, for this type of structure, range from $38.25 to $91.85 per S.F.

Common additives

Description	Unit	$ Cost
Closed Circuit Surveillance, One station		
Camera and monitor	Each	1325
For additional camera station, add	Each	730
Elevators, Hydraulic passenger, 2 stops		
1500# capacity	Each	40,800
2500# capacity	Each	42,100
3500# capacity	Each	45,700
Additional stop, add	Each	3350
Emergency Lighting, 25 watt, battery operated		
Lead battery	Each	350
Nickel cadmium	Each	590
Laundry Equipment		
Dryer, gas, 16 lb. capacity	Each	665
30 lb. capacity	Each	2675
Washer, 4 cycle	Each	740
Commercial	Each	1125

Description	Unit	$ Cost
Sauna, Prefabricated, complete		
6' x 4'	Each	3725
6' x 6'	Each	4350
6' x 9'	Each	5400
8' x 8'	Each	5725
8' x 10'	Each	6800
10' x 12'	Each	8025
Smoke Detectors		
Ceiling type	Each	141
Duct type	Each	335
Swimming Pools, Complete, gunite	S.F.	45 - 54
TV Antenna, Master system, 12 outlet	Outlet	227
30 outlet	Outlet	146
100 outlet	Outlet	140

Model costs calculated for a 3 story building with 9' story height and 73,000 square feet of floor area

Motel, 2-3 Story

				Unit	Unit Cost	Cost Per S.F.	% Of Sub-Total
1.0 Foundations							
.1	Footings & Foundations	Poured concrete; strip and spread footings and 4' foundation wall		S.F. Ground	3.66	1.22	2.5%
.4	Piles & Caissons	N/A		—	—	—	
.9	Excavation & Backfill	Site preparation for slab and trench for foundation wall and footing		S.F. Ground	.89	.30	
2.0 Substructure							
.1	Slab on Grade	4" reinforced concrete with vapor barrier and granular base		S.F. Slab	3.05	1.02	1.7%
.2	Special Substructures	N/A		—	—	—	
3.0 Superstructure							
.1	Columns & Beams	N/A		—	—	—	
.4	Structural Walls	Reinforced concrete block		S.F. Wall	9.26	1.52	
.5	Elevated Floors	Precast concrete plank		S.F. Floor	5.26	3.51	12.5%
.7	Roof	Precast concrete plank		S.F. Roof	5.14	1.71	
.9	Stairs	Concrete filled metal pan		Flight	4925	.81	
4.0 Exterior Closure							
.1	Walls	Face brick with concrete block backup	85% of wall	S.F. Wall	10.74	2.82	
.5	Exterior Wall Finishes	N/A		—	—	—	10.6%
.6	Doors	Aluminum and glass doors and entrance with transom		Each	1107	2.47	
.7	Windows & Glazed Walls	Aluminum sliding	15% of wall	Each	357	1.10	
5.0 Roofing							
.1	Roof Coverings	Built-up tar and gravel with flashing		S.F. Roof	2.04	.68	
.7	Insulation	Perlite/EPS composite		S.F. Roof	1.18	.39	2.0%
.8	Openings & Specialties	Gravel stop, hatches, gutters and downspouts		S.F. Roof	.45	.15	
6.0 Interior Construction							
.1	Partitions	Gypsum board and sound deadening board on wood studs	7 S.F. Floor/L.F. Partition	S.F. Partition	4.76	5.44	
.4	Interior Doors	Wood hollow core	70 S.F. Floor/Door	Each	345	4.93	
.5	Wall Finishes	90% paint, 10% ceramic tile		S.F. Surface	.96	2.19	35.9%
.6	Floor Finishes	85% carpet, 15% vinyl composition tile		S.F. Floor	5.55	5.55	
.7	Ceiling Finishes	Painted gypsum board on furring		S.F. Ceiling	2.83	2.83	
.9	Interior Surface/Exterior Wall	Painted gypsum board on furring	85% of wall	S.F. Wall	2.81	.74	
7.0 Conveying							
.1	Elevators	Two hydraulic passenger elevators		Each	62,050	1.70	2.8%
.2	Special Conveyors	N/A		—	—	—	
8.0 Mechanical							
.1	Plumbing	Toilet and service fixtures, supply and drainage	1 Fixture/180 S.F. Floor	Each	1785	9.92	
.2	Fire Protection	Sprinklers, light hazard		S.F. Floor	1.16	1.16	
.3	Heating	Included in 8.4		—	—	—	23.5%
.4	Cooling	Through the wall electric heating and cooling units		S.F. Floor	3.16	3.16	
.5	Special Systems	N/A		—	—	—	
9.0 Electrical							
.1	Service & Distribution	400 ampere service, panel board and feeders		S.F. Floor	.30	.30	
.2	Lighting & Power	Fluorescent fixtures, receptacles, switches and misc. power		S.F. Floor	4.50	4.50	8.5%
.4	Special Electrical	Alarm systems and emergency lighting		S.F. Floor	.33	.33	
11.0 Special Construction							
.1	Specialties	N/A		—	—	—	0.0%
12.0 Site Work							
.1	Earthwork	N/A		—	—	—	
.3	Utilities	N/A		—	—	—	0.0%
.5	Roads & Parking	N/A		—	—	—	
.7	Site Improvements	N/A		—	—	—	
				Sub-Total		60.45	100%
	GENERAL CONDITIONS (Overhead & Profit)				15%	9.07	
	ARCHITECT FEES				6%	4.18	
				Total Building Cost		73.70	

BUILDING TYPES

COMMERCIAL/INDUSTRIAL/INSTITUTIONAL

M.440 Movie Theater

Costs per square foot of floor area

Exterior Wall	S.F. Area	9000	10000	12000	13000	14000	15000	16000	18000	20000
	L.F. Perimeter	385	410	460	460	480	500	510	547	583
Decorative Concrete Block	Steel Joists	89.25	86.85	83.25	81.00	79.70	78.60	77.35	75.70	74.35
Painted Concrete Block	Steel Joists	84.10	81.95	78.65	76.75	75.55	74.60	73.50	72.05	70.85
Face Brick on Conc. Block	Steel Joists	98.80	96.05	91.85	88.95	87.35	86.05	84.50	82.50	80.85
Precast Concrete Panels	Steel Joists	92.60	90.05	86.30	83.80	82.35	81.20	79.85	78.05	76.60
Tilt-up Panels	Steel Joists	87.80	85.50	82.00	79.85	78.55	77.50	76.30	74.65	73.35
Metal Sandwich Panels	Steel Joists	89.85	87.45	83.85	81.55	80.20	79.10	77.80	76.15	74.75
Perimeter Adj., Add or Deduct	Per 100 L.F.	5.10	4.60	3.85	3.55	3.30	3.05	2.90	2.55	2.30
Story Hgt. Adj., Add or Deduct	Per 1 Ft.	.75	.70	.65	.60	.60	.60	.55	.55	.50

Basement—Not Applicable

The above costs were calculated using the basic specifications shown on the facing page. These costs should be adjusted where necessary for design alternatives and owner's requirements. Reported completed project costs, for this type of structure, range from $45.65 to $119.25 per S.F.

Common additives

Description	Unit	$ Cost
Emergency Lighting, 25 watt, battery operated		
Lead battery	Each	350
Nickel cadmium	Each	590
Seating		
Auditorium chair, all veneer	Each	127
Veneer back, padded seat	Each	152
Upholstered, spring seat	Each	178
Classroom, movable chair & desk	Set	65 - 120
Lecture hall, pedestal type	Each	127 - 370
Smoke Detectors		
Ceiling type	Each	141
Duct type	Each	335
Sound System		
Amplifier, 250 watts	Each	1500
Speaker, ceiling or wall	Each	130
Trumpet	Each	244

Movie Theater

Model costs calculated for a 1 story building with 20' story height and 12,000 square feet of floor area

				Unit	Unit Cost	Cost Per S.F.	% Of Sub-Total
1.0	**Foundations**						
.1	Footings & Foundations	Poured concrete; strip and spread footings and 4' foundation wall		S.F. Ground	2.51	2.51	6.8%
.4	Piles & Caissons	N/A		—	—	—	
.9	Excavation & Backfill	Site preparation for slab and trench for foundation wall and footing		S.F. Ground	2.07	2.07	
2.0	**Substructure**						
.1	Slab on Grade	4" reinforced concrete with vapor barrier and granular base		S.F. Slab	3.05	3.05	4.5%
.2	Special Substructures	N/A		—	—	—	
3.0	**Superstructure**						
.1	Columns & Beams	N/A		—	—	—	
.4	Structural Walls	N/A		—	—	—	
.5	Elevated Floors	Open web steel joists, slab form, concrete	mezzanine 2250 S.F.	S.F. Mezz	5.98	1.12	11.3%
.7	Roof	Metal deck on open web steel joists		S.F. Roof	5.58	5.58	
.9	Stairs	Concrete filled metal pan		Flight	5750	.96	
4.0	**Exterior Closure**						
.1	Walls	Decorative concrete block	100% of wall	S.F. Wall	12.20	9.35	14.9%
.5	Exterior Wall Finishes	N/A		—	—	—	
.6	Doors	Sliding mallfront aluminum and glass and hollow metal		Each	1434	.71	
.7	Windows & Glazed Walls	N/A		—	—	—	
5.0	**Roofing**						
.1	Roof Coverings	Built-up tar and gravel with flashing		S.F. Roof	2.10	2.10	5.5%
.7	Insulation	Perlite/EPS composite		S.F. Roof	1.18	1.18	
.8	Openings & Specialties	Gravel stop, hatches, gutters and downspouts		S.F. Roof	.47	.47	
6.0	**Interior Construction**						
.1	Partitions	Concrete block, toilet partitions	40 S.F. Floor/L.F. Partition	S.F. Partition	6.03	2.41	16.6%
.4	Interior Doors	Single leaf hollow metal	705 S.F. Floor/Door	Each	530	.75	
.5	Wall Finishes	Paint		S.F. Surface	1.08	.86	
.6	Floor Finishes	Carpet	50% of area	S.F. Floor	5.36	2.68	
.7	Ceiling Finishes	Mineral fiber tile on concealed zee runners suspended		S.F. Ceiling	3.06	3.06	
.9	Interior Surface/Exterior Wall	Paint	100% of wall	S.F. Wall	1.93	1.48	
7.0	**Conveying**						
.1	Elevators	N/A		—	—	—	0.0%
.2	Special Conveyors	N/A		—	—	—	
8.0	**Mechanical**						
.1	Plumbing	Toilet and service fixtures, supply and drainage	1 Fixture/500 S.F. Floor	Each	1480	2.96	19.9%
.2	Fire Protection	Wet pipe sprinkler system		S.F. Floor	1.57	1.57	
.3	Heating	Included in 8.4		—	—	—	
.4	Cooling	Single zone rooftop unit, gas heating, electric cooling		S.F. Floor	8.97	8.97	
.5	Special Systems	N/A		—	—	—	
9.0	**Electrical**						
.1	Service & Distribution	400 ampere service, panel board and feeders		S.F. Floor	.61	.61	5.6%
.2	Lighting & Power	Fluorescent fixtures, receptacles, switches, A.C. and misc. power		S.F. Floor	2.70	2.70	
.4	Special Electrical	Alarm systems, sound system and emergency lighting		S.F. Floor	.45	.45	
11.0	**Special Construction**						
.1	Specialties	Projection equipment, screen, seating		S.F. Floor	10.07	10.07	14.9%
12.0	**Site Work**						
.1	Earthwork	N/A		—	—	—	0.0%
.3	Utilities	N/A		—	—	—	
.5	Roads & Parking	N/A		—	—	—	
.7	Site Improvements	N/A		—	—	—	
				Sub-Total		67.67	100%
	GENERAL CONDITIONS (Overhead & Profit)				15%	10.15	
	ARCHITECT FEES				7%	5.43	
				Total Building Cost		**83.25**	

BUILDING TYPES

COMMERCIAL/INDUSTRIAL/INSTITUTIONAL

M.450 Nursing Home

Costs per square foot of floor area

Exterior Wall		S.F. Area	10000	15000	20000	25000	30000	35000	40000	45000	50000
		L.F. Perimeter	286	370	453	457	513	568	624	680	735
Redwood Siding	Wood Frame		87.55	84.35	82.65	80.35	79.55	78.95	78.50	78.15	77.90
Brick Veneer	Wood Frame		94.55	90.35	88.20	84.85	83.70	82.90	82.30	81.85	81.45
Face Brick with Concrete Block Back-up	Wood Joists		97.05	92.50	90.20	86.45	85.20	84.35	83.65	83.15	82.75
	Steel Joists		104.00	99.45	97.15	93.40	92.15	91.30	90.65	90.15	89.75
Stucco on Concrete Block	Wood Joists		89.80	86.25	84.45	81.80	80.85	80.25	79.70	79.35	79.05
	Steel Joists		96.75	93.20	91.40	88.75	87.85	87.20	86.70	86.30	86.00
Perimeter Adj., Add or Deduct	Per 100 L.F.		4.15	2.80	2.10	1.65	1.45	1.20	1.05	.95	.85
Story Hgt. Adj., Add or Deduct	Per 1 Ft.		.80	.70	.65	.55	.50	.45	.45	.40	.40

For Basement, add $17.95 per square foot of basement area

The above costs were calculated using the basic specifications shown on the facing page. These costs should be adjusted where necessary for design alternatives and owner's requirements. Reported completed project costs, for this type of structure, range from $48.65 to $120.85 per S.F.

Common additives

Description	Unit	$ Cost
Beds, Manual	Each	825 - 1450
Doctors In-Out Register, 200 names	Each	12,700
Elevators, Hydraulic passenger, 2 stops		
1500# capacity	Each	40,800
2500# capacity	Each	42,100
3500# capacity	Each	45,700
Emergency Lighting, 25 watt, battery operated		
Lead battery	Each	350
Nickel cadmium	Each	590
Intercom System, 25 station capacity		
Master station	Each	1800
Intercom outlets	Each	113
Handset	Each	299
Kitchen Equipment		
Broiler	Each	3300
Coffee urn, twin 6 gallon	Each	5500
Cooler, 6 ft. long	Each	2600
Dishwasher, 10-12 racks per hr.	Each	2500
Food warmer	Each	715
Freezer, 44 C.F., reach-in	Each	6375

Description	Unit	$ Cost
Kitchen Equipment, cont.		
Ice cube maker, 50 lb. per day	Each	1675
Range with 1 oven	Each	2250
Laundry Equipment		
Dryer, gas, 16 lb. capacity	Each	665
30 lb. capacity	Each	2675
Washer, 4 cycle	Each	740
Commercial	Each	1125
Nurses Call System		
Single bedside call station	Each	199
Pillow speaker	Each	209
Refrigerator, Prefabricated, walk-in		
7'-6" high, 6' x 6'	S.F.	117
10' x 10'	S.F.	91
12' x 14'	S.F.	81
12' x 20'	S.F.	72
TV Antenna, Master system, 12 outlet	Outlet	227
30 outlet	Outlet	146
100 outlet	Outlet	140
Whirlpool Bath, Mobile, 18" x 24" x 60"	Each	3025
X-Ray, Mobile	Each	8800 - 49,500

Location Factors appear in the Reference Section

Model costs calculated for a 2 story building with 10' story height and 25,000 square feet of floor area

Nursing Home

				Unit	Unit Cost	Cost Per S.F.	% Of Sub-Total
1.0 Foundations							
.1	Footings & Foundations	Poured concrete; strip and spread footings and 4' foundation wall		S.F. Ground	2.82	1.41	3.0%
.4	Piles & Caissons	N/A		—	—	—	
.9	Excavation & Backfill	Site preparation for slab and trench for foundation wall and footing		S.F. Ground	.96	.48	
2.0 Substructure							
.1	Slab on Grade	4" reinforced concrete with vapor barrier and granular base		S.F. Slab	3.05	1.52	2.4%
.2	Special Substructures	N/A		—	—	—	
3.0 Superstructure							
.1	Columns & Beams	N/A		—	—	—	
.4	Structural Walls	N/A		—	—	—	5.5%
.5	Elevated Floors	Plywood on wood joists		S.F. Floor	3.62	1.81	
.7	Roof	Plywood on wood joists		S.F. Roof	2.71	1.36	
.9	Stairs	Wood		Flight	1349	.32	
4.0 Exterior Closure							
.1	Walls	Vertical T & G redwood siding	85% of wall	S.F. Wall	6.18	1.92	
.5	Exterior Wall Finishes	N/A		—	—	—	5.6%
.6	Doors	Double aluminum & glass doors, single leaf hollow metal		Each	1429	.28	
.7	Windows & Glazed Walls	Wood double hung	15% of wall	Each	357	1.31	
5.0 Roofing							
.1	Roof Coverings	Built-up tar and gravel with flashing		S.F. Roof	2.06	1.03	
.7	Insulation	Perlite/EPS composite		S.F. Roof	1.18	.59	3.0%
.8	Openings & Specialties	Gravel stop, hatches, gutters and downspouts		S.F. Roof	.50	.25	
6.0 Interior Construction							
.1	Partitions	Gypsum board on wood studs	8 S.F. Floor/L.F. Partition	S.F. Partition	3.04	3.04	
.4	Interior Doors	Single leaf wood	80 S.F. Floor/Door	Each	399	4.99	
.5	Wall Finishes	50% vinyl wall coverings, 45% paint, 5% ceramic tile		S.F. Surface	1.23	2.46	29.3%
.6	Floor Finishes	50% carpet, 45% vinyl tile, 5% ceramic tile		S.F. Floor	4.23	4.23	
.7	Ceiling Finishes	Painted gypsum board on wood furring		S.F. Ceiling	2.83	2.83	
.9	Interior Surface/Exterior Wall	Painted gypsum board on wood furring	85% of wall	S.F. Wall	2.81	.87	
7.0 Conveying							
.1	Elevators	One hydraulic hospital elevator		Each	58,750	2.35	3.7%
.2	Special Conveyors	N/A		—	—	—	
8.0 Mechanical							
.1	Plumbing	Kitchen, toilet and service fixtures, supply and drainage	1 Fixture/230 S.F. Floor	Each	2270	9.87	
.2	Fire Protection	Sprinkler, light hazard		S.F. Floor	2.84	2.84	
.3	Heating	Oil fired hot water, wall fin radiation		S.F. Floor	4.31	4.31	35.9%
.4	Cooling	Split systems with air cooled condensing units		S.F. Floor	5.60	5.60	
.5	Special Systems	N/A		—	—	—	
9.0 Electrical							
.1	Service & Distribution	600 ampere service, panel board and feeders		S.F. Floor	.94	.94	
.2	Lighting & Power	Incandescent fixtures, receptacles, switches, A.C. and misc. power		S.F. Floor	5.43	5.43	11.6%
.4	Special Electrical	Alarm systems and emergency lighting		S.F. Floor	.92	.92	
11.0 Special Construction							
.1	Specialties	N/A		—	—	—	0.0%
12.0 Site Work							
.1	Earthwork	N/A		—	—	—	
.3	Utilities	N/A		—	—	—	0.0%
.5	Roads & Parking	N/A		—	—	—	
.7	Site Improvements	N/A		—	—	—	
				Sub-Total		62.96	100%
	GENERAL CONDITIONS (Overhead & Profit)				15%	9.44	
	ARCHITECT FEES				11%	7.95	
				Total Building Cost		**80.35**	

BUILDING TYPES

COMMERCIAL/INDUSTRIAL/INSTITUTIONAL — M.460 — Office, 2-4 Story

Costs per square foot of floor area

Exterior Wall	S.F. Area	10000	22000	34000	46000	58000	63000	68000	73000	78000
	L.F. Perimeter	246	393	443	543	562	590	603	624	645
Face Brick with Concrete Block Back-up	Wood Joists	91.45	77.90	70.70	68.45	65.55	65.00	64.35	63.90	63.50
	Steel Joists	92.85	79.30	72.10	69.85	66.95	66.40	65.75	65.30	64.90
Glass and Metal Curtain Wall	Steel Frame	88.35	76.55	70.60	68.70	66.35	65.85	65.35	64.95	64.60
	R/Conc. Frame	91.20	79.40	73.45	71.55	69.20	68.75	68.20	67.85	67.50
Wood Siding	Wood Frame	77.45	66.90	61.85	60.15	58.20	57.80	57.35	57.00	56.75
Brick Veneer	Wood Frame	84.95	72.35	65.80	63.75	61.15	60.65	60.05	59.65	59.25
Perimeter Adj., Add or Deduct	Per 100 L.F.	11.40	5.15	3.35	2.45	2.00	1.85	1.70	1.55	1.45
Story Hgt. Adj., Add or Deduct	Per 1 Ft.	2.10	1.50	1.10	1.00	.85	.80	.75	.75	.70

For Basement, add $19.55 per square foot of basement area

The above costs were calculated using the basic specifications shown on the facing page. These costs should be adjusted where necessary for design alternatives and owner's requirements. Reported completed project costs, for this type of structure, range from $38.25 to $107.75 per S.F.

Common additives

Description	Unit	$ Cost
Clock System		
20 room	Each	11,700
50 room	Each	28,300
Closed Circuit Surveillance, One station		
Camera and monitor	Each	1325
For additional camera stations, add	Each	730
Directory Boards, Plastic, glass covered		
30" x 20"	Each	500
36" x 48"	Each	910
Aluminum, 24" x 18"	Each	445
36" x 24"	Each	535
48" x 32"	Each	630
48" x 60"	Each	1425
Elevators, Hydraulic passenger, 2 stops		
1500# capacity	Each	40,800
2500# capacity	Each	42,100
3500# capacity	Each	45,700
Additional stop, add	Each	3350
Emergency Lighting, 25 watt, battery operated		
Lead battery	Each	350
Nickel cadmium	Each	590

Description	Unit	$ Cost
Smoke Detectors		
Ceiling type	Each	141
Duct type	Each	335
Sound System		
Amplifier, 250 watts	Each	1500
Speaker, ceiling or wall	Each	130
Trumpet	Each	244
TV Antenna, Master system, 12 outlet	Outlet	227
30 outlet	Outlet	146
100 outlet	Outlet	140

Model costs calculated for a 3 story building with 12' story height and 58,000 square feet of floor area

Office, 2-4 Story

				Unit	Unit Cost	Cost Per S.F.	% Of Sub-Total
1.0 Foundations							
.1	Footings & Foundations	Poured concrete; strip and spread footings and 4' foundation wall		S.F. Ground	3.36	1.12	2.6%
.4	Piles & Caissons	N/A		—	—	—	
.9	Excavation & Backfill	Site preparation for slab and trench for foundation wall and footing		S.F. Ground	.89	.30	
2.0 Substructure							
.1	Slab on Grade	4" reinforced concrete with vapor barrier and granular base		S.F. Slab	3.05	1.02	1.9%
.2	Special Substructures	N/A		—	—	—	
3.0 Superstructure							
.1	Columns & Beams	Fireproofing; interior columns included in 3.5 and 3.7		L.F. Columns	21	.57	
.4	Structural Walls	N/A		—	—	—	13.2%
.5	Elevated Floors	Open web steel joists, slab form, concrete, columns		S.F. Floor	7.37	4.91	
.7	Roof	Metal deck, open web steel joists, columns		S.F. Roof	3.40	1.13	
.9	Stairs	Concrete filled metal pan		Flight	4925	.59	
4.0 Exterior Closure							
.1	Walls	Face brick with concrete block backup	80% of wall	S.F. Wall	21	5.94	
.5	Exterior Wall Finishes	N/A		—	—	—	13.9%
.6	Doors	Aluminum and glass, hollow metal		Each	2260	.24	
.7	Windows & Glazed Walls	Steel outward projecting	20% of wall	Each	458	1.39	
5.0 Roofing							
.1	Roof Coverings	Built-up tar and gravel with flashing		S.F. Roof	2.04	.68	
.7	Insulation	Perlite/EPS composite		S.F. Roof	1.18	.39	2.0%
.8	Openings & Specialties	N/A		—	—	—	
6.0 Interior Construction							
.1	Partitions	Gypsum board on metal studs, toilet partitions	20 S.F. Floor/L.F. Partition	S.F. Partition	2.91	1.46	
.4	Interior Doors	Single leaf hollow metal	200 S.F. Floor/Door	Each	530	2.65	
.5	Wall Finishes	60% vinyl wall covering, 40% paint		S.F. Surface	1.10	.88	24.7%
.6	Floor Finishes	60% carpet, 30% vinyl composition tile, 10% ceramic tile		S.F. Floor	4.61	4.61	
.7	Ceiling Finishes	Mineral fiber tile on concealed zee bars		S.F. Ceiling	3.06	3.06	
.9	Interior Surface/Exterior Wall	Painted gypsum board on furring	80% of wall	S.F. Wall	2.81	.78	
7.0 Conveying							
.1	Elevators	Two hydraulic passenger elevators		Each	64,090	2.21	4.1%
.2	Special Conveyors	N/A		—	—	—	
8.0 Mechanical							
.1	Plumbing	Toilet and service fixtures, supply and drainage	1 Fixture/1320 S.F. Floor	Each	1966	1.49	
.2	Fire Protection	Standpipes and hose systems		S.F. Floor	.20	.20	
.3	Heating	Included in 8.4		—	—	—	22.8%
.4	Cooling	Multizone unit gas heating, electric cooling		S.F. Floor	10.76	10.76	
.5	Special Systems	N/A		—	—	—	
9.0 Electrical							
.1	Service & Distribution	1000 ampere service, panel board and feeders		S.F. Floor	.91	.91	
.2	Lighting & Power	Fluorescent fixtures, receptacles, switches, A.C. and misc. power		S.F. Floor	6.94	6.94	14.8%
.4	Special Electrical	Alarm systems and emergency lighting		S.F. Floor	.19	.19	
11.0 Special Construction							
.1	Specialties	N/A		—	—	—	0.0%
12.0 Site Work							
.1	Earthwork	N/A		—	—	—	
.3	Utilities	N/A		—	—	—	0.0%
.5	Roads & Parking	N/A		—	—	—	
.7	Site Improvements	N/A		—	—	—	
				Sub-Total		54.42	100%
	GENERAL CONDITIONS (Overhead & Profit)				15%	8.16	
	ARCHITECT FEES				7%	4.37	
				Total Building Cost		**66.95**	

BUILDING TYPES

COMMERCIAL/INDUSTRIAL/INSTITUTIONAL — M.470 — Office, 5-10 Story

Costs per square foot of floor area

Exterior Wall	S.F. Area	50000	60000	70000	80000	90000	100000	110000	120000	130000
	L.F. Perimeter	328	370	378	410	441	450	475	500	510
Precast Concrete Panel	Steel Frame	79.45	77.95	75.70	74.75	73.95	72.80	72.20	71.70	71.00
	R/Conc. Frame	80.90	79.40	77.15	76.15	75.35	74.20	73.55	73.05	72.40
Face Brick with Concrete Block Back-up	Steel Frame	82.35	80.65	78.10	77.00	76.10	74.80	74.10	73.50	72.75
	R/Conc. Frame	83.65	82.00	79.40	78.30	77.45	76.10	75.40	74.80	74.05
Limestone Panel Concrete Block Back-up	Steel Frame	88.10	86.05	82.80	81.50	80.40	78.75	77.85	77.15	76.15
	R/Conc. Frame	89.40	87.40	84.10	82.80	81.70	80.05	79.15	78.45	77.45
Perimeter Adj., Add or Deduct	Per 100 L.F.	5.05	4.20	3.55	3.15	2.80	2.50	2.30	2.10	1.90
Story Hgt. Adj., Add or Deduct	Per 1 Ft.	1.30	1.20	1.05	1.05	1.00	.90	.90	.80	.80

For Basement, add $20.90 per square foot of basement area

The above costs were calculated using the basic specifications shown on the facing page. These costs should be adjusted where necessary for design alternatives and owner's requirements. Reported completed project costs, for this type of structure, range from $43.90 to $129.10 per S.F.

Common additives

Description	Unit	$ Cost
Clock System		
20 room	Each	11,700
50 room	Each	28,300
Closed Circuit Surveillance, One station		
Camera and monitor	Each	1325
For additional camera stations, add	Each	730
Directory Boards, Plastic, glass covered		
30" x 20"	Each	500
36" x 48"	Each	910
Aluminum, 24" x 18"	Each	445
36" x 24"	Each	535
48" x 32"	Each	630
48" x 60"	Each	1425
Elevators, Electric passenger, 5 stops		
2000# capacity	Each	88,800
3500# capacity	Each	94,800
5000# capacity	Each	98,300
Additional stop, add	Each	5225
Emergency Lighting, 25 watt, battery operated		
Lead battery	Each	350
Nickel cadmium	Each	590

Description	Unit	$ Cost
Intercom System, 25 station capacity		
Master station	Each	1800
Intercom outlets	Each	113
Handset	Each	299
Smoke Detectors		
Ceiling type	Each	141
Duct type	Each	335
Sound System		
Amplifier, 250 watts	Each	1500
Speaker, ceiling or wall	Each	130
Trumpet	Each	244
TV Antenna, Master system, 12 oulet	Outlet	227
30 outlet	Outlet	146
100 outlet	Outlet	140

Model costs calculated for an 8 story building with 12' story height and 100,000 square feet of floor area

Office, 5-10 Story

				Unit	Unit Cost	Cost Per S.F.	% Of Sub-Total
1.0 Foundations							
.1	Footings & Foundations	Poured concrete; strip and spread footings and 4' foundation wall		S.F. Ground	8.00	1.00	
.4	Piles & Caissons	N/A		–	–	–	1.9%
.9	Excavation & Backfill	Site preparation for slab and trench for foundation wall and footing		S.F. Ground	.96	.12	
2.0 Substructure							
.1	Slab on Grade	4" reinforced concrete with vapor barrier and granular base		S.F. Slab	3.05	.38	0.6%
.2	Special Substructures	N/A		–	–	–	
3.0 Superstructure							
.1	Columns & Beams	Steel columns w/ fireproofing		L.F. Columns	69	1.89	
.4	Structural Walls	N/A		–	–	–	
.5	Elevated Floors	Concrete slab with metal deck and beams		S.F. Floor	9.56	8.36	19.3%
.7	Roof	Metal deck, open web steel joists, interior columns		S.F. Roof	3.64	.46	
.9	Stairs	Concrete filled metal pan		Flight	4925	.84	
4.0 Exterior Closure							
.1	Walls	Precast concrete panels	80% of wall	S.F. Wall	16.81	5.81	
.5	Exterior Wall Finishes	N/A		–	–	–	13.4%
.6	Doors	Double aluminum and glass doors and entrance with transoms		Each	2521	.13	
.7	Windows & Glazed Walls	Vertical pivoted steel	20% of wall	Each	357	2.06	
5.0 Roofing							
.1	Roof Coverings	Built-up tar and gravel with flashing		S.F. Roof	2.24	.28	
.7	Insulation	Perlite/EPS composite		S.F. Roof	1.18	.15	0.7%
.8	Openings & Specialties	N/A		–	–	–	
6.0 Interior Construction							
.1	Partitions	Gypsum board on metal studs, toilet partitions	30 S.F. Floor/L.F. Partition	S.F. Partition	2.91	1.43	
.4	Interior Doors	Single leaf hollow metal	400 S.F. Floor/Door	Each	530	1.33	
.5	Wall Finishes	60% vinyl wall covering, 40% paint		S.F. Surface	1.10	.73	20.3%
.6	Floor Finishes	60% carpet, 30% vinyl composition tile, 10% ceramic tile		S.F. Floor	4.61	4.61	
.7	Ceiling Finishes	Mineral fiber tile on concealed zee bars		S.F. Ceiling	3.06	3.06	
.9	Interior Surface/Exterior Wall	Painted gypsum board on furring	80% of wall	S.F. Wall	2.81	.97	
7.0 Conveying							
.1	Elevators	Four geared passenger elevators		Each	143,000	5.72	9.6%
.2	Special Conveyors	N/A		–	–	–	
8.0 Mechanical							
.1	Plumbing	Toilet and service fixtures, supply and drainage	1 Fixture/1370 S.F. Floor	Each	1479	1.08	
.2	Fire Protection	Standpipes and hose systems		S.F. Floor	.14	.14	
.3	Heating	Included in 8.4		–	–	–	20.1%
.4	Cooling	Multizone unit gas heating, electric cooling		S.F. Floor	10.76	10.76	
.5	Special Systems	N/A		–	–	–	
9.0 Electrical							
.1	Service & Distribution	1600 ampere service, panel board and feeders		S.F. Floor	.78	.78	
.2	Lighting & Power	Fluorescent fixtures, receptacles, switches, A.C. and misc. power		S.F. Floor	6.96	6.96	14.1%
.4	Special Electrical	Alarm systems and emergency lighting		S.F. Floor	.69	.69	
11.0 Special Construction							
.1	Specialties			–	–	–	0.0%
12.0 Site Work							
.1	Earthwork	N/A		–	–	–	
.3	Utilities	N/A		–	–	–	
.5	Roads & Parking	N/A		–	–	–	0.0%
.7	Site Improvements	N/A		–	–	–	
				Sub-Total		59.74	100%
	GENERAL CONDITIONS (Overhead & Profit)				15%	8.96	
	ARCHITECT FEES				6%	4.10	
				Total Building Cost		72.80	

BUILDING TYPES

COMMERCIAL/INDUSTRIAL/INSTITUTIONAL — M.480 — Office, 11-20 Story

Costs per square foot of floor area

Exterior Wall	S.F. Area	90000	100000	110000	120000	130000	140000	150000	160000	170000
	L.F. Perimeter	320	342	364	360	376	393	400	413	426
Double Glazed Heat Absorbing Tinted Plate Glass Panels	Steel Frame	95.15	93.80	92.70	90.65	89.70	88.90	87.90	87.25	86.65
	R/Conc. Frame	98.75	96.80	95.25	92.75	91.45	90.40	89.10	88.25	87.40
Face Brick with Concret Block Back-up	Steel Frame	99.15	97.25	95.75	93.45	92.20	91.15	90.00	89.15	88.35
	R/Conc. Frame	96.00	94.15	92.65	90.35	89.15	88.15	86.95	86.15	85.35
Precast Concrete Panel With Exposed Aggregate	Steel Frame	96.40	94.60	93.15	91.10	89.95	89.00	87.90	87.15	86.40
	R/Conc. Frame	93.60	91.85	90.40	88.40	87.25	86.30	85.25	84.50	83.75
Perimeter Adj., Add or Deduct	Per 100 L.F.	5.90	5.35	4.85	4.45	4.10	3.80	3.55	3.30	3.15
Story Hgt. Adj., Add or Deduct	Per 1 Ft.	1.80	1.75	1.70	1.55	1.45	1.45	1.35	1.30	1.30

For Basement, add $20.80 per square foot of basement area

The above costs were calculated using the basic specifications shown on the facing page. These costs should be adjusted where necessary for design alternatives and owner's requirements. Reported completed project costs, for this type of structure, range from $55.30 to $134.70 per S.F.

Common additives

Description	Unit	$ Cost
Clock System		
20 room	Each	11,700
50 room	Each	28,300
Directory Boards, Plastic, glass covered		
30" x 20"	Each	500
36" x 48"	Each	910
Aluminum, 24" x 18"	Each	445
36" x 24"	Each	535
48" x 32"	Each	630
48" x 60"	Each	1425
Elevators, Electric passenger, 10 stops		
3000# capacity	Each	196,000
4000# capacity	Each	197,500
5000# capacity	Each	201,500
Additional stop, add	Each	5225
Emergency Lighting, 25 watt, battery operated		
Lead battery	Each	350
Nickel cadmium	Each	590

Description	Unit	$ Cost
Escalators, Metal		
32" wide, 10' story height	Each	92,200
20' story height	Each	105,300
48" wide, 10' story height	Each	97,300
20' story height	Each	110,800
Glass		
32" wide, 10' story height	Each	86,200
20' story height	Each	98,800
48" wide, 10' story height	Each	91,800
20' story height	Each	103,800
Smoke Detectors		
Ceiling type	Each	141
Duct type	Each	335
Sound System		
Amplifier, 250 watts	Each	1500
Speaker, ceiling or wall	Each	130
Trumpet	Each	244
TV Antenna, Master system, 12 oulet	Outlet	227
30 oulet	Outlet	146
100 outlet	Outlet	140

Model costs calculated for a 15 story building with 10' story height and 140,000 square feet of floor area

Office, 11-20 Story

				Unit	Unit Cost	Cost Per S.F.	% Of Sub-Total
1.0 Foundations							
.1	Footings & Foundations	Poured concrete; strip and spread footings and 4' foundation wall		S.F. Ground	19.20	1.28	1.8%
.4	Piles & Caissons	N/A		—	—	—	
.9	Excavation & Backfill	Site preparation for slab and trench for foundation wall and footing		S.F. Ground	.96	.06	
2.0 Substructure							
.1	Slab on Grade	4" reinforced concrete with vapor barrier and granular base		S.F. Slab	3.05	.20	0.3%
.2	Special Substructures	N/A		—	—	—	
3.0 Superstructure							
.1	Columns & Beams	Steel columns with fireproofing		L.F. Column	106	2.76	
.4	Structural Walls	N/A		—	—	—	22.8%
.5	Elevated Floors	Concrete slab, metal deck, beams		S.F. Floor	13.00	12.13	
.7	Roof	Metal deck, open web steel joists, beams, columns		S.F. Roof	4.26	.28	
.9	Stairs	Concrete filled metal pan		Flight	5750	1.44	
4.0 Exterior Closure							
.1	Walls	N/A		—	—	—	
.5	Exterior Wall Finishes	N/A		—	—	—	15.9%
.6	Doors	Double aluminum & glass doors		Each	3896	.78	
.7	Windows & Glazed Walls	Double glazed heat absorbing, tinted plate glass wall panels	100% of wall	S.F. Wall	25	10.82	
5.0 Roofing							
.1	Roof Coverings	Built-up tar and gravel with flashing		S.F. Roof	4.05	.27	
.7	Insulation	Perlite/EPS composite		S.F. Roof	1.18	.08	0.5%
.8	Openings & Specialties	N/A		—	—	—	
6.0 Interior Construction							
.1	Partitions	Gypsum board on metal studs, toilet partitions	30 S.F. Floor/L.F. Partition	S.F. Partition	2.91	1.40	
.4	Interior Doors	Single leaf hollow metal	400 S.F. Floor/Door	Each	530	1.33	
.5	Wall Finishes	60% vinyl wall covering, 40% paint		S.F. Surface	1.11	.59	16.4%
.6	Floor Finishes	60% carpet, 30% vinyl composition tile, 10% ceramic tile		S.F. Floor	4.61	4.61	
.7	Ceiling Finishes	Mineral fiber tile on concealed zee bars		S.F. Ceiling	3.06	3.06	
.9	Interior Surface/Exterior Wall	Painted drywall on metal furring	80% of wall	S.F. Wall	2.81	.95	
7.0 Conveying							
.1	Elevators	Four geared passenger elevators		Each	213,150	6.09	8.3%
.2	Special Conveyors	N/A		—	—	—	
8.0 Mechanical							
.1	Plumbing	Toilet and service fixtures, supply and drainage	1 Fixture/1345 S.F. Floor	Each	1936	1.44	
.2	Fire Protection	Standpipes and hose systems and sprinklers, light hazard		S.F. Floor	2.75	2.75	
.3	Heating	Oil fired hot water		S.F. Floor	2.96	2.96	22.1%
.4	Cooling	Chilled water, fan coil units		S.F. Floor	8.97	8.97	
.5	Special Systems	N/A		—	—	—	
9.0 Electrical							
.1	Service & Distribution	2400 ampere service, panel board and feeders		S.F. Floor	1.09	1.09	
.2	Lighting & Power	Fluorescent fixtures, receptacles, switches, A.C. and misc. power		S.F. Floor	6.94	6.94	11.9%
.4	Special Electrical	Alarm systems and emergency lighting		S.F. Floor	.67	.67	
11.0 Special Construction							
.1	Specialties	N/A		—	—	—	0.0%
12.0 Site Work							
.1	Earthwork	N/A		—	—	—	
.3	Utilities	N/A		—	—	—	0.0%
.5	Roads & Parking	N/A		—	—	—	
.7	Site Improvements	N/A		—	—	—	
				Sub-Total		72.95	100%
	GENERAL CONDITIONS (Overhead & Profit)				15%	10.94	
	ARCHITECT FEES				6%	5.01	
				Total Building Cost		88.90	

BUILDING TYPES

COMMERCIAL/INDUSTRIAL/INSTITUTIONAL

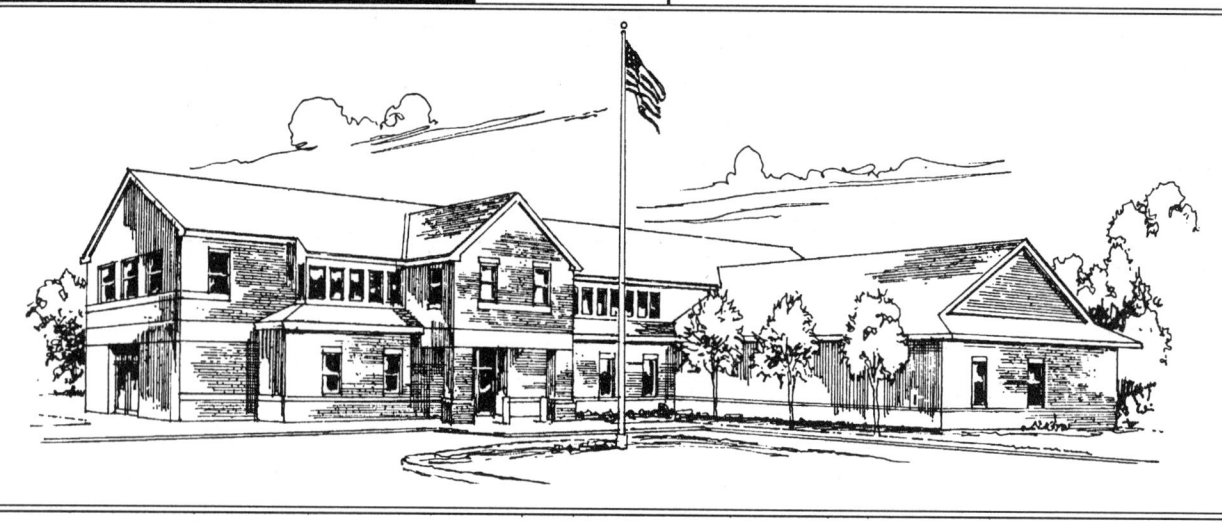

M.490 Police Station

Costs per square foot of floor area

Exterior Wall	S.F. Area	7000	9000	11000	13000	15000	17000	19000	21000	23000
	L.F. Perimeter	240	280	303	325	354	372	397	422	447
Limestone with Concrete Block Back-up	Bearing Walls	132.90	124.35	117.00	111.90	108.70	105.40	103.35	101.60	100.20
	R/Conc. Frame	140.85	132.65	125.85	121.10	118.05	115.05	113.10	111.50	110.20
Face Brick with Concrete Block Back-up	Bearing Walls	119.90	112.45	106.45	102.25	99.55	96.90	95.20	93.75	92.60
	R/Conc. Frame	133.10	125.65	119.65	115.45	112.75	110.10	108.40	106.95	105.80
Decorative Concrete Block	Bearing Walls	112.40	105.65	100.40	96.80	94.40	92.15	90.60	89.35	88.35
	R/Conc. Frame	125.60	118.85	113.60	110.00	107.60	105.35	103.80	102.55	101.55
Perimeter Adj., Add or Deduct	Per 100 L.F.	17.20	13.35	10.95	9.25	8.05	7.05	6.35	5.75	5.20
Story Hgt. Adj., Add or Deduct	Per 1 Ft.	3.05	2.70	2.40	2.20	2.10	1.95	1.85	1.80	1.70

For Basement, add $15.70 per square foot of basement area

The above costs were calculated using the basic specifications shown on the facing page. These costs should be adjusted where necessary for design alternatives and owner's requirements. Reported completed project costs, for this type of structure, range from $63.80 to $163.80 per S.F.

Common additives

Description	Unit	$ Cost
Cells Prefabricated, 5'-6' wide, 7'-8' high, 7'-8' deep	Each	8075
Elevators, Hydraulic passenger, 2 stops		
1500# capacity	Each	40,800
2500# capacity	Each	42,100
3500# capacity	Each	45,700
Emergency Lighting, 25 watt, battery operated		
Lead battery	Each	350
Nickel cadmium	Each	590
Flagpoles, Complete		
Aluminum, 20' high	Each	1050
40' high	Each	2350
70' high	Each	6675
Fiberglass, 23' high	Each	1375
39'-5" high	Each	2400
59' high	Each	6575

Description	Unit	$ Cost
Lockers, Steel, Single tier, 60" to 72"	Opening	134 - 213
2 tier, 60" or 72" total	Opening	83 - 104
5 tier, box lockers	Opening	44 - 56
Locker bench, lam. maple top only	L.F.	18.40
Pedestals, steel pipe	Each	43
Safe, Office type, 4 hour rating		
30" x 18" x 18"	Each	3475
62" x 33" x 20"	Each	7575
Shooting Range, Incl. bullet traps, target provisions, and contols, not incl. structural shell	Each	21,300
Smoke Detectors		
Ceiling type	Each	141
Duct type	Each	335
Sound System		
Amplifier, 250 watts	Each	1500
Speaker, ceiling or wall	Each	130
Trumpet	Each	244

Location Factors appear in the Reference Section

Model costs calculated for a 2 story building with 12' story height and 11,000 square feet of floor area

Police Station

			Unit	Unit Cost	Cost Per S.F.	% Of Sub-Total
1.0 Foundations						
.1	Footings & Foundations	Poured concrete; strip and spread footings and 4' foundation wall	S.F. Ground	5.18	2.59	
.4	Piles & Caissons	N/A	—	—	—	3.3%
.9	Excavation & Backfill	Site preparation for slab and trench for foundation wall and footing	S.F. Ground	.96	.48	
2.0 Substructure						
.1	Slab on Grade	4" reinforced concrete with vapor barrier and granular base	S.F. Slab	3.05	1.52	1.6%
.2	Special Substructures	N/A	—	—	—	
3.0 Superstructure						
.1	Columns & Beams	N/A	—	—	—	
.4	Structural Walls	N/A	—	—	—	
.5	Elevated Floors	Open web steel joists, slab form, concrete	S.F. Floor	6.40	3.20	5.9%
.7	Roof	Metal deck on open web steel joists	S.F. Roof	2.60	1.30	
.9	Stairs	Concrete filled metal pan	Flight	5750	1.05	
4.0 Exterior Closure						
.1	Walls	Limestone with concrete block backup 80% of wall	S.F. Wall	30	16.21	
.5	Exterior Wall Finishes	N/A	—	—	—	24.9%
.6	Doors	Hollow metal	Each	1424	1.04	
.7	Windows & Glazed Walls	Metal horizontal sliding 20% of wall	Each	677	5.97	
5.0 Roofing						
.1	Roof Coverings	Built-up tar and gravel with flashing	S.F. Roof	2.52	1.26	
.7	Insulation	Perlite/EPS composite	S.F. Roof	1.18	.59	2.1%
.8	Openings & Specialties	Gravel stop	L.F. Perimeter	5.55	.15	
6.0 Interior Construction						
.1	Partitions	Concrete block, toilet partitions 20 S.F. Floor/L.F. Partition	S.F. Partition	5.05	2.95	
.4	Interior Doors	Single leaf kalamein fire door 200 S.F. Floor/Door	Each	530	2.65	
.5	Wall Finishes	90% paint, 10% ceramic tile	S.F. Surface	1.32	1.32	15.3%
.6	Floor Finishes	70% vinyl asbestos tile, 20% carpet, 10% ceramic tile	S.F. Floor	3.31	3.31	
.7	Ceiling Finishes	Mineral fiber tile on concealed zee bars	S.F. Ceiling	3.06	3.06	
.9	Interior Surface/Exterior Wall	Paint 80% of wall	S.F. Wall	1.93	1.02	
7.0 Conveying						
.1	Elevators	One hydraulic passenger elevator	Each	49,830	4.53	4.9%
.2	Special Conveyors	N/A	—	—	—	
8.0 Mechanical						
.1	Plumbing	Toilet and service fixtures, supply and drainage 1 Fixture/580 S.F. Floor	Each	2418	4.17	
.2	Fire Protection	Wet pipe sprinkler system	S.F. Floor	1.73	1.73	
.3	Heating	Oil fired hot water, wall fin radiation	S.F. Floor	6.66	6.66	21.5%
.4	Cooling	Split systems with air cooled condensing units	S.F. Floor	7.46	7.46	
.5	Special Systems	N/A	—	—	—	
9.0 Electrical						
.1	Service & Distribution	400 ampere service, panel board and feeders	S.F. Floor	1.10	1.10	
.2	Lighting & Power	Fluorescent fixtures, receptacles, switches, A.C. and misc. power	S.F. Floor	6.58	6.58	8.9%
.4	Special Electrical	Alarm systems and emergency lighting	S.F. Floor	.65	.65	
11.0 Special Construction						
.1	Specialties	Lockers, detention rooms, cells	S.F. Floor	10.80	10.80	11.6%
12.0 Site Work						
.1	Earthwork	N/A	—	—	—	
.3	Utilities	N/A	—	—	—	0.0%
.5	Roads & Parking	N/A	—	—	—	
.7	Site Improvements	N/A	—	—	—	
			Sub-Total		93.35	100%
	GENERAL CONDITIONS (Overhead & Profit)			15%	14.00	
	ARCHITECT FEES			9%	9.65	
			Total Building Cost		117.00	

BUILDING TYPES

COMMERCIAL/INDUSTRIAL/INSTITUTIONAL — M.500 Post Office

Costs per square foot of floor area

Exterior Wall	S.F. Area	5000	7000	9000	11000	13000	15000	17000	19000	21000
	L.F. Perimeter	300	380	420	486	468	513	540	580	620
Face Brick with Concrete Block Back-up	Steel Frame	88.10	83.60	78.80	77.00	72.25	71.10	69.55	68.75	68.10
	Bearing Walls	86.95	82.45	77.65	75.80	71.10	69.90	68.40	67.60	66.95
Limestone with Concrete Block Back-up	Steel Frame	96.00	90.75	84.90	82.80	77.00	75.55	73.75	72.75	72.00
	Bearing Walls	94.40	89.15	83.35	81.20	75.40	74.00	72.15	71.20	70.40
Decorative Concrete Block	Steel Frame	80.45	76.70	72.85	71.35	67.65	66.70	65.50	64.85	64.35
	Bearing Walls	79.30	75.55	71.70	70.20	66.50	65.55	64.35	63.70	63.20
Perimeter Adj., Add or Deduct	Per 100 L.F.	10.60	7.60	5.90	4.80	4.05	3.55	3.10	2.80	2.55
Story Hgt. Adj., Add or Deduct	Per 1 Ft.	1.70	1.55	1.30	1.25	1.00	.95	.90	.85	.85

For Basement, add $16.60 per square foot of basement area

The above costs were calculated using the basic specifications shown on the facing page. These costs should be adjusted where necessary for design alternatives and owner's requirements. Reported completed project costs, for this type of structure, range from $52.15 to $134.60 per S.F.

Common additives

Description	Unit	$ Cost
Closed Circuit Surveillance, One station		
Camera and monitor	Each	1325
For additional camera stations, add	Each	730
Emergency Lighting, 25 watt, battery operated		
Lead battery	Each	350
Nickel cadmium	Each	590
Flagpoles, Complete		
Aluminum, 20' high	Each	1050
40' high	Each	2350
70' high	Each	6675
Fiberglass, 23' high	Each	1375
39'-5" high	Each	2400
59' high	Each	6575

Description	Unit	$ Cost
Mail Boxes, Horizontal, key lock, 15" x 6" x 5"	Each	45
Double 15" x 12" x 5"	Each	77
Quadruple 15" x 12" x 10"	Each	140
Vertical, 6" x 5" x 15", aluminum	Each	37
Bronze	Each	61
Steel, enameled	Each	37
Scales, Dial type, 5 ton cap.		
8' x 6' platform	Each	6850
9' x 7' platform	Each	10,600
Smoke Detectors		
Ceiling type	Each	141
Duct type	Each	335

Post Office

Model costs calculated for a 1 story building with 14' story height and 13,000 square feet of floor area

				Unit	Unit Cost	Cost Per S.F.	% Of Sub-Total
1.0 Foundations							
.1	Footings & Foundations	Poured concrete; strip and spread footings and 4' foundation wall		S.F. Ground	3.38	3.38	
.4	Piles & Caissons	N/A		—	—	—	7.5%
.9	Excavation & Backfill	Site preparation for slab and trench for foundation wall and footing		S.F. Ground	.96	.96	
2.0 Substructure							
.1	Slab on Grade	4" reinforced concrete with vapor barrier and granular base		S.F. Slab	3.05	3.05	5.3%
.2	Special Substructures	N/A		—	—	—	
3.0 Superstructure							
.1	Columns & Beams	Fireproofing, steel columns included in 3.7		L.F. Column	20	.06	
.4	Structural Walls	N/A		—	—	—	
.5	Elevated Floors	N/A		—	—	—	8.5%
.7	Roof	Metal deck, open web steel joists, columns		S.F. Roof	4.84	4.84	
.9	Stairs	N/A		—	—	—	
4.0 Exterior Closure							
.1	Walls	Face brick with concrete block backup	80% of wall	S.F. Wall	21	8.59	
.5	Exterior Wall Finishes	N/A		—	—	—	19.7%
.6	Doors	Double aluminum & glass, single aluminum, hollow metal, steel overhead		Each	1638	.75	
.7	Windows & Glazed Walls	Double strength window glass	20% of wall	Each	458	2.01	
5.0 Roofing							
.1	Roof Coverings	Built-up tar and gravel with flashing		S.F. Roof	2.18	2.18	
.7	Insulation	Perlite/EPS composite		S.F. Roof	1.18	1.18	6.2%
.8	Openings & Specialties	Gravel stop		L.F. Perimeter	5.55	.20	
6.0 Interior Construction							
.1	Partitions	Concrete block, toilet partitions	15 S.F. Floor/L.F. Partition	S.F. Partition	5.05	4.40	
.4	Interior Doors	Single leaf hollow metal	150 S.F. Floor/Door	Each	530	3.53	
.5	Wall Finishes	Paint		S.F. Surface	.90	1.44	
.6	Floor Finishes	50% vinyl tile, 50% paint		S.F. Floor	1.74	1.74	22.0%
.7	Ceiling Finishes	Mineral fiber tile on concealed zee bars	25% of area	S.F. Ceiling	3.06	.77	
.9	Interior Surface/Exterior Wall	Paint	80% of wall	S.F. Wall	1.93	.78	
7.0 Conveying							
.1	Elevators	N/A		—	—	—	0.0%
.2	Special Conveyors	N/A		—	—	—	
8.0 Mechanical							
.1	Plumbing	Toilet and service fixtures, supply and drainage	1 Fixture/1180 S.F. Floor	Each	1899	1.61	
.2	Fire Protection	Wet pipe sprinkler system		S.F. Floor	1.57	1.57	
.3	Heating	Included in 8.4		—	—	—	18.4%
.4	Cooling	Single zone, gas heating, electric cooling		S.F. Floor	7.46	7.46	
.5	Special Systems	N/A		—	—	—	
9.0 Electrical							
.1	Service & Distribution	400 ampere service, panel board and feeders		S.F. Floor	.93	.93	
.2	Lighting & Power	Fluorescent fixtures, receptacles, switches, A.C. and misc. power		S.F. Floor	5.32	5.32	11.5%
.4	Special Electrical	Alarm systems and emergency lighting		S.F. Floor	.38	.38	
11.0 Special Construction							
.1	Specialties	Cabinets, lockers, shelving		S.F. Floor	.52	.52	0.9%
12.0 Site Work							
.1	Earthwork	N/A		—	—	—	
.3	Utilities	N/A		—	—	—	0.0%
.5	Roads & Parking	N/A		—	—	—	
.7	Site Improvements	N/A		—	—	—	
				Sub-Total		57.65	100%
	GENERAL CONDITIONS (Overhead & Profit)				15%	8.65	
	ARCHITECT FEES				9%	5.95	
				Total Building Cost		**72.25**	

BUILDING TYPES

COMMERCIAL/INDUSTRIAL/INSTITUTIONAL   M.510   Racquetball Court

Costs per square foot of floor area

Exterior Wall	S.F. Area	5000	10000	15000	21000	25000	30000	40000	50000	60000
	L.F. Perimeter	287	400	500	600	700	700	834	900	1000
Face Brick with Concrete Block Back-up	Steel Frame	137.35	115.05	106.90	101.35	100.35	95.70	92.90	90.00	88.60
	Bearing Walls	133.10	111.95	104.15	98.85	97.90	93.35	90.65	87.80	86.40
Concrete Block	Steel Frame	115.75	99.75	94.05	90.25	89.40	86.60	84.75	82.95	82.05
Brick Veneer	Steel Frame	130.95	109.55	101.75	96.50	95.50	91.20	88.60	85.90	84.60
Galvanized Steel Siding	Steel Frame	109.05	94.95	90.00	86.70	85.95	83.60	82.05	80.55	79.75
Metal Sandwich Panel	Steel Frame	122.70	104.00	97.25	92.75	91.80	88.25	86.00	83.80	82.65
Perimeter Adj., Add or Deduct	Per 100 L.F.	18.05	9.05	6.00	4.30	3.60	3.00	2.25	1.80	1.55
Story Hgt. Adj., Add or Deduct	Per 1 Ft.	3.75	2.60	2.20	1.85	1.85	1.50	1.35	1.20	1.10

For Basement, add $15.75 per square foot of basement area

The above costs were calculated using the basic specifications shown on the facing page. These costs should be adjusted where necessary for design alternatives and owner's requirements. Reported completed project costs, for this type of structure, range from $50.10 to $126.35 per S.F.

Common additives

Description	Unit	$ Cost
Bar, Front Bar	L.F.	278
Back Bar	L.F.	220
Booth, Upholstered, custom straight	L.F.	127 - 207
"L" or "U" shaped	L.F.	143 - 225
Bleachers, Telescoping, manual		
To 15 tier	Seat	65 - 76
21-30 tier	Seat	88 - 104
Courts		
Ceiling	Court	5200
Floor	Court	9225
Walls	Court	18,300
Emergency Lighting, 25 watt, battery operated		
Lead battery	Each	350
Nickel cadmium	Each	590
Kitchen Equipment		
Broiler	Each	3300
Cooler, 6 ft. long, reach-in	Each	2600
Dishwasher, 10-12 racks per hr.	Each	2500
Food warmer, counter 1.2 KW	Each	715
Freezer, reach-in, 44 C.F.	Each	6375
Ice cube maker, 50 lb. per day	Each	1675

Description	Unit	$ Cost
Lockers, Steel, single tier, 60" or 72"	Opening	134 - 213
2 tier, 60" or 72" total	Opening	83 - 104
5 tier, box lockers	Opening	44 - 56
Locker bench, lam. maple top only	L.F.	18.40
Pedestals, steel pipe	Each	43
Sauna, Prefabricated, complete		
6' x 4'	Each	3725
6' x 9'	Each	5400
8' x 8'	Each	5725
8' x 10'	Each	6800
10' x 12'	Each	8025
Sound System		
Amplifier, 250 watts	Each	1500
Speaker, ceiling or wall	Each	130
Trumpet	Each	244
Steam Bath, Complete, to 140 C.F.	Each	1175
To 300 C.F.	Each	1325
To 800 C.F.	Each	3375
To 2500 C.F.	Each	5525

Location Factors appear in the Reference Section

Model costs calculated for a 2 story building with 12' story height and 30,000 square feet of floor area

Racquetball Court

			Unit	Unit Cost	Cost Per S.F.	% Of Sub-Total
1.0 Foundations						
.1	Footings & Foundations	Poured concrete; strip and spread footings and 4' foundation wall	S.F. Ground	4.12	2.06	3.5%
.4	Piles & Caissons	N/A	—	—	—	
.9	Excavation & Backfill	Site preparation for slab and trench for foundation wall and footing	S.F. Ground	.96	.62	
2.0 Substructure						
.1	Slab on Grade	6" reinforced concrete with vapor barrier and granular base	S.F. Slab	6.28	4.08	5.3%
.2	Special Substructures	N/A				
3.0 Superstructure						
.1	Columns & Beams	Steel columns included in 3.5 and 3.7	—	—	—	
.4	Structural Walls	N/A	—	—	—	
.5	Elevated Floors	Open web steel joists, slab form, concrete, columns 50% of area	S.F. Floor	12.71	4.45	9.3%
.7	Roof	Metal deck on open web steel joists, columns	S.F. Roof	4.47	2.24	
.9	Stairs	Concrete filled metal pan	Flight	4925	.49	
4.0 Exterior Closure						
.1	Walls	Face brick with concrete block backup 95% of wall	S.F. Wall	21	11.41	
.5	Exterior Wall Finishes	N/A	—	—	—	17.4%
.6	Doors	Aluminum and glass and hollow metal	Each	1868	.25	
.7	Windows & Glazed Walls	Storefront 5% of wall	S.F. Window	62	1.74	
5.0 Roofing						
.1	Roof Coverings	Built-up tar and gravel with flashing	S.F. Roof	2.68	1.34	
.7	Insulation	Perlite/EPS composite	S.F. Roof	1.18	.77	3.0%
.8	Openings & Specialties	Gravel stop and hatches	S.F. Roof	.36	.18	
6.0 Interior Construction						
.1	Partitions	Concrete block, gypsum board on metal studs 25 S.F. Floor/L.F. Partition	S.F. Partition	5.05	2.13	
.4	Interior Doors	Single leaf hollow metal 810 S.F. Floor/Door	Each	530	.65	
.5	Wall Finishes	Paint	S.F. Surface	.85	.68	12.1%
.6	Floor Finishes	80% carpet, 20% ceramic tile 50% of floor area	S.F. Floor	4.88	2.44	
.7	Ceiling Finishes	Mineral fiber tile on concealed zee bars 60% of area	S.F. Ceiling	3.06	1.84	
.9	Interior Surface/Exterior Wall	Painted gypsum board on furring 95% of wall	S.F. Wall	2.81	1.57	
7.0 Conveying						
.1	Elevators	N/A	—	—	—	0.0%
.2	Special Conveyors	N/A	—	—	—	
8.0 Mechanical						
.1	Plumbing	Kitchen, bathroom and service fixtures, supply and drainage 1 Fixture/1000 S.F. Floor	Each	3160	3.16	
.2	Fire Protection	Sprinklers, light hazard	S.F. Floor	.14	.14	
.3	Heating	Included in 8.4	—	—	—	29.2%
.4	Cooling	Multizone unit, gas heating, electric cooling	S.F. Floor	19.25	19.25	
.5	Special Systems	N/A	—	—	—	
9.0 Electrical						
.1	Service & Distribution	400 ampere service, panel board and feeders	S.F. Floor	.49	.49	
.2	Lighting & Power	Fluorescent and high intensity discharge fixtures, receptacles, switches, A.C. and misc. power	S.F. Floor	4.01	4.01	6.2%
.4	Special Electrical	Alarm systems and emergency lighting	S.F. Floor	.24	.24	
11.0 Special Construction						
.1	Specialties	Courts, sauna baths	S.F. Floor	10.81	10.81	14.0%
12.0 Site Work						
.1	Earthwork	N/A	—	—	—	
.3	Utilities	N/A	—	—	—	0.0%
.5	Roads & Parking	N/A	—	—	—	
.7	Site Improvements	N/A	—	—	—	
			Sub-Total		77.04	100%
	GENERAL CONDITIONS (Overhead & Profit)			15%	11.56	
	ARCHITECT FEES			8%	7.10	
			Total Building Cost		95.70	

BUILDING TYPES

COMMERCIAL/INDUSTRIAL/INSTITUTIONAL — M.520 Religious Education

Costs per square foot of floor area

Exterior Wall	S.F. Area	5000	6000	7000	8000	9000	10000	11000	12000	13000
	L.F. Perimeter	286	320	353	386	397	425	454	460	486
Face Brick with Concrete Block Back-up	Steel Joists	102.60	100.10	98.25	96.90	94.60	93.65	92.90	91.30	90.75
	Wood Joists	101.55	98.95	97.00	95.60	93.25	92.25	91.45	89.80	89.25
Stucco on Concrete Block	Steel Joists	96.60	94.50	92.95	91.80	90.00	89.20	88.55	87.30	86.85
	Wood Joists	93.10	91.05	89.60	88.50	86.75	86.00	85.35	84.15	83.70
Limestone with Concrete Block Back-up	Steel Joists	111.90	108.75	106.45	104.70	101.75	100.55	99.55	97.50	96.80
	Wood Joists	108.35	105.30	103.05	101.35	98.50	97.35	96.35	94.35	93.70
Perimeter Adj., Add or Deduct	Per 100 L.F.	9.75	8.15	7.00	6.10	5.45	4.90	4.45	4.10	3.75
Story Hgt. Adj., Add or Deduct	Per 1 Ft.	1.70	1.60	1.55	1.50	1.35	1.30	1.25	1.15	1.15

For Basement, add $17.05 per square foot of basement area

The above costs were calculated using the basic specifications shown on the facing page. These costs should be adjusted where necessary for design alternatives and owner's requirements. Reported completed project costs, for this type of structure, range from $43.40 to $99.00 per S.F.

Common additives

Description	Unit	$ Cost
Carrels Hardwood	Each	655 - 850
Emergency Lighting, 25 watt, battery operated		
Lead battery	Each	350
Nickel cadmium	Each	590
Flagpoles, Complete		
Aluminum, 20' high	Each	1050
40' high	Each	2350
70' high	Each	6675
Fiberglass, 23' high	Each	1375
39'-5" high	Each	2400
59' high	Each	6575
Gym Floor, Incl. sleepers and finish, maple	S.F.	8.80
Intercom System, 25 Station capacity		
Master station	Each	1800
Intercom outlets	Each	113
Handset	Each	299

Description	Unit	$ Cost
Lockers, Steel, single tier, 60" to 72"	Opening	134 - 213
2 tier, 60" to 72" total	Opening	83 - 104
5 tier, box lockers	Opening	44 - 56
Locker bench, lam. maple top only	L.F.	18.40
Pedestals, steel pipe	Each	43
Seating		
Auditorium chair, all veneer	Each	127
Veneer back, padded seat	Each	152
Upholstered, spring seat	Each	178
Classroom, movable chair & desk	Set	65 - 120
Lecture hall, pedestal type	Each	127 - 370
Smoke Detectors		
Ceiling type	Each	141
Duct type	Each	335
Sound System		
Amplifier, 250 watts	Each	1500
Speaker, ceiling or wall	Each	130
Trumpet	Each	244
Swimming Pools, Complete, gunite	S.F.	45 - 54

Model costs calculated for a 1 story building with 12' story height and 10,000 square feet of floor area

Religious Education

				Unit	Unit Cost	Cost Per S.F.	% Of Sub-Total
1.0 Foundations							
.1	Footings & Foundations	Poured concrete; strip and spread footings and 4' foundation wall		S.F. Ground	3.49	3.49	
.4	Piles & Caissons	N/A		—	—	—	6.0%
.9	Excavation & Backfill	Site preparation for slab and trench for foundation wall and footing		S.F. Ground	.96	.96	
2.0 Substructure							
.1	Slab on Grade	4" reinforced concrete with vapor barrier and granular base		S.F. Slab	3.05	3.05	4.1%
.2	Special Substructures	N/A		—	—	—	
3.0 Superstructure							
.1	Columns & Beams	Interior columns included in 3.7		—	—	—	
.4	Structural Walls	N/A		—	—	—	
.5	Elevated Floors	N/A		—	—	—	4.9%
.7	Roof	Metal deck, open web steel joists, beams, interior columns		S.F. Roof	3.64	3.64	
.9	Stairs	N/A		—	—	—	
4.0 Exterior Closure							
.1	Walls	Face brick with concrete block backup	85% of wall	S.F. Wall	21	9.23	
.5	Exterior Wall Finishes	N/A		—	—	—	17.5%
.6	Doors	Double aluminum and glass		Each	3962	1.59	
.7	Windows & Glazed Walls	Window wall	15% of wall	S.F. Window	29	2.26	
5.0 Roofing							
.1	Roof Coverings	Built-up tar and gravel with flashing		S.F. Roof	2.16	2.16	
.7	Insulation	Perlite/EPS composite		S.F. Roof	1.18	1.18	4.8%
.8	Openings & Specialties	Gravel stop		L.F. Perimeter	5.55	.24	
6.0 Interior Construction							
.1	Partitions	Concrete block, toilet partitions	8 S.F. Floor/S.F. Partition	S.F. Partition	5.05	7.24	
.4	Interior Doors	Single leaf hollow metal	700 S.F. Floor/Door	Each	530	.76	
.5	Wall Finishes	Paint		S.F. Surface	1.08	2.70	
.6	Floor Finishes	50% vinyl tile, 50% carpet		S.F. Floor	3.88	3.88	24.7%
.7	Ceiling Finishes	Mineral fiber tile on concealed zee bars		S.F. Ceiling	3.06	3.06	
.9	Interior Surface/Exterior Wall	Paint	85% of wall	S.F. Wall	1.93	.84	
7.0 Conveying							
.1	Elevators	N/A		—	—	—	0.0%
.2	Special Conveyors	N/A		—	—	—	
8.0 Mechanical							
.1	Plumbing	Toilet and service fixtures, supply and drainage	1 Fixture/455 S.F. Floor	Each	1797	3.95	
.2	Fire Protection	Sprinkler, light hazard		S.F. Floor	1.57	1.57	
.3	Heating	Oil fired hot water, wall fin radiation		S.F. Floor	6.66	6.66	29.2%
.4	Cooling	Split systems with air cooled condensing units		S.F. Floor	9.65	9.65	
.5	Special Systems	N/A		—	—	—	
9.0 Electrical							
.1	Service & Distribution	200 ampere service, panel board and feeders		S.F. Floor	.60	.60	
.2	Lighting & Power	Fluorescent fixtures, receptacles, switches, A.C. and misc. power		S.F. Floor	5.66	5.66	8.8%
.4	Special Electrical	Alarm systems and emergency lighting		S.F. Floor	.34	.34	
11.0 Special Construction							
.1	Specialties	N/A		—	—	—	0.0%
12.0 Site Work							
.1	Earthwork	N/A		—	—	—	
.3	Utilities	N/A		—	—	—	0.0%
.5	Roads & Parking	N/A		—	—	—	
.7	Site Improvements	N/A		—	—	—	
				Sub-Total		74.71	100%
	GENERAL CONDITIONS (Overhead & Profit)				15%	11.21	
	ARCHITECT FEES				9%	7.73	
				Total Building Cost		93.65	

BUILDING TYPES

COMMERCIAL/INDUSTRIAL/INSTITUTIONAL M.530 Restaurant

Costs per square foot of floor area

Exterior Wall	S.F. Area	2000	2800	3500	4200	5000	5800	6500	7200	8000
	L.F. Perimeter	180	212	240	268	300	314	336	344	368
Wood Siding	Wood Frame	127.85	118.70	114.15	111.10	108.65	105.95	104.60	102.90	101.85
Brick Veneer	Wood Frame	137.20	126.50	121.20	117.70	114.85	111.50	109.90	107.75	106.55
Face Brick with Concrete Block Back-up	Wood Joists	141.00	129.75	124.15	120.45	117.45	113.85	112.15	109.85	108.60
	Steel Joists	143.40	131.10	125.00	120.95	117.70	113.70	111.80	109.25	107.85
Stucco on Concrete Block	Wood Joists	130.30	120.75	116.00	112.85	110.30	107.40	106.00	104.20	103.15
	Steel Joists	126.25	116.70	111.95	108.80	106.25	103.35	101.95	100.15	99.10
Perimeter Adj., Add or Deduct	Per 100 L.F.	15.20	10.85	8.70	7.20	6.05	5.25	4.70	4.20	3.80
Story Hgt. Adj., Add or Deduct	Per 1 Ft.	1.60	1.35	1.25	1.15	1.10	1.00	.95	.90	.85

For Basement, add $19.35 per square foot of basement area

The above costs were calculated using the basic specifications shown on the facing page. These costs should be adjusted where necessary for design alternatives and owner's requirements. Reported completed project costs, for this type of structure, range from $65.90 to $153.95 per S.F.

Common additives

Description	Unit	$ Cost
Bar, Front Bar	L.F.	278
Back bar	L.F.	220
Booth, Upholstered, custom straight	L.F.	127 - 207
"L" or "U" shaped	L.F.	143 - 225
Cupola, Stock unit, redwood		
30" square, 37" high, aluminum roof	Each	455
Copper roof	Each	231
Fiberglass, 5'-0" base, 63" high	Each	2750 - 3350
6'-0" base, 63" high	Each	4150 - 4550
Decorative Wood Beams, Non load bearing		
Rough sawn, 4" x 6"	L.F.	7.40
4" x 8"	L.F.	8.80
4" x 10"	L.F.	10.25
4" x 12"	L.F.	11.75
8" x 8"	L.F.	14.40
Emergency Lighting, 25 watt, battery operated		
Lead battery	Each	350
Nickel cadmium	Each	590

Description	Unit	$ Cost
Fireplace, Brick, not incl. chimney or foundation		
30" x 29" opening	Each	1850
Chimney, standard brick		
Single flue, 16" x 20"	V.L.F.	55
20" x 20"	V.L.F.	63
2 Flue, 20" x 24"	V.L.F.	76
20" x 32"	V.L.F.	92
Kitchen Equipment		
Broiler	Each	3300
Coffee urn, twin 6 gallon	Each	5500
Cooler, 6 ft. long	Each	2600
Dishwasher, 10-12 racks per hr.	Each	2500
Food warmer, counter, 1.2 KW	Each	715
Freezer, 44 C.F., reach-in	Each	6375
Ice cube maker, 50 lb. per day	Each	1675
Range with 1 oven	Each	2250
Refrigerators, Prefabricated, walk-in		
7'-6" high, 6' x 6'	S.F.	117
10' x 10'	S.F.	91
12' x 14'	S.F.	81
12' x 20'	S.F.	72

Location Factors appear in the Reference Section

Restaurant

Model costs calculated for a 1 story building with 12' story height and 5,000 square feet of floor area

				Unit	Unit Cost	Cost Per S.F.	% Of Sub-Total
1.0 Foundations							
.1	Footings & Foundations	Poured concrete; strip and spread footings and 4' foundation wall		S.F. Ground	4.11	4.11	
.4	Piles & Caissons	N/A		—	—	—	5.9%
.9	Excavation & Backfill	Site preparation for slab and trench for foundation wall and footing		S.F. Ground	1.09	1.09	
2.0 Substructure							
.1	Slab on Grade	4" reinforced concrete with vapor barrier and granular base		S.F. Slab	3.05	3.05	3.5%
.2	Special Substructures	N/A		—	—	—	
3.0 Superstructure							
.1	Columns & Beams	Wood columns		S.F. Ground	.19	.19	
.4	Structural Walls	N/A		—	—	—	
.5	Elevated Floors	N/A		—	—	—	2.7%
.7	Roof	Plywood on wood rafters (pitched)		S.F. Roof	1.99	2.22	
.9	Stairs	N/A		—	—	—	
4.0 Exterior Closure							
.1	Walls	Cedar siding on wood studs with insulation	70% of wall	S.F. Wall	7.14	3.60	
.5	Exterior Wall Finishes	N/A		—	—	—	13.9%
.6	Doors	Aluminum and glass doors and entrance with transom		Each	3188	3.19	
.7	Windows & Glazed Walls	Storefront windows	30% of wall	S.F. Window	25	5.48	
5.0 Roofing							
.1	Roof Coverings	Cedar shingles with flashing (pitched)		S.F. Roof	3.13	3.50	
.7	Insulation	Fiberglass sheets		S.F. Roof	.86	.96	5.6%
.8	Openings & Specialties	Gutters and downspouts and skylight		S.F. Roof	.47	.47	
6.0 Interior Construction							
.1	Partitions	Gypsum board on wood studs, toilet partition	25 S.F. Floor/L.F. Partition	S.F. Partition	722	1.80	
.4	Interior Doors	Hollow core wood	250 S.F. Floor/Door	Each	345	1.38	
.5	Wall Finishes	75% paint, 25% ceramic tile		S.F. Surface	1.64	1.31	17.4%
.6	Floor Finishes	65% carpet, 35% quarry tile		S.F. Floor	6.38	6.38	
.7	Ceiling Finishes	Mineral fiber tile on concealed zee bars		S.F. Ceiling	3.06	3.06	
.9	Interior Surface/Exterior Wall	Painted gypsum board on furring	70% of wall	S.F. Wall	2.80	1.41	
7.0 Conveying							
.1	Elevators	N/A		—	—	—	0.0%
.2	Special Conveyors	N/A		—	—	—	
8.0 Mechanical							
.1	Plumbing	Kitchen, bathroom and service fixtures, supply and drainage	1 Fixture/355 S.F. Floor	Each	2815	7.93	
.2	Fire Protection	Sprinklers, light hazard		S.F. Floor	1.57	1.57	
.3	Heating	Included in 8.4		—	—	—	40.5%
.4	Cooling	Multizone unit, gas heating, electric cooling		S.F. Floor	26	26.30	
.5	Special Systems	N/A		—	—	—	
9.0 Electrical							
.1	Service & Distribution	400 ampere service, panel board and feeders		S.F. Floor	2.54	2.54	
.2	Lighting & Power	Fluorescent fixtures, receptacles, switches, A.C. and misc. power		S.F. Floor	6.09	6.09	10.5%
.4	Special Electrical	Alarm systems and emergency lighting		S.F. Floor	.68	.68	
11.0 Special Construction							
.1	Specialties	N/A		—	—	—	0.0%
12.0 Site Work							
.1	Earthwork	N/A		—	—	—	
.3	Utilities	N/A		—	—	—	0.0%
.5	Roads & Parking	N/A		—	—	—	
.7	Site Improvements	N/A		—	—	—	
				Sub-Total		88.31	100%
	GENERAL CONDITIONS (Overhead & Profit)				15%	13.25	
	ARCHITECT FEES				7%	7.09	
				Total Building Cost		108.65	

BUILDING TYPES

COMMERCIAL/INDUSTRIAL/INSTITUTIONAL — M.540 — Restaurant, Fast Food

Costs per square foot of floor area

Exterior Wall	S.F. Area	2000	2800	3500	4000	5000	5800	6500	7200	8000
	L.F. Perimeter	180	212	240	260	300	314	336	344	368
Face Brick with Concrete Block Back-up	Bearing Walls	117.95	109.75	105.65	103.60	100.75	97.80	96.55	94.65	93.65
	Steel Frame	121.05	112.85	108.70	106.65	103.85	100.90	99.60	97.70	96.75
Concrete Block With Stucco	Bearing Walls	109.05	102.25	98.85	97.15	94.80	92.45	91.40	89.90	89.10
	Steel Frame	111.55	104.85	101.50	99.80	97.50	95.15	94.15	92.65	91.90
Wood Siding	Wood Frame	108.85	102.35	99.10	97.50	95.20	93.00	92.00	90.60	89.80
Brick Veneer	Steel Frame	117.65	110.00	106.15	104.25	101.55	98.85	97.70	95.90	95.00
Perimeter Adj., Add or Deduct	Per 100 L.F.	21.10	15.05	12.10	10.60	8.40	7.25	6.50	5.85	5.30
Story Hgt. Adj., Add or Deduct	Per 1 Ft.	2.80	2.35	2.15	2.05	1.85	1.70	1.60	1.50	1.45
Basement—Not Applicable										

The above costs were calculated using the basic specifications shown on the facing page. These costs should be adjusted where necessary for design alternatives and owner's requirements. Reported completed project costs, for this type of structure, range from $65.80 to $128.15 per S.F.

Common additives

Description	Unit	$ Cost
Bar, Front Bar	L.F.	278
Back bar	L.F.	220
Booth, Upholstered, custom straight	L.F.	127 - 207
"L" or "U" shaped	L.F.	143 - 225
Drive-up Window	Each	5950 - 9100
Emergency Lighting, 25 watt, battery operated		
Lead battery	Each	350
Nickel cadmium	Each	590
Kitchen Equipment		
Broiler	Each	3300
Coffee urn, twin 6 gallon	Each	5500
Cooler, 6 ft. long	Each	2600
Dishwasher, 10-12 racks per hr.	Each	2500
Food warmer, counter, 1.2 KW	Each	715
Freezer, 44 C.F., reach-in	Each	6375
Ice cube maker, 50 lb. per day	Each	1675
Range with 1 oven	Each	2250

Description	Unit	$ Cost
Refrigerators, Prefabricated, walk-in		
7'-6" High, 6' x 6'	S.F.	117
10' x 10'	S.F.	91
12' x 14'	S.F.	81
12' x 20'	S.F.	72
Serving		
Counter top (Stainless steel)	L.F.	101
Base cabinets	L.F.	244 - 289
Sound System		
Amplifier, 250 watts	Each	1500
Speaker, ceiling or wall	Each	130
Trumpet	Each	244
Storage		
Shelving	S.F.	11.70
Washing		
Stainless steel counter	L.F.	101

184 — Location Factors appear in the Reference Section

Model costs calculated for a 1 story building with 10' story height and 4,000 square feet of floor area

Restaurant, Fast Food

				Unit	Unit Cost	Cost Per S.F.	% Of Sub-Total
1.0	**Foundations**						
.1	Footings & Foundations	Poured concrete; strip and spread footings and 4' foundation wall		S.F. Ground	4.58	4.58	
.4	Piles & Caissons	N/A		—	—	—	6.8%
.9	Excavation & Backfill	Site preparation for slab and trench for foundation wall and footing		S.F. Ground	1.09	1.09	
2.0	**Substructure**						
.1	Slab on Grade	4" reinforced concrete with vapor barrier and granular base		S.F. Slab	3.05	3.05	3.7%
.2	Special Substructures	N/A		—	—	—	
3.0	**Superstructure**						
.1	Columns & Beams	N/A		—	—	—	
.4	Structural Walls	N/A		—	—	—	
.5	Elevated Floors	N/A		—	—	—	3.6%
.7	Roof	Metal deck on open web steel joists		S.F. Roof	3.03	3.03	
.9	Stairs	N/A		—	—	—	
4.0	**Exterior Closure**						
.1	Walls	Face brick with concrete block backup	70% of wall	S.F. Wall	17.98	9.69	
.5	Exterior Wall Finishes	N/A		—	—	—	24.5%
.6	Doors	Aluminum and glass		Each	2483	4.97	
.7	Windows & Glazed Walls	Window wall	30% of wall	S.F Window	29	5.74	
5.0	**Roofing**						
.1	Roof Coverings	Built-up tar and gravel with flashing		S.F. Roof	2.72	2.72	
.7	Insulation	Perlite/EPS composite		S.F. Roof	1.18	1.18	5.0%
.8	Openings & Specialties	Gravel stop and hatches		S.F. Roof	.27	.27	
6.0	**Interior Construction**						
.1	Partitions	Gypsum board on metal studs	25 S.F. Floor/L.F. Partition	S.F. Partition	3.83	1.38	
.4	Interior Doors	Hollow core wood	1000 S.F. Floor/Door	Each	345	.35	
.5	Wall Finishes	Paint		S.F. Surface	.50	.36	17.2%
.6	Floor Finishes	Quarry tile		S.F. Floor	8.33	8.33	
.7	Ceiling Finishes	Mineral fiber tile on concealed zee bars		S.F. Floor	3.06	3.06	
.9	Interior Surface/Exterior Wall	Paint	70% of wall	S.F. Wall	1.93	.88	
7.0	**Conveying**						
.1	Elevators	N/A		—	—	—	0.0%
.2	Special Conveyors	N/A		—	—	—	
8.0	**Mechanical**						
.1	Plumbing	Kitchen, bathroom and service fixtures, supply and drainage	1 Fixture/400 S.F. Floor	Each	2568	6.42	
.2	Fire Protection	Sprinklers, light hazard		S.F. Floor	2.04	2.04	
.3	Heating	Included in 8.4		—	—	—	22.2%
.4	Cooling	Multizone unit, gas heating, electric cooling		S.F. Floor	10.08	10.08	
.5	Special Systems	N/A		—	—	—	
9.0	**Electrical**						
.1	Service & Distribution	400 ampere service, panel board and feeders		S.F. Floor	2.65	2.65	
.2	Lighting & Power	Fluorescent fixtures, receptacles, switches, A.C. and misc. power		S.F. Floor	6.63	6.63	11.8%
.4	Special Electrical	Alarm systems and emergency lighting		S.F. Floor	.58	.58	
11.0	**Special Construction**						
.1	Specialties	Walk-in refrigerator		Each	17,280	4.32	5.2%
12.0	**Site Work**						
.1	Earthwork	N/A		—	—	—	
.3	Utilities	N/A		—	—	—	0.0%
.5	Roads & Parking	N/A		—	—	—	
.7	Site Improvements	N/A		—	—	—	
				Sub-Total		83.40	100%
	GENERAL CONDITIONS (Overhead & Profit)				15%	12.51	
	ARCHITECT FEES				8%	7.69	
				Total Building Cost		103.60	

BUILDING TYPES

COMMERCIAL/INDUSTRIAL/INSTITUTIONAL — M.550 Rink, Hockey/Indoor Soccer

Costs per square foot of floor area

Exterior Wall	S.F. Area	10000	15000	20000	25000	30000	35000	40000	45000	50000
	L.F. Perimeter	450	500	600	700	740	822	890	920	966
Face Brick with Concrete Block Back-up	Steel Frame	121.50	110.30	106.70	104.60	101.50	100.30	99.10	97.50	96.45
	Lam. Wood Truss	115.30	104.55	101.10	99.05	96.10	94.95	93.80	92.25	91.25
Concrete Block	Steel Frame	98.20	92.00	89.90	88.60	87.00	86.25	85.60	84.75	84.25
	Lam. Wood Truss	99.20	93.00	90.85	89.60	87.95	87.25	86.60	85.75	85.20
Galvanized Steel Siding	Steel Frame	90.20	85.35	83.55	82.55	81.30	80.70	80.20	79.55	79.15
Metal Sandwich Panel	Steel Joists	101.75	93.90	91.30	89.75	87.60	86.75	85.90	84.80	84.10
Perimeter Adj., Add or Deduct	Per 100 L.F.	8.20	5.45	4.10	3.25	2.75	2.35	2.05	1.85	1.65
Story Hgt. Adj., Add or Deduct	Per 1 Ft.	1.30	.95	.90	.80	.70	.70	.65	.60	.55

Basement—Not Applicable

The above costs were calculated using the basic specifications shown on the facing page. These costs should be adjusted where necessary for design alternatives and owner's requirements. Reported completed project costs, for this type of structure, range from $40.35 to $118.20 per S.F.

Common additives

Description	Unit	$ Cost
Bar, Front Bar	L.F.	278
Back bar	L.F.	220
Booth, Upholstered, custom straight	L.F.	127 - 207
"L" or "U" shaped	L.F.	143 - 225
Bleachers, Telescoping, manual		
To 15 tier	Seat	65 - 76
16-20 tier	Seat	72 - 83
21-30 tier	Seat	88 - 104
For power operation, add	Seat	13.40 - 19.40
Emergency Lighting, 25 watt, battery operated		
Lead battery	Each	350
Nickel cadmium	Each	590
Lockers, Steel, single tier, 60" or 72"	Opening	134 - 213
2 tier, 60" or 72" total	Opening	83 - 104
5 tier, box lockers	Opening	44 - 56
Locker bench, lam. maple top only	L.F.	18.40
Pedestals, steel pipe	Each	43

Description	Unit	$ Cost
Rink		
Dasher boards & top guard	Each	110,500
Mats, rubber	S.F.	12.90
Score Board	Each	12,900 - 34,300

Model costs calculated for a 1 story building with 24' story height and 30,000 square feet of floor area

Rink, Hockey/Indoor Soccer

				Unit	Unit Cost	Cost Per S.F.	% Of Sub-Total
1.0 Foundations							
.1	Footings & Foundations	Poured concrete; strip and spread footings and 4' foundation wall		S.F. Ground	2.60	2.60	4.9%
.4	Piles & Caissons	N/A		—	—	—	
.9	Excavation & Backfill	Site preparation for slab and trench for foundation wall and footing		S.F. Ground	.89	.89	
2.0 Substructure							
.1	Slab on Grade	6" reinforced concrete with vapor barrier and granular base		S.F. Slab	3.61	3.61	5.1%
.2	Special Substructures	N/A		—	—	—	
3.0 Superstructure							
.1	Columns & Beams	Wide flange beams and columns		S.F. Ground	8.25	8.25	
.4	Structural Walls	N/A		—	—	—	
.5	Elevated Floors	N/A		—	—	—	18.4%
.7	Roof	Metal deck on steel joist		S.F. Roof	4.74	4.74	
.9	Stairs	N/A		—	—	—	
4.0 Exterior Closure							
.1	Walls	Concrete block	95% of wall	S.F. Wall	7.36	4.14	
.5	Exterior Wall Finishes	N/A		—	—	—	7.8%
.6	Doors	Aluminum and glass, hollow metal, overhead		Each	1849	.49	
.7	Windows & Glazed Walls	Store front	5% of wall	S.F. Window	28	.85	
5.0 Roofing							
.1	Roof Coverings	Elastomeric neoprene membrane with flashing		S.F. Roof	1.71	1.71	
.7	Insulation	Perlite/EPS composite		S.F. Roof	1.18	1.18	4.5%
.8	Openings & Specialties	Hatches		S.F. Roof	.27	.27	
6.0 Interior Construction							
.1	Partitions	Concrete block	140 S.F. Floor/L.F. Partition	S.F. Partition	5.02	.43	
.4	Interior Doors	Hollow metal	2500 S.F. Floor/Door	Each	530	.21	
.5	Wall Finishes	Paint		S.F. Surface	1.11	.19	7.1%
.6	Floor Finishes	80% rubber mat, 20% paint	50% of floor area	S.F. Floor	5.56	2.78	
.7	Ceiling Finishes	Mineral fiber tile on concealed zee bar	10% of area	S.F. Ceiling	3.06	.31	
.9	Interior Surface/Exterior Wall	Paint	95% of wall	S.F. Wall	1.93	1.09	
7.0 Conveying							
.1	Elevators	N/A		—	—	—	0.0%
.2	Special Conveyors	N/A		—	—	—	
8.0 Mechanical							
.1	Plumbing	Toilet and service fixtures, supply and drainage	1 Fixture/1070 S.F. Floor	Each	3702	3.46	
.2	Fire Protection	Standpipes and hose systems		—	—	—	
.3	Heating	Oil fired hot water, unit heaters	10% of area	S.F. Floor	.51	.51	22.2%
.4	Cooling	Single zone, electric cooling	90% of area	S.F. Floor	11.75	11.75	
.5	Special Systems	N/A		—	—	—	
9.0 Electrical							
.1	Service & Distribution	400 ampere service, panel board and feeders		S.F. Floor	.57	.57	
.2	Lighting & Power	High intensity discharge and fluorescent fixtures, receptacles, switches, A.C. and misc. power		S.F. Floor	4.83	4.83	8.8%
.4	Special Electrical	Alarm systems, emergency lighting and public address		S.F. Floor	.81	.81	
11.0 Special Construction							
.1	Specialties	Dasher boards and rink (including ice making system)		S.F. Floor	15.02	15.02	21.2%
12.0 Site Work							
.1	Earthwork	N/A		—	—	—	
.3	Utilities	N/A		—	—	—	0.0%
.5	Roads & Parking	N/A		—	—	—	
.7	Site Improvements	N/A		—	—	—	
				Sub-Total		70.69	100%
	GENERAL CONDITIONS (Overhead & Profit)			15%		10.60	
	ARCHITECT FEES			7%		5.71	
				Total Building Cost		**87.00**	

BUILDING TYPES

COMMERCIAL/INDUSTRIAL/INSTITUTIONAL

M.560 School, Elementary

Costs per square foot of floor area

Exterior Wall	S.F. Area	25000	30000	35000	40000	45000	50000	55000	60000	65000
	L.F. Perimeter	700	740	823	906	922	994	1033	1100	1166
Face Brick with Concrete Block Back-up	Steel Frame	83.45	81.15	80.20	79.40	78.00	77.50	76.75	76.40	76.10
	Bearing Walls	81.05	78.75	77.80	77.00	75.60	75.10	74.35	74.00	73.70
Stucco on Concrete Block	Steel Frame	80.15	78.20	77.40	76.75	75.55	75.15	74.55	74.20	73.95
	Bearing Walls	77.75	75.80	74.95	74.30	73.15	72.75	72.10	71.80	71.55
Decorative Concrete Block	Steel Frame	80.85	78.85	78.00	77.30	76.05	75.65	75.00	74.65	74.40
	Bearing Walls	78.45	76.40	75.55	74.90	73.65	73.25	72.60	72.25	72.00
Perimeter Adj., Add or Deduct	Per 100 L.F.	2.30	1.95	1.65	1.45	1.30	1.15	1.10	1.00	.90
Story Hgt. Adj., Add or Deduct	Per 1 Ft.	.80	.70	.65	.65	.55	.55	.55	.50	.50
For Basement, add $14.60 per square foot of basement area											

The above costs were calculated using the basic specifications shown on the facing page. These costs should be adjusted where necessary for design alternatives and owner's requirements. Reported completed project costs, for this type of structure, range from $44.35 to $118.25 per S.F.

Common additives

Description	Unit	$ Cost
Bleachers, Telescoping, manual		
To 15 tier	Seat	65 - 76
16-20 tier	Seat	72 - 83
21-30 tier	Seat	88 - 104
For power operation, add	Seat	13.40 - 19.40
Carrels Hardwood	Each	655 - 850
Clock System		
20 room	Each	11,700
50 room	Each	28,300
Emergency Lighting, 25 watt, battery operated		
Lead battery	Each	350
Nickel cadmium	Each	590
Flagpoles, Complete		
Aluminum, 20' high	Each	1050
40' high	Each	2350
Fiberglass, 23' high	Each	1375
39'-5" high	Each	2400
Kitchen Equipment		
Broiler	Each	3300
Cooler, 6 ft. long, reach-in	Each	2600

Description	Unit	$ Cost
Kitchen Equipment, cont.		
Dishwasher, 10-12 racks per hr.	Each	2500
Food warmer, counter, 1.2 KW	Each	715
Freezer, 44 C.F., reach-in	Each	6375
Ice cube maker, 50 lb. per day	Each	1675
Range with 1 oven	Each	2250
Lockers, Steel, single tier, 60" to 72"	Opening	134 - 213
2 tier, 60" to 72" total	Opening	83 - 104
5 tier, box lockers	Opening	44 - 56
Locker bench, lam. maple top only	L.F.	18.40
Pedestals, steel pipe	Each	43
Seating		
Auditorium chair, all veneer	Each	127
Veneer back, padded seat	Each	152
Upholstered, spring seat	Each	178
Classroom, movable chair & desk	Set	65 - 120
Lecture hall, pedestal type	Each	127 - 370
Sound System		
Amplifier, 250 watts	Each	1500
Speaker, ceiling or wall	Each	130
Trumpet	Each	244

Model costs calculated for a 1 story building with 12' story height and 45,000 square feet of floor area

School, Elementary

				Unit	Unit Cost	Cost Per S.F.	% Of Sub-Total
1.0 Foundations							
.1	Footings & Foundations	Poured concrete; strip and spread footings and 4' foundation wall		S.F. Ground	3.55	3.55	
.4	Piles & Caissons	N/A		—	—	—	7.2%
.9	Excavation & Backfill	Site preparation for slab and trench for foundation wall and footing		S.F. Ground	.89	.89	
2.0 Substructure							
.1	Slab on Grade	4" reinforced concrete with vapor barrier and granular base		S.F. Slab	3.05	3.05	5.0%
.2	Special Substructures	N/A		—	—	—	
3.0 Superstructure							
.1	Columns & Beams	N/A		—	—	—	
.4	Structural Walls	N/A		—	—	—	
.5	Elevated Floors	N/A		—	—	—	3.9%
.7	Roof	Metal deck on open web steel joists		S.F. Roof	2.37	2.37	
.9	Stairs	N/A		—	—	—	
4.0 Exterior Closure							
.1	Walls	Face brick with concrete block backup	70% of wall	S.F. Wall	21	3.67	
.5	Exterior Wall Finishes	N/A		—	—	—	9.0%
.6	Doors	Metal and glass	5% of wall	Each	2270	.40	
.7	Windows & Glazed Walls	Steel outward projecting	25% of wall	Each	458	1.47	
5.0 Roofing							
.1	Roof Coverings	Built-up tar and gravel with flashing		S.F. Roof	1.89	1.89	
.7	Insulation	Perlite/EPS composite		S.F. Roof	1.18	1.18	5.2%
.8	Openings & Specialties	Gravel stop		L.F. Perimeter	5.55	.11	
6.0 Interior Construction							
.1	Partitions	Concrete block, toilet partitions	20 S.F. Floor/L.F. Partition	S.F. Partition	5.05	3.15	
.4	Interior Doors	Single leaf kalamein fire doors	700 S.F. Floor/Door	Each	530	.76	
.5	Wall Finishes	75% paint, 15% glazed coating, 10% ceramic tile		S.F. Surface	1.41	1.41	21.6%
.6	Floor Finishes	65% vinyl composition tile, 25% carpet, 10% terrazzo		S.F. Floor	4.42	4.42	
.7	Ceiling Finishes	Mineral fiber tile on concealed zee bars		S.F. Ceiling	3.06	3.06	
.9	Interior Surface/Exterior Wall	Painted gypsum board on furring	70% of wall	S.F. Wall	2.81	.48	
7.0 Conveying							
.1	Elevators	N/A		—	—	—	0.0%
.2	Special Conveyors	N/A		—	—	—	
8.0 Mechanical							
.1	Plumbing	Kitchen, bathroom and service fixtures, supply and drainage	1 Fixture/625 S.F. Floor	Each	2237	3.58	
.2	Fire Protection	Sprinklers, light hazard		S.F. Floor	1.41	1.41	
.3	Heating	Oil fired hot water, wall fin radiation		S.F. Floor	6.05	6.05	33.6%
.4	Cooling	Split systems with air cooled condensing units		S.F. Floor	9.67	9.67	
.5	Special Systems	N/A		—	—	—	
9.0 Electrical							
.1	Service & Distribution	600 ampere service, panel board and feeders		S.F. Floor	.62	.62	
.2	Lighting & Power	Fluorescent fixtures, receptacles, switches, A.C. and misc. power		S.F. Floor	6.26	6.26	13.1%
.4	Special Electrical	Alarm systems, communications systems and emergency lighting		S.F. Floor	1.14	1.14	
11.0 Special Construction							
.1	Specialties	Chalkboards		S.F. Floor	.84	.84	1.4%
12.0 Site Work							
.1	Earthwork	N/A		—	—	—	
.3	Utilities	N/A		—	—	—	0.0%
.5	Roads & Parking	N/A		—	—	—	
.7	Site Improvements	N/A		—	—	—	
				Sub-Total		61.43	100%
	GENERAL CONDITIONS (Overhead & Profit)				15%	9.21	
	ARCHITECT FEES				7%	4.96	
				Total Building Cost		75.60	

BUILDING TYPES

COMMERCIAL/INDUSTRIAL/INSTITUTIONAL M.570 School, High, 2-3 Story

Costs per square foot of floor area

Exterior Wall	S.F. Area	50000	70000	90000	110000	130000	150000	170000	190000	210000
	L.F. Perimeter	816	1083	1100	1300	1290	1450	1433	1566	1700
Face Brick with Concrete Block Back-up	Steel Frame	82.85	80.80	77.10	76.25	74.15	73.75	72.40	72.10	71.85
	R/Conc. Frame	85.85	84.05	80.45	79.75	77.70	77.30	76.00	75.70	75.45
Decorative Concrete Block	Steel Frame	78.75	77.15	74.15	73.55	71.90	71.50	70.50	70.25	70.05
	R/Conc. Frame	82.55	80.95	78.00	77.35	75.70	75.35	74.30	74.05	73.85
Limestone with Concrete Block Back-up	Steel Frame	85.40	83.45	79.15	78.35	75.95	75.45	73.95	73.60	73.35
	R/Conc. Frame	89.65	87.70	83.40	82.60	80.15	79.70	78.15	77.85	77.55
Perimeter Adj., Add or Deduct	Per 100 L.F.	1.85	1.35	1.00	.85	.75	.65	.55	.45	.45
Story Hgt. Adj., Add or Deduct	Per 1 Ft.	1.10	1.10	.85	.80	.70	.65	.55	.55	.55

For Basement, add $18.90 per square foot of basement area

The above costs were calculated using the basic specifications shown on the facing page. These costs should be adjusted where necessary for design alternatives and owner's requirements. Reported completed project costs, for this type of structure, range from $54.65 to $126.20 per S.F.

Common additives

Description	Unit	$ Cost
Bleachers, Telescoping, manual		
To 15 tier	Seat	65 - 76
16-20 tier	Seat	72 - 83
21-30 tier	Seat	88 - 104
For power operation, add	Seat	13.40 - 19.40
Carrels Hardwood	Each	655 - 850
Clock System		
20 room	Each	11,700
50 room	Each	28,300
Elevators, Hydraulic passenger, 2 stops		
1500# capacity	Each	40,800
2500# capacity	Each	42,100
Emergency Lighting, 25 watt, battery operated		
Lead battery	Each	350
Nickel cadmium	Each	590
Flagpoles, Complete		
Aluminum, 20' high	Each	1050
40' high	Each	2350
Fiberglass, 23' high	Each	1375
39'-5" high	Each	2400

Description	Unit	$ Cost
Kitchen Equipment		
Broiler	Each	3300
Cooler, 6 ft. long, reach-in	Each	2600
Dishwasher, 10-12 racks per hr.	Each	2500
Food warmer, counter, 1.2 KW	Each	715
Freezer, 44 C.F., reach-in	Each	6375
Lockers, Steel, single tier, 60" or 72"	Opening	134 - 213
2 tier, 60" or 72" total	Opening	83 - 104
5 tier, box lockers	Opening	44 - 56
Locker bench, lam. maple top only	L.F.	18.40
Pedestals, steel pipe	Each	43
Seating		
Auditorium chair, all veneer	Each	127
Veneer back, padded seat	Each	152
Upholstered, spring seat	Each	178
Classroom, movable chair & desk	Set	65 - 120
Lecture hall, pedestal type	Each	127 - 370
Sound System		
Amplifier, 250 watts	Each	1500
Speaker, ceiling or wall	Each	130
Trumpet	Each	244

Model costs calculated for a 2 story building with 12' story height and 130,000 square feet of floor area

School, High, 2-3 Story

				Unit	Unit Cost	Cost Per S.F.	% Of Sub-Total
1.0 Foundations							
.1	Footings & Foundations	Poured concrete; strip and spread footings and 4' foundation wall		S.F. Ground	2.26	1.13	
.4	Piles & Caissons	N/A		—	—	—	2.5%
.9	Excavation & Backfill	Site preparation for slab and trench for foundation wall and footing		S.F. Ground	.89	.44	
2.0 Substructure							
.1	Slab on Grade	4" reinforced concrete with vapor barrier and granular base		S.F. Slab	3.05	1.52	2.4%
.2	Special Substructures	N/A		—	—	—	
3.0 Superstructure							
.1	Columns & Beams	Concrete columns		L.F. Column	78	1.91	
.4	Structural Walls	N/A		—	—	—	
.5	Elevated Floors	Concrete slab without drop panel		S.F. Floor	9.59	4.80	18.3%
.7	Roof	Concrete slab without drop panel		S.F. Roof	9.18	4.59	
.9	Stairs	Concrete filled metal pan		Flight	4925	.23	
4.0 Exterior Closure							
.1	Walls	Face brick with concrete block backup	75% of wall	S.F. Wall	21	3.80	
.5	Exterior Wall Finishes	N/A		—	—	—	10.2%
.6	Doors	Metal and glass		Each	1253	.28	
.7	Windows & Glazed Walls	Window wall	25% of wall	S.F. Window	39	2.38	
5.0 Roofing							
.1	Roof Coverings	Built-up tar and gravel with flashing		S.F. Roof	1.80	.90	
.7	Insulation	Perlite/EPS composite		S.F. Roof	1.18	.59	2.5%
.8	Openings & Specialties	Gravel stop and hatches		S.F. Roof	.18	.09	
6.0 Interior Construction							
.1	Partitions	Concrete block, toilet partitions	25 S.F. Floor/L.F. Partition	S.F. Partition	5.05	2.47	
.4	Interior Doors	Single leaf kalamein fire doors	700 S.F. Floor/Door	Each	530	.76	
.5	Wall Finishes	75% paint, 15% glazed coating, 10% ceramic tile		S.F. Surface	1.41	1.13	19.3%
.6	Floor Finishes	70% vinyl composition tile, 20% carpet, 10% terrazzo		S.F. Floor	4.28	4.28	
.7	Ceiling Finishes	Mineral fiber tile on concealed zee bars		S.F. Ceiling	3.06	3.06	
.9	Interior Surface/Exterior Wall	Painted gypsum board on furring	75% of wall	S.F. Wall	2.81	.50	
7.0 Conveying							
.1	Elevators	One hydraulic passenger elevator		Each	49,400	.38	0.6%
.2	Special Conveyors	N/A		—	—	—	
8.0 Mechanical							
.1	Plumbing	Kitchen, bathroom and service fixtures, supply and drainage	1 Fixture/860 S.F. Floor	Each	2476	2.88	
.2	Fire Protection	Sprinklers, light hazard		S.F. Floor	1.24	1.24	
.3	Heating	Oil fired hot water, wall fin radiation		S.F. Floor	3.09	3.09	28.8%
.4	Cooling	Chilled water, cooling tower systems		S.F. Floor	10.97	10.97	
.5	Special Systems	N/A		—	—	—	
9.0 Electrical							
.1	Service & Distribution	1200 ampere service, panel board and feeders		S.F. Floor	.43	.43	
.2	Lighting & Power	Fluorescent fixtures, receptacles, switches, A.C. and misc. power		S.F. Floor	6.20	6.20	14.3%
.4	Special Electrical	Alarm systems, communications systems and emergency lighting		S.F. Floor	2.38	2.38	
11.0 Special Construction							
.1	Specialties	Chalkboards, laboratory counters, built-in athletic equipment		S.F. Floor	.71	.71	1.1%
12.0 Site Work							
.1	Earthwork	N/A		—	—	—	
.3	Utilities	N/A		—	—	—	0.0%
.5	Roads & Parking	N/A		—	—	—	
.7	Site Improvements	N/A		—	—	—	
				Sub-Total		63.14	100%
	GENERAL CONDITIONS (Overhead & Profit)				15%	9.47	
	ARCHITECT FEES				7%	5.09	
				Total Building Cost		77.70	

BUILDING TYPES

COMMERCIAL/INDUSTRIAL/INSTITUTIONAL

M.580 School, Jr High, 2-3 Story

Costs per square foot of floor area

Exterior Wall	S.F. Area	50000	65000	80000	95000	110000	125000	140000	155000	170000
	L.F. Perimeter	816	1016	1000	1150	1890	2116	2287	2438	2552
Face Brick with Concrete Block Back-up	Steel Frame	80.50	79.65	76.75	76.30	80.60	80.30	79.75	79.20	78.55
	Bearing Walls	77.95	77.05	74.15	73.70	78.05	77.75	77.15	76.60	75.95
Concrete Block Stucco Face	Steel Frame	76.30	75.60	73.50	73.20	76.20	75.95	75.55	75.10	74.65
	Bearing Walls	73.70	73.00	70.95	70.60	73.60	73.35	72.95	72.50	72.05
Decorative Concrete Block	Steel Frame	77.15	76.40	74.15	73.80	77.10	76.85	76.40	75.95	75.45
	Bearing Walls	74.15	73.45	71.20	70.80	74.10	73.90	73.40	72.95	72.45
Perimeter Adj., Add or Deduct	Per 100 L.F.	1.70	1.35	1.10	.90	.80	.70	.60	.55	.50
Story Hgt. Adj., Add or Deduct	Per 1 Ft.	1.00	1.00	.75	.75	1.05	1.05	1.05	.95	.95

For Basement, add $19.05 per square foot of basement area

The above costs were calculated using the basic specifications shown on the facing page. These costs should be adjusted where necessary for design alternatives and owner's requirements. Reported completed project costs, for this type of structure, range from $51.45 to $117.50 per S.F.

Common additives

Description	Unit	$ Cost
Bleachers, Telescoping, manual		
To 15 tier	Seat	65 - 76
16-20 tier	Seat	72 - 83
21-30 tier	Seat	88 - 104
For power operation, add	Seat	13.40 - 19.40
Carrels Hardwood	Each	655 - 850
Clock System		
20 room	Each	11,700
50 room	Each	28,300
Elevators, Hydraulic passenger, 2 stops		
1500# capacity	Each	40,800
2500# capacity	Each	42,100
Emergency Lighting, 25 watt, battery operated		
Lead battery	Each	350
Nickel cadmium	Each	590
Flagpoles, Complete		
Aluminum, 20' high	Each	1050
40' high	Each	2350
Fiberglass, 23' high	Each	1375
39'-5" high	Each	2400

Description	Unit	$ Cost
Kitchen Equipment		
Broiler	Each	3300
Cooler, 6 ft. long, reach-in	Each	2600
Dishwasher, 10-12 racks per hr.	Each	2500
Food warmer, counter, 1.2 KW	Each	715
Freezer, 44 C.F., reach-in	Each	6375
Lockers, Steel, single tier, 60" to 72"	Opening	134 - 213
2 tier, 60" to 72" total	Opening	83 - 104
5 tier, box lockers	Opening	44 - 56
Locker bench, lam. maple top only	L.F.	18.40
Pedestals, steel pipe	Each	43
Seating		
Auditorium chair, all veneer	Each	127
Veneer back, padded seat	Each	152
Upholstered, spring seat	Each	178
Classroom, movable chair & desk	Set	65 - 120
Lecture hall, pedestal type	Each	127 - 370
Sound System		
Amplifier, 250 watts	Each	1500
Speaker, ceiling or wall	Each	130
Trumpet	Each	244

Location Factors appear in the Reference Section

School, Jr High, 2-3 Story

Model costs calculated for a 2 story building with 12' story height and 110,000 square feet of floor area

				Unit	Unit Cost	Cost Per S.F.	% Of Sub-Total
1.0 Foundations							
.1	Footings & Foundations	Poured concrete; strip and spread footings and 4' foundation wall		S.F. Ground	3.12	1.56	3.1%
.4	Piles & Caissons	N/A		—	—	—	
.9	Excavation & Backfill	Site preparation for slab and trench for foundation wall and footing		S.F. Ground	.89	.44	
2.0 Substructure							
.1	Slab on Grade	4" reinforced concrete with vapor barrier and granular base		S.F. Slab	3.05	1.52	2.3%
.2	Special Substructures	N/A		—	—	—	
3.0 Superstructure							
.1	Columns & Beams	Fireproofing, steel columns included in 3.5 and 3.7		L.F. Column	20	.29	
.4	Structural Walls	N/A		—	—	—	
.5	Elevated Floors	Open web steel joists, slab form, concrete, columns		S.F. Floor	13.16	6.58	14.9%
.7	Roof	Metal deck, open web steel joists, columns		S.F. Roof	5.28	2.64	
.9	Stairs	Concrete filled metal pan		Flight	4925	.27	
4.0 Exterior Closure							
.1	Walls	Face brick with concrete block backup	75% of wall	S.F. Wall	21	6.59	
.5	Exterior Wall Finishes	N/A		—	—	—	15.2%
.6	Doors	Double aluminum & glass		Each	1337	.35	
.7	Windows & Glazed Walls	Window wall	25% of wall	S.F. Window	29	3.04	
5.0 Roofing							
.1	Roof Coverings	Built-up tar and gravel with flashing		S.F. Roof	2.14	1.07	
.7	Insulation	Perlite/EPS composite		S.F. Roof	1.18	.59	2.7%
.8	Openings & Specialties	Gravel stop and hatches		S.F. Roof	.24	.12	
6.0 Interior Construction							
.1	Partitions	Concrete block, toilet partitions	20 S.F. Floor/L.F. Partition	S.F. Partition	5.05	2.95	
.4	Interior Doors	Single leaf kalamein fire doors	750 S.F. Floor/Door	Each	530	.71	
.5	Wall Finishes	50% paint, 40% glazed coatings, 10% ceramic tile		S.F. Surface	1.55	1.55	22.9%
.6	Floor Finishes	50% vinyl composition tile, 30% carpet, 20% terrrazzo		S.F. Floor	5.87	5.87	
.7	Ceiling Finishes	Mineral fiberboard on concealed zee bars		S.F. Ceiling	3.06	3.06	
.9	Interior Surface/Exterior Wall	Painted gypsum board on furring	75% of wall	S.F. Wall	2.81	.87	
7.0 Conveying							
.1	Elevators	One hydraulic passenger elevator		Each	49,500	.45	0.7%
.2	Special Conveyors	N/A		—	—	—	
8.0 Mechanical							
.1	Plumbing	Kitchen, toilet and service fixtures, supply and drainage	1 Fixture/1170 S.F. Floor	Each	2726	2.33	
.2	Fire Protection	Sprinklers, light hazard	10% of area	S.F. Floor	.25	.25	
.3	Heating	Included in 8.4		—	—	—	23.9%
.4	Cooling	Multizone unit, gas heating, electric cooling		S.F. Floor	13.05	13.05	
.5	Special Systems	N/A		—	—	—	
9.0 Electrical							
.1	Service & Distribution	1000 ampere service, panel board and feeders		S.F. Floor	.47	.47	
.2	Lighting & Power	Fluorescent fixtures, receptacles, switches, A.C. and misc. power		S.F. Floor	6.28	6.28	13.2%
.4	Special Electrical	Alarm systems, communications systems and emergency lighting		S.F. Floor	1.93	1.93	
11.0 Special Construction							
.1	Specialties	Chalkboards, laboratory counters		S.F. Floor	.69	.69	1.1%
12.0 Site Work							
.1	Earthwork	N/A		—	—	—	
.3	Utilities	N/A		—	—	—	0.0%
.5	Roads & Parking	N/A		—	—	—	
.7	Site Improvements	N/A		—	—	—	

	Sub-Total	65.52	100%
GENERAL CONDITIONS (Overhead & Profit)	15%	9.83	
ARCHITECT FEES	7%	5.25	
Total Building Cost		80.60	

COMMERCIAL/INDUSTRIAL/INSTITUTIONAL

M.590 School, Vocational

Costs per square foot of floor area

Exterior Wall	S.F. Area	20000	30000	40000	50000	60000	70000	80000	90000	100000
	L.F. Perimeter	400	500	685	700	800	900	1000	1100	1200
Face Brick with Concrete Block Back-up	Steel Frame	85.60	82.20	82.25	79.45	78.80	78.30	77.95	77.65	77.45
	Bearing Walls	83.85	80.45	80.55	77.75	77.10	76.60	76.25	75.95	75.70
Decorative Concrete Block	Steel Frame	79.75	77.35	77.30	75.40	74.90	74.55	74.35	74.15	73.95
	Bearing Walls	78.05	75.65	75.60	73.70	73.20	72.85	72.60	72.40	72.25
Steel Siding on Steel Studs	Steel Frame	77.15	75.15	75.05	73.55	73.15	72.90	72.70	72.50	72.35
Metal Sandwich Panel	Steel Frame	81.30	78.60	78.60	76.45	75.95	75.55	75.30	75.05	74.90
Perimeter Adj., Add or Deduct	Per 100 L.F.	4.25	2.85	2.15	1.70	1.45	1.25	1.05	.95	.85
Story Hgt. Adj., Add or Deduct	Per 1 Ft.	1.20	1.05	1.05	.85	.85	.75	.75	.75	.75

For Basement, add $19.00 per square foot of basement area

The above costs were calculated using the basic specifications shown on the facing page. These costs should be adjusted where necessary for design alternatives and owner's requirements. Reported completed project costs, for this type of structure, range from $46.15 to $131.70 per S.F.

Common additives

Description	Unit	$ Cost
Carrels Hardwood	Each	655 - 850
Clock System		
20 room	Each	11,700
50 room	Each	28,300
Directory Boards, Plastic, glass covered		
30" x 20"	Each	500
36" x 48"	Each	910
Aluminum, 24" x 18"	Each	445
36" x 24"	Each	535
48" x 32"	Each	630
48" x 60"	Each	1425
Elevators, Hydraulic passenger, 2 stops		
1500# capacity	Each	40,800
2500# capacity	Each	42,100
3500# capacity	Each	45,700
Emergency Lighting, 25 watt, battery operated		
Lead battery	Each	350
Nickel cadmium	Each	590

Description	Unit	$ Cost
Flagpoles, Complete		
Aluminum, 20' high	Each	1050
40' high	Each	2350
Fiberglass, 23' high	Each	1375
39'-5" high	Each	2400
Seating		
Auditorium chair, all veneer	Each	127
Veneer back, padded seat	Each	152
Upholstered, spring seat	Each	178
Classroom, movable chair & desk	Set	65 - 120
Lecture hall, pedestal type	Each	127 - 370
Shops & Workroom:		
Benches, metal	Each	380
Parts bins 6'-3" high, 3' wide, 12" deep, 72 bins	Each	540
Shelving, metal 1' x 3'	S.F.	9.00
Wide span 6' wide x 24" deep	S.F.	11.70
Sound System		
Amplifier, 250 watts	Each	1500
Speaker, ceiling or wall	Each	130
Trumpet	Each	244

Model costs calculated for a 2 story building with 12' story height and 40,000 square feet of floor area

School, Vocational

				Unit	Unit Cost	Cost Per S.F.	% Of Sub-Total
1.0 Foundations							
.1	Footings & Foundations	Poured concrete; strip and spread footings and 4' foundation wall		S.F. Ground	3.24	1.62	3.1%
.4	Piles & Caissons	N/A		—	—	—	
.9	Excavation & Backfill	Site preparation for slab and trench for foundation wall and footing		S.F. Ground	.89	.44	
2.0 Substructure							
.1	Slab on Grade	5" reinforced concrete with vapor barrier and granular base		S.F. Slab	6.28	3.14	4.8%
.2	Special Substructures	N/A		—	—	—	
3.0 Superstructure							
.1	Columns & Beams	Interior steel columns included in 3.5 and 3.7		—	—	—	
.4	Structural Walls	N/A		—	—	—	
.5	Elevated Floors	Open web steel joists, slab form, concrete, beams, columns		S.F. Floor	9.40	4.70	10.4%
.7	Roof	Metal deck, open web steel joists, beams, columns		S.F. Roof	3.23	1.62	
.9	Stairs	Concrete filled metal pan		Flight	4925	.49	
4.0 Exterior Closure							
.1	Walls	Face brick with concrete block backup	85% of wall	S.F. Wall	21	7.44	14.6%
.5	Exterior Wall Finishes	N/A		—	—	—	
.6	Doors	Metal and glass doors		Each	1511	.31	
.7	Windows & Glazed Walls	Steel outward projecting	15% of wall	S.F. Wall	29	1.81	
5.0 Roofing							
.1	Roof Coverings	Built-up tar and gravel with flashing		S.F. Roof	2.14	1.07	2.8%
.7	Insulation	Perlite/EPS composite		S.F. Roof	1.18	.59	
.8	Openings & Specialties	Gravel stop and hatches		S.F. Roof	.36	.18	
6.0 Interior Construction							
.1	Partitions	Concrete block, toilet partitions	20 S.F. Floor/L.F. Partition	S.F. Partition	5.05	3.11	19.3%
.4	Interior Doors	Single leaf kalamein fire doors	600 S.F. Floor/Door	Each	530	.88	
.5	Wall Finishes	50% paint, 40% glazed coating, 10% ceramic tile		S.F. Surface	1.55	1.55	
.6	Floor Finishes	70% vinyl composition tile, 20% carpet, 10% terrazzo		S.F. Floor	4.28	4.28	
.7	Ceiling Finishes	Mineral fiber tile on concealed zee bars	60% of area	S.F. Ceiling	3.06	1.84	
.9	Interior Surface/Exterior Wall	Painted gypsum board on furring	85% of wall	S.F. Wall	2.81	.98	
7.0 Conveying							
.1	Elevators	One hydraulic passenger elevator		Each	49,600	1.24	1.9%
.2	Special Conveyors	N/A		—	—	—	
8.0 Mechanical							
.1	Plumbing	Toilet and service fixtures, supply and drainage	1 Fixture/700 S.F. Floor	Each	2254	3.22	29.3%
.2	Fire Protection	Sprinklers, light hazard		S.F. Floor	.81	.81	
.3	Heating	Oil fired hot water, wall fin radiation and unit heaters		S.F. Floor	11.17	5.68	
.4	Cooling	Chilled water, cooling tower systems		S.F. Floor	9.42	9.42	
.5	Special Systems	N/A		—	—	—	
9.0 Electrical							
.1	Service & Distribution	600 ampere service, panel board and feeders		S.F. Floor	.71	.71	12.8%
.2	Lighting & Power	Fluorescent fixtures, receptacles, switches, A.C. and misc. power		S.F. Floor	6.29	6.29	
.4	Special Electrical	Alarm systems, communications systems and emergency lighting		S.F. Floor	1.41	1.41	
11.0 Special Construction							
.1	Specialties	Chalkboards		S.F. Floor	.64	.64	1.0%
12.0 Site Work							
.1	Earthwork	N/A		—	—	—	0.0%
.3	Utilities	N/A		—	—	—	
.5	Roads & Parking	N/A		—	—	—	
.7	Site Improvements	N/A		—	—	—	
				Sub-Total		65.47	100%
	GENERAL CONDITIONS (Overhead & Profit)				15%	9.82	
	ARCHITECT FEES				7%	5.26	
				Total Building Cost		80.55	

BUILDING TYPES

COMMERCIAL/INDUSTRIAL/INSTITUTIONAL

M.600 Store, Convenience

Costs per square foot of floor area

Exterior Wall	S.F. Area	1000	2000	3000	4000	6000	8000	10000	12000	15000
	L.F. Perimeter	126	179	219	253	310	358	400	438	490
Wood Siding	Wood Frame	74.45	66.75	63.30	61.25	58.75	57.35	56.30	55.60	54.75
Face Brick Veneer	Wood Frame	93.90	80.30	74.25	70.60	66.30	63.75	62.00	60.70	59.25
Stucco on Concrete Block	Steel Frame	83.90	73.75	69.20	66.50	63.25	61.35	60.05	59.05	58.00
	Bearing Walls	82.55	72.45	67.85	65.15	61.95	60.05	58.70	57.75	56.70
Metal Sandwich Panel	Steel Frame	86.30	75.60	70.75	67.90	64.50	62.45	61.05	60.05	58.90
Precast Concrete	Steel Frame	96.45	82.55	76.25	72.55	68.10	65.45	63.65	62.35	60.85
Perimeter Adj., Add or Deduct	Per 100 L.F.	21.10	10.55	7.05	5.25	3.55	2.65	2.10	1.75	1.40
Story Hgt. Adj., Add or Deduct	Per 1 Ft.	1.55	1.10	.90	.75	.65	.55	.50	.45	.40

For Basement, add $14.25 per square foot of basement area

The above costs were calculated using the basic specifications shown on the facing page. These costs should be adjusted where necessary for design alternatives and owner's requirements. Reported completed project costs, for this type of structure, range from $38.25 to $105.50 per S.F.

Common additives

Description	Unit	$ Cost
Check Out Counter		
Single belt	Each	2175
Double belt	Each	3675
Emergency Lighting, 25 watt, battery operated		
Lead battery	Each	350
Nickel cadmium	Each	590
Refrigerators, Prefabricated, walk-in		
7'-6" high, 6' x 6'	S.F.	117
10' x 10'	S.F.	91
12' x 14'	S.F.	81
12' x 20'	S.F.	72
Refrigerated Food Cases		
Dairy, multi deck, 12' long	Each	6975
Delicatessen case, single deck, 12' long	Each	4925
Multi deck, 18 S.F. shelf display	Each	7425
Freezer, self-contained chest type, 30 C.F.	Each	2950
Glass door upright, 78 C.F.	Each	6525

Description	Unit	$ Cost
Refrigerated Food Cases, cont.		
Frozen food, chest type, 12' long	Each	4775
Glass door reach-in, 5 door	Each	9025
Island case 12' long, single deck	Each	5400
Multi deck	Each	11,100
Meat cases, 12' long, single deck	Each	3925
Multi deck	Each	6700
Produce, 12' long single deck	Each	5175
Multi deck	Each	5775
Safe, Office type, 4 hour rating		
30" x 18" x 18"	Each	3475
62" x 33" x 20"	Each	7575
Smoke Detectors		
Ceiling type	Each	141
Duct type	Each	335
Sound System		
Amplifier, 250 watts	Each	1500
Speaker, ceiling or wall	Each	130
Trumpet	Each	244

Model costs calculated for a 1 story building with 12' story height and 4,000 square feet of floor area

Store, Convenience

				Unit	Unit Cost	Cost Per S.F.	% Of Sub-Total
1.0 Foundations							
.1	Footings & Foundations	Poured concrete; strip and spread footings and 4' foundation wall		S.F. Ground	3.42	3.42	
.4	Piles & Caissons	N/A		—	—	—	9.1%
.9	Excavation & Backfill	Site preparation for slab and trench for foundation wall and footing		S.F. Ground	1.09	1.09	
2.0 Substructure							
.1	Slab on Grade	4" reinforced concrete with vapor barrier and granular base		S.F. Slab	3.05	3.05	6.1%
.2	Special Substructures	N/A		—	—	—	
3.0 Superstructure							
.1	Columns & Beams	N/A		—	—	—	
.4	Structural Walls	N/A		—	—	—	
.5	Elevated Floors	N/A		—	—	—	8.7%
.7	Roof	Wood truss with plywood sheathing		S.F. Ground	4.35	4.35	
.9	Stairs	N/A		—	—	—	
4.0 Exterior Closure							
.1	Walls	Wood siding on wood studs, insulated	80% of wall	S.F. Wall	6.32	3.84	
.5	Exterior Wall Finishes	N/A		—	—	—	16.4%
.6	Doors	Double aluminum and glass, solid core wood		Each	1920	1.44	
.7	Windows & Glazed Walls	Storefront	20% of wall	S.F. Window	28	2.87	
5.0 Roofing							
.1	Roof Coverings	Asphalt shingles		S.F. Roof	1.64	1.64	
.7	Insulation	Fiberglass sheets		S.F. Roof	.86	.86	5.2%
.8	Openings & Specialties	Gutters and downspouts		S.F. Roof	.10	.10	
6.0 Interior Construction							
.1	Partitions	Gypsum board on wood studs	60 S.F. Floor/L.F. Partition	S.F. Partition	3.06	.51	
.4	Interior Doors	Single leaf wood, hollow metal	1300 S.F. Floor/Door	Each	607	.46	
.5	Wall Finishes	Paint		S.F. Surface	.72	.24	12.2%
.6	Floor Finishes	Asphalt tile		S.F. Floor	1.87	1.87	
.7	Ceiling Finishes	Mineral fiber tile on wood furring		S.F. Ceiling	2.12	2.12	
.9	Interior Surface/Exterior Wall	Painted gypsum board on furring	80% of wall	S.F. Wall	1.47	.89	
7.0 Conveying							
.1	Elevators	N/A		—	—	—	0.0%
.2	Special Conveyors	N/A		—	—	—	
8.0 Mechanical							
.1	Plumbing	Toilet and service fixtures, supply and drainage	1 Fixture/1000 S.F. Floor	Each	1880	1.88	
.2	Fire Protection	Sprinkler, ordinary hazard		S.F. Floor	2.04	2.04	
.3	Heating	Included in 8.4		—	—	—	20.6%
.4	Cooling	Single zone rooftop unit, gas heating, electric cooling		S.F. Floor	6.29	6.29	
.5	Special Systems	N/A		—	—	—	
9.0 Electrical							
.1	Service & Distribution	200 ampere service, panel board and feeders		S.F. Floor	1.65	1.65	
.2	Lighting & Power	Fluorescent fixtures, receptacles, switches, A.C. and misc. power		S.F. Floor	6.37	6.37	21.7%
.4	Special Electrical	Alarm systems and emergency lighting		S.F. Floor	2.78	2.78	
11.0 Special Construction							
.1	Specialties	N/A		—	—	—	0.0%
12.0 Site Work							
.1	Earthwork	N/A		—	—	—	
.3	Utilities	N/A		—	—	—	0.0%
.5	Roads & Parking	N/A		—	—	—	
.7	Site Improvements	N/A		—	—	—	
				Sub-Total		49.76	100%
	GENERAL CONDITIONS (Overhead & Profit)				15%	7.46	
	ARCHITECT FEES				7%	4.03	
				Total Building Cost		61.25	

BUILDING TYPES

COMMERCIAL/INDUSTRIAL/INSTITUTIONAL   M.610   Store, Department, 1 Story

Costs per square foot of floor area

Exterior Wall	S.F. Area	50000	65000	80000	95000	110000	125000	140000	155000	170000
	L.F. Perimeter	920	1065	1167	1303	1333	1433	1533	1633	1733
Face Brick with Concrete Block Back-up	R/Conc. Frame	61.45	59.90	58.60	57.90	56.90	56.40	56.05	55.75	55.50
	Steel Frame	51.90	50.35	49.05	48.35	47.35	46.85	46.50	46.20	45.95
Decorative Concrete Block	R/Conc. Frame	57.95	56.75	55.80	55.30	54.60	54.25	53.95	53.75	53.55
	Steel Joists	48.90	47.65	46.65	46.15	45.35	45.00	44.70	44.45	44.30
Precast Concrete Panels	R/Conc. Frame	58.25	57.10	56.15	55.60	54.90	54.55	54.30	54.05	53.90
	Steel Joists	48.35	47.15	46.20	45.70	44.95	44.65	44.35	44.15	43.95
Perimeter Adj., Add or Deduct	Per 100 L.F.	1.10	.80	.70	.55	.45	.45	.35	.35	.30
Story Hgt. Adj., Add or Deduct	Per 1 Ft.	.60	.50	.45	.40	.40	.35	.35	.30	.30

For Basement, add $13.70 per square foot of basement area

The above costs were calculated using the basic specifications shown on the facing page. These costs should be adjusted where necessary for design alternatives and owner's requirements. Reported completed project costs, for this type of structure, range from $32.35 to $71.45 per S.F.

Common additives

Description	Unit	$ Cost
Closed Circuit Surveillance, One station		
Camera and monitor	Each	1325
For additional camera stations, add	Each	730
Directory Boards, Plastic, glass covered		
30" x 20"	Each	500
36" x 48"	Each	910
Aluminum, 24" x 18"	Each	445
36" x 24"	Each	535
48" x 32"	Each	630
48" x 60"	Each	1425
Emergency Lighting, 25 watt, battery operated		
Lead battery	Each	350
Nickel cadmium	Each	590
Safe, Office type, 4 hour rating		
30" x 18" x 18"	Each	3475
62" x 33" x 20"	Each	7575
Sound System		
Amplifier, 250 watts	Each	1500
Speaker, ceiling or wall	Each	130
Trumpet	Each	244

Model costs calculated for a 1 story building with 14' story height and 110,000 square feet of floor area

Store, Department, 1 Story

				Unit	Unit Cost	Cost Per S.F.	% Of Sub-Total
1.0 Foundations							
.1	Footings & Foundations	Poured concrete; strip and spread footings and 4' foundation wall		S.F. Ground	1.08	1.08	
.4	Piles & Caissons	N/A		—	—	—	4.1%
.9	Excavation & Backfill	Site preparation for slab and trench for foundation wall and footing		S.F. Ground	.85	.85	
2.0 Substructure							
.1	Slab on Grade	4" reinforced concrete with vapor barrier and granular base		S.F. Slab	3.05	3.05	6.5%
.2	Special Substructures	N/A		—	—	—	
3.0 Superstructure							
.1	Columns & Beams	Concrete columns		L.F. Column	45	.58	
.4	Structural Walls	N/A		—	—	—	
.5	Elevated Floors	N/A		—	—	—	26.8%
.7	Roof	Precast concrete beam and plank		S.F. Roof	11.95	11.95	
.9	Stairs	N/A		—	—	—	
4.0 Exterior Closure							
.1	Walls	Face brick with concrete block backup	90% of wall	S.F. Wall	21	3.28	
.5	Exterior Wall Finishes	N/A		—	—	—	8.9%
.6	Doors	Sliding electric operated entrance, hollow metal		Each	4385	.24	
.7	Windows & Glazed Walls	Storefront	10% of wall	S.F. Window	37	.64	
5.0 Roofing							
.1	Roof Coverings	Built-up tar and gravel with flashing		S.F. Roof	1.70	1.70	
.7	Insulation	Perlite/EPS composite		S.F. Roof	1.18	1.18	6.4%
.8	Openings & Specialties	Gravel stop and hatches		S.F. Roof	.10	.10	
6.0 Interior Construction							
.1	Partitions	Gypsum board on metal studs	60 S.F. Floor/L.F. Partition	S.F. Partition	3.84	.64	
.4	Interior Doors	Single leaf hollow metal	600 S.F. Floor/Door	Each	530	.88	
.5	Wall Finishes	Paint		S.F. Surface	.51	.17	15.4%
.6	Floor Finishes	50% vinyl tile, 50% carpet		S.F. Floor	3.88	3.88	
.7	Ceiling Finishes	Fiberglass board on exposed grid system, suspended		S.F. Ceiling	1.18	1.18	
.9	Interior Surface/Exterior Wall	Painted gypsum board on furring	90% of wall	S.F. Wall	2.81	.43	
7.0 Conveying							
.1	Elevators	N/A		—	—	—	0.0%
.2	Special Conveyors	N/A		—	—	—	
8.0 Mechanical							
.1	Plumbing	Toilet and service fixtures, supply and drainage	1 Fixture/4075 S.F. Floor	Each	2689	.66	
.2	Fire Protection	Sprinklers, light hazard		S.F. Floor	1.41	1.41	
.3	Heating	Included in 8.4		—	—	—	19.3%
.4	Cooling	Single zone unit gas heating, electric cooling		S.F. Floor	6.87	6.87	
.5	Special Systems	N/A		—	—	—	
9.0 Electrical							
.1	Service & Distribution	1200 ampere service, panel board and feeders		S.F. Floor	.50	.50	
.2	Lighting & Power	Fluorescent fixtures, receptacles, switches, A.C. and misc. power		S.F. Floor	5.10	5.10	12.6%
.4	Special Electrical	Alarm systems and emergency lighting		S.F. Floor	.30	.30	
11.0 Special Construction							
.1	Specialties	N/A		—	—	—	0.0%
12.0 Site Work							
.1	Earthwork	N/A		—	—	—	
.3	Utilities	N/A		—	—	—	0.0%
.5	Roads & Parking	N/A		—	—	—	
.7	Site Improvements	N/A		—	—	—	
				Sub-Total		46.67	100%
	GENERAL CONDITIONS (Overhead & Profit)				15%	7.00	
	ARCHITECT FEES				6%	3.23	
				Total Building Cost		**56.90**	

BUILDING TYPES

COMMERCIAL/INDUSTRIAL/INSTITUTIONAL

M.620 Store, Department, 3 Story

Costs per square foot of floor area

Exterior Wall	S.F. Area	50000	65000	80000	95000	110000	125000	140000	155000	170000
	L.F. Perimeter	533	593	670	715	778	840	871	923	976
Face Brick with Concrete Block Back-up	Steel Frame	75.60	72.35	70.65	68.90	67.95	67.20	66.25	65.75	65.30
	R/Conc. Frame	78.05	74.75	73.05	71.35	70.40	69.65	68.70	68.20	67.75
Face Brick on Steel Studs	Steel Frame	73.15	70.30	68.75	67.25	66.35	65.70	64.85	64.40	64.00
	R/Conc. Frame	76.65	73.75	72.20	70.70	69.80	69.15	68.30	67.85	67.45
Precast Concrete Panels Exposed Aggregate	Steel Frame	72.05	69.30	67.85	66.45	65.65	65.00	64.25	63.80	63.45
	R/Conc. Frame	75.20	72.45	71.00	69.60	68.80	68.15	67.35	66.95	66.60
Perimeter Adj., Add or Deduct	Per 100 L.F.	3.00	2.35	1.85	1.60	1.40	1.25	1.10	1.00	.90
Story Hgt. Adj., Add or Deduct	Per 1 Ft.	.95	.80	.75	.65	.60	.60	.55	.50	.50

For Basement, add $23.95 per square foot of basement area

The above costs were calculated using the basic specifications shown on the facing page. These costs should be adjusted where necessary for design alternatives and owner's requirements. Reported completed project costs, for this type of structure, range from $33.45 to $89.60 per S.F.

Common additives

Description	Unit	$ Cost
Closed Circuit Surveillance, One station		
Camera and monitor	Each	1325
For additional camera stations, add	Each	730
Directory Boards, Plastic, glass covered		
30" x 20"	Each	500
36" x 48"	Each	910
Aluminum, 24" x 18"	Each	445
36" x 24"	Each	535
48" x 32"	Each	630
48" x 60"	Each	1425
Elevators, Hydraulic passenger, 2 stops		
1500# capacity	Each	40,800
2500# capacity	Each	42,100
3500# capacity	Each	45,700
Additional stop, add	Each	3350
Emergency Lighting, 25 watt, battery operated		
Lead battery	Each	350
Nickel cadmium	Each	590

Description	Unit	$ Cost
Escalators, Metal		
32" wide, 10' story height	Each	92,200
20' story height	Each	105,300
48" wide, 10' story height	Each	97,300
20' story height	Each	110,800
Glass		
32" wide, 10' story height	Each	86,200
20' story height	Each	98,800
48" wide, 10' story height	Each	91,800
20' story height	Each	103,800
Safe, Office type, 4 hour rating		
30" x 18" x 18"	Each	3475
62" x 33" x 20"	Each	7575
Sound System		
Amplifier, 250 watts	Each	1500
Speaker, ceiling or wall	Each	130
Trumpet	Each	244

Location Factors appear in the Reference Section

Model costs calculated for a 3 story building with 16' story height and 95,000 square feet of floor area

Store, Department, 3 Story

			Unit	Unit Cost	Cost Per S.F.	% Of Sub-Total
1.0 Foundations						
.1	Footings & Foundations	Poured concrete; strip and spread footings and 4' foundation wall	S.F. Ground	3.36	1.12	
.4	Piles & Caissons	N/A	—	—	—	2.5%
.9	Excavation & Backfill	Site preparation for slab and trench for foundation wall and footing	S.F. Ground	.89	.30	
2.0 Substructure						
.1	Slab on Grade	4" reinforced concrete with vapor barrier and granular base	S.F. Slab	3.05	1.02	1.8%
.2	Special Substructures	N/A	—	—	—	
3.0 Superstructure						
.1	Columns & Beams	Steel columns with sprayed fiber fireproofing	L.F. Column	98	1.17	
.4	Structural Walls	N/A	—	—	—	
.5	Elevated Floors	Concrete slab with metal deck and beams	S.F. Floor	13.35	8.90	22.2%
.7	Roof	Metal deck, open web steel joists, beams, columns	S.F. Roof	5.28	1.76	
.9	Stairs	Concrete filled metal pan	Flight	6625	.70	
4.0 Exterior Closure						
.1	Walls	Face brick with concrete block backup 90% of wall	S.F. Wall	21	6.97	
.5	Exterior Wall Finishes	N/A	—	—	—	15.9%
.6	Doors	Revolving and sliding panel, mall-front	Each	6849	1.01	
.7	Windows & Glazed Walls	Storefront 10% of wall	S.F. Window	28	1.02	
5.0 Roofing						
.1	Roof Coverings	Built-up tar and gravel with flashing	S.F. Roof	1.95	.65	
.7	Insulation	Perlite/EPS composite	S.F. Roof	1.18	.39	2.0%
.8	Openings & Specialties	Gravel stop and hatches	S.F. Roof	.24	.08	
6.0 Interior Construction						
.1	Partitions	Gypsum board on metal studs 60 S.F. Floor/L.F. Partition	S.F. Partition	3.84	.64	
.4	Interior Doors	Single leaf hollow metal 600 S.F. Floor/Door	Each	530	.88	
.5	Wall Finishes	70% paint, 20% vinyl wall covering, 10% ceramic tile	S.F. Surface	1.17	.39	19.3%
.6	Floor Finishes	50% carpet, 40% vinyl tile, 10% terrazzo	S.F. Floor	5.30	5.30	
.7	Ceiling Finishes	Mineral fiber tile on concealed zee bars	S.F. Ceiling	3.06	3.06	
.9	Interior Surface/Exterior Wall	Paint 90% of wall	S.F. Wall	1.93	.63	
7.0 Conveying						
.1	Elevators	One hydraulic passenger, one hydraulic freight	Each	133,950	1.41	9.3%
.2	Special Conveyors	Four escalators	Each	—	3.87	
8.0 Mechanical						
.1	Plumbing	Toilet and service fixtures, supply and drainage 1 Fixture/2570 S.F. Floor	Each	1542	.60	
.2	Fire Protection	Sprinklers, light hazard	S.F. Floor	1.16	1.16	
.3	Heating	Included in 8.4	—	—	—	15.3%
.4	Cooling	Multizone rooftop unit, gas heating, electric cooling	S.F. Floor	6.87	6.87	
.5	Special Systems	N/A	—	—	—	
9.0 Electrical						
.1	Service & Distribution	1200 ampere service, panel board and feeders	S.F. Floor	.75	.75	
.2	Lighting & Power	Fluorescent fixtures, receptacles, switches, A.C. and misc. power	S.F. Floor	5.37	5.37	11.7%
.4	Special Electrical	Alarm systems and emergency lighting	S.F. Floor	.52	.52	
11.0 Special Construction						
.1	Specialties	N/A	—	—	—	0.0%
12.0 Site Work						
.1	Earthwork	N/A	—	—	—	
.3	Utilities	N/A	—	—	—	0.0%
.5	Roads & Parking	N/A	—	—	—	
.7	Site Improvements	N/A	—	—	—	
			Sub-Total		56.54	100%
	GENERAL CONDITIONS (Overhead & Profit)			15%	8.48	
	ARCHITECT FEES			6%	3.88	
			Total Building Cost		68.90	

BUILDING TYPES

COMMERCIAL/INDUSTRIAL/INSTITUTIONAL

M.630 Store, Retail

Costs per square foot of floor area

Exterior Wall	S.F. Area	4000	6000	8000	10000	12000	15000	18000	20000	22000
	L.F. Perimeter	260	340	360	410	440	490	540	565	594
Split Face Concrete Block	Steel Joists	77.55	70.90	64.95	62.45	60.15	58.00	56.55	55.70	55.05
Stucco on Concrete Block	Steel Joists	74.25	68.05	62.65	60.35	58.25	56.30	55.00	54.20	53.60
Painted Concrete Block	Steel Joists	72.20	66.00	60.65	58.35	56.30	54.30	53.05	52.25	51.65
Face Brick on Concrete Block	Steel Joists	88.35	80.30	72.40	69.20	66.20	63.35	61.50	60.35	59.45
Painted Reinforced Concrete	Steel Joists	82.05	74.85	68.05	65.25	62.65	60.20	58.60	57.60	56.85
Tilt-up Concrete Panels	Steel Joists	73.40	67.30	62.05	59.80	57.75	55.85	54.60	53.85	53.25
Perimeter Adj., Add or Deduct	Per 100 L.F.	8.85	5.90	4.45	3.55	2.95	2.35	1.95	1.75	1.60
Story Hgt. Adj., Add or Deduct	Per 1 Ft.	1.15	1.00	.80	.70	.65	.60	.50	.50	.45

For Basement, add $19.50 per square foot of basement area

The above costs were calculated using the basic specifications shown on the facing page. These costs should be adjusted where necessary for design alternatives and owner's requirements. Reported completed project costs, for this type of structure, range from $33.05 to $87.10 per S.F.

Common additives

Description	Unit	$ Cost
Emergency Lighting, 25 watt, battery operated		
Lead battery	Each	350
Nickel cadmium	Each	590
Safe, Office type, 4 hour rating		
30" x 18" x 18"	Each	3475
62" x 33" x 20"	Each	7575
Smoke Detectors		
Ceiling type	Each	141
Duct type	Each	335
Sound System		
Amplifier, 250 watts	Each	1500
Speaker, ceiling or wall	Each	130
Trumpet	Each	244

Location Factors appear in the Reference Section

Model costs calculated for a 1 story building with 14' story height and 8,000 square feet of floor area

Store, Retail

				Unit	Unit Cost	Cost Per S.F.	% Of Sub-Total
1.0 Foundations							
.1	Footings & Foundations	Poured concrete; strip and spread footings and 4' foundation wall		S.F. Ground	3.03	3.03	7.6%
.4	Piles & Caissons	N/A		–	–	–	
.9	Excavation & Backfill	Site preparation for slab and trench for foundation wall and footing		S.F. Ground	.96	.96	
2.0 Substructure							
.1	Slab on Grade	4" reinforced concrete with vapor barrier and granular base		S.F. Slab	3.05	3.05	5.8%
.2	Special Substructures	N/A		–	–	–	
3.0 Superstructure							
.1	Columns & Beams	Interior columns included in 3.7		–	–	–	
.4	Structural Walls	N/A		–	–	–	
.5	Elevated Floors	N/A		–	–	–	6.7%
.7	Roof	Metal deck, open web steel joists, beams, interior columns		S.F. Roof	3.51	3.51	
.9	Stairs	N/A		–	–	–	
4.0 Exterior Closure							
.1	Walls	Decorative concrete block	90% of wall	S.F. Wall	10.74	6.09	
.5	Exterior Wall Finishes	N/A		–	–	–	15.8%
.6	Doors	Sliding entrance door and hollow metal service doors		Each	1593	.40	
.7	Windows & Glazed Walls	Storefront windows	10% of wall	S.F. Window	28	1.78	
5.0 Roofing							
.1	Roof Coverings	Built-up tar and gravel with flashing		S.F. Roof	2.20	2.20	
.7	Insulation	Perlite/EPS composite		S.F. Roof	1.18	1.18	7.1%
.8	Openings & Specialties	Gravel stop and hatches		S.F. Roof	.31	.31	
6.0 Interior Construction							
.1	Partitions	Gypsum board on metal studs	60 S.F. Floor/L.F. Partition	S.F. Partition	3.84	.64	
.4	Interior Doors	Single leaf hollow metal	600 S.F. Floor/Door	Each	530	.88	
.5	Wall Finishes	Paint		S.F. Surface	.51	.17	15.8%
.6	Floor Finishes	Vinyl tile		S.F. Floor	2.40	2.40	
.7	Ceiling Finishes	Mineral fiber tile on concealed zee bars		S.F. Ceiling	3.06	3.06	
.9	Interior Surface/Exterior Wall	Paint	90% of wall	S.F. Wall	1.93	1.09	
7.0 Conveying							
.1	Elevators	N/A		–	–	–	0.0%
.2	Special Conveyors	N/A		–	–	–	
8.0 Mechanical							
.1	Plumbing	Toilet and service fixtures, supply and drainage	1 Fixture/890 S.F. Floor	Each	3853	4.33	
.2	Fire Protection	Wet pipe sprinkler system		S.F. Floor	2.09	2.09	
.3	Heating	Included in 8.4		–	–	–	25.4%
.4	Cooling	Single zone unit, gas heating, electric cooling		S.F. Floor	6.87	6.87	
.5	Special Systems	N/A		–	–	–	
9.0 Electrical							
.1	Service & Distribution	400 ampere service, panel board and feeders		S.F. Floor	1.53	1.53	
.2	Lighting & Power	Fluorescent fixtures, receptacles, switches, A.C. and misc. power		S.F. Floor	6.26	6.26	15.8%
.4	Special Electrical	Alarm systems and emergency lighting		S.F. Floor	.47	.47	
11.0 Special Construction							
.1	Specialties	N/A		–	–	–	0.0%
12.0 Site Work							
.1	Earthwork	N/A		–	–	–	
.3	Utilities	N/A		–	–	–	0.0%
.5	Roads & Parking	N/A		–	–	–	
.7	Site Improvements	N/A		–	–	–	
				Sub-Total		52.30	100%
	GENERAL CONDITIONS (Overhead & Profit)				15%	7.85	
	ARCHITECT FEES				8%	4.80	
				Total Building Cost		64.95	

BUILDING TYPES

COMMERCIAL/INDUSTRIAL/INSTITUTIONAL M.640 Supermarket

Costs per square foot of floor area

Exterior Wall	S.F. Area	6000	8000	12000	16000	20000	25000	32000	38000	45000
	L.F. Perimeter	310	360	460	520	600	656	773	806	900
Face Brick with Concrete Block Back-up	Steel Frame	88.75	82.35	75.95	71.00	68.75	65.75	63.90	61.70	60.70
	Bearing Walls	87.35	80.95	74.50	69.60	67.35	64.30	62.50	60.25	59.30
Stucco on Concrete Block	Steel Frame	77.55	72.60	67.60	63.95	62.25	60.05	58.65	57.10	56.35
	Bearing Walls	78.05	73.10	68.10	64.45	62.75	60.55	59.15	57.60	56.85
Precast Concrete Panels	Steel Frame	83.00	77.35	71.70	67.40	65.40	62.85	61.25	59.35	58.50
Metal Sandwich Panels	Steel Frame	80.50	75.15	69.80	65.80	63.95	61.55	60.05	58.30	57.50
Perimeter Adj., Add or Deduct	Per 100 L.F.	11.40	8.55	5.70	4.25	3.40	2.75	2.15	1.80	1.50
Story Hgt. Adj., Add or Deduct	Per 1 Ft.	1.60	1.40	1.20	1.00	.95	.85	.80	.65	.65

For Basement, add $14.35 per square foot of basement area

The above costs were calculated using the basic specifications shown on the facing page. These costs should be adjusted where necessary for design alternatives and owner's requirements. Reported completed project costs, for this type of structure, range from $38.05 to $83.00 per S.F.

Common additives

Description	Unit	$ Cost
Check Out Counter		
Single belt	Each	2175
Double belt	Each	3675
Scanner, registers, guns & memory 2 lanes	Each	13,200
10 lanes	Each	125,500
Power take away	Each	4575
Emergency Lighting, 25 watt, battery operated		
Lead battery	Each	350
Nickel cadmium	Each	590
Refrigerators, Prefabricated, walk-in		
7'-6" High, 6' x 6'	S.F.	117
10' x 10'	S.F.	91
12' x 14'	S.F.	81
12' x 20'	S.F.	72
Refrigerated Food Cases		
Dairy, multi deck, 12' long	Each	6975
Delicatessen case, single deck, 12' long	Each	4925
Multi deck, 18 S.F. shelf display	Each	7425
Freezer, self-contained chest type, 30 C.F.	Each	2950
Glass door upright, 78 C.F.	Each	6525

Description	Unit	$ Cost
Refrigerated Food Cases, cont.		
Frozen food, chest type, 12' long	Each	4775
Glass door reach in, 5 door	Each	9025
Island case 12' long, single deck	Each	5400
Multi deck	Each	11,100
Meat case 12' long, single deck	Each	3925
Multi deck	Each	6700
Produce, 12' long single deck	Each	5175
Multi deck	Each	5775
Safe, Office type, 4 hour rating		
30" x 18" x 18"	Each	3475
62" x 33" x 20"	Each	7575
Smoke Detectors		
Ceiling type	Each	141
Duct type	Each	335
Sound System		
Amplifier, 250 watts	Each	1500
Speaker, ceiling or wall	Each	130
Trumpet	Each	244

Model costs calculated for a 1 story building with 18' story height and 20,000 square feet of floor area

Supermarket

				Unit	Unit Cost	Cost Per S.F.	% Of Sub-Total
1.0 Foundations							
.1	Footings & Foundations	Poured concrete; strip and spread footings and 4' foundation wall		S.F. Ground	2.35	2.35	
.4	Piles & Caissons	N/A		—	—	—	5.9%
.9	Excavation & Backfill	Site preparation for slab and trench for foundation wall and footing		S.F. Ground	.89	.89	
2.0 Substructure							
.1	Slab on Grade	4" reinforced concrete with vapor barrier and granular base		S.F. Slab	3.05	3.05	5.6%
.2	Special Substructures	N/A		—	—	—	
3.0 Superstructure							
.1	Columns & Beams	Interior columns included in 3.7		—	—	—	
.4	Structural Walls	N/A		—	—	—	
.5	Elevated Floors	N/A		—	—	—	7.1%
.7	Roof	Metal deck, open web steel joists, beams, interior columns		S.F. Roof	3.87	3.87	
.9	Stairs	N/A		—	—	—	
4.0 Exterior Closure							
.1	Walls	Face brick with concrete block backup	85% of wall	S.F. Wall	21	9.78	
.5	Exterior Wall Finishes	N/A		—	—	—	27.0%
.6	Doors	Sliding entrance doors with electrical operator, hollow metal		Each	3074	1.69	
.7	Windows & Glazed Walls	Storefront windows	15% of wall	S.F. Window	40	3.29	
5.0 Roofing							
.1	Roof Coverings	Built-up tar and gravel with flashing		S.F. Roof	1.97	1.97	
.7	Insulation	Perlite/EPS composite		S.F. Roof	1.18	1.18	6.3%
.8	Openings & Specialties	Gravel stop and hatches		S.F. Roof	.31	.31	
6.0 Interior Construction							
.1	Partitions	50% concrete block, 50% gypsum board on metal studs	50 S.F. Floor/L.F. Partition	S.F. Partition	4.46	1.07	
.4	Interior Doors	Single leaf hollow metal	2000 S.F. Floor/Door	Each	530	.27	
.5	Wall Finishes	Paint		S.F. Surface	.71	.34	14.7%
.6	Floor Finishes	Vinyl composition tile		S.F. Floor	2.40	2.40	
.7	Ceiling Finishes	Mineral fiber tile on concealed zee bars		S.F. Ceiling	3.06	3.06	
.9	Interior Surface/Exterior Wall	Paint	85% of wall	S.F. Wall	1.93	.89	
7.0 Conveying							
.1	Elevators	N/A		—	—	—	0.0%
.2	Special Conveyors	N/A		—	—	—	
8.0 Mechanical							
.1	Plumbing	Toilet and service fixtures, supply and drainage	1 Fixture/1820 S.F. Floor	Each	3112	1.71	
.2	Fire Protection	Sprinklers, light hazard		S.F. Floor	1.57	1.57	
.3	Heating	Included in 8.4		—	—	—	18.1%
.4	Cooling	Single zone rooftop unit, gas heating, electric cooling		S.F. Floor	6.67	6.67	
.5	Special Systems	N/A		—	—	—	
9.0 Electrical							
.1	Service & Distribution	400 ampere service, panel board and feeders		S.F. Floor	.84	.84	
.2	Lighting & Power	Fluorescent fixtures, receptacles, switches, A.C. and misc. power		S.F. Floor	7.17	7.17	15.3%
.4	Special Electrical	Alarm systems and emergency lighting		S.F. Floor	.35	.35	
11.0 Special Construction							
.1	Specialties	N/A		—	—	—	0.0%
12.0 Site Work							
.1	Earthwork	N/A		—	—	—	
.3	Utilities	N/A		—	—	—	0.0%
.5	Roads & Parking	N/A		—	—	—	
.7	Site Improvements	N/A		—	—	—	

	Sub-Total	54.72	100%
GENERAL CONDITIONS (Overhead & Profit)	15%	8.21	
ARCHITECT FEES	7%	4.42	
Total Building Cost		**67.35**	

BUILDING TYPES

COMMERCIAL/INDUSTRIAL/INSTITUTIONAL — M.650 Swimming Pool, Enclosed

Costs per square foot of floor area

Exterior Wall	S.F. Area	16000	18000	20000	22000	24000	26000	28000	30000	32000
	L.F. Perimeter	560	560	600	640	680	673	706	740	737
Face Brick with Concrete Block Back-up	Wood Truss	115.80	111.60	110.25	109.10	108.20	105.60	104.75	104.10	102.40
	Precast Conc.	133.95	129.45	128.00	126.80	125.85	123.05	122.20	121.50	119.65
Metal Sandwich Panel	Wood Truss	125.00	121.45	120.30	119.30	118.50	116.35	115.65	115.05	113.60
Wood Siding	Wood Frame	120.85	117.90	116.90	116.00	115.35	113.50	112.90	112.40	111.20
Painted Concrete Block	Wood Frame	119.80	116.95	115.95	115.15	114.50	112.75	112.15	111.70	110.55
	Precast Conc.	121.70	118.55	117.50	116.65	115.90	114.00	113.35	112.85	111.60
Perimeter Adj., Add or Deduct	Per 100 L.F.	6.20	5.50	4.95	4.50	4.10	3.80	3.55	3.30	3.10
Story Hgt. Adj., Add or Deduct	Per 1 Ft.	1.05	.95	.90	.85	.85	.75	.75	.75	.70

For Basement, add $21.35 per square foot of basement area

The above costs were calculated using the basic specifications shown on the facing page. These costs should be adjusted where necessary for design alternatives and owner's requirements. Reported completed project costs, for this type of structure, range from $47.85 to $140.65 per S.F.

Common additives

Description	Unit	$ Cost
Bleachers, Telescoping, manual		
To 15 tier	Seat	65 - 76
16-20 tier	Seat	72 - 83
21-30 tier	Seat	88 - 104
For power operation, add	Seat	13.40 - 19.40
Emergency Lighting, 25 watt, battery operated		
Lead battery	Each	350
Nickel cadmium	Each	590
Lockers, Steel, single tier, 60" or 72"	Opening	134 - 213
2 tier, 60" or 72" total	Opening	83 - 104
5 tier, box lockers	Opening	44 - 56
Locker bench, lam. maple top only	L.F.	18.40
Pedestal, steel pipe	Each	43
Pool Equipment		
Diving stand, 3 meter	Each	7950
1 meter	Each	4725
Diving board, 16' aluminum	Each	1525
Fiberglass	Each	885
Lifeguard chair, fixed	Each	1550
Portable	Each	1300
Lights, underwater, 12 volt, 300 watt	Each	1175

Description	Unit	$ Cost
Sauna, Prefabricated, complete		
6' x 4'	Each	3725
6' x 6'	Each	4350
6' x 9'	Each	5400
8' x 8'	Each	5725
8' x 10'	Each	6800
10' x 12'	Each	8025
Sound System		
Amplifier, 250 watts	Each	1500
Speaker, ceiling or wall	Each	130
Trumpet	Each	244
Steam Bath, Complete, to 140 C.F.	Each	1175
To 300 C.F.	Each	1325
To 800 C.F.	Each	3375
To 2500 C.F.	Each	5525

Model costs calculated for a 1 story building with 24' story height and 20,000 square feet of floor area

Swimming Pool, Enclosed

			Unit	Unit Cost	Cost Per S.F.	% Of Sub-Total
1.0 Foundations						
.1	Footings & Foundations	Poured concrete; strip and spread footings and 4' foundation wall	S.F. Ground	4.07	4.07	11.2%
.4	Piles & Caissons	N/A	–	–	–	
.9	Excavation & Backfill	Site preparation for slab and trench for foundation wall and footing	S.F. Ground	5.85	5.85	
2.0 Substructure						
.1	Slab on Grade	4" reinforced concrete with vapor barrier and granular base	S.F. Slab	3.05	3.05	6.2%
.2	Special Substructures	Pool walls	S.F. Ground	2.41	2.41	
3.0 Superstructure						
.1	Columns & Beams	Wood columns included in 3.7	–	–	–	
.4	Structural Walls	N/A	–	–	–	
.5	Elevated Floors	N/A	–	–	–	13.2%
.7	Roof	Wood deck on laminated wood truss	S.F. Ground	11.75	11.75	
.9	Stairs	N/A	–	–	–	
4.0 Exterior Closure						
.1	Walls	Face brick with concrete block backup 80% of wall	S.F. Wall	21	12.36	
.5	Exterior Wall Finishes	N/A	–	–	–	18.9%
.6	Doors	Double aluminum and glass, hollow metal	Each	1429	.36	
.7	Windows & Glazed Walls	Outward projecting steel 20% of wall	S.F. Window	28	4.08	
5.0 Roofing						
.1	Roof Coverings	Asphalt shingles with flashing	S.F. Ground	1.01	1.01	
.7	Insulation	Perlite/EPS composite	S.F. Ground	–	–	1.3%
.8	Openings & Specialties	Gutters and downspouts	S.F. Ground	.15	.15	
6.0 Interior Construction						
.1	Partitions	Concrete block, toilet partitions 100 S.F. Floor/L.F. Partition	S.F. Partition	5.05	.65	
.4	Interior Doors	Single leaf hollow metal 1000 S.F. Floor/Door	Each	345	.35	
.5	Wall Finishes	Acrylic glazed coating	S.F. Surface	1.45	.29	17.4%
.6	Floor Finishes	70% terrazzo, 30% ceramic tile	S.F. Floor	12.76	12.76	
.7	Ceiling Finishes	Mineral fiber tile on concealed zee bars 20% of area	S.F. Ceiling	1.18	.24	
.9	Interior Surface/Exterior Wall	Paint 80% of wall	S.F. Wall	1.93	1.11	
7.0 Conveying						
.1	Elevators	N/A	–	–	–	0.0%
.2	Special Conveyors	N/A	–	–	–	
8.0 Mechanical						
.1	Plumbing	Toilet and service fixtures, supply and drainage 1 Fixture/210 S.F. Floor	Each	2005	9.55	
.2	Fire Protection	N/A	–	–	–	
.3	Heating	Terminal unit heaters 10% of area	–	.51	.51	24.5%
.4	Cooling	Single zone unit gas heating, electric cooling	S.F. Floor	11.75	11.75	
.5	Special Systems	N/A	–	–	–	
9.0 Electrical						
.1	Service & Distribution	200 ampere service, panel board and feeders	S.F. Floor	.63	.63	
.2	Lighting & Power	Fluorescent fixtures, receptacles, switches, A.C. and misc. power	S.F. Floor	5.54	5.54	7.3%
.4	Special Electrical	Alarm systems and emergency lighting	S.F. Floor	.29	.29	
11.0 Special Construction						
.1	Specialties	N/A	–	–	–	0.0%
12.0 Site Work						
.1	Earthwork	N/A	–	–	–	
.3	Utilities	N/A	–	–	–	0.0%
.5	Roads & Parking	N/A	–	–	–	
.7	Site Improvements	N/A	–	–	–	
				Sub-Total	88.76	100%
	GENERAL CONDITIONS (Overhead & Profit)			15%	13.31	
	ARCHITECT FEES			8%	8.18	
				Total Building Cost	110.25	

BUILDING TYPES

COMMERCIAL/INDUSTRIAL/INSTITUTIONAL — M.660 Telephone Exchange

Costs per square foot of floor area

Exterior Wall	S.F. Area	2000	3000	4000	5000	6000	7000	8000	9000	10000
	L.F. Perimeter	180	220	260	286	320	353	368	397	425
Face Brick with Concrete Block Back-up	Steel Frame	113.90	101.70	95.55	90.35	87.60	85.55	82.80	81.50	80.40
	Bearing Walls	112.45	100.25	94.10	88.90	86.15	84.10	81.35	80.05	78.90
Limestone with Concrete Block Back-up	Steel Frame	124.25	110.10	103.00	96.90	93.70	91.35	88.05	86.55	85.25
	Bearing Walls	122.80	108.65	101.55	95.45	92.25	89.90	86.60	85.10	83.80
Decorative Concrete Block	Steel Frame	103.90	93.50	88.30	84.00	81.65	79.95	77.65	76.60	75.65
	Bearing Walls	102.45	92.05	86.85	82.55	80.20	78.50	76.20	75.10	74.20
Perimeter Adj., Add or Deduct	Per 100 L.F.	27.60	18.40	13.80	11.05	9.20	7.90	6.90	6.15	5.55
Story Hgt. Adj., Add or Deduct	Per 1 Ft.	3.25	2.65	2.35	2.05	1.95	1.85	1.65	1.60	1.50

For Basement, add $20.45 per square foot of basement area

The above costs were calculated using the basic specifications shown on the facing page. These costs should be adjusted where necessary for design alternatives and owner's requirements. Reported completed project costs, for this type of structure, range from $52.70 to $186.00 per S.F.

Common additives

Description	Unit	$ Cost
Emergency Lighting, 25 watt, battery operated		
Lead battery	Each	350
Nickel cadmium	Each	590
Smoke Detectors		
Ceiling type	Each	141
Duct type	Each	335
Emergency Generators, complete system, gas		
15 kw	Each	11,800
85 kw	Each	31,400
170 kw	Each	93,500
Diesel, 50 kw	Each	24,000
150 kw	Each	42,400
350 kw	Each	64,000

Model costs calculated for a 1 story building with 12' story height and 5,000 square feet of floor area

Telephone Exchange

				Unit	Unit Cost	Cost Per S.F.	% Of Sub-Total
1.0 Foundations							
.1	Footings & Foundations	Poured concrete; strip and spread footings and 4' foundation wall		S.F. Ground	4.38	4.38	7.7%
.4	Piles & Caissons	N/A		—	—	—	
.9	Excavation & Backfill	Site preparation for slab and trench for foundation wall and footing		S.F. Ground	1.09	1.09	
2.0 Substructure							
.1	Slab on Grade	4" reinforced concrete with vapor barrier and granular base		S.F. Slab	3.05	3.05	4.3%
.2	Special Substructures	N/A		—	—	—	
3.0 Superstructure							
.1	Columns & Beams	Steel columns included in 3.7		—	—	—	
.4	Structural Walls	N/A		—	—	—	
.5	Elevated Floors	N/A		—	—	—	6.0%
.7	Roof	Metal deck, open web steel joists, beams, columns		S.F. Roof	4.26	4.26	
.9	Stairs	N/A		—	—	—	
4.0 Exterior Closure							
.1	Walls	Face brick with concrete block backup	80% of wall	S.F. Wall	21	11.70	
.5	Exterior Wall Finishes	N/A		—	—	—	27.3%
.6	Doors	Single aluminum glass with transom		Each	2258	.90	
.7	Windows & Glazed Walls	Outward projecting steel	20% of wall	Each	985	6.76	
5.0 Roofing							
.1	Roof Coverings	Built-up tar and gravel with flashing		S.F. Roof	2.39	2.39	
.7	Insulation	Perlite/EPS composite		S.F. Roof	1.18	1.18	5.5%
.8	Openings & Specialties	Gravel stop and hatches		S.F. Roof	.32	.32	
6.0 Interior Construction							
.1	Partitions	Double layer gypsum board on metal studs, toilet partitions	15 S.F. Floor/L.F. Partition	S.F. Partition	5.23	3.49	
.4	Interior Doors	Single leaf hollow metal	150 S.F. Floor/Door	Each	530	3.53	
.5	Wall Finishes	Paint		S.F. Surface	.50	.67	22.5%
.6	Floor Finishes	90% carpet, 10% terrazzo		S.F. Floor	5.29	5.29	
.7	Ceiling Finishes	Fiberglass board on exposed grid system, suspended		S.F. Ceiling	1.87	1.87	
.9	Interior Surface/Exterior Wall	Paint	80% of wall	S.F. Floor	1.93	1.06	
7.0 Conveying							
.1	Elevators	N/A		—	—	—	0.0%
.2	Special Conveyors	N/A		—	—	—	
8.0 Mechanical							
.1	Plumbing	Kitchen, toilet and service fixtures, supply and drainage	1 Fixture/715 S.F. Floor	Each	2824	3.95	
.2	Fire Protection	Wet pipe sprinkler system		S.F. Floor	2.19	2.19	
.3	Heating	Included in 8.4		—	—	—	18.7%
.4	Cooling	Single zone unit, gas heating, electric cooling		S.F. Floor	7.05	7.05	
.5	Special Systems	N/A		—	—	—	
9.0 Electrical							
.1	Service & Distribution	200 ampere service, panel board and feeders		S.F. Floor	1.05	1.05	
.2	Lighting & Power	Fluorescent fixtures, receptacles, switches, A.C. and misc. power		S.F. Floor	3.60	3.60	8.0%
.4	Special Electrical	Alarm systems and emergency lighting		S.F. Floor	1.01	1.01	
11.0 Special Construction							
.1	Specialties	N/A		—	—	—	0.0%
12.0 Site Work							
.1	Earthwork	N/A		—	—	—	
.3	Utilities	N/A		—	—	—	0.0%
.5	Roads & Parking	N/A		—	—	—	
.7	Site Improvements	N/A		—	—	—	
				Sub-Total		70.79	100%
	GENERAL CONDITIONS (Overhead & Profit)				15%	10.62	
	ARCHITECT FEES				11%	8.94	
				Total Building Cost		**90.35**	

BUILDING TYPES

COMMERCIAL/INDUSTRIAL/INSTITUTIONAL — M.670 Town Hall, 1 Story

Costs per square foot of floor area

Exterior Wall	S.F. Area	5000	6500	8000	9500	11000	14000	17500	21000	24000
	L.F. Perimeter	300	360	386	396	435	510	550	620	680
Face Brick with Concrete Block Back-up	Steel Joists	90.55	87.05	82.90	79.30	77.90	75.85	73.10	72.00	71.30
	Wood Joists	89.75	86.20	81.90	78.20	76.75	74.70	71.85	70.70	69.95
Stone with Concrete Block Back-up	Steel Joists	90.50	87.05	82.90	79.30	77.85	75.85	73.10	72.00	71.30
	Wood Joists	89.70	86.15	81.90	78.15	76.70	74.65	71.85	70.70	69.95
Brick Veneer	Wood Frame	87.25	84.00	80.15	76.80	75.50	73.60	71.10	70.10	69.40
Wood Siding	Wood Frame	81.80	78.95	75.80	73.05	71.90	70.35	68.25	67.40	66.85
Perimeter Adj., Add or Deduct	Per 100 L.F.	9.20	7.10	5.80	4.85	4.20	3.30	2.70	2.20	1.90
Story Hgt. Adj., Add or Deduct	Per 1 Ft.	1.70	1.60	1.40	1.20	1.15	1.05	.90	.80	.80

For Basement, add $17.10 per square foot of basement area

The above costs were calculated using the basic specifications shown on the facing page. These costs should be adjusted where necessary for design alternatives and owner's requirements. Reported completed project costs, for this type of structure, range from $46.85 to $132.15 per S.F.

Common additives

Description	Unit	$ Cost
Directory Boards, Plastic, glass covered		
30" x 20"	Each	500
36" x 48"	Each	910
Aluminum, 24" x 18"	Each	445
36" x 24"	Each	535
48" x 32"	Each	630
48" x 60"	Each	1425
Emergency Lighting, 25 watt, battery operated		
Lead battery	Each	350
Nickel cadmium	Each	590
Flagpoles, Complete		
Aluminum, 20' high	Each	1050
40' high	Each	2350
70' high	Each	6675
Fiberglass, 23' high	Each	1375
39'-5" high	Each	2400
59' high	Each	6575
Safe, Office type, 4 hour rating		
30" x 18" x 18"	Each	3475
62" x 33" x 20"	Each	7575

Description	Unit	$ Cost
Smoke Detectors		
Ceiling type	Each	141
Duct type	Each	335
Vault Front, Door & frame		
1 Hour test, 32" x 78"	Opening	3225
2 Hour test, 32" door	Opening	3625
40" door	Opening	4150
4 Hour test, 32" door	Opening	3800
40" door	Opening	4475
Time lock movement; two movement	Each	1575

Model costs calculated for a 1 story building with 12' story height and 11,000 square feet of floor area

Town Hall, 1 Story

				Unit	Unit Cost	Cost Per S.F.	% Of Sub-Total
1.0 Foundations							
.1	Footings & Foundations	Poured concrete; strip and spread footings and 4' foundation wall		S.F. Ground	2.89	2.89	6.2%
.4	Piles & Caissons	N/A		—	—	—	
.9	Excavation & Backfill	Site preparation for slab and trench for foundation wall and footing		S.F. Ground	.96	.96	
2.0 Substructure							
.1	Slab on Grade	4" reinforced concrete with vapor barrier and granular base		S.F. Slab	3.05	3.05	4.9%
.2	Special Substructures	N/A		—	—	—	
3.0 Superstructure							
.1	Columns & Beams	Steel columns included in 3.7		—	—	—	
.4	Structural Walls	N/A		—	—	—	
.5	Elevated Floors	N/A		—	—	—	6.0%
.7	Roof	Metal deck, open web steel joists, beams, interior columns		S.F. Roof	3.70	3.70	
.9	Stairs	N/A		—	—	—	
4.0 Exterior Closure							
.1	Walls	Face brick with concrete block backup	70% of wall	S.F. Wall	21	7.08	
.5	Exterior Wall Finishes	N/A		—	—	—	16.9%
.6	Doors	Metal and glass with transom		Each	1643	.60	
.7	Windows & Glazed Walls	Metal outward projecting	30% of wall	Each	458	2.83	
5.0 Roofing							
.1	Roof Coverings	Built-up tar and gravel with flashing		S.F. Roof	2.25	2.25	
.7	Insulation	Perlite/EPS composite		S.F. Roof	1.18	1.18	5.9%
.8	Openings & Specialties	Gravel stop and hatches		S.F. Roof	.22	.22	
6.0 Interior Construction							
.1	Partitions	Gypsum board on metal studs, toilet partition	20 S.F. Floor/L.F. Partition	S.F. Partition	3.84	2.31	
.4	Interior Doors	Wood solid core	200 S.F. Floor/Door	Each	399	2.00	
.5	Wall Finishes	90% paint, 10% ceramic tile		S.F. Surface	.96	.96	25.5%
.6	Floor Finishes	70% carpet, 15% terrazzo, 15% vinyl composition tile		S.F. Floor	6.59	6.59	
.7	Ceiling Finishes	Mineral fiber tile on concealed zee bars		S.F. Ceiling	3.06	3.06	
.9	Interior Surface/Exterior Wall	Painted gypsum board on furring	70% of wall	S.F. Wall	2.81	.93	
7.0 Conveying							
.1	Elevators	N/A		—	—	—	0.0%
.2	Special Conveyors	N/A		—	—	—	
8.0 Mechanical							
.1	Plumbing	Kitchen, toilet and service fixtures, supply and drainage	1 Fixture/500 S.F. Floor	Each	2365	4.73	
.2	Fire Protection	Wet pipe sprinkler system		S.F. Floor	1.57	1.57	
.3	Heating	Included in 8.4		—	—	—	22.1%
.4	Cooling	Multizone unit, gas heating, electric cooling		S.F. Floor	7.46	7.46	
.5	Special Systems	N/A		—	—	—	
9.0 Electrical							
.1	Service & Distribution	400 ampere service, panel board and feeders		S.F. Floor	1.10	1.10	
.2	Lighting & Power	Fluorescent fixtures, receptacles, switches, A.C. and misc. power		S.F. Floor	6.11	6.11	12.5%
.4	Special Electrical	Alarm systems and emergency lighting		S.F. Floor	.55	.55	
11.0 Special Construction							
.1	Specialties	N/A		—	—	—	0.0%
12.0 Site Work							
.1	Earthwork	N/A		—	—	—	
.3	Utilities	N/A		—	—	—	0.0%
.5	Roads & Parking	N/A		—	—	—	
.7	Site Improvements	N/A		—	—	—	
				Sub-Total		62.13	100%
	GENERAL CONDITIONS (Overhead & Profit)				15%	9.32	
	ARCHITECT FEES				9%	6.45	
				Total Building Cost		77.90	

BUILDING TYPES

211

COMMERCIAL/INDUSTRIAL/INSTITUTIONAL

M.680 Town Hall, 2-3 Story

Costs per square foot of floor area

Exterior Wall	S.F. Area	8000	10000	12000	15000	18000	24000	28000	35000	40000
	L.F. Perimeter	206	233	260	300	320	360	393	451	493
Face Brick with Concrete Block Back-up	Steel Frame	107.15	102.00	98.55	95.15	91.50	87.05	85.35	83.40	82.40
	R/Conc. Frame	108.55	103.40	99.95	96.55	92.90	88.45	86.75	84.80	83.80
Stone with Concrete Block Back-up	Steel Frame	101.00	96.65	93.75	90.80	87.75	83.95	82.50	80.80	79.95
	R/Conc. Frame	108.50	103.40	99.90	96.50	92.90	88.45	86.75	84.75	83.75
Limestone with Concrete Block Back-up	Steel Frame	114.75	108.90	104.95	101.05	96.75	91.50	89.50	87.20	86.05
	R/Conc. Frame	116.15	110.30	106.35	102.45	98.15	92.90	90.90	88.60	87.45
Perimeter Adj., Add or Deduct	Per 100 L.F.	14.40	11.50	9.60	7.60	6.40	4.80	4.10	3.25	2.90
Story Hgt. Adj., Add or Deduct	Per 1 Ft.	2.20	2.00	1.85	1.70	1.55	1.30	1.20	1.10	1.05

For Basement, add $16.45 per square foot of basement area

The above costs were calculated using the basic specifications shown on the facing page. These costs should be adjusted where necessary for design alternatives and owner's requirements. Reported completed project costs, for this type of structure, range from $46.85 to $132.15 per S.F.

Common additives

Description	Unit	$ Cost
Directory Boards, Plastic, glass covered		
30" x 20"	Each	500
36" x 48"	Each	910
Aluminum, 24" x 18"	Each	445
36" x 24"	Each	535
48" x 32"	Each	630
48" x 60"	Each	1425
Elevators, Hydraulic passenger, 2 stops		
1500# capacity	Each	40,800
2500# capacity	Each	42,100
3500# capacity	Each	45,700
Additional stop, add	Each	3350
Emergency Lighting, 25 watt, battery operated		
Lead battery	Each	350
Nickel cadmium	Each	590

Description	Unit	$ Cost
Flagpoles, Complete		
Aluminum, 20' high	Each	1050
40' high	Each	2350
70' high	Each	6675
Fiberglass, 23' high	Each	1375
39'-5" high	Each	2400
59' high	Each	6575
Safe, Office type, 4 hour rating		
30" x 18" x 18"	Each	3475
62" x 33" x 20"	Each	7575
Smoke Detectors		
Ceiling type	Each	141
Duct type	Each	335
Vault Front, Door & frame		
1 Hour test, 32" x 78"	Opening	3225
2 Hour test, 32" door	Opening	3625
40" door	Opening	4150
4 Hour test, 32" door	Opening	3800
40" door	Opening	4475
Time lock movement; two movement	Each	1575

Model costs calculated for a 3 story building with 12′ story height and 18,000 square feet of floor area

Town Hall, 2-3 Story

				Unit	Unit Cost	Cost Per S.F.	% Of Sub-Total
1.0	**Foundations**						
.1	Footings & Foundations	Poured concrete; strip and spread footings and 4′ foundation wall		S.F. Ground	5.28	1.76	
.4	Piles & Caissons	N/A		—	—	—	2.9%
.9	Excavation & Backfill	Site preparation for slab and trench for foundation wall and footing		S.F. Ground	.89	.30	
2.0	**Substructure**						
.1	Slab on Grade	4″ reinforced concrete with vapor barrier and granular base		S.F. Slab	3.05	1.02	1.5%
.2	Special Substructures	N/A		—	—	—	
3.0	**Superstructure**						
.1	Columns & Beams	Wide flange steel columns with gypsum board fireproofing		L.F. Column	46	1.41	
.4	Structural Walls	N/A		—	—	—	
.5	Elevated Floors	Open web steel joists, slab form, concrete		S.F. Floor	11.36	7.57	17.6%
.7	Roof	Metal deck, open web steel joists, beams, interior columns		S.F. Roof	4.33	1.44	
.9	Stairs	Concrete filled metal pan		Flight	5750	1.92	
4.0	**Exterior Closure**						
.1	Walls	Stone with concrete block backup	70% of wall	S.F. Wall	21	9.52	
.5	Exterior Wall Finishes	N/A		—	—	—	16.3%
.6	Doors	Metal and glass with transoms		Each	1517	.43	
.7	Windows & Glazed Walls	Metal outward projecting	10% of wall	Each	631	1.46	
5.0	**Roofing**						
.1	Roof Coverings	Built-up tar and gravel with flashing		S.F. Roof	2.49	.83	
.7	Insulation	Perlite/EPS composite		S.F. Roof	1.18	.39	1.7%
.8	Openings & Specialties	N/A		—	—	—	
6.0	**Interior Construction**						
.1	Partitions	Gypsum board on metal studs, toilet partitions	20 S.F. Floor/L.F. Partition	S.F. Partition	3.84	2.16	
.4	Interior Doors	Wood solid core	200 S.F. Floor/Door	Each	399	2.00	
.5	Wall Finishes	90% paint, 10% ceramic tile		S.F. Surface	.96	.96	22.9%
.6	Floor Finishes	70% carpet, 15% terrazzo, 15% vinyl composition tile		S.F. Floor	6.59	6.59	
.7	Ceiling Finishes	Mineral fiber tile on concealed zee bars		S.F. Ceiling	3.06	3.06	
.9	Interior Surface/Exterior Wall	Painted gypsum board on furring	70% of wall	S.F. Wall	2.81	1.26	
7.0	**Conveying**						
.1	Elevators	Two hydraulic elevators		Each	64,170	7.13	10.2%
.2	Special Conveyors	N/A		—	—	—	
8.0	**Mechanical**						
.1	Plumbing	Toilet and service fixtures, supply and drainage	1 Fixture/1385 S.F. Floor	Each	2119	1.53	
.2	Fire Protection	Sprinklers, light hazard		S.F. Floor	1.50	1.50	
.3	Heating	Included in 8.4		—	—	—	15.1%
.4	Cooling	Multizone unit, gas heating, electric cooling		S.F. Floor	7.46	7.46	
.5	Special Systems	N/A		—	—	—	
9.0	**Electrical**						
.1	Service & Distribution	400 ampere service, panel board and feeders		S.F. Floor	.94	.94	
.2	Lighting & Power	Fluorescent fixtures, receptacles, switches, A.C. and misc. power		S.F. Floor	6.77	6.77	11.8%
.4	Special Electrical	Alarm systems and emergency lighting		S.F. Floor	.58	.58	
11.0	**Special Construction**						
.1	Specialties	N/A		—	—	—	0.0%
12.0	**Site Work**						
.1	Earthwork	N/A		—	—	—	
.3	Utilities	N/A		—	—	—	0.0%
.5	Roads & Parking	N/A		—	—	—	
.7	Site Improvements	N/A		—	—	—	
				Sub-Total		69.99	100%
	GENERAL CONDITIONS (Overhead & Profit)				15%	10.50	
	ARCHITECT FEES				9%	7.26	
				Total Building Cost		**87.75**	

BUILDING TYPES

COMMERCIAL/INDUSTRIAL/INSTITUTIONAL — M.690 Warehouse

Costs per square foot of floor area

Exterior Wall	S.F. Area	10000	15000	20000	25000	30000	35000	40000	50000	60000
	L.F. Perimeter	410	500	600	700	700	766	833	966	1000
Brick with Concrete Block Back-up	Steel Frame	73.55	65.75	62.25	60.10	56.10	54.70	53.70	52.25	50.00
	Bearing Walls	73.00	65.00	61.40	59.25	55.15	53.75	52.70	51.20	48.90
Concrete Block	Steel Frame	57.85	52.90	50.65	49.30	47.10	46.25	45.60	44.75	43.50
	Bearing Walls	57.00	52.00	49.70	48.30	46.05	45.20	44.60	43.70	42.45
Galvanized Steel Siding	Steel Frame	58.60	53.95	51.75	50.45	48.40	47.60	47.00	46.20	45.05
Metal Sandwich Panels	Steel Frame	60.00	54.65	52.20	50.75	48.30	47.40	46.70	45.75	44.40
Perimeter Adj., Add or Deduct	Per 100 L.F.	7.80	5.20	3.90	3.15	2.60	2.25	1.95	1.55	1.30
Story Hgt. Adj., Add or Deduct	Per 1 Ft.	1.10	.90	.80	.75	.65	.60	.55	.55	.45

For Basement, add $15.75 per square foot of basement area

The above costs were calculated using the basic specifications shown on the facing page. These costs should be adjusted where necessary for design alternatives and owner's requirements. Reported completed project costs, for this type of structure, range from $19.90 to $79.60 per S.F.

Common additives

Description	Unit	$ Cost
Dock Leveler, 10 ton cap.		
6' x 8'	Each	3875
7' x 8'	Each	4100
Emergency Lighting, 25 watt, battery operated		
Lead battery	Each	350
Nickel cadmium	Each	590
Fence, Chain link, 6' high		
9 ga. wire	L.F.	13.30
6 ga. wire	L.F.	17.20
Gate	Each	241
Flagpoles, Complete		
Aluminum, 20' high	Each	1050
40' high	Each	2350
70' high	Each	6675
Fiberglass, 23' high	Each	1375
39'-5" high	Each	2400
59' high	Each	6575
Paving, Bituminous		
Wearing course plus base course	S.Y.	6.95
Sidewalks, Concrete 4" thick	S.F.	2.56

Description	Unit	$ Cost
Sound System		
Amplifier, 250 watts	Each	1500
Speaker, ceiling or wall	Each	130
Trumpet	Each	244
Yard Lighting, 20' aluminum pole with 400 watt high pressure sodium fixture.	Each	2165

Model costs calculated for a 1 story building with 24' story height and 30,000 square feet of floor area

Warehouse

				Unit	Unit Cost	Cost Per S.F.	% Of Sub-Total
1.0 Foundations							
.1	Footings & Foundations	Poured concrete; strip and spread footings and 4' foundation wall		S.F. Ground	2.31	2.31	
.4	Piles & Caissons	N/A		—	—	—	8.5%
.9	Excavation & Backfill	Site preparation for slab and trench for foundation wall and footing		S.F. Ground	.89	.89	
2.0 Substructure							
.1	Slab on Grade	5" reinforced concrete with vapor barrier and granular base		S.F. Slab	6.28	6.28	16.8%
.2	Special Substructures	N/A		—	—	—	
3.0 Superstructure							
.1	Columns & Beams	Steel columns included in 3.5 and 3.7		—	—	—	
.4	Structural Walls	N/A		—	—	—	
.5	Elevated Floors	Open web steel joists, slab form, concrete beams, columns, (above office)	10% of area	S.F. Floor	8.59	1.12	12.9%
.7	Roof	Metal deck, open web steel joists, beams, columns		S.F. Roof	3.40	3.40	
.9	Stairs	Steel gate with rails		Flight	4650	.31	
4.0 Exterior Closure							
.1	Walls	Concrete block	95% of wall	S.F. Wall	7.31	3.89	
.5	Exterior Wall Finishes	N/A		—	—	—	12.0%
.6	Doors	Steel overhead, hollow metal	5% of wall	Each	1592	.59	
.7	Windows & Glazed Walls	N/A		—	—	—	
5.0 Roofing							
.1	Roof Coverings	Built-up tar and gravel with flashing		S.F. Roof	1.87	1.87	
.7	Insulation	Perlite/EPS composite		S.F. Roof	1.18	1.18	9.0%
.8	Openings & Specialties	Gravel stop, hatches and skylight		S.F. Roof	.31	.31	
6.0 Interior Construction							
.1	Partitions	Concrete block (office and washrooms)	100 S.F. Floor/L.F. Partition	S.F. Partition	5.00	.40	
.4	Interior Doors	Single leaf hollow metal	5000 S.F. Floor/Door	Each	530	.11	
.5	Wall Finishes	Paint		S.F. Surface	.88	.14	8.1%
.6	Floor Finishes	90% hardener, 10% vinyl composition tile		S.F. Floor	1.03	1.03	
.7	Ceiling Finishes	Suspended mineral tile on zee channels in office area	10% of area	S.F. Ceiling	3.06	.31	
.9	Interior Surface/Exterior Wall	Paint	95% of wall	S.F. Wall	1.93	1.03	
7.0 Conveying							
.1	Elevators	N/A		—	—	—	0.0%
.2	Special Conveyors	N/A		—	—	—	
8.0 Mechanical							
.1	Plumbing	Toilet and service fixtures, supply and drainage	1 Fixture/2500 S.F. Floor	Each	3625	1.45	
.2	Fire Protection	Sprinklers, ordinary hazard		S.F. Floor	1.81	1.81	
.3	Heating	Oil fired hot water, unit heaters	90% of area	S.F. Floor	3.21	3.21	19.2%
.4	Cooling	Single zone unit gas, heating, electric cooling	10% of area	S.F. Floor	.74	.74	
.5	Special Systems	N/A		—	—	—	
9.0 Electrical							
.1	Service & Distribution	200 ampere service, panel board and feeders		S.F. Floor	.29	.29	
.2	Lighting & Power	Fluorescent fixtures, receptacles, switches, A.C. and misc. power		S.F. Floor	2.93	2.93	9.3%
.4	Special Electrical	Alarm systems		S.F. Floor	.27	.27	
11.0 Special Construction							
.1	Specialties	Dock boards, dock levelers		S.F. Floor	1.57	1.57	4.2%
12.0 Site Work							
.1	Earthwork	N/A		—	—	—	
.3	Utilities	N/A		—	—	—	0.0%
.5	Roads & Parking	N/A		—	—	—	
.7	Site Improvements	N/A		—	—	—	
				Sub-Total		37.44	100%
	GENERAL CONDITIONS (Overhead & Profit)				15%	5.62	
	ARCHITECT FEES				7%	2.99	
				Total Building Cost		46.05	

BUILDING TYPES

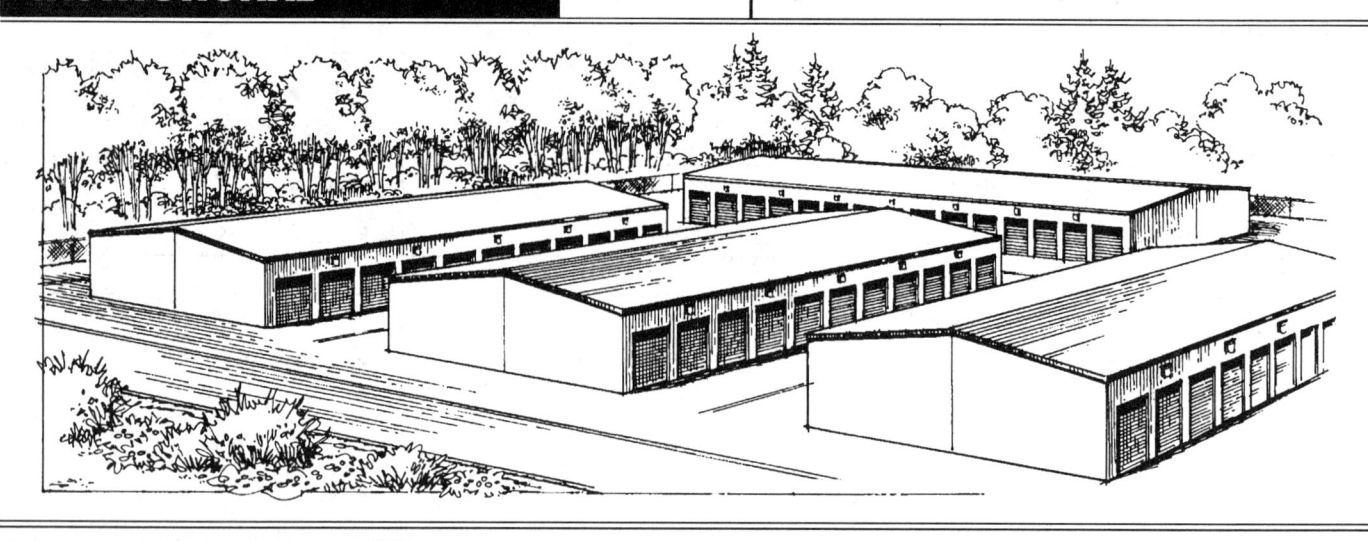

COMMERCIAL/INDUSTRIAL/INSTITUTIONAL — M.700 Warehouse, Mini

Costs per square foot of floor area

Exterior Wall	S.F. Area	2000	3000	5000	8000	12000	20000	30000	50000	100000
	L.F. Perimeter	179	220	284	350	438	566	693	894	1266
Concrete Block	Steel Frame	83.75	75.85	69.10	62.00	56.15	50.75	45.45	40.95	37.20
	R/Conc. Frame	89.60	82.20	75.25	70.35	67.70	64.85	63.15	61.55	59.95
Metal Sandwich Panel	Steel Frame	81.90	74.10	66.85	61.65	58.80	55.85	54.05	52.30	50.65
Tilt-up Concrete Panel	R/Conc. Frame	89.00	81.50	74.50	69.55	66.80	63.95	62.25	60.55	59.00
Precast Concrete Panel	Steel Frame	81.85	73.95	66.60	61.30	58.40	55.35	53.55	51.75	50.05
	Bearing Wall	93.85	85.30	77.30	71.50	68.35	65.00	62.95	60.95	59.05
Perimeter Adj., Add or Deduct	Per 100 L.F.	12.30	8.20	4.90	3.10	2.05	1.20	.80	.50	.25
Story Hgt. Adj., Add or Deduct	Per 1 Ft.	.70	.55	.45	.35	.30	.20	.20	.15	.10

Basement—Not Applicable

The above costs were calculated using the basic specifications shown on the facing page. These costs should be adjusted where necessary for design alternatives and owner's requirements. Reported completed project costs, for this type of structure, range from $16.55 to $70.25 per S.F.

Common additives

Description	Unit	$ Cost
Dock Leveler, 10 ton cap.		
6' x 8'	Each	3875
7' x 8'	Each	4100
Emergency Lighting, 25 watt, battery operated		
Lead battery	Each	350
Nickel cadmium	Each	590
Fence, Chain link, 6' high		
9 ga. wire	L.F.	13.30
6 ga. wire	L.F.	17.20
Gate	Each	241
Flagpoles, Complete		
Aluminum, 20' high	Each	1050
40' high	Each	2350
70' high	Each	6675
Fiberglass, 23' high	Each	1375
39'-5" high	Each	2400
59' high	Each	6575
Paving, Bituminous		
Wearing course plus base course	S.Y.	6.95
Sidewalks, Concrete 4" thick	S.F.	2.56

Description	Unit	$ Cost
Sound System		
Amplifier, 250 watts	Each	1500
Speaker, ceiling or wall	Each	130
Trumpet	Each	244
Yard Lighting, 20' aluminum pole with 400 watt high pressure sodium fixture	Each	2165

Model costs calculated for a 1 story building with 12' story height and 20,000 square feet of floor area

Warehouse, Mini

				Unit	Unit Cost	Cost Per S.F.	% Of Sub-Total
1.0	**Foundations**						
.1	Footings & Foundations	Poured concrete; strip and spread footings and 4' foundation wall		S.F. Ground	.71	.71	4.0%
.4	Piles & Caissons	N/A		–	–	–	
.9	Excavation & Backfill	Site preparation for slab and trench for foundation wall and footing		S.F. Ground	.96	.96	
2.0	**Substructure**						
.1	Slab on Grade	4" reinforced concrete with vapor barrier and granular base		S.F. Slab	5.30	5.30	12.9%
.2	Special Substructures	N/A		–	–	–	
3.0	**Superstructure**						
.1	Columns & Beams	Column fireproofing, steel columns included in 3.7		S.F. Floor	2.37	2.37	
.4	Structural Walls	N/A		–	–	–	
.5	Elevated Floors	N/A		–	–	–	18.5%
.7	Roof	Metal deck, open web steel joists, beams, columns		S.F. Roof	5.28	5.28	
.9	Stairs	N/A		–	–	–	
4.0	**Exterior Closure**						
.1	Walls	Concrete block	70% of wall	S.F. Wall	8.96	2.13	
.5	Exterior Wall Finishes	N/A		–	–	–	21.5%
.6	Doors	Steel overhead, hollow metal	25% of wall	Each	805	6.52	
.7	Windows & Glazed Walls	Aluminum projecting	5% of wall	Each	582	.23	
5.0	**Roofing**						
.1	Roof Coverings	Built-up tar and gravel with flashing		S.F. Roof	1.89	1.89	
.7	Insulation	Perlite/EPS composite		S.F. Roof	.78	.78	6.5%
.8	Openings & Specialties	N/A		–	–	–	
6.0	**Interior Construction**						
.1	Partitions	Concrete block, gypsum board on metal studs	10.65 S.F. Floor/L.F. Partition	S.F. Partition	7.81	5.56	
.4	Interior Doors	Single leaf hollow metal	1835 S.F. Floor/Door	Each	66	.22	
.5	Wall Finishes	Paint		S.F. Floor	.39	.39	15.6%
.6	Floor Finishes	Carpet, asphalt tile, ceramic tile		S.F. Floor	.25	.25	
.7	Ceiling Finishes	N/A		–	–	–	
.9	Interior Surface/Exterior Wall	N/A		–	–	–	
7.0	**Conveying**						
.1	Elevators	N/A		–	–	–	0.0%
.2	Special Conveyors	N/A		–	–	–	
8.0	**Mechanical**						
.1	Plumbing	Toilet and service fixtures, supply and drainage	1 Fixture/5000 S.F. Floor	Each	5000	1.00	
.2	Fire Protection	Wet pipe sprinkler system		S.F. Floor	2.09	2.09	
.3	Heating	Oil fired hot water, unit heaters		S.F. Floor	11,892	.59	8.9%
.4	Cooling	N/A		–	–	–	
.5	Special Systems	N/A		–	–	–	
9.0	**Electrical**						
.1	Service & Distribution	200 ampere service, panel board and feeders		S.F. Floor	.93	.93	
.2	Lighting & Power	Fluorescent fixtures, receptacles, switches and misc. power		S.F. Floor	3.23	3.23	10.7%
.4	Special Electrical	Alarm systems		S.F. Floor	.25	.25	
11.0	**Special Construction**						
.1	Specialties	Bathroom access, appliances, cabinets		S.F. Floor	.56	.56	1.4%
12.0	**Site Work**						
.1	Earthwork	N/A		–	–	–	
.3	Utilities	N/A		–	–	–	0.0%
.5	Roads & Parking	N/A		–	–	–	
.7	Site Improvements	N/A		–	–	–	
				Sub-Total		41.24	100%
	GENERAL CONDITIONS (Overhead & Profit)				15%	6.19	
	ARCHITECT FEES				7%	3.32	
				Total Building Cost		50.75	

BUILDING TYPES

COMMERCIAL/INDUSTRIAL/INSTITUTIONAL

Adjustments: Depreciation

Deterioration occurs within the structure itself and is determined by the observation of both materials and equipment.

Curable deterioration can be remedied either by maintenance, repair or replacement, within prudent economic limits.

Incurable deterioration that has progressed to the point of actually affecting the structural integrity of the structure, making repair or replacement not economically feasible.

Actual Versus Observed Age

The observed age of a structure refers to the age that the structure appears to be. Periodic maintenance, remodeling and renovation all tend to reduce the amount of deterioration that has taken place, thereby decreasing the observed age. Actual age on the other hand relates solely to the year that the structure was built.

The Depreciation Table shown here relates to the observed age.

Obsolescence arises from conditions either occurring within the structure (functional) or caused by factors outside the limits of the structure (economic).

Functional obsolescence is any inadequacy caused by outmoded design, dated construction materials or oversized or undersized areas, all of which cause excessive operation costs.

Incurable is so costly as not to economically justify the capital expenditure required to correct the deficiency.

Economic obsolescence is caused by factors outside the limits of the structure. The prime causes of economic obsolescence are:
- zoning and environmental laws
- government legislation
- negative neighborhood influences
- business climate
- proximity to transportation facilities

Depreciation Table
Commercial/Industrial/Institutional

Observed Age (Years)	Building Material		
	Frame	Masonry On Wood	Masonry On Masonry or Steel
1	1%	0%	0%
2	2	1	0
3	3	2	1
4	4	3	2
5	6	5	3
10	20	15	8
15	25	20	15
20	30	25	20
25	35	30	25
30	40	35	30
35	45	40	35
40	50	45	40
45	55	50	45
50	60	55	50
55	65	60	55
60	70	65	60

Assemblies Section

Table of Contents

Table No.		Page
	Introduction	221
	How to Use Assemblies	222
	Example	224

Foundations — 233

Footings & Foundations
- 1.1-120 Spread Footings ... 234
- 1.1-140 Strip Footings ... 234
- 1.1-210 Walls—Cast in Place ... 235
- 1.1-292 Foundation Dampproofing ... 235

Excavation & Backfill
- 1.9-100 Building Excavation & Backfill ... 236

Substructures — 237

Slab on Grade
- 2.1-200 Plain & Reinforced ... 238

Superstructures — 239

Columns, Beams & Joists
- 3.1-114 C.I.P. Columns—Square Tied ... 240
- 3.1-120 Precast Concrete Columns ... 242
- 3.1-130 Steel Columns ... 243
- 3.1-140 Wood Columns ... 246
- 3.1-190 Steel Column Fireproofing ... 247
- 3.1-224 "T" Shaped Precast Beams ... 247
- 3.1-226 "L" Shaped Precast Beams ... 247

Structural Walls
- 3.4-300 Metal Siding Support ... 248

Floors
- 3.5-110 C.I.P. Slabs—One Way ... 249
- 3.5-120 C.I.P. Beam & Slab—One Way ... 249
- 3.5-130 C.I.P. Beam & Slab—Two Way ... 250
- 3.5-140 C.I.P. Flat Slab with Drop Panels ... 250
- 3.5-150 C.I.P. Flat Plate ... 251
- 3.5-160 C.I.P. Multispan Joist Slab ... 251
- 3.5-170 C.I.P. Waffle Slab ... 252
- 3.5-210 Precast Plank ... 253
- 3.5-230 Precast Double "T" Beams ... 254
- 3.5-242 Precast Beam & Plank—No Topping ... 255
- 3.5-244 Precast Beam & Plank—2" Topping ... 255
- 3.5-254 Precast Double "T" & 2" Topping ... 256
- 3.5-310 Wide Flnge. Bms. & Gdrs. ... 257
- 3.5-360 Light Gauge Steel Floor Systems ... 258
- 3.5-420 Deck & Joists on Bearing Walls ... 259
- 3.5-440 Steel Joists on Beam & Wall ... 260
- 3.5-460 Steel Joists, Beams & Slab on Columns ... 261
- 3.5-520 Composite Beam & C.I.P. Slab ... 263
- 3.5-530 Wide Flange, Composite Deck & Slab ... 264
- 3.5-540 Composite Beams, Deck & Slab ... 265
- 3.5-580 Metal Deck/Concrete Fill ... 266
- 3.5-710 Wood Joist ... 266
- 3.5-720 Wood Beam & Joist ... 266

Roofs
- 3.7-410 Steel Joists, Beams & Deck on Columns & Walls ... 267
- 3.7-420 Steel Joists, Beams & Deck on Columns ... 268
- 3.7-430 Steel Joists & Deck on Bearing Walls ... 269
- 3.7-440 Steel Joists & Joist Girders on Columns & Walls ... 270
- 3.7-450 Steel Joists & Joist Girders on Columns ... 271
- 3.7-510 Wood/Flat or Pitched ... 272

Stairs
- 3.9-100 Stairs ... 273

Exterior Closure — 275

Walls
- 4.1-110 Cast in Place Concrete ... 276
- 4.1-140 Flat Precast Concrete ... 276
- 4.1-140 Fluted Window or Mullion Precast Concrete ... 277
- 4.1-140 Precast Concrete Specialties ... 278
- 4.1-140 Ribbed Precast Concrete ... 279
- 4.1-160 Tilt Up Concrete Panel ... 280
- 4.1-211 Concrete Block Wall ... 281
- 4.1-212 Split Ribbed Block Wall ... 284
- 4.1-213 Split Face Block Wall ... 285
- 4.1-231 Solid Brick Walls ... 287
- 4.1-242 Stone Veneer ... 289
- 4.1-252 Brick Veneer/Wood Stud Backup ... 291
- 4.1-252 Brick Veneer/Metal Stud Backup ... 292
- 4.1-272 Brick Face Composite Wall ... 294
- 4.1-273 Brick Face Cavity Wall ... 297
- 4.1-282 Glass Block ... 301
- 4.1-384 Metal Siding Panel ... 302
- 4.1-412 Wood & Other Siding ... 303

Exterior Wall Finishes
- 4.5-110 Stucco Wall ... 305

Doors
- 4.6-100 Wood, Steel & Aluminum ... 306

Windows & Glazed Walls
- 4.7-110 Wood, Steel & Aluminum ... 309
- 4.7-582 Tubular Alum. Framing ... 311
- 4.7-584 Curtain Wall Panels ... 312

Roofing — 313

Roof Covers
- 5.1-103 Built Up ... 314
- 5.1-220 Single Ply Membrane ... 315
- 5.1-310 Preformed Metal ... 315
- 5.1-330 Formed Metal ... 315
- 5.1-410 Shingle & Tile ... 316
- 5.1-520 Roof Edges ... 316
- 5.1-620 Flashings ... 317

Insulation
- 5.7-101 Roof Deck Rigid Insulation ... 318

Openings & Specialties
- 5.8-100 Hatches ... 319
- 5.8-100 Skylights ... 319
- 5.8-400 Gutters ... 320
- 5.8-500 Downspouts ... 320
- 5.8-500 Gravel Stop ... 320

Interior Construction — 321

Partitions
- 6.1-210 Concrete Block Partitions ... 322
- 6.1-270 Tile Partitions ... 324
- 6.1-510 Drywall Partitions ... 325
- 6.1-580 Drywall Components ... 327
- 6.1-610 Plaster Partitions ... 328
- 6.1-680 Plaster Partitions/Components ... 330
- 6.1-820 Folding Partitions ... 331
- 6.1-870 Toilet Partitions ... 332

Table of Contents

Table No.	Page
Doors	
6.4-100 Special Doors	333
Wall Finishes	
6.5-100 Paint & Covering	334
6.5-100 Paint Trim	335
Floor Finishes	
6.6-100 Tile & Covering	336
Ceiling Finishes	
6.7-100 Plaster Ceilings	338
6.7-100 Drywall Ceilings	338
6.7-100 Acoustical Ceilings	339
6.7-810 Acoustical Ceilings	339
6.7-820 Plaster Ceilings	340
Conveying Systems	341
Elevators	
7.1-100 Hydraulic	342
7.1-200 Traction Geared Elevators	342
7.1-300 Traction Gearless Elevators	343
Mechanical	345
Plumbing	
8.1-040 Piping—Installed—Unit Costs	346
8.1-120 Gas Fired Water Heaters—Residential	349
8.1-130 Oil Fired Water Heaters—Residential	349
8.1-160 Electric Water Heaters—Commercial	350
8.1-170 Gas Fired Water Heaters—Commercial	350
8.1-180 Oil Fired Water Heaters—Commercial	352
8.1-310 Roof Drain Systems	353
Reference Min. Plumbing Fixture Requirements	354
8.1-410 Bathtub Systems	355
8.1-420 Drinking Fountain Systems	355
8.1-431 Kitchen Sink Systems	356
8.1-432 Laundry Sink Systems	356
8.1-433 Lavatory Systems	357
8.1-434 Laboratory Sink Systems	357
8.1-434 Service Sink Systems	357
8.1-440 Shower Systems	358
8.1-450 Urinal Systems	358
8.1-460 Water Cooler Systems	358
8.1-470 Water Closet Systems	359

Table No.	Page
8.1-510 Water Closets—Group Systems	359
8.1-560 Group Wash Fountain Systems	359
8.1-620 Two Fixture Bathroom	360
8.1-630 Three Fixture Bathroom	360
8.1-640 Four Fixture Bathroom	361
8.1-650 Five Fixture Bathroom	361
Fire Protection	
Reference Sprinkler System Types	362
Reference Sprinkler System Classification	363
8.2-110 Wet Pipe Sprinkler Systems	364
8.2-120 Dry Pipe Sprinkler Systems	366
8.2-130 Preaction Sprinkler Systems	368
8.2-140 Deluge Sprinkler Systems	370
8.2-150 Firecycle Sprinkler Systems	372
8.2-310 Wet Standpipe Risers	374
8.2-320 Dry Standpipe Risers	375
8.2-390 Standpipe Equipment	376
8.2-810 FM200 Fire Suppression	377
Heating	
8.3-110 Small Hydronic Electric Boilers	378
8.3-120 Large Hydronic Electric Boilers	378
8.3-130 Boilers—Hot Water & Steam	379
8.3-141 Hydronic Heating—Fossil Fuel Unit Heaters	381
8.3-142 Hydronic Heating—Fossil Fuel Fin Tube Radiation	382
8.3-151 Apartment Bldg. Heating—Fin Tube Radiation	383
8.3-161 Commercial Bldg. Heating—Fin Tube Radiation	383
8.3-162 Comm. Bldg. Heating—Terminal Unit Heaters	383
Cooling	
8.4-110 Chilled Water—Air Cooled Condenser	384
8.4-120 Chilled Water—Cooling Tower	386
8.4-210 Rooftop Single Zone Units	388
8.4-220 Rooftop Multizone Units	390
8.4-230 Self-Contained Water Cooled	392
8.4-240 Self-Contained Air Cooled	394
8.4-250 Split System/Air Cooled Condensing Units	396
Special Systems	
8.5-110 Garage Exhaust System	398

Table No.	Page
Electrical	399
Service & Distribution	
9.1-210 Electric Service	400
9.1-310 Feeder Installation	400
9.1-410 Switchgear	400
Lighting & Power	
9.2-213 Fluorescent Fixtures (by Wattage)	401
9.2-223 Incandescent Fixtures (by Wattage)	401
9.2-235 H.I.D. Fixture—High Bay (by Wattage)	401
9.2-239 H.I.D. Fixture—High Bay (by Wattage)	402
9.2-242 H.I.D. Fixture—Low Bay (by Wattage)	403
9.2-244 H.I.D. Fixture—Low Bay (by Wattage)	404
9.2-252 Light Poles (Installed)	405
9.2-522 Receptacle (by Wattage)	406
9.2-524 Receptacles	406
9.2-542 Wall Switch by Square Foot	407
9.2-582 Miscellaneous Power	407
9.2-610 Central A.C. Power (by Wattage)	407
9.2-710 Motor Installation	408
9.2-720 Motor Feeder	409
Special Electrical	
9.4-100 Communication & Alarm Systems	410
9.4-310 Generators (by KW)	411
Special Construction	413
Specialties	
11.1-100 Architectural Specialties	414
11.1-200 Architectural Equipment	415
11.1-500 Furnishings	418
11.1-700 Special Construction	420
Site Work	423
Utilities	
12.3-110 Trenching	424
12.3-310 Pipe Bedding	424
12.3-710 Manholes & Catch Basins	424
Roads & Parking	
12.5-110 Roadway Pavement	425
12.5-510 Parking Lots	425

Introduction to the Assemblies Section

This section of the manual contains the technical description and installed cost of all components used in the commercial/industrial/institutional section. In most cases, there is an illustration of the component to aid in identification.

All Costs Include:

Materials purchased in lot sizes typical of normal construction projects with discounts appropriate to established contractors.

Installation work performed by union labor working under standard conditions at a normal pace.

Installing Contractor's Overhead & Profit

Standard installing contractor's overhead & profit are included in the assemblies costs.

These assemblies costs contain no provisions for General Contractor's overhead and profit or Architect's fees, nor do they include premiums for labor or materials or savings which may be realized under certain economic situations.

The assemblies tables are arranged by physical size (dimensions) of the component wherever possible.

How to Use the Assemblies Cost Tables

The following is a detailed explanation of a sample Assemblies Cost Table. Most Assembly Tables are separated into two parts:
1) an illustration of the system to be estimated with descriptive notes; and 2) the costs for similar systems with dimensional and/or size variations. Next to each bold number below is the item being described with the appropriate component of the sample entry following in parenthesis. In most cases, if the work is to be subcontracted, the general contractor will need to add an additional markup (R.S. Means suggests using 10%) to the "Total" figures.

1 System/Line Numbers (A2.1-200-2240)

Each Assemblies Cost Line has been assigned a unique identification number based on the UniFormat classification system.

2.1 200 2240

- UniFormat Division
- Means Subdivision
- Means Major Classification
- Means Individual Line Number

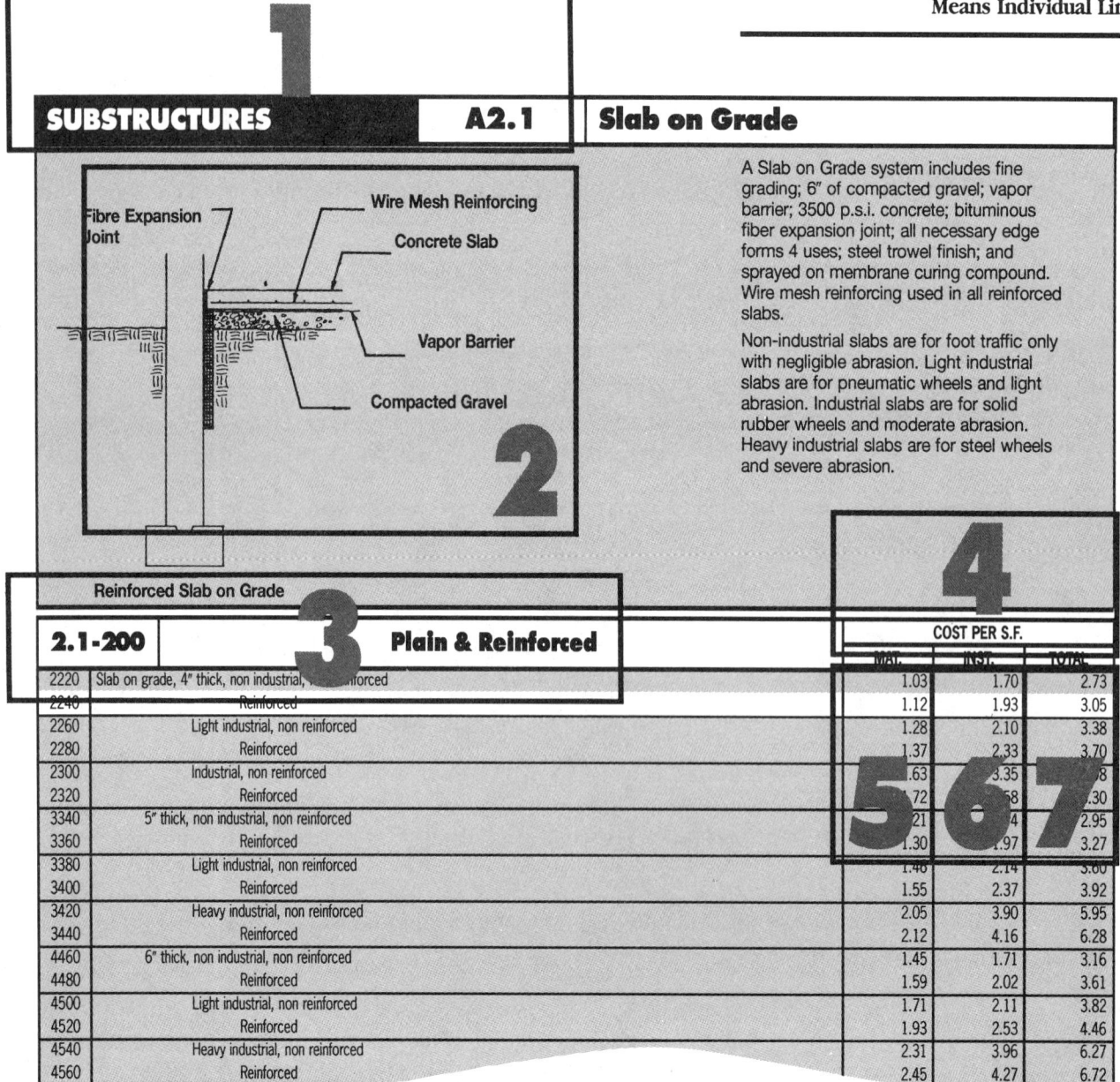

SUBSTRUCTURES — A2.1 — Slab on Grade

A Slab on Grade system includes fine grading; 6" of compacted gravel; vapor barrier; 3500 p.s.i. concrete; bituminous fiber expansion joint; all necessary edge forms 4 uses; steel trowel finish; and sprayed on membrane curing compound. Wire mesh reinforcing used in all reinforced slabs.

Non-industrial slabs are for foot traffic only with negligible abrasion. Light industrial slabs are for pneumatic wheels and light abrasion. Industrial slabs are for solid rubber wheels and moderate abrasion. Heavy industrial slabs are for steel wheels and severe abrasion.

Reinforced Slab on Grade

2.1-200	Plain & Reinforced	MAT.	INST.	TOTAL
2220	Slab on grade, 4" thick, non industrial, non reinforced	1.03	1.70	2.73
2240	Reinforced	1.12	1.93	3.05
2260	Light industrial, non reinforced	1.28	2.10	3.38
2280	Reinforced	1.37	2.33	3.70
2300	Industrial, non reinforced	1.63	3.35	
2320	Reinforced	1.72	3.58	5.30
3340	5" thick, non industrial, non reinforced	1.21	1.74	2.95
3360	Reinforced	1.30	1.97	3.27
3380	Light industrial, non reinforced	1.46	2.14	3.60
3400	Reinforced	1.55	2.37	3.92
3420	Heavy industrial, non reinforced	2.05	3.90	5.95
3440	Reinforced	2.12	4.16	6.28
4460	6" thick, non industrial, non reinforced	1.45	1.71	3.16
4480	Reinforced	1.59	2.02	3.61
4500	Light industrial, non reinforced	1.71	2.11	3.82
4520	Reinforced	1.93	2.53	4.46
4540	Heavy industrial, non reinforced	2.31	3.96	6.27
4560	Reinforced	2.45	4.27	6.72
	7" thick, non industrial,		1.76	

2. Illustration
At the top of most assembly pages is an illustration, a brief description, and the design criteria used to develop the cost.

3. System Description
The components of a typical system are listed in the description to show what has been included in the development of the total system price. The rest of the table contains prices for other similar systems with dimensional and/or size variations.

4. Unit of Measure for Each System (S.F.)
Costs shown in the three right hand columns have been adjusted by the component quantity and unit of measure for the entire system. In this example, "Cost per S.F." is the unit of measure for this system or "assembly."

5. Materials (1.12)
This column contains the Materials Cost of each component. These cost figures are bare costs plus 10% for profit.

6. Installation (1.93)
Installation includes labor and equipment plus the installing contractor's overhead and profit. Equipment costs are the bare rental costs plus 10% for profit. The labor overhead and profit is defined on the inside back cover of this book.

7. Total (3.05)
The figure in this column is the sum of the material and installation costs.

Material Cost	+	Installation Cost	=	Total
$1.12	+	$1.93	=	$3.05

223

Example

Assemblies costs can be used to calculate the cost for each component of a building and, accumulated with appropriate markups, produce a cost for the complete structure. Components from a model building with similar characteristics in conjunction with Assemblies components may be used to calculate costs for a complete structure.

Example:
Outline Specifications

General	Building size—60' x 100', 8 suspended floors, 12' floor to floor, 4' high parapet above roof, full basement 11'-8" floor to floor, bay size 25' x 30', ceiling heights - 9' in office area and 8' in core area.
1.0 Foundations	Concrete, spread and strip footings and concrete walls waterproofed.
2.0 Substructures	4" concrete slab on grade.
3.0 Superstructures	Columns, wide flange; 3 hr. fire rated; Floors, composite steel frame & deck with concrete slab; Roof, steel beams, open web joists and deck.
4.0 Exterior Closure	Walls; North, East & West, brick & lightweight concrete block with 2" cavity insulation, 25% window; South, 8" lightweight concrete block insulated, 10% window. Doors, aluminum & glass. Windows, aluminum, 3'-0" x 5'-4", insulating glass.
5.0 Roofing	Tar & gravel, 2" rigid insulation, R 12.5.
6.0 Interior Construction	Core—6" lightweight concrete block, full height with plaster on exposed faces to ceiling height.
	Corridors—1st & 2nd floor; 3-5/8" steel studs with F.R. gypsum board, full height.
	Exterior Wall—plaster, ceiling height.
	Doors—hollow metal.
	Wall Finishes—lobby, mahogany paneling on furring, remainder plaster & gypsum board, paint.
	Floor Finishes—1st floor lobby, corridors & toilet rooms, terrazzo; remainder, concrete, tenant developed. 2nd thru 8th, toilet rooms, ceramic tile; office & corridor, carpet.
	Ceiling Finishes—24" x 48" fiberglass board on Tee grid.
7.0 Conveying Systems	2-2500 lb. capacity, 200 F.P.M. geared elevators, 9 stops.
8.0 Mechanical	Fixtures, see sketch.
	Roof Drains—2-4" C.I. pipe.
	Fire Protection—4" standpipe, 9 hose cabinets.
	Heating—fin tube radiation, forced hot water.
	Air Conditioning—chilled water with water cooled condenser.
9.0 Electrical	Lighting, 1st thru 8th, 15 fluorescent fixtures/1000 S.F., 3 Watts/S.F.
	Basement, 10 fluorescent fixtures/1000 S.F., 2 Watts/S.F.
	Receptacles, 1st thru 8th, 16.5/1000 S.F., 2 Watts/S.F.
	Basement, 10 receptacles/1000 S.F., 1.2 Watts/S.F.
	Air Conditioning, 4 Watts/S.F.
	Miscellaneous Connections, 1.2 Watts/S.F.
	Elevator Power, 2-10 H.P., 230 volt motors.
	Wall Switches, 2/1000 S.F.
	Service, panel board & feeder, 2000 Amp.
	Fire Detection System, pull stations, signals, smoke and heat detectors.
	Emergency Lighting Generator, 30 KW.
10.0 General Conditions & Profit	See Summary Sheet.
11.0 Special Construction	Toilet Accessories & Directory Boards.
12.0 Site Work	NA.

Front Elevation

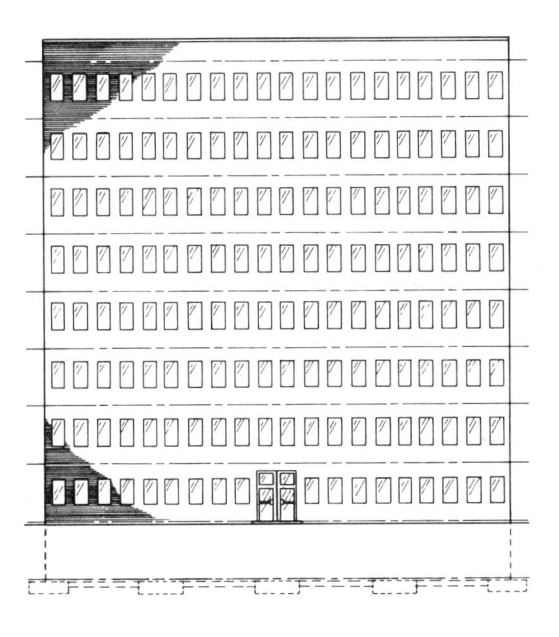

Basement Plan

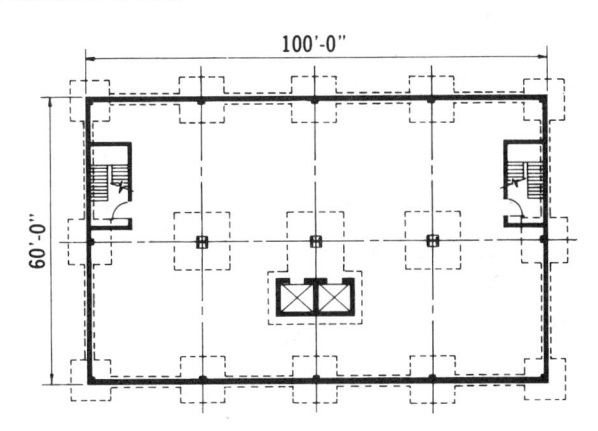

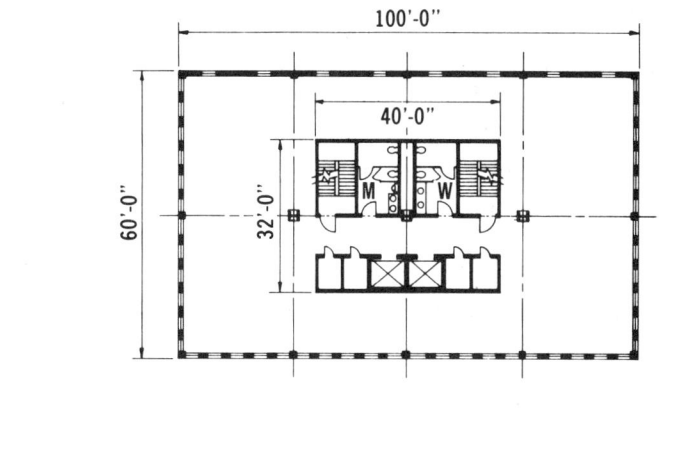

Ground Floor Plan

Typical Floor Plan

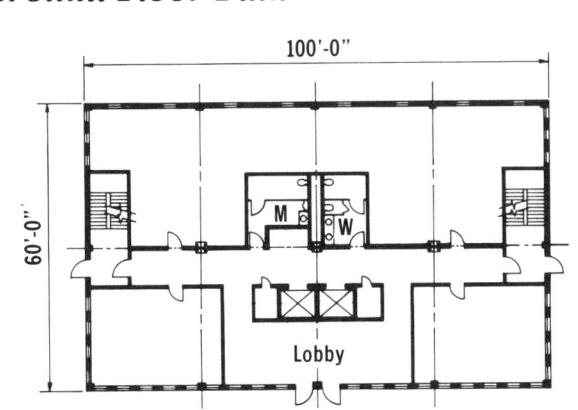

Example

If spread footing & column sizes are unknown, develop approximate loads as follows. Enter tables with these loads to determine costs.

Superimposed Load Ranges

Apartments & Residential Structures	65 to	75 psf
Assembly Areas & Retail Stores	110 to	125 psf
Commercial & Manufacturing	150 to	250 psf
Offices	75 to	100 psf

Approximate loads/S.F. for roof & floors.
Roof. Assume 40 psf superimposed load.
Steel joists, beams & deck.
Table 3.7-420—Line 3900

3.7-420 Steel Joists, Beams, & Deck on Columns

	BAY SIZE (FT.)	SUPERIMPOSED LOAD (P.S.F.)	DEPTH (IN.)	TOTAL LOAD (P.S.F.)	COLUMN ADD	MAT.	INST.	TOTAL
3500	25x30	20	22	40		2.61	.99	3.60
3600					columns	.52	.17	.69
3900		40	25	60		3.16	1.17	4.33
4000					columns	.62	.21	.83

3.5-540 Composite Beams, Deck & Slab

	BAY SIZE (FT.)	SUPERIMPOSED LOAD (P.S.F.)	SLAB THICKNESS (IN.)	TOTAL DEPTH (FT. - IN.)	TOTAL LOAD (P.S.F.)	MAT.	INST.	TOTAL
3400	25x30	40	5-1/2	1 - 11-1/2	83	5.80	3.65	9.45
3600		75	5-1/2	1 - 11-1/2	119	6.25	3.69	9.94
3900		125	5-1/2	1 - 11-1/2	170	7.20	4.16	11.36
4000		200	6-1/4	2 - 6-1/4	252	8.65	4.71	13.36

Floors—Total load, 119 psf.
Interior foundation load.
Roof

[(25' × 30' × 60 psf) + 8 floors × (25' × 30' × 119 psf)] × 1/1000 lb./Kip =	759 Kips
Approximate Footing Loads, Interior footing =	759 Kips
Exterior footing (1/2 bay) 759 k × .6 =	455 Kips
Corner footing (1/4 bay) 759 k × .45 =	342 Kips
[Factors to convert Interior load to Exterior & Corner loads]	
Approximate average Column load 759 k/2 =	379 Kips

Means Forms
PRELIMINARY ESTIMATE

PROJECT **Office Building** TOTAL SITE AREA

BUILDING TYPE OWNER

LOCATION ARCHITECT

DATE OF CONSTRUCTION ESTIMATED CONSTRUCTION PERIOD

BRIEF DESCRIPTION **Building size: 60' x 100' - 8 structural floors - 12' floor to floor 4' high parapet above roof, full basement, bay size 25' x 30', ceiling heights - 9' in office area + 8' in core area**

TYPE OF PLAN TYPE OF CONSTRUCTION

QUALITY BUILDING CAPACITY

Floor			Wall Areas				
Below Grade Levels			Foundation Walls	L.F.		Ht.	S.F.
Area		S.F.	Frost Walls	L.F.		Ht.	S.F.
Area		S.F.	Exterior Closure		Total		S.F.
Total Area		S.F.	Comment				
Ground Floor			Fenestration		%		S.F.
Area		S.F.			%		S.F.
Area		S.F.	Exterior Wall		%		S.F.
Total Area		S.F.			%		S.F.
Supported Levels			Site Work				
Area			Parking		S.F. (For		Cars)
Area		S.F.	Access Roads		L.F. (X		Ft. Wide)
Area		S.F.	Sidewalk		L.F. (X		Ft. Wide)
Area		S.F.	Landscaping		S.F. (		% Unbuilt Site)
Area		S.F.	Building Codes				
Total Area		S.F.	City		Country		
Miscellaneous			National		Other		
Area		S.F.	Loading				
Area		S.F.	Roof	psf	Ground Floor		psf
Area		S.F.	Supported Floors	psf	Corridor		psf
Area		S.F.	Balcony	psf	Partition, allow		psf
Total Area		S.F.	Miscellaneous				psf
Net Finished Area		S.F.	Live Load Reduction				
Net Floor Area		S.F.	Wind				
Gross Floor Area	**54,000**	S.F.	Earthquake		Zone		
Roof			Comment				
Total Area		S.F.	Soil Type				
Comments			Bearing Capacity				K.S.F.
			Frost Depth				Ft.
Volume			Frame				
Depth of Floor System			Type		Bay Spacing		
Minimum		In.	Foundation, Standard				
Maximum		In.	Special				
Foundation Wall Height		Ft.	Substructure				
Floor to Floor Height		Ft.	Comment				
Floor to Ceiling Height		Ft.	Superstructure, Vertical		Horizontal		
Subgrade Volume		C.F.	Fireproofing	☐ Columns			Hrs.
Above Grade Volume		C.F.	☐ Girders	Hrs.	☐ Beams		Hrs.
Total Building Volume	**648,000**	C.F.	☐ Floor	Hrs.	☐ None		

Means Forms

ASSEMBLY NUMBER		QTY.	UNIT	TOTAL COST UNIT	TOTAL COST TOTAL	COST PER S.F.
1.0	**Foundations**					
1.1-120-7900	Corner Footings 8'-6" SQ. x 27"	4	Ea.	1170	4,680	
-8010	Exterior 9'-6" SQ. x 30"	8		1560	12,480	
-8300	Interior 12" SQ.	3	↓	2825	8,475	
1.1-140-2700	Strip 2' Wide x 1' Thick					
	320 L.F. [(4 x 8.5) + (8 x 9.5)] =	210	L.F.	25.25	5,303	
1.1-210-7262	Foundation Wall 12' High, 1' Thick	↓	↓	142.50	29,925	
1.1-292-2800	Foundation Waterproofing	↓	↓	11.57	2,430	
1.9-100-3440	Building Excavation + Backfill	6000	S.F.	3.91	23,460	
-3500	(Interpolated ; 12' Between					
-4620	8' and 16' ; 6,000 Between					
-4680	4,000 and 10,000 S.F.)					
	Total				86,753	1.61
2.0	**Slab on Grade**					
2.1-200-2240	4", non-industrial, reinf.	6000	S.F.	3.05	18,300	
	Total				18,300	.34
3.0	**Superstructure**					
3.1-130-5800	Columns: Load 379K → Use 400K					
	Exterior 12 x 96' + Interior 3 x 108'	1476	V.L.F.	66.55	98,228	
3.1-190-3650	Column Fireproofing - Interior 4 sides					
	Exterior 1/2 (Interpolated for 12")	900	V.L.F	21.04	18,936	
3.5-540-3600	Floors: Composite steel + lt. wt. conc.	48,000	S.F.	9.94	477,120	
3.7-420-3900	Roof: Open web joists, beams + decks	6000	S.F.	4.33	25,980	
3.9-100-0760	Stairs: Steel w/conc. fill	18	Flt.	5750	103,500	
	Total				723,764	13.40
4.0	**Exterior Closure**					
4.1-273-1200	4" Brick + 6" Block - Insulated					
	75% NE + W Walls 220' x 100' x .75	16,500	S.F.	19.95	329,175	
4.1-211-3410	8" Block - Insulated 90%					
	100' x 100' x .9	9000	S.F.	6.39	57,510	
4.6-100-6950	Double Aluminum + Glass Door	1	Pr.	3525	3,525	
-6300	Single	2	Ea.	1580	3,160	
4.7-110-8800	Aluminum windows - Insul. glass					
	(5,500 S.F. + 1,000 S.F.)/3 x 5.33	406	Ea.	620.50	251,923	
4.1-140-6776	Precast concrete coping	320	L.F.	22.45	7,184	
	Total				652,477	12.08
5.0	**Roofing**					
5.7-101-4020	Insulation: Perlite/Polyisocyanurate composite	6000	S.F.	1.19	7,140	
5.1-103-1400	Roof: 4 Ply T & G	6000	↓	1.51	9,060	
5.1-620-0300	Flashing: Aluminum - fabric backed	640	↓	2.04	1,306	
5.8-100-0500	Roof hatch	1	Ea.	749	749	
	Total				18,255	.34
			Page Total			

Means Forms

ASSEMBLY NUMBER		QTY.	UNIT	TOTAL COST UNIT	TOTAL COST TOTAL	COST PER S.F.
6.0	**Interior Construction**					
6.1-210-5500	Core partitions: 6" Lt. Wt. concrete					
	block 288 L.F. x 11.5' x 8 Floors	26,449	S.F.	5.05	133,567	
6.1-680-0920	Plaster: [(196 L.F. x 8') + (144 L.F. x 9')] 8					
	+ Ext. Wall [(320 x 9 x 8 Floors) - 6,500]	39,452	S.F.	1.93	76,142	
6.1-510-5400	Corridor partitions 1st + 2nd floors					
	steel studs + F.R. gypsum board	2530	S.F.	2.83	7,160	
6.4-100-2600	Interior doors	66	Ea.	530.00	34,980	
	Wall finishes - Lobby: Mahogany w/furring					
6.1-580-0652	Furring: 150 L.F. x 9' + Ht.	1350	S.F.	.97	1,310	
6.5-100-1662	Mahogany paneling	1350	S.F.	3.74	5,049	
	Paint, plaster + gypsum board					
6.5-100-0080	39,452 + [(220' x 9' x 2) - 1,350] + 72	42,134	S.F.	.72	30,336	
	Floor Finishes					
6.6-100-1100	1st floor lobby + terrazzo	2175	S.F.	7.41	16,117	
	Remainder: Concrete - Tenant finished					
6.6-100-0060	2nd - 8th: Carpet - 4,725 S.F. x 7	33,075	S.F.	3.31	109,478	
6.6-100-1720	Ceramic Tile - 300 S.F. x 7	2100	S.F.	6.76	14,196	
6.7-810-2780/3260	Suspended Ceiling - 5,200 S.F. x 8	41,600	S.F.	2.00	83,200	
6.1-870-0700	Toilet partitions	31	Ea.	649	20,119	
-0760	Handicapped addition	16	Ea.	263	4,208	
	Total				535,862	9.92
7.0	**Conveying**					
	2 Elevators - 2500# 200'/min. Geared					
7.1-200-1600	5 Floors $91,800					
-1800	15 Floors $ 174,000					
	Diff. $82,200/10 = 8,220/Flr.					
	$91,800 + (4 x 8,220) = $124,680/Ea.	2	Ea.	124,680	249,360	
	Total				249,360	4.62
8.0	**Mechanical System**					
8.1-433-2120	Plumbing - Lavatories	31	Ea.	860	26,660	
-434-4340	-Service Sink	8	Ea.	1450	11,600	
-450-2000	-Urinals	8	Ea.	1035	8,280	
-470-2080	-Water Closets	31	Ea.	1065	33,015	
	Water Control, Waste Vent Piping	45%	OF TOTAL ABOVE		35,800	
-310-4200	- Roof Drains, 4" C.I.	2	Ea.	795	1,590	
-310-4240	- Pipe, 9 Flr. x 12' x 12' Ea.	216	L.F.	21.05	4,547	
	Fire Protection					
8.2-310-0560	Wet Stand Pipe: 4" x 10' 1st floor	12/10	Ea.	3850	4,620	
-310-0580	Additional Floors: (12'/10 x 8)	9.6	Ea.	1170	11,232	
-390-8400	Cabinet Assembly	9	Ea.	892	8,028	
	Heating - Fin Tube Radiation					
				Page Total		

Means Forms

ASSEMBLY NUMBER		QTY.	UNIT	TOTAL COST UNIT	TOTAL COST TOTAL	COST PER S.F.
8.3-161-2000	10,000 S.F. @ $6.05/S.F. ⎤ Interpolate					
-2040	100,000 S.F. @ $2.69/S.F. ⎦	48,000	S.F.	4.63	222,240	
	Cooling - Chilled Water, Air Cool, Cond.					
8.4-120-4000	40,000 S.F. @8.75/S.F. ⎤ Interpolate	48,000	S.F.	8.90	427,200	
-4040	60,000 S.F. @8.97/S.F. ⎦					
	Total				794,812	14.72
9.0	Electrical					
9.2-213-0280	Office Lighting, 15/1000 S.F. - 3 Watts/S.F.	48,000	S.F.	4.87	233,760	
-0240	Basement Lighting, 10/1000 S.F. - 2 Watts/S.F.	6000	S.F.	3.24	19,440	
-522-0640	Office Receptacles 16.5/1000 S.F. - 2 Watts/S.F.	48,000	S.F.	2.63	126,240	
-0560	Basement Receptacles 10/1000 S.F. - 1.2 Watts/S.F.	6000	S.F.	2.01	12,060	
-610-0280	Central A.C. - 4 Watts/S.F.	48,000	S.F.	.34	16,320	
-582-0320	Misc. Connections - 1.2 Watts/S.F.	48,000	S.F.	.18	8,640	
-710-0680	Elevator Motor Power - 10 H.P.	2	Ea.	1815	3,630	
-542-0280	Wall Switches - 2/1000 S.F.	54,000	S.F.	.26	14,040	
9.1-210-0560	2000 Amp Service	1	Ea.	22,825	22,825	
-410-0400	Switchgear	1	Ea.	35,700	35,700	
-310-0560	Feeder	50	L.F.	296	14,800	
9.4-100-0400	Fire Detection System - 50 Detectors	1	Ea.	19,725	19,725	
-310-0320	Emergency Generator - 30 kW	30	kW	546.50	16,395	
	Total				543,575	10.07
10.0	General Conditions					
11.0	Special Construction					
11.0-100-1100	Towel Dispenser	16	Ea.	62.50	1,000	
-1200	Grab Bar	32	Ea.	50.20	1,606	
-1300	Mirror	31	Ea.	115.45	3,579	
-1400	Toilet Tissue Dispenser	31	Ea.	21.50	667	
-2100	Directory Board	8	Ea.	178.50	1,428	
	Total				8,280	.15
12.0	Site Work					
13.0	Miscellaneous					
		Page Total				

Means Forms

PRELIMINARY ESTIMATE (Cost Summary)

PROJECT	Office Building	TOTAL AREA	54,000 S.F.	SHEET NO.	
LOCATION		TOTAL VOLUME	648,000 C.F.	ESTIMATE NO.	
ARCHITECT		COST PER S.F.		DATE	
OWNER		COST PER C.F.		NO. OF STORIES	
QUANTITIES BY		PRICES BY	EXTENSIONS BY	CHECKED BY	

NO.	DESCRIPTION	SUBTOTAL COST	COST/S.F.	%
1.0	Foundation	86,753	1.61	
2.0	Substructure	18,300	.34	
3.0	Superstructure	723,764	13.40	
4.0	Exterior Closure	652,477	12.08	
5.0	Roofing	18,255	.34	
6.0	Interior Construction	535,862	9.92	
7.0	Conveying	249,360	4.62	
8.0	Mechanical System	794,812	14.72	
9.0	Electrical	543,575	10.07	
10.0	General Conditions			
11.0	Special Construction	8,280	.15	
12.0	Site Work			

Building Subtotal 3,631,438 → $ 3,631,438

Sales Tax N/A % x Subtotal $ N/A /2 $ _____

General Conditions (%) 5 % x Subtotal $ 3,631,438 = 181,572

General Conditions $ 181,572
Subtotal "A" $ 3,813,010

Overhead 7 % x Subtotal "A" $ 3,813,010 $ 266,911
Subtotal "B" $ 4,079,921

Profit 3 % x Subtotal "B" $ 4,079,921 $ 122,398
Subtotal "C" $ 4,202,319

Location Factor N/A % x Subtotal "C" $ N/A
Adjusted Building Cost $ 4,202,319

Architect's Fee 6.5 % x Adjusted Building Cost 4,202,319 = $ 273,151
Contingency N/A % x Adjusted Building Cost N/A = $
Total Cost 4,475,470

Square Foot Cost $ 4,475,470 / 54,000 S.F. 82.88 $/S.F.
Cubic Foot Cost $ 4,475,470 / 648,000 C.F. 6.91 $/S.F.

Division 1
Foundations

FOUNDATIONS — A1.1 — Footings & Foundations

Waterproofing — Spread Footing — Strip Footing

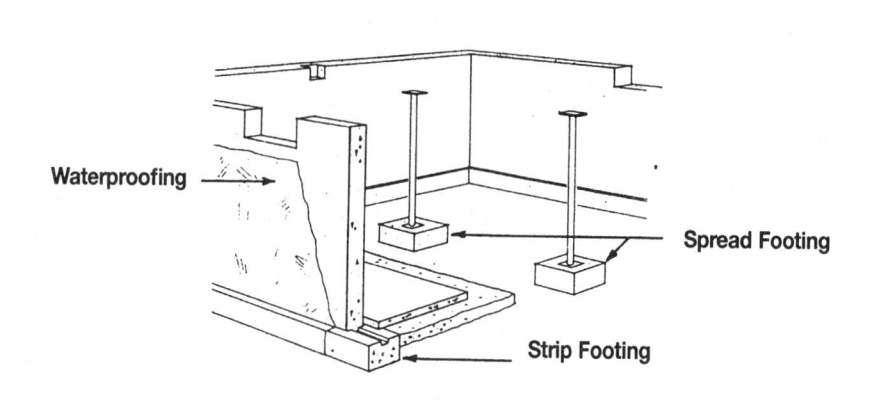

1.1-120 Spread Footings

		MAT.	INST.	TOTAL
7090	Spread footings, 3000 psi concrete, chute delivered			
7100	Load 25K, soil capacity 3 KSF, 3'-0" sq. x 12" deep	42	73.50	115.50
7150	Load 50K, soil capacity 3 KSF, 4'-6" sq. x 12" deep	83.50	128	211.50
7200	Load 50K, soil capacity 6 KSF, 3'-0" sq. x 12" deep	42	73.50	115.50
7250	Load 75K, soil capacity 3 KSF, 5'-6" sq. x 13" deep	128	181	309
7300	Load 75K, soil capacity 6 KSF, 4'-0" sq. x 12" deep	69	109	178
7350	Load 100K, soil capacity 3 KSF, 6'-0" sq. x 14" deep	160	216	376
7410	Load 100K, soil capacity 6 KSF, 4'-6" sq. x 15" deep	102	150	252
7450	Load 125K, soil capacity 3 KSF, 7'-0" sq. x 17" deep	250	310	560
7500	Load 125K, soil capacity 6 KSF, 5'-0" sq. x 16" deep	130	180	310
7550	Load 150K, soil capacity 3 KSF 7'-6" sq. x 18" deep	299	365	664
7610	Load 150K, soil capacity 6 KSF, 5'-6" sq. x 18" deep	171	227	398
7650	Load 200K, soil capacity 3 KSF, 8'-6" sq. x 20" deep	420	485	905
7700	Load 200K, soil capacity 6 KSF, 6'-0" sq. x 20" deep	221	280	501
7750	Load 300K, soil capacity 3 KSF, 10'-6" sq. x 25" deep	755	785	1,540
7810	Load 300K, soil capacity 6 KSF, 7'-6" sq. x 25" deep	410	470	880
7850	Load 400K, soil capacity 3 KSF, 12'-6" sq. x 28" deep	1,175	1,150	2,325
7900	Load 400K, soil capacity 6 KSF, 8'-6" sq. x 27" deep	560	610	1,170
8010	Load 500K, soil capacity 6 KSF, 9'-6" sq. x 30" deep	760	800	1,560
8100	Load 600K, soil capacity 6 KSF, 10'-6" sq. x 33" deep	1,025	1,025	2,050
8200	Load 700K, soil capacity 6 KSF, 11'-6" sq. x 36" deep	1,300	1,275	2,575
8300	Load 800K, soil capacity 6 KSF, 12'-0" sq. x 37" deep	1,450	1,375	2,825
8400	Load 900K, soil capacity 6 KSF, 13'-0" sq. x 39" deep	1,775	1,650	3,425
8500	Load 1000K, soil capacity 6 KSF, 13'-6" sq. x 41" deep	2,000	1,850	3,850

1.1-140 Strip Footings

		MAT.	INST.	TOTAL
2100	Strip footing, load 2.6KLF, soil capacity 3KSF, 16"wide x 8"deep plain	5.20	9	14.20
2300	Load 3.9 KLF, soil capacity, 3 KSF, 24"wide x 8"deep, plain	6.20	9.95	16.15
2500	Load 5.1KLF, soil capacity 3 KSF, 24"wide x 12"deep, reinf.	10.15	15.10	25.25
2700	Load 11.1KLF, soil capacity 6 KSF, 24"wide x 12"deep, reinf.	10.15	15.10	25.25
2900	Load 6.8 KLF, soil capacity 3 KSF, 32"wide x 12"deep, reinf.	12.05	16.50	28.55
3100	Load 14.8 KLF, soil capacity 6 KSF, 32"wide x 12"deep, reinf.	12.05	16.50	28.55
3300	Load 9.3 KLF, soil capacity 3 KSF, 40"wide x 12"deep, reinf.	13.85	17.90	31.75
3500	Load 18.4 KLF, soil capacity 6 KSF, 40"wide x 12"deep, reinf.	13.95	18.05	32
4500	Load 10KLF, soil capacity 3 KSF, 48"wide x 16"deep, reinf.	18.95	22	40.95
4700	Load 22KLF, soil capacity 6 KSF, 48"wide, 16"deep, reinf.	19.35	22.50	41.85
5700	Load 15KLF, soil capacity 3 KSF, 72"wide x 20"deep, reinf.	31.50	31.50	63
5900	Load 33KLF, soil capacity 6 KSF, 72"wide x 20"deep, reinf.	33.50	33.50	67

Important: See the Reference Section for critical supporting data - Location Factors & Historical Cost Indexes

FOUNDATIONS — A1.1 — Footings & Foundations

1.1-210 Walls, Cast in Place

	WALL HEIGHT (FT.)	PLACING METHOD	CONCRETE (C.Y./L.F.)	REINFORCING (LBS./L.F.)	WALL THICKNESS (IN.)	MAT.	INST.	TOTAL
1500	4'	direct chute	.074	3.3	6	9.15	28	37.15
1520			.099	4.8	8	11.10	29	40.10
1540			.123	6.0	10	12.90	29.50	42.40
1561			.148	7.2	12	14.75	30.50	45.25
1580			.173	8.1	14	16.55	31	47.55
1600			.197	9.44	16	18.40	32	50.40
3000	6'	direct chute	.111	4.95	6	13.70	42.50	56.20
3020			.149	7.20	8	16.65	44	60.65
3040			.184	9.00	10	19.30	44.50	63.80
3061			.222	10.8	12	22	45.50	67.50
5000	8'	direct chute	.148	6.6	6	18.25	57	75.25
5020			.199	9.6	8	22	58.50	80.50
5040			.250	12	10	26	59.50	85.50
5061			.296	14.39	12	29.50	61	90.50
6020	10'	direct chute	.248	12	8	27.50	73	100.50
6040			.307	14.99	10	32	74	106
6061			.370	17.99	12	37	76.50	113.50
7220	12'	pumped	.298	14.39	8	33.50	92.50	126
7240			.369	17.99	10	38.50	94	132.50
7262			.444	21.59	12	44.50	98	142.50
9220	16'	pumped	.397	19.19	8	44.50	123	167.50
9240			.492	23.99	10	51.50	126	177.50
9260			.593	28.79	12	59	130	189

1.1-292 Foundation Dampproofing

	Description	MAT.	INST.	TOTAL
1000	Foundation dampproofing, bituminous, 1 coat, 4' high	.28	2.69	2.97
1400	8' high	.56	5.40	5.96
1800	12' high	.84	8.35	9.19
2000	2 coats, 4' high	.44	3.33	3.77
2400	8' high	.88	6.65	7.53
2800	12' high	1.32	10.25	11.57
3000	Asphalt with fibers, 1/16" thick, 4' high	.68	3.33	4.01
3400	8' high	1.36	6.65	8.01
3800	12' high	2.04	10.25	12.29
4000	1/8" thick, 4' high	1.20	3.97	5.17
4400	8' high	2.40	7.95	10.35
4800	12' high	3.60	12.15	15.75
5000	Asphalt coated board and mastic, 1/4" thick, 4' high	2.20	3.61	5.81
5400	8' high	4.40	7.20	11.60
5800	12' high	6.60	11.10	17.70
6000	1/2" thick, 4' high	3.30	5.05	8.35
6400	8' high	6.60	10.05	16.65
6800	12' high	9.90	15.35	25.25
7000	Metallic coating, on walls, w/ 1/8" thick cement. coating, 4' high	7.20	26.50	33.70
7400	8' high	14.40	53.50	67.90
7800	12' high	21.50	80	101.50
8000	3 coat system on slabs, no protection course, 2' wide	.44	1.70	2.14
8400	4' wide	.88	3.40	4.28
8800	6' wide	1.32	5.10	6.42

For expanded coverage of these items see *Means Assemblies Cost Data 1997*

FOUNDATIONS — A1.9 Excavation & Backfill

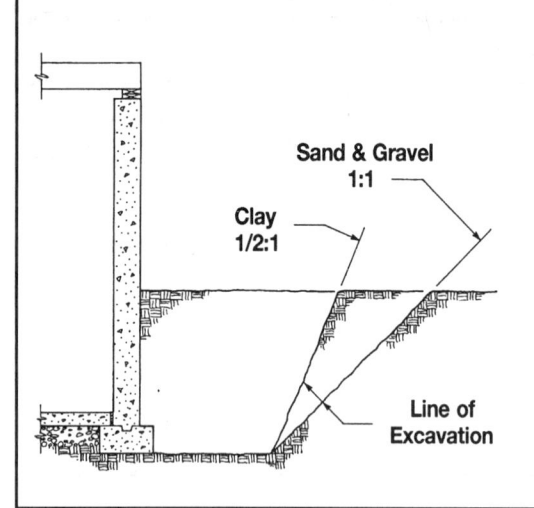

In general, the following items are accounted for in the table below.

Costs:
1) Excavation for building or other structure to depth and extent indicated.
2) Backfill compacted in place
3) Haul of excavated waste
4) Replacement of unsuitable backfill material with bank run gravel.

1.9-100	Building Excavation & Backfill	MAT.	INST.	TOTAL
2220	Excav & fill, 1000 S.F. 4' sand, gravel, or common earth, on site storage		1.43	1.43
2240	Off site storage		2.86	2.86
2260	Clay excavation, bank run gravel borrow for backfill	.83	2.74	3.57
2280	8' deep, sand, gravel, or common earth, on site storage		3.42	3.42
2300	Off site storage		7.35	7.35
2320	Clay excavation, bank run gravel borrow for backfill	2	6.25	8.25
2340	16' deep, sand, gravel, or common earth, on site storage		9.70	9.70
2350	Off site storage		18.25	18.25
2360	Clay excavation, bank run gravel borrow for backfill	5.55	16.15	21.70
3380	4000 S.F., 4' deep, sand, gravel, or common earth, on site storage		1.09	1.09
3400	Off site storage		1.75	1.75
3420	Clay excavation, bank run gravel borrow for backfill	.38	1.69	2.07
3440	8' deep, sand, gravel, or common earth, on site storage		2.38	2.38
3460	Off site storage		4.12	4.12
3480	Clay excavation, bank run gravel borrow for backfill	.91	3.72	4.63
3500	16' deep, sand, gravel, or common earth, on site storage		5.90	5.90
3520	Off site storage		11.40	11.40
3540	Clay, excavation, bank run gravel borrow for backfill	2.44	8.85	11.29
4560	10,000 S.F., 4' deep, sand gravel, or common earth, on site storage		.96	.96
4580	Off site storage		1.36	1.36
4600	Clay excavation, bank run gravel borrow for backfill	.24	1.34	1.58
4620	8' deep, sand, gravel, or common earth, on site storage		2.07	2.07
4640	Off site storage		3.11	3.11
4660	Clay excavation, bank run gravel borrow for backfill	.55	2.87	3.42
4680	16' deep, sand, gravel, or common earth, on site storage		4.81	4.81
4700	Off site storage		8	8
4720	Clay excavation, bank run gravel borrow for backfill	1.46	6.60	8.06
5740	30,000 S.F., 4' deep, sand, gravel, or common earth, on site storage		.89	.89
5760	Off site storage		1.12	1.12
5780	Clay excavation, bank run gravel borrow for backfill	.13	1.10	1.23
5860	16' deep, sand, gravel, or common earth, on site storage		4.09	4.09
5880	Off site storage		5.85	5.85
5900	Clay excavation, bank run gravel borrow for backfill	.81	5.05	5.86
6910	100,000 S.F., 4' deep, sand, gravel, or common earth, on site storage		.85	.85
6940	8' deep, sand, gravel, or common earth, on site storage		1.73	1.73
6970	16' deep, sand, gravel, or common earth, on site storage		3.69	3.69
6980	Off site storage		4.61	4.61
6990	Clay excavation, bank run gravel borrow for backfill	.44	4.19	4.63

For information about Means Estimating Seminars, see yellow pages 11 and 12 in back of book

Division 2
Substructures

SUBSTRUCTURES A2.1 Slab on Grade

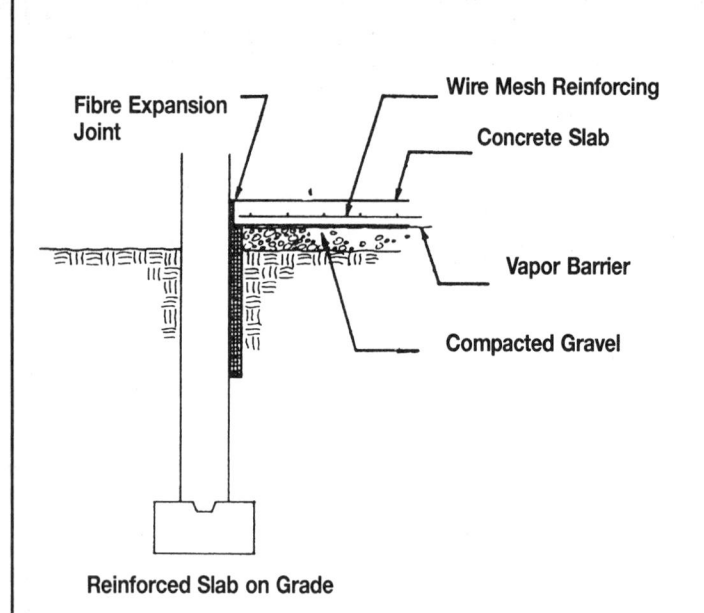

A Slab on Grade system includes fine grading; 6" of compacted gravel; vapor barrier; 3500 p.s.i. concrete; bituminous fiber expansion joint; all necessary edge forms 4 uses; steel trowel finish; and sprayed on membrane curing compound. Wire mesh reinforcing used in all reinforced slabs.

Non-industrial slabs are for foot traffic only with negligible abrasion. Light industrial slabs are for pneumatic wheels and light abrasion. Industrial slabs are for solid rubber wheels and moderate abrasion. Heavy industrial slabs are for steel wheels and severe abrasion.

Reinforced Slab on Grade

2.1-200	Plain & Reinforced	MAT.	INST.	TOTAL
2220	Slab on grade, 4" thick, non industrial, non reinforced	1.03	1.70	2.73
2240	Reinforced	1.12	1.93	3.05
2260	Light industrial, non reinforced	1.28	2.10	3.38
2280	Reinforced	1.37	2.33	3.70
2300	Industrial, non reinforced	1.63	3.35	4.98
2320	Reinforced	1.72	3.58	5.30
3340	5" thick, non industrial, non reinforced	1.21	1.74	2.95
3360	Reinforced	1.30	1.97	3.27
3380	Light industrial, non reinforced	1.46	2.14	3.60
3400	Reinforced	1.55	2.37	3.92
3420	Heavy industrial, non reinforced	2.05	3.90	5.95
3440	Reinforced	2.12	4.16	6.28
4460	6" thick, non industrial, non reinforced	1.45	1.71	3.16
4480	Reinforced	1.59	2.02	3.61
4500	Light industrial, non reinforced	1.71	2.11	3.82
4520	Reinforced	1.93	2.53	4.46
4540	Heavy industrial, non reinforced	2.31	3.96	6.27
4560	Reinforced	2.45	4.27	6.72
5580	7" thick, non industrial, non reinforced	1.65	1.76	3.41
5600	Reinforced	1.85	2.09	3.94
5620	Light industrial, non reinforced	1.91	2.16	4.07
5640	Reinforced	2.11	2.49	4.60
5660	Heavy industrial, non reinforced	2.50	3.89	6.39
5680	Reinforced	2.64	4.20	6.84
6700	8" thick, non industrial, non reinforced	1.83	1.79	3.62
6720	Reinforced	2	2.07	4.07
6740	Light industrial, non reinforced	2.10	2.19	4.29
6760	Reinforced	2.27	2.47	4.74
6780	Heavy industrial, non reinforced	2.71	3.95	6.66
6800	Reinforced	2.93	4.25	7.18

For information about Means Estimating Seminars, see yellow pages 11 and 12 in back of book

Division 3
Superstructures

SUPERSTRUCTURES — A3.1 — Columns, Beams & Joists

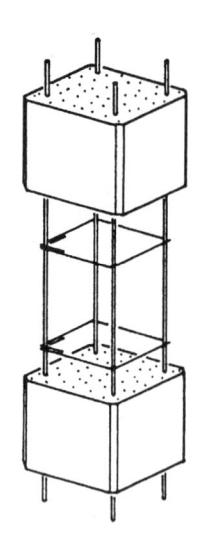

General: It is desirable for purposes of consistency and simplicity to maintain constant column sizes throughout building height. To do this, concrete strength may be varied (higher strength concrete at lower stories and lower strength concrete at upper stories), as well as varying the amount of reinforcing.

The table provides probably minimum column sizes with related costs and weight per lineal foot of story height.

3.1-114 C.I.P. Column, Square Tied

	LOAD (KIPS)	STORY HEIGHT (FT.)	COLUMN SIZE (IN.)	COLUMN WEIGHT (P.L.F.)	CONCRETE STRENGTH (PSI)	MAT.	INST.	TOTAL
0640	100	10	10	96	4000	9.15	28.50	37.65
0680		12	10	97	4000	9	28.50	37.50
0720		14	12	142	4000	11.05	34.50	45.55
0840	200	10	12	140	4000	11.15	35	46.15
0860		12	12	142	4000	11.10	34.50	45.60
0900		14	14	196	4000	20.50	40.50	61
0920	300	10	14	192	4000	20.50	41	61.50
0960		12	14	194	4000	20.50	40.50	61
0980		14	16	253	4000	23	44.50	67.50
1020	400	10	16	248	4000	24	46.50	70.50
1060		12	16	251	4000	24	46	70
1080		14	16	253	4000	24	46	70
1200	500	10	18	315	4000	21.50	55.50	77
1250		12	20	394	4000	22	57.50	79.50
1300		14	20	397	4000	22	57.50	79.50
1350	600	10	20	388	4000	25	63	88
1400		12	20	394	4000	25	62.50	87.50
1600		14	20	397	4000	24.50	62	86.50
3400	900	10	24	560	4000	34.50	80	114.50
3800		12	24	567	4000	34	79.50	113.50
4000		14	24	571	4000	33.50	79	112.50
7300	300	10	14	192	6000	20.50	40.50	61
7500		12	14	194	6000	20.50	40.50	61
7600		14	14	196	6000	20.50	40.50	61
8000	500	10	16	248	6000	24	46.50	70.50
8050		12	16	251	6000	24	46	70
8100		14	16	253	6000	24	46	70
8200	600	10	18	315	6000	28	53.50	81.50
8300		12	18	319	6000	27.50	53.50	81
8400		14	18	321	6000	27.50	53	80.50
8800	800	10	20	388	6000	22	58	80
8900		12	20	394	6000	22	57.50	79.50
9000		14	20	397	6000	22	57.50	79.50

Important: See the Reference Section for critical supporting data - Location Factors & Historical Cost Indexes

SUPERSTRUCTURES | A3.1 | Columns, Beams & Joists

3.1-114 C.I.P. Column, Square Tied

	LOAD (KIPS)	STORY HEIGHT (FT.)	COLUMN SIZE (IN.)	COLUMN WEIGHT (P.L.F.)	CONCRETE STRENGTH (PSI)	COST PER V.L.F. MAT.	COST PER V.L.F. INST.	COST PER V.L.F. TOTAL
9100	900	10	20	388	6000	30	72	102
9300		12	20	394	6000	29.50	71	100.50
9600		14	20	397	6000	29.50	70.50	100

3.1-114 C.I.P. Column, Square Tied-Minimum Reinforcing

	LOAD (KIPS)	STORY HEIGHT (FT.)	COLUMN SIZE (IN.)	COLUMN WEIGHT (P.L.F.)	CONCRETE STRENGTH (PSI)	COST PER V.L.F. MAT.	COST PER V.L.F. INST.	COST PER V.L.F. TOTAL
9912	150	10-14	12	135	4000	10.90	34.50	45.40
9918	300	10-14	16	240	4000	23	44	67
9924	500	10-14	20	375	4000	21.50	57	78.50
9930	700	10-14	24	540	4000	29.50	71.50	101
9936	1000	10-14	28	740	4000	36.50	85	121.50
9942	1400	10-14	32	965	4000	49	99	148
9948	1800	10-14	36	1220	4000	58	113	171

For expanded coverage of these items see *Means Assemblies Cost Data 1997*

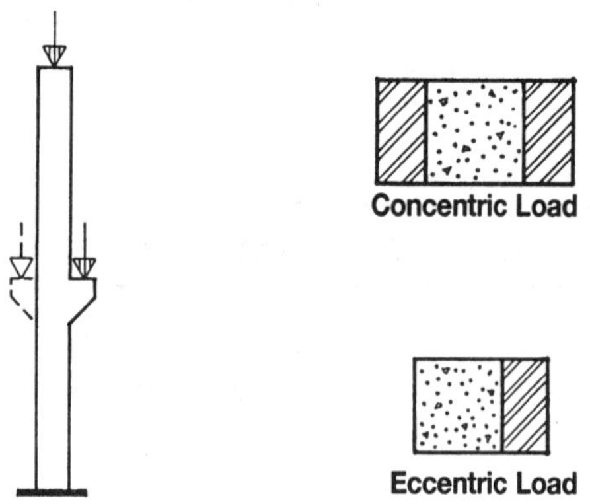

SUPERSTRUCTURES — A3.1 — Columns, Beams & Joists

General: Data presented here is for plant produced members transported 50 miles to 100 miles to the site and erected.

Design and pricing assumptions:
Normal wt. concrete, f'c = 5 KSI

Main reinforcement, fy = 60 KSI
Ties, fy = 40 KSI

Minimum design eccentricity, 0.1t.

Concrete encased structural steel haunches are assumed where practical; otherwise galvanized rebar haunches are assumed.

Base plates are integral with columns.

Foundation anchor bolts, nuts and washers are included in price.

3.1-120 Tied, Concentric Loaded Precast Concrete Columns

	LOAD (KIPS)	STORY HEIGHT (FT.)	COLUMN SIZE (IN.)	COLUMN WEIGHT (P.L.F.)	LOAD LEVELS	MAT.	INST.	TOTAL
0560	100	10	12x12	164	2	34	8	42
0570		12	12x12	162	2	33	6.70	39.70
0580		14	12x12	161	2	32	6.70	38.70
0590	150	10	12x12	166	3	32.50	6.70	39.20
0600		12	12x12	169	3	31	6	37
0610		14	12x12	162	3	31	6	37
0620	200	10	12x12	168	4	34	7.35	41.35
0630		12	12x12	170	4	34	6.70	40.70
0640		14	14x14	220	4	36.50	6.70	43.20

3.1-120 Tied, Eccentric Loaded Precast Concrete Columns

	LOAD (KIPS)	STORY HEIGHT (FT.)	COLUMN SIZE (IN.)	COLUMN WEIGHT (P.L.F.)	LOAD LEVELS	MAT.	INST.	TOTAL
1130	100	10	12x12	161	2	30.50	8	38.50
1140		12	12x12	159	2	32.50	6.70	39.20
1150		14	12x12	159	2	32	6.70	38.70
1390	600	10	18x18	385	4	60.50	7.35	67.85
1400		12	18x18	380	4	58.50	6.70	65.20
1410		14	18x18	375	4	58	6.70	64.70
1480	800	10	20x20	490	4	77	7.35	84.35
1490		12	20x20	480	4	76.50	6.70	83.20
1500		14	20x20	475	4	76	6.70	82.70

SUPERSTRUCTURES — A3.1 — Columns, Beams & Joists

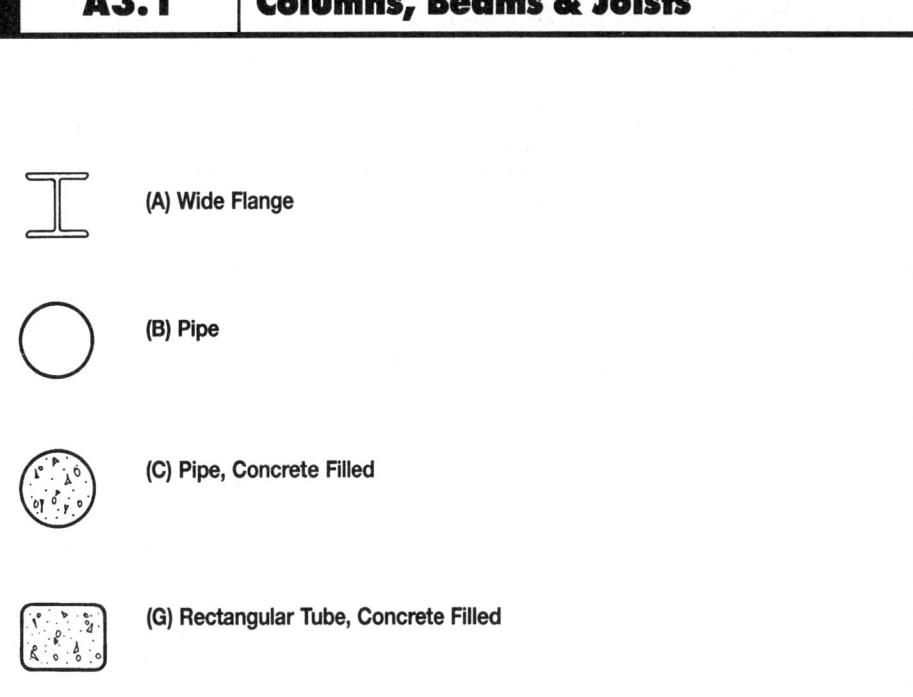

- (A) Wide Flange
- (B) Pipe
- (C) Pipe, Concrete Filled
- (G) Rectangular Tube, Concrete Filled

3.1-130 Steel Columns

	LOAD (KIPS)	UNSUPPORTED HEIGHT (FT.)	WEIGHT (P.L.F.)	SIZE (IN.)	TYPE	COST PER V.L.F. MAT.	INST.	TOTAL
1000	25	10	13	4	A	9.90	6.55	16.45
1020			7.58	3	B	9	6.55	15.55
1040			15	3-1/2	C	10.90	6.55	17.45
1120			20	4x3	G	12.10	6.55	18.65
1200		16	16	5	A	11.30	4.91	16.21
1220			10.79	4	B	11.85	4.91	16.76
1240			36	5-1/2	C	16.40	4.91	21.31
1320			64	8x6	G	24.50	4.91	29.41
1600	50	10	16	5	A	12.20	6.55	18.75
1620			14.62	5	B	17.35	6.55	23.90
1640			24	4-1/2	C	13	6.55	19.55
1720			28	6x3	G	16.05	6.55	22.60
1800		16	24	8	A	16.95	4.91	21.86
1840			36	5-1/2	C	16.40	4.91	21.31
1920			64	8x6	G	24.50	4.91	29.41
2000		20	28	8	A	18.70	4.91	23.61
2040			49	6-5/8	C	20	4.91	24.91
2120			64	8x6	G	23	4.91	27.91
2200	75	10	20	6	A	15.25	6.55	21.80
2240			36	4-1/2	C	33	6.55	39.55
2320			35	6x4	G	18.10	6.55	24.65
2400		16	31	8	A	22	4.91	26.91
2440			49	6-5/8	C	21.50	4.91	26.41
2520			64	8x6	G	24.50	4.91	29.41
2600		20	31	8	A	20.50	4.91	25.41
2640			81	8-5/8	C	30.50	4.91	35.41
2720			64	8x6	G	23	4.91	27.91
2800	100	10	24	8	A	18.30	6.55	24.85
2840			35	4-1/2	C	33	6.55	39.55
2920			46	8x4	G	22	6.55	28.55

For expanded coverage of these items see *Means Assemblies Cost Data 1997*

SUPERSTRUCTURES A3.1 Columns, Beams & Joists

3.1-130 Steel Columns

	LOAD (KIPS)	UNSUPPORTED HEIGHT (FT.)	WEIGHT (P.L.F.)	SIZE (IN.)	TYPE	COST PER V.L.F. MAT.	COST PER V.L.F. INST.	COST PER V.L.F. TOTAL
3000	100	16	31	8	A	22	4.91	26.91
3040			56	6-5/8	C	32	4.91	36.91
3120			64	8x6	G	24.50	4.91	29.41
3200		20	40	8	A	26.50	4.91	31.41
3240			81	8-5/8	C	30.50	4.91	35.41
3320			70	8x6	G	33	4.91	37.91
3400	125	10	31	8	A	23.50	6.55	30.05
3440			81	8	C	34.50	6.55	41.05
3520			64	8x6	G	26	6.55	32.55
3600	125	16	40	8	A	28	4.91	32.91
3640			81	8	C	32	4.91	36.91
3720			64	8x6	G	24.50	4.91	29.41
3800		20	48	8	A	32	4.91	36.91
3840			81	8	C	30.50	4.91	35.41
3920			60	8x6	G	33	4.91	37.91
4000	150	10	35	8	A	26.50	6.55	33.05
4040			81	8-5/8	C	34.50	6.55	41.05
4120			64	8x6	G	26	6.55	32.55
4200		16	45	10	A	32	4.91	36.91
4240			81	8-5/8	C	32	4.91	36.91
4320			70	8x6	G	35	4.91	39.91
4400		20	49	10	A	32.50	4.91	37.41
4440			123	10-3/4	C	43.50	4.91	48.41
4520			86	10x6	G	32.50	4.91	37.41
4600	200	10	45	10	A	34.50	6.55	41.05
4640			81	8-5/8	C	34.50	6.55	41.05
4720			70	8x6	G	37.50	6.55	44.05
4800		16	49	10	A	34.50	4.91	39.41
4840			123	10-3/4	C	46	4.91	50.91
4920			85	10x6	G	40.50	4.91	45.41
5200	300	10	61	14	A	46.50	6.55	53.05
5240			169	12-3/4	C	60.50	6.55	67.05
5320			86	10x6	G	56.50	6.55	63.05
5400		16	72	12	A	51	4.91	55.91
5440			169	12-3/4	C	56.50	4.91	61.41
5600		20	79	12	A	53	4.91	57.91
5640			169	12-3/4	C	53.50	4.91	58.41
5800	400	10	79	12	A	60	6.55	66.55
5840			178	12-3/4	C	79.50	6.55	86.05
6000		16	87	12	A	61.50	4.91	66.41
6040			178	12-3/4	C	73.50	4.91	78.41
6400	500	10	99	14	A	75.50	6.55	82.05
6600		16	109	14	A	77	4.91	81.91
6800		20	120	12	A	80	4.91	84.91
7000	600	10	120	12	A	91.50	6.55	98.05
7200		16	132	14	A	93	4.91	97.91
7400		20	132	14	A	88	4.91	92.91
7600	700	10	136	12	A	104	6.55	110.55
7800		16	145	14	A	102	4.91	106.91
8000		20	145	14	A	97	4.91	101.91
8200	800	10	145	14	A	111	6.55	117.55
8300		16	159	14	A	112	4.91	116.91
8400		20	176	14	A	118	4.91	122.91

Important: See the Reference Section for critical supporting data - Location Factors & Historical Cost Indexes

SUPERSTRUCTURES — A3.1 Columns, Beams & Joists

3.1-130 Steel Columns

	LOAD (KIPS)	UNSUPPORTED HEIGHT (FT.)	WEIGHT (P.L.F.)	SIZE (IN.)	TYPE	COST PER V.L.F. MAT.	COST PER V.L.F. INST.	COST PER V.L.F. TOTAL
8800	900	10	159	14	A	121	6.55	127.55
8900		16	176	14	A	124	4.91	128.91
9000		20	193	14	A	129	4.91	133.91
9100	1000	10	176	14	A	134	6.55	140.55
9200		16	193	14	A	136	4.91	140.91
9300		20	211	14	A	141	4.91	145.91

For expanded coverage of these items see *Means Assemblies Cost Data 1997*

SUPERSTRUCTURES   A3.1   Columns, Beams & Joists

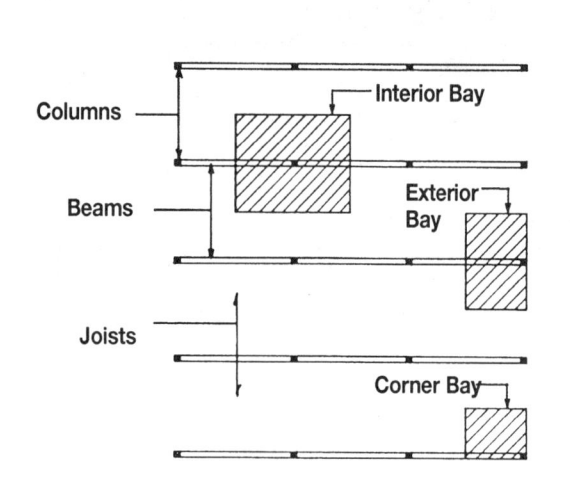

Description: Table below lists costs per S.F. of bay size for wood columns of various sizes and unsupported heights and the maximum allowable total load per S.F. per bay size.

3.1-140 Wood Columns

	NOMINAL COLUMN SIZE (IN.)	BAY SIZE (FT.)	UNSUPPORTED HEIGHT (FT.)	MATERIAL (BF/M.S.F.)	TOTAL LOAD (P.S.F.)	COST PER S.F. MAT.	COST PER S.F. INST.	COST PER S.F. TOTAL
1000	4 x 4	10 x 8	8	133	100	.10	.15	.25
1050			10	167	60	.12	.19	.31
1200		10 x 10	8	106	80	.08	.12	.20
1250			10	133	50	.10	.15	.25
1400		10 x 15	8	71	50	.05	.08	.13
1450			10	88	30	.06	.10	.16
1600		15 x 15	8	47	30	.03	.05	.08
1650			10	59	15	.04	.06	.10
2000	6 x 6	10 x 15	8	160	230	.12	.17	.29
2050			10	200	210	.15	.21	.36
2200		15 x 15	8	107	150	.08	.11	.19
2250			10	133	140	.10	.13	.23
2400		15 x 20	8	80	110	.06	.08	.14
2450			10	100	100	.07	.10	.17
2600		20 x 20	8	60	80	.04	.06	.10
2650			10	75	70	.06	.08	.14
2800		20 x 25	8	48	60	.04	.05	.09
2850			10	60	50	.04	.06	.10
3400	8 x 8	20 x 20	8	107	160	.09	.10	.19
3450			10	133	160	.11	.13	.24
3600		20 x 25	8	85	130	.06	.09	.15
3650			10	107	130	.08	.12	.20
3800		25 x 25	8	68	100	.06	.06	.12
3850			10	85	100	.08	.07	.15
4200	10 x 10	20 x 25	8	133	210	.13	.12	.25
4250			10	167	210	.16	.16	.32
4400		25 x 25	8	107	160	.10	.09	.19
4450			10	133	160	.13	.12	.25
4700	12 x 12	20 x 25	8	192	310	.19	.17	.36
4750			10	240	310	.23	.21	.44
4900		25 x 25	8	154	240	.15	.13	.28
4950			10	192	240	.19	.17	.36

Important: See the Reference Section for critical supporting data - Location Factors & Historical Cost Indexes

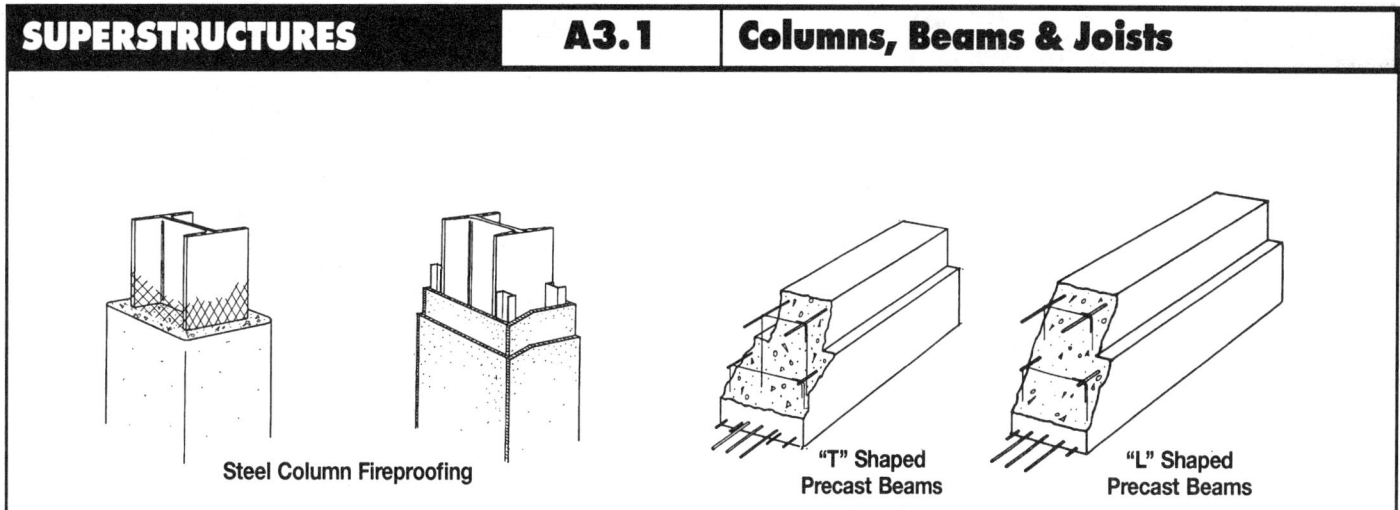

SUPERSTRUCTURES A3.1 Columns, Beams & Joists

Steel Column Fireproofing | **"T" Shaped Precast Beams** | **"L" Shaped Precast Beams**

3.1-190 Steel Column Fireproofing

	ENCASEMENT SYSTEM	COLUMN SIZE (IN.)	THICKNESS (IN.)	FIRE RATING (HRS.)	WEIGHT (P.L.F.)	MAT.	INST.	TOTAL
3000	Concrete	8	1	1	110	5.50	22	27.50
3300		14	1	1	258	8.55	32	40.55
3400			2	3	325	10.15	36	46.15
3450	Gypsum board	8	1/2	2	8	2.35	12.70	15.05
3550	1 layer	14	1/2	2	18	2.58	13.55	16.13
3600	Gypsum board	8	1	3	14	3.25	16.20	19.45
3650	1/2" fire rated	10	1	3	17	3.52	17.20	20.72
3700	2 layers	14	1	3	22	3.65	17.70	21.35
3750	Gypsum board	8	1-1/2	3	23	4.35	20.50	24.85
3800	1/2" fire rated	10	1-1/2	3	27	4.92	22.50	27.42
3850	3 layers	14	1-1/2	3	35	5.50	25	30.50
3900	Sprayed fiber	8	1-1/2	2	6.3	2.92	4.95	7.87
3950	Direct application		2	3	8.3	4.02	6.85	10.87
4050		10	1-1/2	2	7.9	3.52	6	9.52
4200		14	1-1/2	2	10.8	4.38	7.45	11.83

3.1-224 "T" Shaped Precast Beams

	SPAN (FT.)	SUPERIMPOSED LOAD (K.L.F.)	SIZE W X D (IN.)	BEAM WEIGHT (P.L.F.)	TOTAL LOAD (K.L.F.)	MAT.	INST.	TOTAL
2300	15	2.8	12x16	260	3.06	92.50	14.25	106.75
2500		8.37	12x28	515	8.89	117	14	131
8900	45	3.34	12x60	1165	4.51	206	13.35	219.35
9900		6.5	24x60	1915	8.42	260	13.35	273.35

3.1-226 "L" Shaped Precast Beams

	SPAN (FT.)	SUPERIMPOSED LOAD (K.L.F.)	SIZE W X D (IN.)	BEAM WEIGHT (P.L.F.)	TOTAL LOAD (K.L.F.)	MAT.	INST.	TOTAL
2250	15	2.58	12x16	230	2.81	69.50	14.25	83.75
2400		5.92	12x24	370	6.29	83.50	13.35	96.85
4000	25	2.64	12x28	435	3.08	92	8	100
4450		6.44	18x36	790	7.23	123	10.05	133.05
5300	30	2.80	12x36	565	3.37	110	8	118
6400		8.66	24x44	1245	9.90	169	14.70	183.70

For expanded coverage of these items see *Means Assemblies Cost Data 1997*

SUPERSTRUCTURES — A3.4 Structural Walls

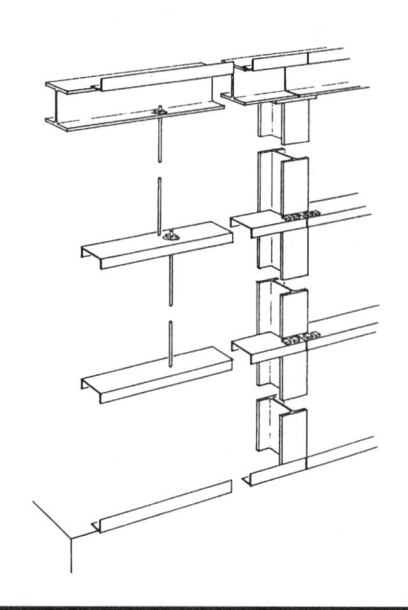

Description: The table below lists costs, $/S.F., for channel girts with sag rods and connector angles top and bottom for various column spacings, building heights and wind loads. Additive costs are shown for wind columns.

3.4-300 Metal Siding Support

	BLDG. HEIGHT (FT.)	WIND LOAD (P.S.F.)	COL. SPACING (FT.)		INTERMEDIATE COLUMNS	MAT.	INST.	TOTAL
3000	18	20	20			1.27	.71	1.98
3100					wind cols.	.54	.18	.72
3200		20	25			1.37	.74	2.11
3300					wind cols.	.43	.14	.57
3400		20	30			1.49	.78	2.27
3500					wind cols.	.36	.12	.48
3600		20	35			1.62	.82	2.44
3700					wind cols.	.31	.10	.41
3800		30	20			1.38	.74	2.12
3900					wind cols.	.54	.18	.72
4000		30	25			1.49	.78	2.27
4100					wind cols.	.43	.14	.57
4200		30	30			1.62	.83	2.45
4300					wind cols.	.49	.16	.65
4600	30	20	20			1.16	.57	1.73
4700					wind cols.	.76	.25	1.01
4800		20	25			1.29	.61	1.90
4900					wind cols.	.61	.20	.81
5000		20	30			1.43	.66	2.09
5100					wind cols.	.60	.20	.80
5200		20	35			1.58	.72	2.30
5300					wind cols.	.59	.20	.79
5400		30	20			1.30	.62	1.92
5500					wind cols.	.90	.30	1.20
5600		30	25			1.43	.66	2.09
5700					wind cols.	.86	.29	1.15
5800		30	30			1.59	.72	2.31
5900					wind cols.	.81	.27	1.08

SUPERSTRUCTURES — A3.5 Floors

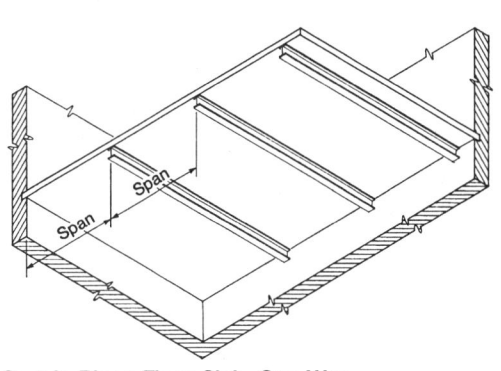

Cast in Place Floor Slab, One Way
General: Solid concrete slabs of uniform depth reinforced for flexure in one direction and for temperature and shrinkage in the other direction.

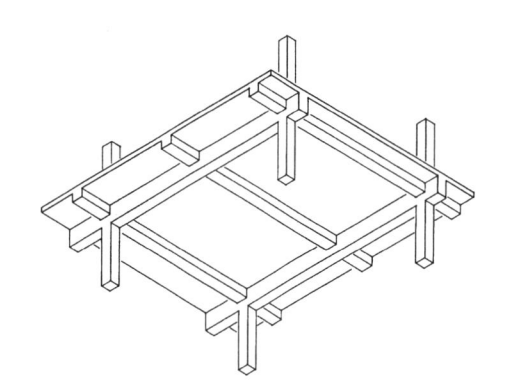

Cast in Place Beam & Slab, One Way
General: Solid concrete one way slab cast monolithically with reinforced concrete beams and girders.

3.5-110 Cast in Place Slabs, One Way

	SLAB DESIGN & SPAN (FT.)	SUPERIMPOSED LOAD (P.S.F.)	THICKNESS (IN.)	TOTAL LOAD (P.S.F.)	MAT.	INST.	TOTAL
2500	Single 8	40	4	90	1.98	4.99	6.97
2600		75	4	125	2.05	5.40	7.45
2700		125	4-1/2	181	2.14	5.45	7.59
2800		200	5	262	2.39	5.65	8.04
3000	Single 10	40	4	90	2.12	5.35	7.47
3100		75	4	125	2.12	5.35	7.47
3200		125	5	188	2.36	5.50	7.86
3300		200	7-1/2	293	3.21	6.10	9.31
3500	Single 15	40	5-1/2	90	2.54	5.50	8.04
3600		75	6-1/2	156	2.86	5.70	8.56
3700		125	7-1/2	219	3.12	5.85	8.97
3800		200	8-1/2	306	3.40	5.95	9.35
4000	Single 20	40	7-1/2	115	3.10	5.75	8.85
4100		75	9	200	3.40	5.80	9.20
4200		125	10	250	3.77	6	9.77
4300		200	10	324	4.06	6.25	10.31

3.5-120 Cast in Place Beam & Slab, One Way

	BAY SIZE (FT.)	SUPERIMPOSED LOAD (P.S.F.)	MINIMUM COL. SIZE (IN.)	SLAB THICKNESS (IN.)	TOTAL LOAD (P.S.F.)	MAT.	INST.	TOTAL
5000	20x25	40	12	5-1/2	121	3.18	6.85	10.03
5200		125	16	5-1/2	215	3.75	7.85	11.60
5500	25x25	40	12	6	129	3.33	6.75	10.08
5700		125	18	6	227	4.22	8.25	12.47
7500	30x35	40	16	8	158	4.12	7.60	11.72
7600		75	18	8	196	4.37	7.80	12.17
7700		125	22	8	254	4.91	8.70	13.61
7800		200	26	8	332	5.30	9.05	14.35
8000	35x35	40	16	9	169	4.56	7.85	12.41
8200		75	20	9	213	4.94	8.55	13.49
8400		125	24	9	272	5.40	8.85	14.25
8600		200	26	9	355	5.90	9.50	15.40

For expanded coverage of these items see *Means Assemblies Cost Data 1997*

SUPERSTRUCTURES A3.5 Floors

General: Solid concrete two way slab cast monolithically with reinforced concrete support beams and girders.

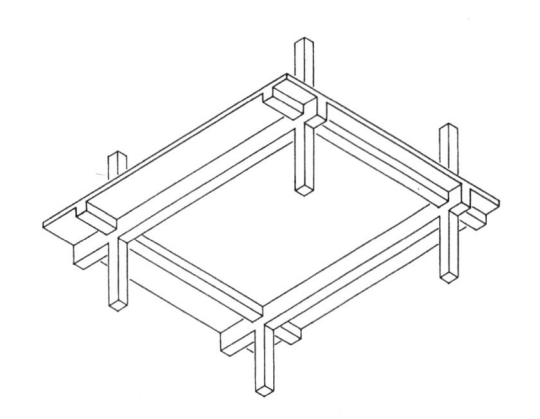

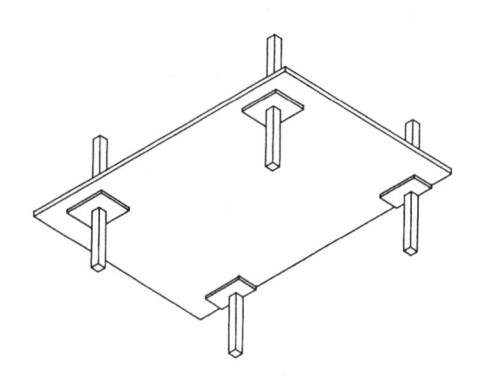

General: Flat Slab: Solid uniform depth concrete two way slabs with drop panels at columns and no column capitals.

3.5-130 Cast in Place Beam & Slab, Two Way

	BAY SIZE (FT.)	SUPERIMPOSED LOAD (P.S.F.)	MINIMUM COL. SIZE (IN.)	SLAB THICKNESS (IN.)	TOTAL LOAD (P.S.F.)	COST PER S.F. MAT.	COST PER S.F. INST.	COST PER S.F. TOTAL
4000	20 x 25	40	12	7	141	3.64	6.95	10.59
4300		75	14	7	181	4.15	7.55	11.70
4500		125	16	7	236	4.22	7.75	11.97
5100	25 x 25	40	12	7-1/2	149	3.78	7	10.78
5200		75	16	7-1/2	185	4.11	7.50	11.61
5300		125	18	7-1/2	250	4.46	8.15	12.61
7600	30 x 35	40	16	10	188	4.93	7.90	12.83
7700		75	18	10	225	5.20	8.20	13.40
8000		125	22	10	282	5.80	8.90	14.70
8500	35 x 35	40	16	10-1/2	193	5.25	8.10	13.35
8600		75	20	10-1/2	233	5.50	8.40	13.90
9000		125	24	10-1/2	287	6.15	9.05	15.20

3.5-140 Cast in Place Flat Slab with Drop Panels

	BAY SIZE (FT.)	SUPERIMPOSED LOAD (P.S.F.)	MINIMUM COL. SIZE (IN.)	SLAB & DROP (IN.)	TOTAL LOAD (P.S.F.)	COST PER S.F. MAT.	COST PER S.F. INST.	COST PER S.F. TOTAL
1960	20 x 20	40	12	7 - 3	132	3.14	5.50	8.64
1980		75	16	7 - 4	168	3.31	5.65	8.96
2000		125	18	7 - 6	221	3.67	5.85	9.52
3200	25 x 25	40	12	8-1/2 - 5-1/2	154	3.66	5.80	9.46
4000		125	20	8-1/2 - 8-1/2	243	4.12	6.25	10.37
4400		200	24	9 - 8-1/2	329	4.32	6.40	10.72
5000	25 x 30	40	14	9-1/2 - 7	168	3.96	6	9.96
5200		75	18	9-1/2 - 7	203	4.23	6.25	10.48
5600		125	22	9-1/2 - 8	256	4.42	6.40	10.82
6400	30 x 30	40	14	10-1/2 - 7-1/2	182	4.27	6.15	10.42
6600		75	18	10-1/2 - 7-1/2	217	4.55	6.40	10.95
6800		125	22	10-1/2 - 9	269	4.76	6.55	11.31
7400	30 x 35	40	16	11-1/2 - 9	196	4.65	6.45	11.10
7900		75	20	11-1/2 - 9	231	4.96	6.70	11.66
8000		125	24	11-1/2 - 11	284	5.15	6.85	12
9000	35 x 35	40	16	12 - 9	202	4.77	6.50	11.27
9400		75	20	12 - 11	240	5.15	6.80	11.95
9600		125	24	12 - 11	290	5.30	6.95	12.25

Important: See the Reference Section for critical supporting data - Location Factors & Historical Cost Indexes

SUPERSTRUCTURES A3.5 Floors

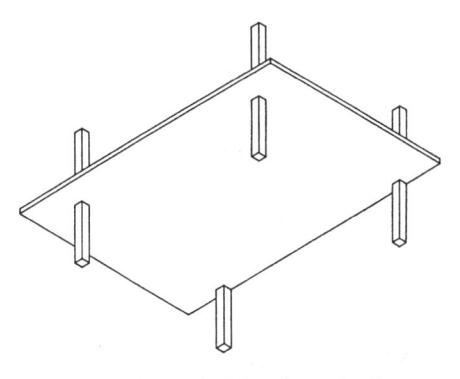

General: Flat Plates: Solid uniform depth concrete two way slab without drops or interior beams. Primary design limit is shear at columns.

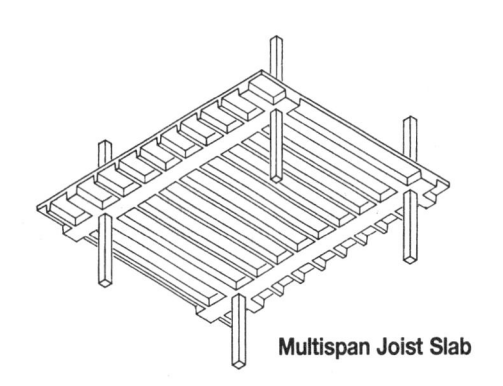

General: Combination of thin concrete slab and monolithic ribs at uniform spacing to reduce dead weight and increase rigidity.

Multispan Joist Slab

3.5-150 Cast in Place Flat Plate

	BAY SIZE (FT.)	SUPERIMPOSED LOAD (P.S.F.)	MINIMUM COL. SIZE (IN.)	SLAB THICKNESS (IN.)	TOTAL LOAD (P.S.F.)	MAT.	INST.	TOTAL
3000	15 x 20	40	14	7	127	3.01	5.35	8.36
3400		75	16	7-1/2	169	3.19	5.45	8.64
3600		125	22	8-1/2	231	3.49	5.55	9.04
3800		175	24	8-1/2	281	3.52	5.60	9.12
4200	20 x 20	40	16	7	127	3.01	5.30	8.31
4400		75	20	7-1/2	175	3.21	5.45	8.66
4600		125	24	8-1/2	231	3.49	5.55	9.04
5000		175	24	8-1/2	281	3.52	5.60	9.12
5600	20 x 25	40	18	8-1/2	146	3.48	5.55	9.03
6000		75	20	9	188	3.58	5.60	9.18
6400		125	26	9-1/2	244	3.85	5.80	9.65
6600		175	30	10	300	4	5.90	9.90
7000	25 x 25	40	20	9	152	3.58	5.60	9.18
7400		75	24	9-1/2	194	3.79	5.80	9.59
7600		125	30	10	250	4.02	5.90	9.92

3.5-160 Cast in Place Multispan Joist Slab

	BAY SIZE (FT.)	SUPERIMPOSED LOAD (P.S.F.)	MINIMUM COL. SIZE (IN.)	RIB DEPTH (IN.)	TOTAL LOAD (P.S.F.)	MAT.	INST.	TOTAL
2000	15 x 15	40	12	8	115	2.94	6.45	9.39
2100		75	12	8	150	2.95	6.50	9.45
2200		125	12	8	200	3.03	6.55	9.58
2300		200	14	8	275	3.16	6.80	9.96
2600	15 x 20	40	12	8	115	2.98	6.50	9.48
2800		75	12	8	150	3.06	6.60	9.66
3000		125	14	8	200	3.21	7	10.21
3300		200	16	8	275	3.38	7.05	10.43
3600	20 x 20	40	12	10	120	3.05	6.40	9.45
3900		75	14	10	155	3.22	6.75	9.97
4000		125	16	10	205	3.26	6.90	10.16
4100		200	18	10	280	3.46	7.15	10.61
6200	30 x 30	40	14	14	131	3.42	6.65	10.07
6400		75	18	14	166	3.57	6.90	10.47
6600		125	20	14	216	3.81	7.25	11.06
6700		200	24	16	297	4.12	7.55	11.67

For expanded coverage of these items see *Means Assemblies Cost Data 1997*

SUPERSTRUCTURES — A3.5 Floors

Waffle Slab

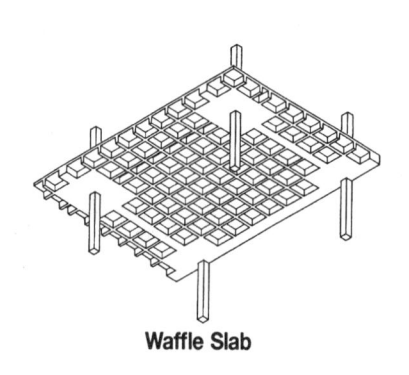

3.5-170 Cast in Place Waffle Slab

	BAY SIZE (FT.)	SUPERIMPOSED LOAD (P.S.F.)	MINIMUM COL. SIZE (IN.)	RIB DEPTH (IN.)	TOTAL LOAD (P.S.F.)	MAT.	INST.	TOTAL
3900	20 x 20	40	12	8	144	4.56	6.40	10.96
4000		75	12	8	179	4.64	6.50	11.14
4100		125	16	8	229	4.74	6.60	11.34
4200		200	18	8	304	5	6.80	11.80
4400	20 x 25	40	12	8	146	4.65	6.45	11.10
4500		75	14	8	181	4.76	6.55	11.31
4600		125	16	8	231	4.86	6.65	11.51
4700		200	18	8	306	5.10	6.85	11.95
4900	25 x 25	40	12	10	150	4.77	6.55	11.32
5000		75	16	10	185	4.92	6.65	11.57
5300		125	18	10	235	5.05	6.80	11.85
5500		200	20	10	310	5.20	6.95	12.15
5700	25 x 30	40	14	10	154	4.86	6.55	11.41
5800		75	16	10	189	5	6.70	11.70
5900		125	18	10	239	5.15	6.80	11.95
6000		200	20	12	329	5.65	7.20	12.85
6400	30 x 30	40	14	12	169	5.15	6.75	11.90
6500		75	18	12	204	5.30	6.85	12.15
6600		125	20	12	254	5.40	6.95	12.35
6700		200	24	12	329	5.85	7.35	13.20
6900	30 x 35	40	16	12	169	5.25	6.80	12.05
7000		75	18	12	204	5.25	6.80	12.05
7100		125	22	12	254	5.55	7.05	12.60
7200		200	26	14	334	6.15	7.50	13.65
7400	35 x 35	40	16	14	174	5.55	6.95	12.50
7500		75	20	14	209	5.70	7.10	12.80
7600		125	24	14	259	5.80	7.20	13
7700		200	26	16	346	6.25	7.60	13.85
8000	35 x 40	40	18	14	176	5.70	7.10	12.80
8300		75	22	14	211	5.90	7.30	13.20
8500		125	26	16	271	6.20	7.45	13.65
8750		200	30	20	372	6.80	7.85	14.65
9200	40 x 40	40	18	14	176	5.90	7.30	13.20
9400		75	24	14	211	6.15	7.50	13.65
9500		125	26	16	271	6.30	7.55	13.85
9700	40 x 45	40	20	16	186	6.10	7.40	13.50
9800		75	24	16	221	6.35	7.60	13.95
9900		125	28	16	271	6.45	7.70	14.15

Important: See the Reference Section for critical supporting data - Location Factors & Historical Cost Indexes

SUPERSTRUCTURES   A3.5   Floors

Precast Plank with No Topping

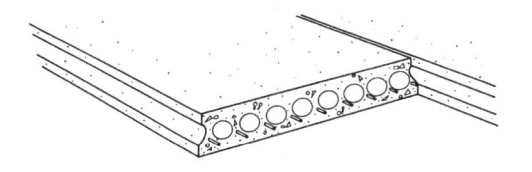

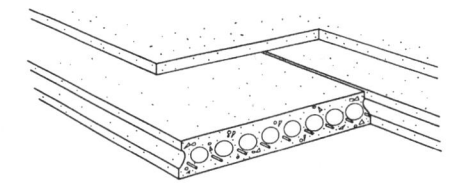

Precast Plank with 2" Concrete Topping

3.5-210 Precast Plank with No Topping

	SPAN (FT.)	SUPERIMPOSED LOAD (P.S.F.)	TOTAL DEPTH (IN.)	DEAD LOAD (P.S.F.)	TOTAL LOAD (P.S.F.)	COST PER S.F. MAT.	COST PER S.F. INST.	COST PER S.F. TOTAL
0720	10	40	4	50	90	3.96	1.10	5.06
0750		75	6	50	125	4.23	.91	5.14
0770		100	6	50	150	4.23	.91	5.14
0800	15	40	6	50	90	4.23	.91	5.14
0820		75	6	50	125	4.23	.91	5.14
0850		100	6	50	150	4.23	.91	5.14
0950	25	40	6	50	90	4.23	.91	5.14
0970		75	8	55	130	4.51	.75	5.26
1000		100	8	55	155	4.51	.75	5.26
1200	30	40	8	55	95	4.51	.75	5.26
1300		75	8	55	130	4.51	.75	5.26
1400		100	10	70	170	4.95	.60	5.55
1500	40	40	10	70	110	4.95	.60	5.55
1600		75	12	70	145	5.20	.56	5.76
1700	45	40	12	70	110	5.20	.56	5.76

3.5-210 Precast Plank with 2" Concrete Topping

	SPAN (FT.)	SUPERIMPOSED LOAD (P.S.F.)	TOTAL DEPTH (IN.)	DEAD LOAD (P.S.F.)	TOTAL LOAD (P.S.F.)	COST PER S.F. MAT.	COST PER S.F. INST.	COST PER S.F. TOTAL
2000	10	40	6	75	115	4.49	2.35	6.84
2100		75	8	75	150	4.76	2.16	6.92
2200		100	8	75	175	4.76	2.16	6.92
2500	15	40	8	75	115	4.76	2.16	6.92
2600		75	8	75	150	4.76	2.16	6.92
2700		100	8	75	175	4.76	2.16	6.92
3100	25	40	8	75	115	4.76	2.16	6.92
3200		75	8	75	150	4.76	2.16	6.92
3300		100	10	80	180	5.05	2	7.05
3400	30	40	10	80	120	5.05	2	7.05
3500		75	10	80	155	5.05	2	7.05
3600		100	10	80	180	5.05	2	7.05
4000	40	40	12	95	135	5.50	1.85	7.35
4500		75	14	95	170	5.75	1.81	7.56
5000	45	40	14	95	135	5.75	1.81	7.56

For expanded coverage of these items see *Means Assemblies Cost Data 1997*

SUPERSTRUCTURES   A3.5   Floors

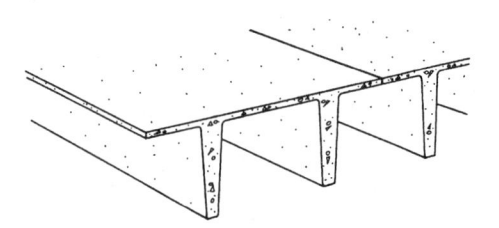

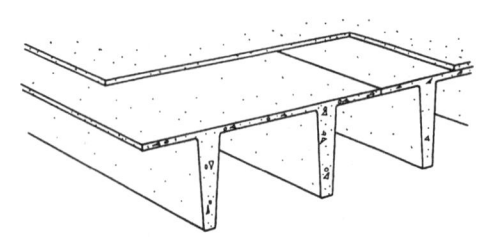

Most widely used for moderate span floors and moderate and long span roofs. At shorter spans, they tend to be competitive with hollow core slabs. They are also used as wall panels.

Precast Double "T" Beams with No Topping

Precast Double "T" Beams with 2" Topping

3.5-230 Precast Double "T" Beams with No Topping

	SPAN (FT.)	SUPERIMPOSED LOAD (P.S.F.)	DBL. "T" SIZE D (IN.) W (FT.)	CONCRETE "T" TYPE	TOTAL LOAD (P.S.F.)	MAT.	INST.	TOTAL
4300	50	30	20x8	Lt. Wt.	66	4.95	1.02	5.97
4400		40	20x8	Lt. Wt.	76	5.25	.93	6.18
4500		50	20x8	Lt. Wt.	86	5.40	1.11	6.51
4600		75	20x8	Lt. Wt.	111	5.55	1.27	6.82
5600	70	30	32x10	Lt. Wt.	78	5.70	.70	6.40
5750		40	32x10	Lt. Wt.	88	5.85	.85	6.70
5900		50	32x10	Lt. Wt.	98	6	.97	6.97
6000		75	32x10	Lt. Wt.	123	6.20	1.17	7.37
6100		100	32x10	Lt. Wt.	148	6.60	1.58	8.18
6200	80	30	32x10	Lt. Wt.	78	6	.97	6.97
6300		40	32x10	Lt. Wt.	88	6.20	1.17	7.37
6400		50	32x10	Lt. Wt.	98	6.40	1.38	7.78

3.5-230 Precast Double "T" Beams With 2" Topping

	SPAN (FT.)	SUPERIMPOSED LOAD (P.S.F.)	DBL. "T" SIZE D (IN.) W (FT.)	CONCRETE "T" TYPE	TOTAL LOAD (P.S.F.)	MAT.	INST.	TOTAL
7100	40	30	18x8	Reg. Wt.	120	4.53	2.06	6.59
7200		40	20x8	Reg. Wt.	130	4.43	1.97	6.40
7300		50	20x8	Reg. Wt.	140	4.61	2.15	6.76
7400		75	20x8	Reg. Wt.	165	4.69	2.24	6.93
7500		100	20x8	Reg. Wt.	190	4.95	2.49	7.44
7550	50	30	24x8	Reg. Wt.	120	4.84	2.06	6.90
7600		40	24x8	Reg. Wt.	130	4.90	2.13	7.03
7750		50	24x8	Reg. Wt.	140	4.92	2.14	7.06
7800		75	24x8	Reg. Wt.	165	5.15	2.40	7.55
7900		100	32x10	Reg. Wt.	189	5.65	1.96	7.61

Important: See the Reference Section for critical supporting data - Location Factors & Historical Cost Indexes

SUPERSTRUCTURES A3.5 Floors

Precast Beam and Plank with No Topping

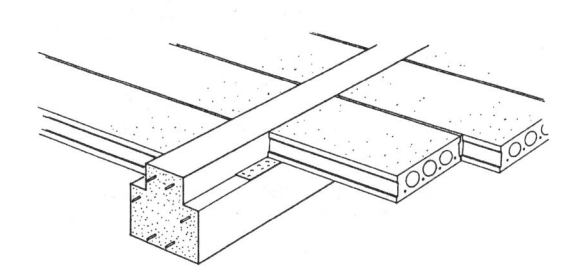

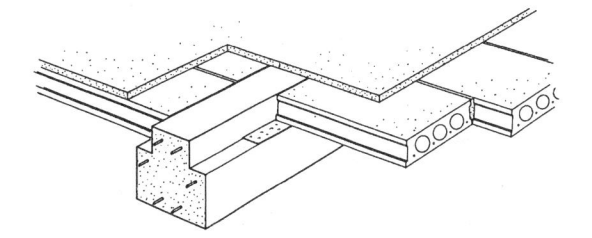

Precast Beam and Plank with 2" Topping

3.5-242 Precast Beam & Plank with No Topping

	BAY SIZE (FT.)	SUPERIMPOSED LOAD (P.S.F.)	PLANK THICKNESS (IN.)	TOTAL DEPTH (IN.)	TOTAL LOAD (P.S.F.)	COST PER S.F. MAT.	INST.	TOTAL
6200	25x25	40	8	28	118	9.90	2.09	11.99
6400		75	8	36	158	10.40	2.09	12.49
6500		100	8	36	183	10.40	2.09	12.49
7000	25x30	40	8	36	110	9.45	2.09	11.54
7200		75	10	36	159	10.15	1.94	12.09
7400		100	12	36	188	10.85	1.90	12.75
7600	30x30	40	8	36	121	9.80	2.09	11.89
8000		75	10	44	140	10.35	2.09	12.44
8250		100	12	52	206	11.80	1.90	13.70
8500	30x35	40	12	44	135	10.05	1.90	11.95
8750		75	12	52	176	10.70	1.90	12.60
9000	35x35	40	12	52	141	10.75	1.90	12.65
9250		75	12	60	181	11.35	1.90	13.25
9500	35x40	40	12	52	137	10.25	2.57	12.82

3.5-244 Precast Beam & Plank with 2" Topping

	BAY SIZE (FT.)	SUPERIMPOSED LOAD (P.S.F.)	PLANK THICKNESS (IN.)	TOTAL DEPTH (IN.)	TOTAL LOAD (P.S.F.)	COST PER S.F. MAT.	INST.	TOTAL
4300	20x20	40	6	22	135	10.15	3.49	13.64
4400		75	6	24	173	10.75	3.49	14.24
4500		100	6	28	200	11.10	3.49	14.59
4600	20x25	40	6	26	134	9.50	3.48	12.98
5000		75	8	30	177	10.20	3.32	13.52
5200		100	8	30	202	10.20	3.32	13.52
5400	25x25	40	6	38	143	10.65	3.46	14.11
5600		75	8	38	183	10.65	3.46	14.11
6000		100	8	46	216	11.80	3.30	15.10
6200	25x30	40	8	38	144	9.95	3.28	13.23
6400		75	10	46	200	11.10	3.13	14.23
6600		100	10	46	225	11.10	3.13	14.23
7000	30x30	40	8	46	150	10.85	3.27	14.12
7200		75	10	54	181	11.65	3.27	14.92
7600		100	10	54	231	11.65	3.27	14.92
7800	30x35	40	10	54	166	11	3.12	14.12
8000		75	12	54	200	11.05	3.08	14.13

For expanded coverage of these items see *Means Assemblies Cost Data 1997*

SUPERSTRUCTURES A3.5 Floors

General: Beams and double tees priced here are for plant produced prestressed members transported to the site and erected.

The 2" structural topping is applied after the beams and double tees are in place and is reinforced with W.W.F.

Note: Deduct from prices 20% for Southern States. Add to prices 10% for Western States.

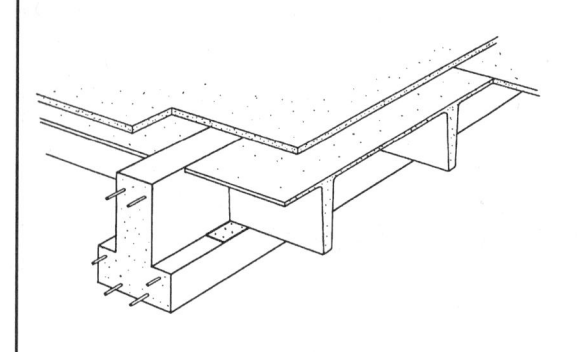

With Topping

3.5-254 Precast Double "T" & 2" Topping on Precast Beams

	BAY SIZE (FT.)	SUPERIMPOSED LOAD (P.S.F.)	DEPTH (IN.)		TOTAL LOAD (P.S.F.)	COST PER S.F. MAT.	INST.	TOTAL
3000	25x30	40	38		130	9.90	3.64	13.54
3100		75	38		168	9.90	3.64	13.54
3300		100	46		196	10.35	3.64	13.99
3600	30x30	40	46		150	10.75	3.62	14.37
3750		75	46		174	10.75	3.62	14.37
4000		100	54		203	11.30	3.62	14.92
4100	30x40	40	46		136	8.75	3.43	12.18
4300		75	54		173	9.15	3.43	12.58
4400		100	62		204	9.70	3.43	13.13
4600	30x50	40	54		138	8.35	3.34	11.69
4800		75	54		181	8.80	3.34	12.14
5000		100	54		219	9.65	3.13	12.78
5200	30x60	40	62		151	8.80	3.13	11.93
5400		75	62		192	9.25	3.13	12.38
5600		100	62		215	9.25	3.13	12.38
5800	35x40	40	54		139	9.35	3.43	12.78
6000		75	62		179	10.30	3.35	13.65
6250		100	62		212	10.50	3.35	13.85
6500	35x50	40	62		142	8.95	3.34	12.29
6750		75	62		186	9.40	3.34	12.74
7300		100	62		231	10.90	3.13	14.03
7600	35x60	40	54		154	10	3.06	13.06
7750		75	54		179	10.30	3.06	13.36
8000		100	62		224	10.85	3.06	13.91
8250	40x40	40	62		145	10.40	4.09	14.49
8400		75	62		187	10.85	4.10	14.95
8750		100	62		223	11.70	4.10	15.80
9000	40x50	40	62		151	9.70	3.99	13.69
9300		75	62		193	10.30	3.37	13.67
9800	40x60	40	62		164	12.50	3.48	15.98

Important: See the Reference Section for critical supporting data - Location Factors & Historical Cost Indexes

SUPERSTRUCTURES | A3.5 | Floors

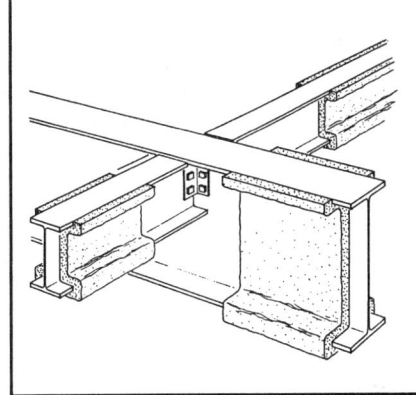

General: The following table is based upon structural wide flange (WF) beam and girder framing. Non-composite action is assumed between WF framing and decking. Deck costs not included.

The deck spans the short direction. The steel beams and girders are fireproofed with sprayed fiber fireproofing.

No columns included in price.

3.5-310 W Shape Beams & Girders

	BAY SIZE (FT.) BEAM X GIRD	SUPERIMPOSED LOAD (P.S.F.)	STEEL FRAMING DEPTH (IN.)	FIREPROOFING (S.F./S.F.)	TOTAL LOAD (P.S.F.)	MAT.	INST.	TOTAL
1350	15x20	40	12	.535	50	2.28	1.22	3.50
1400		40	16	.65	90	2.97	1.56	4.53
1450		75	18	.694	125	3.88	1.96	5.84
1500		125	24	.796	175	5.35	2.76	8.11
2000	20x20	40	12	.55	50	2.54	1.34	3.88
2050		40	14	.579	90	3.45	1.72	5.17
2100		75	16	.672	125	4.13	2.05	6.18
2150		125	16	.714	175	4.92	2.55	7.47
2900	20x25	40	16	.53	50	2.79	1.43	4.22
2950		40	18	.621	96	4.37	2.12	6.49
3000		75	18	.651	131	5	2.39	7.39
3050		125	24	.77	200	6.60	3.31	9.91
5200	25x25	40	18	.486	50	3.03	1.49	4.52
5300		40	18	.592	96	4.47	2.15	6.62
5400		75	21	.668	131	5.40	2.57	7.97
5450		125	24	.738	191	7.10	3.51	10.61
6300	25x30	40	21	.568	50	3.44	1.72	5.16
6350		40	21	.694	90	4.65	2.27	6.92
6400		75	24	.776	125	5.95	2.86	8.81
6450		125	30	.904	175	7.15	3.63	10.78
7700	30x30	40	21	.52	50	3.68	1.78	5.46
7750		40	24	.629	103	5.65	2.64	8.29
7800		75	30	.715	138	6.70	3.13	9.83
7850		125	36	.822	206	8.80	4.30	13.10
8250	30x35	40	21	.508	50	4.19	1.98	6.17
8300		40	24	.651	109	6.15	2.86	9.01
8350		75	33	.732	150	7.50	3.45	10.95
8400		125	36	.802	225	9.30	4.51	13.81
9550	35x35	40	27	.560	50	4.34	2.07	6.41
9600		40	36	.706	109	7.85	3.58	11.43
9650		75	36	.750	150	8.15	3.72	11.87
9820		125	36	.797	225	10.85	5.15	16

For expanded coverage of these items see *Means Assemblies Cost Data 1997*

SUPERSTRUCTURES A3.5 Floors

Description: Table below lists costs for light gage CEE or PUNCHED DOUBLE joists to suit the span and loading with the minimum thickness subfloor required by the joist spacing.

3.5-360 Light Gauge Steel Floor Systems

	SPAN (FT.)	SUPERIMPOSED LOAD (P.S.F.)	FRAMING DEPTH (IN.)	FRAMING SPAC. (IN.)	TOTAL LOAD (P.S.F.)	COST PER S.F. MAT.	COST PER S.F. INST.	COST PER S.F. TOTAL
1500	15	40	8	16	54	1.86	1.32	3.18
1550			8	24	54	2	1.36	3.36
1600		65	10	16	80	2.42	1.61	4.03
1650			10	24	80	2.44	1.57	4.01
1700		75	10	16	90	2.42	1.61	4.03
1750			10	24	90	2.44	1.57	4.01
1800		100	10	16	116	3.14	2.03	5.17
1850			10	24	116	2.44	1.57	4.01
1900		125	10	16	141	3.14	2.03	5.17
1950			10	24	141	2.44	1.57	4.01
2500	20	40	8	16	55	2.13	1.49	3.62
2550			8	24	55	2	1.36	3.36
2600		65	8	16	80	2.47	1.69	4.16
2650			10	24	80	2.45	1.59	4.04
2700		75	10	16	90	2.28	1.52	3.80
2750			12	24	90	2.72	1.39	4.11
2800		100	10	16	115	2.28	1.52	3.80
2850			10	24	116	3.28	1.76	5.04
2900		125	12	16	142	3.55	1.75	5.30
2950			12	24	141	3.55	1.89	5.44
3500	25	40	10	16	55	2.42	1.61	4.03
3550			10	24	55	2.44	1.57	4.01
3600		65	10	16	81	3.14	2.03	5.17
3650			12	24	81	3.01	1.51	4.52
3700		75	12	16	92	3.55	1.75	5.30
3750			12	24	91	3.55	1.89	5.44
3800		100	12	16	117	3.99	1.93	5.92
3850		125	12	16	143	4.83	2.51	7.34
4500	30	40	12	16	57	3.55	1.75	5.30
4550			12	24	56	2.99	1.51	4.50
4600		65	12	16	82	3.97	1.92	5.89
4650		75	12	16	92	3.97	1.92	5.89

Important: See the Reference Section for critical supporting data - Location Factors & Historical Cost Indexes

SUPERSTRUCTURES — A3.5 Floors

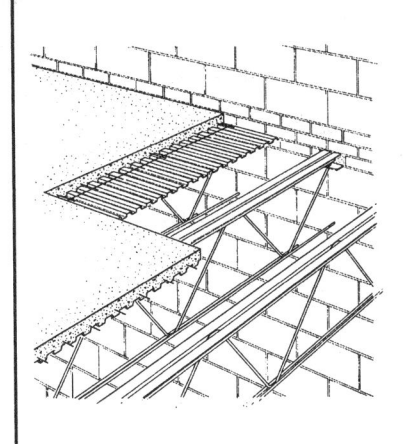

Description: The table below lists costs per S.F. for a floor system on bearing walls using open web steel joists, galvanized steel slab form and 2-1/2" concrete slab reinforced with welded wire fabric. Costs of the bearing walls are not included.

3.5-420 Deck & Joists on Bearing Walls

	SPAN (FT.)	SUPERIMPOSED LOAD (P.S.F.)	JOIST SPACING FT./IN.	DEPTH (IN.)	TOTAL LOAD (P.S.F.)	MAT.	INST.	TOTAL
1050	20	40	2-0	14-1/2	83	2.76	2.21	4.97
1070		65	2-0	16-1/2	109	3.02	2.34	5.36
1100		75	2-0	16-1/2	119	3.02	2.34	5.36
1120		100	2-0	18-1/2	145	3.04	2.35	5.39
1150		125	1-9	18-1/2	170	3.56	2.81	6.37
1170	25	40	2-0	18-1/2	84	3.21	2.62	5.83
1200		65	2-0	20-1/2	109	3.33	2.68	6.01
1220		75	2-0	20-1/2	119	3.39	2.50	5.89
1250		100	2-0	22-1/2	145	3.45	2.53	5.98
1270		125	1-9	22-1/2	170	3.97	2.78	6.75
1300	30	40	2-0	22-1/2	84	3.48	2.29	5.77
1320		65	2-0	24-1/2	110	3.80	2.40	6.20
1350		75	2-0	26-1/2	121	3.95	2.45	6.40
1370		100	2-0	26-1/2	146	4.24	2.54	6.78
1400		125	2-0	24-1/2	172	4.63	3.31	7.94
1420	35	40	2-0	26-1/2	85	4.37	3.17	7.54
1450		65	2-0	28-1/2	111	4.52	3.24	7.76
1470		75	2-0	28-1/2	121	4.52	3.24	7.76
1500		100	1-11	28-1/2	147	4.93	3.46	8.39
1520		125	1-8	28-1/2	172	5.45	3.73	9.18
1550 1560	5/8" gyp. fireproof. On metal furring, add					.52	1.81	2.33

For expanded coverage of these items see *Means Assemblies Cost Data 1997*

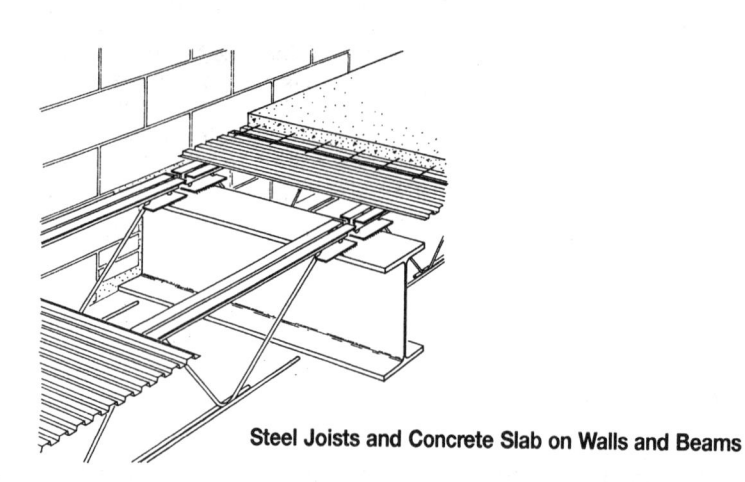

SUPERSTRUCTURES — A3.5 Floors

The table below lists costs for a floor system on exterior bearing walls and interior columns and beams using open web steel joists, galvanized steel slab form, 2-1/2" concrete slab reinforced with welded wire fabric.

Columns are sized to accommodate one floor plus roof loading but costs for columns are from suspended floor to slab on grade only. Costs of the bearing walls are not included.

Steel Joists and Concrete Slab on Walls and Beams

3.5-440 Steel Joists on Beam & Wall

	BAY SIZE (FT.)	SUPERIMPOSED LOAD (P.S.F.)	DEPTH (IN.)	TOTAL LOAD (P.S.F.)	COLUMN ADD	MAT.	INST.	TOTAL
1720	25x25	40	23	84		3.95	2.88	6.83
1730					columns	.18	.08	.26
1750	25x25	65	29	110		4.16	2.99	7.15
1760					columns	.18	.08	.26
1770	25x25	75	26	120		4.46	2.91	7.37
1780					columns	.21	.09	.30
1800	25x25	100	29	145		5.05	3.18	8.23
1810					columns	.21	.09	.30
1820	25x25	125	29	170		5.30	3.29	8.59
1830					columns	.23	.09	.32
2020	30x30	40	29	84		4.42	2.63	7.05
2030					columns	.16	.07	.23
2050	30x30	65	29	110		5	2.86	7.86
2060					columns	.16	.07	.23
2070	30x30	75	32	120		5.15	2.91	8.06
2080					columns	.19	.08	.27
2100	30x30	100	35	145		5.75	3.12	8.87
2110					columns	.22	.09	.31
2120	30x30	125	35	172		6.30	3.95	10.25
2130					columns	.26	.10	.36
2170	30x35	40	29	85		5.05	2.85	7.90
2180					columns	.16	.06	.22
2200	30x35	65	29	111		5.70	3.69	9.39
2210					columns	.18	.08	.26
2220	30x35	75	32	121		5.70	3.69	9.39
2230					columns	.18	.08	.26
2250	30x35	100	35	148		6.15	3.25	9.40
2260					columns	.22	.09	.31
2320	35x35	40	32	85		5.20	2.91	8.11
2330					columns	.16	.07	.23
2350	35x35	65	35	111		6	3.83	9.83
2360					columns	.19	.08	.27
2370	35x35	75	35	121		6.10	3.88	9.98
2380					columns	.19	.08	.27
2400	35x35	100	38	148		6.40	4.02	10.42
2410					columns	.24	.09	.33
2460	5/8 gyp. fireproof.							
2475	On metal furring, add					.52	1.81	2.33

Important: See the Reference Section for critical supporting data - Location Factors & Historical Cost Indexes

SUPERSTRUCTURES — A3.5 Floors

The table below lists costs per S.F. for a floor system on steel columns and beams using open web steel joists, galvanized steel slab form, 2-1/2" concrete slab reinforced with welded wire fabric.

Columns are sized to accommodate one floor plus roof loading but costs for columns are from floor to grade only.

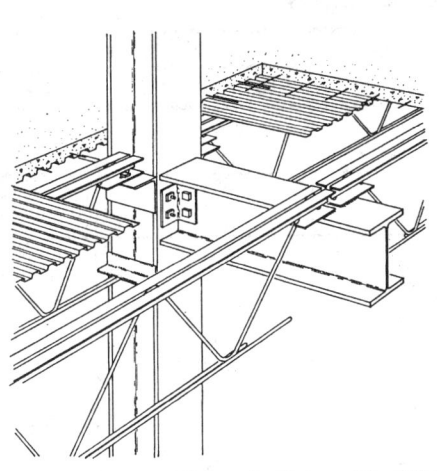

Steel Joists and Concrete Slab on Steel Columns and Beams

3.5-460 Steel Joists, Beams & Slab on Columns

	BAY SIZE (FT.)	SUPERIMPOSED LOAD (P.S.F.)	DEPTH (IN.)	TOTAL LOAD (P.S.F.)	COLUMN ADD	MAT.	INST.	TOTAL
2350	15x20	40	17	83		3.90	2.64	6.54
2400					column	.60	.25	.85
2450	15x20	65	19	108		4.30	2.80	7.10
2500					column	.60	.25	.85
2550	15x20	75	19	119		4.50	2.89	7.39
2600					column	.65	.27	.92
2650	15x20	100	19	144		4.78	3.02	7.80
2700					column	.65	.27	.92
2750	15x20	125	19	170		5.20	3.41	8.61
2800					column	.87	.36	1.23
2850	20x20	40	19	83		4.21	2.76	6.97
2900					column	.49	.20	.69
2950	20x20	65	23	109		4.65	2.96	7.61
3000					column	.65	.27	.92
3100	20x20	75	26	119		4.90	3.06	7.96
3200					column	.65	.27	.92
3400	20x20	100	23	144		5.10	3.14	8.24
3450					column	.65	.27	.92
3500	20x20	125	23	170		5.70	3.40	9.10
3600					column	.78	.32	1.10
3700	20x25	40	44	83		4.70	3.19	7.89
3800					column	.52	.22	.74
3900	20x25	65	26	110		5.15	3.37	8.52
4000					column	.52	.22	.74
4100	20x25	75	26	120		5.20	3.20	8.40
4200					column	.62	.26	.88
4300	20x25	100	26	145		5.50	3.34	8.84
4400					column	.62	.26	.88
4500	20x25	125	29	170		6.15	3.62	9.77
4600					column	.73	.30	1.03
4700	25x25	40	23	84		5.05	3.31	8.36
4800					column	.50	.21	.71
4900	25x25	65	29	110		5.35	3.46	8.81
5000					column	.50	.21	.71
5100	25x25	75	26	120		5.75	3.43	9.18
5200					column	.58	.24	.82

For expanded coverage of these items see *Means Assemblies Cost Data 1997*

SUPERSTRUCTURES		A3.5	Floors				
3.5-460		Steel Joists, Beams & Slab on Columns					

	BAY SIZE (FT.)	SUPERIMPOSED LOAD (P.S.F.)	DEPTH (IN.)	TOTAL LOAD (P.S.F.)	COLUMN ADD	COST PER S.F.		
						MAT.	INST.	TOTAL
5300	25x25	100	29	145		6.40	3.72	10.12
5400					column	.58	.24	.82
5500	25x25	125	32	170		6.75	3.88	10.63
5600					column	.65	.26	.91
5700	25x30	40	29	84		5.45	3.46	8.91
5800					column	.49	.20	.69
5900	25x30	65	29	110		5.70	3.61	9.31
6000					column	.49	.20	.69
6050	25x30	75	29	120		6.10	3.29	9.39
6100					column	.54	.22	.76
6150	25x30	100	29	145		6.65	3.48	10.13
6200					column	.54	.22	.76
6250	25x30	125	32	170		7.15	4.26	11.41
6300					column	.62	.26	.88
6350	30x30	40	29	84		5.65	3.13	8.78
6400					column	.45	.18	.63
6500	30x30	65	29	110		6.45	3.44	9.89
6600					column	.45	.18	.63
6700	30x30	75	32	120		6.60	3.48	10.08
6800					column	.52	.21	.73
6900	30x30	100	35	145		7.35	3.77	11.12
7000					column	.60	.25	.85
7100	30x30	125	35	172		8.05	4.66	12.71
7200					column	.67	.27	.94
7300	30x35	40	29	85		6.45	3.41	9.86
7400					column	.38	.16	.54
7500	30x35	65	29	111		7.20	4.30	11.50
7600					column	.50	.21	.71
7700	30x35	75	32	121		7.20	4.30	11.50
7800					column	.51	.21	.72
7900	30x35	100	35	148		7.80	3.91	11.71
8000					column	.62	.26	.88
8100	30x35	125	38	173		8.70	4.23	12.93
8200					column	.63	.26	.89
8300	35x35	40	32	85		6.60	3.49	10.09
8400					column	.44	.18	.62
8500	35x35	65	35	111		7.55	4.46	12.01
8600					column	.53	.22	.75
9300	35x35	75	38	121		7.75	4.54	12.29
9400					column	.53	.22	.75
9500	35x35	100	38	148		8.35	4.81	13.16
9600					column	.65	.27	.92
9750	35x35	125	41	173		9	4.39	13.39
9800					column	.67	.27	.94
9810	5/8 gyp. fireproof.							
9815	On metal furring, add					.52	1.81	2.33

SUPERSTRUCTURES   A3.5   Floors

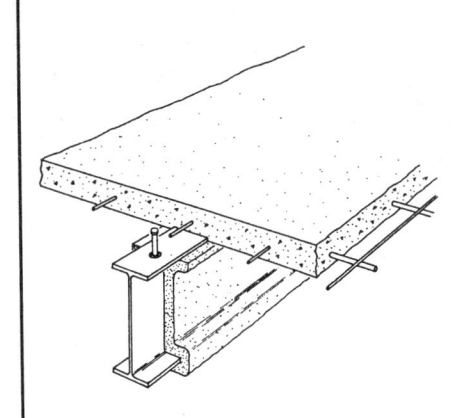

General: Composite construction of wide flange beams and concrete slabs is most efficiently used when loads are heavy and spans are moderately long. It is stiffer with less deflection than non-composite construction of similar depth and spans.

Composite Beam & Cast in Place Slab

3.5-520 Composite Beam & Cast in Place Slab

	BAY SIZE (FT.)	SUPERIMPOSED LOAD (P.S.F.)	SLAB THICKNESS (IN.)	TOTAL DEPTH (FT.-IN.)	TOTAL LOAD (P.S.F.)	MAT.	INST.	TOTAL
3800	20x25	40	4	1 - 8	94	4.73	6.45	11.18
3900		75	4	1 - 8	130	5.50	6.85	12.35
4000		125	4	1 - 10	181	6.40	7.35	13.75
4100		200	5	2 - 2	272	8.30	8.35	16.65
4200	25x25	40	4-1/2	1 - 8-1/2	99	4.94	6.50	11.44
4300		75	4-1/2	1 - 10-1/2	136	5.95	7.05	13
4400		125	5-1/2	2 - 0-1/2	200	7.05	7.65	14.70
4500		200	5-1/2	2 - 2-1/2	278	8.75	8.50	17.25
4600	25x30	40	4-1/2	1 - 8-1/2	100	5.50	6.80	12.30
4700		75	4-1/2	1 - 10-1/2	136	6.45	7.25	13.70
4800		125	5-1/2	2 - 2-1/2	202	7.90	8	15.90
4900		200	5-1/2	2 - 5-1/2	279	9.40	8.80	18.20
5000	30x30	40	4	1 - 8	95	5.50	6.80	12.30
5200		75	4	2 - 1	131	6.35	7.15	13.50
5400		125	4	2 - 4	183	7.85	7.95	15.80
5600		200	5	2 - 10	274	10	9.05	19.05
5800	30x35	40	4	2 - 1	95	5.85	6.90	12.75
6000		75	4	2 - 4	139	7	7.50	14.50
6250		125	4	2 - 4	185	9.05	8.45	17.50
6500		200	5	2 - 11	276	11.05	9.50	20.55
8000	35x35	40	4	2 - 1	96	6.15	7.05	13.20
8250		75	4	2 - 4	133	7.30	7.65	14.95
8500		125	4	2 - 10	185	8.75	8.35	17.10
8750		200	5	3 - 5	276	11.35	9.60	20.95
9000	35x40	40	4	2 - 4	97	6.80	7.40	14.20
9250		75	4	2 - 4	134	7.95	7.85	15.80
9500		125	4	2 - 7	186	9.60	8.75	18.35
9750		200	5	3 - 5	278	12.45	10.05	22.50

For expanded coverage of these items see *Means Assemblies Cost Data 1997*

SUPERSTRUCTURES A3.5 Floors

The table below lists costs per S.F. for floors using steel beams and girders, composite steel deck, concrete slab reinforced with W.W.F. and sprayed fiber fireproofing (non-asbestos) on the steel beams and girders and on the steel deck.

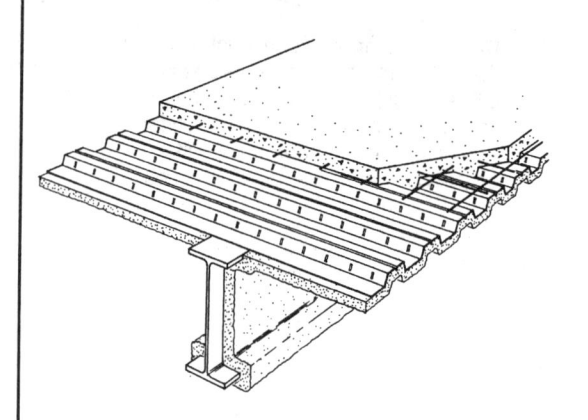

Wide Flange, Composite Deck & Slab

3.5-530 W Shape, Composite Deck, & Slab

	BAY SIZE (FT.) BEAM X GIRD	SUPERIMPOSED LOAD (P.S.F.)	SLAB THICKNESS (IN.)	TOTAL DEPTH (FT.-IN.)	TOTAL LOAD (P.S.F.)	COST PER S.F. MAT.	INST.	TOTAL
0540	15x20	40	5	1-7	89	5.65	4.32	9.97
0560		75	5	1-9	125	6.15	4.56	10.71
0580		125	5	1-11	176	7.15	4.95	12.10
0600		200	5	2-2	254	8.95	5.60	14.55
0700	20x20	40	5	1-9	90	6.10	4.49	10.59
0720		75	5	1-9	126	6.85	4.85	11.70
0740		125	5	1-11	177	7.70	5.25	12.95
0760		200	5	2-5	255	9.45	5.85	15.30
0780	20x25	40	5	1-9	90	6.45	4.49	10.94
0800		75	5	1-9	127	7.85	5.10	12.95
0820		125	5	1-11	180	9.10	5.55	14.65
0840		200	5	2-5	256	10	6.25	16.25
0940	25x25	40	5	1-11	91	7.20	4.77	11.97
0960		75	5	2-5	178	8.20	5.20	13.40
0980		125	5	2-5	181	9.70	5.80	15.50
1000		200	5-1/2	2-8-1/2	263	10.85	6.65	17.50
1400	25x30	40	5	2-5	91	6.95	4.90	11.85
1500		75	5	2-5	128	8.55	5.40	13.95
1600		125	5	2-8	180	9.40	6	15.40
1700		200	5	2-11	259	11.90	6.85	18.75
2200	30x30	40	5	2-2	92	7.65	5.20	12.85
2300		75	5	2-5	129	8.85	5.75	14.60
2400		125	5	2-11	182	10.45	6.40	16.85
2500		200	5	3-2	263	14	7.95	21.95
2600	30x35	40	5	2-5	94	8.30	5.50	13.80
2700		75	5	2-11	131	9.55	6	15.55
2800		125	5	3-2	183	11.15	6.70	17.85
2900		200	5-1/2	3-5-1/2	268	13.80	7.65	21.45
3800	35x35	40	5	2-8	94	8.75	5.45	14.20
3900		75	5	2-11	131	10.05	6.05	16.10
4000		125	5	3-5	184	11.95	6.85	18.80
4100		200	5-1/2	3-5-1/2	270	15	8.05	23.05

SUPERSTRUCTURES — A3.5 Floors

The table below lists costs per S.F. for floors using composite steel beams with welded shear studs, composite steel deck and light weight concrete slab reinforced with W.W.F.

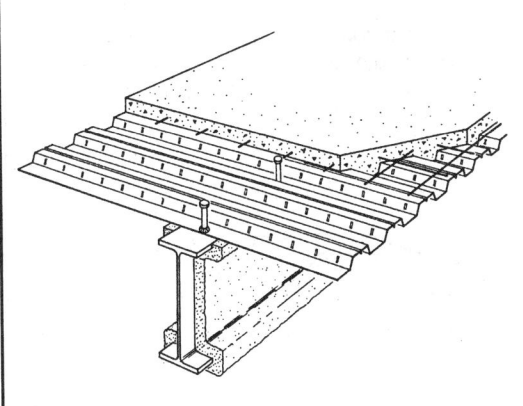

Composite Beam, Deck & Slab

3.5-540 Composite Beams, Deck & Slab

	BAY SIZE (FT.)	SUPERIMPOSED LOAD (P.S.F.)	SLAB THICKNESS (IN.)	TOTAL DEPTH (FT. - IN.)	TOTAL LOAD (P.S.F.)	MAT.	INST.	TOTAL
2400	20x25	40	5-1/2	1 - 5-1/2	80	5.50	3.85	9.35
2500		75	5-1/2	1 - 9-1/2	115	5.70	3.86	9.56
2750		125	5-1/2	1 - 9-1/2	167	7.05	4.53	11.58
2900		200	6-1/4	1 - 11-1/2	251	7.95	4.90	12.85
3000	25x25	40	5-1/2	1 - 9-1/2	82	5.70	3.69	9.39
3100		75	5-1/2	1 - 11-1/2	118	6.30	3.73	10.03
3200		125	5-1/2	2 - 2-1/2	169	6.60	4.06	10.66
3300		200	6-1/4	2 - 6-1/4	252	8.60	4.69	13.29
3400	25x30	40	5-1/2	1 - 11-1/2	83	5.80	3.65	9.45
3600		75	5-1/2	1 - 11-1/2	119	6.25	3.69	9.94
3900		125	5-1/2	1 - 11-1/2	170	7.20	4.16	11.36
4000		200	6-1/4	2 - 6-1/4	252	8.65	4.71	13.36
4200	30x30	40	5-1/2	1 - 11-1/2	81	5.55	3.77	9.32
4400		75	5-1/2	2 - 2-1/2	116	6.05	3.92	9.97
4500		125	5-1/2	2 - 5-1/2	168	7.30	4.41	11.71
4700		200	6-1/4	2 - 9-1/4	252	8.75	5.10	13.85
4900	30x35	40	5-1/2	2 - 2-1/2	82	5.80	3.88	9.68
5100		75	5-1/2	2 - 5-1/2	117	6.35	3.99	10.34
5300		125	5-1/2	2 - 5-1/2	169	7.55	4.49	12.04
5500		200	6-1/4	2 - 9-1/4	254	8.75	5.10	13.85
5750	35x35	40	5-1/2	2 - 5-1/2	84	6.45	3.90	10.35
6000		75	5-1/2	2 - 5-1/2	121	7.35	4.17	11.52
7000		125	5-1/2	2 - 8-1/2	170	8.60	4.75	13.35
7200		200	5-1/2	2 - 11-1/2	254	9.65	5.25	14.90
7400	35x40	40	5-1/2	2 - 5-1/2	85	7.10	4.18	11.28
7600		75	5-1/2	2 - 5-1/2	121	7.70	4.31	12.01
8000		125	5-1/2	2 - 5-1/2	171	8.80	4.81	13.61
9000		200	5-1/2	2 - 11-1/2	255	10.40	5.45	15.85

For expanded coverage of these items see *Means Assemblies Cost Data 1997*

SUPERSTRUCTURES — A3.5 Floors

Description: Table 3.5-580 lists S.F. costs for steel deck and concrete slabs for various spans.

Description: Table 3.5-710 lists the S.F. costs for wood joists and a minimum thickness plywood subfloor.

Description: Table 3.5-720 lists the S.F. costs, total load, and member sizes, for various bay sizes and loading conditions.

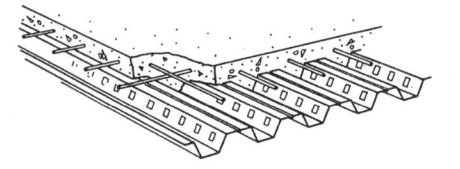

Metal Deck/Concrete Fill

Wood Joist

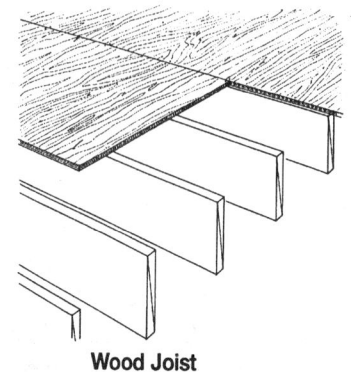

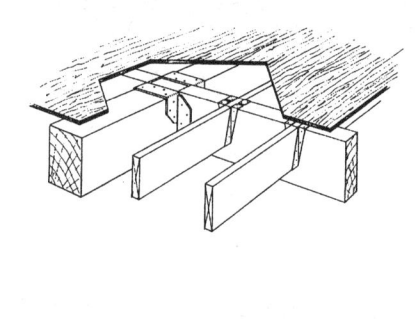

Wood Beam and Joist

3.5-580 Metal Deck/Concrete Fill

	SUPERIMPOSED LOAD (P.S.F.)	DECK SPAN (FT.)	DECK GAGE DEPTH	SLAB THICKNESS (IN.)	TOTAL LOAD (P.S.F.)	MAT.	INST.	TOTAL
0900	125	6	22 1-1/2	4	164	1.46	1.64	3.10
0920		7	20 1-1/2	4	164	1.50	1.72	3.22
0950		8	20 1-1/2	4	165	1.50	1.72	3.22
0970		9	18 1-1/2	4	165	1.72	1.72	3.44
1000		10	18 2	4	165	1.78	1.80	3.58
1020		11	18 3	5	169	2.38	1.95	4.33

3.5-710 Wood Joist

		MAT.	INST.	TOTAL
2902	Wood joist, 2" x 8", 12" O.C.	1.48	1.15	2.63
2950	16" O.C.	1.27	.99	2.26
3000	24" O.C.	1.19	.92	2.11
3300	2"x10", 12" O.C.	1.87	1.31	3.18
3350	16" O.C.	1.56	1.10	2.66
3400	24" O.C.	1.37	.99	2.36
3700	2"x12", 12" O.C.	2.29	1.33	3.62
3750	16" O.C.	1.87	1.11	2.98
3800	24" O.C.	1.59	1	2.59
4100	2"x14", 12" O.C.	2.58	1.43	4.01
4150	16" O.C.	2.10	1.19	3.29
4200	24" O.C.	1.74	1.07	2.81
7101	Note: Subfloor cost is included in these prices			

3.5-720 Wood Beam & Joist

	BAY SIZE (FT.)	SUPERIMPOSED LOAD (P.S.F.)	GIRDER BEAM (IN.)	JOISTS (IN.)	TOTAL LOAD (P.S.F.)	MAT.	INST.	TOTAL
2000	15x15	40	8 x 12 4 x 12	2 x 6 @ 16	53	4.39	2.64	7.03
2050		75	8 x 16 4 x 16	2 x 8 @ 16	90	6.05	2.89	8.94
2100		125	12 x 16 6 x 16	2 x 8 @ 12	144	9.45	3.71	13.16
2150		200	14 x 22 12 x 16	2 x 10 @ 12	227	17.45	5.65	23.10
3000	20x20	40	10 x 14 10 x 12	2 x 8 @ 16	63	6.55	2.66	9.21
3050		75	12 x 16 8 x 16	2 x 10 @ 16	102	8.35	2.99	11.34
3100		125	14 x 22 12 x 16	2 x 10 @ 12	163	14.10	4.74	18.84

SUPERSTRUCTURES A3.7 Roofs

The table below lists the cost per S.F. for a roof system with steel columns, beams and deck using open web steel joists and 1-1/2" galvanized metal deck. Perimeter of system is supported on bearing walls.

Fireproofing is not included. Costs/S.F. are based on a building 4 bays long and 4 bays wide.

Column costs are additive. Costs for the bearing walls are not included.

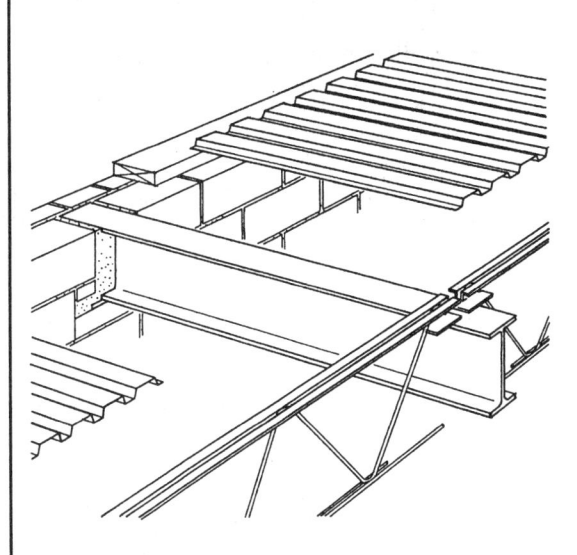

Steel Joists, Beams and Deck on Bearing Walls

3.7-410 Steel Joists, Beams & Deck on Columns & Walls

	BAY SIZE (FT.)	SUPERIMPOSED LOAD (P.S.F.)	DEPTH (IN.)	TOTAL LOAD (P.S.F.)	COLUMN ADD	MAT.	INST.	TOTAL
3000	25x25	20	18	40		2.01	.86	2.87
3100					columns	.17	.05	.22
3200		30	22	50		2.22	1.02	3.24
3300					columns	.22	.08	.30
3400		40	20	60		2.39	1.01	3.40
3500					columns	.22	.08	.30
3600	25x30	20	22	40		2.17	.83	3
3700					columns	.19	.06	.25
3800		30	20	50		2.35	1.08	3.43
3900					columns	.19	.06	.25
4000		40	25	60		2.55	.96	3.51
4100					columns	.22	.08	.30
4200	30x30	20	25	42		2.36	.90	3.26
4300					columns	.16	.05	.21
4400		30	22	52		2.59	.98	3.57
4500					columns	.19	.06	.25
4600		40	28	62		2.68	1.02	3.70
4700					columns	.19	.06	.25
4800	30x35	20	22	42		2.47	.94	3.41
4900					columns	.16	.05	.21
5000		30	28	52		2.61	.98	3.59
5100					columns	.16	.05	.21
5200		40	25	62		2.82	1.05	3.87
5300					columns	.19	.06	.25
5400	35x35	20	28	42		2.49	.95	3.44
5500					columns	.14	.04	.18

For expanded coverage of these items see *Means Assemblies Cost Data 1997*

SUPERSTRUCTURES A3.7 Roofs

Description: The table below lists the cost per S.F. for a roof system with steel columns, beams, and deck, using open web steel joists and 1-1/2" galvanized metal deck.

Column costs are additive. Fireproofing is not included.

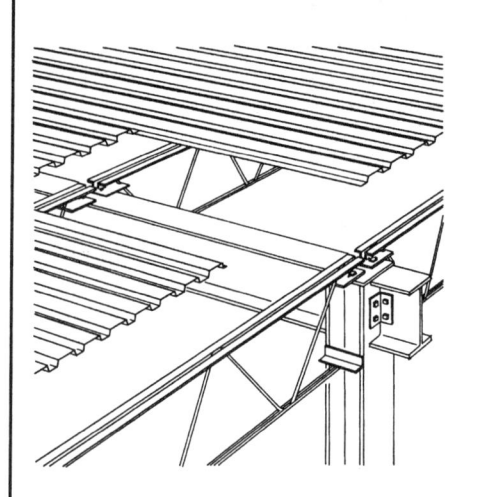

Steel Joists, Beams and Deck on Columns

3.7-420 Steel Joists, Beams, & Deck on Columns

	BAY SIZE (FT.)	SUPERIMPOSED LOAD (P.S.F.)	DEPTH (IN.)	TOTAL LOAD (P.S.F.)	COLUMN ADD	MAT.	INST.	TOTAL
1100	15x20	20	16	40		1.99	.85	2.84
1200					columns	.97	.33	1.30
1500		40	18	60		2.22	.96	3.18
1600					columns	.97	.33	1.30
2300	20x25	20	18	40		2.26	.95	3.21
2400					columns	.58	.20	.78
2700		40	20	60		2.56	1.08	3.64
2800					columns	.78	.26	1.04
2900	25x25	20	18	40		2.63	1.08	3.71
3000					columns	.47	.16	.63
3100		30	22	50		2.79	1.23	4.02
3200					columns	.62	.21	.83
3300		40	20	60		3.03	1.23	4.26
3400					columns	.62	.21	.83
3500	25x30	20	22	40		2.61	.99	3.60
3600					columns	.52	.17	.69
3900		40	25	60		3.16	1.17	4.33
4000					columns	.62	.21	.83
4100	30x30	20	25	42		2.95	1.09	4.04
4200					columns	.43	.14	.57
4300		30	22	52		3.25	1.19	4.44
4400					columns	.52	.17	.69
4500		40	28	62		3.39	1.24	4.63
4600					columns	.52	.17	.69
5300	35x35	20	28	42		3.23	1.19	4.42
5400					columns	.38	.13	.51
5500		30	25	52		3.58	1.31	4.89
5600					columns	.45	.15	.60
5700		40	28	62		3.87	1.41	5.28
5800					columns	.49	.17	.66

Important: See the Reference Section for critical supporting data - Location Factors & Historical Cost Indexes

SUPERSTRUCTURES — A3.7 Roofs

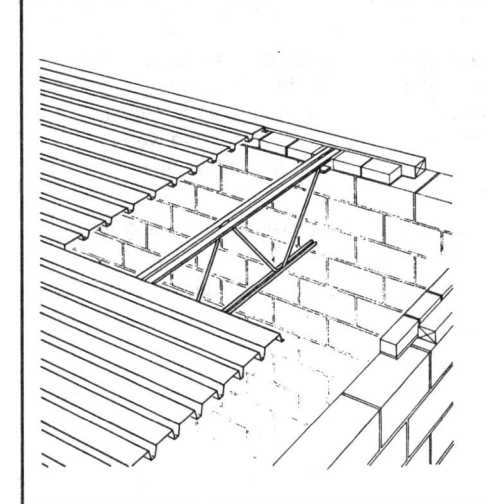

Description: The table below lists the cost per S.F. for a roof system using open web steel joists and 1-1/2" galvanized metal deck. The system is assumed supported on bearing walls or other suitable support. Costs for the supports are not included.

3.7-430 Steel Joists & Deck on Bearing Walls

	BAY SIZE (FT.)	SUPERIMPOSED LOAD (P.S.F.)	DEPTH (IN.)	TOTAL LOAD (P.S.F.)	MAT.	INST.	TOTAL
1100	20	20	13-1/2	40	1.32	.63	1.95
1200		30	15-1/2	50	1.36	.64	2
1300		40	15-1/2	60	1.46	.69	2.15
1400	25	20	17-1/2	40	1.46	.69	2.15
1500		30	17-1/2	50	1.58	.74	2.32
1600		40	19-1/2	60	1.61	.76	2.37
1700	30	20	19-1/2	40	1.60	.75	2.35
1800		30	21-1/2	50	1.64	.77	2.41
1900		40	23-1/2	60	1.77	.83	2.60
2000	35	20	23-1/2	40	1.77	.70	2.47
2100		30	25-1/2	50	1.83	.73	2.56
2200		40	25-1/2	60	1.95	.77	2.72
2300	40	20	25-1/2	41	1.98	.78	2.76
2400		30	25-1/2	51	2.11	.82	2.93
2500		40	25-1/2	61	2.18	.85	3.03
2600	45	20	27-1/2	41	2.22	1.14	3.36
2700		30	31-1/2	51	2.35	1.20	3.55
2800		40	31-1/2	61	2.48	1.28	3.76
2900	50	20	29-1/2	42	2.48	1.28	3.76
3000		30	31-1/2	52	2.72	1.40	4.12
3100		40	31-1/2	62	2.89	1.49	4.38
3200	60	20	37-1/2	42	2.80	1.32	4.12
3300		30	37-1/2	52	3.10	1.46	4.56
3400		40	37-1/2	62	3.10	1.46	4.56
3500	70	20	41-1/2	42	3.10	1.46	4.56
3600		30	41-1/2	52	3.30	1.55	4.85
3700		40	41-1/2	64	4	1.88	5.88
3800	80	20	45-1/2	44	4.31	2.10	6.41
3900		30	45-1/2	54	4.31	2.10	6.41
4000		40	45-1/2	64	4.77	2.33	7.10
4400	100	20	57-1/2	44	3.88	1.73	5.61
4500		30	57-1/2	54	4.63	2.06	6.69
4600		40	57-1/2	65	5.10	2.27	7.37
4700	125	20	69-1/2	44	5.50	2.38	7.88
4800		30	69-1/2	56	6.40	2.76	9.16
4900		40	69-1/2	67	7.35	3.15	10.50

For expanded coverage of these items see *Means Assemblies Cost Data 1997*

SUPERSTRUCTURES A3.7 | Roofs

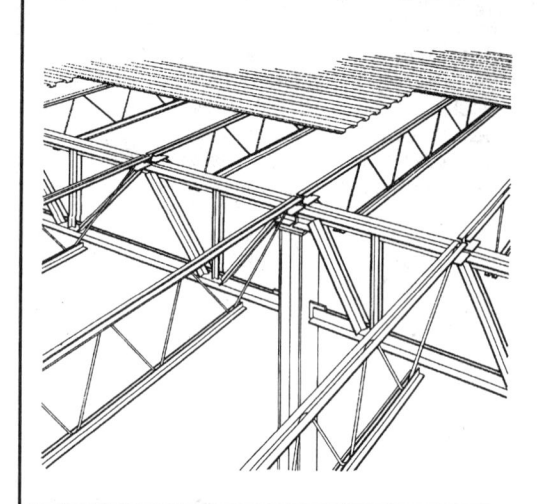

Description: The table below lists costs for a roof system supported on exterior bearing walls and interior columns. Costs include bracing, joist girders, open web steel joists and 1-1/2" galvanized metal deck.

Column costs are additive. Fireproofing is not included.

Costs/S.F. are based on a building 4 bays long and 4 bays wide. Costs for bearing walls are not included.

3.7-440 Steel Joists, Joist Girders & Deck on Columns & Walls

	BAY SIZE (FT.) GIRD X JOISTS	SUPERIMPOSED LOAD (P.S.F.)	DEPTH (IN.)	TOTAL LOAD (P.S.F.)	COLUMN ADD	MAT.	INST.	TOTAL
2350	30x35	20	32-1/2	40		1.96	.94	2.90
2400					columns	.16	.05	.21
2550		40	36-1/2	60		2.20	1.03	3.23
2600					columns	.19	.06	.25
3000	35x35	20	36-1/2	40		2.12	1.23	3.35
3050					columns	.14	.04	.18
3200		40	36-1/2	60		2.40	1.34	3.74
3250					columns	.18	.06	.24
3300	35x40	20	36-1/2	40		2.16	1.07	3.23
3350					columns	.14	.05	.19
3500		40	36-1/2	60		2.44	1.19	3.63
3550					columns	.15	.05	.20
3900	40x40	20	40-1/2	41		2.41	1.34	3.75
3950					columns	.13	.04	.17
4100		40	40-1/2	61		2.70	1.45	4.15
4150					columns	.13	.04	.17
5100	45x50	20	52-1/2	41		2.72	1.54	4.26
5150					columns	.10	.03	.13
5300		40	52-1/2	61		3.31	1.85	5.16
5350					columns	.14	.05	.19
5400	50x45	20	56-1/2	41		2.63	1.50	4.13
5450					columns	.10	.03	.13
5600		40	56-1/2	61		3.39	1.89	5.28
5650					columns	.13	.04	.17
5700	50x50	20	56-1/2	42		2.97	1.69	4.66
5750					columns	.10	.03	.13
5900		40	59	64		3.54	1.89	5.43
5950					columns	.14	.04	.18
6300	60x50	20	62-1/2	43		2.98	1.74	4.72
6350					columns	.15	.05	.20
6500		40	71	65		3.62	1.96	5.58
6550					columns	.20	.07	.27

Important: See the Reference Section for critical supporting data - Location Factors & Historical Cost Indexes

SUPERSTRUCTURES   A3.7   Roofs

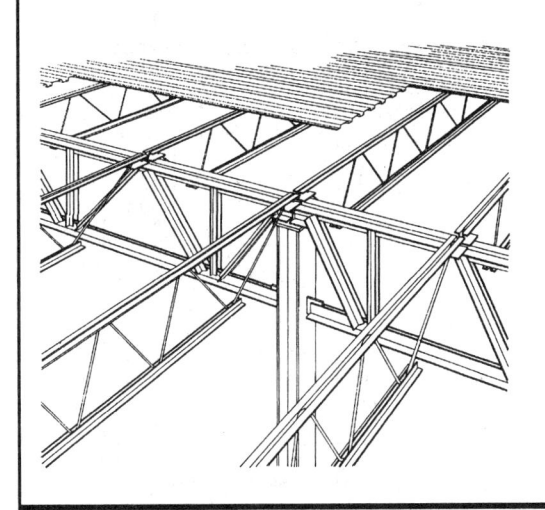

Description: The table below lists the cost per S.F. for a roof system supported on columns. Costs include joist girders, open web steel joists and 1-1/2" galvanized metal deck. Column costs are additive.

Fireproofing is not included. Costs/S.F. are based on a building 4 bays long and 4 bays wide.

Costs for wind columns are not included.

3.7-450 Steel Joists & Joist Girders on Columns

	BAY SIZE (FT.) GIRD X JOISTS	SUPERIMPOSED LOAD (P.S.F.)	DEPTH (IN.)	TOTAL LOAD (P.S.F.)	COLUMN ADD	MAT.	INST.	TOTAL
2000	30x30	20	17-1/2	40		1.96	1.13	3.09
2050					columns	.52	.17	.69
2100		30	17-1/2	50		2.15	1.20	3.35
2150					columns	.52	.17	.69
2200		40	21-1/2	60		2.20	1.24	3.44
2250					columns	.52	.17	.69
2300	30x35	20	32-1/2	40		2.10	1.07	3.17
2350					columns	.44	.15	.59
2500		40	36-1/2	60		2.38	1.19	3.57
2550					columns	.52	.17	.69
3200	35x40	20	36-1/2	40		2.62	1.68	4.30
3250					columns	.39	.13	.52
3400		40	36-1/2	60		2.96	1.85	4.81
3450					columns	.43	.14	.57
3800	40x40	20	40-1/2	41		2.67	1.66	4.33
3850					columns	.37	.13	.50
4000		40	40-1/2	61		3	1.82	4.82
4050					columns	.37	.13	.50
4100	40x45	20	40-1/2	41		2.87	1.92	4.79
4150					columns	.33	.12	.45
4200		30	40-1/2	51		3.12	2.04	5.16
4250					columns	.33	.12	.45
4300		40	40-1/2	61		3.48	2.22	5.70
4350					columns	.38	.13	.51
5000	45x50	20	52-1/2	41		3.03	1.95	4.98
5050					columns	.27	.09	.36
5200		40	52-1/2	61		3.71	2.29	6
5250					columns	.34	.12	.46
5300	50x45	20	56-1/2	41		2.95	1.89	4.84
5350					columns	.27	.09	.36
5500		40	56-1/2	61		3.81	2.34	6.15
5550					columns	.34	.12	.46
5600	50x50	20	56-1/2	42		3.29	2.09	5.38
5650					columns	.27	.09	.36
5800		40	59	64		3.11	1.91	5.02
5850					columns	.38	.13	.51

For expanded coverage of these items see *Means Assemblies Cost Data 1997*

SUPERSTRUCTURES | A3.7 | Roofs

The table below lists prices per S.F. for roof rafters and sheathing by nominal size and spacing. Sheathing is 5/16" CDX for 12" and 16" spacing and 3/8" CDX for 24" spacing.

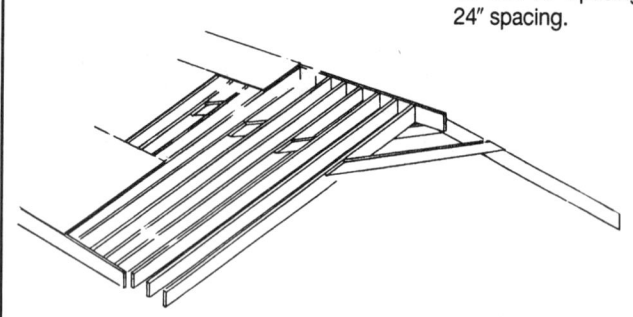

Factors for Converting Inclined to Horizontal

Roof Slope	Approx. Angle	Factor	Roof Slope	Approx. Angle	Factor
Flat	0°	1.000	12 in 12	45.0°	1.414
1 in 12	4.8°	1.003	13 in 12	47.3°	1.474
2 in 12	9.5°	1.014	14 in 12	49.4°	1.537
3 in 12	14.0°	1.031	15 in 12	51.3°	1.601
4 in 12	18.4°	1.054	16 in 12	53.1°	1.667
5 in 12	22.6°	1.083	17 in 12	54.8°	1.734
6 in 12	26.6°	1.118	18 in 12	56.3°	1.803
7 in 12	30.3°	1.158	19 in 12	57.7°	1.873
8 in 12	33.7°	1.202	20 in 12	59.0°	1.943
9 in 12	36.9°	1.250	21 in 12	60.3°	2.015
10 in 12	39.8°	1.302	22 in 12	61.4°	2.088
11 in 12	42.5°	1.357	23 in 12	62.4°	2.162

3.7-510 Wood/Flat or Pitched

		MAT.	INST.	TOTAL
2500	Flat rafter, 2"x4", 12" O.C.	.80	1.04	1.84
2550	16" O.C.	.70	.88	1.58
2600	24" O.C.	.62	.76	1.38
2900	2"x6", 12" O.C.	.99	1.04	2.03
2950	16" O.C.	.84	.89	1.73
3000	24" O.C.	.71	.76	1.47
3300	2"x8", 12" O.C.	1.24	1.12	2.36
3350	16" O.C.	1.03	.96	1.99
3400	24" O.C.	.85	.80	1.65
3700	2"x10", 12" O.C.	1.63	1.28	2.91
3750	16" O.C.	1.32	1.07	2.39
3800	24" O.C.	1.03	.87	1.90
4100	2"x12", 12" O.C.	2.05	1.30	3.35
4150	16" O.C.	1.63	1.08	2.71
4200	24" O.C.	1.25	.88	2.13
4500	2"x14", 12" O.C.	2.34	1.90	4.24
4550	16" O.C.	1.86	1.54	3.40
4600	24" O.C.	1.40	1.19	2.59
4900	3"x6", 12" O.C.	2.44	1.10	3.54
4950	16" O.C.	1.93	.94	2.87
5000	24" O.C.	1.45	.79	2.24
5300	3"x8", 12" O.C.	3.13	1.23	4.36
5350	16" O.C.	2.44	1.03	3.47
5400	24" O.C.	1.79	.85	2.64
5700	3"x10", 12" O.C.	3.82	1.39	5.21
5750	16" O.C.	2.97	1.15	4.12
5800	24" O.C.	2.14	.94	3.08
6100	3"x12", 12" O.C.	4.51	1.68	6.19
6150	16" O.C.	3.48	1.38	4.86
6200	24" O.C.	2.47	1.08	3.55
7001	Wood truss, 4 in 12 slope, 24" O.C., 24' to 29' span	2.61	1.57	4.18
7100	30' to 43' span	2.78	1.57	4.35
7200	44' to 60' span	2.94	1.57	4.51

SUPERSTRUCTURES A3.9 Stairs

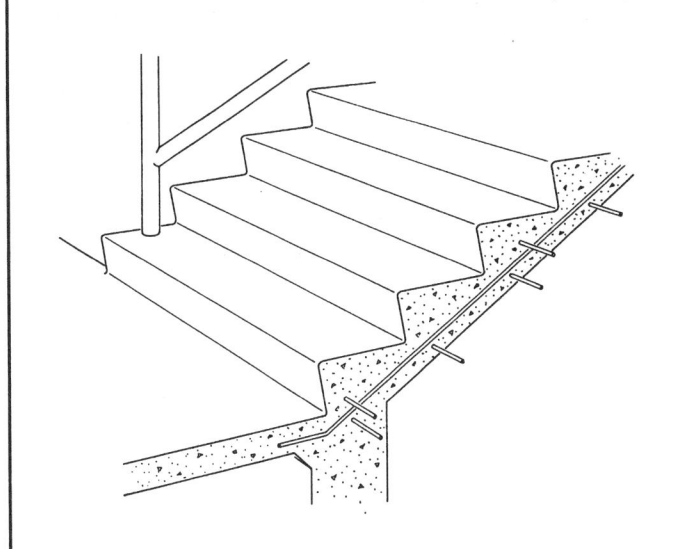

The table below lists the cost per flight for 4'-0" wide stairs. Side walls are not included. Railings are included in the prices.

3.9-100	Stairs	MAT.	INST.	TOTAL
0470	Stairs, C.I.P. concrete, w/o landing, 12 risers, w/o nosing	625	1,400	2,025
0480	With nosing	960	1,600	2,560
0550	W/landing, 12 risers, w/o nosing	700	1,725	2,425
0560	With nosing	1,025	1,925	2,950
0570	16 risers, w/o nosing	870	2,175	3,045
0580	With nosing	1,325	2,425	3,750
0590	20 risers, w/o nosing	1,050	2,625	3,675
0600	With nosing	1,600	2,950	4,550
0610	24 risers, w/o nosing	1,225	3,050	4,275
0620	With nosing	1,875	3,450	5,325
0630	Steel, grate type w/nosing & rails, 12 risers, w/o landing	1,575	510	2,085
0640	With landing	2,450	790	3,240
0660	16 risers, with landing	3,000	955	3,955
0680	20 risers, with landing	3,525	1,125	4,650
0700	24 risers, with landing	4,050	1,300	5,350
0701				
0710	Cement fill metal pan & picket rail, 12 risers, w/o landing	2,025	510	2,535
0720	With landing	3,200	880	4,080
0740	16 risers, with landing	3,875	1,050	4,925
0760	20 risers, with landing	4,550	1,200	5,750
0780	24 risers, with landing	5,225	1,400	6,625
0790	Cast iron tread & pipe rail, 12 risers, w/o landing	2,075	510	2,585
0800	With landing	3,225	880	4,105
1120	Wood, prefab box type, oak treads, wood rails 3'-6" wide, 14 risers	1,075	274	1,349
1150	Prefab basement type, oak treads, wood rails 3'-0" wide, 14 risers	790	219	1,009

For information about Means Estimating Seminars, see yellow pages 11 and 12 in back of book

For expanded coverage of these items see *Means Assemblies Cost Data 1997*

Division 4
Exterior Closure

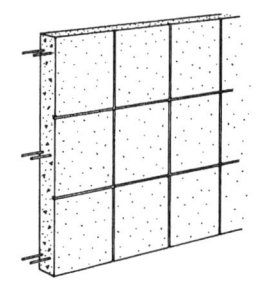

EXTERIOR CLOSURE — A4.1 Walls

Table 4.1-110 describes a concrete wall system for exterior closure. There are several types of wall finishes priced from plain finish to a finish with 3/4" rustication strip.

In Table 4.1-140, precast concrete wall panels are either solid or insulated. Prices below are based on delivery within 50 miles of a plant. Small jobs can double the prices below. For large, highly repetitive jobs, deduct up to 15%.

Cast in Place Concrete Wall

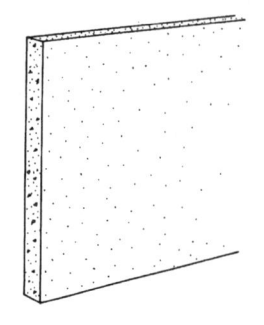

Flat Precast Concrete Wall

4.1-110 Cast In Place Concrete

		MAT.	INST.	TOTAL
2100	Conc wall reinforced, 8' high, 6" thick, plain finish, 3000 PSI	3	9.75	12.75
2700	Aged wood liner, 3000 PSI	4.72	10.20	14.92
3000	Sand blast light 1 side, 3000 PSI	3.15	10.60	13.75
3700	3/4" bevel rustication strip, 3000 PSI	3.15	11.95	15.10
4000	8" thick, plain finish, 3000 PSI	3.46	10.05	13.51
4100	4000 PSI	3.49	10.05	13.54
4300	Rub concrete 1 side, 3000 PSI	3.54	11.75	15.29
4550	8" thick, aged wood liner, 3000 PSI	5.20	10.50	15.70
4750	Sand blast light 1 side, 3000 PSI	3.61	10.85	14.46
5300	3/4" bevel rustication strip, 3000 PSI	3.61	12.25	15.86
5600	10" thick, plain finish, 3000 PSI	3.91	10.35	14.26
5900	Rub concrete 1 side, 3000 PSI	3.99	12.05	16.04
6200	Aged wood liner, 3000 PSI	5.65	10.75	16.40
6500	Sand blast light 1 side, 3000 PSI	4.06	11.20	15.26
7100	3/4" bevel rustication strip, 3000 PSI	4.06	12.50	16.56
7400	12" thick, plain finish, 3000 PSI	4.43	10.65	15.08
7700	Rub concrete 1 side, 3000 PSI	4.51	12.35	16.86
8000	Aged wood liner, 3000 PSI	6.15	11.10	17.25
8300	Sand blast light 1 side, 3000 PSI	4.58	11.50	16.08
8900	3/4" bevel rustication strip, 3000 PSI	4.58	12.80	17.38

4.1-140 Flat Precast Concrete

	THICKNESS (IN.)	PANEL SIZE (FT.)	FINISHES	RIGID INSULATION (IN)	TYPE	MAT.	INST.	TOTAL
3000	4	5x18	smooth gray	none	low rise	5.10	5.10	10.20
3050		6x18				4.26	4.28	8.54
3100		8x20				6.25	1.61	7.86
3150		12x20				5.90	1.52	7.42
3200	6	5x18	smooth gray	2	low rise	5.85	5.45	11.30
3250		6x18				5.05	4.61	9.66
3300		8x20				7.10	2	9.10
3350		12x20				6.50	1.83	8.33
3400	8	5x18	smooth gray	2	low rise	9.10	2.50	11.60
3450		6x18				8.65	2.38	11.03
3500		8x20				7.90	2.20	10.10
3550		12x20				7.25	2.03	9.28
3600	4	4x8	white face	none	low rise	15.55	2.96	18.51
3650		8x8				11.65	2.21	13.86
3700		10x10				10.20	1.95	12.15
3750		20x10				9.25	1.76	11.01

Important: See the Reference Section for critical supporting data - Location Factors & Historical Cost Indexes

EXTERIOR CLOSURE — A4.1 Walls

4.1-140 Flat Precast Concrete

	THICKNESS (IN.)	PANEL SIZE (FT.)	FINISHES	RIGID INSULATION (IN)	TYPE	COST PER S.F. MAT.	INST.	TOTAL
3800	5	4x8	white face	none	low rise	15.85	3.02	18.87
3850		8x8				12	2.28	14.28
3900		10x10				10.60	2.02	12.62
3950		20x20				9.70	1.85	11.55
4000	6	4x8	white face	none	low rise	16.45	3.12	19.57
4050		8x8				12.50	2.37	14.87
4100		10x10				11	2.10	13.10
4150		20x10				10.10	1.92	12.02
4200	6	4x8	white face	2	low rise	17.25	3.49	20.74
4250		8x8				13.30	2.74	16.04
4300		10x10				11.85	2.47	14.32
4350		20x10				10.10	1.92	12.02
4400	7	4x8	white face	none	low rise	16.85	3.21	20.06
4450		8x8				12.95	2.46	15.41
4500		10x10				11.60	2.21	13.81
4550		20x10				10.60	2.02	12.62
4600	7	4x8	white face	2	low rise	17.65	3.58	21.23
4650		8x8				13.75	2.83	16.58
4700		10x10				12.40	2.58	14.98
4750		20x10				11.45	2.39	13.84
4800	8	4x8	white face	none	low rise	17.20	3.28	20.48
4850		8x8				13.25	2.52	15.77
4900		10x10				11.95	2.27	14.22
4950		20x10				11	2.09	13.09
5000	8	4x8	white face	2	low rise	18.05	3.65	21.70
5050		8x8				14.10	2.89	16.99
5100		10x10				12.75	2.64	15.39
5150		20x10				11.80	2.46	14.26

4.1-140 Fluted Window or Mullion Precast Concrete

	THICKNESS (IN.)	PANEL SIZE (FT.)	FINISHES	RIGID INSULATION (IN)	TYPE	COST PER S.F. MAT.	INST.	TOTAL
5200	4	4x8	smooth gray	none	high rise	11.65	11.75	23.40
5250		8x8				8.35	8.35	16.70
5300		10x10				11.90	3.07	14.97
5350		20x10				10.45	2.70	13.15
5400	5	4x8	smooth gray	none	high rise	11.85	11.95	23.80
5450		8x8				8.60	8.65	17.25
5500		10x10				12.40	3.20	15.60
5550		20x10				11	2.83	13.83
5600	6	4x8	smooth gray	none	high rise	12.15	12.25	24.40
5650		8x8				8.90	8.95	17.85
5700		10x10				12.75	3.30	16.05
5750		20x10				11.35	2.92	14.27
5800	6	4x8	smooth gray	2	high rise	13	12.60	25.60
5850		8x8				9.75	9.35	19.10
5900		10x10				13.60	3.67	17.27
5950		20x10				12.15	3.29	15.44
6000	7	4x8	smooth gray	none	high rise	12.40	12.55	24.95
6050		8x8				9.15	9.25	18.40
6100		10x10				13.35	3.44	16.79
6150		20x10				11.70	3.02	14.72

For expanded coverage of these items see *Means Assemblies Cost Data 1997*

EXTERIOR CLOSURE — A4.1 Walls

4.1-140 Fluted Window or Mullion Precast Concrete

	THICKNESS (IN.)	PANEL SIZE (FT.)	FINISHES	RIGID INSULATION (IN)	TYPE	COST PER S.F. MAT.	COST PER S.F. INST.	COST PER S.F. TOTAL
6200	7	4x8	smooth gray	2	high rise	13.25	12.90	26.15
6250		8x8				10	9.60	19.60
6300		10x10				14.15	3.81	17.96
6350		20x10				12.55	3.39	15.94
6400	8	4x8	smooth gray	none	high rise	12.60	12.70	25.30
6450		8x8				9.40	9.45	18.85
6500		10x10				13.75	3.55	17.30
6550		20x10				12.30	3.17	15.47
6600	8	4x8	smooth gray	2	high rise	13.45	13.05	26.50
6650		8x8				10.20	9.80	20
6700		10x10				14.55	3.92	18.47
6750		20x10				13.10	3.54	16.64

4.1-140 Precast Concrete Specialties

	TYPE	SIZE	COST PER L.F. MAT.	COST PER L.F. INST.	COST PER L.F. TOTAL
6773	Coping, precast	6" wide	10.45	5.85	16.30
6774	Stock units	10" wide	11	6.45	17.45
6775		12" wide	9.90	6.85	16.75
6776		14" wide	15.15	7.30	22.45
6777	Window sills	6" wide	10.75	8.30	19.05
6778	Precast	10" wide	10.10	9.70	19.80
6779		14" wide	13.15	11.65	24.80

EXTERIOR CLOSURE — A4.1 Walls

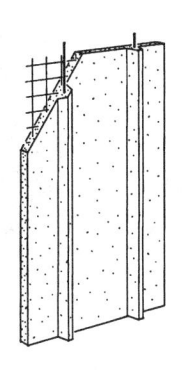

Ribbed Precast Panel

4.1-140 Ribbed Precast Concrete

	THICKNESS (IN.)	PANEL SIZE (FT.)	FINISHES	RIGID INSULATION (IN.)	TYPE	MAT.	INST.	TOTAL
6800	4	4x8	aggregate	none	high rise	12.65	11.70	24.35
6850		8x8				9.25	8.55	17.80
6900		10x10				13.15	3.10	16.25
6950		20x10				11.65	2.74	14.39
7000	5	4x8	aggregate	none	high rise	12.85	11.95	24.80
7050		8x8				9.50	8.80	18.30
7100		10x10				13.65	3.22	16.87
7150		20x10				12.20	2.88	15.08
7200	6	4x8	aggregate	none	high rise	13.15	12.15	25.30
7250		8x8				9.75	9.05	18.80
7300		10x10				14.10	3.32	17.42
7350		20x10				12.65	2.98	15.63
7400	6	4x8	aggregate	2	high rise	13.95	12.50	26.45
7450		8x8				10.60	9.45	20.05
7500		10x10				14.95	3.69	18.64
7550		20x10				13.45	3.35	16.80
7600	7	4x8	aggregate	none	high rise	13.45	12.45	25.90
7650		8x8				10	9.25	19.25
7700		10x10				14.55	3.42	17.97
7750		20x10				13	3.07	16.07
7800	7	4x8	aggregate	2	high rise	14.25	12.80	27.05
7850		8x8				10.85	9.65	20.50
7900		10x10				15.40	3.79	19.19
7950		20x10				13.85	3.44	17.29
8000	8	4x8	aggregate	none	high rise	13.65	12.65	26.30
8050		8x8				10.30	9.55	19.85
8100		10x10				15.05	3.55	18.60
8150		20x10				13.60	3.20	16.80
8200	8	4x8	aggregate	2	high rise	14.50	13	27.50
8250		8x8				11.15	9.95	21.10
8300		10x10				15.90	3.92	19.82
8350		20x10				14.45	3.57	18.02

For expanded coverage of these items see *Means Assemblies Cost Data 1997*

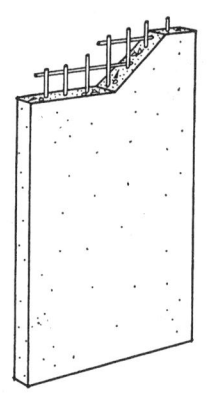

EXTERIOR CLOSURE — A4.1 Walls

The advantage of tilt up construction is in the low cost of forms and placing of concrete and reinforcing. Tilt up has been used for several types of buildings, including warehouses, stores, offices, and schools. The panels are cast in forms on the ground, or floor slab. Most jobs use 5-1/2" thick solid reinforced concrete panels.

Tilt Up Concrete Panel

4.1-160	Tilt-Up Concrete Panel	MAT.	INST.	TOTAL
		COST PER S.F.		
3200	Tilt up conc panels, broom finish, 5-1/2" thick, 3000 PSI	2.28	2.92	5.20
3250	5000 PSI	2.32	2.85	5.17
3300	6" thick, 3000 PSI	2.51	3.01	5.52
3350	5000 PSI	2.55	2.94	5.49
3400	7-1/2" thick, 3000 PSI	3.11	3.16	6.27
3450	5000 PSI	3.19	3.09	6.28
3500	8" thick, 3000 PSI	3.34	3.25	6.59
3550	5000 PSI	3.42	3.18	6.60
3700	Steel trowel finish, 5-1/2" thick, 3000 PSI	2.27	3.03	5.30
3750	5000 PSI	2.32	2.96	5.28
3800	6" thick, 3000 PSI	2.51	3.12	5.63
3850	5000 PSI	2.55	3.05	5.60
3900	7-1/2" thick, 3000 PSI	3.11	3.27	6.38
3950	5000 PSI	3.19	3.20	6.39
4000	8" thick, 3000 PSI	3.34	3.36	6.70
4050	5000 PSI	3.42	3.29	6.71
4200	Exp. aggregate finish, 5-1/2" thick, 3000 PSI	2.73	2.95	5.68
4250	5000 PSI	2.77	2.88	5.65
4300	6" thick, 3000 PSI	2.96	3.04	6
4350	5000 PSI	3.01	2.97	5.98
4400	7-1/2" thick, 3000 PSI	3.57	3.19	6.76
4450	5000 PSI	3.64	3.12	6.76
4500	8" thick, 3000 PSI	3.80	3.28	7.08
4550	5000 PSI	3.88	3.21	7.09
4600	Exposed aggregate & vert. rustication 5-1/2" thick, 3000 PSI	4.05	5.45	9.50
4650	5000 PSI	4.09	5.35	9.44
4700	6" thick, 3000 PSI	4.28	5.50	9.78
4750	5000 PSI	4.33	5.45	9.78
4800	7-1/2" thick, 3000 PSI	4.89	5.65	10.54
4850	5000 PSI	4.96	5.60	10.56
4900	8" thick, 3000 PSI	5.10	5.75	10.85
4950	5000 PSI	5.20	5.70	10.90
5000	Vertical rib & light sandblast, 5-1/2" thick, 3000 PSI	4.25	3.74	7.99
5100	6" thick, 3000 PSI	4.48	3.83	8.31
5200	7-1/2" thick, 3000 PSI	5.10	3.98	9.08
5300	8" thick, 3000 PSI	5.30	4.07	9.37
6000	Broom finish w/2" urethane insulation, 6" thick, 3000 PSI	2.28	3.78	6.06
6100	Broom finish 2" fiberplank insulation, 6" thick, 3000 PSI	2.77	3.74	6.51
6200	Exposed aggregate w/2" urethane insulation, 6" thick, 3000 PSI	2.68	3.81	6.49
6300	Exposed aggregate 2" fiberplank insulation, 6" thick, 3000 PSI	3.17	3.77	6.94

EXTERIOR CLOSURE — A4.1 Walls

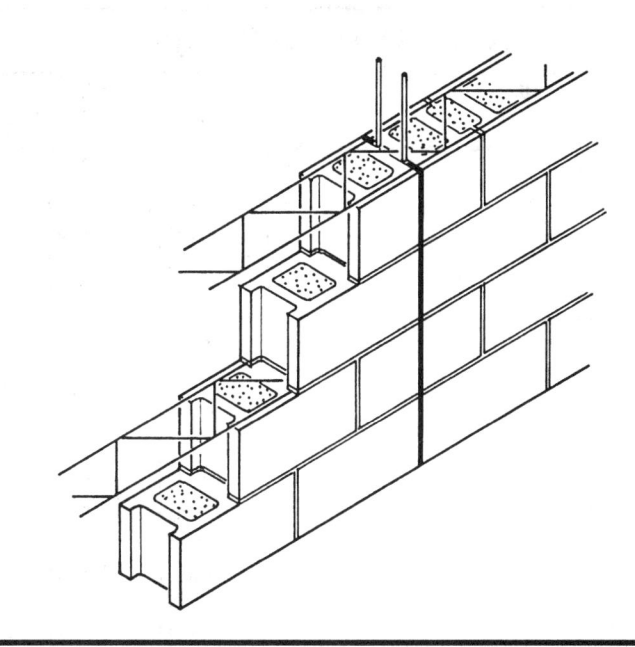

Exterior concrete block walls are defined in the following terms; structural reinforcement, weight, percent solid, size, strength and insulation. Within each of these categories, two to four variations are shown. No costs are included for brick shelf or relieving angles.

4.1-211 Concrete Block Wall - Regular Weight

	TYPE	SIZE (IN.)	STRENGTH (P.S.I.)	CORE FILL	MAT.	INST.	TOTAL
1200	Hollow	4x8x16	2,000	none	.94	3.59	4.53
1250			4,500	none	1.36	3.59	4.95
1400		8x8x16	2,000	perlite	1.96	4.36	6.32
1410				styrofoam	2.38	4.12	6.50
1440				none	1.57	4.12	5.69
1450			4,500	perlite	2.45	4.36	6.81
1460				styrofoam	2.87	4.12	6.99
1490				none	2.06	4.12	6.18
2000	75% solid	4x8x16	2,000	none	1.43	3.64	5.07
2100		6x8x16	2,000	perlite	1.85	4	5.85
2500	Solid	4x8x16	2,000	none	1.37	3.72	5.09
2700		8x8x16	2,000	none	2.21	4.29	6.50

4.1-211 Concrete Block Wall - Lightweight

	TYPE	SIZE (IN.)	WEIGHT (P.C.F.)	CORE FILL	MAT.	INST.	TOTAL
3100	Hollow	8x4x16	105	perlite	2.66	4.11	6.77
3110				styrofoam	3.08	3.87	6.95
3200		4x8x16	105	none	1.05	3.52	4.57
3250			85	none	1.28	3.44	4.72
3300		6x8x16	105	perlite	1.55	3.95	5.50
3310				styrofoam	2.10	3.76	5.86
3340				none	1.29	3.76	5.05
3400		8x8x16	105	perlite	1.95	4.26	6.21
3410				styrofoam	2.37	4.02	6.39
3440				none	1.56	4.02	5.58
3450			85	perlite	2.13	4.16	6.29
4000	75% solid	4x8x16	105	none	1.57	3.55	5.12
4050			85	none	2.13	3.48	5.61
4100		6x8x16	105	perlite	2.27	3.91	6.18
4500	Solid	4x8x16	105	none	1.41	3.68	5.09
4700		8x8x16	105	none	2.30	4.23	6.53

For expanded coverage of these items see *Means Assemblies Cost Data 1997*

EXTERIOR CLOSURE — A4.1 Walls

4.1-211 Reinforced Concrete Block Wall - Regular Weight

	TYPE	SIZE (IN.)	STRENGTH (P.S.I.)	VERT. REINF & GROUT SPACING	MAT.	INST.	TOTAL
5200	Hollow	4x8x16	2,000	#4 @ 48"	1.06	3.97	5.03
5300		6x8x16	2,000	#4 @ 48"	1.33	4.22	5.55
5330				#5 @ 32"	1.46	4.36	5.82
5340				#5 @ 16"	1.79	4.87	6.66
5350			4,500	#4 @ 28"	1.82	4.22	6.04
5390				#5 @ 16"	2.28	4.87	7.15
5400	Hollow	8x8x16	2,000	#4 @ 48"	1.86	4.49	6.35
5430				#5 @ 32"	1.98	4.73	6.71
5440				#5 @ 16"	2.39	5.35	7.74
5450		8x8x16	4,500	#4 @ 48"	2.32	4.55	6.87
5490				#5 @ 16"	2.88	5.35	8.23
5500		12x8x16	2,000	#4 @ 48"	2.53	5.75	8.28
5540				#5 @ 16"	3.35	6.55	9.90
6100	75% solid	6x8x16	2,000	#4 @ 48"	1.82	4.23	6.05
6140				#5 @ 16"	2.08	4.80	6.88
6150			4,500	#4 @ 48"	2.13	4.23	6.36
6190				#5 @ 16"	2.39	4.80	7.19
6200		8x8x16	2,000	#4 @ 48"	2.16	4.56	6.72
6230				#5 @ 32"	2.25	4.71	6.96
6240				#5 @ 16"	2.46	5.25	7.71
6250			4,500	#4 @ 48"	2.52	4.56	7.08
6280				#5 @ 32"	2.61	4.71	7.32
6290				#5 @ 16"	2.82	5.25	8.07
6500	Solid-double Wythe	2-4x8x16	2,000	#4 @ 48" E.W.	3.47	8.35	11.82
6530				#5 @ 16" E.W.	3.79	8.90	12.69
6550			4,500	#4 @ 48" E.W.	4.09	8.30	12.39
6580				#5 @ 16" E.W.	4.41	8.80	13.21

4.1-211 Reinforced Concrete Block Wall - Lightweight

	TYPE	SIZE (IN.)	WEIGHT (P.C.F.)	VERT REINF. & GROUT SPACING	MAT.	INST.	TOTAL
7100	Hollow	8x4x16	105	#4 @ 48"	2.53	4.30	6.83
7140				#5 @ 16"	3.09	5.10	8.19
7150			85	#4 @ 48"	3.07	4.21	7.28
7190				#5 @ 16"	3.63	4.99	8.62
7400		8x8x16	105	#4 @ 48"	1.82	4.45	6.27
7440				#5 @ 16"	2.38	5.25	7.63
7450		8x8x16	85	#4 @ 48"	2	4.35	6.35
7490				#5 @ 16"	2.56	5.15	7.71
7800		8x8x24	105	#4 @ 48"	2.31	4.78	7.09
7840				#5 @ 16"	2.87	5.55	8.42
7850			85	#4 @ 48"	3.33	4.09	7.42
7890				#5 @ 16"	3.89	4.87	8.76
8100	75% solid	6x8x16	105	#4 @ 48"	2.24	4.14	6.38
8130				#5 @ 32"	2.32	4.26	6.58
8150			85	#4 @ 48"	2.54	4.05	6.59
8180				#5 @ 32"	2.62	4.17	6.79
8200		8x8x16	105	#4 @ 48"	2.72	4.45	7.17
8230				#5 @ 32"	2.81	4.60	7.41
8250			85	#4 @ 48"	3.43	4.35	7.78
8280				#5 @ 32"	3.52	4.50	8.02

Important: See the Reference Section for critical supporting data - Location Factors & Historical Cost Indexes

EXTERIOR CLOSURE		A4.1	Walls				

4.1-211 Reinforced Concrete Block Wall - Lightweight

	TYPE	SIZE (IN.)	WEIGHT (P.C.F.)	VERT REINF. & GROUT SPACING		COST PER S.F.	
					MAT.	INST.	TOTAL
8500	Solid-double	2-4x8x16	105	#4 @ 48"	3.55	8.30	11.85
8530	Wythe		105	#5 @ 16"	3.87	8.80	12.67
8600		2-6x8x16	105	#4 @ 48"	4.78	8.85	13.63
8630				#5 @ 16"	5.10	9.35	14.45
8650			85	#4 @ 48"	6.05	8.45	14.50

For expanded coverage of these items see *Means Assemblies Cost Data 1997*

EXTERIOR CLOSURE — A4.1 Walls

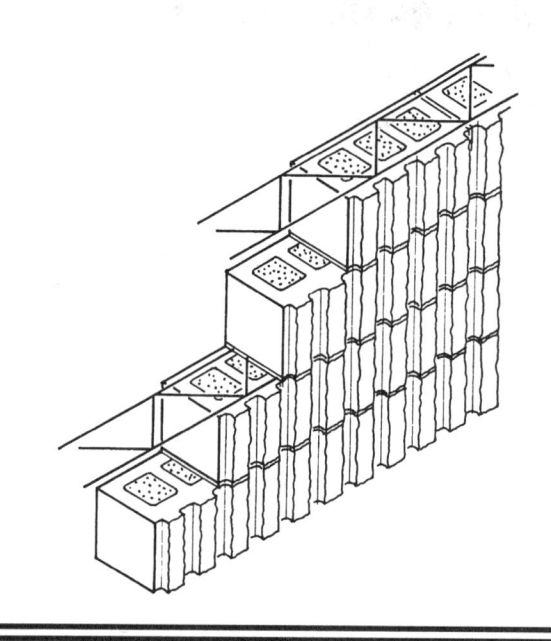

Exterior split ribbed block walls are defined in the following terms; structural reinforcement, weight, percent solid, size, number of ribs and insulation. Within each of these categories two to four variations are shown. No costs are included for brick shelf or relieving angles. Costs include control joints every 20' and horizontal reinforcing.

4.1-212 Split Ribbed Block Wall - Regular Weight

	TYPE	SIZE (IN.)	RIBS	CORE FILL	MAT.	INST.	TOTAL
1220	Hollow	4x8x16	4	none	2.05	4.45	6.50
1250			8	none	2.39	4.45	6.84
1280			16	none	1.94	4.51	6.45
1430		8x8x16	8	perlite	4.06	5.30	9.36
1440				styrofoam	4.48	5.05	9.53
1450				none	3.67	5.05	8.72
1530		12x8x16	8	perlite	5.05	7	12.05
1540				styrofoam	5.45	6.50	11.95
1550				none	4.43	6.50	10.93
2120	75% solid	4x8x16	4	none	2.57	4.51	7.08
2150			8	none	3	4.51	7.51
2180			16	none	2.46	4.58	7.04
2520	Solid	4x8x16	4	none	3.10	4.58	7.68
2550			8	none	3.63	4.58	8.21
2580			16	none	2.93	4.65	7.58

4.1-212 Reinforced Split Ribbed Block Wall - Regular Weight

	TYPE	SIZE (IN.)	RIBS	VERT. REINF. & GROUT SPACING	MAT.	INST.	TOTAL
5200	Hollow	4x8x16	4	#4 @ 48"	2.17	4.83	7
5230			8	#4 @ 48"	2.51	4.83	7.34
5260			16	#4 @ 48"	2.06	4.89	6.95
5430		8x8x16	8	#4 @ 48"	3.93	5.50	9.43
5440				#5 @ 32"	4.08	5.65	9.73
5450				#5 @ 16"	4.49	6.25	10.74
5530		12x8x16	8	#4 @ 48"	4.81	6.95	11.76
5540				#5 @ 32"	5.05	7.15	12.20
5550				#5 @ 16"	5.65	7.75	13.40
6230	75% solid	6x8x16	8	#4 @ 48"	3.82	5.10	8.92
6240				#5 @ 32"	3.90	5.25	9.15
6250				#5 @ 16"	4.08	5.70	9.78
6330		8x8x16	8	#4 @ 48"	4.79	5.50	10.29
6340				#5 @ 32"	4.88	5.65	10.53
6350				#5 @ 16"	5.10	6.15	11.25

Important: See the Reference Section for critical supporting data - Location Factors & Historical Cost Indexes

EXTERIOR CLOSURE   A4.1   Walls

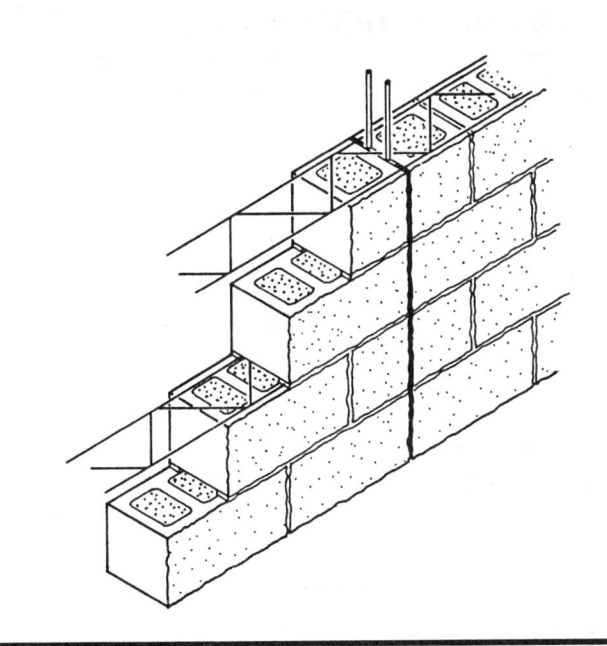

Exterior split face block walls are defined in the following terms; structural reinforcement, weight, percent solid, size, scores and insulation. Within each of these categories two to four variations are shown. No costs are included for brick shelf or relieving angles. Costs include control joints every 20' and horizontal reinforcing.

4.1-213   Split Face Block Wall - Regular Weight

	TYPE	SIZE (IN.)	SCORES	CORE FILL	MAT.	INST.	TOTAL
1200	Hollow	8x4x16	0	perlite	5.90	5.85	11.75
1210				styrofoam	6.35	5.60	11.95
1240				none	5.50	5.60	11.10
1250			1	perlite	6.30	5.85	12.15
1260				styrofoam	6.75	5.60	12.35
1290				none	5.90	5.60	11.50
1300		12x4x16	0	perlite	7.65	6.65	14.30
1310				styrofoam	8.05	6.15	14.20
1340				none	7.05	6.15	13.20
1350			1	perlite	8.05	6.65	14.70
1360				styrofoam	8.45	6.15	14.60
1390				none	7.45	6.15	13.60
1400		4x8x16	0	none	2.05	4.39	6.44
1450			1	none	2.30	4.45	6.75
1500		6x8x16	0	perlite	2.89	5.05	7.94
1510				styrofoam	3.44	4.86	8.30
1540				none	2.63	4.86	7.49
1550			1	perlite	3.15	5.15	8.30
1560				styrofoam	3.70	4.94	8.64
1590				none	2.89	4.94	7.83
1600		8x8x16	0	perlite	3.49	5.45	8.94
1610				styrofoam	3.91	5.20	9.11
1640				none	3.10	5.20	8.30
1650			1	perlite	3.70	5.55	9.25
1660				styrofoam	4.12	5.30	9.42
1690				none	3.31	5.30	8.61
1700		12x8x16	0	perlite	4.47	7.10	11.57
1710				styrofoam	4.83	6.60	11.43
1740				none	3.83	6.60	10.43
1750			1	perlite	4.67	7.25	11.92
1760				styrofoam	5.05	6.75	11.80
1790				none	4.03	6.75	10.78

For expanded coverage of these items see *Means Assemblies Cost Data 1997*

EXTERIOR CLOSURE — A4.1 Walls

4.1-213 Split Face Block Wall - Regular Weight

	TYPE	SIZE (IN.)	SCORES	CORE FILL	MAT.	INST.	TOTAL
1800	75% solid	8x4x16	0	perlite	7.15	5.80	12.95
1840				none	6.95	5.70	12.65
1852			1	perlite	7.55	5.80	13.35
1890				none	7.35	5.70	13.05
2000		4x8x16	0	none	2.57	4.45	7.02
2050			1	none	2.78	4.51	7.29
2400	Solid	8x4x16	0	none	8.45	5.80	14.25
2450			1	none	8.90	5.80	14.70
2900		12x8x16	0	none	5.85	6.85	12.70
2950			1	none	6.05	7	13.05

4.1-213 Reinforced Split Face Block Wall - Regular Weight

	TYPE	SIZE (IN.)	SCORES	VERT. REINF. & GROUT SPACING	MAT.	INST.	TOTAL
5200	Hollow	8x4x16	0	#4 @ 48"	5.80	6.05	11.85
5210				#5 @ 32"	5.95	6.20	12.15
5240				#5 @ 16"	6.35	6.80	13.15
5250			1	#4 @ 48"	6.20	6.05	12.25
5260				#5 @ 32"	6.35	6.20	12.55
5290				#5 @ 16"	6.75	6.80	13.55
5700		12x8x16	0	#4 @ 48"	4.21	7.05	11.26
5710				#5 @ 32"	4.43	7.25	11.68
5740				#5 @ 16"	5.05	7.85	12.90
5750			1	#4 @ 48"	4.41	7.20	11.61
5760				#5 @ 32"	4.63	7.40	12.03
5790				#5 @ 16"	5.25	8	13.25
6000	75% solid	8x4x16	0	#4 @ 48"	7.10	6.05	13.15
6010				#5 @ 32"	7.20	6.20	13.40
6040				#5 @ 16"	7.40	6.75	14.15
6050			1	#4 @ 48"	7.50	6.05	13.55
6060				#5 @ 32"	7.60	6.20	13.80
6090				#5 @ 16"	7.80	6.75	14.55
6100		12x4x16	0	#4 @ 48"	9.15	6.65	15.80
6110				#5 @ 32"	9.25	6.80	16.05
6140				#5 @ 16"	9.55	7.20	16.75
6150			1	#4 @ 48"	9.50	6.65	16.15
6160				#5 @ 32"	9.60	6.80	16.40
6190				#5 @ 16"	9.95	7.30	17.25
6700	Solid-double Wythe	2-4x8x16	0	#4 @ 48" E.W.	6.95	9.95	16.90
6710				#5 @ 32" E.W.	7.05	10.05	17.10
6740				#5 @ 16" E.W.	7.25	10.45	17.70
6750			1	#4 @ 48" E.W.	7.35	10.10	17.45
6760				#5 @ 32" E.W.	7.45	10.15	17.60
6790				#5 @ 16" E.W.	7.65	10.60	18.25
6800		2-6x8x16	0	#4 @ 48" E.W.	8.90	11	19.90
6810				#5 @ 32" E.W.	9.05	11.05	20.10
6840				#5 @ 16" E.W.	9.25	11.50	20.75
6850			1	#4 @ 48" E.W.	9.30	11.15	20.45
6860				#5 @ 32" E.W.	9.45	11.25	20.70
6890				#5 @ 16" E.W.	9.65	11.65	21.30

Important: See the Reference Section for critical supporting data - Location Factors & Historical Cost Indexes

EXTERIOR CLOSURE — A4.1 Walls

Exterior solid brick walls are defined in the following terms; structural reinforcement, type of brick, thickness, and bond. Nine different types of face bricks are presented, with single wythes shown in four different bonds. Shelf angles are included in system components. These walls do not include ties and as such are not tied to a backup wall.

4.1-231 Solid Brick Walls - Single Wythe

	TYPE	THICKNESS (IN.)	BOND			MAT.	INST.	TOTAL
1000	Common	4	running			3.11	8.25	11.36
1010			common			3.60	9.40	13
1050			Flemish			3.92	11.40	15.32
1100			English			4.35	12.05	16.40
1150	Standard	4	running			5.70	8.55	14.25
1160			common			4.27	9.85	14.12
1200			Flemish			4.68	11.70	16.38
1250			English			5.20	12.40	17.60
1300	Glazed	4	running			9.10	8.85	17.95
1310			common			10.80	10.30	21.10
1350			Flemish			11.90	12.40	24.30
1400			English			13.35	13.25	26.60
1600	Economy	4	running			3.96	6.60	10.56
1610			common			4.62	7.50	12.12
1650			Flemish			5.05	8.85	13.90
1700			English			5.60	9.40	15
1900	Fire	4-1/2	running			10.40	7.50	17.90
1910			common			12.35	8.55	20.90
1950			Flemish			13.65	10.30	23.95
2000			English			15.30	10.80	26.10
2050	King	3-1/2	running			3.49	6.75	10.24
2060			common			4.04	7.75	11.79
2100			Flemish			4.84	9.20	14.04
2150			English			5.30	9.60	14.90
2200	Roman	4	running			5.75	7.75	13.50
2210			common			6.75	8.85	15.60
2250			Flemish			7.45	10.55	18
2300			English			8.35	11.40	19.75
2350	Norman	4	running			9	6.45	15.45
2360			common			5.65	7.30	12.95
2400			Flemish			6.20	8.70	14.90
2450			English			6.90	9.20	16.10
2500	Norwegian	4	running			4.07	5.75	9.82
2510			common			4.76	6.50	11.26
2550			Flemish			5.20	7.75	12.95
2600			English			5.80	8.10	13.90

For expanded coverage of these items see *Means Assemblies Cost Data 1997*

EXTERIOR CLOSURE — A4.1 Walls

4.1-231 Solid Brick Walls - Double Wythe

	TYPE	THICKNESS (IN.)	COLLAR JOINT THICKNESS (IN.)			MAT.	INST.	TOTAL
4100	Common	8	3/4			5.85	13.75	19.60
4150	Standard	8	1/2			10.35	14.20	24.55
4200	Glazed	8	1/2			17.85	14.70	32.55
4250	Engineer	8	1/2			7.50	12.55	20.05
4300	Economy	8	3/4			7.40	10.80	18.20
4350	Double	8	3/4			6.45	8.75	15.20
4400	Fire	8	3/4			20.50	12.55	33.05
4450	King	7	3/4			6.60	11.05	17.65
4500	Roman	8	1			11.10	12.90	24
4550	Norman	8	3/4			9.25	10.55	19.80
4600	Norwegian	8	3/4			7.80	9.35	17.15
4650	Utility	8	3/4			7.10	8.10	15.20
4700	Triple	8	3/4			4.43	7.30	11.73
4750	SCR	12	3/4			10.35	10.80	21.15
4800	Norwegian	12	3/4			8.70	9.55	18.25
4850	Jumbo	12	3/4			9.90	8.35	18.25

4.1-231 Reinforced Brick Walls

	TYPE	THICKNESS (IN.)	WYTHE	REINF. & SPACING		MAT.	INST.	TOTAL
7200	Common	4"	1	#4 @ 48"vert		3.29	8.55	11.84
7220				#5 @ 32"vert		3.35	8.65	12
7230				#5 @ 16"vert		3.51	9.05	12.56
7700	King	9"	2	#4 @ 48" E.W.		7	11.55	18.55
7720				#5 @ 32" E.W.		7.10	11.60	18.70
7730				#5 @ 16" E.W.		7.30	11.95	19.25
7800	Common	10"	2	#4 @ 48" E.W.		6.25	14.25	20.50
7820				#5 @ 32" E.W.		6.35	14.30	20.65
7830				#5 @ 16" E.W.		6.55	14.65	21.20
7900	Standard	10"	2	#4 @ 48" E.W.		10.75	14.70	25.45
7920				#5 @ 32" E.W.		10.85	14.75	25.60
7930				#5 @ 16" E.W.		11.05	15.10	26.15
8100	Economy	10"	2	#4 @ 48" E.W.		7.80	11.30	19.10
8120				#5 @ 32" E.W.		7.90	11.35	19.25
8130				#5 @ 16" E.W.		8.10	11.70	19.80
8200	Double	10"	2	#4 @ 48" E.W.		6.85	9.25	16.10
8220				#5 @ 32" E.W.		6.95	9.30	16.25
8230				#5 @ 16" E.W.		7.15	9.65	16.80
8300	Fire	10"	2	#4 @ 48" E.W.		21	13.05	34.05
8320				#5 @ 32" E.W.		21	13.10	34.10
8330				#5 @ 16" E.W.		21	13.45	34.45
8400	Roman	10"	2	#4 @ 48" E.W.		11.50	13.40	24.90
8420				#5 @ 32" E.W.		11.60	13.45	25.05
8430				#5 @ 16" E.W.		11.80	13.80	25.60
8500	Norman	10"	2	#4 @ 48" E.W.		9.65	11.05	20.70
8520				#5 @ 32" E.W.		9.75	11.10	20.85
8530				#5 @ 16" E.W.		9.95	11.45	21.40
8600	Norwegian	10"	2	#4 @ 48" E.W.		8.20	9.85	18.05
8620				#5 @ 32" E.W.		8.30	9.90	18.20
8630				#5 @ 16" E.W.		8.50	10.25	18.75
8800	Triple	10"	2	#4 @ 48" E.W.		4.81	7.80	12.61
8820				#5 @ 32" E.W.		4.93	7.85	12.78
8830				#5 @ 16" E.W.		5.15	8.20	13.35

Important: See the Reference Section for critical supporting data - Location Factors & Historical Cost Indexes

EXTERIOR CLOSURE — A4.1 Walls

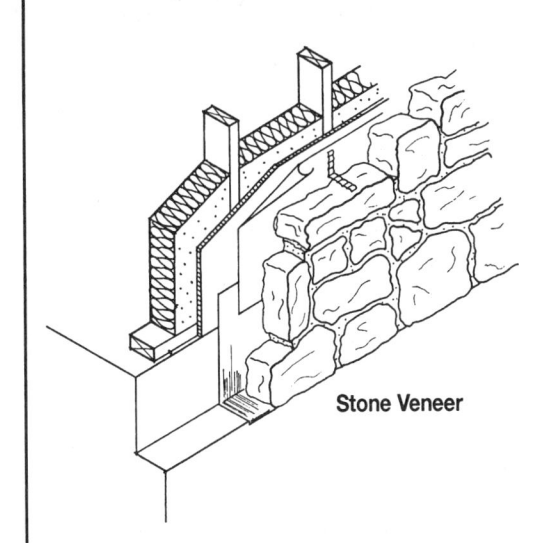

The table below lists costs per S.F. for stone veneer walls on various backup using different stone.

Stone Veneer

4.1-242	Stone Veneer	MAT.	INST.	TOTAL
2000	Ashlar veneer, 4", 2"x4" stud backup, 16" O.C., 8' high, low priced stone	6.55	11.20	17.75
2100	Metal stud backup, 8' high, 16" O.C.	7.10	12.25	19.35
2150	24" O.C.	6.95	11.90	18.85
2200	Conc. block backup, 4" thick	6.90	13.75	20.65
2300	6" thick	7.05	13.90	20.95
2350	8" thick	7.20	14.05	21.25
2400	10" thick	7.60	14.15	21.75
2500	12" thick	7.70	15.05	22.75
3100	High priced stone, wood stud backup, 10' high, 16" O.C.	11.30	12.60	23.90
3200	Metal stud backup, 10' high, 16" O.C.	11.75	13.65	25.40
3250	24" O.C.	11.60	13.30	24.90
3300	Conc. block backup, 10' high, 4" thick	11.55	15.15	26.70
3350	6" thick	11.70	15.20	26.90
3400	8" thick	11.85	15.45	27.30
3450	10" thick	12.25	15.55	27.80
3500	12" thick	12.35	16.45	28.80
4000	Indiana limestone 2" thick, sawn finish, wood stud backup, 10' high, 16" O.C.	13.10	9.55	22.65
4100	Metal stud backup, 10' high, 16" O.C.	13.65	10.60	24.25
4150	24" O.C.	13.50	10.25	23.75
4200	Conc. block backup, 4" thick	13.45	12.10	25.55
4250	6" thick	13.60	12.25	25.85
4300	8" thick	13.75	12.40	26.15
4350	10" thick	14.15	12.50	26.65
4400	12" thick	14.25	13.40	27.65
4450	2" thick, smooth finish, wood stud backup, 8' high, 16" O.C.	13.10	9.55	22.65
4550	Metal stud backup, 8' high, 16" O.C.	13.65	10.50	24.15
4600	24" O.C.	13.50	10.15	23.65
4650	Conc. block backup, 4" thick	13.45	12	25.45
4700	6" thick	13.60	12.15	25.75
4750	8" thick	13.75	12.30	26.05
4800	10" thick	14.15	12.50	26.65
4850	12" thick	14.25	13.40	27.65
5350	4" thick, smooth finish, wood stud backup, 8' high, 16" O.C.	17.60	9.55	27.15
5450	Metal stud backup, 8' high, 16" O.C.	18.15	10.60	28.75
5500	24" O.C.	18	10.25	28.25
5550	Conc. block backup, 4" thick	17.95	12.10	30.05
5600	6" thick	18.10	12.25	30.35
5650	8" thick	18.25	12.40	30.65
5700	10" thick	18.65	12.50	31.15
5750	12" thick	18.75	13.40	32.15

For expanded coverage of these items see *Means Assemblies Cost Data 1997*

EXTERIOR CLOSURE — A4.1 Walls

4.1-242 Stone Veneer

		COST PER S.F.		
		MAT.	INST.	TOTAL
6000	Granite, grey or pink, 2" thick, wood stud backup, 8' high, 16" O.C.	18.50	15.80	34.30
6100	Metal studs, 8' high, 16" O.C.	19.05	16.85	35.90
6150	24" O.C.	18.90	16.50	35.40
6200	Conc. block backup, 4" thick	18.85	18.35	37.20
6250	6" thick	19	18.50	37.50
6300	8" thick	19.15	18.65	37.80
6350	10" thick	19.55	18.75	38.30
6400	12" thick	19.65	19.65	39.30
6900	4" thick, wood stud backup, 8' high, 16" O.C.	30	17.85	47.85
7000	Metal studs, 8' high, 16" O.C.	30.50	18.90	49.40
7050	24" O.C.	30.50	18.55	49.05
7100	Conc. block backup, 4" thick	30.50	20.50	51
7150	6" thick	30.50	20.50	51
7200	8" thick	30.50	20.50	51
7250	10" thick	31	21	52
7300	12" thick	31	21.50	52.50

Important: See the Reference Section for critical supporting data - Location Factors & Historical Cost Indexes

EXTERIOR CLOSURE — A4.1 Walls

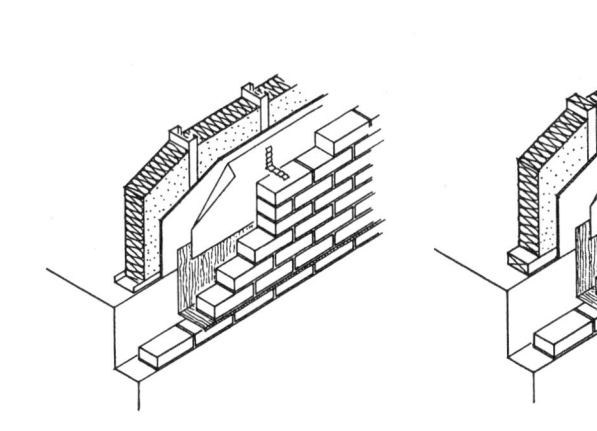

Exterior brick veneer/stud backup walls are defined in the following terms; type of brick and studs, stud spacing and bond. All systems include a brick shelf, ties to the backup and all necessary dampproofing and insulation.

4.1-252 Brick Veneer/Wood Stud Backup

	FACE BRICK	STUD BACKUP	STUD SPACING (IN.)	BOND	MAT.	INST.	TOTAL
1100	Standard	2x4-wood	16	running	7.05	10.15	17.20
1120				common	5.60	11.45	17.05
1140				Flemish	6	13.30	19.30
1160				English	6.50	14	20.50
1400		2x6-wood	16	running	7.20	10.20	17.40
1420				common	5.75	11.50	17.25
1440				Flemish	6.15	13.35	19.50
1460				English	6.70	14.05	20.75
1700	Glazed	2x4-wood	16	running	10.40	10.45	20.85
1720				common	12.10	11.90	24
1740				Flemish	13.20	14	27.20
1760				English	14.65	14.85	29.50
2300	Engineer	2x4-wood	16	running	5.25	9.10	14.35
2320				common	5.90	10.15	16.05
2340				Flemish	6.35	11.90	18.25
2360				English	6.90	12.40	19.30
2900	Roman	2x4-wood	16	running	7.05	9.35	16.40
2920				common	8.05	10.45	18.50
2940				Flemish	8.75	12.15	20.90
2960				English	9.65	13	22.65
4100	Norwegian	2x4-wood	16	running	5.40	7.30	12.70
4120				common	6.05	8.05	14.10
4140				Flemish	6.50	9.35	15.85
4160				English	7.10	9.70	16.80

For expanded coverage of these items see *Means Assemblies Cost Data 1997*

EXTERIOR CLOSURE — A4.1 Walls

4.1-252 Brick Veneer/Metal Stud Backup

	FACE BRICK	STUD BACKUP	STUD SPACING (IN.)	BOND		MAT.	INST.	TOTAL
5100	Standard	25ga.x6"NLB	24	running		6.65	10.10	16.75
5120				common		5.20	11.40	16.60
5140				Flemish		5.35	12.65	18
5160				English		6.15	13.95	20.10
5200		20ga.x3-5/8"NLB	16	running		7	10.15	17.15
5220				common		5.55	11.45	17
5240				Flemish		5.95	13.30	19.25
5260				English		6.45	14	20.45
5400		16ga.x3-5/8"LB	16	running		7.35	11.35	18.70
5420				common		5.90	12.65	18.55
5440				Flemish		6.30	14.50	20.80
5460				English		6.80	15.20	22
5700	Glazed	25ga.x6"NLB	24	running		10.05	10.40	20.45
5720				common		11.75	11.85	23.60
5740				Flemish		12.85	13.95	26.80
5760				English		14.30	14.80	29.10
5800		20ga.x3-5/8"NLB	24	running		10.25	10.35	20.60
5820				common		11.95	11.80	23.75
5840				Flemish		13.05	13.90	26.95
5860				English		14.50	14.75	29.25
6000		16ga.x3-5/8"LB	16	running		10.70	11.65	22.35
6020				common		12.40	13.10	25.50
6040				Flemish		13.50	15.20	28.70
6060				English		14.95	16.05	31
6300	Engineer	25ga.x6"NLB	24	running		4.86	9.05	13.91
6320				common		5.50	10.10	15.60
6340				Flemish		5.95	11.85	17.80
6360				English		6.50	12.35	18.85
6400		20ga.x3-5/8"NLB	16	running		5.20	9.10	14.30
6420				common		5.85	10.15	16
6440				Flemish		6.30	11.90	18.20
6460				English		6.85	12.40	19.25
6900	Roman	25ga.x6"NLB	24	running		6.70	9.30	16
6920				common		7.45	10.20	17.65
6940				Flemish		8.40	12.10	20.50
6960				English		9.30	12.95	22.25
7000		20ga.x3-5/8"NLB	16	running		7	9.35	16.35
7020				common		8	10.45	18.45
7040				Flemish		8.70	12.15	20.85
7060				English		9.60	13	22.60
7500	Norman	25ga.x6"NLB	24	running		9.95	7.95	17.90
7520				common		6.60	8.85	15.45
7540				Flemish		7.15	10.25	17.40
7560				English		7.85	10.75	18.60
7600		20ga.x3-5/8"NLB	24	running		10.15	7.95	18.10
7620				common		6.80	8.80	15.60
7640				Flemish		7.35	10.20	17.55
7660				English		8.05	10.70	18.75
8100	Norwegian	25ga.x6"NLB	24	running		5	7.25	12.25
8120				common		5.70	8.05	13.75
8140				Flemish		5.85	9.15	15
8160				English		6.70	9.65	16.35

Important: See the Reference Section for critical supporting data - Location Factors & Historical Cost Indexes

EXTERIOR CLOSURE | A4.1 Walls

4.1-252 Brick Veneer/Metal Stud Backup

	FACE BRICK	STUD BACKUP	STUD SPACING (IN.)	BOND	MAT.	INST.	TOTAL
8400	Norwegian	16ga.x3-5/8"LB	16	running	5.70	8.50	14.20
8420				common	6.35	9.30	15.65
8440				Flemish	6.80	10.55	17.35
8460				English	7.40	10.90	18.30

(COST PER S.F.)

For expanded coverage of these items see Means Assemblies Cost Data 1997

EXTERIOR CLOSURE — A4.1 Walls

Brick Face Composite Wall

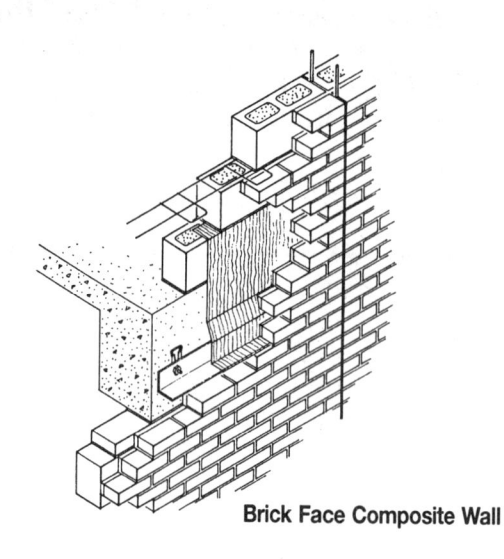

4.1-272 Brick Face Composite Wall - Double Wythe

	FACE BRICK	BACKUP MASONRY	BACKUP THICKNESS (IN.)	BACKUP CORE FILL		MAT.	INST.	TOTAL
1000	Standard	common brick	4	none		8.40	15.50	23.90
1040		SCR brick	6	none		10.65	13.85	24.50
1080		conc. block	4	none		7	12.50	19.50
1120			6	perlite		7.45	12.95	20.40
1160				styrofoam		8	12.75	20.75
1200			8	perlite		8.05	13.25	21.30
1240				styrofoam		8.45	13	21.45
1520		glazed block	4	none		11.80	13.35	25.15
1560			6	perlite		12.75	13.75	26.50
1600				styrofoam		13.30	13.55	26.85
1640			8	perlite		13.25	14.10	27.35
1680				styrofoam		13.65	13.85	27.50
1720		clay tile	4	none		10.30	12	22.30
1760			6	none		10.40	12.35	22.75
1800			8	none		11.35	12.75	24.10
1840		glazed tile	4	none		13.65	15.80	29.45
1880								
2000	Glazed	common brick	4	none		11.80	15.80	27.60
2040		SCR brick	6	none		14.05	14.15	28.20
2080		conc. block	4	none		10.40	12.80	23.20
2200			8	perlite		11.45	13.55	25
2240				styrofoam		11.85	13.30	25.15
2280		L.W. block	4	none		10.50	12.70	23.20
2400			8	perlite		11.40	13.45	24.85
2520		glazed block	4	none		15.20	13.65	28.85
2560			6	perlite		16.15	14.05	30.20
2600				styrofoam		16.70	13.85	30.55
2640			8	perlite		16.65	14.40	31.05
2680				styrofoam		17.05	14.15	31.20
2720		clay tile	4	none		13.70	12.30	26
2760			6	none		13.75	12.65	26.40
2800			8	none		14.75	13.05	27.80
2840		glazed tile	4	none		17.05	16.10	33.15
3000	Engineer	common brick	4	none		6.60	14.45	21.05
3040		SCR brick	6	none		8.85	12.80	21.65

Important: See the Reference Section for critical supporting data - Location Factors & Historical Cost Indexes

EXTERIOR CLOSURE — A4.1 Walls

4.1-272 Brick Face Composite Wall - Double Wythe

	FACE BRICK	BACKUP MASONRY	BACKUP THICKNESS (IN.)	BACKUP CORE FILL		MAT.	INST.	TOTAL
3080	Engineer	conc. block	4	none		5.20	11.45	16.65
3200			8	perlite		6.25	12.20	18.45
3280		L.W. block	4	none		4.50	8.15	12.65
3322			6	perlite		5.85	11.80	17.65
3520		glazed block	4	none		10	12.30	22.30
3560			6	perlite		10.95	12.70	23.65
3600				styrofoam		11.50	12.50	24
3640			8	perlite		11.45	13.05	24.50
3680				styrofoam		11.85	12.80	24.65
3720		clay tile	4	none		8.50	10.95	19.45
3800			8	none		9.55	11.70	21.25
3840		glazed tile	4	none		11.85	14.75	26.60
4000	Roman	common brick	4	none		8.45	14.70	23.15
4040		SCR brick	6	none		10.70	13.05	23.75
4080		conc. block	4	none		7.05	11.70	18.75
4120			6	perlite		7.50	12.15	19.65
4200			8	perlite		8.05	12.45	20.50
5000	Norman	common brick	4	none		11.70	13.40	25.10
5040		SCR brick	6	none		13.95	11.70	25.65
5280		L.W. block	4	none		10.40	10.30	20.70
5320			6	perlite		10.95	10.85	21.80
5720		clay tile	4	none		13.60	9.90	23.50
5760			6	none		13.65	10.20	23.85
5840		glazed tile	4	none		16.95	13.70	30.65
6000	Norwegian	common brick	4	none		6.75	12.70	19.45
6040		SCR brick	6	none		9	11	20
6080		conc. block	4	none		5.35	9.65	15
6120			6	perlite		5.80	10.10	15.90
6160				styrofoam		6.35	9.95	16.30
6200			8	perlite		6.40	10.40	16.80
6520		glazed block	4	none		10.15	10.55	20.70
6560			6	perlite		11.10	10.90	22
7000	Utility	common brick	4	none		6.40	12.05	18.45
7040		SCR brick	6	none		8.65	10.35	19
7080		conc. block	4	none		5	9	14
7120			6	perlite		5.50	9.45	14.95
7160				styrofoam		6.05	9.25	15.30
7200			8	perlite		6.05	9.75	15.80
7240				styrofoam		6.45	9.50	15.95
7280		L.W. block	4	none		5.15	8.95	14.10
7320			6	perlite		5.65	9.35	15
7520		glazed block	4	none		9.80	9.85	19.65
7560			6	perlite		10.75	10.25	21
7720		clay tile	4	none		8.30	8.55	16.85
7760			6	none		8.40	8.85	17.25
7840		glazed tile	4	none		11.65	12.35	24

4.1-272 Brick Face Composite Wall - Triple Wythe

	FACE BRICK	MIDDLE WYTHE	INSIDE MASONRY	TOTAL THICKNESS (IN.)		MAT.	INST.	TOTAL
8000	Standard	common brick	standard brick	12		13.50	22	35.50
8100		4" conc. brick	standard brick	12		13.65	22.50	36.15
8120			common brick	12		11.05	22.50	33.55

For expanded coverage of these items see *Means Assemblies Cost Data 1997*

EXTERIOR CLOSURE		A4.1	Walls				
4.1-272	Brick Face Composite Wall - Triple Wythe						

	FACE BRICK	MIDDLE WYTHE	INSIDE MASONRY	TOTAL THICKNESS (IN.)		COST PER S.F.	
						MAT.	INST.	TOTAL
8200	Glazed	common brick	standard brick	12		16.90	22.50	39.40
8300		4" conc. brick	standard brick	12		17.05	23	40.05
8320			glazed brick	12		14.45	22.50	36.95

Important: See the Reference Section for critical supporting data - Location Factors & Historical Cost Indexes

EXTERIOR CLOSURE — A4.1 Walls

Brick Face Cavity Wall

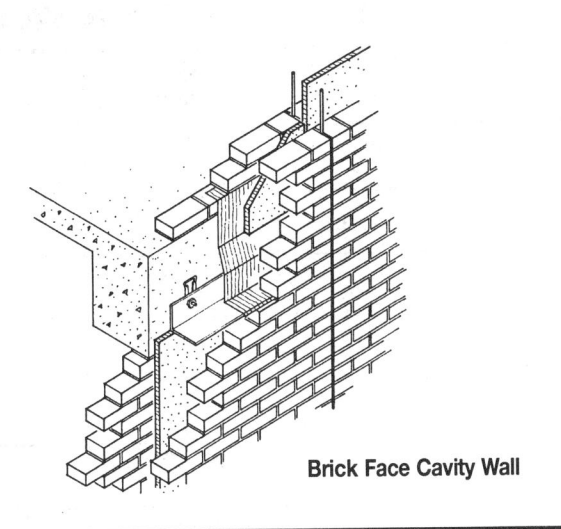

4.1-273 Brick Face Cavity Wall

	FACE BRICK	BACKUP MASONRY	TOTAL THICKNESS (IN.)	CAVITY INSULATION		MAT.	INST.	TOTAL
1000	Standard	4" common brick	10	polystyrene		8.20	15.60	23.80
1020				none		8	15.20	23.20
1040		6" SCR brick	12	polystyrene		10.50	13.90	24.40
1060				none		10.30	13.50	23.80
1080		4" conc. block	10	polystyrene		6.80	12.55	19.35
1100				none		6.60	12.15	18.75
1120		6" conc. block	12	polystyrene		7.05	12.80	19.85
1140				none		6.85	12.40	19.25
1160		4" L.W. block	10	polystyrene		6.95	12.50	19.45
1180				none		6.75	12.10	18.85
1200		6" L.W. block	12	polystyrene		7.20	12.75	19.95
1220				none		7	12.30	19.30
1240		4" glazed block	10	polystyrene		11.60	13.40	25
1260				none		11.40	13	24.40
1280		6" glazed block	12	polystyrene		11.65	13.40	25.05
1300				none		11.45	13	24.45
1320		4" clay tile	10	polystyrene		10.10	12.10	22.20
1340				none		9.90	11.65	21.55
1360		4" glazed tile	10	polystyrene		13.45	15.90	29.35
1380				none		13.25	15.50	28.75
1500	Glazed	4" common brick	10	polystyrene		11.60	15.90	27.50
1520				none		11.40	15.50	26.90
1580		4" conc. block	10	polystyrene		10.20	12.85	23.05
1600				none		10	12.45	22.45
1660		4" L.W. block	10	polystyrene		10.30	12.80	23.10
1680				none		10.10	12.40	22.50
1740		4" glazed block	10	polystyrene		15	13.70	28.70
1760				none		14.80	13.30	28.10
1820		4" clay tile	10	polystyrene		13.50	12.40	25.90
1840				none		13.30	11.95	25.25
1860		4" glazed tile	10	polystyrene		13.60	12.70	26.30
1880				none		13.40	12.30	25.70
2000	Engineer	4" common brick	10	polystyrene		6.40	14.55	20.95
2020				none		6.20	14.15	20.35
2080		4" conc. block	10	polystyrene		5.05	11.50	16.55
2100				none		4.83	11.10	15.93
2162		4" L.W. block	10	polystyrene		5.15	11.45	16.60
2180				none		4.94	11.05	15.99

For expanded coverage of these items see *Means Assemblies Cost Data 1997*

EXTERIOR CLOSURE — A4.1 Walls

4.1-273 Brick Face Cavity Wall

	FACE BRICK	BACKUP MASONRY	TOTAL THICKNESS (IN.)	CAVITY INSULATION		MAT.	INST.	TOTAL
2240	Engineer	4" glazed block	10	polystyrene		9.80	12.35	22.15
2260				none		9.60	11.95	21.55
2320		4" clay tile	10	polystyrene		8.30	11.05	19.35
2340				none		8.10	10.60	18.70
2360		4" glazed tile	10	polystyrene		11.65	14.85	26.50
2380				none		11.45	14.45	25.90
2500	Roman	4" common brick	10	polystyrene		8.25	14.80	23.05
2520				none		8.05	14.40	22.45
2580		4" conc. block	10	polystyrene		6.85	11.75	18.60
2600				none		6.65	11.35	18
2660		4" L.W. block	10	polystyrene		6.95	11.70	18.65
2680				none		6.75	11.30	18.05
2740		4" glazed block	10	polystyrene		11.65	12.60	24.25
2760				none		11.45	12.20	23.65
2820		4" clay tile	10	polystyrene		10.15	11.30	21.45
2840				none		9.95	10.85	20.80
2860		4" glazed tile	10	polystyrene		13.50	15.10	28.60
2880				none		13.30	14.70	28
3000	Norman	4" common brick	10	polystyrene		11.50	13.45	24.95
3020				none		11.30	13.05	24.35
3080		4" conc. block	10	polystyrene		10.10	10.45	20.55
3100				none		9.90	10	19.90
3160		4" L.W. block	10	polystyrene		10.20	10.35	20.55
3180				none		10	9.95	19.95
3240		4" glazed block	10	polystyrene		14.90	11.30	26.20
3260				none		14.70	10.90	25.60
3320		4" clay tile	10	polystyrene		13.40	9.95	23.35
3340				none		13.20	9.55	22.75
3360		4" glazed tile	10	polystyrene		16.75	13.75	30.50
3380				none		16.55	13.35	29.90
3500	Norwegian	4" common brick	10	polystyrene		6.55	12.75	19.30
3520				none		6.35	12.35	18.70
3580		4" conc. block	10	polystyrene		5.15	9.75	14.90
3600				none		4.97	9.35	14.32
3660		4" L.W. block	10	polystyrene		5.30	9.65	14.95
3680				none		5.10	9.25	14.35
3740		4" glazed block	10	polystyrene		9.95	10.60	20.55
3760				none		9.75	10.20	19.95
3820		4" clay tile	10	polystyrene		8.45	9.25	17.70
3840				none		8.25	8.85	17.10
3860		4" glazed tile	10	polystyrene		11.80	13.05	24.85
3880				none		11.60	12.65	24.25
4000	Utility	4" common brick	10	polystyrene		6.20	12.10	18.30
4020				none		6	11.70	17.70
4080		4" conc. block	10	polystyrene		4.83	9.10	13.93
4100				none		4.63	8.65	13.28
4160		4" L.W. block	10	polystyrene		4.94	9	13.94
4180				none		4.74	8.60	13.34
4240		4" glazed block	10	polystyrene		9.60	9.95	19.55
4260				none		9.40	9.55	18.95
4320		4" clay tile	10	polystyrene		8.10	8.60	16.70
4340				none		7.90	8.20	16.10
4360		4" glazed tile	10	polystyrene		11.45	12.40	23.85
4380				none		11.25	12	23.25

Important: See the Reference Section for critical supporting data - Location Factors & Historical Cost Indexes

EXTERIOR CLOSURE — A4.1 Walls

4.1-273 Brick Face Cavity Wall - Insulated Backup

	FACE BRICK	BACKUP MASONRY	TOTAL THICKNESS (IN.)	BACKUP CORE FILL		MAT.	INST.	TOTAL
5100	Standard	6" conc. block	10	perlite		7.10	12.60	19.70
5120				styrofoam		7.65	12.40	20.05
5180		6" L.W. block	10	perlite		7.25	12.50	19.75
5200				styrofoam		7.80	12.30	20.10
5260		6" glazed block	10	perlite		12.35	13.40	25.75
5280				styrofoam		12.90	13.20	26.10
5340		6" clay tile	10	none		10	12	22
5360		8" clay tile	12	none		11	12.40	23.40
5600	Glazed	6" conc. block	10	perlite		10.45	12.90	23.35
5620				styrofoam		11	12.70	23.70
5680		6" L.W. block	10	perlite		10.65	12.80	23.45
5700				styrofoam		11.20	12.60	23.80
5760		6" glazed block	10	perlite		15.75	13.70	29.45
5780				styrofoam		16.30	13.50	29.80
5840		6" clay tile	10	none		13.40	12.30	25.70
5860		8"clay tile	8	none		14.35	12.70	27.05
6100	Engineer	6" conc. block	10	perlite		5.30	11.55	16.85
6120				styrofoam		5.85	11.35	17.20
6180		6" L.W. block	10	perlite		5.45	11.45	16.90
6200				styrofoam		6	11.25	17.25
6260		6" glazed block	10	perlite		10.60	12.35	22.95
6280				styrofoam		11.15	12.15	23.30
6340		6" clay tile	10	none		8.20	10.95	19.15
6360		8" clay tile	12	none		9.20	11.35	20.55
6600	Roman	6" conc. block	10	perlite		7.10	11.80	18.90
6620				styrofoam		7.65	11.60	19.25
6680		6" L.W. block	10	perlite		7.25	11.70	18.95
6700				styrofoam		7.80	11.50	19.30
6760		6" glazed block	10	perlite		12.40	12.60	25
6780				styrofoam		12.95	12.40	25.35
6840		6" clay tile	10	none		10	11.20	21.20
6860		8" clay tile	12	none		11	11.60	22.60
7100	Norman	6" conc. block	10	perlite		10.35	10.45	20.80
7120				styrofoam		10.90	10.30	21.20
7180		6" L.W. block	10	perlite		10.55	10.40	20.95
7200				styrofoam		11.10	10.20	21.30
7260		6" glazed block	10	perlite		15.65	11.25	26.90
7280				styrofoam		16.20	11.10	27.30
7340		6" clay tile	10	none		13.30	9.85	23.15
7360		8" clay tile	12	none		14.25	10.30	24.55
7600	Norwegian	6" conc. block	10	perlite		5.45	9.80	15.25
7620				styrofoam		6	9.60	15.60
7680		6" L.W. block	10	perlite		5.60	9.70	15.30
7700				styrofoam		6.15	9.50	15.65
7760		6" glazed block	10	perlite		10.70	10.60	21.30
7780				styrofoam		11.25	10.40	21.65
7840		6" clay tile	10	none		8.35	9.20	17.55
7860		8" clay tile	12	none		9.35	9.60	18.95
8100	Utility	6" conc. block	10	perlite		5.10	9.10	14.20
8120				styrofoam		5.65	8.95	14.60
8180		6" L.W. block	10	perlite		5.25	9.05	14.30
8200				styrofoam		5.80	8.85	14.65
8260		6" glazed block	10	perlite		10.40	9.90	20.30
8280				styrofoam		10.95	9.75	20.70

For expanded coverage of these items see *Means Assemblies Cost Data 1997*

EXTERIOR CLOSURE — A4.1 Walls

4.1-273 Brick Face Cavity Wall - Insulated Backup

	FACE BRICK	BACKUP MASONRY	TOTAL THICKNESS (IN.)	BACKUP CORE FILL		COST PER S.F. MAT.	INST.	TOTAL
8340	Utility	6" clay tile	10	none		8	8.50	16.50
8360		8" clay tile	12	none		9	8.95	17.95

EXTERIOR CLOSURE — A4.1 Walls

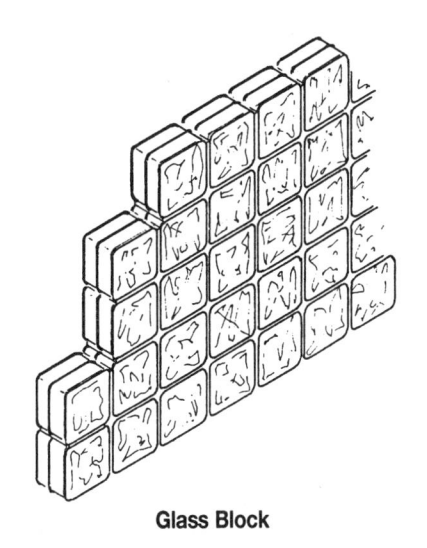

Glass Block

4.1-282	Glass Block	MAT.	INST.	TOTAL
2300	Glass block 4" thick, 6"x6" plain, under 1,000 S.F.	14.50	14.50	29
2400	1,000 to 5,000 S.F.	14.20	12.55	26.75
2500	Over 5,000 S.F.	13.65	11.80	25.45
2600	Solar reflective, under 1,000 S.F.	20.50	19.70	40.20
2700	1,000 to 5,000 S.F.	19.85	16.95	36.80
2800	Over 5,000 S.F.	19.10	15.90	35
3500	8"x8" plain, under 1,000 S.F.	10.20	10.85	21.05
3600	1,000 to 5,000 S.F.	10.50	9.35	19.85
3700	Over 5,000 S.F.	10.10	8.45	18.55
3800	Solar reflective, under 1,000 S.F.	14.25	14.60	28.85
3900	1,000 to 5,000 S.F.	14.65	12.50	27.15
4000	Over 5,000 S.F.	14.10	11.20	25.30
5000	12"x12" plain, under 1,000 S.F.	12.35	10.05	22.40
5100	1,000 to 5,000 S.F.	12.70	8.45	21.15
5200	Over 5,000 S.F.	12.25	7.70	19.95
5300	Solar reflective, under 1,000 S.F.	17.25	13.45	30.70
5400	1,000 to 5,000 S.F.	17.75	11.20	28.95
5600	Over 5,000 S.F.	17.10	10.15	27.25
5800	3" thinline, 6"x6" plain, under 1,000 S.F.	16.35	14.50	30.85
5900	Over 5,000 S.F.	16.35	14.50	30.85
6000	Solar reflective, under 1,000 S.F.	23	19.70	42.70
6100	Over 5,000 S.F.	21	15.90	36.90
6200	8"x8" plain, under 1,000 S.F.	12.05	10.85	22.90
6300	Over 5,000 S.F.	10.65	8.45	19.10
6400	Solar reflective, under 1,000 S.F.	16.85	14.60	31.45
6500	Over 5,000 S.F.	14.90	11.20	26.10

For expanded coverage of these items see *Means Assemblies Cost Data 1997*

EXTERIOR CLOSURE — A4.1 Walls

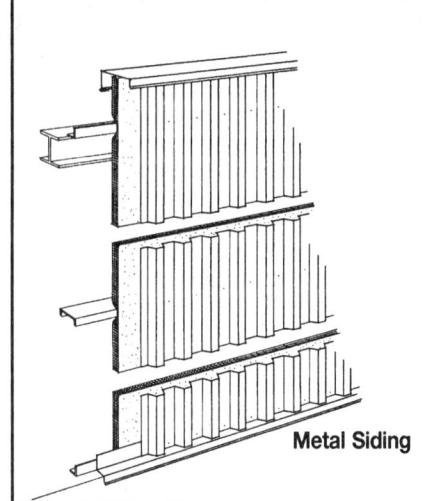

The table below lists costs for metal siding of various descriptions, not including the steel frame, or the structural steel, of a building. Costs are per S.F. including all accessories and insulation.

Metal Siding

4.1-384	Metal Siding Panel	MAT.	INST.	TOTAL
1400	Metal siding aluminum panel, corrugated, .024" thick, natural	1.38	2.40	3.78
1450	Painted	1.59	2.40	3.99
1500	.032" thick, natural	1.77	2.40	4.17
1550	Painted	2.13	2.40	4.53
1600	Ribbed 4" pitch, .032" thick, natural	1.80	2.40	4.20
1650	Painted	2.07	2.40	4.47
1700	.040" thick, natural	2.16	2.40	4.56
1750	Painted	2.47	2.40	4.87
1800	.050" thick, natural	2.45	2.40	4.85
1850	Painted	2.73	2.40	5.13
1900	8" pitch panel, .032" thick, natural	1.72	2.31	4.03
1950	Painted	1.98	2.32	4.30
2000	.040" thick, natural	2.08	2.32	4.40
2050	Painted	2.37	2.34	4.71
2100	.050" thick, natural	2.36	2.33	4.69
2150	Painted	2.66	2.35	5.01
3000	Steel, corrugated or ribbed, 29 Ga. .0135" thick, galvanized	1.17	2.15	3.32
3050	Colored	1.49	2.19	3.68
3100	26 Ga. .0179" thick, galvanized	1.19	2.16	3.35
3150	Colored	1.53	2.20	3.73
3200	24 Ga. .0239" thick, galvanized	1.27	2.17	3.44
3250	Colored	1.71	2.21	3.92
3300	22 Ga. .0299" thick, galvanized	1.45	2.18	3.63
3350	Colored	2.04	2.22	4.26
3400	20 Ga. .0359" thick, galvanized	1.45	2.18	3.63
3450	Colored	2.16	2.34	4.50
4100	Sandwich panels, factory fab., 1" polystyrene, steel core, 26 Ga., galv.	2.96	3.40	6.36
4200	Colored, 1 side	3.72	3.40	7.12
4300	2 sides	4.85	3.40	8.25
4400	2" polystyrene, steel core, 26 Ga., galvanized	3.53	3.40	6.93
4500	Colored, 1 side	4.29	3.40	7.69
4600	2 sides	5.40	3.40	8.80
4700	22 Ga., baked enamel exterior	7.20	3.58	10.78
4800	Polyvinyl chloride exterior	7.60	3.58	11.18
5100	Field assem. panels; 18Ga.galv. steel int., 16Ga.alum ext., 1.5" F.G. insul	6.20	6.60	12.80
5200	Porcelain face, on aluminum	7.05	6.60	13.65

Important: See the Reference Section for critical supporting data - Location Factors & Historical Cost Indexes

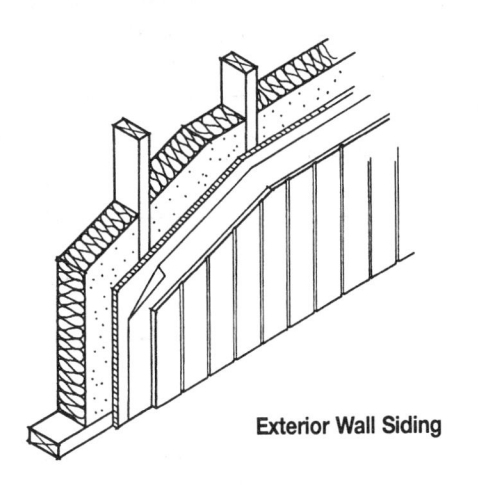

EXTERIOR CLOSURE — A4.1 Walls

The table below lists costs per S.F. for exterior walls with wood siding. A variety of systems are presented using both wood and metal studs at 16" and 24" O.C.

Exterior Wall Siding

4.1-412	Wood & Other Siding	MAT.	INST.	TOTAL
1400	Wood siding w/2"x4"studs, 16"O.C., insul. wall, 5/8"text 1-11 fir plywood	2.52	3.45	5.97
1450	5/8" text 1-11 cedar plywood	3.02	3.34	6.36
1500	1" x 4" vert T.&G. redwood	4.07	4.71	8.78
1600	1" x 8" vert T.&G. redwood	3.71	4.26	7.97
1650	1" x 5" rabbetted cedar bev. siding	3.55	3.77	7.32
1700	1" x 6" cedar drop siding	3.59	3.80	7.39
1750	1" x 12" rough sawn cedar	2.69	3.48	6.17
1800	1" x 12" sawn cedar, 1" x 4" battens	3.44	3.71	7.15
1850	1" x 10" redwood shiplap siding	3.65	3.60	7.25
1900	18" no. 1 red cedar shingles, 5-1/2" exposed	3.38	4.49	7.87
1950	6" exposed	3.24	4.35	7.59
2000	6-1/2" exposed	3.11	4.20	7.31
2100	7" exposed	2.97	4.06	7.03
2150	7-1/2" exposed	2.83	3.91	6.74
3000	8" wide aluminum siding	2.67	3.08	5.75
3150	8" plain vinyl siding	2.04	3.05	5.09
3250	8" insulated vinyl siding	2.20	3.05	5.25
3300				
3400	2" x 6" studs, 16" O.C., insul. wall, w/ 5/8" text 1-11 fir plywood	2.76	3.56	6.32
3500	5/8" text 1-11 cedar plywood	3.26	3.56	6.82
3600	1" x 4" vert T.&G. redwood	4.31	4.82	9.13
3700	1" x 8" vert T.&G. redwood	3.95	4.37	8.32
3800	1" x 5" rabbetted cedar bev siding	3.79	3.88	7.67
3900	1" x 6" cedar drop siding	3.83	3.91	7.74
4000	1" x 12" rough sawn cedar	2.93	3.59	6.52
4200	1" x 12" sawn cedar, 1" x 4" battens	3.68	3.82	7.50
4500	1" x 10" redwood shiplap siding	3.89	3.71	7.60
4550	18" no. 1 red cedar shingles, 5-1/2" exposed	3.62	4.60	8.22
4600	6" exposed	3.48	4.46	7.94
4650	6-1/2" exposed	3.35	4.31	7.66
4700	7" exposed	3.21	4.17	7.38
4750	7-1/2" exposed	3.07	4.02	7.09
4800	8" wide aluminum siding	2.91	3.19	6.10
4850	8" plain vinyl siding	2.28	3.16	5.44
4900	8" insulated vinyl siding	2.44	3.16	5.60
4910				
5000	2" x 6" studs, 24" O.C., insul. wall, 5/8" text 1-11, fir plywood	2.62	3.35	5.97
5050	5/8" text 1-11 cedar plywood	3.12	3.35	6.47
5100	1" x 4" vert T.&G. redwood	4.17	4.61	8.78
5150	1" x 8" vert T.&G. redwood	3.81	4.16	7.97
5200	1" x 5" rabbetted cedar bev siding	3.65	3.67	7.32
5250	1" x 6" cedar drop siding	3.69	3.70	7.39

For expanded coverage of these items see *Means Assemblies Cost Data 1997*

EXTERIOR CLOSURE — A4.1 Walls

4.1-412 Wood & Other Siding

		COST PER S.F.		
		MAT.	INST.	TOTAL
5300	1" x 12" rough sawn cedar	2.79	3.38	6.17
5400	1" x 12" sawn cedar, 1" x 4" battens	3.54	3.61	7.15
5450	1" x 10" redwood shiplap siding	3.75	3.50	7.25
5500	18" no. 1 red cedar shingles, 5-1/2" exposed	3.48	4.39	7.87
5550	6" exposed	3.34	4.25	7.59
5650	7" exposed	3.07	3.96	7.03
5700	7-1/2" exposed	2.93	3.81	6.74
5750	8" wide aluminum siding	2.77	2.98	5.75
5800	8" plain vinyl siding	2.14	2.95	5.09
5850	8" insulated vinyl siding	2.30	2.95	5.25
5900	3-5/8" metal studs, 16 Ga., 16" OC insul.wall, 5/8"text 1-11 fir plywood	3.05	4.52	7.57
5950	5/8" text 1-11 cedar plywood	3.55	4.52	8.07
6000	1" x 4" vert T.&G. redwood	4.60	5.80	10.40
6050	1" x 8" vert T.&G. redwood	4.24	5.35	9.59
6100	1" x 5" rabbetted cedar bev siding	4.08	4.84	8.92
6150	1" x 6" cedar drop siding	4.12	4.87	8.99
6200	1" x 12" rough sawn cedar	3.22	4.55	7.77
6250	1" x 12" sawn cedar, 1" x 4" battens	3.97	4.78	8.75
6300	1" x 10" redwood shiplap siding	4.18	4.67	8.85
6350	18" no. 1 red cedar shingles, 5-1/2" exposed	3.91	5.55	9.46
6500	6" exposed	3.77	5.40	9.17
6550	6-1/2" exposed	3.64	5.25	8.89
6600	7" exposed	3.50	5.10	8.60
6650	7-1/2" exposed	3.50	5	8.50
6700	8" wide aluminum siding	3.20	4.15	7.35
6750	8" plain vinyl siding	2.57	4.01	6.58
6800	8" insulated vinyl siding	2.73	4.01	6.74
7000	3-5/8" metal studs, 16 Ga. 24" OC insul wall, 5/8" text 1-11 fir plywood	2.91	4.05	6.96
7050	5/8" text 1-11 cedar plywood	3.41	4.05	7.46
7100	1" x 4" vert T.&G. redwood	4.46	5.45	9.91
7150	1" x 8" vert T.&G. redwood	4.10	4.97	9.07
7200	1" x 5" rabbetted cedar bev siding	3.94	4.48	8.42
7250	1" x 6" cedar drop siding	3.98	4.51	8.49
7300	1" x 12" rough sawn cedar	3.08	4.19	7.27
7350	1" x 12" sawn cedar 1" x 4" battens	3.83	4.42	8.25
7400	1" x 10" redwood shiplap siding	4.04	4.31	8.35
7450	18" no. 1 red cedar shingles, 5-1/2" exposed	3.91	5.35	9.26
7500	6" exposed	3.77	5.20	8.97
7550	6-1/2" exposed	3.63	5.05	8.68
7600	7" exposed	3.50	4.91	8.41
7650	7-1/2" exposed	3.22	4.62	7.84
7700	8" wide aluminum siding	3.06	3.79	6.85
7750	8" plain vinyl siding	2.43	3.76	6.19
7800	8" insul. vinyl siding	2.59	3.76	6.35

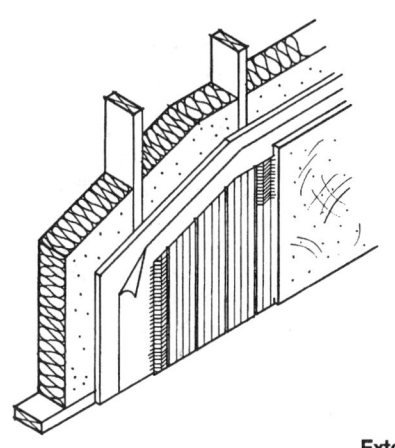

EXTERIOR CLOSURE — A4.5 Exterior Wall Finishes

The table below lists costs for some typical stucco walls including all the components as demonstrated in the component block below. Prices are presented for backup walls using wood studs, metal studs and CMU.

Exterior Stucco Wall

4.5-110	Stucco Wall	MAT.	INST.	TOTAL
2100	Cement stucco, 7/8" th., plywood sheathing, stud wall, 2" x 4", 16" O.C.	2.23	3.85	6.08
2200	24" O.C.	2.14	3.67	5.81
2300	2" x 6", 16" O.C.	2.47	3.96	6.43
2400	24" O.C.	2.33	3.75	6.08
2500	No sheathing, metal lath on stud wall, 2" x 4", 16" O.C.	1.64	3.11	4.75
2600	24" O.C.	1.55	2.93	4.48
2700	2" x 6", 16" O.C.	1.88	3.22	5.10
2800	24" O.C.	1.74	3.01	4.75
2900	1/2" gypsum sheathing, 3-5/8" metal studs, 16" O.C.	2.20	4.20	6.40
2950	24" O.C.	2.06	3.84	5.90
3000	Cement stucco, 5/8" thick, 2 coats on std. CMU block, 8"x 16", 8" thick	2.02	5.40	7.42
3100	10" Thick	2.91	5.65	8.56
3200	12" Thick	2.97	6.65	9.62
3300	Std. light Wt. block 8" x 16", 8" Thick	2.51	5.20	7.71
3400	10" Thick	3.12	5.35	8.47
3500	12" Thick	3.44	6.30	9.74
3600	3 coat stucco, self furring metal lath 3.4 Lb/SY, on 8" x 16", 8" thick	2.07	5.55	7.62
3700	10" Thick	2.96	5.75	8.71
3800	12" Thick	3.02	6.75	9.77
3900	Lt. Wt. block, 8" Thick	2.56	5.30	7.86
4000	10" Thick	3.17	5.45	8.62
4100	12" Thick	3.49	6.40	9.89

COST PER S.F.

For expanded coverage of these items see *Means Assemblies Cost Data 1997*

EXTERIOR CLOSURE — A4.6 Doors

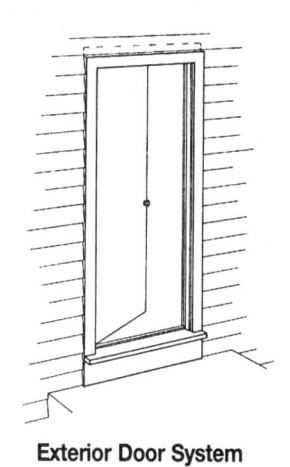

The table below lists exterior door systems by material, type and size. Prices between sizes listed can be interpolated with reasonable accuracy. Prices are per opening for a complete door system including frame and required hardware.

Wood doors in this table are designed with wood frames and metal doors with hollow metal frames. Depending upon quality the total material and installation cost of a wood door frame is about the same as a hollow metal frame.

Exterior Door System

4.6-100 Wood, Steel & Aluminum

	MATERIAL	TYPE	DOORS	SPECIFICATION	OPENING	MAT.	INST.	TOTAL
2350	Birch	solid core	single door	hinged	2'-6" x 6'-8"	710	182	892
2400					2'-6" x 7'-0"	715	183	898
2450					2'-8" x 7'-0"	720	183	903
2500					3'-0" x 7'-0"	720	186	906
2550			double door	hinged	2'-6" x 6'-8"	1,350	330	1,680
2600					2'-6" x 7'-0"	1,350	335	1,685
2650					2'-8" x 7'-0"	1,350	335	1,685
2700					3'-0" x 7'-0"	1,350	340	1,690
2750	Wood	combination	storm & screen	hinged	3'-0" x 6'-8"	229	45.50	274.50
2800					3'-0" x 7'-0"	252	50	302
2850		overhead	panels, H.D.	manual oper.	8'-0" x 8'-0"	415	330	745
2900					10'-0" x 10'-0"	635	365	1,000
2950					12'-0" x 12'-0"	885	440	1,325
3000					14'-0" x 14'-0"	1,525	505	2,030
3050					20'-0" x 16'-0"	3,175	1,100	4,275
3100				electric oper.	8'-0" x 8'-0"	820	495	1,315
3150					10'-0" x 10'-0"	1,050	530	1,580
3200					12'-0" x 12'-0"	1,300	605	1,905
3250					14'-0" x 14'-0"	1,925	670	2,595
3300					20'-0" x 16'-0"	3,600	1,425	5,025
3350	Steel 18 Ga.	hollow metal	1 door w/frame	no label	2'-6" x 7'-0"	815	201	1,016
3400					2'-8" x 7'-0"	815	201	1,016
3450					3'-0" x 7'-0"	815	201	1,016
3500					3'-6" x 7'-0"	875	210	1,085
3550		hollow metal	1 door w/frame	no label	4'-0" x 8'-0"	1,075	212	1,287
3600			2 doors w/frame	no label	5'-0" x 7'-0"	1,575	375	1,950
3650					5'-4" x 7'-0"	1,575	375	1,950
3700					6'-0" x 7'-0"	1,575	375	1,950
3750					7'-0" x 7'-0"	1,700	390	2,090
3800					8'-0" x 8'-0"	2,100	395	2,495
3850			1 door w/frame	"A" label	2'-6" x 7'-0"	955	237	1,192
3900					2'-8" x 7'-0"	965	240	1,205
3950					3'-0" x 7'-0"	965	240	1,205
4000					3'-6" x 7'-0"	1,025	246	1,271
4050					4'-0" x 8'-0"	1,050	259	1,309
4100			2 doors w/frame	"A" label	5'-0" x 7'-0"	1,850	440	2,290
4150					5'-4" x 7'-0"	1,850	445	2,295
4200					6'-0" x 7'-0"	1,850	445	2,295

Important: See the Reference Section for critical supporting data - Location Factors & Historical Cost Indexes

EXTERIOR CLOSURE — A4.6 Doors

4.6-100 Wood, Steel & Aluminum

	MATERIAL	TYPE	DOORS	SPECIFICATION	OPENING	MAT.	INST.	TOTAL
4250	Steel 18 Ga.				7'-0" x 7'-0"	1,975	460	2,435
4300					8'-0" x 8'-0"	1,975	460	2,435
4350	Steel 24 Ga.	overhead	sectional	manual oper.	8'-0" x 8'-0"	370	330	700
4400					10'-0" x 10'-0"	480	365	845
4450					12'-0" x 12'-0"	650	440	1,090
4500					20'-0" x 14'-0"	1,625	940	2,565
4550				electric oper.	8'-0" x 8'-0"	775	495	1,270
4600					10'-0" x 10'-0"	885	530	1,415
4650					12'-0" x 12'-0"	1,050	605	1,655
4700					20'-0" x 14'-0"	2,050	1,275	3,325
4750	Steel	overhead	rolling	manual oper.	8'-0" x 8'-0"	990	535	1,525
4800					10'-0" x 10'-0"	1,000	610	1,610
4850					12'-0" x 12'-0"	1,275	715	1,990
4900					14'-0" x 14'-0"	1,625	1,075	2,700
4950					20'-0" x 12'-0"	2,800	950	3,750
5000					20'-0" x 16'-0"	3,800	1,425	5,225
5050				electric oper.	8'-0" x 8'-0"	1,925	705	2,630
5100					10'-0" x 10'-0"	1,950	780	2,730
5150					12'-0" x 12'-0"	2,225	885	3,110
5200					14'-0" x 14'-0"	2,575	1,250	3,825
5250					20'-0" x 12'-0"	3,625	1,125	4,750
5300					20'-0" x 16'-0"	4,625	1,600	6,225
5350				fire rated	10'-0" x 10'-0"	1,950	780	2,730
5400			rolling grill	manual oper.	10'-0" x 10'-0"	1,600	855	2,455
5450					15'-0" x 8'-0"	1,875	1,075	2,950
5500		vertical lift	1 door w/frame	motor operator	16'-0" x 16'-0"	16,100	3,400	19,500
5550					32'-0" x 24'-0"	29,900	2,250	32,150
5600	St. Stl. & glass	revolving	stock unit	manual oper.	6'-0" x 7'-0"	26,400	5,700	32,100
5650				auto Cntrls.	6'-10" x 7'-0"	33,600	6,050	39,650
5700	Bronze	revolving	stock unit	manual oper.	6'-10" x 7'-0"	30,800	11,400	42,200
5750				auto Cntrls.	6'-10" x 7'-0"	38,000	11,800	49,800
5800	St. Stl. & glass	balanced	standard	economy	3'-0" x 7'-0"	5,700	965	6,665
5850				premium	3'-0" x 7'-0"	9,875	1,250	11,125
6000	Aluminum	combination	storm & screen	hinged	3'-0" x 6'-8"	262	48.50	310.50
6050					3'-0" x 7'-0"	288	53.50	341.50
6100		overhead	rolling grill	manual oper.	12'-0" x 12'-0"	2,575	1,500	4,075
6150				motor oper.	12'-0" x 12'-0"	3,375	1,675	5,050
6200	Alum. & Fbrgls.	overhead	heavy duty	manual oper.	12'-0" x 12'-0"	1,300	440	1,740
6250				electric oper.	12'-0" x 12'-0"	1,700	605	2,305
6300	Alum. & glass	w/o transom	narrow stile	w/panic Hrdwre.	3'-0" x 7'-0"	945	635	1,580
6350				dbl. door, Hrdwre.	6'-0" x 7'-0"	1,700	1,050	2,750
6400			wide stile	hdwre.	3'-0" x 7'-0"	1,350	620	1,970
6450				dbl. door, Hdwre.	6'-0" x 7'-0"	2,700	1,250	3,950
6500			full vision	hdwre.	3'-0" x 7'-0"	1,400	985	2,385
6550				dbl. door, Hdwre.	6'-0" x 7'-0"	2,100	1,400	3,500
6600			non-standard	hdwre.	3'-0" x 7'-0"	1,525	620	2,145
6650				dbl. door, Hdwre.	6'-0" x 7'-0"	3,025	1,250	4,275
6700			bronze fin.	hdwre.	3'-0" x 7'-0"	1,550	620	2,170
6750				dbl. door, Hrdwre.	6'-0" x 7'-0"	3,075	1,250	4,325
6800			black fin.	hdwre.	3'-0" x 7'-0"	1,600	620	2,220
6850				dbl. door, Hdwre.	6'-0" x 7'-0"	3,175	1,250	4,425
6900		w/transom	narrow stile	hdwre.	3'-0" x 10'-0"	1,325	710	2,035
6950				dbl. door, Hdwre.	6'-0" x 10'-0"	2,275	1,250	3,525
7000			wide stile	hdwre.	3'-0" x 10'-0"	1,575	855	2,430

For expanded coverage of these items see *Means Assemblies Cost Data 1997*

EXTERIOR CLOSURE — A4.6 Doors

4.6-100 Wood, Steel & Aluminum

	MATERIAL	TYPE	DOORS	SPECIFICATION	OPENING	MAT.	INST.	TOTAL
7050	Alum. & glass			dbl. door, Hdwre.	6'-0" x 10'-0"	2,625	1,500	4,125
7100			full vision	hdwre.	3'-0" x 10'-0"	1,725	950	2,675
7150				dbl. door, Hdwre.	6'-0" x 10'-0"	2,875	1,650	4,525
7200			non-standard	hdwre.	3'-0" x 10'-0"	1,550	665	2,215
7250				dbl. door, Hdwre.	6'-0" x 10'-0"	3,100	1,325	4,425
7300			bronze fin.	hdwre.	3'-0" x 10'-0"	1,575	665	2,240
7350				dbl. door, Hdwre.	6'-0" x 10'-0"	3,150	1,325	4,475
7400			black fin.	hdwre.	3'-0" x 10'-0"	1,625	665	2,290
7450				dbl. door, Hdwre.	6'-0" x 10'-0"	3,250	1,325	4,575
7500		revolving	stock design	minimum	6'-10" x 7'-0"	14,100	2,275	16,375
7550				average	6'-0" x 7'-0"	17,600	2,850	20,450
7600				maximum	6'-10" x 7'-0"	23,000	3,800	26,800
7650				min., automatic	6'-10" x 7'-0"	21,300	2,625	23,925
7700				avg., automatic	6'-10" x 7'-0"	24,800	3,200	28,000
7750				max., automatic	6'-10" x 7'-0"	30,200	4,150	34,350
7800		balanced	standard	economy	3'-0" x 7'-0"	3,600	950	4,550
7850				premium	3'-0" x 7'-0"	4,975	1,225	6,200
7900		mall front	sliding panels	alum. fin.	16'-0" x 9'-0"	2,200	475	2,675
7950					24'-0" x 9'-0"	3,200	885	4,085
8000				bronze fin.	16'-0" x 9'-0"	2,575	555	3,130
8050					24'-0" x 9'-0"	3,725	1,025	4,750
8100			fixed panels	alum. fin.	48'-0" x 9'-0"	5,950	690	6,640
8150				bronze fin.	48'-0" x 9'-0"	6,925	805	7,730
8200		sliding entrance	5' x 7' door	electric oper.	12'-0" x 7'-6"	5,600	885	6,485
8250		sliding patio	temp. glass	economy	6'-0" x 7'-0"	680	164	844
8300			temp. glass	economy	12'-0" x 7'-0"	2,275	219	2,494
8350				premium	6'-0" x 7'-0"	1,025	246	1,271
8400					12'-0" x 7'-0"	3,425	330	3,755

EXTERIOR CLOSURE — A4.7 Windows & Glazed Walls

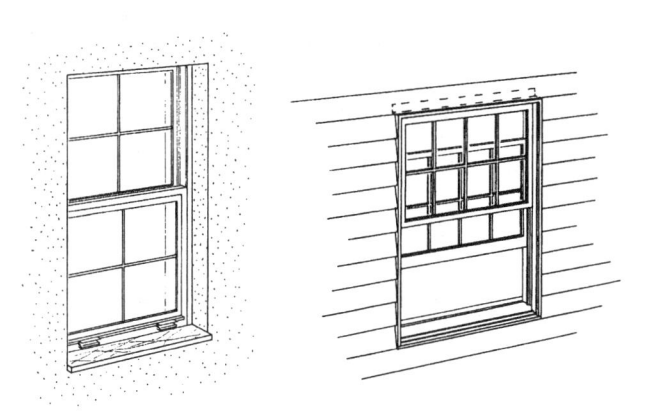

The table below lists window systems by material, type and size. Prices between sizes listed can be interpolated with reasonable accuracy. Prices include frame, hardware, and casing as illustrated in the component block below.

4.7-110 Wood, Steel & Aluminum

	MATERIAL	TYPE	GLAZING	SIZE	DETAIL	MAT.	INST.	TOTAL
3000	Wood	double hung	std. glass	2'-8" x 4'-6"		150	144	294
3050				3'-0" x 5'-6"		174	166	340
3100			insul. glass	2'-8" x 4'-6"		182	144	326
3150				3'-0" x 5'-6"		226	166	392
3200		sliding	std. glass	3'-4" x 2'-7"		154	119	273
3250				4'-4" x 3'-3"		214	132	346
3300				5'-4" x 6'-0"		279	159	438
3350			insul. glass	3'-4" x 2'-7"		186	139	325
3400				4'-4" x 3'-3"		261	152	413
3450				5'-4" x 6'-0"		340	181	521
3500		awning	std. glass	2'-10" x 1'-9"		186	72.50	258.50
3600				4'-4" x 2'-8"		221	70.50	291.50
3700			insul. glass	2'-10" x 1'-9"		228	83.50	311.50
3800				4'-4" x 2'-8"		270	77	347
3900		casement	std. glass	1'-10" x 3'-2"	1 lite	231	95.50	326.50
3950				4'-2" x 4'-2"	2 lite	360	127	487
4000				5'-11" x 5'-2"	3 lite	610	163	773
4050				7'-11" x 6'-3"	4 lite	770	196	966
4100				9'-11" x 6'-3"	5 lite	965	228	1,193
4150			insul. glass	1'-10" x 3'-2"	1 lite	261	95.50	356.50
4200				4'-2" x 4'-2"	2 lite	455	127	582
4250				5'-11" x 5'-2"	3 lite	800	163	963
4300				7'-11" x 6'-3"	4 lite	1,025	196	1,221
4350				9'-11" x 6'-3"	5 lite	1,275	228	1,503
4400		picture	std. glass	4'-6" x 4'-6"		219	163	382
4450				5'-8" x 4'-6"		268	181	449
4500		picture	insul. glass	4'-6" x 4'-6"		265	190	455
4550				5'-8" x 4'-6"		325	210	535
4600		fixed bay	std. glass	8' x 5'		620	298	918
4650				9'-9" x 5'-4"		730	420	1,150
4700			insul. glass	8' x 5'		775	298	1,073
4750				9'-9" x 5'-4"		935	420	1,355
4800		casement bay	std. glass	8' x 5'		660	340	1,000
4850			insul. glass	8' x 5'		930	340	1,270
4900		vert. bay	std. glass	8' x 5'		1,175	340	1,515
4950			insul. glass	8' x 5'		1,425	340	1,765
5000	Steel	double hung	1/4" tempered	2'-8" x 4'-6"		515	113	628
5050				3'-4" x 5'-6"		790	173	963

For expanded coverage of these items see *Means Assemblies Cost Data 1997*

EXTERIOR CLOSURE — A4.7 Windows & Glazed Walls

4.7-110 Wood, Steel & Aluminum

	MATERIAL	TYPE	GLAZING	SIZE	DETAIL	MAT.	INST.	TOTAL
5100	Steel		insul. glass	2'-8" x 4'-6"		535	130	665
5150				3'-4" x 5'-6"		815	198	1,013
5202		horiz. pivoted	std. glass	2' x 2'		122	37.50	159.50
5250				3' x 3'		275	85	360
5300				4' x 4'		490	151	641
5350				6' x 4'		735	226	961
5400			insul. glass	2' x 2'		129	43.50	172.50
5450				3' x 3'		289	97.50	386.50
5500				4' x 4'		515	173	688
5550				6' x 4'		770	260	1,030
5600		picture window	std. glass	3' x 3'		169	85	254
5650				6' x 4'		450	226	676
5700			insul. glass	3' x 3'		183	97.50	280.50
5750				6' x 4'		485	260	745
5800		industrial security	std. glass	2'-9" x 4'-1"		500	106	606
5850				4'-1" x 5'-5"		985	209	1,194
5900			insul. glass	2'-9" x 4'-1"		520	122	642
5950				4'-1" x 5'-5"		1,025	239	1,264
6000		comm. projected	std. glass	3'-9" x 5'-5"		735	192	927
6050				6'-9" x 4'-1"		995	260	1,255
6100			insul. glass	3'-9" x 5'-5"		765	220	985
6150				6'-9" x 4'-1"		1,025	298	1,323
6200		casement	std. glass	4'-2" x 4'-2"	2 lite	730	164	894
6250			insul. glass	4'-2" x 4'-2"		755	188	943
6300			std. glass	5'-11" x 5'-2"	3 lite	1,225	288	1,513
6350			insul. glass	5'-11" x 5'-2"		1,275	330	1,605
6400	Aluminum	projecting	std. glass	3'-1" x 3'-2"		195	85.50	280.50
6450				4'-5" x 5'-3"		275	107	382
6500			insul. glass	3'-1" x 3'-2"		234	103	337
6550				4'-5" x 5'-3"		330	128	458
6600		sliding	std. glass	3' x 2'		147	85.50	232.50
6650				5' x 3'		187	95	282
6700				8' x 4'		270	143	413
6750				9' x 5'		410	214	624
6800			insul. glass	3' x 2'		163	85.50	248.50
6850				5' x 3'		262	95	357
6900				8' x 4'		435	143	578
6950				9' x 5'		650	214	864
7000		single hung	std. glass	2' x 3'		131	85.50	216.50
7050				2'-8" x 6'-8"		278	107	385
7100				3'-4" x 5'0"		179	95	274
7150			insul. glass	2' x 3'		159	85.50	244.50
7200				2'-8" x 6'-8"		360	107	467
7250				3'-4" x 5'		252	95	347
7300		double hung	std. glass	2' x 3'		187	56.50	243.50
7350				2'-8" x 6'-8"		555	168	723
7400				3'-4" x 5'		520	157	677
7450			insul. glass	2' x 3'		196	65	261
7500				2'-8" x 6'-8"		580	193	773
7550				3'-4" x 5'-0"		545	180	725
7600		casements	std. glass	3'-1" x 3'-2"		140	92	232
7650				4'-5" x 5'-3"		340	223	563
7700			insul. glass	3'-1" x 3'-2"		155	106	261
7750				4'-5" x 5'-3"		375	256	631

Important: See the Reference Section for critical supporting data - Location Factors & Historical Cost Indexes

EXTERIOR CLOSURE — A4.7 Windows & Glazed Walls

4.7-110 Wood, Steel & Aluminum

	MATERIAL	TYPE	GLAZING	SIZE	DETAIL	MAT.	INST.	TOTAL
7800	Aluminum	hinged swing	std. glass	3' x 4'		330	113	443
7850				4' x 5'		550	189	739
7900			insul. glass	3' x 4'		350	130	480
7950				4' x 5'		585	217	802
8002		folding type	std. glass	3'-0" x 4'-0"		197	92.50	289.50
8050				4'-0" x 5'-0"		275	92.50	367.50
8100			insul. glass	3'-0" x 4'-0"		247	92.50	339.50
8150				4'-0" x 5'-0"		360	92.50	452.50
8200		picture unit	std. glass	2'-0" x 3'-0"		95	56.50	151.50
8250				2'-8" x 6'-8"		282	168	450
8300				3'-4" x 5'-0"		264	157	421
8350			insul. glass	2'-0" x 3'-0"		104	65	169
8400				2'-8" x 6'-8"		310	193	503
8450				3'-4" x 5'-0"		290	180	470
8500		awning type	std. glass	3'-0" x 3'-0"	2 lite	320	61	381
8550				3'-0" x 4'-0"	3 lite	375	85.50	460.50
8600				3'-0" x 5'-4"	4 lite	455	85.50	540.50
8650				4'-0" x 5'-4"	4 lite	500	95	595
8700			insul. glass	3'-0" x 3'-0"	2 lite	345	61	406
8750				3'-0" x 4'-0"	3 lite	435	85.50	520.50
8800				3'-0" x 5'-4"	4 lite	535	85.50	620.50
8850				4'-0" x 5'-4"	4 lite	595	95	690
8900		jalousie type	std. glass	1'-7" x 3'-2"		118	85.50	203.50
8950				2'-3" x 4'-0"		169	85.50	254.50
9000				3'-1" x 2'-0"		116	85.50	201.50
9050				3'-1" x 5'-3"		241	85.50	326.50
9051								
9052								

4.7-582 Tubular Aluminum Framing

COST/S.F. OPNG.

		MAT.	INST.	TOTAL
1100	Alum flush tube frame, for 1/4"glass, 1-3/4"x4", 5'x6'opng, no inter horizntls	8.50	7.60	16.10
1150	One intermediate horizontal	11.55	8.95	20.50
1200	Two intermediate horizontals	14.60	10.25	24.85
1250	5' x 20' opening, three intermediate horizontals	7.65	6.35	14
1400	1-3/4" x 4-1/2", 5' x 6' opening, no intermediate horizontals	9.65	7.60	17.25
1450	One intermediate horizontal	12.90	8.95	21.85
1500	Two intermediate horizontals	16.05	10.25	26.30
1550	5' x 20' opening, three intermediate horizontals	8.60	6.35	14.95
1700	For insulating glass, 2"x4-1/2", 5'x6' opening, no intermediate horizontals	10.85	8	18.85
1750	One intermediate horizontal	14.20	9.40	23.60
1800	Two intermediate horizontals	17.50	10.75	28.25
1850	5' x 20' opening, three intermediate horizontals	9.50	6.65	16.15
2000	Thermal break frame, 2-1/4"x4-1/2", 5'x6'opng, no intermediate horizontals	11.65	8.10	19.75
2050	One intermediate horizontal	15.65	9.75	25.40
2100	Two intermediate horizontals	19.60	11.35	30.95
2150	5' x 20' opening, three intermediate horizontals	10.65	6.95	17.60

For expanded coverage of these items see *Means Assemblies Cost Data 1997*

EXTERIOR CLOSURE — A4.7 — Windows & Glazed Walls

The table below lists costs of curtain wall and spandrel panels per S.F. Costs do not include structural framing used to hang the panels.

Glazing Panel

Plate Glass

Spandrel Glass

Polycarbonate

Sandwich Panel

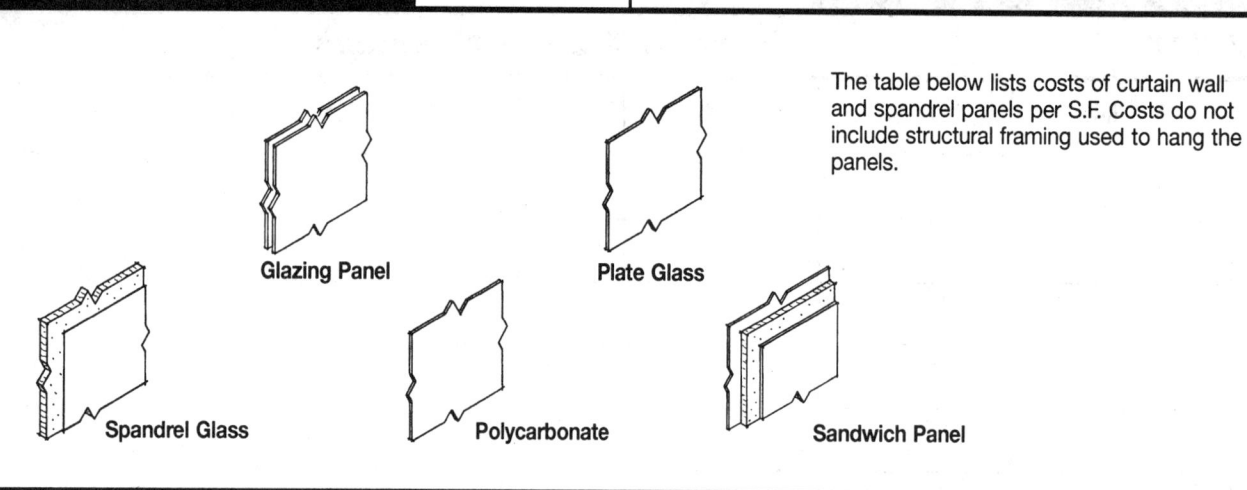

4.7-584	Curtain Wall Panels	MAT.	INST.	TOTAL
1000	Glazing panel, insulating, 1/2" thick, 2 lites 1/8" float, clear	5.65	6.55	12.20
1100	Tinted	6.85	6.55	13.40
1200	5/8" thick units, 2 lites 3/16" float, clear	6.35	6.90	13.25
1300	Tinted	7.10	6.90	14
1400	1" thick units, 2 lites, 1/4" float, clear	9.70	8.25	17.95
1500	Tinted	9	8.25	17.25
1600	Heat reflective film inside	13.75	7.30	21.05
1700	Light and heat reflective glass, tinted	15	7.30	22.30
2000	Plate glass, 1/4" thick, clear	3.85	5.15	9
2050	Tempered	4.11	5.15	9.26
2100	Tinted	4.03	5.15	9.18
2200	3/8" thick, clear	6.60	8.25	14.85
2250	Tempered	9	8.25	17.25
2300	Tinted	7.80	8.25	16.05
2400	1/2" thick, clear	12.55	11.25	23.80
2450	Tempered	14.45	11.25	25.70
2500	Tinted	13.35	11.25	24.60
2600	3/4" thick, clear	17.35	17.70	35.05
2650	Tempered	20	17.70	37.70
3000	Spandrel glass, panels, 1/4" plate glass insul w/fiberglass, 1" thick	8.45	5.15	13.60
3100	2" thick	9.85	5.15	15
3200	Galvanized steel backing, add	2.92		2.92
3300	3/8" plate glass, 1" thick	14.15	5.15	19.30
3400	2" thick	15.55	5.15	20.70
4000	Polycarbonate, masked, clear or colored, 1/8" thick	5.20	3.65	8.85
4100	3/16" thick	6.05	3.76	9.81
4200	1/4" thick	6.90	4	10.90
4300	3/8" thick	11.80	4.13	15.93
5000	Sandwich panel, 1-1/2" fiberglass, 16 Ga. aluminum exterior	6.20	6.60	12.80
5100	16 Ga. porcelainized aluminum exterior	7.05	6.60	13.65
5200	18 Ga. galvanized steel exterior	5.45	6.60	12.05
5300	20 Ga. protected metal exterior	6	6.60	12.60
5400	20 Ga stainless steel exterior	7.90	6.60	14.50
5500	22 Ga. galv., both sides 2" insulation, enamel exterior	7.20	3.58	10.78
5600	Polyvinylidene floride exterior finish	7.60	3.58	11.18
5700	26 Ga., galv. both sides, 1" insulation, colored 1 side	3.72	3.40	7.12
5800	Colored 2 sides	4.85	3.40	8.25

For information about Means Estimating Seminars, see yellow pages 11 and 12 in back of book

Division 5
Roofing

ROOFING — A5.1 Roof Covers

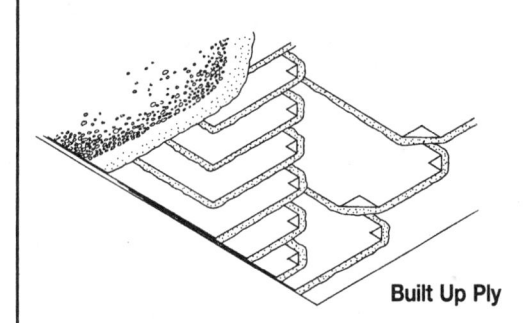

Multiple ply roofing is the most popular covering for minimum pitch roofs.

Built Up Ply

5.1-103	Built-Up	MAT.	INST.	TOTAL
1200	Asphalt flood coat w/gravel; not incl. insul, flash., nailers			
1300				
1400	Asphalt base sheets & 3 plies #15 asphalt felt, mopped	.37	1.14	1.51
1500	On nailable deck	.41	1.19	1.60
1600	4 plies #15 asphalt felt, mopped	.52	1.25	1.77
1700	On nailable deck	.48	1.31	1.79
1800	Coated glass base sheet, 2 plies glass (type IV), mopped	.41	1.14	1.55
1900	For 3 plies	.48	1.25	1.73
2000	On nailable deck	.46	1.31	1.77
2300	4 plies glass fiber felt (type IV), mopped	.58	1.25	1.83
2400	On nailable deck	.54	1.31	1.85
2500	Organic base sheet & 3 plies #15 organic felt, mopped	.39	1.26	1.65
2600	On nailable deck	.42	1.31	1.73
2700	4 plies #15 organic felt, mopped	.50	1.14	1.64
2750				
2800	Asphalt flood coat, smooth surface			
2900	Asphalt base sheet & 3 plies #15 asphalt felt, mopped	.37	1.05	1.42
3000	On nailable deck	.35	1.08	1.43
3100	Coated glass fiber base sheet & 2 plies glass fiber felt, mopped	.34	1	1.34
3200	On nailable deck	.33	1.05	1.38
3300	For 3 plies, mopped	.42	1.08	1.50
3400	On nailable deck	.40	1.14	1.54
3700	4 plies glass fiber felt (type IV), mopped	.49	1.08	1.57
3800	On nailable deck	.47	1.14	1.61
3900	Organic base sheet & 3 plies #15 organic felt, mopped	.37	1.05	1.42
4000	On nailable decks	.35	1.08	1.43
4100	4 plies #15 organic felt, mopped	.44	1.14	1.58
4200	Coal tar pitch with gravel surfacing			
4300	4 plies #15 tarred felt, mopped	1	1.19	2.19
4400	3 plies glass fiber felt (type IV), mopped	.83	1.31	2.14
4500	Coated glass fiber base sheets 2 plies glass fiber felt, mopped	.83	1.31	2.14
4600	On nailable decks	.74	1.39	2.13
4800	3 plies glass fiber felt (type IV), mopped	1.13	1.19	2.32
4900	On nailable decks	1.03	1.25	2.28

Important: See the Reference Section for critical supporting data - Location Factors & Historical Cost Indexes

ROOFING — A5.1 Roof Covers

Full Adhered Single Ply Membrane

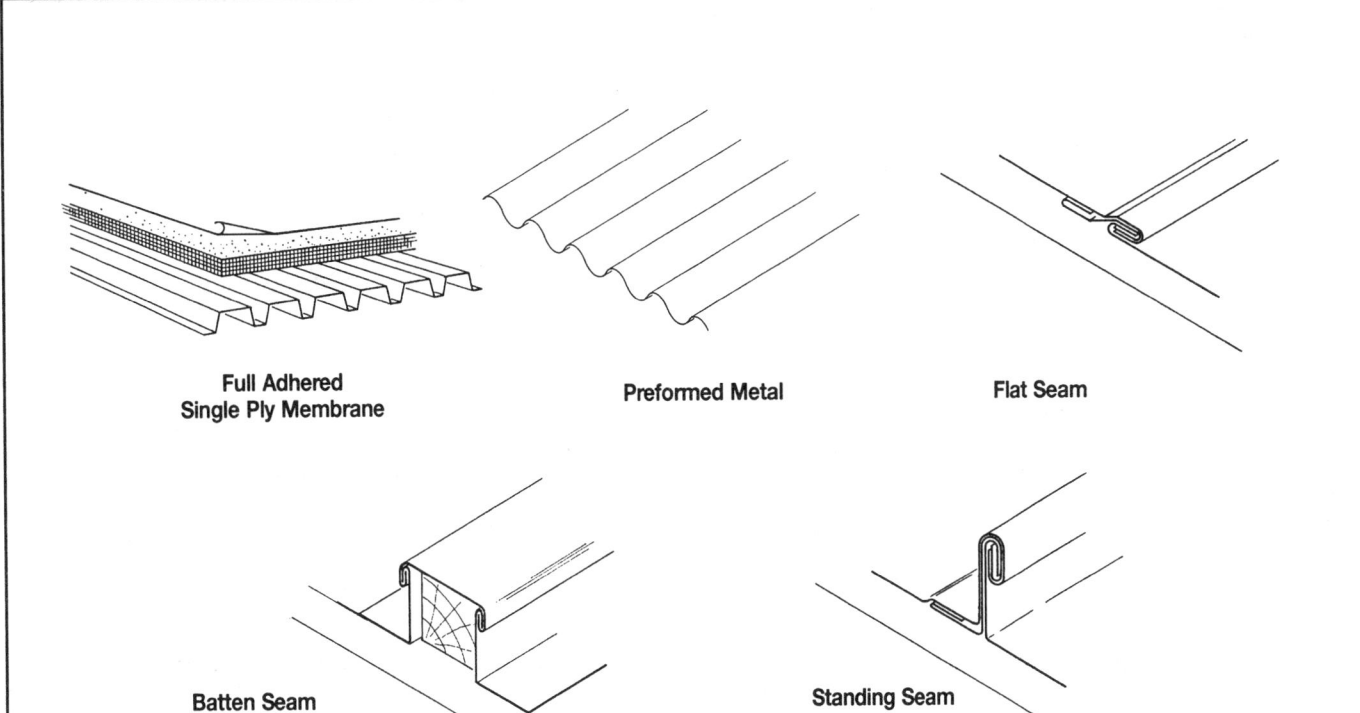

Preformed Metal

Flat Seam

Batten Seam

Standing Seam

5.1-220	Single Ply Membrane	MAT.	INST.	TOTAL
1000	CSPE (Chlorosulfonated polyethylene), 35 mils, fully adhered	1.32	.62	1.94
2000	EPDM (Ethylene propylene diene monomer), 45 mils, fully adhered	.87	.62	1.49
3000	55 mils, fully adhered	1.01	.62	1.63
4000	Modified bit., SBS modified, granule surface cap sheet, mopped, 150 mils	.42	1.25	1.67
4500	APP modified, granule surface cap sheet, torched, 180 mils	.42	.81	1.23
6000	Reinforced PVC, 48 mils, loose laid and ballasted with stone	1	.31	1.31
6200	Fully adhered with adhesive	1.18	.62	1.80

5.1-310	Preformed Metal Roofing	MAT.	INST.	TOTAL
0200	Corrugated roofing, aluminum, mill finish, .0175" thick, .272 P.S.F.	.71	1.08	1.79
0250	.0215 thick, .334 P.S.F.	.91	1.08	1.99

5.1-330	Formed Metal	MAT.	INST.	TOTAL
1000	Batten seam, formed copper roofing, 3"min slope, 16 oz., 1.2 P.S.F.	4.86	3.41	8.27
1100	18 oz., 1.35 P.S.F.	5.40	3.76	9.16
2000	Zinc copper alloy, 3" min slope, .020" thick, .88 P.S.F.	5.70	3.16	8.86
3000	Flat seam, copper, 1/4" min. slope, 16 oz., 1.2 P.S.F.	4.31	3.16	7.47
3100	18 oz., 1.35 P.S.F.	4.86	3.26	8.12
5000	Standing seam, copper, 2-1/2" min. slope, 16 oz., 1.25 P.S.F.	4.66	2.91	7.57
5100	18 oz., 1.40 P.S.F.	5.20	3.16	8.36
6000	Zinc copper alloy, 2-1/2" min. slope, .020" thick, .87 P.S.F.	5.60	3.16	8.76
6100	.032" thick, 1.39 P.S.F.	7.60	3.41	11.01

For expanded coverage of these items see *Means Assemblies Cost Data 1997*

ROOFING — A5.1 Roof Covers

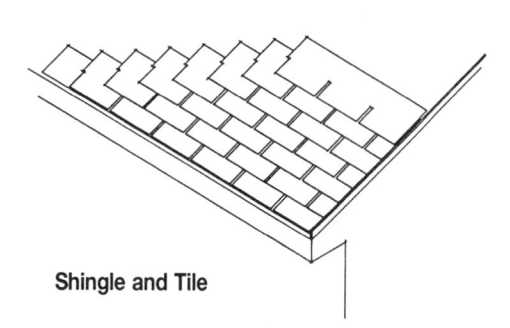

Shingle and Tile

Shingles and tiles are practical in applications where the roof slope is more than 3-1/2" per foot of rise. Lines 1100 through 6000 list the various materials and the weight per square foot.

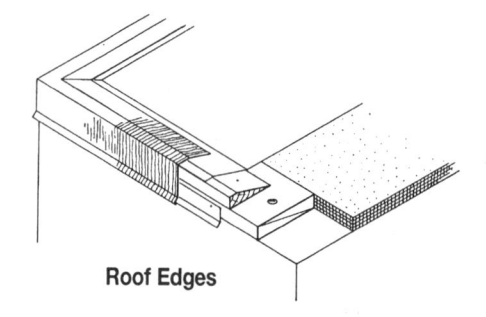

Roof Edges

The table below lists the costs for various types of roof perimeter edge treatments.

Roof edge systems include the cost per L.F. for a 2" x 8" treated wood nailer fastened at 4'-0" O.C. and a diagonally cut 4" x 6" treated wood cant.

Roof edge and base flashing are assumed to be made from the same material.

5.1-410 Shingle & Tile

Line	Description	MAT.	INST.	TOTAL
1095	Asphalt roofing			
1100	Strip shingles, 4" slope, inorganic class A 210-235 lb/Sq.	.31	.67	.98
1150	Organic, class C, 235-240 lb./sq.	.42	.73	1.15
1200	Premium laminated multi-layered, class A, 260-300 lb./Sq.	.55	1.01	1.56
1545	Metal roofing			
1550	Alum., shingles, colors, 3" min slope, .019" thick, 0.4 PSF	1.92	4.16	6.08
1850	Steel, colors, 3" min slope, 26 gauge, 1.0 PSF	2.01	1.52	3.53
2795	Slate roofing			
2800	4" min. slope, shingles, 3/16" thick, 8.0 PSF	5.70	1.89	7.59
3495	Wood roofing			
3500	4" min slope, cedar shingles, 16" x 5", 5" exposure 1.6 PSF	1.71	1.42	3.13
4000	Shakes, 18", 8-1/2" exposure, 2.8 PSF	1.57	1.74	3.31
5095	Tile roofing			
5100	Aluminum, mission, 3" min slope, .019" thick, 0.65 PSF	3.76	1.38	5.14
6000	Clay, Americana, 3" minimum slope, 8 PSF	5.25	2	7.25

5.1-520 Roof Edges

Line	EDGE TYPE	DESCRIPTION	SPECIFICATION	FACE HEIGHT	MAT.	INST.	TOTAL
1000	Aluminum	mill finish	.050" thick	4"	7	6.30	13.30
1100				6"	7.55	6.50	14.05
1300		duranodic	.050" thick	4"	8.05	6.30	14.35
1400				6"	8.80	6.50	15.30
1600		painted	.050" thick	4"	8.70	6.30	15
1700				6"	9.55	6.50	16.05
2000	Copper	plain	16 oz.	4"	5.65	6.30	11.95
2100				6"	6.50	6.50	13
2300			20 oz.	4"	6.20	7.15	13.35
2400				6"	6.90	6.95	13.85
2700	Sheet Metal	galvanized	20 Ga.	4"	6.35	7.45	13.80
2800				6"	7.55	7.45	15
3000			24 Ga.	4"	5.65	6.30	11.95
3100				6"	6.50	6.30	12.80

ROOFING — A5.1 Roof Covers

5.1-620 Flashing

	MATERIAL	BACKING	SIDES	SPECIFICATION	QUANTITY	MAT.	INST.	TOTAL
0040	Aluminum	none		.019"		.85	2.56	3.41
0050				.032"		1.01	2.56	3.57
0300		fabric	2	.004"		.92	1.12	2.04
0400		mastic		.004"		.88	1.12	2
0700	Copper	none		16 oz.	<500 lbs.	3.50	3.22	6.72
0800				24 oz.	<500 lbs.	5.25	3.53	8.78
2000	Copper lead	fabric	1	2 oz.		1.54	1.12	2.66
3500	PVC black	none		.010"		.14	1.12	1.26
3700				.030"		.33	1.12	1.45
4200	Neoprene			1/16"		1.62	1.12	2.74
4500	Stainless steel	none		.015"	<500 lbs.	3.30	3.22	6.52
4600	Copper clad				>2000 lbs.	3.28	2.39	5.67
5000	Plain			32 ga.		2.37	2.39	4.76

Cost per S.F.

For expanded coverage of these items see *Means Assemblies Cost Data 1997*

ROOFING | A5.7 | Insulation

5.7-101 Roof Deck Rigid Insulation

		COST PER S.F.		
		MAT.	INST.	TOTAL
0100	Fiberboard low density, 1/2" thick, R1.39			
0150	1" thick R2.78	.35	.40	.75
0300	1 1/2" thick R4.17	.53	.40	.93
0350	2" thick R5.56	.70	.40	1.10
0370	Fiberboard high density, 1/2" thick R1.3	.20	.32	.52
0380	1" thick R2.5	.37	.40	.77
0390	1 1/2" thick R3.8	.61	.40	1.01
0410	Fiberglass, 3/4" thick R2.78	.42	.32	.74
0450	15/16" thick R3.70	.54	.32	.86
0550	1-5/16" thick R5.26	.95	.32	1.27
0650	2 7/16" thick R10	1.14	.40	1.54
1510	Polyisocyanurate 2#/CF density, 1" thick R7.14	.34	.23	.57
1550	1 1/2" thick R10.87	.37	.26	.63
1600	2" thick R14.29	.47	.29	.76
1650	2 1/2" thick R16.67	.58	.30	.88
1700	3" thick R21.74	.68	.32	1
1750	3 1/2" thick R25	.88	.32	1.20
1800	Tapered for drainage	.44	.23	.67
1810	Expanded polystyrene, 1#/CF density, 3/4" thick R2.89	.20	.21	.41
1820	2" thick R7.69	.33	.26	.59
1830	Extruded polystyrene, 15 PSI compressive strength, 1" thick R5	.21	.21	.42
1835	2" thick R10	.35	.26	.61
1840	3" thick R15	.80	.32	1.12
2550	40 PSI compressive strength, 1" thick R5	.40	.21	.61
2600	2" thick R10	.75	.26	1.01
2650	3" thick R15	1.12	.32	1.44
2700	4" thick R20	1.50	.32	1.82
2750	Tapered for drainage	.53	.23	.76
2810	60 PSI compressive strength, 1" thick R5	.46	.22	.68
2850	2" thick R10	.86	.27	1.13
2900	Tapered for drainage	.64	.23	.87
2910	115 PSI compressive strength, 1" thick R5	.99	.23	1.22
2950	2" thick R10	1.96	.28	2.24
3000	Tapered for drainage	1.07	.23	1.30
3050	Composites with 2" EPS			
3060	1" Fiberboard	.75	.34	1.09
3070	7/16" oriented strand board	.88	.40	1.28
3080	1/2" plywood	.94	.40	1.34
3090	1" perlite	.78	.40	1.18
4000	Composites with 1-1/2" polyisocyanurate			
4010	1" fiberboard	.79	.40	1.19
4020	1" perlite	.81	.38	1.19
4030	7/16" oriented strand board	.92	.40	1.32

ROOFING — A5.8 Openings & Specialties

Roof Hatch   Smoke Hatch   Skylight

5.8-100	Hatches	MAT.	INST.	TOTAL
0200	Roof hatches, with curb, and 1" fiberglass insulation, 2'-6"x3'-0", aluminum	435	129	564
0300	Galvanized steel 165 lbs.	365	129	494
0400	Primed steel 164 lbs.	325	129	454
0500	2'-6"x4'-6" aluminum curb and cover, 150 lbs.	605	144	749
0600	Galvanized steel 220 lbs.	515	144	659
0650	Primed steel 218 lbs.	495	144	639
0800	2'x6"x8'-0" aluminum curb and cover, 260 lbs.	1,075	195	1,270
0900	Galvanized steel, 360 lbs.	990	195	1,185
0950	Primed steel 358 lbs.	975	195	1,170
1200	For plexiglass panels, add to the above	375		375
2100	Smoke hatches, unlabeled not incl. hand winch operator, 2'-6"x3', galv	440	156	596
2200	Plain steel, 160 lbs.	395	156	551
2400	2'-6"x8'-0", galvanized steel, 360 lbs.	1,075	214	1,289
2500	Plain steel, 350 lbs.	1,075	214	1,289
3000	4'-0"x8'-0", double leaf low profile, aluminum cover, 359 lb.	1,575	161	1,736
3100	Galvanized steel 475 lbs.	1,375	161	1,536
3200	High profile, aluminum cover, galvanized curb, 361 lbs.	1,425	161	1,586

5.8-100	Skylights	MAT.	INST.	TOTAL
5100	Skylights, plastic domes, insul curbs, nom. size to 10 S.F., single glaze	24.50	10.95	35.45
5200	Double glazing	24.50	12.05	36.55
5300	10 S.F. to 20 S.F., single glazing	27	13.45	40.45
5400	Double glazing	21	5	26
5500	20 S.F. to 30 S.F., single glazing	17.40	3.75	21.15
5600	Double glazing	20.50	3.99	24.49
5700	30 S.F. to 65 S.F., single glazing	17.40	2.87	20.27
5800	Double glazing	19.10	3.38	22.48
6000	Sandwich panels fiberglass, 9-1/16"thick, 2 S.F. to 10 S.F.	14.60	6.45	21.05
6100	10 S.F. to 18 S.F.	13.10	4.87	17.97
6200	2-3/4" thick, 25 S.F. to 40 S.F.	21	4.37	25.37
6300	40 S.F. to 70 S.F.	17.25	3.91	21.16
6301				

For expanded coverage of these items see *Means Assemblies Cost Data 1997*

ROOFING — A5.8 Openings & Specialties

5.8-400 Gutters

	SECTION	MATERIAL	THICKNESS	SIZE	FINISH	COST PER L.F. MAT.	INST.	TOTAL
0050	Box	aluminum	.027"	5"	enameled	1.07	3.09	4.16
0100					mill	.96	3.09	4.05
0200			.032"	5"	enameled	1.32	3.09	4.41
0500		copper	16 Oz.	4"	lead coated	7.65	3.09	10.74
0600					mill	3.91	3.09	7
1000		steel galv.	28 Ga.	5"	enameled	1	3.09	4.09
1200			26 Ga.	5"	mill	.90	3.09	3.99
1800		vinyl		4"	colors	.97	2.99	3.96
1900				5"	colors	1.14	2.99	4.13
2300		hemlock or fir		4"x5"	treated	7.75	3.29	11.04
3000	Half round	copper	16 Oz.	4"	lead coated	5.30	3.09	8.39
3100					mill	3.66	3.09	6.75
3600		steel galv.	28 Ga.	5"	enameled	1	3.09	4.09
4102		stainless steel		5"	mill	5.50	3.09	8.59
5000		vinyl		4"	white	.69	2.99	3.68

5.8-500 Downspouts

	MATERIALS	SECTION	SIZE	FINISH	THICKNESS	COST PER V.L.F. MAT.	INST.	TOTAL
0100	Aluminum	rectangular	2"x3"	embossed mill	.020"	.68	1.95	2.63
0150				enameled	.020"	.71	1.95	2.66
0250			3"x4"	enameled	.024"	1.57	2.65	4.22
0300		round corrugated	3"	enameled	.020"	.94	1.95	2.89
0350			4"	enameled	.025"	1.77	2.65	4.42
0500	Copper	rectangular corr.	2"x3"	mill	16 Oz.	4.42	1.95	6.37
0600		smooth		mill	16 Oz.	5.20	1.95	7.15
0700		rectangular corr.	3"x4"	mill	16 Oz.	5.90	2.56	8.46
1300	Steel	rectangular corr.	2"x3"	galvanized	28 Ga.	.58	1.95	2.53
1350				epoxy coated	24 Ga.	1.12	1.95	3.07
1400		smooth		galvanized	28 Ga.	.87	1.95	2.82
1450		rectangular corr.	3"x4"	galvanized	28 Ga.	1.63	2.56	4.19
1500				epoxy coated	24 Ga.	1.88	2.56	4.44
1550		smooth		galvanized	28 Ga.	1.33	2.56	3.89
1652		round corrugated	3"	galvanized	28 Ga.	.78	1.95	2.73
1700			4"	galvanized	28 Ga.	1.01	2.56	3.57
1750			5"	galvanized	28 Ga.	1.35	2.85	4.20
2000	Steel pipe	round	4"	black	X.H.	10.15	18.55	28.70
2552	S.S. tubing sch.5	rectangular	3"x4"	mill		17.95	2.56	20.51

5.8-500 Gravel Stop

	MATERIALS	SECTION	SIZE	FINISH	THICKNESS	COST PER L.F. MAT.	INST.	TOTAL
5100	Aluminum	extruded	4"	mill	.050"	2.99	2.56	5.55
5200			4"	duranodic	.050"	4.06	2.56	6.62
5300			8"	mill	.050"	4.37	2.97	7.34
5400			8"	duranodic	.050"	5.50	2.97	8.47
6000			12"-2 pc.	duranodic	.050"	6.90	3.71	10.61
6100	Stainless	formed	6"	mill	24 Ga.	7.85	2.75	10.60

For information about Means Estimating Seminars, see yellow pages 11 and 12 in back of book

Important: See the Reference Section for critical supporting data - Location Factors & Historical Cost Indexes

Division 6
Interior Construction

INTERIOR CONSTR. A6.1 Block Partitions

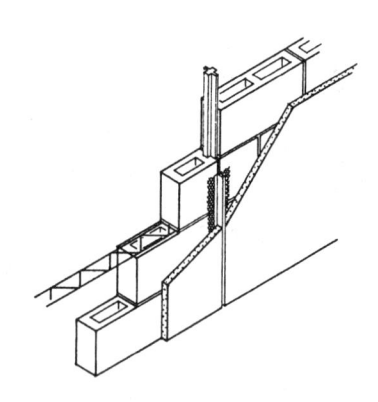

The Concrete Block Partition Systems are defined by weight and type of block, thickness, type of finish and number of sides finished. System components include joint reinforcing on alternate courses and vertical control joints.

6.1-210 Concrete Block Partitions - Regular Weight

	TYPE	THICKNESS (IN.)	TYPE FINISH	SIDES FINISHED	MAT.	INST.	TOTAL
1000	Hollow	4	none	0	.94	3.59	4.53
1010			gyp. plaster 2 coat	1	1.24	5.20	6.44
1020				2	1.54	6.85	8.39
1200			portland - 3 coat	1	1.14	5.45	6.59
1400			5/8" drywall	1	1.27	4.80	6.07
1500		6	none	0	1.13	3.85	4.98
1510			gyp. plaster 2 coat	1	1.43	5.50	6.93
1520				2	1.73	7.10	8.83
1700			portland - 3 coat	1	1.33	5.70	7.03
1900			5/8" drywall	1	1.46	5.05	6.51
1910				2	1.79	6.25	8.04
2000		8	none	0	1.57	4.12	5.69
2010			gyp. plaster 2 coat	1	1.87	5.75	7.62
2020			gyp. plaster 2 coat	2	2.17	7.35	9.52
2200			portland - 3 coat	1	1.77	6	7.77
2400			5/8" drywall	1	1.90	5.35	7.25
2410				2	2.23	6.55	8.78
2500		10	none	0	1.95	4.31	6.26
2510			gyp. plaster 2 coat	1	2.25	5.95	8.20
2520				2	2.55	7.55	10.10
2700			portland - 3 coat	1	2.15	6.15	8.30
2900			5/8" drywall	1	2.28	5.50	7.78
2910				2	2.61	6.75	9.36
3000	Solid	2	none	0	1.18	3.55	4.73
3010			gyp. plaster	1	1.48	5.20	6.68
3020				2	1.78	6.80	8.58
3200			portland - 3 coat	1	1.38	5.40	6.78
3400			5/8" drywall	1	1.51	4.76	6.27
3410				2	1.84	5.95	7.79
3500		4	none	0	1.37	3.72	5.09
3510			gyp. plaster	1	1.74	5.35	7.09
3520				2	1.97	6.95	8.92
3700			portland - 3 coat	1	1.57	5.60	7.17
3900			5/8" drywall	1	1.70	4.93	6.63
3910				2	2.03	6.15	8.18

Important: See the Reference Section for critical supporting data - Location Factors & Historical Cost Indexes

INTERIOR CONSTR.   A6.1   Block Partitions

6.1-210 Concrete Block Partitions - Regular Weight

	TYPE	THICKNESS (IN.)	TYPE FINISH	SIDES FINISHED		MAT.	INST.	TOTAL
4002	Solid	6	none	0		1.83	4	5.83
4010			gyp. plaster	1		2.13	5.65	7.78
4020				2		2.43	7.25	9.68
4200			portland - 3 coat	1		2.03	5.85	7.88
4400			5/8" drywall	1		2.16	5.20	7.36
4410				2		2.49	6.40	8.89

6.1-210 Concrete Block Partitions - Lightweight

	TYPE	THICKNESS (IN.)	TYPE FINISH	SIDES FINISHED		MAT.	INST.	TOTAL
5000	Hollow	4	none	0		1.05	3.52	4.57
5010			gyp. plaster	1		1.35	5.15	6.50
5020				2		1.65	6.75	8.40
5200			portland - 3 coat	1		1.25	5.40	6.65
5400			5/8" drywall	1		1.38	4.73	6.11
5410				2		1.71	5.95	7.66
5500		6	none	0		1.29	3.76	5.05
5510			gyp. plaster	1		1.59	5.40	6.99
5520			gyp. plaster	2		1.89	7	8.89
5700			portland - 3 coat	1		1.49	5.60	7.09
5900			5/8" drywall	1		1.62	4.97	6.59
5910				2		1.95	6.20	8.15
6000		8	none	0		1.56	4.02	5.58
6010			gyp. plaster	1		1.86	5.65	7.51
6020				2		2.16	7.25	9.41
6200			portland - 3 coat	1		1.76	5.90	7.66
6400			5/8" drywall	1		1.89	5.25	7.14
6410				2		2.22	6.45	8.67
6500		10	none	0		2.09	4.20	6.29
6510			gyp. plaster	1		2.39	5.85	8.24
6520				2		2.69	7.45	10.14
6700			portland - 3 coat	1		2.29	6.05	8.34
6900			5/8" drywall	1		2.42	5.40	7.82
6910				2		2.75	6.60	9.35
7000	Solid	4	none	0		1.41	3.68	5.09
7010			gyp. plaster	1		1.71	5.30	7.01
7020				2		2.01	6.95	8.96
7200			portland - 3 coat	1		1.61	5.55	7.16
7400			5/8" drywall	1		1.74	4.89	6.63
7410				2		2.07	6.10	8.17
7500		6	none	0		2.01	3.95	5.96
7510			gyp. plaster	1		2.40	5.60	8
7520				2		2.61	7.20	9.81
7700			portland - 3 coat	1		2.21	5.80	8.01
7900			5/8" drywall	1		2.34	5.15	7.49
7910				2		2.67	6.35	9.02
8000		8	none	0		2.30	4.23	6.53
8010			gyp. plaster	1		2.60	5.85	8.45
8020				2		2.90	7.50	10.40
8200			portland - 3 coat	1		2.50	6.10	8.60
8400			5/8" drywall	1		2.63	5.45	8.08
8410				2		2.96	6.65	9.61

For expanded coverage of these items see *Means Assemblies Cost Data 1997*

INTERIOR CONSTR. — A6.1 Partitions

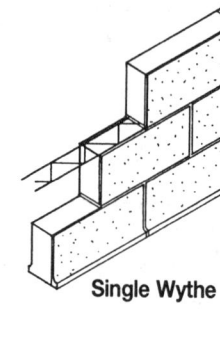

Structural facing tile
8W series
8" x 16"

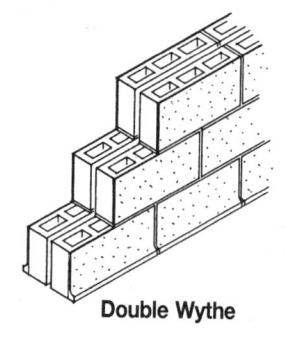

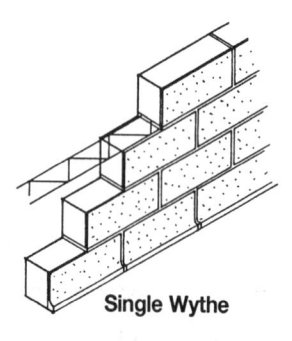

Single Wythe

Double Wythe

Structural facing tile
6T series
5-1/3" x 12"

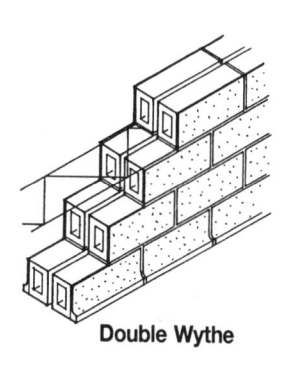

Single Wythe

Double Wythe

6.1-270	Tile Partitions	MAT.	INST.	TOTAL
1000	8W series 8"x16", 4" thick wall, reinf every 2 courses, glazed 1 side	6.75	4.33	11.08
1100	Glazed 2 sides	10.80	4.60	15.40
1200	Glazed 2 sides, using 2 wythes of 2" thick tile	12.50	8.30	20.80
1300	6" thick wall, horizontal reinf every 2 courses, glazed 1 side	9.55	4.53	14.08
1400	Glazed 2 sides, each face different color, 2" and 4" tile	13	8.50	21.50
1500	8" thick wall, glazed 2 sides using 2 wythes of 4" thick tile	13.50	8.65	22.15
1600	10" thick wall, glazed 2 sides using 1 wythe of 4" tile & 1 wythe of 6" tile	16.30	8.85	25.15
1700	Glazed 2 sides cavity wall, using 2 wythes of 4" thick tile	13.50	8.65	22.15
1800	12" thick wall, glazed 2 sides using 2 wythes of 6" thick tile	19.10	9.05	28.15
1900	Glazed 2 sides cavity wall, using 2 wythes of 4" thick tile	13.50	8.65	22.15
2100	6T series 5-1/3"x12" tile, 4" thick, non load bearing glazed one side,	7.40	6.80	14.20
2200	Glazed two sides	11.80	7.65	19.45
2300	Glazed two sides, using two wythes of 2" thick tile	12.30	13.30	25.60
2400	6" thick, glazed one side	10.95	7.10	18.05
2500	Glazed two sides	17.85	8.10	25.95
2600	Glazed two sides using 2" thick tile and 4" thick tile	13.55	13.45	27
2700	8" thick, glazed one side	14.10	8.30	22.40
2800	Glazed two sides using two wythes of 4" thick tile	14.80	13.60	28.40
2900	Glazed two sides using 6" thick tile and 2" thick tile	17.10	13.75	30.85
3000	10" thick cavity wall, glazed two sides using two wythes of 4" tile	14.80	13.60	28.40
3100	12" thick, glazed two sides using 4" thick tile and 8" thick tile	21.50	15.10	36.60
3200	2" thick facing tile, glazed one side, on 6" concrete block	7.20	7.90	15.10
3300	On 8" concrete block	7.65	8.15	15.80
3400	On 10" concrete block	8	8.30	16.30

Important: See the Reference Section for critical supporting data - Location Factors & Historical Cost Indexes

INTERIOR CONSTR. — A6.1 Partitions

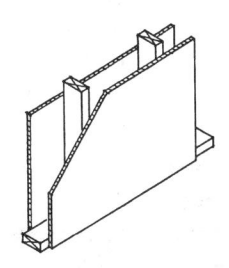

Gypsum board, single layer each side on wood studs.

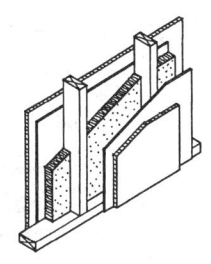

Gypsum board, sound deadening board each side, with 1-1/2" insulation on wood studs.

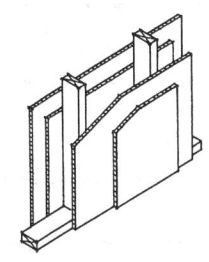

Gypsum board, two layers each side on wood studs.

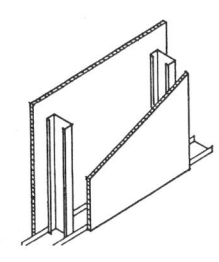

Gypsum board, single layer each side on metal studs.

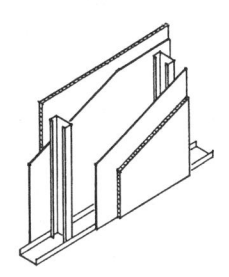

Gypsum board, sound deadening board each side on metal studs.

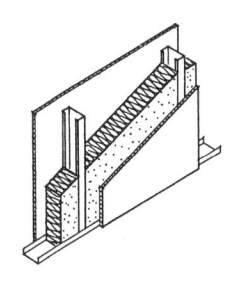

Gypsum board two layers one side, single layer opposite side, with 3-1/2" insulation on metal studs.

6.1-510 Drywall Partitions/Wood Stud Framing

	FACE LAYER	BASE LAYER	FRAMING	OPPOSITE FACE	INSULATION	MAT.	INST.	TOTAL
1200	5/8" FR drywall	none	2 x 4, @ 16" O.C.	same	0	1.04	2	3.04
1250				5/8" reg. drywall	0	1.03	2	3.03
1300				nothing	0	.72	1.34	2.06
1400		1/4" SD gypsum	2 x 4 @ 16" O.C.	same	1-1/2" fiberglass	1.69	3.07	4.76
1450				5/8" FR drywall	1-1/2" fiberglass	1.53	2.70	4.23
1500				nothing	1-1/2" fiberglass	1.21	2.04	3.25
1600		resil. channels	2 x 4 @ 16", O.C.	same	1-1/2" fiberglass	1.61	3.88	5.49
1650				5/8" FR drywall	1-1/2" fiberglass	1.49	3.10	4.59
1700				nothing	1-1/2" fiberglass	1.17	2.44	3.61
1800		5/8" FR drywall	2 x 4 @ 24" O.C.	same	0	1.39	2.53	3.92
1850				5/8" FR drywall	0	1.17	2.20	3.37
1900				nothing	0	.85	1.54	2.39
2200		5/8" FR drywall	2 rows-2 x 4 16"O.C.	same	2" fiberglass	2.26	3.67	5.93
2250				5/8" FR drywall	2" fiberglass	2.04	3.34	5.38
2300				nothing	2" fiberglass	1.72	2.68	4.40
2400	5/8" WR drywall	none	2 x 4, @ 16" O.C.	same	0	1.12	2	3.12
2450				5/8" FR drywall	0	1.08	2	3.08
2500				nothing	0	.76	1.34	2.10
2600		5/8" FR drywall	2 x 4, @ 24" O.C.	same	0	1.47	2.53	4
2650				5/8" FR drywall	0	1.21	2.20	3.41
2700				nothing	0	.89	1.54	2.43
2800	5/8 VF drywall	none	2 x 4, @ 16" O.C.	same	0	1.72	2.14	3.86
2850				5/8" FR drywall	0	1.38	2.07	3.45
2900				nothing	0	1.06	1.41	2.47
3000		5/8" FR drywall	2 x 4, 24" O.C.	same	0	2.07	2.67	4.74
3050				5/8" FR drywall	0	1.51	2.27	3.78
3100				nothing	0	1.19	1.61	2.80

For expanded coverage of these items see *Means Assemblies Cost Data 1997*

INTERIOR CONSTR. — A6.1 Partitions

6.1-510 Drywall Partitions/Wood Stud Framing

	FACE LAYER	BASE LAYER	FRAMING	OPPOSITE FACE	INSULATION	MAT.	INST.	TOTAL
3200	1/2" reg drywall	3/8" reg drywall	2 x 4, @ 16" O.C.	same	0	1.22	2.66	3.88
3252				5/8" FR drywall	0	1.13	2.33	3.46
3300				nothing	0	.81	1.67	2.48

6.1-510 Drywall Partitions/Metal Stud Framing

	FACE LAYER	BASE LAYER	FRAMING	OPPOSITE FACE	INSULATION	MAT.	INST.	TOTAL
5200	5/8" FR drywall	none	1-5/8" @ 24" O.C.	same	0	.80	1.95	2.75
5250				5/8" reg. drywall	0	.79	1.95	2.74
5300				nothing	0	.48	1.29	1.77
5400			3-5/8" @ 24" O.C.	same	0	.85	1.98	2.83
5450				5/8" reg. drywall	0	.84	1.98	2.82
5500				nothing	0	.53	1.32	1.85
5600		1/4" SD gypsum	1-5/8" @ 24" O.C.	same	0	1.12	2.69	3.81
5650				5/8" FR drywall	0	.96	2.32	3.28
5700				nothing	0	.64	1.66	2.30
5800			2-1/2" @ 24" O.C.	same	0	1.14	2.70	3.84
5850				5/8" FR drywall	0	.98	2.33	3.31
5900				nothing	0	.66	1.67	2.33
6000		5/8" FR drywall	2-1/2" @ 16" O.C.	same	0	1.59	2.73	4.32
6050				5/8" FR drywall	0	1.37	2.40	3.77
6100				nothing	0	1.05	1.74	2.79
6200			3-5/8" @ 24" O.C.	same	0	1.29	2.64	3.93
6250				5/8" FR drywall	3-1/2" fiberglass	1.35	2.52	3.87
6300				nothing	0	.75	1.65	2.40
6400	5/8" WR drywall	none	1-5/8" @ 24" O.C.	same	0	.88	1.95	2.83
6450				5/8" FR drywall	0	.84	1.95	2.79
6500				nothing	0	.52	1.29	1.81
6600			3-5/8" @ 24" O.C.	same	0	.93	1.98	2.91
6650				5/8" FR drywall	0	.89	1.98	2.87
6700				nothing	0	.57	1.32	1.89
6800		5/8" FR drywall	2-1/2" @ 16" O.C.	same	0	1.67	2.73	4.40
6850				5/8" FR drywall	0	1.41	2.40	3.81
6900				nothing	0	1.09	1.74	2.83
7000			3-5/8" @ 24" O.C.	same	0	1.37	2.64	4.01
7050				5/8" FR drywall	3-1/2" fiberglass	1.39	2.52	3.91
7100				nothing	0	.79	1.65	2.44
7200	5/8" VF drywall	none	1-5/8" @ 24" O.C.	same	0	1.48	2.09	3.57
7250				5/8" FR drywall	0	1.14	2.02	3.16
7300				nothing	0	.82	1.36	2.18
7400			3-5/8" @ 24" O.C.	same	0	1.53	2.12	3.65
7450				5/8" FR drywall	0	1.19	2.05	3.24
7500				nothing	0	.87	1.39	2.26
7600		5/8" FR drywall	2-1/2" @ 16" O.C.	same	0	2.27	2.87	5.14
7650				5/8" FR drywall	0	1.71	2.47	4.18
7700				nothing	0	1.39	1.81	3.20
7800			3-5/8" @ 24" O.C.	same	0	1.97	2.78	4.75
7850				5/8" FR drywall	3-1/2" fiberglass	1.69	2.59	4.28
7900				nothing	0	1.09	1.72	2.81

Important: See the Reference Section for critical supporting data - Location Factors & Historical Cost Indexes

INTERIOR CONSTR. | A6.1 | Partitions

6.1-580 Drywall Components

		COST PER S.F.		
		MAT.	INST.	TOTAL
0060	Metal studs, 24" O.C. including track, load bearing, 20 gage, 2-1/2"	.55	1.37	1.92
0080	3-5/8"	.63	1.43	2.06
0100	4"	.67	1.49	2.16
0120	6"	.81	1.57	2.38
0140	Metal studs, 24" O.C. including track, load bearing, 18 gage, 2-1/2"	.55	1.37	1.92
0160	3-5/8"	.63	1.43	2.06
0180	4"	.67	1.49	2.16
0200	6"	.81	1.57	2.38
0220	16 gage, 2-1/2"	.59	1.49	2.08
0240	3-5/8"	.82	1.57	2.39
0260	4"	.82	1.64	2.46
0280	6"	1	1.73	2.73
0300	Non load bearing, 25 gage, 1-5/8"	.16	.63	.79
0340	3-5/8"	.21	.66	.87
0360	4"	.22	.67	.89
0380	6"	.28	.68	.96
0400	20 gage, 2-1/2"	.40	.64	1.04
0420	3-5/8"	.48	.66	1.14
0440	4"	.51	.67	1.18
0460	6"	.65	.68	1.33
0540	Wood studs including blocking, shoe and double top plate, 2"x4", 12"O.C.	.50	.85	1.35
0560	16" O.C.	.40	.68	1.08
0580	24" O.C.	.31	.55	.86
0600	2"x6", 12" O.C.	.72	.97	1.69
0620	16" O.C.	.58	.76	1.34
0640	24" O.C.	.44	.59	1.03
0642	Furring one side only, steel channels, 3/4", 12" O.C.	.19	1.30	1.49
0644	16" O.C.	.17	1.16	1.33
0646	24" O.C.	.11	.88	.99
0647	1-1/2" , 12" O.C.	.27	1.46	1.73
0648	16" O.C.	.24	1.28	1.52
0649	24" O.C.	.16	1	1.16
0650	Wood strips, 1" x 3", on wood, 12" O.C.	.31	.60	.91
0651	16" O.C.	.23	.45	.68
0652	On masonry, 12" O.C.	.31	.66	.97
0653	16" O.C.	.23	.50	.73
0654	On concrete, 12" O.C.	.31	1.26	1.57
0655	16" O.C.	.23	.95	1.18
0665	Gypsum board, one face only, exterior sheathing, 1/2"	.19	.60	.79
0680	Fire resistant, 1/2"	.22	.33	.55
0700	5/8"	.22	.33	.55
0720	Sound deadening board 1/4"	.16	.37	.53
0740	Standard drywall 3/8"	.16	.33	.49
0760	1/2"	.15	.33	.48
0780	5/8"	.21	.33	.54
0800	Tongue & groove coreboard 1"	.55	1.37	1.92
0820	Water resistant, 1/2"	.24	.33	.57
0840	5/8"	.26	.33	.59
0860	Add for the following:, foil backing	.08		.08
0880	Fiberglass insulation, 3-1/2"	.28	.21	.49
0900	6"	.33	.24	.57
0920	Rigid insulation 1"	.34	.33	.67
0940	Resilient furring @ 16" O.C.	.16	1.03	1.19
0960	Taping and finishing	.10	.33	.43
0980	Texture spray	.13	.38	.51
1000	Thin coat plaster	.13	.40	.53
1040	2"x4" staggered studs 2"x6" plates & blocking	.58	.74	1.32

For expanded coverage of these items see *Means Assemblies Cost Data 1997*

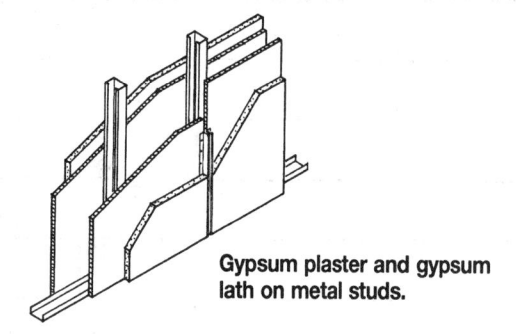

INTERIOR CONSTR. — A6.1 Partitions

Gypsum plaster and gypsum lath on metal studs.

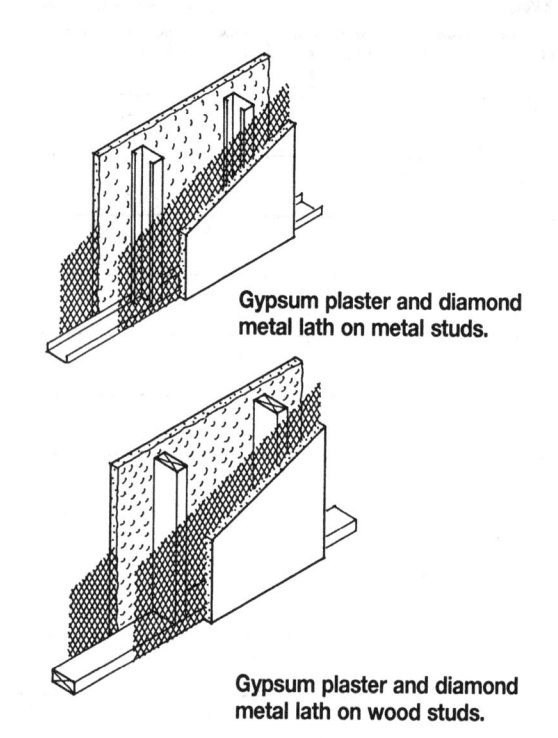

Gypsum plaster and diamond metal lath on metal studs.

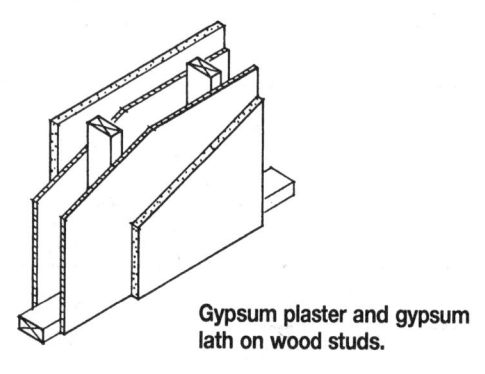

Gypsum plaster and gypsum lath on wood studs.

Gypsum plaster and diamond metal lath on wood studs.

6.1-610 Plaster Partitions/Metal Stud Framing

	TYPE	FRAMING	LATH	OPPOSITE FACE	MAT.	INST.	TOTAL
1000	2 coat gypsum	2-1/2" @ 16"O.C.	3/8" gypsum	same	1.93	4.87	6.80
1010				nothing	1.11	2.86	3.97
1100		3-1/4" @ 24"O.C.	1/2" gypsum	same	2.03	4.84	6.87
1110				nothing	1.15	2.81	3.96
1500	2 coat vermiculite	2-1/2" @ 16"O.C.	3/8" gypsum	same	2.09	5.30	7.39
1510				nothing	1.18	3.07	4.25
1600		3-1/4" @ 24"O.C.	1/2" gypsum	same	2.19	5.25	7.44
1610				nothing	1.22	3.02	4.24
2000	3 coat gypsum	2-1/2" @ 16"O.C.	3/8" gypsum	same	1.86	5.50	7.36
2010				nothing	1.07	3.18	4.25
2020			3.4lb. diamond	same	1.55	5.50	7.05
2030				nothing	.92	3.18	4.10
2040			2.75lb. ribbed	same	1.40	5.50	6.90
2050				nothing	.84	3.18	4.02
2100		3-1/4" @ 24"O.C.	1/2" gypsum	same	1.96	5.45	7.41
2110				nothing	1.11	3.13	4.24
2120			3.4lb. ribbed	same	1.75	5.45	7.20
2130				nothing	1.01	3.13	4.14
3500	3 coat gypsum W/med. Keenes	2-1/2" @ 16"O.C.	3/8" gypsum	same	2.30	7	9.30
3510				nothing	1.29	3.93	5.22
3520			3.4lb. diamond	same	1.99	7	8.99
3530				nothing	1.14	3.93	5.07
3540			2.75lb. ribbed	same	1.84	7	8.84
3550				nothing	1.06	3.93	4.99
3600		3-1/4" @ 24"O.C.	1/2" gypsum	same	2.40	7	9.40
3610				nothing	1.33	3.88	5.21
3620			3.4lb. ribbed	same	2.19	7	9.19
3630				nothing	1.23	3.88	5.11

Important: See the Reference Section for critical supporting data - Location Factors & Historical Cost Indexes

INTERIOR CONSTR.   A6.1   Partitions

6.1-610   Plaster Partitions/Metal Stud Framing

	TYPE	FRAMING	LATH	OPPOSITE FACE		MAT.	INST.	TOTAL
4000	3 coat gypsum W/hard Keenes	2-1/2" @ 16"O.C.	3/8" gypsum	same		2.30	7.80	10.10
4010				nothing		1.29	4.30	5.59
4022		2-1/2" @ 16"O.C.	3.4 lb. diamond	same		1.99	7.80	9.79
4032				nothing		1.14	4.30	5.44
4040			2.75lb. ribbed	same		1.84	7.80	9.64
4050				nothing		1.06	4.30	5.36
4100		3-1/4" @ 24"O.C.	1/2" gypsum	same		2.40	7.75	10.15
4110				nothing		1.33	4.25	5.58
4120			3.4lb. ribbed	same		2.19	7.75	9.94
4130				nothing		1.23	4.25	5.48

6.1-610   Plaster Partitions/Wood Stud Framing

	TYPE	FRAMING	LATH	OPPOSITE FACE		MAT.	INST.	TOTAL
5000	2 coat gypsum	2"x4" @ 16"O.C.	3/8" gypsum	same		2.08	4.69	6.77
5010				nothing		1.28	2.74	4.02
5100		2"x4" @ 24"O.C.	1/2" gypsum	same		2.04	4.61	6.65
5110				nothing		1.21	2.64	3.85
5500	2 coat vermiculite	2"x4" @ 16"O.C.	3/8" gypsum	same		2.24	5.15	7.39
5510				nothing		1.35	2.95	4.30
5600		2"x4" @ 24"O.C.	1/2" gypsum	same		2.20	5.05	7.25
5610				nothing		1.28	2.85	4.13
6000	3 coat gypsum	2"x4" @ 16"O.C.	3/8" gypsum	same		2.01	5.35	7.36
6010				nothing		1.24	3.06	4.30
6020			3.4lb. diamond	same		1.73	5.40	7.13
6030				nothing		1.10	3.09	4.19
6040			2.75lb. ribbed	same		1.57	5.40	6.97
6050				nothing		1.02	3.10	4.12
6100		2"x4" @ 24"O.C.	1/2" gypsum	same		1.97	5.25	7.22
6110				nothing		1.17	2.96	4.13
6120			3.4lb. ribbed	same		1.49	5.30	6.79
6130				nothing		.93	2.98	3.91
7500	3 coat gypsum W/med Keenes	2"x4" @ 16"O.C.	3/8" gypsum	same		2.45	6.85	9.30
7510				nothing		1.46	3.81	5.27
7520			3.4lb. diamond	same		2.17	6.90	9.07
7530				nothing		1.32	3.84	5.16
7540			2.75lb. ribbed	same		2.01	6.95	8.96
7550				nothing		1.24	3.85	5.09
7600		2"x4" @ 24"O.C.	1/2" gypsum	same		2.41	6.80	9.21
7610				nothing		1.39	3.71	5.10
7620			3.4lb. ribbed	same		2.29	6.90	9.19
7630				nothing		1.33	3.77	5.10
8000	3 coat gypsum W/hard Keenes	2"x4" @ 16"O.C.	3/8" gypsum	same		2.45	7.60	10.05
8010				nothing		1.46	4.18	5.64
8020			3.4lb. diamond	same		2.17	7.65	9.82
8030				nothing		1.32	4.21	5.53
8040			2.75lb. ribbed	same		2.01	7.65	9.66
8050				nothing		1.24	4.22	5.46
8100		2"x4" @ 24"O.C.	1/2" gypsum	same		2.41	7.50	9.91
8110				nothing		1.39	4.08	5.47
8120			3.4lb. ribbed	same		2.29	7.60	9.89
8130				nothing		1.33	4.14	5.47

For expanded coverage of these items see *Means Assemblies Cost Data 1997*

INTERIOR CONSTR. — A6.1 Partitions

6.1-680 Plaster Partition Components

		COST PER S.F.		
		MAT.	INST.	TOTAL
0060	Metal studs, 16" O.C., including track, non load bearing, 25 gage, 1-5/8"	.22	.75	.97
0080	2-1/2"	.22	.75	.97
0100	3-1/4"	.25	.76	1.01
0120	3-5/8"	.25	.76	1.01
0140	4"	.28	.78	1.06
0160	6"	.35	.80	1.15
0180	Load bearing, 20 gage, 2-1/2"	.68	1.64	2.32
0200	3-5/8"	.78	1.73	2.51
0220	4"	.83	1.83	2.66
0240	6"	1.02	1.93	2.95
0260	16 gage 2-1/2"	.74	1.83	2.57
0280	3-5/8"	.96	1.93	2.89
0300	4"	1.02	2.05	3.07
0320	6"	1.25	2.19	3.44
0340	Wood studs, including blocking, shoe and double plate, 2"x4", 12" O.C.	.50	.85	1.35
0360	16" O.C.	.40	.68	1.08
0380	24" O.C.	.31	.55	.86
0400	2"x6", 12" O.C.	.72	.97	1.69
0420	16" O.C.	.58	.76	1.34
0440	24" O.C.	.44	.59	1.03
0460	Furring one face only, steel channels, 3/4", 12" O.C.	.19	1.30	1.49
0480	16" O.C.	.17	1.16	1.33
0500	24" O.C.	.11	.88	.99
0520	1-1/2", 12" O.C.	.27	1.46	1.73
0540	16" O.C.	.24	1.28	1.52
0560	24" O.C.	.16	1	1.16
0580	Wood strips 1"x3", on wood., 12" O.C.	.31	.60	.91
0600	16" O.C.	.23	.45	.68
0620	On masonry, 12" O.C.	.31	.66	.97
0640	16" O.C.	.23	.50	.73
0660	On concrete, 12" O.C.	.31	1.26	1.57
0680	16" O.C.	.23	.95	1.18
0700	Gypsum lath. plain or perforated, nailed to studs, 3/8" thick	.42	.40	.82
0720	1/2" thick	.44	.43	.87
0740	Clipped to studs, 3/8" thick	.43	.45	.88
0760	1/2" thick	.48	.49	.97
0780	Metal lath, diamond painted, nailed to wood studs, 2.5 lb.	.21	.40	.61
0800	3.4 lb.	.28	.43	.71
0820	Screwed to steel studs, 2.5 lb.	.21	.43	.64
0840	3.4 lb.	.25	.45	.70
0860	Rib painted, wired to steel, 2.75 lb	.20	.45	.65
0880	3.4 lb	.38	.49	.87
0900	4.0 lb	.33	.52	.85
0910				
0920	Gypsum plaster, 2 coats	.38	1.55	1.93
0940	3 coats	.53	1.87	2.40
0960	Perlite or vermiculite plaster, 2 coats	.47	1.76	2.23
0980	3 coats	.75	2.20	2.95
1000	Stucco, 3 coats, 1" thick, on wood framing	.77	1.46	2.23
1020	On masonry	.33	.81	1.14
1100	Metal base galvanized and painted 2-1/2" high	.46	1.28	1.74

INTERIOR CONSTR. — A6.1 Partitions

Folding Accordion

Folding Leaf

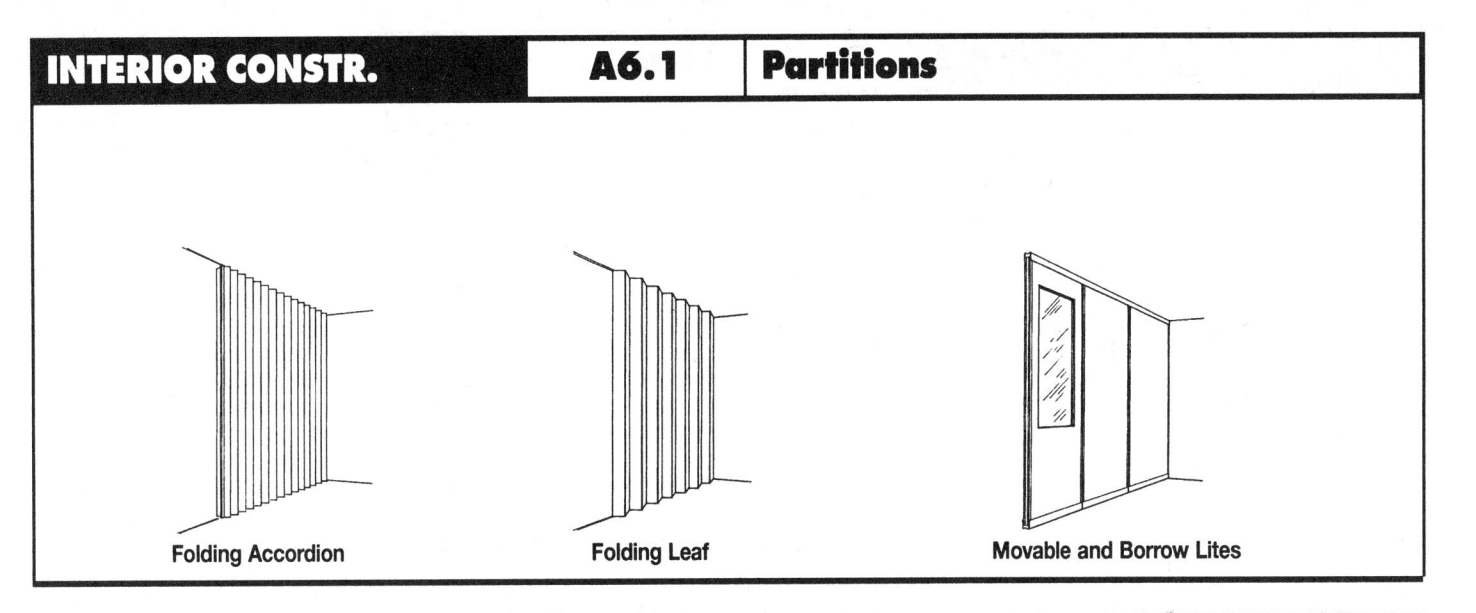

Movable and Borrow Lites

6.1-820	Partitions	MAT.	INST.	TOTAL
0360	Folding accordion, vinyl covered, acoustical, 3 lb. S.F., 17 ft max. hgt	18.55	6.60	25.15
0380	5 lb. per S.F. 27 ft max height	27.50	6.90	34.40
0400	5.5 lb. per S.F., 17 ft. max height	31	7.30	38.30
0420	Commercial, 1.75 lb per S.F., 8 ft. max height	12.25	2.92	15.17
0440	2.0 Lb per S.F., 17 ft. max height	14.85	4.38	19.23
0460	Industrial, 4.0 lb. per S.F. 27 ft max height	25	8.75	33.75
0480	Vinyl clad wood or steel, electric operation 6 psf	31	4.11	35.11
0500	Wood, non acoustic, birch or mahogany	16.25	2.19	18.44
0560	Folding leaf, aluminum framed acoustical 12 ft.high.,5.5 lb per S.F.,min.	26	10.95	36.95
0580	Maximum	31.50	22	53.50
0600	6.5 lb. per S.F., minimum	27.50	10.95	38.45
0620	Maximum	34	22	56
0640	Steel acoustical, 7.5 per S.F., vinyl faced, minimum	39	10.95	49.95
0660	Maximum	47.50	22	69.50
0680	Wood acoustic type, vinyl faced to 18' high 6 psf, minimum	36.50	10.95	47.45
0700	Average	43.50	14.60	58.10
0720	Maximum	56	22	78
0740	Formica or hardwood faced, minimum	37.50	10.95	48.45
0760	Maximum	40	22	62
0780	Wood, low acoustical type to 12 ft. high 4.5 psf	27.50	13.15	40.65
0840	Demountable, trackless wall, cork finish, semi acous, 1-5/8"th, min	18.65	2.02	20.67
0860	Maximum	22.50	3.46	25.96
0880	Acoustic, 2" thick, minimum	19.50	2.16	21.66
0900	Maximum	28.50	2.92	31.42
0920	In-plant modular office system, w/prehung steel door			
0940	3" thick honeycomb core panels			
0960	12' x 12', 2 wall	9.75	.41	10.16
0970	4 wall	10	.54	10.54
0980	16' x 16', 2 wall	9.65	.29	9.94
0990	4 wall	6.95	.29	7.24
1000	Gypsum, demountable, 3" to 3-3/4" thick x 9' high, vinyl clad	2.66	1.52	4.18
1020	Fabric clad	8	1.66	9.66
1040	1.75 system, vinyl clad hardboard, paper honeycomb core panel			
1060	1-3/4" to 2-1/2" thick x 9' high	5.45	1.52	6.97
1080	Unitized gypsum panel system, 2" to 2-1/2" thick x 9' high			
1100	Vinyl clad gypsum	7.25	1.52	8.77
1120	Fabric clad gypsum	13.10	1.66	14.76
1140	Movable steel walls, modular system			
1160	Unitized panels, 48" wide x 9' high			
1180	Baked enamel, pre-finished	8.50	1.22	9.72
1200	Fabric clad	13	1.30	14.30

For expanded coverage of these items see *Means Assemblies Cost Data 1997*

INTERIOR CONSTR. | A6.1 | Partitions

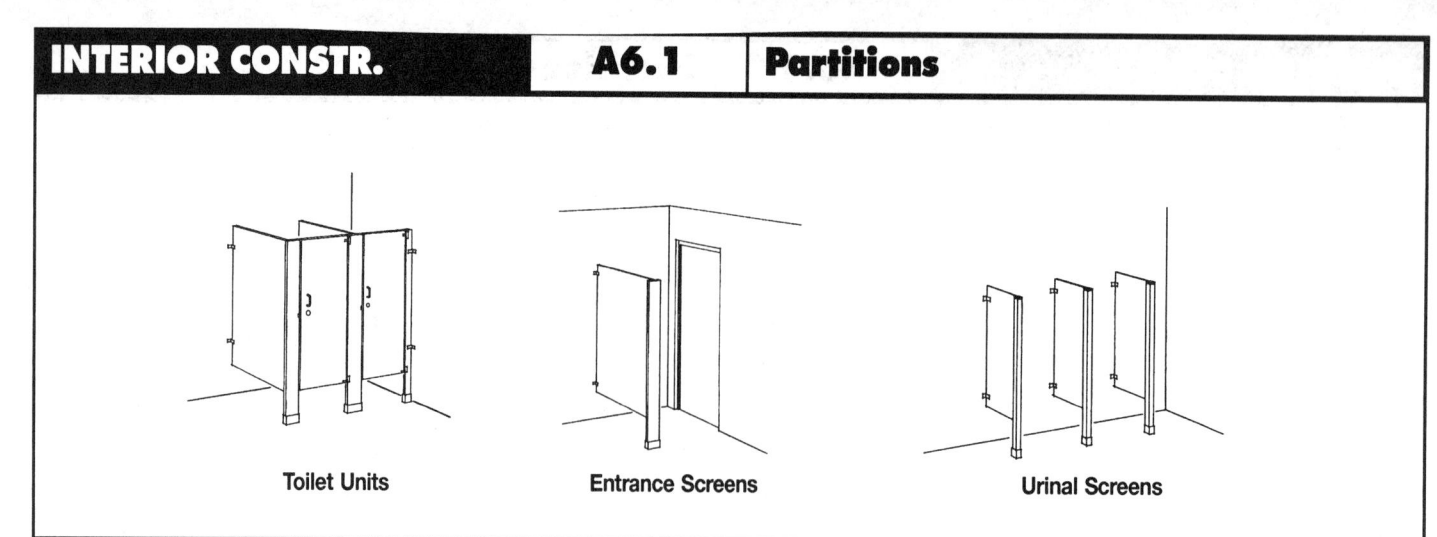

Toilet Units | Entrance Screens | Urinal Screens

6.1-870	Toilet Partitions	COST PER UNIT		
		MAT.	INST.	TOTAL
0380	Toilet partitions, cubicles, ceiling hung, marble	1,350	325	1,675
0400	Painted metal	415	164	579
0420	Plastic laminate	555	164	719
0440	Porcelain enamel	835	164	999
0460	Stainless steel	1,000	164	1,164
0480	Handicap addition	263		263
0520	Floor and ceiling anchored, marble	1,475	261	1,736
0540	Painted metal	420	132	552
0560	Plastic laminate	555	132	687
0580	Porcelain enamel	860	132	992
0600	Stainless steel	1,175	132	1,307
0620	Handicap addition	263		263
0660	Floor mounted marble	875	217	1,092
0680	Painted metal	375	94	469
0700	Plastic laminate	555	94	649
0720	Porcelain enamel	840	94	934
0740	Stainless steel	1,150	94	1,244
0760	Handicap addition	263		263
0780	Juvenile deduction	37.50		37.50
0820	Floor mounted with handrail marble	1,025	217	1,242
0840	Painted metal	385	110	495
0860	Plastic laminate	540	110	650
0880	Porcelain enamel	850	110	960
0900	Stainless steel	1,150	110	1,260
0920	Handicap addition	263		263
0960	Wall hung, painted metal	475	94	569
1000	Porcelain enamel	855	94	949
1020	Stainless steel	1,125	94	1,219
1040	Handicap addition	263		263
1080	Entrance screens, floor mounted, 54" high, marble	555	72.50	627.50
1100	Painted metal	173	44	217
1120	Porcelain enamel	340	44	384
1140	Stainless steel	625	44	669
1300	Urinal screens, floor mounted, 24" wide, laminated plastic	251	82	333
1320	Marble	520	97	617
1340	Painted metal	180	82	262
1360	Porcelain enamel	340	82	422
1380	Stainless steel	520	82	602
1428	Wall mounted wedge type, painted metal	203	66	269
1440	Porcelain enamel	293	66	359
1460	Stainless steel	420	66	486

Important: See the Reference Section for critical supporting data - Location Factors & Historical Cost Indexes

INTERIOR CONSTR. | A6.4 | Doors

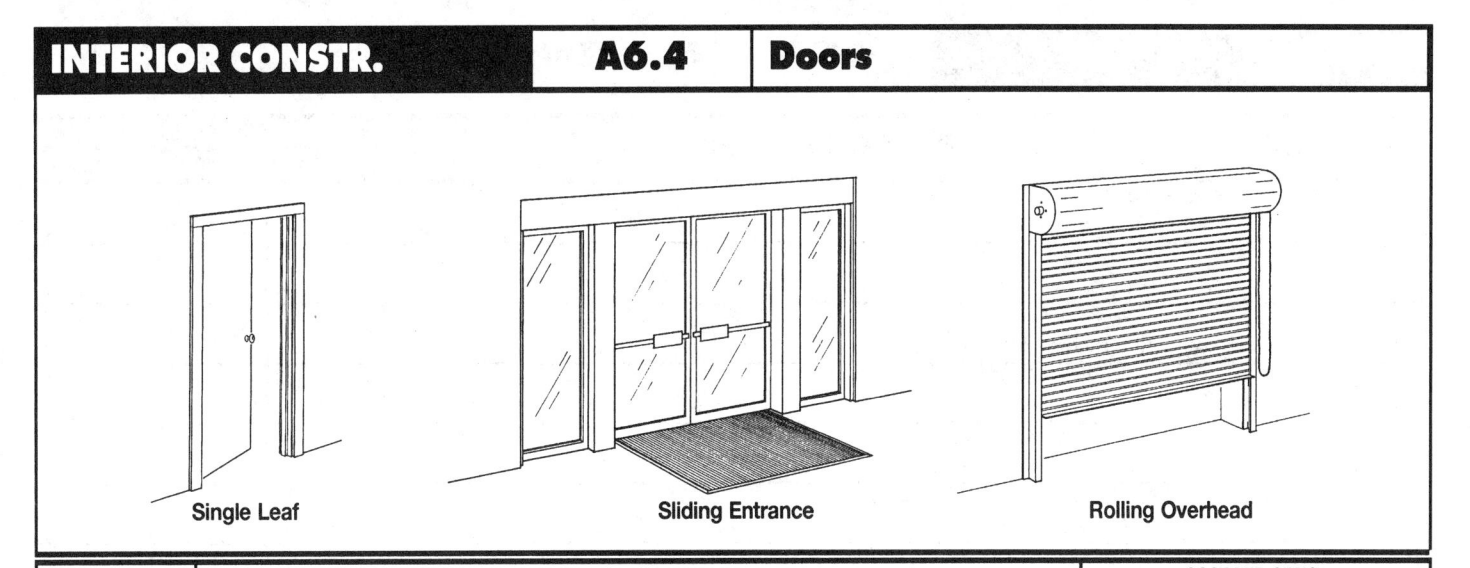

Single Leaf | Sliding Entrance | Rolling Overhead

6.4-100	Special Doors	MAT.	INST.	TOTAL
2500	Single leaf, wood, 3'-0"x7'-0"x1 3/8", birch, solid core	272	127	399
2510	Hollow core	225	120	345
2530	Hollow core, lauan	215	120	335
2540	Louvered pine	325	120	445
2550	Paneled pine	330	120	450
2600	Hollow metal, comm. quality, flush, 3'-0"x7'-0"x1-3/8"	395	135	530
2650	3'-0"x10'-0" openings with panel	625	174	799
2700	Metal fire, comm. quality, 3'-0"x7'-0"x1-3/8"	430	138	568
2800	Kalamein fire, comm. quality, 3'-0"x7'-0"x1-3/4"	360	248	608
3200	Double leaf, wood, hollow core, 2 - 3'-0"x7'-0"x1-3/8"	335	230	565
3300	Hollow metal, comm. quality, B label, 2'-3'-0"x7'-0"x1-3/8"	805	290	1,095
3400	6'-0"x10'-0" opening, with panel	1,200	365	1,565
3500	Double swing door system, 12'-0"x7'-0", mill finish	5,225	1,875	7,100
3700	Black finish	5,725	1,925	7,650
3800	Sliding entrance door and system mill finish	6,225	1,575	7,800
3900	Bronze finish	6,850	1,700	8,550
4000	Black finish	7,150	1,750	8,900
4100	Sliding panel mall front, 16'x9' opening, mill finish	2,200	475	2,675
4200	Bronze finish	2,850	620	3,470
4300	Black finish	3,525	760	4,285
4400	24'x9' opening mill finish	3,200	885	4,085
4500	Bronze finish	4,150	1,150	5,300
4600	Black finish	5,125	1,425	6,550
4700	48'x9' opening mill finish	5,950	690	6,640
4800	Bronze finish	7,725	895	8,620
4900	Black finish	9,525	1,100	10,625
5000	Rolling overhead steel door, manual, 8' x 8' high	990	535	1,525
5100	10' x 10' high	1,000	610	1,610
5200	20' x 10' high	2,500	855	3,355
5300	12' x 12' high	1,275	715	1,990
5400	Motor operated, 8' x 8' high	1,925	705	2,630
5500	10' x 10' high	1,950	780	2,730
5600	20' x 10' high	3,450	1,025	4,475
5700	12' x 12' high	2,225	885	3,110
5800	Roll up grille, aluminum, manual, 10' x 10' high, mill finish	1,800	1,050	2,850
5900	Bronze anodized	2,850	1,050	3,900
6000	Motor operated, 10' x 10' high, mill finish	2,600	1,225	3,825
6100	Bronze anodized	3,650	1,225	4,875
6200	Steel, manual, 10' x 10' high	1,600	855	2,455
6300	15' x 8' high	1,875	1,075	2,950
6400	Motor operated, 10' x 10' high	2,400	1,025	3,425
6500	15' x 8' high	2,675	1,250	3,925

For expanded coverage of these items see *Means Assemblies Cost Data 1997*

INTERIOR CONSTR. — A6.5 Wall Finishes

6.5-100 Paint & Covering

		COST PER S.F.		
		MAT.	INST.	TOTAL
0060	Painting, interior on plaster and drywall, brushwork, primer & 1 coat	.08	.45	.53
0080	Primer & 2 coats	.12	.60	.72
0100	Primer & 3 coats	.17	.73	.90
0120	Walls & ceilings, roller work, primer & 1 coat	.08	.30	.38
0140	Primer & 2 coats	.12	.38	.50
0160	Woodwork incl. puttying, brushwork, primer & 1 coat	.08	.65	.73
0180	Primer & 2 coats	.12	.86	.98
0200	Primer & 3 coats	.17	1.17	1.34
0260	Cabinets and casework, enamel, primer & 1 coat	.08	.73	.81
0280	Primer & 2 coats	.13	.90	1.03
0300	Masonry or concrete, latex, brushwork, primer & 1 coat	.15	.61	.76
0320	Primer & 2 coats	.21	.87	1.08
0340	Addition for block filler	.10	.75	.85
0380	Fireproof paints, intumescent, 1/8" thick 3/4 hour	1.63	.60	2.23
0400	3/16" thick 1 hour	3.63	.90	4.53
0420	7/16" thick 2 hour	4.68	2.08	6.76
0440	1-1/16" thick 3 hour	7.65	4.17	11.82
0480	Miscellaneous metal brushwork, exposed metal, primer & 1 coat	.06	.73	.79
0500	Gratings, primer & 1 coat	.14	.91	1.05
0600	Pipes over 12" diameter	.39	2.92	3.31
0700	Structural steel, brushwork, light framing 300-500 S.F./Ton	.05	1.20	1.25
0720	Heavy framing 50-100 S.F./Ton	.05	.60	.65
0740	Spraywork, light framing 300-500 S.F./Ton	.06	.27	.33
0760	Heavy framing 50-100 S.F./Ton	.06	.30	.36
0800	Varnish, interior wood trim, no sanding sealer & 1 coat	.06	.73	.79
0820	Hardwood floor, no sanding 2 coats	.11	.15	.26
0840	Wall coatings, acrylic glazed coatings, minimum	.23	.56	.79
0860	Maximum	.51	.96	1.47
0880	Epoxy coatings, minimum	.31	.56	.87
0900	Maximum	.96	1.72	2.68
0940	Exposed epoxy aggregate, troweled on, 1/16" to 1/4" aggregate, minimum	.48	1.24	1.72
0960	Maximum	1.03	2.24	3.27
0980	1/2" to 5/8" aggregate, minimum	.95	2.24	3.19
1000	Maximum	1.62	3.65	5.27
1020	1" aggregate, minimum	1.65	3.24	4.89
1040	Maximum	2.53	5.30	7.83
1060	Sprayed on, minimum	.45	.99	1.44
1080	Maximum	.83	2.01	2.84
1100	High build epoxy 50 mil, minimum	.52	.75	1.27
1120	Maximum	.90	3.07	3.97
1140	Laminated epoxy with fiberglass minimum	.57	.99	1.56
1160	Maximum	1.02	2.01	3.03
1180	Sprayed perlite or vermiculite 1/16" thick, minimum	.20	.10	.30
1200	Maximum	.58	.46	1.04
1260	Wall coatings, vinyl plastic, minimum	.25	.40	.65
1280	Maximum	.65	1.22	1.87
1300	Urethane on smooth surface, 2 coats, minimum	.19	.26	.45
1320	Maximum	.44	.44	.88
1340	3 coats, minimum	.25	.35	.60
1360	Maximum	.58	.62	1.20
1380	Ceramic-like glazed coating, cementitious, minimum	.39	.66	1.05
1400	Maximum	.64	.85	1.49
1420	Resin base, minimum	.25	.46	.71
1440	Maximum	.43	.88	1.31
1460	Wall coverings, aluminum foil	.81	1.08	1.89
1480	Copper sheets, .025" thick, phenolic backing	5.55	1.23	6.78
1500	Vinyl backing	4.29	1.23	5.52
1520	Cork tiles, 12"x12", light or dark, 3/16" thick	2.75	1.23	3.98
1540	5/16" thick	2.86	1.26	4.12
1560	Basketweave, 1/4" thick	4.40	1.23	5.63

INTERIOR CONSTR. — A6.5 Wall Finishes

6.5-100 Paint & Covering

		COST PER S.F.		
		MAT.	INST.	TOTAL
1580	Natural, non-directional, 1/2" thick	4.32	1.23	5.55
1600	12"x36", granular, 3/16" thick	.95	.77	1.72
1620	1" thick	1.23	.80	2.03
1640	12"x12", polyurethane coated, 3/16" thick	2.97	1.23	4.20
1660	5/16" thick	4.21	1.26	5.47
1661	Paneling, prefinished plywood, birch	1.10	1.62	2.72
1662	Mahogany, African	2.04	1.70	3.74
1663	Philippine (lauan)	.88	1.36	2.24
1664	Oak or cherry	2.86	1.70	4.56
1665	Rosewood	4.07	2.12	6.19
1666	Teak	2.86	1.70	4.56
1667	Chestnut	4.23	1.81	6.04
1668	Pecan	1.82	1.70	3.52
1669	Walnut	4.62	1.70	6.32
1670	Wood board, knotty pine, finished	1.55	2.82	4.37
1671	Rough sawn cedar	1.93	2.82	4.75
1672	Redwood	4.36	2.82	7.18
1673	Aromatic cedar	3.26	3.03	6.29
1680	Cork wallpaper, paper backed, natural	1.68	.62	2.30
1700	Color	2.09	.62	2.71
1720	Gypsum based, fabric backed, minimum	.62	.37	.99
1740	Average	.92	.41	1.33
1760	Maximum	1.03	.46	1.49
1780	Vinyl wall covering, fabric back, light weight	.70	.46	1.16
1800	Medium weight	.88	.62	1.50
1820	Heavy weight	1.20	.68	1.88
1840	Wall paper, double roll, solid pattern, avg. workmanship	.29	.46	.75
1860	Basic pattern, avg. workmanship	.50	.55	1.05
1880	Basic pattern, quality workmanship	.95	.68	1.63
1900	Grass cloths with lining paper, minimum	.61	.74	1.35
1920	Maximum	1.95	.85	2.80
1940	Ceramic tile, thin set, 4-1/4" x 4-1/4"	2.19	2.88	5.07
1941				
1942				

6.5-100 Paint Trim

		COST PER L.F.		
		MAT.	INST.	TOTAL
2040	Painting, wood trim, to 6" wide, enamel, primer & 1 coat	.08	.36	.44
2060	Primer & 2 coats	.13	.46	.59
2080	Misc. metal brushwork, ladders	.29	3.65	3.94
2100	Pipes, to 4" dia.	.05	.77	.82
2120	6" to 8" dia.	.11	1.53	1.64
2140	10" to 12" dia.	.31	2.30	2.61
2160	Railings, 2" pipe	.10	1.82	1.92
2180	Handrail, single	.08	.73	.81

For expanded coverage of these items see *Means Assemblies Cost Data 1997*

INTERIOR CONSTR. — A6.6 Floor Finishes

6.6-100 Tile & Covering

		COST PER S.F.		
		MAT.	INST.	TOTAL
0060	Carpet tile, nylon, fusion bonded, 18" x 18" or 24" x 24", 24 oz.	2.89	.42	3.31
0080	35 oz.	3.39	.42	3.81
0100	42 oz.	4.33	.42	4.75
0120				
0140	Carpet, tufted, nylon, roll goods, 12' wide, 26 oz.	2.66	.68	3.34
0160	36 oz.	4.05	.68	4.73
0180	Woven, wool, 36 oz.	4.50	.69	5.19
0200	42 oz.	5.75	.69	6.44
0220	Padding, add to above, minimum	.38	.23	.61
0240	Maximum	.65	.23	.88
0260	Composition flooring, acrylic, 1/4" thick	1.15	3.25	4.40
0280	3/8" thick	1.54	3.75	5.29
0300	Epoxy, minimum	2.09	2.50	4.59
0320	Maximum	2.53	3.45	5.98
0340	Epoxy terrazzo, minimum	4.29	3.32	7.61
0360	Maximum	7.20	4.44	11.64
0380	Mastic, hot laid, 1-1/2" thick, minimum	3.03	2.45	5.48
0400	Maximum	3.91	3.25	7.16
0420	Neoprene 1/4" thick, minimum	2.97	3.10	6.07
0440	Maximum	4.13	3.93	8.06
0460	Polyacrylate with ground granite 1/4", minimum	2.29	2.30	4.59
0480	Maximum	4.71	3.52	8.23
0500	Polyester with colored quart 2 chips 1/16", minimum	2.09	1.59	3.68
0520	Maximum	3.74	2.50	6.24
0540	Polyurethane with vinyl chips, minimum	5.85	1.59	7.44
0560	Maximum	8.35	1.97	10.32
0600	Concrete topping, granolithic concrete, 1/2" thick	.17	1.59	1.76
0620	1" thick	.33	1.63	1.96
0640	2" thick	.66	1.88	2.54
0650				
0660	Heavy duty 3/4" thick, minimum	.23	2.47	2.70
0680	Maximum	.23	2.94	3.17
0700	For colors, add to above, minimum	.46	.57	1.03
0720	Maximum	2.05	.63	2.68
0740	Exposed aggregate finish, minimum	.40	.47	.87
0760	Maximum	1.25	.64	1.89
0780	Abrasives, .25 P.S.F. add to above, minimum	.10	.40	.50
0800	Maximum	.15	.40	.55
0820	Dust on coloring, add, minimum	.56	.26	.82
0840	Maximum	1.96	.54	2.50
0860	1/2" integral, minimum	.23	2.12	2.35
0880	Maximum	.17	1.59	1.76
0900	Dustproofing, add, minimum	.06	.16	.22
0920	Maximum	.10	.23	.33
0930	Paint	.21	.87	1.08
0940	Hardeners, metallic add, minimum	.23	.40	.63
0960	Maximum	.53	.60	1.13
0980	Non-metallic, minimum	.06	.40	.46
1000	Maximum	.15	.60	.75
1020	Integral topping, 3/16" thick	.12	.94	1.06
1040	1/2" thick	.17	.99	1.16
1060	3/4" thick	.25	1.10	1.35
1080	1" thick	.33	1.25	1.58
1100	Terrazzo, minimum	2.31	5.10	7.41
1120	Maximum	4.28	11.05	15.33
1280	Resilient, asphalt tile, 1/8" thick on concrete, minimum	1.01	.76	1.77
1300	Maximum	1.11	.76	1.87
1320	On wood, add for felt underlay	.20		.20
1340	Cork tile, minimum	2.61	.97	3.58
1360	Maximum	9.90	.97	10.87

Important: See the Reference Section for critical supporting data - Location Factors & Historical Cost Indexes

INTERIOR CONSTR. — A6.6 Floor Finishes

6.6-100 Tile & Covering

		COST PER S.F.		
		MAT.	INST.	TOTAL
1380	Polyethylene, in rolls, minimum	2.37	1.11	3.48
1400	Maximum	4.59	1.11	5.70
1420	Polyurethane, thermoset, minimum	3.63	3.04	6.67
1440	Maximum	4.31	6.10	10.41
1460	Rubber, sheet goods, minimum	3.25	2.53	5.78
1480	Maximum	5.35	3.38	8.73
1500	Tile, minimum	3.27	.76	4.03
1520	Maximum	6.35	1.11	7.46
1580	Vinyl, composition tile, minimum	.74	.61	1.35
1600	Maximum	1.79	.61	2.40
1620	Tile, minimum	1.66	.61	2.27
1640	Maximum	2.61	.61	3.22
1660	Sheet goods, minimum	1.35	1.22	2.57
1680	Maximum	5.15	1.52	6.67
1720	Tile, ceramic natural clay	3.77	2.99	6.76
1730	Marble, synthetic 12"x12"x5/8"	7.40	9.10	16.50
1740	Porcelain type, minimum	4.07	2.99	7.06
1760	Maximum	4.29	2.88	7.17
1800	Quarry tile, mud set, minimum	2.92	3.90	6.82
1820	Maximum	4.86	4.97	9.83
1840	Thin set, deduct		.78	.78
1850				
1860	Terrazzo precast, minimum	6.75	4.05	10.80
1880	Maximum	7.65	4.05	11.70
1900	Non-slip, minimum	15.40	20	35.40
1920	Maximum	16.60	28	44.60
1960	Wood, block, end grain factory type, creosoted, 2" thick	2.80	1.11	3.91
1980	2-1/2" thick	2.90	2.63	5.53
2000	3" thick	2.82	2.63	5.45
2020	Natural finish, 2" thick	2.93	2.63	5.56
2040	Fir, vertical grain, 1"x4", no finish, minimum	2.38	1.29	3.67
2060	Maximum	2.53	1.29	3.82
2080	Prefinished white oak, prime grade, 2-1/4" wide	6.45	1.93	8.38
2100	3-1/4" wide	8	1.78	9.78
2120	Maple strip, sanded and finished, minimum	3.39	2.94	6.33
2140	Maximum	3.97	2.94	6.91
2160	Oak strip, sanded and finished, minimum	4.19	2.94	7.13
2180	Maximum	3.62	2.94	6.56
2200	Parquetry, sanded and finished, minimum	2.24	3.06	5.30
2220	Maximum	5.70	4.30	10
2260	Add for sleepers on concrete, treated, 24" O.C., 1"x2"	1.05	1.74	2.79
2280	1"x3"	1.05	1.38	2.43
2300	2"x4"	.83	.69	1.52
2340	Underlayment, plywood, 3/8" thick	.46	.45	.91
2350	1/2" thick	.57	.47	1.04
2360	5/8" thick	.68	.49	1.17
2370	3/4" thick	.83	.53	1.36
2380	Particle board, 3/8" thick	.34	.45	.79
2390	1/2" thick	.36	.47	.83
2400	5/8" thick	.39	.49	.88
2410	3/4" thick	.45	.53	.98
2420	Hardboard, 4' x 4', .215" thick	.50	.45	.95

For expanded coverage of these items see *Means Assemblies Cost Data 1997*

INTERIOR CONSTR. — A6.7 Ceiling Finishes

Plaster Partitions are defined as follows: type of plaster, type and spacing of framing, type of lath and treatment of opposite face.

Included in the system components are expansion joints. Metal studs are assumed to be nonload bearing.

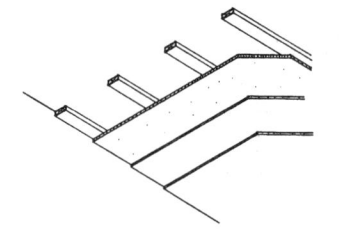

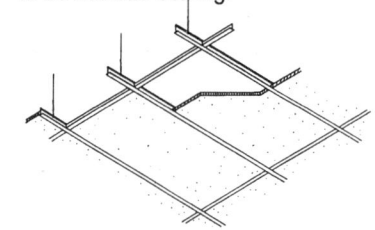

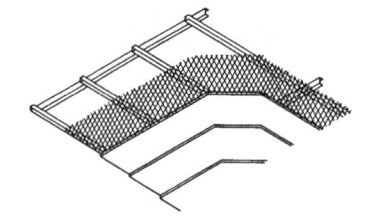

2 Coats of Plaster on Gypsum Lath on Wood Furring

Fiberglass Board on Exposed Suspended Grid System

Plaster and Metal Lath on Metal Furring

6.7-100 Plaster Ceilings

	TYPE	LATH	FURRING	SUPPORT	MAT.	INST.	TOTAL
2400	2 coat gypsum	3/8" gypsum	1"x3" wood, 16" O.C.	wood	1.11	3.33	4.44
2500	Painted			masonry	1.11	3.39	4.50
2600				concrete	1.11	3.80	4.91
2700	3 coat gypsum	3.4# metal	1"x3" wood, 16" O.C.	wood	1.12	3.58	4.70
2800	Painted			masonry	1.12	3.64	4.76
2900				concrete	1.12	4.05	5.17
3000	2 coat perlite	3/8" gypsum	1"x3" wood, 16" O.C.	wood	1.20	3.46	4.66
3100	Painted			masonry	1.20	3.52	4.72
3200				concrete	1.20	3.93	5.13
3300	3 coat perlite	3.4# metal	1"x3" wood, 16" O.C.	wood	1.06	3.55	4.61
3400	Painted			masonry	1.06	3.61	4.67
3500				concrete	1.06	4.02	5.08
3600	2 coat gypsum	3/8" gypsum	3/4" CRC, 12" O.C.	1-1/2" CRC, 48"O.C.	1.07	4.08	5.15
3700	Painted		3/4" CRC, 16" O.C.	1-1/2" CRC, 48"O.C.	1.05	3.68	4.73
3800			3/4" CRC, 24" O.C.	1-1/2" CRC, 48"O.C.	.99	3.35	4.34
3900	2 coat perlite	3/8" gypsum	3/4" CRC, 12" O.C.	1-1/2" CRC, 48"O.C	1.16	4.37	5.53
4000	Painted		3/4" CRC, 16" O.C.	1-1/2" CRC, 48"O.C.	1.14	3.97	5.11
4100			3/4" CRC, 24" O.C.	1-1/2" CRC, 48"O.C.	1.08	3.64	4.72
4200	3 coat gypsum	3.4# metal	3/4" CRC, 12" O.C.	1-1/2" CRC, 36" O.C.	1.34	5.30	6.64
4300	Painted		3/4" CRC, 16" O.C.	1-1/2" CRC, 36" O.C.	1.32	4.91	6.23
4400			3/4" CRC, 24" O.C.	1-1/2" CRC, 36" O.C.	1.26	4.58	5.84
4500	3 coat perlite	3.4# metal	3/4" CRC, 12" O.C.	1-1/2" CRC,36" O.C.	1.56	5.85	7.41
4600	Painted		3/4" CRC, 16" O.C.	1-1/2" CRC, 36" O.C.	1.54	5.45	6.99
4700			3/4" CRC, 24" O.C.	1-1/2" CRC, 36" O.C.	1.48	5.10	6.58

6.7-100 Drywall Ceilings

	TYPE	FINISH	FURRING	SUPPORT	MAT.	INST.	TOTAL
4800	1/2" F.R. drywall	painted and textured	1"x3" wood, 16" O.C.	wood	.76	2.07	2.83
4900				masonry	.76	2.13	2.89
5000				concrete	.76	2.54	3.30
5100	5/8" F.R. drywall	painted and textured	1"x3" wood, 16" O.C.	wood	.76	2.07	2.83
5200				masonry	.76	2.13	2.89
5300				concrete	.76	2.54	3.30
5400	1/2" F.R. drywall	painted and textured	7/8"resil. channels	24" O.C.	.63	2.01	2.64
5500			1"x2" wood	stud clips	.75	1.94	2.69
5602	1/2" F.R. drywall	painted	1-5/8" metal studs	24" O.C.	.69	1.99	2.68

Important: See the Reference Section for critical supporting data - Location Factors & Historical Cost Indexes

INTERIOR CONSTR. — A6.7 Ceiling Finishes

6.7-100 Drywall Ceilings

	TYPE	FINISH	FURRING	SUPPORT	MAT.	INST.	TOTAL
5700 5702	5/8" F.R. drywall	painted and textured	1-5/8" metal studs	24" O.C.	.69	1.99	2.68

6.7-100 Acoustical Ceilings

	TYPE	TILE	GRID	SUPPORT	MAT.	INST.	TOTAL
5800	5/8" fiberglass board	24" x 48"	tee	suspended	.81	.95	1.76
5900		24" x 24"	tee	suspended	.90	1.05	1.95
6000	3/4" fiberglass board	24" x 48"	tee	suspended	.90	.97	1.87
6100		24" x 24"	tee	suspended	.99	1.07	2.06
6500	5/8" mineral fiber	12" x 12"	1"x3" wood, 12" O.C.	wood	.85	1.27	2.12
6600				masonry	.85	1.36	2.21
6700				concrete	.85	1.90	2.75
6800	3/4" mineral fiber	12" x 12"	1"x3" wood, 12" O.C.	wood	1.12	1.27	2.39
6900				masonry	1.12	1.27	2.39
7000				concrete	1.12	1.27	2.39
7100 7102	3/4" mineral fiber on 5/8" F.R. drywall	12" x 12"	25 ga. channels	runners	1.28	1.74	3.02
7200 7201 7202	5/8" plastic coated Mineral fiber	12" x 12"		adhesive backed	.85	.33	1.18
7300 7301 7302	3/4" plastic coated Mineral fiber	12" x 12"		adhesive backed	1.12	.33	1.45
7400 7401 7402	3/4" mineral fiber	12" x 12"	conceal 2" bar & channels	suspended	1.39	1.67	3.06

6.7-810 Acoustical Ceilings

		MAT.	INST.	TOTAL
2480	Ceiling boards, eggcrate, acrylic, 1/2" x 1/2" x 1/2" cubes	1.32	.66	1.98
2500	Polystyrene, 3/8" x 3/8" x 1/2" cubes	1.10	.64	1.74
2520	1/2" x 1/2" x 1/2" cubes	1.49	.66	2.15
2540	Fiberglass boards, plain, 5/8" thick	.48	.53	1.01
2560	3/4" thick	.57	.55	1.12
2580	Grass cloth faced, 3/4" thick	1.60	.66	2.26
2600	1" thick	1.81	.68	2.49
2620	Luminous panels, prismatic, acrylic	1.60	.82	2.42
2640	Polystyrene	.81	.82	1.63
2660	Flat or ribbed, acrylic	2.78	.82	3.60
2680	Polystyrene	1.89	.82	2.71
2700	Drop pan, white, acrylic	4.07	.82	4.89
2720	Polystyrene	3.41	.82	4.23
2740	Mineral fiber boards, 5/8" thick, standard	.56	.49	1.05
2760	Plastic faced	.88	.82	1.70
2780	2 hour rating	.78	.49	1.27
2800	Perforated aluminum sheets, .024 thick, corrugated painted	1.63	.67	2.30
2820	Plain	1.46	.66	2.12
3080	Mineral fiber, plastic coated, 12" x 12" or 12" x 24", 5/8" thick	.54	.33	.87
3100	3/4" thick	.81	.33	1.14
3120	Fire rated, 3/4" thick, plain faced	.84	.33	1.17
3140	Mylar faced	.91	.33	1.24
3160	Add for ceiling primer	.12		.12
3180	Add for ceiling cement	.31		.31
3240	Suspension system, furring, 1" x 3" wood 12" O.C.	.31	.94	1.25
3260	T bar suspension system, 2' x 4' grid	.32	.41	.73

For expanded coverage of these items see *Means Assemblies Cost Data 1997*

INTERIOR CONSTR.   A6.7   Ceiling Finishes

6.7-810 Acoustical Ceilings

		COST PER S.F.		
		MAT.	INST.	TOTAL
3280	2' x 2' grid	.41	.51	.92
3300	Concealed Z bar suspension system 12" module	.36	.63	.99
3320	Add to above for 1-1/2" carrier channels 4' O.C.	.18	.70	.88
3340	Add to above for carrier channels for recessed lighting	.35	.71	1.06

6.7-820 Plaster Ceilings

		COST PER S.F.		
		MAT.	INST.	TOTAL
0060	Plaster, gypsum incl. finish, 2 coats			
0080	3 coats	.53	2.08	2.61
0100	Perlite, incl. finish, 2 coats	.47	2.05	2.52
0120	3 coats	.75	2.58	3.33
0140	Thin coat on drywall	.13	.40	.53
0200	Lath, gypsum, 3/8" thick	.42	.56	.98
0220	1/2" thick	.44	.59	1.03
0240	5/8" thick	.49	.68	1.17
0260	Metal, diamond, 2.5 lb.	.21	.45	.66
0280	3.4 lb.	.28	.49	.77
0300	Flat rib, 2.75 lb.	.20	.45	.65
0320	3.4 lb.	.38	.49	.87
0440	Furring, steel channels, 3/4" galvanized, 12" O.C.	.19	1.46	1.65
0460	16" O.C.	.17	1.06	1.23
0480	24" O.C.	.11	.73	.84
0500	1-1/2" galvanized, 12" O.C.	.27	1.61	1.88
0520	16" O.C.	.24	1.18	1.42
0540	24" O.C.	.16	.79	.95
0560	Wood strips, 1"x3", on wood, 12" O.C.	.31	.94	1.25
0580	16" O.C.	.23	.71	.94
0600	24" O.C.	.16	.47	.63
0620	On masonry, 12" O.C.	.31	1.03	1.34
0640	16" O.C.	.23	.77	1
0660	24" O.C.	.16	.52	.68
0680	On concrete, 12" O.C.	.31	1.57	1.88
0700	16" O.C.	.23	1.18	1.41
0720	24" O.C.	.16	.79	.95
0722				
0740				
0940	Paint on plaster or drywall, roller work, primer + 1 coat	.08	.30	.38
0960	Primer + 2 coats	.12	.38	.50

For information about Means Estimating Seminars, see yellow pages 11 and 12 in back of book

Division 7
Conveying Systems

CONVEYING — A7.1 Elevators

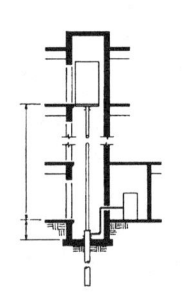

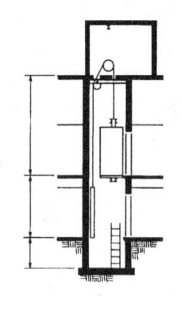

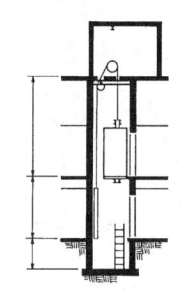

Hydraulic

Traction Geared

Traction Gearless

7.1-100	Hydraulic	MAT.	INST.	TOTAL
1300	Pass. elev., 1500 lb., 2 Floors, 100 FPM	33,100	7,700	40,800
1400	5 Floors, 100 FPM	44,700	20,500	65,200
1600	2000 lb., 2 Floors, 100 FPM	34,000	7,700	41,700
1700	5 floors, 100 FPM	45,500	20,500	66,000
1900	2500 lb., 2 Floors, 100 FPM	34,400	7,700	42,100
2000	5 floors, 100 FPM	46,000	20,500	66,500
2200	3000 lb., 2 Floors, 100 FPM	35,900	7,700	43,600
2300	5 floors, 100 FPM	47,500	20,500	68,000
2500	3500 lb., 2 Floors, 100 FPM	38,000	7,700	45,700
2600	5 floors, 100 FPM	49,600	20,500	70,100
2800	4000 lb., 2 Floors, 100 FPM	39,000	7,700	46,700
2900	5 floors, 100 FPM	50,500	20,500	71,000
3100	4500 lb., 2 Floors, 100 FPM	39,900	7,700	47,600
3200	5 floors, 100 FPM	51,500	20,500	72,000
4000	Hospital elevators, 3500 lb., 2 Floors, 100 FPM	46,100	7,700	53,800
4100	5 floors, 100 FPM	69,500	20,500	90,000
4300	4000 lb., 2 Floors, 100 FPM	46,100	7,700	53,800
4400	5 floors, 100 FPM	69,500	20,500	90,000
4600	4500 lb., 2 Floors, 100 FPM	51,500	7,700	59,200
4800	5 floors, 100 FPM	75,000	20,500	95,500
4900	5000 lb., 2 Floors, 100 FPM	54,000	7,700	61,700
5000	5 floors, 100 FPM	77,500	20,500	98,000
6700	Freight elevators (Class "B"), 3000 lb., 2 Floors, 50 FPM	42,600	9,800	52,400
6800	5 floors, 100 FPM	60,500	25,800	86,300
7000	4000 lb., 2 Floors, 50 FPM	45,500	9,800	55,300
7100	5 floors, 100 FPM	63,000	25,800	88,800
7500	10,000 lb., 2 Floors, 50 FPM	56,500	9,800	66,300
7600	5 floors, 100 FPM	74,000	25,800	99,800
8100	20,000 lb., 2 Floors, 50 FPM	68,000	9,800	77,800
8200	5 Floors, 100 FPM	86,000	25,800	111,800

7.1-200	Traction Geared Elevators	MAT.	INST.	TOTAL
1300	Passenger, 2000 Lb., 5 floors, 200 FPM	69,500	19,300	88,800
1500	15 floors, 350 FPM	112,500	58,500	171,000
1600	2500 Lb., 5 floors, 200 FPM	72,500	19,300	91,800
1800	15 floors, 350 FPM	115,500	58,500	174,000
1900	3000 Lb., 5 floors, 200 FPM	73,500	19,300	92,800
2100	15 floors, 350 FPM	116,500	58,500	175,000
2200	3500 Lb., 5 floors, 200 FPM	75,500	19,300	94,800
2400	15 floors, 350 FPM	118,000	58,500	176,500
2500	4000 Lb., 5 floors, 200 FPM	75,500	19,300	94,800
2700	15 floors, 350 FPM	118,000	58,500	176,500
2800	4500 Lb., 5 floors, 200 FPM	77,500	19,300	96,800
3000	15 floors, 350 FPM	120,000	58,500	178,500

Important: See the Reference Section for critical supporting data - Location Factors & Historical Cost Indexes

CONVEYING — A7.1 Elevators

7.1-200 Traction Geared Elevators

		MAT.	INST.	TOTAL
3100	5000 Lb., 5 floors, 200 FPM	79,000	19,300	98,300
3300	15 floors, 350 FPM	122,000	58,500	180,500
4000	Hospital, 3500 Lb., 5 floors, 200 FPM	85,500	19,300	104,800
4200	15 floors, 350 FPM	148,000	58,500	206,500
4300	4000 Lb., 5 floors, 200 FPM	85,500	19,300	104,800
4500	15 floors, 350 FPM	148,000	58,500	206,500
4600	4500 Lb., 5 floors, 200 FPM	90,000	19,300	109,300
4800	15 floors, 350 FPM	152,000	58,500	210,500
4900	5000 Lb., 5 floors, 200 FPM	91,000	19,300	110,300
5100	15 floors, 350 FPM	153,500	58,500	212,000
6000	Freight, 4000 Lb., 5 floors, 50 FPM class 'B'	71,000	20,000	91,000
6200	15 floors, 200 FPM class 'B'	124,000	69,500	193,500
6300	8000 Lb., 5 floors, 50 FPM class 'B'	85,500	20,000	105,500
6500	15 floors, 200 FPM class 'B'	138,000	69,500	207,500
7000	10,000 Lb., 5 floors, 50 FPM class 'B'	101,500	20,000	121,500
7200	15 floors, 200 FPM class 'B'	288,500	69,500	358,000
8000	20,000 Lb., 5 floors, 50 FPM class 'B'	111,500	20,000	131,500
8200	15 floors, 200 FPM class 'B'	298,500	69,500	368,000

7.1-300 Traction Gearless Elevators

		MAT.	INST.	TOTAL
1700	Passenger, 2500 Lb., 10 floors, 200 FPM	144,500	50,500	195,000
1900	30 floors, 600 FPM	255,000	128,500	383,500
2000	3000 Lb., 10 floors, 200 FPM	145,500	50,500	196,000
2200	30 floors, 600 FPM	256,000	128,500	384,500
2300	3500 Lb., 10 floors, 200 FPM	147,000	50,500	197,500
2500	30 floors, 600 FPM	257,500	128,500	386,000
2700	50 floors, 800 FPM	340,000	206,500	546,500
2800	4000 lb., 10 floors, 200 FPM	147,000	50,500	197,500
3000	30 floors, 600 FPM	257,500	128,500	386,000
3200	50 floors, 800 FPM	340,000	206,500	546,500
3300	4500 lb., 10 floors, 200 FPM	149,000	50,500	199,500
3500	30 floors, 600 FPM	259,500	128,500	388,000
3700	50 floors, 800 FPM	342,000	206,500	548,500
3800	5000 lb., 10 floors, 200 FPM	151,000	50,500	201,500
4000	30 floors, 600 FPM	261,500	128,500	390,000
4200	50 floors, 800 FPM	344,000	206,500	550,500
6000	Hospital, 3500 Lb., 10 floors, 200 FPM	167,000	50,500	217,500
6200	30 floors, 600 FPM	316,500	128,500	445,000
6400	4000 Lb., 10 floors, 200 FPM	167,000	50,500	217,500
6600	30 floors, 600 FPM	316,500	128,500	445,000
6800	4500 Lb., 10 floors, 200 FPM	171,500	50,500	222,000
7000	30 floors, 600 FPM	320,500	128,500	449,000
7200	5000 Lb., 10 floors, 200 FPM	172,500	50,500	223,000
7400	30 floors, 600 FPM	322,000	128,500	450,500

For information about Means Estimating Seminars, see yellow pages 11 and 12 in back of book

For expanded coverage of these items see *Means Assemblies Cost Data 1997*

Division 8
Mechanical

MECHANICAL — A8.1 Plumbing

8.1-040 Piping - Installed - Unit Costs

		COST PER L.F.		
		MAT.	INST.	TOTAL
0840	Cast iron, soil, B & S, service weight, 2" diameter	4.93	10.70	15.63
0860	3" diameter	6.80	11.20	18
0880	4" diameter	8.80	12.25	21.05
0900	5" diameter	11.70	13.80	25.50
0920	6" diameter	14.25	14.35	28.60
0940	8" diameter	22.50	24	46.50
0960	10" diameter	36	26.50	62.50
0980	12" diameter	50.50	29.50	80
1040	No hub, 1-1/2" diameter	5.80	9.50	15.30
1060	2" diameter	5.80	10.05	15.85
1080	3" diameter	7.70	10.50	18.20
1100	4" diameter	9.75	11.60	21.35
1120	5" diameter	12.70	12.60	25.30
1140	6" diameter	17.10	13.25	30.35
1160	8" diameter	27	20.50	47.50
1180	10" diameter	44	23.50	67.50
1220	Copper tubing, hard temper, solder, type K, 1/2" diameter	1.91	4.79	6.70
1260	3/4" diameter	3.12	5.05	8.17
1280	1" diameter	3.94	5.65	9.59
1300	1-1/4" diameter	4.85	6.70	11.55
1320	1-1/2" diameter	6.35	7.50	13.85
1340	2" diameter	9.85	9.35	19.20
1360	2-1/2" diameter	14.35	11.20	25.55
1380	3" diameter	19.85	12.45	32.30
1400	4" diameter	30	17.70	47.70
1480	5" diameter	55.50	21	76.50
1500	6" diameter	113	27.50	140.50
1520	8" diameter	141	31	172
1560	Type L, 1/2" diameter	1.65	4.62	6.27
1600	3/4" diameter	2.27	4.92	7.19
1620	1" diameter	3.06	5.50	8.56
1640	1-1/4" diameter	4.18	6.45	10.63
1660	1-1/2" diameter	5.20	7.20	12.40
1680	2" diameter	8.25	8.90	17.15
1700	2-1/2" diameter	11.70	10.85	22.55
1720	3" diameter	15.80	12	27.80
1740	4" diameter	26.50	17.25	43.75
1760	5" diameter	50.50	19.80	70.30
1780	6" diameter	63.50	26	89.50
1800	8" diameter	109	29	138
1840	Type M, 1/2" diameter	1.32	4.45	5.77
1880	3/4" diameter	1.84	4.79	6.63
1900	1" diameter	2.48	5.35	7.83
1920	1-1/4" diameter	3.48	6.25	9.73
1940	1-1/2" diameter	4.63	6.95	11.58
1960	2" diameter	7.35	8.50	15.85
1980	2-1/2" diameter	10.30	10.50	20.80
2000	3" diameter	13.85	11.60	25.45
2020	4" diameter	22.50	16.85	39.35
2060	6" diameter	62	25	87
2080	8" diameter	102	27.50	129.50
2120	Type DWV, 1-1/4" diameter	3.41	6.25	9.66
2160	1-1/2" diameter	4.15	6.95	11.10
2180	2" diameter	5.65	8.50	14.15
2200	3" diameter	9.15	11.60	20.75
2220	4" diameter	16.25	16.85	33.10
2240	5" diameter	35	18.70	53.70
2260	6" diameter	49.50	25	74.50

Important: See the Reference Section for critical supporting data - Location Factors & Historical Cost Indexes

MECHANICAL | A8.1 | Plumbing

8.1-040 Piping - Installed - Unit Costs

		COST PER L.F.		
		MAT.	INST.	TOTAL
2280	8" diameter	113	27.50	140.50
2800	Plastic, PVC, DWV, schedule 40, 1-1/4" diameter	2.19	8.90	11.09
2820	1-1/2" diameter	2.22	10.40	12.62
2830	2" diameter	2.46	11.40	13.86
2840	3" diameter	3.92	12.70	16.62
2850	4" diameter	5.05	14.05	19.10
2890	6" diameter	8.70	17.25	25.95
3010	Pressure pipe 200 PSI, 1/2" diameter	1.77	6.95	8.72
3030	3/4" diameter	1.83	7.35	9.18
3040	1" diameter	2.02	8.15	10.17
3050	1-1/4" diameter	2.23	8.90	11.13
3060	1-1/2" diameter	2.35	10.40	12.75
3070	2" diameter	2.63	11.40	14.03
3080	2-1/2" diameter	3.92	12	15.92
3090	3" diameter	4.40	12.70	17.10
3100	4" diameter	5.80	14.05	19.85
3110	6" diameter	10.25	17.25	27.50
3120	8" diameter	17.25	22	39.25
4000	Steel, schedule 40, threaded, black, 1/2" diameter	1.65	5.95	7.60
4020	3/4" diameter	1.89	6.15	8.04
4030	1" diameter	2.60	7.05	9.65
4040	1-1/4" diameter	3.15	7.55	10.70
4050	1-1/2" diameter	3.58	8.40	11.98
4060	2" diameter	4.73	10.50	15.23
4070	2-1/2" diameter	7.40	13.45	20.85
4080	3" diameter	9.55	15.65	25.20
4090	4" diameter	13.95	18.70	32.65
4100	5" diameter	18.90	26	44.90
4110	6" diameter	24.50	34	58.50
4120	8" diameter	38.50	39	77.50
4130	10" diameter	69.50	45.50	115
4140	12" diameter	86.50	58	144.50
4200	Galvanized, 1/2" diameter	1.91	5.95	7.86
4220	3/4" diameter	2.20	6.15	8.35
4230	1" diameter	3.35	7.05	10.40
4240	1-1/4" diameter	3.71	7.55	11.26
4250	1-1/2" diameter	4.27	8.40	12.67
4260	2" diameter	5.60	10.50	16.10
4270	2-1/2" diameter	8.30	13.45	21.75
4280	3" diameter	10.55	15.65	26.20
4290	4" diameter	14.85	18.70	33.55
4300	5" diameter	24	26	50
4310	6" diameter	31.50	34	65.50
4320	8" diameter	45.50	39	84.50
4330	10" diameter	67.50	45.50	113
4340	12" diameter	89.50	58	147.50
5010	Flanged, black, 1" diameter	7.55	10.40	17.95
5020	1-1/4" diameter	8.35	11.35	19.70
5030	1-1/2" diameter	8.70	12.55	21.25
5040	2" diameter	10.05	16.20	26.25
5050	2-1/2" diameter	12.20	20.50	32.70
5060	3" diameter	13.85	22.50	36.35
5070	4" diameter	17.85	28	45.85
5080	5" diameter	28	34.50	62.50
5090	6" diameter	32.50	44	76.50
5100	8" diameter	49.50	58	107.50
5110	10" diameter	96	69	165
5120	12" diameter	123	79	202
5720	Grooved joints, black, 3/4" diameter	2.29	5.25	7.54
5730	1" diameter	2.55	5.95	8.50

For expanded coverage of these items see *Means Mechanical or Plumbing Cost Data 1997*

MECHANICAL — A8.1 Plumbing

8.1-040 Piping - Installed - Unit Costs

		COST PER L.F.		
		MAT.	INST.	TOTAL
5740	1-1/4" diameter	3.18	6.45	9.63
5750	1-1/2" diameter	3.60	7.35	10.95
5760	2" diameter	4.44	9.35	13.79
5770	2-1/2" diameter	6.30	11.80	18.10
5900	3" diameter	7.80	13.45	21.25
5910	4" diameter	10.55	14.95	25.50
5920	5" diameter	16.45	18.20	34.65
5930	6" diameter	21	25	46
5940	8" diameter	31.50	28.50	60
5950	10" diameter	59.50	34	93.50
5960	12" diameter	71	39	110

Important: See the Reference Section for critical supporting data - Location Factors & Historical Cost Indexes

MECHANICAL — A8.1 Plumbing

Installation includes piping and fittings within 10' of heater. Gas heaters require vent piping (not included with these units).

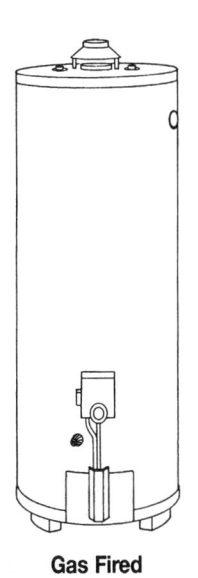

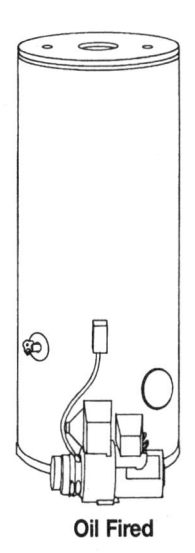

Gas Fired

Oil Fired

8.1-120 Gas Fired Water Heaters - Residential Systems

Line	Description	MAT.	INST.	TOTAL
2200	Gas fired water heater, residential, 100° F rise			
2220	20 gallon tank, 25 GPH	520	760	1,280
2260	30 gallon tank, 32 GPH	525	770	1,295
2300	40 gallon tank, 32 GPH	610	865	1,475
2340	50 gallon tank, 63 GPH	695	865	1,560
2380	75 gallon tank, 63 GPH	1,050	970	2,020
2420	100 gallon tank, 63 GPH	1,500	1,025	2,525
2422				

8.1-130 Oil Fired Water Heaters - Residential Systems

Line	Description	MAT.	INST.	TOTAL
2200	Oil fired water heater, residential, 100° F rise			
2220	30 gallon tank, 103 GPH	985	715	1,700
2260	50 gallon tank, 145 GPH	1,325	800	2,125
2300	70 gallon tank, 164 GPH	1,600	905	2,505
2340	85 gallon tank, 181 GPH	2,175	925	3,100

For expanded coverage of these items see *Means Mechanical or Plumbing Cost Data 1997*

MECHANICAL | A8.1 | Plumbing

Systems below include piping and fittings within 10' of heater. Electric water heaters do not require venting. Gas fired heaters require vent piping (not included in these prices).

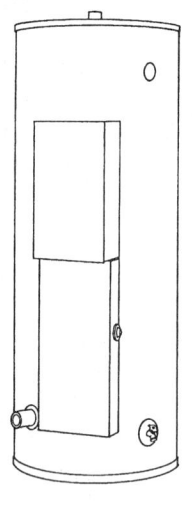

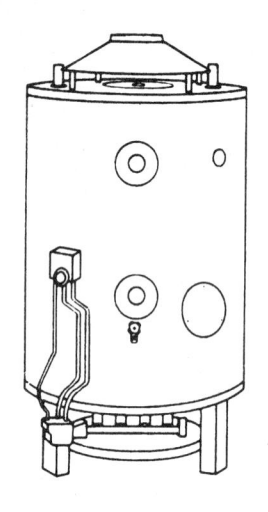

Electric

Gas Fired

8.1-160	Electric Water Heaters - Commercial Systems	MAT.	INST.	TOTAL
1800	Electric water heater, commercial, 100° F rise			
1820	50 gallon tank, 9 KW 37 GPH	2,925	645	3,570
1860	80 gal, 12 KW 49 GPH	3,600	800	4,400
1900	36 KW 147 GPH	4,900	865	5,765
1940	120 gal, 36 KW 147 GPH	5,200	930	6,130
1980	150 gal, 120 KW 490 GPH	18,300	1,000	19,300
2020	200 gal, 120 KW 490 GPH	19,200	1,025	20,225
2060	250 gal, 150 KW 615 GPH	21,400	1,200	22,600
2100	300 gal, 180 KW 738 GPH	31,300	1,275	32,575
2140	350 gal, 30 KW 123 GPH	16,300	1,350	17,650
2180	180 KW 738 GPH	23,900	1,350	25,250
2220	500 gal, 30 KW 123 GPH	21,100	1,600	22,700
2260	240 KW 984 GPH	31,500	1,600	33,100
2300	700 gal, 30 KW 123 GPH	25,500	1,825	27,325
2340	300 KW 1230 GPH	40,600	1,825	42,425
2380	1000 gal, 60 KW 245 GPH	31,000	2,550	33,550
2420	480 KW 1970 GPH	52,000	2,575	54,575
2460	1500 gal, 60 KW 245 GPH	45,800	3,175	48,975
2500	480 KW 1970 GPH	65,000	3,175	68,175

8.1-170	Gas Fired Water Heaters - Commercial Systems	MAT.	INST.	TOTAL
1760	Gas fired water heater, commercial, 100° F rise			
1780	75.5 MBH input, 63 GPH	1,575	995	2,570
1820	95 MBH input, 86 GPH	2,450	995	3,445
1860	100 MBH input, 91 GPH	2,650	1,050	3,700
1900	115 MBH input, 110 GPH	2,050	1,075	3,125
1980	155 MBH input, 150 GPH	2,375	1,200	3,575
2020	175 MBH input, 168 GPH	2,625	1,300	3,925
2060	200 MBH input, 192 GPH	3,075	1,475	4,550
2100	240 MBH input, 230 GPH	3,650	1,600	5,250
2140	300 MBH input, 278 GPH	4,775	1,825	6,600
2180	390 MBH input, 374 GPH	5,400	1,825	7,225
2220	500 MBH input, 480 GPH	7,100	1,975	9,075

Important: See the Reference Section for critical supporting data - Location Factors & Historical Cost Indexes

MECHANICAL — A8.1 Plumbing

8.1-170 Gas Fired Water Heaters - Commercial Systems

		COST EACH		
		MAT.	INST.	TOTAL
2260	600 MBH input, 576 GPH	8,050	2,125	10,175
2300	800 MBH input, 768 GPH	9,425	2,350	11,775
2340	1000 MBH input, 960 GPH	11,200	2,425	13,625
2380	1200 MBH input, 1150 GPH	13,900	3,000	16,900
2420	1500 MBH input, 1440 GPH	16,900	3,075	19,975
2460	1800 MBH input, 1730 GPH	18,900	3,425	22,325
2500	2450 MBH input, 2350 GPH	24,800	3,975	28,775
2540	3000 MBH input, 2880 GPH	29,900	4,850	34,750
2580	3750 MBH input, 3600 GPH	37,400	4,975	42,375

For expanded coverage of these items see *Means Mechanical or Plumbing Cost Data 1997*

MECHANICAL — A8.1 Plumbing

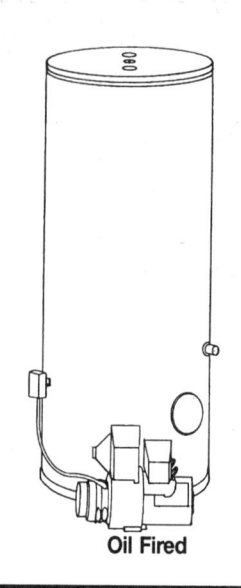

Oil fired water heater systems include piping and fittings within 10' of heater. Oil fired heaters require vent piping (not included in these systems).

Oil Fired

8.1-180	Oil Fired Water Heaters - Commercial Systems	MAT.	INST.	TOTAL
1800	Oil fired water heater, commercial, 100° F rise			
1820	103 MBH output, 116 GPH	2,000	875	2,875
1900	134 MBH output, 161 GPH	2,300	1,125	3,425
1940	161 MBH output, 192 GPH	2,750	1,275	4,025
1980	187 MBH output, 224 GPH	2,975	1,475	4,450
2060	262 MBH output, 315 GPH	3,725	1,700	5,425
2100	341 MBH output, 409 GPH	4,925	1,800	6,725
2140	420 MBH output, 504 GPH	5,575	1,975	7,550
2180	525 MBH output, 630 GPH	7,075	2,225	9,300
2220	630 MBH output, 756 GPH	7,825	2,225	10,050
2260	735 MBH output, 880 GPH	9,625	2,550	12,175
2300	840 MBH output, 1000 GPH	10,900	2,550	13,450
2340	1050 MBH output, 1260 GPH	13,000	2,650	15,650
2380	1365 MBH output, 1640 GPH	16,600	3,000	19,600
2420	1680 MBH output, 2000 GPH	19,700	3,775	23,475
2460	2310 MBH output, 2780 GPH	25,000	4,650	29,650
2500	2835 MBH output, 3400 GPH	30,800	4,650	35,450
2540	3150 MBH output, 3780 GPH	34,400	6,375	40,775

COST EACH

Important: See the Reference Section for critical supporting data - Location Factors & Historical Cost Indexes

MECHANICAL — A8.1 Plumbing

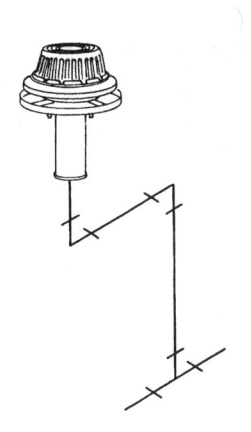

Design Assumptions: Vertical conductor size is based on a maximum rate of rainfall of 4" per hour. To convert roof area to other rates multiply "Max. S.F. Roof Area" shown by four and divide the result by desired local rate. The answer is the local roof area that may be handled by the indicated pipe diameter.

Basic cost is for roof drain, 10' of vertical leader and 10' of horizontal, plus connection to the main.

Pipe Dia.	Max. S.F. Roof Area	Gallons per Min.
2"	544	23
3"	1610	67
4"	3460	144
5"	6280	261
6"	10,200	424
8"	22,000	913

8.1-310 Roof Drain Systems

Line	Description	MAT.	INST.	TOTAL
1880	Roof drain, DWV PVC, 2" diam., piping, 10' high	110	390	500
1920	For each additional foot add	2.46	11.40	13.86
1960	3" diam., 10' high	150	470	620
2000	For each additional foot add	3.92	12.70	16.62
2040	4" diam., 10' high	186	530	716
2080	For each additional foot add	5.05	14.05	19.10
2120	5" diam., 10' high	380	570	950
2160	For each additional foot add	7.80	15.65	23.45
2200	6" diam., 10' high	415	675	1,090
2240	For each additional foot add	8.70	17.25	25.95
2280	8" diam., 10' high	900	1,075	1,975
2320	For each additional foot add	14.40	22	36.40
3940	C.I., soil, single hub, service wt., 2" diam. piping, 10' high	194	410	604
3980	For each additional foot add	4.93	10.70	15.63
4120	3" diam., 10' high	252	445	697
4160	For each additional foot add	6.80	11.20	18
4200	4" diam., 10' high	310	485	795
4240	For each additional foot add	8.80	12.25	21.05
4280	5" diam., 10' high	455	535	990
4320	For each additional foot add	11.70	13.80	25.50
4360	6" diam., 10' high	530	575	1,105
4400	For each additional foot add	14.25	14.35	28.60
4440	8" diam., 10' high	1,100	1,175	2,275
4480	For each additional foot add	22.50	24	46.50
6040	Steel galv. sch 40 threaded, 2" diam. piping, 10' high	233	400	633
6080	For each additional foot add	5.60	10.50	16.10
6120	3" diam., 10' high	435	570	1,005
6160	For each additional foot add	10.55	15.65	26.20
6200	4" diam., 10' high	630	735	1,365
6240	For each additional foot add	14.85	18.70	33.55
6280	5" diam., 10' high	880	935	1,815
6320	For each additional foot add	24	26	50
6360	6" diam, 10' high	1,325	1,150	2,475
6400	For each additional foot add	31.50	34	65.50
6440	8" diam., 10' high	2,425	1,550	3,975
6480	For each additional foot add	45.50	39	84.50

For expanded coverage of these items see *Means Mechanical or Plumbing Cost Data 1997*

MECHANICAL — A8.1 Plumbing

Reference—Minimum Plumbing Fixture Requirements

Type of Building/Use	Water Closets Persons	Water Closets Fixtures	Urinals Persons	Urinals Fixtures	Lavatories Persons	Lavatories Fixtures	Bathtubs or Showers Persons	Bathtubs or Showers Fixtures	Drinking Fountain Fixtures	Other Fixtures
Assembly Halls Auditoriums, Theater, Public assembly	1-100; 101-200; 201-400; Over 400 add 1 fixt. for ea. 500 men; 1 fixt. for ea. 300 women	1; 2; 3	1-200; 201-400; 401-600; Over 600 add 1 fixture for each 300 men	1; 2; 3	1-200; 201-400; 401-750; Over 750 add 1 fixture for each 500 persons	1; 2; 3			1 for each 1000 persons	1 service sink
Assembly Public Worship	300 men; 150 women	1; 1	300 men	1	men; women	1; 1			1	
Dormitories	Men: 1 for each 10 persons; Women: 1 for each 8 persons		1 for each 25 men; over 150 add 1 fixture for each 50 men		1 for ea. 12 persons; 1 separate dental lav. for each 50 persons recom.		1 for ea. 8 persons; For women add 1 additional for each 30. Over 150 persons add 1 for each 20.		1 for each 75 persons	Laundry trays 1 for each 50 serv. sink 1 for ea. 100
Dwellings Apartments and homes	1 fixture for each unit				1 fixture for each unit		1 fixture for each unit			
Hospitals Indiv. Room, Ward, Waiting room	8 persons	1; 1; 1			10 persons	1; 1; 1	20 persons	1; 1	1 for 100 patients	1 service sink per floor
Industrial Mfg. plants, Warehouses	1-10; 11-25; 26-50; 51-75; 76-100; 1 fixture for each additional 30 persons	1; 2; 3; 4; 5	0-30; 31-80; 81-160; 161-240	1; 2; 3; 4	1-100; over 100	1 for ea. 10; 1 for ea. 15	1 Shower for each 15 persons subject to excessive heat or occupational hazard		1 for each 75 persons	
Public Buildings Businesses, Offices	1-15; 16-35; 36-55; 56-80; 81-110; 111-150; 1 fixture for ea. additional 40 persons	1; 2; 3; 4; 5; 6	Urinals may be provided in place of water closets but may not replace more than 1/3 required number of men's water closets		1-15; 16-35; 36-60; 61-90; 91-125; 1 fixture for ea. additional 45 persons	1; 2; 3; 4; 5			1 for each 75 persons	1 service sink per floor.
Schools Elementary	1 for ea. 30 boys; 1 for ea. 25 girls		1 for ea. 25 boys		1 for ea. 35 boys; 1 for ea. 35 girls		For gym or pool shower room 1/5 of a class		1 for each 40 pupils	
Schools Secondary	1 for ea. 40 boys; 1 for ea. 30 girls		1 for ea. 25 boys		1 for ea. 40 boys; 1 for ea. 40 girls		For gym or pool shower room 1/5 of a class		1 for each 50 pupils	

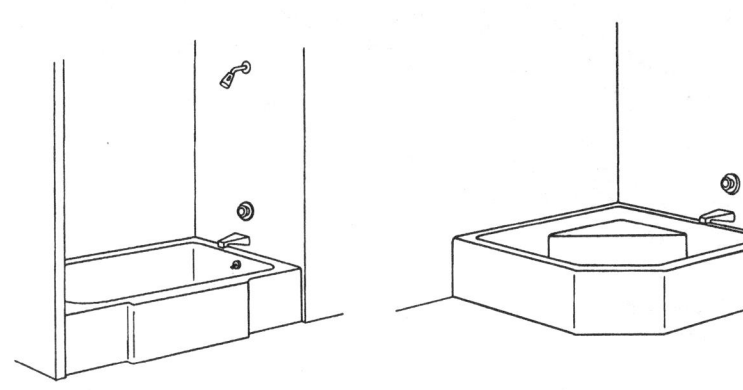

MECHANICAL — A8.1 Plumbing

Recessed Bathtub

Corner Bathtub

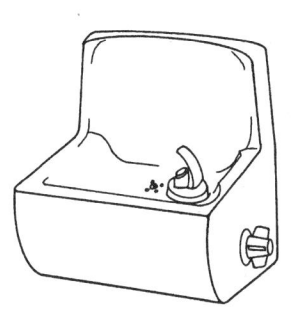

Wall Mounted, Low Back

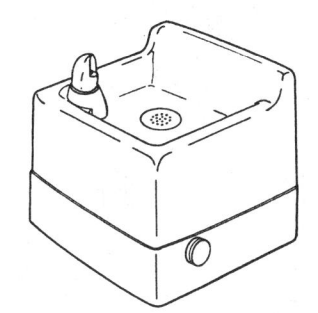

Wall Mounted, No Back

Systems are complete with trim and rough-in (supply, waste and vent) to connect to supply branches and waste mains.

8.1-410	Bathtub Systems	MAT.	INST.	TOTAL
1960	Bathtub, recessed, P.E. on Cl., 42" x 37"	975	460	1,435
2000	48" x 42"	1,400	495	1,895
2040	72" x 36"	1,475	550	2,025
2080	Mat bottom, 5' long	570	480	1,050
2120	5'-6" long	1,175	495	1,670
2160	Corner, 48" x 42"	1,575	480	2,055
2200	Formed steel, enameled, 4'-6" long	445	440	885
2240	5' long	425	450	875

8.1-420	Drinking Fountain Systems	MAT.	INST.	TOTAL
1740	Drinking fountain, one bubbler, wall mounted			
1760	Non recessed			
1800	Bronze, no back	975	262	1,237
1840	Cast iron, enameled, low back	465	262	727
1880	Fiberglass, 12" back	535	262	797
1920	Stainless steel, no back	860	262	1,122
1960	Semi-recessed, poly marble	635	262	897
2040	Stainless steel	670	262	932
2080	Vitreous china	490	262	752
2120	Full recessed, poly marble	710	262	972
2200	Stainless steel	615	262	877
2240	Floor mounted, pedestal type, aluminum	455	355	810
2320	Bronze	1,075	355	1,430
2360	Stainless steel	1,075	355	1,430

For expanded coverage of these items see *Means Mechanical or Plumbing Cost Data 1997*

MECHANICAL — A8.1 Plumbing

Systems are complete with trim and rough-in (supply, waste and vent) to connect to supply branches and waste mains.

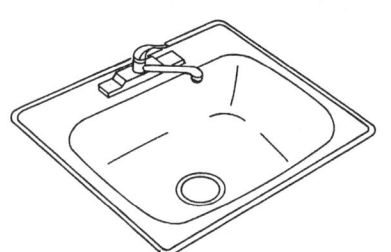

Counter Top Single Bowl

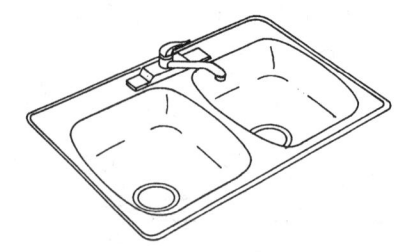

Counter Top Double Bowl

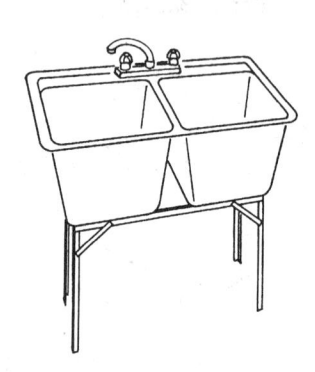

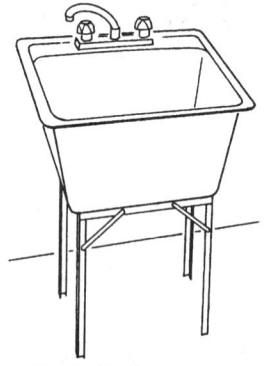

Single Compartment Sink

Double Compartment Sink

8.1-431 Kitchen Sink Systems

		MAT.	INST.	TOTAL
1720	Kitchen sink w/trim, countertop, PE on CI, 24"x21", single bowl	340	435	775
1760	30" x 21" single bowl	415	435	850
1800	32" x 21" double bowl	460	470	930
1840	42" x 21" double bowl	615	475	1,090
1880	Stainless steel, 19" x 18" single bowl	375	435	810
1920	25" x 22" single bowl	405	435	840
1960	33" x 22" double bowl	545	470	1,015
2000	43" x 22" double bowl	600	475	1,075
2040	44" x 22" triple bowl	765	495	1,260
2080	44" x 24" corner double bowl	550	475	1,025
2120	Steel, enameled, 24" x 21" single bowl	221	435	656
2160	32" x 21" double bowl	248	470	718
2240	Raised deck, PE on CI, 32" x 21", dual level, double bowl	465	595	1,060
2280	42" x 21" dual level, triple bowl	730	650	1,380

8.1-432 Laundry Sink Systems

		MAT.	INST.	TOTAL
1740	Laundry sink w/trim, PE on CI, black iron frame			
1760	24" x 20", single compartment	430	430	860
1840	48" x 21" double compartment	895	465	1,360
1920	Molded stone, on wall, 22" x 21" single compartment	236	430	666
1960	45"x 21" double compartment	340	465	805
2040	Plastic, on wall or legs, 18" x 23" single compartment	197	420	617
2080	20" x 24" single compartment	226	420	646
2120	36" x 23" double compartment	255	450	705
2160	40" x 24" double compartment	315	450	765
2200	Stainless steel, countertop, 22" x 17" single compartment	365	430	795
2240	19" x 22" single compartment	430	430	860
2280	33" x 22" double compartment	415	465	880

MECHANICAL — A8.1 Plumbing

Systems are complete with trim and rough-in (supply, waste and vent) to connect to supply branches and waste mains.

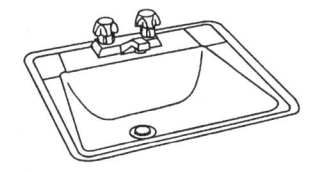

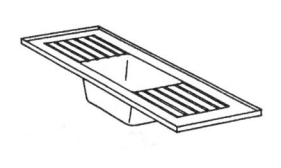

Vanity Top — **Wall Hung** — **Laboratory Sink** — **Service Sink**

8.1-433 Lavatory Systems

		MAT.	INST.	TOTAL
1560	Lavatory w/trim, vanity top, PE on CI, 20" x 18", Vanity top by others.	275	400	675
1600	19" x 16" oval	279	400	679
1640	18" round	293	400	693
1680	Cultured marble, 19" x 17"	191	400	591
1720	25" x 19"	208	400	608
1760	Stainless, self-rimming, 25" x 22"	256	400	656
1800	17" x 22"	250	400	650
1840	Steel enameled, 20" x 17"	192	410	602
1880	19" round	190	410	600
1920	Vitreous china, 20" x 16"	287	420	707
1960	19" x 16"	287	420	707
2000	22" x 13"	298	420	718
2040	Wall hung, PE on CI, 18" x 15"	585	440	1,025
2080	19" x 17"	565	440	1,005
2120	20" x 18"	420	440	860
2160	Vitreous china, 18" x 15"	410	450	860
2200	19" x 17"	350	450	800
2240	24" x 20"	485	450	935
2241				
2243				

8.1-434 Laboratory Sink Systems

		MAT.	INST.	TOTAL
1580	Laboratory sink w/trim, polyethylene, single bowl,			
1600	Double drainboard, 54" x 24" O.D.	915	560	1,475
1640	Single drainboard, 47" x 24" O.D.	950	560	1,510
1680	70" x 24" O.D.	970	560	1,530
1760	Flanged, 14-1/2" x 14-1/2" O.D.	350	500	850
1800	18-1/2" x 18-1/2" O.D.	315	500	815
1840	23-1/2" x 20-1/2" O.D.	370	500	870
1920	Polypropylene, cup sink, oval, 7" x 4" O.D.	165	445	610
1960	10" x 4-1/2" O.D.	176	445	621
1961				

8.1-434 Service Sink Systems

		MAT.	INST.	TOTAL
4260	Service sink w/trim, PE on CI, corner floor, 28" x 28", w/rim guard	835	560	1,395
4300	Wall hung w/rim guard, 22" x 18"	760	650	1,410
4340	24" x 20"	800	650	1,450
4380	Vitreous china, wall hung 22" x 20"	815	650	1,465

For expanded coverage of these items see *Means Mechanical or Plumbing Cost Data 1997*

PLUMBING — A8.1 Plumbing

Systems are complete with trim and rough-in (supply, waste and vent) to connect to supply branches and waste mains.

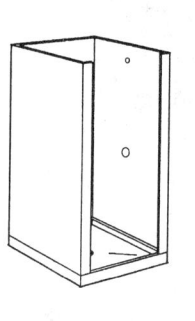

Square Shower Stall

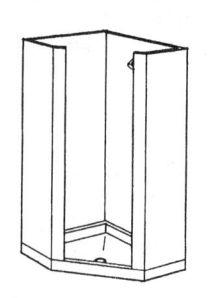

Corner Angle Shower Stall

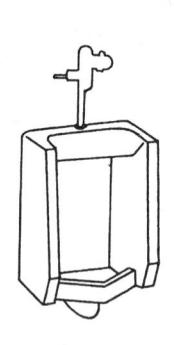

Wall Hung Urinal

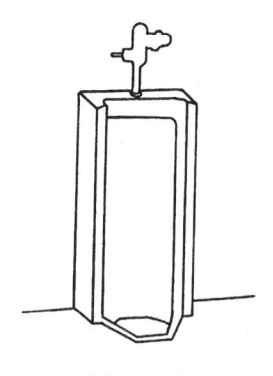

Stall Type Urinal

Wall Hung Water Cooler

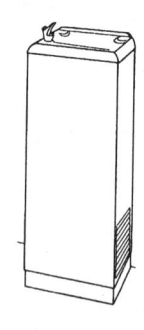

Floor Mounted Water Cooler

8.1-440	Shower Systems	MAT.	INST.	TOTAL
1560	Shower, stall, baked enamel, molded stone receptor, 30" square	465	665	1,130
1600	32" square	485	665	1,150
1640	Terrazzo receptor, 32" square	920	665	1,585
1680	36" square	980	705	1,685
1720	36" corner angle	900	705	1,605
1800	Fiberglass one piece, three walls, 32" square	520	610	1,130
1840	36" square	575	610	1,185
1880	Polypropylene, molded stone receptor, 30" square	475	665	1,140
1920	32" square	485	665	1,150
1960	Built-in head, arm, bypass, stops and handles	84	173	257

8.1-450	Urinal Systems	MAT.	INST.	TOTAL
2000	Urinal, vitreous china, wall hung	555	480	1,035
2040	Stall type	940	530	1,470

8.1-460	Water Cooler Systems	MAT.	INST.	TOTAL
1840	Water cooler, electric, wall hung, 8.2 GPH	560	335	895
1880	Dual height, 14.3 GPH	815	345	1,160
1920	Wheelchair type, 7.5 G.P.H.	1,250	335	1,585
1960	Semi recessed, 8.1 G.P.H.	750	335	1,085
2000	Full recessed, 8 G.P.H.	1,150	360	1,510
2040	Floor mounted, 14.3 G.P.H.	590	294	884
2080	Dual height, 14.3 G.P.H.	815	355	1,170
2120	Refrigerated compartment type, 1.5 G.P.H.	1,050	294	1,344

MECHANICAL — A8.1 Plumbing

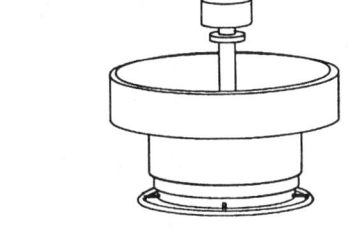

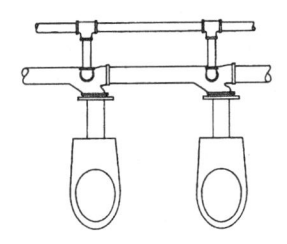

One Piece Wall Hung Water Closet

Circular Wash Fountain

Side By Side Water Closet Group

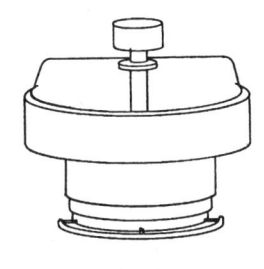

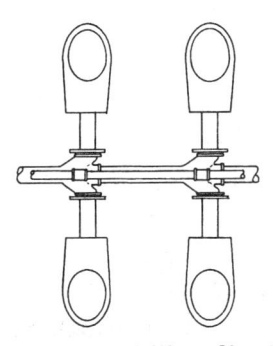

Floor Mount Water Closet

Semi-Circular Wash Fountain

Back to Back Water Closet Group

8.1-470	Water Closet Systems	MAT.	INST.	TOTAL
1800	Water closet, vitreous china, elongated			
1840	Tank type, wall hung, one piece	1,000	390	1,390
1880	Close coupled two piece	650	390	1,040
1920	Floor mount, one piece	735	425	1,160
1960	One piece low profile	1,025	425	1,450
2000	Two piece close coupled	315	425	740
2040	Bowl only with flush valve			
2080	Wall hung	620	445	1,065
2120	Floor mount	510	430	940
2122				

8.1-510	Water Closets, Group	MAT.	INST.	TOTAL
1760	Water closets, battery mount, wall hung, side by side, first closet	655	455	1,110
1800	Each additional water closet, add	625	430	1,055
3000	Back to back, first pair of closets	1,175	615	1,790
3100	Each additional pair of closets, back to back	1,175	605	1,780

8.1-560	Group Wash Fountain Systems	MAT.	INST.	TOTAL
1740	Group wash fountain, precast terrazzo			
1760	Circular, 36" diameter	1,925	720	2,645
1800	54" diameter	2,375	790	3,165
1840	Semi-circular, 36" diameter	1,800	720	2,520
1880	54" diameter	2,000	790	2,790
1960	Stainless steel, circular, 36" diameter	2,125	670	2,795
2000	54" diameter	2,750	745	3,495
2040	Semi-circular, 36" diameter	1,900	670	2,570
2080	54" diameter	2,375	745	3,120
2160	Thermoplastic, circular, 36" diameter	1,725	545	2,270
2200	54" diameter	2,000	630	2,630
2240	Semi-circular, 36" diameter	1,625	545	2,170
2280	54" diameter	1,950	630	2,580

For expanded coverage of these items see *Means Mechanical or Plumbing Cost Data 1997*

MECHANICAL — A8.1 Plumbing

Two Fixture

Three Fixture

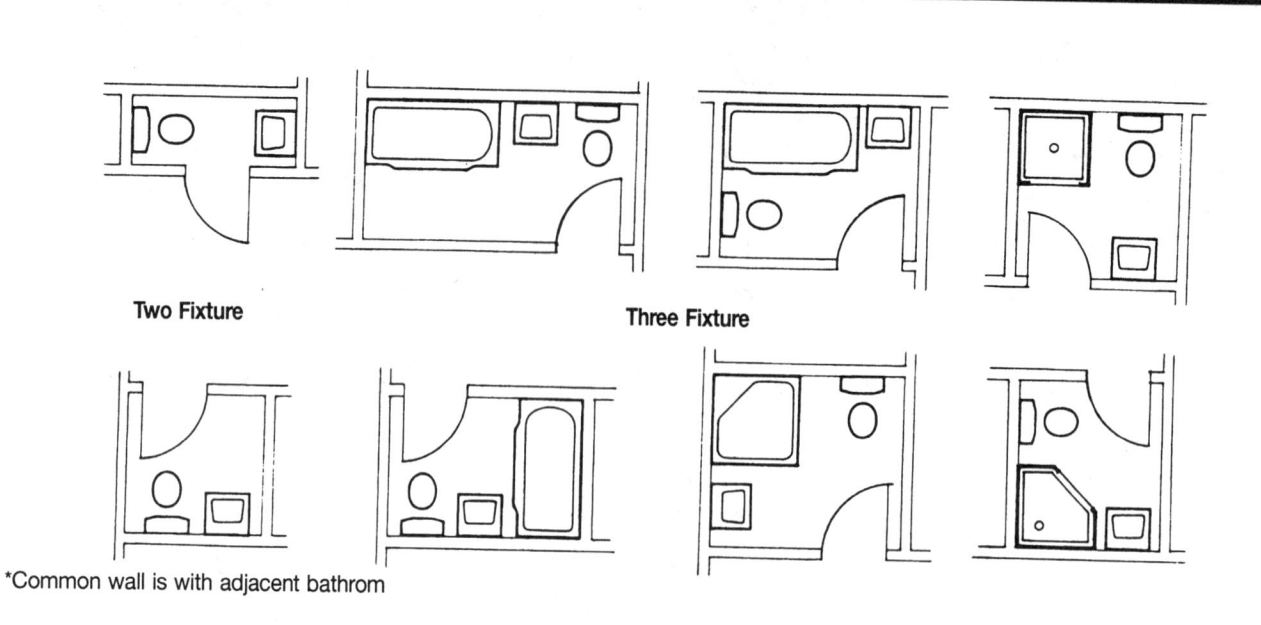

*Common wall is with adjacent bathroom

8.1-620	Two Fixture Bathroom, Two Wall Plumbing	MAT.	INST.	TOTAL
1180	Bathroom, lavatory & water closet, 2 wall plumbing, stand alone	860	1,100	1,960
1200	Share common plumbing wall*	790	945	1,735

8.1-620	Two Fixture Bathroom, One Wall Plumbing	MAT.	INST.	TOTAL
2220	Bathroom, lavatory & water closet, one wall plumbing, stand alone	810	985	1,795
2240	Share common plumbing wall*	715	835	1,550

8.1-630	Three Fixture Bathroom, One Wall Plumbing	MAT.	INST.	TOTAL
1150	Bathroom, three fixture, one wall plumbing			
1160	Lavatory, water closet & bathtub			
1170	Stand alone	1,275	1,275	2,550
1180	Share common plumbing wall *	1,100	925	2,025

8.1-630	Three Fixture Bathroom, Two Wall Plumbing	MAT.	INST.	TOTAL
2130	Bathroom, three fixture, two wall plumbing			
2140	Lavatory, water closet & bathtub			
2160	Stand alone	1,275	1,300	2,575
2180	Long plumbing wall common *	1,150	1,025	2,175
3610	Lavatory, bathtub & water closet			
3620	Stand alone	1,375	1,475	2,850
3640	Long plumbing wall common *	1,300	1,325	2,625
4660	Water closet, corner bathtub & lavatory			
4680	Stand alone	2,300	1,325	3,625
4700	Long plumbing wall common *	2,150	990	3,140
6100	Water closet, stall shower & lavatory			
6120	Stand alone	1,300	1,675	2,975
6140	Long plumbing wall common *	1,250	1,575	2,825
7060	Lavatory, corner stall shower & water closet			
7080	Stand alone	1,625	1,550	3,175
7100	Short plumbing wall common *	1,475	1,100	2,575

Important: See the Reference Section for critical supporting data - Location Factors & Historical Cost Indexes

MECHANICAL — A8.1 Plumbing

Four Fixture

Five Fixture

*Common wall is with adjacent bathrom

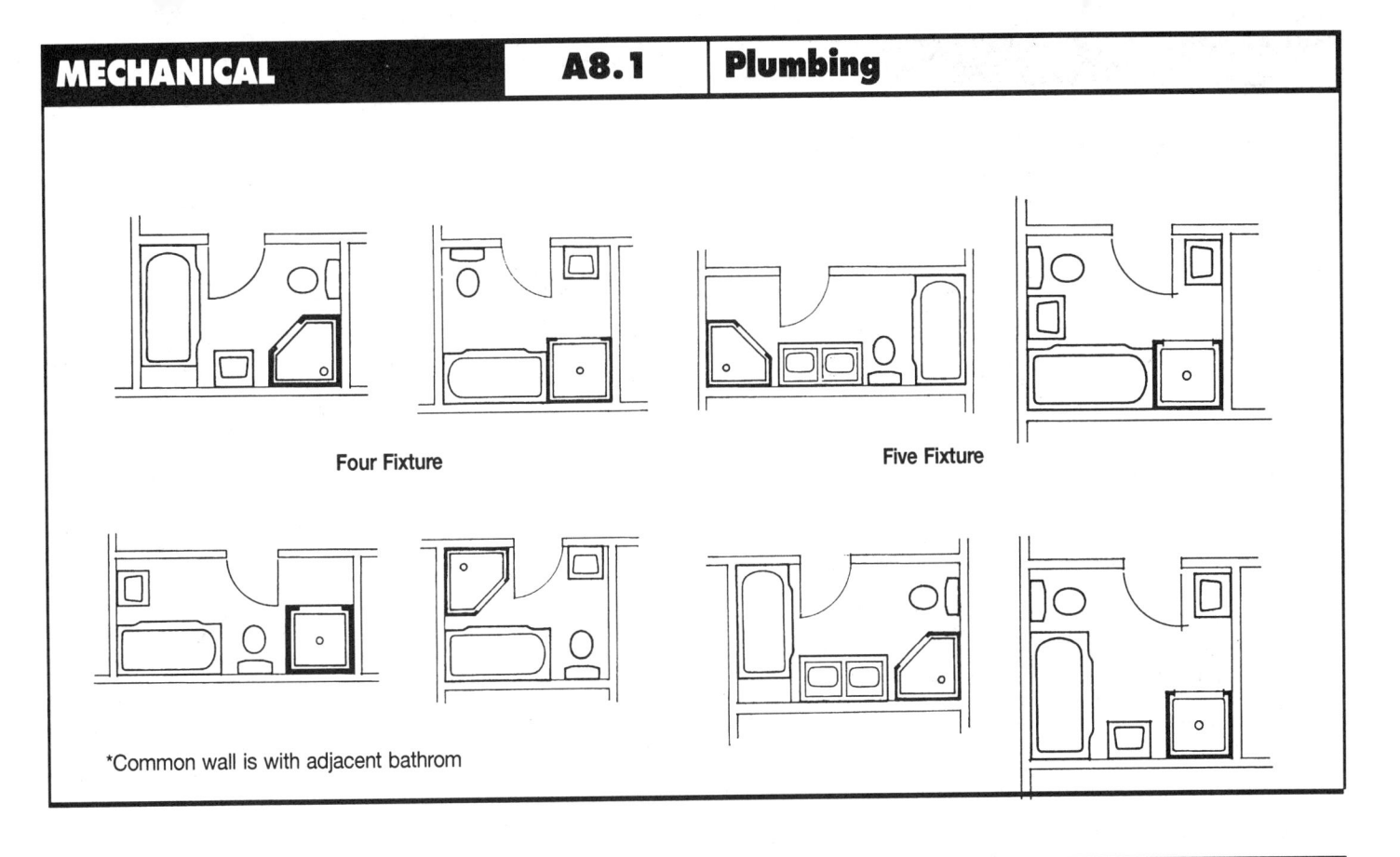

8.1-640	Four Fixture Bathroom, Two Wall Plumbing	MAT.	INST.	TOTAL
1140	Bathroom, four fixture, two wall plumbing			
1150	Bathtub, water closet, stall shower & lavatory			
1160	Stand alone	1,575	1,625	3,200
1180	Long plumbing wall common *	1,425	1,300	2,725
2260	Bathtub, lavatory, corner stall shower & water closet			
2280	Stand alone	2,000	1,650	3,650
2320	Long plumbing wall common *	1,850	1,325	3,175
3620	Bathtub, stall shower, lavatory & water closet			
3640	Stand alone	1,775	1,975	3,750
3661	Long plumbing wall (opposite door) common*	1,600	1,650	3,250
4680	Bathroom, four fixture, three wall plumbing			
4700	Bathtub, stall shower, lavatory & water closet			
4720	Stand alone	2,275	2,200	4,475
4761	Long plumbing wall (opposite door) common*	2,200	2,050	4,250

8.1-650	Five Fixture Bathroom, Two Wall Plumbing	MAT.	INST.	TOTAL
1320	Bathroom, five fixture, two wall plumbing			
1340	Bathtub, water closet, stall shower & two lavatories			
1360	Stand alone	2,200	2,475	4,675
1400	One short plumbing wall common *	2,025	2,150	4,175
2360	Bathroom, five fixture, three wall plumbing			
2380	Water closet, bathtub, two lavatories & stall shower			
2400	Stand alone	2,600	2,500	5,100
2440	One short plumbing wall common *	2,450	2,175	4,625
4080	Bathroom, five fixture, one wall plumbing			
4100	Bathtub, two lavatories, corner stall shower & water closet			
4120	Stand alone	2,475	2,275	4,750
4160	Share common wall *	2,100	1,550	3,650

For expanded coverage of these items see *Means Mechanical or Plumbing Cost Data 1997*

MECHANICAL — A8.2 — Fire Protection

Reference—Sprinkler Systems (Automatic)

Sprinkler systems may be classified by type as follows:

1. **Wet Pipe System.** A system employing automatic sprinklers attached to a piping system containing water and connected to a water supply so that water discharges immediately from sprinklers opened by a fire.
2. **Dry Pipe System.** A system employing automatic sprinklers attached to a piping system containing air under pressure, the release of which as from the opening of sprinklers permits the water pressure to open a valve known as a "dry pipe valve". The water then flows into the piping system and out the opened sprinklers.
3. **Pre-Action System.** A system employing automatic sprinklers attached to a piping system containing air that may or may not be under pressure, with a supplemental heat responsive system of generally more sensitive characteristics than the automatic sprinklers themselves, installed in the same areas as the sprinklers; actuation of the heat responsive system, as from a fire, opens a valve which permits water to flow into the sprinkler piping system and to be discharged from any sprinklers which may be open.
4. **Deluge System.** A system employing open sprinklers attached to a piping system connected to a water supply through a valve which is opened by the operation of a heat responsive system installed in the same areas as the sprinklers. When this valve opens, water flows into the piping system and discharges from all sprinklers attached thereto.
5. **Combined Dry Pipe and Pre-Action Sprinkler System.** A system employing automatic sprinklers attached to a piping system containing air under pressure with a supplemental heat responsive system of generally more sensitive characteristics than the automatic sprinklers themselves, installed in the same areas as the sprinklers; operation of the heat responsive system, as from a fire, actuates tripping devices which open dry pipe valves simultaneously and without loss of air pressure in the system. Operation of the heat responsive system also opens approved air exhaust valves at the end of the feed main which facilitates the filling of the system with water which usually precedes the opening of sprinklers. The heat responsive system also serves as an automatic fire alarm system.
6. **Limited Water Supply System.** A system employing automatic sprinklers and conforming to these standards but supplied by a pressure tank of limited capacity.
7. **Chemical Systems.** Systems using halon, carbon dioxide, dry chemical or high expansion foam as selected for special requirements. Agent may extinguish flames by chemically inhibiting flame propagation, suffocate flames by excluding oxygen, interrupting chemical action of oxygen uniting with fuel or sealing and cooling the combustion center.
8. **Firecycle System.** Firecycle is a fixed fire protection sprinkler system utilizing water as its extinguishing agent. It is a time delayed, recycling, preaction type which automatically shuts the water off when heat is reduced below the detector operating temperature and turns the water back on when that temperature is exceeded. The sytem senses a fire condition through a closed circuit electrical detector system which controls water flow to the fire automatically. Batteries supply up to 90 hour emergency power supply for system operation. The piping system is dry (until water is required) and is monitored with pressurized air. Should any leak in the system piping occur, an alarm will sound, but water will not enter the system until heat is sensed by a firecycle detector.

Area coverage sprinkler systems may be laid out and fed from the supply in any one of several patterns as shown below. It is desirable, if possible, to utilize a central feed and achieve a shorter flow path from the riser to the furthest sprinkler. This permits use of the smallest sizes of pipe possible with resulting savings.

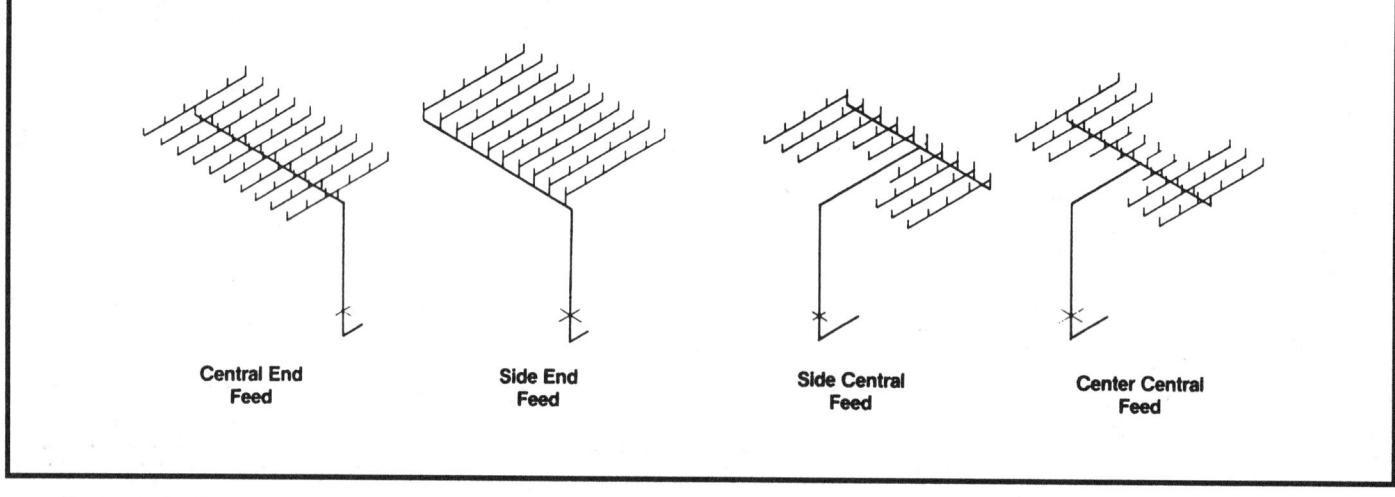

Central End Feed   Side End Feed   Side Central Feed   Center Central Feed

MECHANICAL | A8.2 | Fire Protection

Reference—System Classification

System Classification
Rules for installation of sprinkler systems vary depending on the classification of occupancy falling into one of three categories as follows:

Light Hazard Occupancy
The protection area allotted per sprinkler should not exceed 200 S.F. with the maximum distance between lines and sprinklers on lines being 15'. The sprinklers do not need to be staggered. Branch lines should not exceed eight sprinklers on either side of a cross main. Each large area requiring more than 100 sprinklers and without a sub-dividing partition should be supplied by feed mains or risers sized for ordinary hazard occupancy.

Included in this group are:
- Auditoriums
- Churches
- Clubs
- Educational
- Hospitals
- Institutional
- Libraries (except large stack rooms)
- Museums
- Nursing Homes
- Offices
- Residential
- Restaurants
- Schools
- Theaters

Ordinary Hazard Occupancy
The protection area allotted per sprinkler shall not exceed 130 S.F. of noncombustible ceiling and 120 S.F. of combustible ceiling. The maximum allowable distance between sprinkler lines and sprinklers on line is 15'. Sprinklers shall be staggered if the distance between heads exceeds 12'. Branch lines should not exceed eight sprinklers on either side of a cross main.

Included in this group are:
- Automotive garages
- Bakeries
- Beverage manufacturing
- Bleacheries
- Boiler houses
- Canneries
- Cement plants
- Clothing factories
- Cold storage warehouses
- Dairy products manufacturing
- Distilleries
- Dry cleaning
- Electric generating stations
- Feed mills
- Grain elevators
- Ice manufacturing
- Laundries
- Machine shops
- Mercantiles
- Paper mills
- Printing and Publishing
- Shoe factories
- Warehouses
- Wood product assembly

Extra Hazard Occupancy
The protection area allotted per sprinkler shall not exceed 90 S.F. of noncombustible ceiling and 80 S.F. of combustible ceiling. The maximum allowable distance between lines and between sprinklers on lines is 12'. Sprinklers on alternate lines shall be staggered if the distance between sprinklers on lines exceeds 8'. Branch lines should not exceed six sprinklers on either side of a cross main.

Included in this group are:
- Aircraft hangars
- Chemical works
- Explosives manufacturing
- Linoleum manufacturing
- Linseed oil mills
- Oil refineries
- Paint shops
- Shade cloth manufacturing
- Solvent extracting
- Varnish works
- Volatile flammable liquid manufacturing & use

For expanded coverage of these items see *Means Mechanical or Plumbing Cost Data 1997*

MECHANICAL — A8.2 Fire Protection

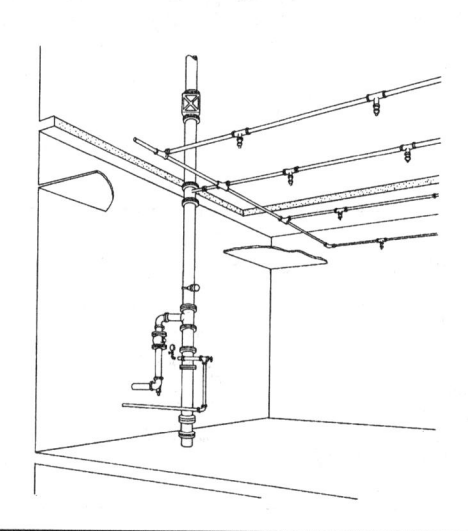

Wet Pipe System. A system employing automatic sprinklers attached to a piping system containing water and connected to a water supply so that water discharges immediately from sprinklers opened by heat from a fire.

All areas are assumed to be open.

8.2-110 Wet Pipe Sprinkler Systems

		MAT.	INST.	TOTAL
0520	Wet pipe sprinkler systems, steel, black, sch. 40 pipe			
0530	Light hazard, one floor, 500 S.F.	1.08	1.71	2.79
0560	1000 S.F.	1.49	1.78	3.27
0580	2000 S.F.	1.56	1.81	3.37
0600	5000 S.F.	.79	1.25	2.04
0620	10,000 S.F.	.52	1.05	1.57
0640	50,000 S.F.	.43	.98	1.41
0660	Each additional floor, 500 S.F.	.60	1.45	2.05
0680	1000 S.F.	.51	1.35	1.86
0700	2000 S.F.	.47	1.25	1.72
0720	5000 S.F.	.37	1.04	1.41
0740	10,000 S.F.	.32	.95	1.27
0760	50,000 S.F.	.29	.77	1.06
1000	Ordinary hazard, one floor, 500 S.F.	1.18	1.83	3.01
1020	1000 S.F.	1.45	1.73	3.18
1040	2000 S.F.	1.64	1.89	3.53
1060	5000 S.F.	.86	1.33	2.19
1080	10,000 S.F.	.68	1.41	2.09
1100	50,000 S.F.	.64	1.39	2.03
1140	Each additional floor, 500 S.F.	.75	1.63	2.38
1160	1000 S.F.	.47	1.32	1.79
1180	2000 S.F.	.55	1.34	1.89
1200	5000 S.F.	.53	1.26	1.79
1220	10,000 S.F.	.48	1.31	1.79
1240	50,000 S.F.	.48	1.21	1.69
1500	Extra hazard, one floor, 500 S.F.	2.74	2.86	5.60
1520	1000 S.F.	1.83	2.46	4.29
1540	2000 S.F.	1.75	2.57	4.32
1560	5000 S.F.	1.20	2.19	3.39
1580	10,000 S.F.	1.13	2.09	3.22
1600	50,000 S.F.	1.22	1.98	3.20
1660	Each additional floor, 500 S.F.	.77	2.04	2.81
1680	1000 S.F.	.75	1.92	2.67
1700	2000 S.F.	.70	1.96	2.66
1720	5000 S.F.	.60	1.71	2.31
1740	10,000 S.F.	.68	1.55	2.23
1760	50,000 S.F.	.71	1.48	2.19
2020	Grooved steel, black sch. 40 pipe, light hazard, one floor, 2000 S.F.	1.57	1.54	3.11
2060	10,000 S.F.	.64	.96	1.60
2100	Each additional floor, 2000 S.F.	.49	1	1.49
2150	10,000 S.F.	.34	.82	1.16
2200	Ordinary hazard, one floor, 2000 S.F.	1.58	1.66	3.24

MECHANICAL — A8.2 Fire Protection

8.2-110 Wet Pipe Sprinkler Systems

		COST PER S.F.		
		MAT.	INST.	TOTAL
2250	10,000 S.F.	.63	1.18	1.81
2300	Each additional floor, 2000 S.F.	.49	1.11	1.60
2350	10,000 S.F.	.43	1.08	1.51
2400	Extra hazard, one floor, 2000 S.F.	1.75	2.11	3.86
2450	10,000 S.F.	.87	1.56	2.43
2500	Each additional floor, 2000 S.F.	.72	1.61	2.33
2550	10,000 S.F.	.61	1.38	1.99
3050	Grooved steel black sch. 10 pipe, light hazard, one floor, 2000 S.F.	1.55	1.52	3.07
3100	10,000 S.F.	.53	.90	1.43
3150	Each additional floor, 2000 S.F.	.47	.98	1.45
3200	10,000 S.F.	.33	.80	1.13
3250	Ordinary hazard, one floor, 2000 S.F.	1.57	1.63	3.20
3300	10,000 S.F.	.61	1.16	1.77
3350	Each additional floor, 2000 S.F.	.48	1.08	1.56
3400	10,000 S.F.	.41	1.06	1.47
3450	Extra hazard, one floor, 2000 S.F.	1.74	2.08	3.82
3500	10,000 S.F.	.82	1.54	2.36
3550	Each additional floor, 2000 S.F.	.71	1.58	2.29
3600	10,000 S.F.	.58	1.36	1.94
4050	Copper tubing, type M, light hazard, one floor, 2000 S.F.	1.60	1.51	3.11
4100	10,000 S.F.	.63	.92	1.55
4150	Each additional floor, 2000 S.F.	.52	.98	1.50
4200	10,000 S.F.	.43	.82	1.25
4250	Ordinary hazard, one floor, 2000 S.F.	1.67	1.70	3.37
4300	10,000 S.F.	.74	1.09	1.83
4350	Each additional floor, 2000 S.F.	.60	1.10	1.70
4400	10,000 S.F.	.53	.97	1.50
4450	Extra hazard, one floor, 2000 S.F.	1.85	2.10	3.95
4500	10,000 S.F.	1.29	1.67	2.96
4550	Each additional floor, 2000 S.F.	.82	1.60	2.42
4600	10,000 S.F.	.87	1.48	2.35
5050	Copper tubing, type M, T-drill system, light hazard, one floor			
5060	2000 S.F.	1.61	1.41	3.02
5100	10,000 S.F.	.61	.77	1.38
5150	Each additional floor, 2000 S.F.	.53	.88	1.41
5200	10,000 S.F.	.41	.67	1.08
5250	Ordinary hazard, one floor, 2000 S.F.	1.62	1.44	3.06
5300	10,000 S.F.	.72	.97	1.69
5350	Each additional floor, 2000 S.F.	.53	.89	1.42
5400	10,000 S.F.	.52	.87	1.39
5450	Extra hazard, one floor, 2000 S.F.	1.70	1.73	3.43
5500	10,000 S.F.	1.06	1.22	2.28
5550	Each additional floor, 2000 S.F.	.70	1.26	1.96
5600	10,000 S.F.	.64	1.03	1.67

For expanded coverage of these items see *Means Mechanical or Plumbing Cost Data 1997*

MECHANICAL — A8.2 Fire Protection

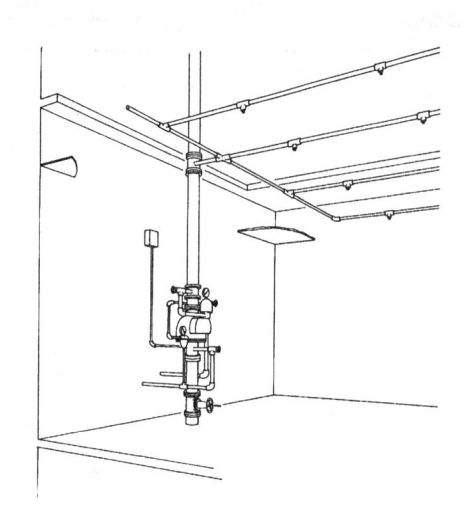

Dry Pipe System: A system employing automatic sprinklers attached to a piping system containing air under pressure, the release of which as from the opening of sprinklers permits the water pressure to open a valve known as a "dry pipe valve". The water then flows into the piping system and out the opened sprinklers.

All areas are assumed to be open.

8.2-120	Dry Pipe Sprinkler Systems	MAT.	INST.	TOTAL
0520	Dry pipe sprinkler systems, steel, black, sch. 40 pipe			
0530	Light hazard, one floor, 500 S.F.	4.03	3.39	7.42
0560	1000 S.F.	2.16	1.96	4.12
0580	2000 S.F.	1.88	2	3.88
0600	5000 S.F.	.98	1.32	2.30
0620	10,000 S.F.	.67	1.10	1.77
0640	50,000 S.F.	.54	1	1.54
0660	Each additional floor, 500 S.F.	.86	1.61	2.47
0680	1000 S.F.	.62	1.33	1.95
0700	2000 S.F.	.59	1.26	1.85
0720	5000 S.F.	.49	1.05	1.54
0740	10,000 S.F.	.43	.97	1.40
0760	50,000 S.F.	.41	.89	1.30
1000	Ordinary hazard, one floor, 500 S.F.	4.08	3.43	7.51
1020	1000 S.F.	2.17	1.98	4.15
1040	2000 S.F.	1.96	2.08	4.04
1060	5000 S.F.	1.09	1.41	2.50
1080	10,000 S.F.	.88	1.47	2.35
1100	50,000 S.F.	.82	1.43	2.25
1140	Each additional floor, 500 S.F.	.91	1.65	2.56
1160	1000 S.F.	.69	1.49	2.18
1180	2000 S.F.	.71	1.36	2.07
1200	5000 S.F.	.62	1.16	1.78
1220	10,000 S.F.	.56	1.16	1.72
1240	50,000 S.F.	.56	1.05	1.61
1500	Extra hazard, one floor, 500 S.F.	5.25	4.30	9.55
1520	1000 S.F.	3.07	3.05	6.12
1540	2000 S.F.	2.13	2.70	4.83
1560	5000 S.F.	1.28	1.99	3.27
1580	10,000 S.F.	1.28	1.90	3.18
1600	50,000 S.F.	1.39	1.81	3.20
1660	Each additional floor, 500 S.F.	1	2.07	3.07
1680	1000 S.F.	.98	1.95	2.93
1700	2000 S.F.	.93	1.99	2.92
1720	5000 S.F.	.80	1.72	2.52
1740	10,000 S.F.	.91	1.55	2.46
1760	50,000 S.F.	.96	1.48	2.44
2020	Grooved steel, black, sch. 40 pipe, light hazard, one floor, 2000 S.F.	1.85	1.72	3.57
2060	10,000 S.F.	.69	.97	1.66
2100	Each additional floor, 2000 S.F.	.61	1.01	1.62
2150	10,000 S.F.	.45	.84	1.29
2200	Ordinary hazard, one floor, 2000 S.F.	1.90	1.85	3.75

Important: See the Reference Section for critical supporting data - Location Factors & Historical Cost Indexes

MECHANICAL — A8.2 Fire Protection

8.2-120 Dry Pipe Sprinkler Systems

		\multicolumn{3}{c	}{COST PER S.F.}
		MAT.	INST.	TOTAL
2250	10,000 S.F.	.83	1.24	2.07
2300	Each additional floor, 2000 S.F.	.65	1.13	1.78
2350	10,000 S.F.	.59	1.11	1.70
2400	Extra hazard, one floor, 2000 S.F.	2.14	2.31	4.45
2450	10,000 S.F.	1.15	1.62	2.77
2500	Each additional floor, 2000 S.F.	.95	1.64	2.59
2550	10,000 S.F.	.85	1.41	2.26
3050	Grooved steel black sch. 10 pipe, light hazard, one floor, 2000 S.F.	1.83	1.70	3.53
3100	10,000 S.F.	.68	.95	1.63
3150	Each additional floor, 2000 S.F.	.59	.99	1.58
3200	10,000 S.F.	.44	.82	1.26
3250	Ordinary hazard, one floor, 2000 S.F.	1.89	1.82	3.71
3300	10,000 S.F.	.81	1.22	2.03
3350	Each additional floor, 2000 S.F.	.64	1.10	1.74
3400	10,000 S.F.	.57	1.09	1.66
3450	Extra hazard, one floor, 2000 S.F.	2.13	2.28	4.41
3500	10,000 S.F.	1.10	1.60	2.70
3550	Each additional floor, 2000 S.F.	.94	1.61	2.55
3600	10,000 S.F.	.82	1.39	2.21
4050	Copper tubing, type M, light hazard, one floor, 2000 S.F.	1.88	1.69	3.57
4100	10,000 S.F.	.78	.97	1.75
4150	Each additional floor, 2000 S.F.	.64	.99	1.63
4200	10,000 S.F.	.54	.84	1.38
4250	Ordinary hazard, one floor, 2000 S.F.	1.99	1.89	3.88
4300	10,000 S.F.	.94	1.15	2.09
4350	Each additional floor, 2000 S.F.	.85	1.16	2.01
4400	10,000 S.F.	.69	1	1.69
4450	Extra hazard, one floor, 2000 S.F.	2.24	2.30	4.54
4500	10,000 S.F.	1.58	1.74	3.32
4550	Each additional floor, 2000 S.F.	1.05	1.63	2.68
4600	10,000 S.F.	1.11	1.51	2.62
5050	Copper tubing, type M, T-drill system, light hazard, one floor			
5060	2000 S.F.	1.89	1.59	3.48
5100	10,000 S.F.	.76	.82	1.58
5150	Each additional floor, 2000 S.F.	.65	.89	1.54
5200	10,000 S.F.	.52	.69	1.21
5250	Ordinary hazard, one floor, 2000 S.F.	1.94	1.63	3.57
5300	10,000 S.F.	.92	1.03	1.95
5350	Each additional floor, 2000 S.F.	.69	.91	1.60
5400	10,000 S.F.	.64	.85	1.49
5450	Extra hazard, one floor, 2000 S.F.	2.09	1.93	4.02
5500	10,000 S.F.	1.35	1.29	2.64
5550	Each additional floor, 2000 S.F.	.90	1.26	2.16
5600	10,000 S.F.	.88	1.06	1.94

For expanded coverage of these items see *Means Mechanical or Plumbing Cost Data 1997*

MECHANICAL — A8.2 Fire Protection

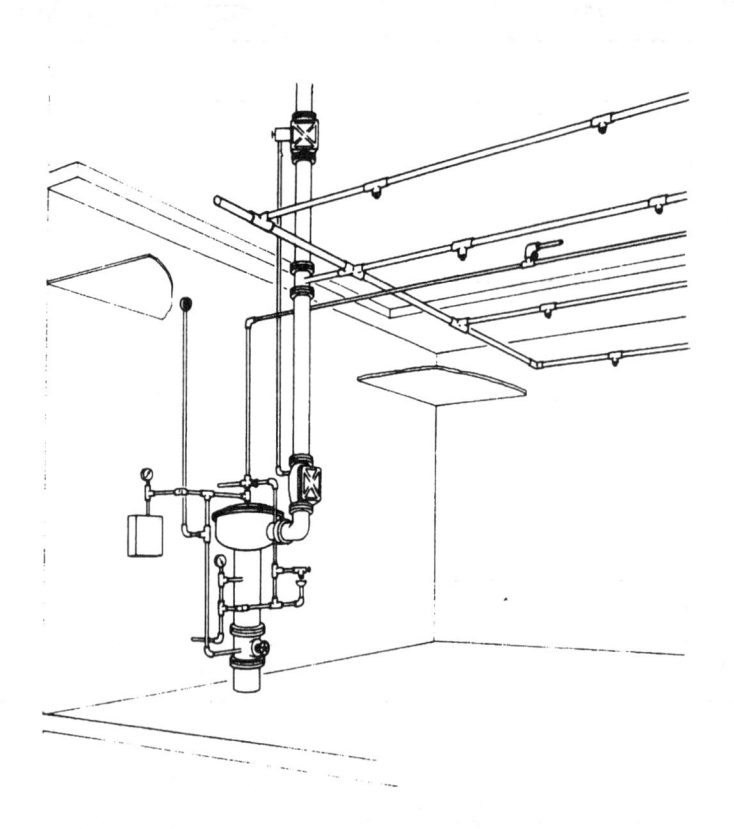

Preaction System: A system employing automatic sprinklers attached to a piping system containing air that may or may not be under pressure, with a supplemental heat responsive system of generally more sensitive characteristics than the automatic sprinklers themselves, installed in the same areas as the sprinklers. Actuation of the heat responsive system, as from a fire, opens a valve which permits water to flow into the sprinkler piping system and to be discharged from those sprinklers which were opened by heat from the fire.

All areas are assumed to be opened.

8.2-130	Preaction Sprinkler Systems	MAT.	INST.	TOTAL
0520	Preaction sprinkler systems, steel, black, sch. 40 pipe			
0530	Light hazard, one floor, 500 S.F.	3.92	2.71	6.63
0560	1000 S.F.	2.19	2.01	4.20
0580	2000 S.F.	1.87	2	3.87
0600	5000 S.F.	.97	1.32	2.29
0620	10,000 S.F.	.66	1.09	1.75
0640	50,000 S.F.	.53	.99	1.52
0660	Each additional floor, 500 S.F.	.88	1.45	2.33
0680	1000 S.F.	.64	1.33	1.97
0700	2000 S.F.	.61	1.26	1.87
0720	5000 S.F.	.48	1.05	1.53
0740	10,000 S.F.	.42	.96	1.38
0760	50,000 S.F.	.45	.92	1.37
1000	Ordinary hazard, one floor, 500 S.F.	1.53	1.95	3.48
1020	1000 S.F.	2.15	1.96	4.11
1040	2000 S.F.	2.01	2.08	4.09
1060	5000 S.F.	1.04	1.40	2.44
1080	10,000 S.F.	.82	1.45	2.27
1100	50,000 S.F.	.75	1.42	2.17
1140	Each additional floor, 500 S.F.	1.04	1.66	2.70
1160	1000 S.F.	.61	1.33	1.94
1180	2000 S.F.	.62	1.35	1.97
1200	5000 S.F.	.63	1.25	1.88
1220	10,000 S.F.	.58	1.32	1.90
1240	50,000 S.F.	.57	1.22	1.79
1500	Extra hazard, one floor, 500 S.F.	5.15	3.78	8.93
1520	1000 S.F.	2.91	2.77	5.68
1540	2000 S.F.	2.04	2.68	4.72
1560	5000 S.F.	1.29	2.14	3.43
1580	10,000 S.F.	1.24	2.11	3.35

MECHANICAL — A8.2 Fire Protection

8.2-130 Preaction Sprinkler Systems

		COST PER S.F.		
		MAT.	INST.	TOTAL
1600	50,000 S.F.	1.32	2	3.32
1660	Each additional floor, 500 S.F.	1.06	2.07	3.13
1680	1000 S.F.	.89	1.93	2.82
1700	2000 S.F.	.84	1.97	2.81
1720	5000 S.F.	.71	1.72	2.43
1740	10,000 S.F.	.78	1.56	2.34
1760	50,000 S.F.	.80	1.46	2.26
2020	Grooved steel, black, sch. 40 pipe, light hazard, one floor, 2000 S.F.	1.87	1.72	3.59
2060	10,000 S.F.	.68	.96	1.64
2100	Each additional floor of 2000 S.F.	.63	1.01	1.64
2150	10,000 S.F.	.44	.83	1.27
2200	Ordinary hazard, one floor, 2000 S.F.	1.88	1.84	3.72
2250	10,000 S.F.	.77	1.22	1.99
2300	Each additional floor, 2000 S.F.	.63	1.12	1.75
2350	10,000 S.F.	.53	1.09	1.62
2400	Extra hazard, one floor, 2000 S.F.	2.05	2.29	4.34
2450	10,000 S.F.	1.02	1.61	2.63
2500	Each additional floor, 2000 S.F.	.86	1.62	2.48
2550	10,000 S.F.	.71	1.39	2.10
3050	Grooved steel, black, sch. 10 pipe light hazard, one floor, 2000 S.F.	1.85	1.70	3.55
3100	10,000 S.F.	.67	.94	1.61
3150	Each additional floor, 2000 S.F.	.61	.99	1.60
3200	10,000 S.F.	.43	.81	1.24
3250	Ordinary hazard, one floor, 2000 S.F.	1.85	1.70	3.55
3300	10,000 S.F.	.68	1.19	1.87
3350	Each additional floor, 2000 S.F.	.62	1.09	1.71
3400	10,000 S.F.	.51	1.07	1.58
3450	Extra hazard, one floor, 2000 S.F.	2.04	2.26	4.30
3500	10,000 S.F.	.96	1.58	2.54
3550	Each additional floor, 2000 S.F.	.85	1.59	2.44
3600	10,000 S.F.	.68	1.37	2.05
4050	Copper tubing, type M, light hazard, one floor, 2000 S.F.	1.90	1.69	3.59
4100	10,000 S.F.	.77	.96	1.73
4150	Each additional floor, 2000 S.F.	.66	.99	1.65
4200	10,000 S.F.	.46	.82	1.28
4250	Ordinary hazard, one floor, 2000 S.F.	1.97	1.88	3.85
4300	10,000 S.F.	.88	1.13	2.01
4350	Each additional floor, 2000 S.F.	.66	1.02	1.68
4400	10,000 S.F.	.56	.90	1.46
4450	Extra hazard, one floor, 2000 S.F.	2.15	2.28	4.43
4500	10,000 S.F.	1.43	1.71	3.14
4550	Each additional floor, 2000 S.F.	.96	1.61	2.57
4600	10,000 S.F.	.97	1.49	2.46
5050	Copper tubing, type M, T-drill system, light hazard, one floor			
5060	2000 S.F.	1.91	1.59	3.50
5100	10,000 S.F.	.75	.81	1.56
5150	Each additional floor, 2000 S.F.	.67	.89	1.56
5200	10,000 S.F.	.51	.68	1.19
5250	Ordinary hazard, one floor, 2000 S.F.	1.92	1.62	3.54
5300	10,000 S.F.	.86	1.01	1.87
5350	Each additional floor, 2000 S.F.	.68	.90	1.58
5400	10,000 S.F.	.62	.88	1.50
5450	Extra hazard, one floor, 2000 S.F.	2	1.91	3.91
5500	10,000 S.F.	1.20	1.26	2.46
5550	Each additional floor, 2000 S.F.	.81	1.24	2.05
5600	10,000 S.F.	.74	1.04	1.78

For expanded coverage of these items see *Means Mechanical or Plumbing Cost Data 1997*

MECHANICAL — A8.2 — Fire Protection

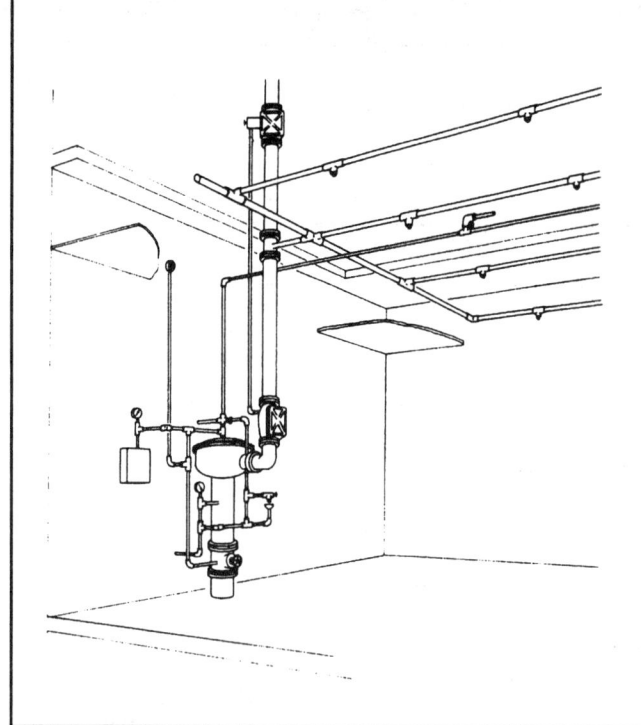

Deluge System: A system employing open sprinklers attached to a piping system connected to a water supply through a valve which is opened by the operation of a heat responsive system installed in the same areas as the sprinklers. When this valve opens, water flows into the piping system and discharges from all sprinklers attached thereto.

8.2-140	Deluge Sprinkler Systems	MAT.	INST.	TOTAL
0520	Deluge sprinkler systems, steel, black, sch. 40 pipe			
0530	Light hazard, one floor, 500 S.F.	5.60	2.74	8.34
0560	1000 S.F.	3.01	1.94	4.95
0580	2000 S.F.	2.39	2	4.39
0600	5000 S.F.	1.18	1.32	2.50
0620	10,000 S.F.	.76	1.09	1.85
0640	50,000 S.F.	.55	.99	1.54
0660	Each additional floor, 500 S.F.	.88	1.45	2.33
0680	1000 S.F.	.64	1.33	1.97
0700	2000 S.F.	.61	1.26	1.87
0720	5000 S.F.	.48	1.05	1.53
0740	10,000 S.F.	.42	.96	1.38
0760	50,000 S.F.	.45	.92	1.37
1000	Ordinary hazard, one floor, 500 S.F.	6	3.13	9.13
1020	1000 S.F.	2.99	1.98	4.97
1040	2000 S.F.	2.54	2.09	4.63
1060	5000 S.F.	1.25	1.40	2.65
1080	10,000 S.F.	.92	1.45	2.37
1100	50,000 S.F.	.80	1.45	2.25
1140	Each additional floor, 500 S.F.	1.04	1.66	2.70
1160	1000 S.F.	.61	1.33	1.94
1180	2000 S.F.	.62	1.35	1.97
1200	5000 S.F.	.57	1.15	1.72
1220	10,000 S.F.	.54	1.17	1.71
1240	50,000 S.F.	.53	1.12	1.65
1500	Extra hazard, one floor, 500 S.F.	6.85	3.81	10.66
1520	1000 S.F.	3.92	2.89	6.81
1540	2000 S.F.	2.57	2.69	5.26
1560	5000 S.F.	1.38	1.97	3.35
1580	10,000 S.F.	1.27	1.92	3.19
1600	50,000 S.F.	1.35	1.85	3.20
1660	Each additional floor, 500 S.F.	1.06	2.07	3.13
1680	1000 S.F.	.89	1.93	2.82
1700	2000 S.F.	.84	1.97	2.81

Important: See the Reference Section for critical supporting data - Location Factors & Historical Cost Indexes

MECHANICAL — A8.2 Fire Protection

8.2-140 Deluge Sprinkler Systems

		COST PER S.F.		
		MAT.	INST.	TOTAL
1720	5000 S.F.	.71	1.72	2.43
1740	10,000 S.F.	.81	1.62	2.43
1760	50,000 S.F.	.86	1.56	2.42
2000	Grooved steel, black, sch. 40 pipe, light hazard, one floor			
2020	2000 S.F.	2.41	1.73	4.14
2060	10,000 S.F.	.79	.97	1.76
2100	Each additional floor, 2,000 S.F.	.63	1.01	1.64
2150	10,000 S.F.	.44	.83	1.27
2200	Ordinary hazard, one floor, 2000 S.F.	1.88	1.84	3.72
2250	10,000 S.F.	.87	1.22	2.09
2300	Each additional floor, 2000 S.F.	.63	1.12	1.75
2350	10,000 S.F.	.53	1.09	1.62
2400	Extra hazard, one floor, 2000 S.F.	2.58	2.30	4.88
2450	10,000 S.F.	1.13	1.61	2.74
2500	Each additional floor, 2000 S.F.	.86	1.62	2.48
2550	10,000 S.F.	.71	1.39	2.10
3000	Grooved steel, black, sch. 10 pipe, light hazard, one floor			
3050	2000 S.F.	2.15	1.63	3.78
3100	10,000 S.F.	.77	.94	1.71
3150	Each additional floor, 2000 S.F.	.61	.99	1.60
3200	10,000 S.F.	.43	.81	1.24
3250	Ordinary hazard, one floor, 2000 S.F.	2.40	1.82	4.22
3300	10,000 S.F.	.78	1.19	1.97
3350	Each additional floor, 2000 S.F.	.62	1.09	1.71
3400	10,000 S.F.	.51	1.07	1.58
3450	Extra hazard, one floor, 2000 S.F.	2.57	2.27	4.84
3500	10,000 S.F.	1.06	1.58	2.64
3550	Each additional floor, 2000 S.F.	.85	1.59	2.44
3600	10,000 S.F.	.68	1.37	2.05
4000	Copper tubing, type M, light hazard, one floor			
4050	2000 S.F.	2.43	1.70	4.13
4100	10,000 S.F.	.87	.96	1.83
4150	Each additional floor, 2000 S.F.	.66	.99	1.65
4200	10,000 S.F.	.46	.82	1.28
4250	Ordinary hazard, one floor, 2000 S.F.	2.50	1.89	4.39
4300	10,000 S.F.	.98	1.13	2.11
4350	Each additional floor, 2000 S.F.	.66	1.02	1.68
4400	10,000 S.F.	.56	.90	1.46
4450	Extra hazard, one floor, 2000 S.F.	2.68	2.29	4.97
4500	10,000 S.F.	1.55	1.72	3.27
4550	Each additional floor, 2000 S.F.	.96	1.61	2.57
4600	10,000 S.F.	.97	1.49	2.46
5000	Copper tubing, type M, T-drill system, light hazard, one floor			
5050	2000 S.F.	2.44	1.60	4.04
5100	10,000 S.F.	.85	.81	1.66
5150	Each additional floor, 2000 S.F.	.67	.90	1.57
5200	10,000 S.F.	.51	.68	1.19
5250	Ordinary hazard, one floor, 2000 S.F.	2.45	1.63	4.08
5300	10,000 S.F.	.96	1.01	1.97
5350	Each additional floor, 2000 S.F.	.67	.90	1.57
5400	10,000 S.F.	.62	.88	1.50
5450	Extra hazard, one floor, 2000 S.F.	2.53	1.92	4.45
5500	10,000 S.F.	1.30	1.26	2.56
5550	Each additional floor, 2000 S.F.	.81	1.24	2.05
5600	10,000 S.F.	.74	1.04	1.78

For expanded coverage of these items see *Means Mechanical or Plumbing Cost Data 1997*

MECHANICAL — A8.2 — Fire Protection

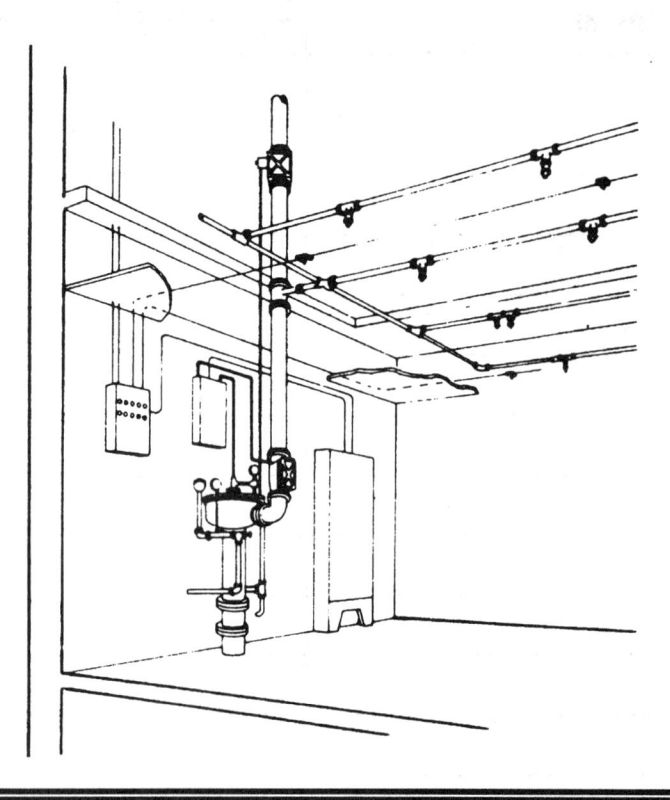

Firecycle is a fixed fire protection sprinkler system utilizing water as its extinguishing agent. It is a time delayed, recycling, preaction type which automatically shuts the water off when heat is reduced below the detector operating temperature and turns the water back on when that temperature is exceeded.

The system senses a fire condition through a closed circuit electrical detector system which controls water flow to the fire automatically. Batteries supply up to 90 hours emergency power supply for system operation. The piping system is dry (until water is required) and is monitored with pressurized air. Should any leak in the system piping occur, an alarm will sound, but water will not enter the system until heat is sensed by a Firecycle detector.

8.2-150	Firecycle Sprinkler Systems	MAT.	INST.	TOTAL
0520	Firecycle sprinkler systems, steel black sch. 40 pipe			
0530	Light hazard, one floor, 500 S.F.	15.20	5.40	20.60
0560	1000 S.F.	7.85	3.36	11.21
0580	2000 S.F.	4.68	2.47	7.15
0600	5000 S.F.	2.13	1.51	3.64
0620	10,000 S.F.	1.26	1.18	2.44
0640	50,000 S.F.	.69	1	1.69
0660	Each additional floor of 500 S.F.	.98	1.46	2.44
0680	1000 S.F.	.70	1.34	2.04
0700	2000 S.F.	.57	1.26	1.83
0720	5000 S.F.	.53	1.05	1.58
0740	10,000 S.F.	.48	.96	1.44
0760	50,000 S.F.	.50	.92	1.42
1000	Ordinary hazard, one floor, 500 S.F.	15.40	5.60	21
1020	1000 S.F.	7.80	3.31	11.11
1040	2000 S.F.	4.76	2.54	7.30
1060	5000 S.F.	2.20	1.59	3.79
1080	10,000 S.F.	1.42	1.54	2.96
1100	50,000 S.F.	1.05	1.60	2.65
1140	Each additional floor, 500 S.F.	1.14	1.67	2.81
1160	1000 S.F.	.67	1.34	2.01
1180	2000 S.F.	.70	1.24	1.94
1200	5000 S.F.	.62	1.15	1.77
1220	10,000 S.F.	.56	1.14	1.70
1240	50,000 S.F.	.56	1.06	1.62
1500	Extra hazard, one floor, 500 S.F.	16.40	6.45	22.85
1520	1000 S.F.	8.50	4.12	12.62
1540	2000 S.F.	4.86	3.15	8.01
1560	5000 S.F.	2.33	2.16	4.49
1580	10,000 S.F.	1.82	2.18	4
1600	50,000 S.F.	1.65	2.37	4.02
1660	Each additional floor, 500 S.F.	1.16	2.08	3.24

Cost per S.F.

MECHANICAL — A8.2 Fire Protection

8.2-150 Firecycle Sprinkler Systems

		MAT.	INST.	TOTAL
1680	1000 S.F.	.95	1.94	2.89
1700	2000 S.F.	.90	1.98	2.88
1720	5000 S.F.	.76	1.72	2.48
1740	10,000 S.F.	.84	1.56	2.40
1760	50,000 S.F.	.89	1.51	2.40
2020	Grooved steel, black, sch. 40 pipe, light hazard, one floor			
2030	2000 S.F.	4.69	2.19	6.88
2060	10,000 S.F.	1.38	1.47	2.85
2100	Each additional floor, 2000 S.F.	.69	1.02	1.71
2150	10,000 S.F.	.50	.83	1.33
2200	Ordinary hazard, one floor, 2000 S.F.	4.70	2.31	7.01
2250	10,000 S.F.	1.46	1.38	2.84
2300	Each additional floor, 2000 S.F.	.69	1.13	1.82
2350	10,000 S.F.	.59	1.09	1.68
2400	Extra hazard, one floor, 2000 S.F.	4.87	2.76	7.63
2450	10,000 S.F.	1.60	1.68	3.28
2500	Each additional floor, 2000 S.F.	.92	1.63	2.55
2550	10,000 S.F.	.77	1.39	2.16
3050	Grooved steel, black, sch. 10 pipe light hazard, one floor,			
3060	2000 S.F.	4.67	2.17	6.84
3100	10,000 S.F.	1.27	1.03	2.30
3150	Each additional floor, 2000 S.F.	.67	1	1.67
3200	10,000 S.F.	.49	.81	1.30
3250	Ordinary hazard, one floor, 2000 S.F.	4.69	2.28	6.97
3300	10,000 S.F.	1.35	1.29	2.64
3350	Each additional floor, 2000 S.F.	.68	1.10	1.78
3400	10,000 S.F.	.57	1.07	1.64
3450	Extra hazard, one floor, 2000 S.F.	4.86	2.73	7.59
3500	10,000 S.F.	1.55	1.66	3.21
3550	Each additional floor, 2000 S.F.	.91	1.60	2.51
3600	10,000 S.F.	.74	1.37	2.11
4060	Copper tubing, type M, light hazard, one floor, 2000 S.F.	4.72	2.16	6.88
4100	10,000 S.F.	1.37	1.05	2.42
4150	Each additional floor, 2000 S.F.	.72	1	1.72
4200	10,000 S.F.	.59	.83	1.42
4250	Ordinary hazard, one floor, 2000 S.F.	4.79	2.35	7.14
4300	10,000 S.F.	1.48	1.22	2.70
4350	Each additional floor, 2000 S.F.	.72	1.03	1.75
4400	10,000 S.F.	.61	.89	1.50
4450	Extra hazard, one floor, 2000 S.F.	4.97	2.75	7.72
4500	10,000 S.F.	2.07	1.82	3.89
4550	Each additional floor, 2000 S.F.	1.02	1.62	2.64
4600	10,000 S.F.	1.03	1.49	2.52
5060	Copper tubing, type M, T-drill system, light hazard, one floor 2000 S.F.	4.73	2.06	6.79
5100	10,000 S.F.	1.35	.90	2.25
5150	Each additional floor, 2000 S.F.	.79	.95	1.74
5200	10,000 S.F.	.57	.68	1.25
5250	Ordinary hazard, one floor, 2000 S.F.	4.74	2.09	6.83
5300	10,000 S.F.	1.46	1.10	2.56
5350	Each additional floor, 2000 S.F.	.73	.91	1.64
5400	10,000 S.F.	.68	.88	1.56
5450	Extra hazard, one floor, 2000 S.F.	4.82	2.38	7.20
5500	10,000 S.F.	1.79	1.34	3.13
5550	Each additional floor, 2000 S.F.	.87	1.25	2.12
5600	10,000 S.F.	.80	1.04	1.84

For expanded coverage of these items see *Means Mechanical or Plumbing Cost Data 1997*

MECHANICAL — A8.2 — Fire Protection

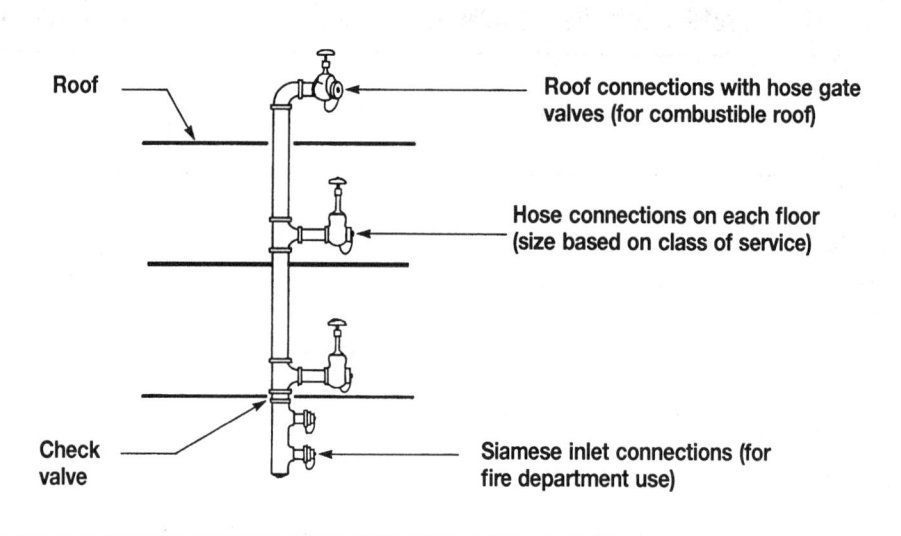

Roof — Roof connections with hose gate valves (for combustible roof)

Hose connections on each floor (size based on class of service)

Check valve — Siamese inlet connections (for fire department use)

8.2-310	Wet Standpipe Risers, Class I	MAT.	INST.	TOTAL
0550	Wet standpipe risers, Class I, steel black sch. 40, 10' height			
0560	4" diameter pipe, one floor	1,925	1,925	3,850
0580	Additional floors	565	605	1,170
0600	6" diameter pipe, one floor	3,275	3,350	6,625
0620	Additional floors	935	955	1,890
0640	8" diameter pipe, one floor	4,800	4,000	8,800
0660	Additional floors	1,250	1,150	2,400

8.2-310	Wet Standpipe Risers, Class II	MAT.	INST.	TOTAL
1030	Wet standpipe risers, Class II, steel black sch. 40, 10' height			
1040	2" diameter pipe, one floor	730	675	1,405
1060	Additional floors	272	262	534
1080	2-1/2" diameter pipe, one floor	1,025	995	2,020
1100	Additional floors	310	305	615

8.2-310	Wet Standpipe Risers, Class III	MAT.	INST.	TOTAL
1530	Wet standpipe risers, Class III, steel black sch. 40, 10' height			
1540	4" diameter pipe, one floor	2,000	1,925	3,925
1560	Additional floors	470	510	980
1580	6" diameter pipe, one floor	3,350	3,350	6,700
1600	Additional floors	970	955	1,925
1620	8" diameter pipe, one floor	4,850	4,000	8,850
1640	Additional floors	1,275	1,150	2,425

Important: See the Reference Section for critical supporting data - Location Factors & Historical Cost Indexes

MECHANICAL — A8.2 — Fire Protection

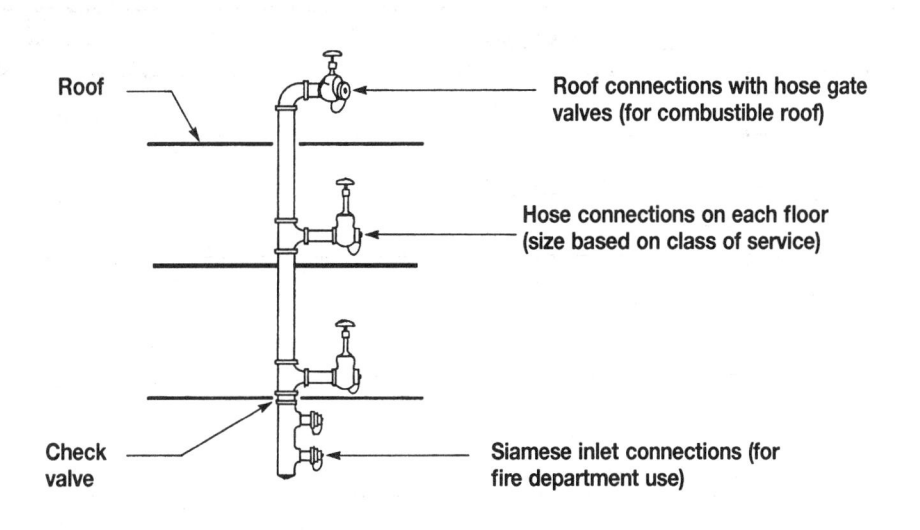

Diagram: Dry standpipe riser showing Roof, Roof connections with hose gate valves (for combustible roof), Hose connections on each floor (size based on class of service), Check valve, and Siamese inlet connections (for fire department use).

8.2-320	Dry Standpipe Risers, Class I	COST PER FLOOR		
		MAT.	INST.	TOTAL
0530	Dry standpipe riser, Class I, steel black sch. 40, 10' height			
0540	4" diameter pipe, one floor	1,300	1,525	2,825
0560	Additional floors	470	550	1,020
0580	6" diameter pipe, one floor	2,550	2,650	5,200
0600	Additional floors	845	900	1,745
0620	8" diameter pipe, one floor	3,750	3,200	6,950
0640	Additional floors	1,150	1,075	2,225

8.2-320	Dry Standpipe Risers, Class II	COST PER FLOOR		
		MAT.	INST.	TOTAL
1030	Dry standpipe risers, Class II, steel black sch. 40, 10' height			
1040	2" diameter pipe, one floor	685	710	1,395
1060	Additional floor	239	230	469
1080	2-1/2" diameter pipe, one floor	845	825	1,670
1100	Additional floors	276	274	550

8.2-320	Dry Standpipe Risers, Class III	COST PER FLOOR		
		MAT.	INST.	TOTAL
1530	Dry standpipe risers, Class III, steel black sch. 40, 10' height			
1540	4" diameter pipe, one floor	1,325	1,525	2,850
1560	Additional floors	385	495	880
1580	6" diameter pipe, one floor	2,575	2,650	5,225
1600	Additional floors	875	900	1,775
1620	8" diameter pipe, one floor	3,775	3,200	6,975
1640	Additional floor	1,200	1,075	2,275

For expanded coverage of these items see *Means Mechanical or Plumbing Cost Data 1997*

MECHANICAL | A8.2 | Fire Protection

8.2-390 Standpipe Equipment

		MAT.	INST.	TOTAL
0100	Adapters, reducing, 1 piece, FxM, hexagon, cast brass, 2-1/2" x 1-1/2"	33.50		33.50
0200	Pin lug, 1-1/2" x 1"	11.55		11.55
0250	3" x 2-1/2"	42		42
0300	For polished chrome, add 75% mat.			
0400	Cabinets, D.S. glass in door, recessed, steel box, not equipped			
0500	Single extinguisher, steel door & frame	67	87	154
0550	Stainless steel door & frame	156	87	243
0600	Valve, 2-1/2" angle, steel door & frame	87.50	58	145.50
0650	Aluminum door & frame	124	58	182
0700	Stainless steel door & frame	173	58	231
0750	Hose rack assy, 2-1/2" x 1-1/2" valve & 100' hose, steel door & frame	150	116	266
0800	Aluminum door & frame	197	116	313
0850	Stainless steel door & frame	360	116	476
0900	Hose rack assy,& extinguisher,2-1/2"x1-1/2" valve & hose,steel door & frame	188	139	327
0950	Aluminum	330	139	469
1000	Stainless steel	415	139	554
1550	Compressor, air, dry pipe system, automatic, 200 gal., 1/3 H.P.	740	298	1,038
1600	520 gal., 1 H.P.	780	298	1,078
1650	Alarm, electric pressure switch (circuit closer)	124	14.90	138.90
2500	Couplings, hose, rocker lug, cast brass, 1-1/2"	26		26
2550	2-1/2"	54		54
3000	Escutcheon plate, for angle valves, polished brass, 1-1/2"	12.65		12.65
3050	2-1/2"	31.50		31.50
3500	Fire pump, electric, w/controller, fittings, relief valve			
3550	4" pump, 30 H.P., 500 G.P.M.	13,000	2,175	15,175
3600	5" pump, 40 H.P., 1000 G.P.M.	19,000	2,450	21,450
3650	5" pump, 100 H.P., 1000 G.P.M.	21,300	2,725	24,025
3700	For jockey pump system, add	2,325	350	2,675
5000	Hose, per linear foot, synthetic jacket, lined,			
5100	300 lb. test, 1-1/2" diameter	2.11		2.11
5150	2-1/2" diameter	3.49		3.49
5200	500 lb. test, 1-1/2" diameter	2.31		2.31
5250	2-1/2" diameter	4		4
5500	Nozzle, plain stream, polished brass, 1-1/2" x 10"	26		26
5550	2-1/2" x 15" x 13/16" or 1-1/2"	99		99
5600	Heavy duty combination adjustable fog and straight stream w/handle 1-1/2"	253		253
5650	2-1/2" direct connection	360		360
6000	Rack, for 1-1/2" diameter hose 100 ft. long, steel	33.50	35	68.50
6050	Brass	53	35	88
6500	Reel, steel, for 50 ft. long 1-1/2" diameter hose	80.50	49.50	130
6550	For 75 ft. long 2-1/2" diameter hose	130	49.50	179.50
7050	Siamese, w/plugs & chains, polished brass, sidewalk, 4" x 2-1/2" x 2-1/2"	310	278	588
7100	6" x 2-1/2" x 2-1/2"	455	350	805
7200	Wall type, flush, 4" x 2-1/2" x 2-1/2"	345	139	484
7250	6" x 2-1/2" x 2-1/2"	435	151	586
7300	Projecting, 4" x 2-1/2" x 2-1/2"	305	139	444
7350	6" x 2-1/2" x 2-1/2"	505	151	656
7400	For chrome plate, add 15% mat.			
8000	Valves, angle, wheel handle, 300 Lb., rough brass, 1-1/2"	32.50	32	64.50
8050	2-1/2"	60.50	55.50	116
8100	Combination pressure restricting, 1-1/2"	50.50	32	82.50
8150	2-1/2"	114	55.50	169.50
8200	Pressure restricting, adjustable, satin brass, 1-1/2"	63	32	95
8250	2-1/2"	105	55.50	160.50
8300	Hydrolator, vent and drain, rough brass, 1-1/2"	33.50	32	65.50
8350	2-1/2"	93.50	55.50	149
8400	Cabinet assy, incls. 2-1/2" valve, adapter, rack, hose, nozzle & hydrolator	630	262	892

MECHANICAL — A8.2 Fire Protection

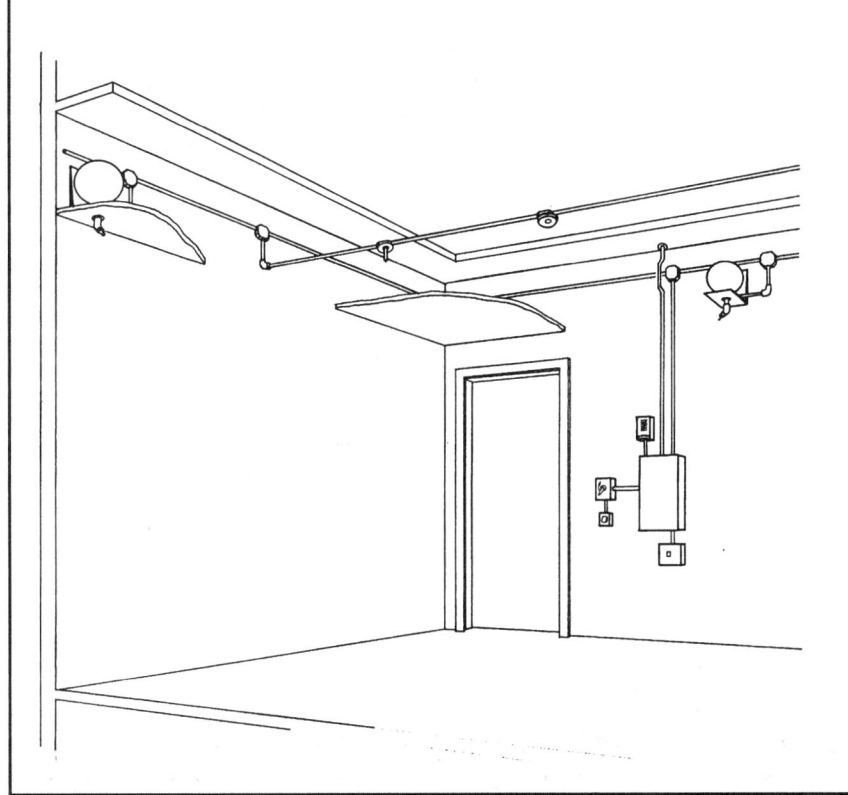

General: Automatic fire protection (suppression) systems other than water sprinklers may be desired for special environments, high risk areas, isolated locations or unusual hazards. Some typical applications would include:

1. Paint dip tanks
2. Securities vaults
3. Electronic data processing
4. Tape and data storage
5. Transformer rooms
6. Spray booths
7. Petroleum storage
8. High rack storage

Piping and wiring costs are highly variable and are not included.

8.2-810 FM200 Fire Suppression

		COST EACH		
		MAT.	INST.	TOTAL
0020	Detectors with brackets			
0040	Fixed temperature heat detector	31	45	76
0060	Rate of temperature rise detector	37	45	82
0080	Ion detector (smoke) detector	82.50	58	140.50
0200	Extinguisher agent			
0240	200 lb FM200, container	6,600	168	6,768
0280	75 lb carbon dioxide cylinder	1,050	112	1,162
0320	Dispersion nozzle			
0340	FM200 1-1/2" dispersion nozzle	60.50	26.50	87
0380	Carbon dioxide 3" x 5" dispersion nozzle	55	21	76
0420	Control station			
0440	Single zone control station with batteries	1,800	360	2,160
0470	Multizone (4) control station with batteries	2,700	720	3,420
0500	Electric mechanical release	132	184	316
0550	Manual pull station	49.50	62.50	112
0640	Battery standby power 10" x 10" x 17"	760	90	850
0740	Bell signalling device	54.50	45	99.50

8.2-810 FM200 Systems

		COST PER C.F.		
		MAT.	INST.	TOTAL
0820	Average FM200 system, minimum			1.10
0840	Maximum			2.20

For expanded coverage of these items see *Means Mechanical or Plumbing Cost Data 1997*

MECHANICAL | A8.3 | Heating

Small Electric Boiler Systems Considerations:

1. Terminal units are fin tube baseboard radiation rated at 720 BTU/hr with 200° water temperature or 820 BTU/hr steam.
2. Primary use being for residential or smaller supplementary areas, the floor levels are based on 7-1/2" ceiling heights.
3. All distribution piping is copper for boilers through 205 MBH. All piping for larger systems is steel pipe.

Large Electric Boiler System Considerations:

1. Terminal units are all unit heaters of the same size. Quantities are varied to accommodate total requirements.
2. All air is circulated through the heaters a minimum of three times per hour.
3. As the capacities are adequate for commercial use, gross output rating by floor levels are based on 10' ceiling height.
4. All distribution piping is black steel pipe.

8.3-110 Small Heating Systems, Hydronic, Electric Boilers

		MAT.	INST.	TOTAL
1100	Small heating systems, hydronic, electric boilers			
1120	Steam, 1 floor, 1480 S.F., 61 M.B.H.	8.45	4.67	13.12
1160	3,000 S.F., 123 M.B.H.	5.05	4.10	9.15
1200	5,000 S.F., 205 M.B.H.	3.86	3.76	7.62
1240	2 floors, 12,400 S.F., 512 M.B.H.	2.82	3.74	6.56
1280	3 floors, 24,800 S.F., 1023 M.B.H.	2.49	3.68	6.17
1320	34,750 S.F., 1,433 M.B.H.	2.23	3.60	5.83
1360	Hot water, 1 floor, 1,000 S.F., 41 M.B.H.	7.65	2.60	10.25
1400	2,500 S.F., 103 M.B.H.	4.89	4.65	9.54
1440	2 floors, 4,850 S.F., 205 M.B.H.	4.58	5.60	10.18
1480	3 floors, 9,700 S.F., 410 M.B.H.	4.61	5.80	10.41

8.3-120 Large Heating Systems, Hydronic, Electric Boilers

		MAT.	INST.	TOTAL
1230	Large heating systems, hydronic, electric boilers			
1240	9,280 S.F., 150 K.W., 510 M.B.H., 1 floor	3.16	1.75	4.91
1280	14,900 S.F., 240 K.W., 820 M.B.H., 2 floors	3.34	2.86	6.20
1320	18,600 S.F., 300 K.W., 1,024 M.B.H., 3 floors	3.54	3.14	6.68
1360	26,100 S.F., 420 K.W., 1,432 M.B.H., 4 floors	3.41	3.06	6.47
1400	39,100 S.F., 630 K.W., 2,148 M.B.H., 4 floors	3	2.53	5.53
1440	57,700 S.F., 900 K.W., 3,071 M.B.H., 5 floors	2.84	2.48	5.32
1480	111,700 S.F., 1,800 K.W., 6,148 M.B.H., 6 floors	2.57	2.15	4.72
1520	149,000 S.F., 2,400 K.W., 8,191 M.B.H., 8 floors	2.49	2.20	4.69
1560	223,300 S.F., 3,600 K.W., 12,283 M.B.H., 14 floors	2.73	2.52	5.25

MECHANICAL — A8.3 Heating

Boiler Selection: The maximum allowable working pressures are limited by ASME "Code for Heating Boilers" to 15 PSI for steam and 160 PSI for hot water heating boilers, with a maximum temperature limitation of 250° F. Hot water boilers are generally rated for a working pressure of 30 PSI. High pressure boilers are governed by the ASME "Code for Power Boilers" which is used almost universally for boilers operating over 15 PSIG. High pressure boilers used for a combination of heating/process loads are usually designed for 150 PSIG.

Boiler ratings are usually indicated as either Gross or Net Output. The Gross Load is equal to the Net Load plus a piping and pickup allowance. When this allowance cannot be determined, divide the gross output rating by 1.25 for a value equal to or greater than the next heat loss requirement of the building.

Table below lists installed cost per boiler and includes insulating jacket, standard controls, burner and safety controls. Costs do not include piping or boiler base pad. Outputs are Gross.

8.3-130	Boilers, Hot Water & Steam	MAT.	INST.	TOTAL
0600	Boiler, electric, steel, hot water, 12 K.W., 41 M.B.H.	3,550	800	4,350
0620	30 K.W., 103 M.B.H.	3,900	865	4,765
0640	60 K.W., 205 M.B.H.	4,950	945	5,895
0660	120 K.W., 410 M.B.H.	6,300	1,150	7,450
0680	210 K.W., 716 M.B.H.	10,500	1,725	12,225
0700	510 K.W., 1,739 M.B.H.	19,800	3,225	23,025
0720	720 K.W., 2,452 M.B.H.	24,900	3,625	28,525
0740	1,200 K.W., 4,095 M.B.H.	35,200	4,150	39,350
0760	2,100 K.W., 7,167 M.B.H.	55,000	5,250	60,250
0780	3,600 K.W., 12,283 M.B.H.	75,500	8,850	84,350
0820	Steam, 6 K.W., 20.5 M.B.H.	8,450	865	9,315
0840	24 K.W., 81.8 M.B.H.	8,800	945	9,745
0860	60 K.W., 205 M.B.H.	10,200	1,050	11,250
0880	150 K.W., 512 M.B.H.	13,200	1,600	14,800
0900	510 K.W., 1,740 M.B.H.	21,500	3,925	25,425
0920	1,080 K.W., 3,685 M.B.H.	33,000	5,650	38,650
0940	2,340 K.W., 7,984 M.B.H.	62,500	8,850	71,350
0980	Gas, cast iron, hot water, 80 M.B.H.	1,350	985	2,335
1000	100 M.B.H.	1,550	1,075	2,625
1020	163 M.B.H.	2,125	1,425	3,550
1040	280 M.B.H.	2,950	1,600	4,550
1060	544 M.B.H.	5,025	2,400	7,425
1080	1,088 M.B.H.	8,450	2,875	11,325
1100	2,000 M.B.H.	14,500	3,775	18,275
1120	2,856 M.B.H.	20,600	4,350	24,950
1140	4,720 M.B.H.	44,200	7,975	52,175
1160	6,970 M.B.H.	62,000	18,000	80,000
1180	For steam systems under 2,856 M.B.H., add 8%			
1240	Steel, hot water, 72 M.B.H.	1,850	525	2,375
1260	101 M.B.H.	2,125	585	2,710
1280	132 M.B.H.	2,425	620	3,045
1300	150 M.B.H.	2,825	700	3,525
1320	240 M.B.H.	4,325	810	5,135
1340	400 M.B.H.	6,175	1,325	7,500
1360	640 M.B.H.	8,375	1,750	10,125
1380	800 M.B.H.	9,775	2,100	11,875
1400	960 M.B.H.	12,100	2,350	14,450
1420	1,440 M.B.H.	16,900	3,000	19,900
1440	2,400 M.B.H.	27,300	5,275	32,575
1460	3,000 M.B.H.	33,700	7,025	40,725

For expanded coverage of these items see *Means Mechanical or Plumbing Cost Data 1997*

MECHANICAL | A8.3 Heating

8.3-130 Boilers, Hot Water & Steam

		COST EACH		
		MAT.	INST.	TOTAL
1520	Oil, cast iron, hot water, 109 M.B.H.	1,650	1,200	2,850
1540	173 M.B.H.	2,150	1,425	3,575
1560	236 M.B.H.	2,475	1,700	4,175
1580	1,084 M.B.H.	8,850	3,425	12,275
1600	1,600 M.B.H.	11,800	4,625	16,425
1620	2,480 M.B.H.	16,600	5,750	22,350
1640	3,550 M.B.H.	22,000	6,525	28,525
1660	Steam systems same price as hot water			
1700	Steel, hot water, 103 M.B.H.	2,225	555	2,780
1720	137 M.B.H.	2,350	660	3,010
1740	225 M.B.H.	3,350	755	4,105
1760	315 M.B.H.	4,925	960	5,885
1780	420 M.B.H.	5,550	1,325	6,875
1800	630 M.B.H.	7,150	1,750	8,900
1820	735 M.B.H.	9,550	1,925	11,475
1840	1,050 M.B.H.	13,200	2,500	15,700
1860	1,365 M.B.H.	16,900	2,850	19,750
1880	1,680 M.B.H.	19,900	3,200	23,100
1900	2,310 M.B.H.	25,600	4,400	30,000
1920	2,835 M.B.H.	31,700	6,200	37,900
1940	3,150 M.B.H.	35,600	8,100	43,700

MECHANICAL — A8.3 Heating

Unit Heater

Fossil Fuel Boiler System Considerations:

1. Terminal units are horizontal unit heaters. Quantities are varied to accommodate total heat loss per building.
2. Unit heater selection was determined by their capacity to circulate the building volume a minimum of three times per hour in addition to the BTU output.
3. Systems shown are forced hot water. Steam boilers cost slightly more than hot water boilers. However, this is compensated for by the smaller size or fewer terminal units required with steam.
4. Floor levels are based on 10' story heights.
5. MBH requirements are gross boiler output.

8.3-141 Heating Systems, Unit Heaters

Line	Description	MAT.	INST.	TOTAL
1260	Heating systems, hydronic, fossil fuel, terminal unit heaters,			
1280	Cast iron boiler, gas, 80 M.B.H., 1,070 S.F. bldg.	6.65	5.55	12.20
1320	163 M.B.H., 2,140 S.F. bldg.	4.56	3.76	8.32
1360	544 M.B.H., 7,250 S.F. bldg.	3.08	2.62	5.70
1400	1,088 M.B.H., 14,500 S.F. bldg.	2.74	2.50	5.24
1440	3,264 M.B.H., 43,500 S.F. bldg.	2.32	1.84	4.16
1480	5,032 M.B.H., 67,100 S.F. bldg.	2.54	1.95	4.49
1520	Oil, 109 M.B.H., 1,420 S.F. bldg.	6.50	4.96	11.46
1560	235 M.B.H., 3,150 S.F. bldg.	4.30	3.57	7.87
1600	940 M.B.H., 12,500 S.F. bldg.	3.26	2.27	5.53
1640	1,600 M.B.H., 21,300 S.F. bldg.	3.15	2.18	5.33
1680	2,480 M.B.H., 33,100 S.F. bldg.	3.10	1.96	5.06
1720	3,350 M.B.H., 44,500 S.F. bldg.	2.75	2.01	4.76
1760	Coal, 148 M.B.H., 1,975 S.F. bldg.	4.94	3.23	8.17
1800	300 M.B.H., 4,000 S.F. bldg.	3.88	2.52	6.40
1840	2,360 M.B.H., 31,500 S.F. bldg.	2.76	2.04	4.80
1880	Steel boiler, gas, 72 M.B.H., 1,020 S.F. bldg.	6	4.11	10.11
1920	240 M.B.H., 3,200 S.F. bldg.	4.26	3.26	7.52
1960	480 M.B.H., 6,400 S.F. bldg.	3.48	2.42	5.90
2000	800 M.B.H., 10,700 S.F. bldg.	3.05	2.15	5.20
2040	1,960 M.B.H., 26,100 S.F. bldg.	2.75	1.92	4.67
2080	3,000 M.B.H., 40,000 S.F. bldg.	2.73	1.97	4.70
2120	Oil, 97 M.B.H., 1,300 S.F. bldg.	6.80	4.56	11.36
2160	315 M.B.H., 4,550 S.F. bldg.	4.06	2.34	6.40
2200	525 M.B.H., 7,000 S.F. bldg.	4.09	2.37	6.46
2240	1,050 M.B.H., 14,000 S.F. bldg.	3.58	2.39	5.97
2280	2,310 M.B.H. 30,800 S.F. bldg.	3.40	2.07	5.47
2320	3,150 M.B.H., 42,000 S.F. bldg.	3.26	2.11	5.37

For expanded coverage of these items see Means Mechanical or Plumbing Cost Data 1997

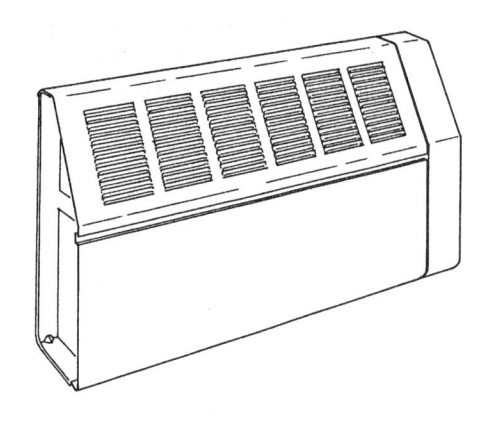

MECHANICAL — A8.3 Heating

Fossil Fuel Boiler System Considerations:

1. Terminal units are commercial steel fin tube radiation. Quantities are varied to accommodate total heat loss per building.
2. Systems shown are forced hot water. Steam boilers cost slightly more than hot water boilers. However, this is compensated for by the smaller size or fewer terminal units required with steam.
3. Floor levels are based on 10' story heights.
4. MBH requirements are gross boiler output.

Fin Tube Radiator

8.3-142	Heating System, Fin Tube Radiation	MAT.	INST.	TOTAL
3230	Heating systems, hydronic, fossil fuel, fin tube radiation			
3240	Cast iron boiler, gas, 80 MBH, 1,070 S.F. bldg.	8.75	8.40	17.15
3280	169 M.B.H., 2,140 S.F. bldg.	5.55	5.30	10.85
3320	544 M.B.H., 7,250 S.F. bldg.	4.59	4.51	9.10
3360	1,088 M.B.H., 14,500 S.F. bldg.	4.34	4.44	8.78
3400	3,264 M.B.H., 43,500 S.F. bldg.	4.04	3.87	7.91
3440	5,032 M.B.H., 67,100 S.F. bldg.	4.27	3.98	8.25
3480	Oil, 109 M.B.H., 1,420 S.F. bldg.	10.10	9.15	19.25
3520	235 M.B.H., 3,150 S.F. bldg.	5.80	5.40	11.20
3560	940 M.B.H., 12,500 S.F. bldg.	4.88	4.23	9.11
3600	1,600 M.B.H., 21,300 S.F. bldg.	4.87	4.21	9.08
3640	2,480 M.B.H., 33,100 S.F. bldg.	4.83	3.98	8.81
3680	3,350 M.B.H., 44,500 S.F. bldg.	4.47	4.03	8.50
3720	Coal, 148 M.B.H., 1,975 S.F. bldg.	6.40	5.05	11.45
3760	300 M.B.H., 4,000 S.F. bldg.	5.30	4.35	9.65
3800	2,360 M.B.H., 31,500 S.F. bldg.	4.44	4.03	8.47
3840	Steel boiler, gas, 72 M.B.H., 1,020 S.F. bldg.	9.50	7.95	17.45
3880	240 M.B.H., 3,200 S.F. bldg.	5.85	5.20	11.05
3920	480 M.B.H., 6,400 S.F. bldg.	5	4.31	9.31
3960	800 M.B.H., 10,700 S.F. bldg.	5.15	4.58	9.73
4000	1,960 M.B.H., 26,100 S.F. bldg.	4.42	3.91	8.33
4040	3,000 M.B.H., 40,000 S.F. bldg.	4.38	3.96	8.34
4080	Oil, 97 M.B.H., 1,300 S.F. bldg.	8.35	7.40	15.75
4120	315 M.B.H., 4,550 S.F. bldg.	5.45	4.07	9.52
4160	525 M.B.H., 7,000 S.F. bldg.	5.75	4.35	10.10
4200	1,050 M.B.H., 14,000 S.F. bldg.	5.25	4.37	9.62
4240	2,310 M.B.H., 30,800 S.F. bldg.	5.05	4.05	9.10
4280	3,150 M.B.H., 42,000 S.F. bldg.	4.93	4.10	9.03

COST PER S.F.

MECHANICAL — A8.3 Heating

Forced Hot Water Heating System

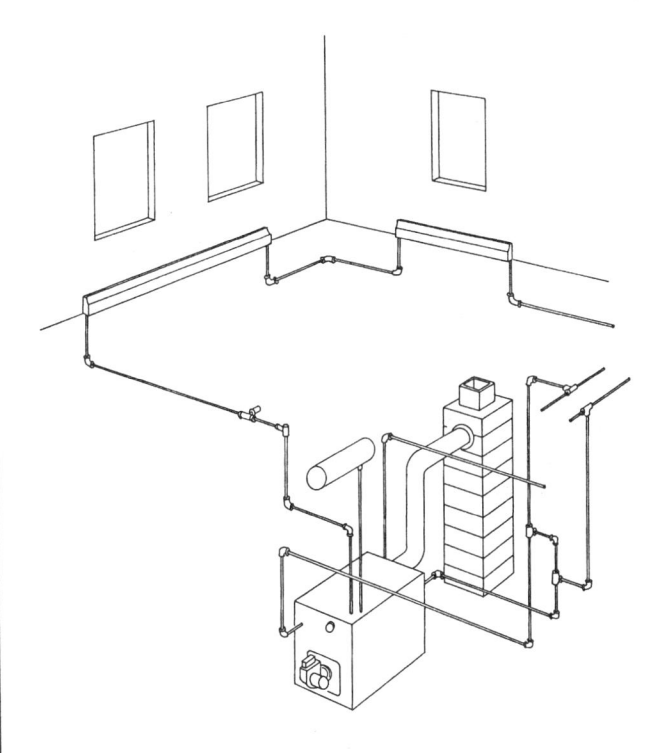

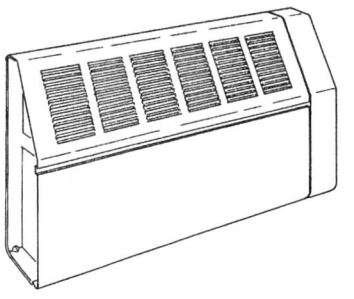

Fin Tube Radiation

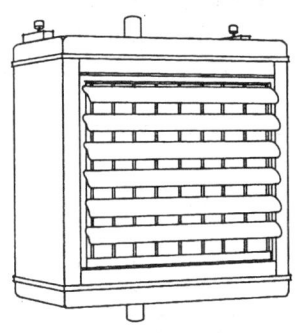

Terminal Unit Heater

8.3-151 Apartment Building Heating - Fin Tube Radiation

		MAT.	INST.	TOTAL
1740	Heating systems, fin tube radiation, forced hot water			
1760	1,000 S.F. area, 10,000 C.F. volume	4.54	3.55	8.09
1800	10,000 S.F. area, 100,000 C.F. volume	1.52	2.19	3.71
1840	20,000 S.F. area, 200,000 C.F. volume	1.61	2.49	4.10
1880	30,000 S.F. area, 300,000 C.F. volume	1.61	2.38	3.99

COST PER S.F.

8.3-161 Commercial Building Heating - Fin Tube Radiation

		MAT.	INST.	TOTAL
1940	Heating systems, fin tube radiation, forced hot water			
1960	1,000 S.F. bldg, one floor	9.90	8.20	18.10
2000	10,000 S.F., 100,000 C.F., total two floors	2.90	3.15	6.05
2040	100,000 S.F., 1,000,000 C.F., total three floors	1.29	1.40	2.69
2080	1,000,000 S.F., 10,000,000 C.F., total five floors	.60	.52	1.12

8.3-162 Commercial Bldg. Heating - Terminal Unit Heaters

		MAT.	INST.	TOTAL
1860	Heating systems, terminal unit heaters, forced hot water			
1880	1,000 S.F. bldg., one floor	9.70	7.50	17.20
1920	10,000 S.F. bldg., 100,000 C.F. total two floors	2.51	2.61	5.12
1960	100,000 S.F. bldg., 1,000,000 C.F. total three floors	1.30	1.27	2.57
2000	1,000,000 S.F. bldg., 10,000,000 C.F. total five floors	.84	.59	1.43

For expanded coverage of these items see *Means Mechanical or Plumbing Cost Data 1997*

MECHANICAL — A8.4 Cooling

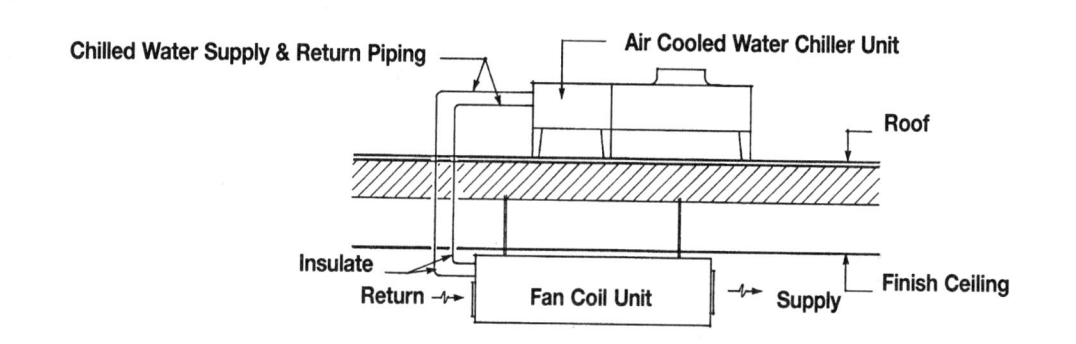

*Cooling requirements would lead to a choice of multiple chillers.

8.4-110 Chilled Water, Air Cooled Condenser Systems

Line	Description	MAT.	INST.	TOTAL
1180	Packaged chiller, air cooled, with fan coil unit			
1200	Apartment corridors, 3,000 S.F., 5.50 ton	4.68	4.57	9.25
1240	6,000 S.F., 11.00 ton	3.68	3.67	7.35
1280	10,000 S.F., 18.33 ton	3.17	2.79	5.96
1320	20,000 S.F., 36.66 ton	2.56	2.07	4.63
1360	40,000 S.F., 73.33 ton	3.03	2.19	5.22
1440	Banks and libraries, 3,000 S.F., 12.50 ton	6.80	5.35	12.15
1480	6,000 S.F., 25.00 ton	6.25	4.36	10.61
1520	10,000 S.F., 41.66 ton	5.05	3.17	8.22
1560	20,000 S.F., 83.33 ton	5.45	3.10	8.55
1600	40,000 S.F., 167 ton*			
1680	Bars and taverns, 3,000 S.F., 33.25 ton	12.50	6.70	19.20
1720	6,000 S.F., 66.50 ton	11.65	5.80	17.45
1760	10,000 S.F., 110.83 ton	9.25	2.66	11.91
1800	20,000 S.F., 220 ton*			
1840	40,000 S.F., 440 ton*			
1920	Bowling alleys, 3,000 S.F., 17.00 ton	8.50	6	14.50
1960	6,000 S.F., 34.00 ton	6.90	4.43	11.33
2000	10,000 S.F., 56.66 ton	6.25	3.52	9.77
2040	20,000 S.F., 113.33 ton	6.05	3.25	9.30
2080	40,000 S.F., 227 ton*			
2160	Department stores, 3,000 S.F., 8.75 ton	6.50	5.10	11.60
2200	6,000 S.F., 17.50 ton	4.81	3.94	8.75
2240	10,000 S.F., 29.17 ton	3.97	2.94	6.91
2280	20,000 S.F., 58.33 ton	3.52	2.34	5.86
2320	40,000 S.F., 116.66 ton	3.83	2.43	6.26
2400	Drug stores, 3,000 S.F., 20.00 ton	9.70	6.20	15.90
2440	6,000 S.F., 40.00 ton	8.10	4.85	12.95
2480	10,000 S.F., 66.66 ton	8.35	4.51	12.86
2520	20,000 S.F., 133.33 ton	7.30	3.67	10.97
2560	40,000 S.F., 267 ton*			
2640	Factories, 2,000 S.F., 10.00 ton	5.95	5.15	11.10
2680	6,000 S.F., 20.00 ton	5.40	4.16	9.56
2720	10,000 S.F., 33.33 ton	4.33	3.02	7.35
2760	20,000 S.F., 66.66 ton	4.67	2.91	7.58
2800	40,000 S.F., 133.33 ton	4.19	2.53	6.72
2880	Food supermarkets, 3,000 S.F., 8.50 ton	6.40	5.05	11.45
2920	6,000 S.F., 17.00 ton	4.72	3.91	8.63
2960	10,000 S.F., 28.33 ton	3.81	2.85	6.66
3000	20,000 S.F., 56.66 ton	3.40	2.28	5.68
3040	40,000 S.F., 113.33 ton	3.71	2.39	6.10
3120	Medical centers, 3,000 S.F., 7.00 ton	5.70	4.93	10.63
3160	6,000 S.F., 14.00 ton	4.20	3.79	7.99
3200	10,000 S.F., 23.33 ton	3.74	2.89	6.63

Important: See the Reference Section for critical supporting data - Location Factors & Historical Cost Indexes

MECHANICAL — A8.4 Cooling

8.4-110 Chilled Water, Air Cooled Condenser Systems

		COST PER S.F.		
		MAT.	INST.	TOTAL
3240	20,000 S.F., 46.66 ton	3.08	2.20	5.28
3280	40,000 S.F., 93.33 ton	3.32	2.28	5.60
3360	Offices, 3,000 S.F., 9.50 ton	5.60	5	10.60
3400	6,000 S.F., 19.00 ton	5.20	4.12	9.32
3440	10,000 S.F., 31.66 ton	4.18	2.98	7.16
3480	20,000 S.F., 63.33 ton	4.62	2.93	7.55
3520	40,000 S.F., 126.66 ton	4.09	2.52	6.61
3600	Restaurants, 3,000 S.F., 15.00 ton	7.65	5.60	13.25
3640	6,000 S.F., 30.00 ton	6.50	4.41	10.91
3680	10,000 S.F., 50.00 ton	5.85	3.52	9.37
3720	20,000 S.F., 100.00 ton	6.05	3.37	9.42
3760	40,000 S.F., 200 ton*			
3840	Schools and colleges, 3,000 S.F., 11.50 ton	6.45	5.30	11.75
3880	6,000 S.F., 23.00 ton	5.90	4.29	10.19
3920	10,000 S.F., 38.33 ton	4.77	3.11	7.88
3960	20,000 S.F., 76.66 ton	5.25	3.11	8.36
4000	40,000 S.F., 153 ton*			

For expanded coverage of these items see Means Mechanical or Plumbing Cost Data 1997

MECHANICAL — A8.4 Cooling

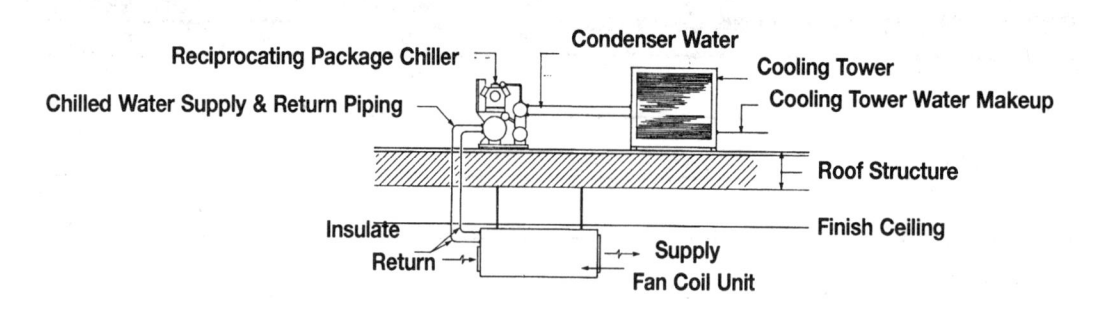

*Cooling requirements would lead to a choice of multiple chillers

8.4-120	Chilled Water, Cooling Tower Systems	MAT.	INST.	TOTAL
1300	Packaged chiller, water cooled, with fan coil unit			
1320	Apartment corridors, 4,000 S.F., 7.33 ton	5.70	4.55	10.25
1360	6,000 S.F., 11.00 ton	4.47	3.88	8.35
1400	10,000 S.F., 18.33 ton	3.66	2.92	6.58
1440	20,000 S.F., 26.66 ton	2.88	2.19	5.07
1480	40,000 S.F., 73.33 ton	3.28	2.31	5.59
1520	60,000 S.F., 110.00 ton	3.22	2.37	5.59
1600	Banks and libraries, 4,000 S.F., 16.66 ton	8.15	5.10	13.25
1640	6,000 S.F., 25.00 ton	6.60	4.42	11.02
1680	10,000 S.F., 41.66 ton	5.20	3.33	8.53
1720	20,000 S.F., 83.33 ton	5.95	3.35	9.30
1760	40,000 S.F., 166.66 ton	5.80	4.07	9.87
1800	60,000 S.F., 250.00 ton	5.70	4.41	10.11
1880	Bars and taverns, 4,000 S.F., 44.33 ton	12.50	6.55	19.05
1920	6,000 S.F., 66.50 ton	12.80	6.80	19.60
1960	10,000 S.F., 110.83 ton	12.75	5.85	18.60
2000	20,000 S.F., 221.66 ton	11.95	6.35	18.30
2040	40,000 S.F., 440 ton			
2080	60,000 S.F., 660 ton			
2160	Bowling alleys, 4,000 S.F., 22.66 ton	9.45	5.55	15
2200	6,000 S.F., 34.00 ton	8	4.93	12.93
2240	10,000 S.F., 56.66 ton	6.45	3.70	10.15
2280	20,000 S.F., 113.33 ton	7	3.65	10.65
2320	40,000 S.F., 226.66 ton	6.80	4.37	11.17
2360	60,000 S.F., 340 ton			
2440	Department stores, 4,000 S.F., 11.66 ton	6.30	4.86	11.16
2480	6,000 S.F., 17.50 ton	6	4.13	10.13
2520	10,000 S.F., 29.17 ton	4.36	3.07	7.43
2560	20,000 S.F., 58.33 ton	3.37	2.32	5.69
2600	40,000 S.F., 116.66 ton	4.10	2.52	6.62
2640	60,000 S.F., 175.00 ton	4.71	3.96	8.67
2720	Drug stores, 4,000 S.F., 26.66 ton	9.85	5.70	15.55
2760	6,000 S.F., 40.00 ton	8.25	4.97	13.22
2800	10,000 S.F., 66.66 ton	8.10	4.55	12.65
2840	20,000 S.F., 133.33 ton	8	3.94	11.94
2880	40,000 S.F., 266.67 ton	7.80	4.97	12.77
2920	60,000 S.F., 400 ton			
3000	Factories, 4,000 S.F., 13.33 ton	6.95	4.83	11.78
3040	6,000 S.F., 20.00 ton	6.05	4.14	10.19
3080	10,000 S.F., 33.33 ton	4.78	3.18	7.96
3120	20,000 S.F., 66.66 ton	4.49	2.89	7.38
3160	40,000 S.F., 133.33 ton	4.53	2.67	7.20
3200	60,000 S.F., 200.00 ton	5.05	4.15	9.20
3280	Food supermarkets, 4,000 S.F., 11.33 ton	6.20	4.83	11.03
3320	6,000 S.F., 17.00 ton	5.35	4.01	9.36
3360	10,000 S.F., 28.33 ton	4.26	3.04	7.30

MECHANICAL — A8.4 Cooling

8.4-120 Chilled Water, Cooling Tower Systems

		COST PER S.F.		
		MAT.	INST.	TOTAL
3400	20,000 S.F., 56.66 ton	3.45	2.33	5.78
3440	40,000 S.F., 113.33 ton	4.04	2.51	6.55
3480	60,000 S.F., 170.00 ton	4.64	3.95	8.59
3560	Medical centers, 4,000 S.F., 9.33 ton	5.35	4.43	9.78
3600	6,000 S.F., 14.00 ton	5.15	3.94	9.09
3640	10,000 S.F., 23.33 ton	3.94	2.92	6.86
3680	20,000 S.F., 46.66 ton	3.06	2.26	5.32
3720	40,000 S.F., 93.33 ton	3.70	2.41	6.11
3760	60,000 S.F., 140.00 ton	4.28	3.86	8.14
3840	Offices, 4,000 S.F., 12.66 ton	6.70	4.78	11.48
3880	6,000 S.F., 19.00 ton	5.90	4.23	10.13
3920	10,000 S.F., 31.66 ton	4.70	3.20	7.90
3960	20,000 S.F., 63.33 ton	4.41	2.90	7.31
4000	40,000 S.F., 126.66 ton	4.90	3.85	8.75
4040	60,000 S.F., 190.00 ton	4.88	4.09	8.97
4120	Restaurants, 4,000 S.F., 20.00 ton	8.50	5.15	13.65
4160	6,000 S.F., 30.00 ton	7.20	4.61	11.81
4200	10,000 S.F., 50.00 ton	5.90	3.54	9.44
4240	20,000 S.F., 100.00 ton	6.75	3.57	10.32
4280	40,000 S.F., 200.00 ton	6.10	4.14	10.24
4320	60,000 S.F., 300.00 ton	6.40	4.65	11.05
4400	Schools and colleges, 4,000 S.F., 15.33 ton	7.70	4.98	12.68
4440	6,000 S.F., 23.00 ton	6.25	4.34	10.59
4480	10,000 S.F., 38.33 ton	4.91	3.26	8.17
4520	20,000 S.F., 76.66 ton	5.65	3.30	8.95
4560	40,000 S.F., 153.33 ton	5.45	3.97	9.42
4600	60,000 S.F., 230.00 ton	5.30	4.24	9.54

For expanded coverage of these items see *Means Mechanical or Plumbing Cost Data 1997*

MECHANICAL — A8.4 Cooling

Rooftop, Single Zone System

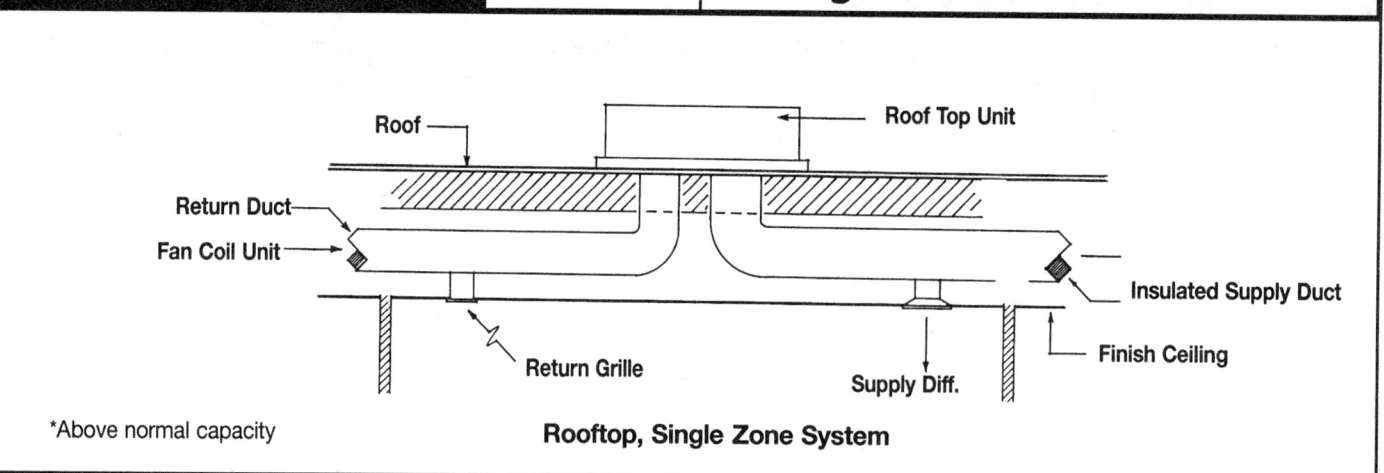

*Above normal capacity

8.4-210	Rooftop Single Zone Unit Systems	MAT.	INST.	TOTAL
1260	Rooftop, single zone, air conditioner			
1280	Apartment corridors, 500 S.F., .92 ton	4.83	1.76	6.59
1320	1,000 S.F., 1.83 ton	4.80	1.75	6.55
1360	1500 S.F., 2.75 ton	2.94	1.55	4.49
1400	3,000 S.F., 5.50 ton	2.58	1.71	4.29
1440	5,000 S.F., 9.17 ton	2.61	1.69	4.30
1480	10,000 S.F., 18.33 ton	2.75	1.68	4.43
1560	Banks or libraries, 500 S.F., 2.08 ton	10.90	3.97	14.87
1600	1,000 S.F., 4.17 ton	6.70	3.54	10.24
1640	1,500 S.F., 6.25 ton	5.85	3.90	9.75
1680	3,000 S.F., 12.50 ton	5.90	3.86	9.76
1720	5,000 S.F., 20.80 ton	6.25	3.83	10.08
1760	10,000 S.F., 41.67 ton	5.90	3.83	9.73
1840	Bars and taverns, 500 S.F. 5.54 ton	13.15	6.50	19.65
1880	1,000 S.F., 11.08 ton	13.30	6.40	19.70
1920	1,500 S.F., 16.62 ton	12	6.35	18.35
1960	3,000 S.F., 33.25 ton	13.55	6.25	19.80
2000	5,000 S.F., 55.42 ton	13	6.30	19.30
2040	10,000 S.F., 110.83 ton*			
2080	Bowling alleys, 500 S.F., 2.83 ton	9.10	4.80	13.90
2120	1,000 S.F., 5.67 ton	7.95	5.30	13.25
2160	1,500 S.F., 8.50 ton	8.05	5.25	13.30
2200	3,000 S.F., 17.00 ton	7.40	5.20	12.60
2240	5,000 S.F., 28.33 ton	8.15	5.15	13.30
2280	10,000 S.F., 56.67 ton	7.85	5.20	13.05
2360	Department stores, 500 S.F., 1.46 ton	7.65	2.79	10.44
2400	1,000 S.F., 2.92 ton	4.68	2.48	7.16
2480	3,000 S.F., 8.75 ton	4.14	2.70	6.84
2520	5,000 S.F., 14.58 ton	3.80	2.69	6.49
2560	10,000 S.F., 29.17 ton	4.21	2.66	6.87
2640	Drug stores, 500 S.F., 3.33 ton	10.70	5.65	16.35
2680	1,000 S.F., 6.67 ton	9.35	6.25	15.60
2720	1,500 S.F., 10.00 ton	9.45	6.15	15.60
2760	3,000 S.F., 20.00 ton	10	6.15	16.15
2800	5,000 S.F., 33.33 ton	9.60	6.10	15.70
2840	10,000 S.F., 66.67 ton	9.25	6.10	15.35
2920	Factories, 500 S.F., 1.67 ton	8.70	3.18	11.88
3000	1,500 S.F., 5.00 ton	4.68	3.12	7.80
3040	3,000 S.F., 10.00 ton	4.73	3.09	7.82
3080	5,000 S.F., 16.67 ton	4.34	3.08	7.42
3120	10,000 S.F., 33.33 ton	4.81	3.04	7.85
3200	Food supermarkets, 500 S.F., 1.42 ton	7.45	2.71	10.16
3240	1,000 S.F., 2.83 ton	4.50	2.40	6.90
3280	1,500 S.F., 4.25 ton	3.98	2.65	6.63

MECHANICAL — A8.4 Cooling

8.4-210 Rooftop Single Zone Unit Systems

		COST PER S.F.		
		MAT.	INST.	TOTAL
3320	3,000 S.F., 8.50 ton	4.02	2.62	6.64
3360	5,000 S.F., 14.17 ton	3.68	2.61	6.29
3400	10,000 S.F., 28.33 ton	4.08	2.59	6.67
3480	Medical centers, 500 S.F., 1.17 ton	6.10	2.24	8.34
3520	1,000 S.F., 2.33 ton	6.10	2.23	8.33
3560	1,500 S.F., 3.50 ton	3.75	1.98	5.73
3640	5,000 S.F., 11.67 ton	3.32	2.16	5.48
3680	10,000 S.F., 23.33 ton	3.50	2.15	5.65
3760	Offices, 500 S.F., 1.58 ton	8.30	3.02	11.32
3800	1,000 S.F., 3.17 ton	5.10	2.69	7.79
3840	1,500 S.F., 4.75 ton	4.45	2.96	7.41
3880	3,000 S.F., 9.50 ton	4.50	2.93	7.43
3920	5,000 S.F., 15.83 ton	4.13	2.92	7.05
3960	10,000 S.F., 31.67 ton	4.57	2.89	7.46
4000	Restaurants, 500 S.F., 2.50 ton	13.10	4.78	17.88
4040	1,000 S.F., 5.00 ton	7.05	4.68	11.73
4080	1,500 S.F., 7.50 ton	7.10	4.63	11.73
4120	3,000 S.F., 15.00 ton	6.50	4.61	11.11
4160	5,000 S.F., 25.00 ton	7.50	4.61	12.11
4200	10,000 S.F., 50.00 ton	6.95	4.58	11.53
4240	Schools and colleges, 500 S.F., 1.92 ton	10.05	3.67	13.72
4280	1,000 S.F., 3.83 ton	6.15	3.25	9.40
4360	3,000 S.F., 11.50 ton	5.45	3.54	8.99
4400	5,000 S.F., 19.17 ton	5.75	3.52	9.27

For expanded coverage of these items see *Means Mechanical or Plumbing Cost Data 1997*

MECHANICAL — A8.4 Cooling

*Note A: Small single zone unit recommended.

*Note B: A combination of multizone units recommended.

8.4-220	Rooftop Multizone Unit Systems	MAT.	INST.	TOTAL
1240	Rooftop, multizone, air conditioner			
1260	Apartment corridors, 1,500 S.F., 2.75 ton. See Note A.			
1280	3,000 S.F., 5.50 ton	7.45	3.39	10.84
1320	10,000 S.F., 18.30 ton	5.55	3.05	8.60
1360	15,000 S.F., 27.50 ton	4.90	3.02	7.92
1400	20,000 S.F., 36.70 ton	4.96	3.02	7.98
1440	25,000 S.F., 45.80 ton	4.78	3.02	7.80
1520	Banks or libraries, 1,500 S.F., 6.25 ton	16.90	7.70	24.60
1560	3,000 S.F., 12.50 ton	14.60	7.35	21.95
1600	10,000 S.F., 41.67 ton	10.85	6.85	17.70
1640	15,000 S.F., 62.50 ton	7.70	6.45	14.15
1680	20,000 S.F., 83.33 ton	7.70	6.45	14.15
1720	25,000 S.F., 104.00 ton	7.70	6.45	14.15
1800	Bars and taverns, 1,500 S.F., 16.62 ton	34.50	12.15	46.65
1840	3,000 S.F., 33.24 ton	26	10.90	36.90
1880	10,000 S.F., 110.83 ton	16.45	9.85	26.30
1920	15,000 S.F., 165 ton, See Note B			
1960	20,000 S.F., 220 ton, See Note B			
2000	25,000 S.F., 275 ton, See Note B			
2080	Bowling alleys, 1,500 S.F., 8.50 ton	23	10.45	33.45
2120	3,000 S.F., 17.00 ton	19.85	9.95	29.80
2160	10,000 S.F., 56.70 ton	14.75	9.35	24.10
2200	15,000 S.F., 85.00 ton	10.50	8.80	19.30
2240	20,000 S.F., 113.00 ton	10.45	8.80	19.25
2280	25,000 S.F., 140.00 ton see Note B			
2360	Department stores, 1,500 S.F., 4.37 ton, See Note A.			
2400	3,000 S.F., 8.75 ton	11.80	5.40	17.20
2440	10,000 S.F., 29.17 ton	7.90	4.80	12.70
2520	20,000 S.F., 58.33 ton	5.40	4.53	9.93
2560	25,000 S.F., 72.92 ton	5.40	4.52	9.92
2640	Drug stores, 1,500 S.F., 10.00 ton	27	12.30	39.30
2680	3,000 S.F., 20.00 ton	20	11.10	31.10
2720	10,000 S.F., 66.66 ton	12.35	10.35	22.70
2760	15,000 S.F., 100.00 ton	12.35	10.35	22.70
2800	20,000 S.F., 135 ton, See Note B			
2840	25,000 S.F., 165 ton, See Note B			
2920	Factories, 1,500 S.F., 5 ton, See Note A			
3000	10,000 S.F., 33.33 ton	9	5.50	14.50
3040	15,000 S.F., 50.00 ton	8.70	5.50	14.20
3080	20,000 S.F., 66.66 ton	6.20	5.20	11.40
3120	25,000 S.F., 83.33 ton	6.20	5.15	11.35
3200	Food supermarkets, 1,500 S.F., 4.25 ton, See Note A			

Important: See the Reference Section for critical supporting data - Location Factors & Historical Cost Indexes

MECHANICAL — A8.4 Cooling

8.4-220 Rooftop Multizone Unit Systems

		COST PER S.F.		
		MAT.	INST.	TOTAL
3240	3,000 S.F., 8.50 ton	11.50	5.25	16.75
3280	10,000 S.F., 28.33 ton	7.55	4.66	12.21
3320	15,000 S.F., 42.50 ton	7.40	4.67	12.07
3360	20,000 S.F., 56.67 ton	5.25	4.40	9.65
3400	25,000 S.F., 70.83 ton	5.25	4.40	9.65
3480	Medical centers, 1,500 S.F., 3.5 ton, See Note A			
3520	3,000 S.F., 7.00 ton	9.45	4.31	13.76
3560	10,000 S.F., 23.33 ton	6.45	3.83	10.28
3600	15,000 S.F., 35.00 ton	6.30	3.83	10.13
3640	20,000 S.F., 46.66 ton	6.10	3.85	9.95
3680	25,000 S.F., 58.33 ton	4.33	3.62	7.95
3760	Offices, 1,500 S.F., 4.75 ton, See Note A			
3800	3,000 S.F., 9.50 ton	12.85	5.85	18.70
3840	10,000 S.F., 31.66 ton	8.55	5.20	13.75
3920	20,000 S.F., 63.33 ton	5.85	4.92	10.77
3960	25,000 S.F., 79.16 ton	5.85	4.91	10.76
4000	Restaurants, 1,500 S.F., 7.50 ton	20.50	9.25	29.75
4040	3,000 S.F., 15.00 ton	17.50	8.80	26.30
4080	10,000 S.F., 50.00 ton	13	8.25	21.25
4120	15,000 S.F., 75.00 ton	9.25	7.75	17
4160	20,000 S.F., 100.00 ton	9.25	7.75	17
4200	25,000 S.F., 125 ton, See Note B			
4240	Schools and colleges, 1,500 S.F., 5.75 ton	15.55	7.10	22.65
4320	10,000 S.F., 38.33 ton	10	6.30	16.30
4360	15,000 S.F., 57.50 ton	7.10	5.95	13.05
4400	20,000 S.F., 76.66 ton	7.10	5.95	13.05
4440	25,000 S.F., 95.83 ton	7.10	5.95	13.05

For expanded coverage of these items see *Means Mechanical or Plumbing Cost Data 1997*

MECHANICAL — A8.4 Cooling

Self-Contained Water Cooled System

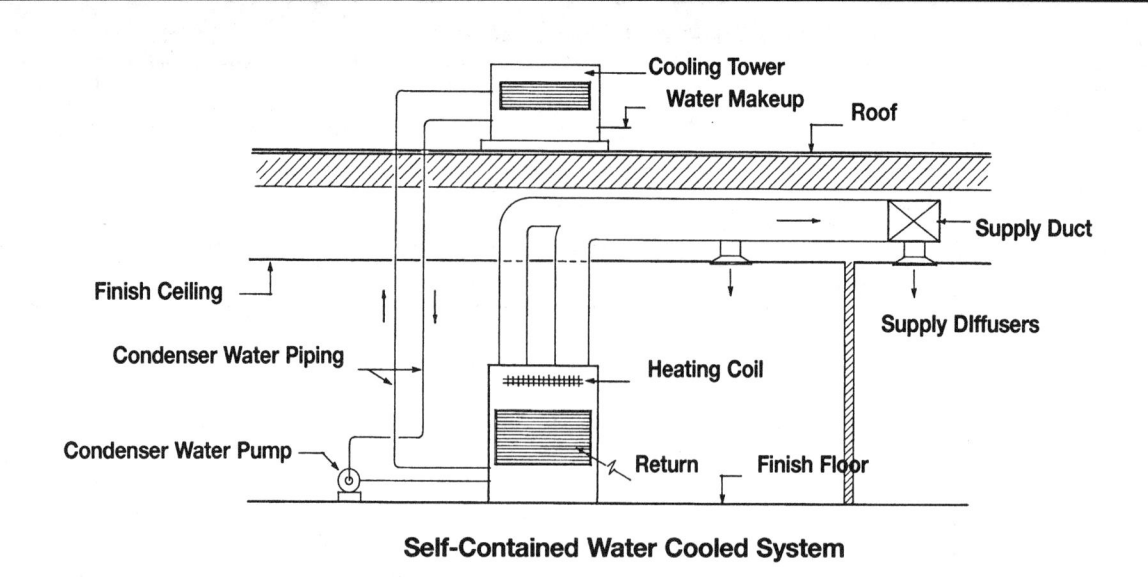

8.4-230	Self-contained, Water Cooled Unit Systems	MAT.	INST.	TOTAL
1280	Self-contained, water cooled unit	3.09	1.76	4.85
1300	Apartment corridors, 500 S.F., .92 ton	3.09	1.75	4.84
1320	1,000 S.F., 1.83 ton	3.08	1.74	4.82
1360	3,000 S.F., 5.50 ton	2.47	1.46	3.93
1400	5,000 S.F., 9.17 ton	2.50	1.36	3.86
1440	10,000 S.F., 18.33 ton	2.34	1.22	3.56
1520	Banks or libraries, 500 S.F., 2.08 ton	5.25	1.76	7.01
1560	1,000 S.F., 4.17 ton	5.65	3.33	8.98
1600	3,000 S.F., 12.50 ton	5.70	3.10	8.80
1640	5,000 S.F., 20.80 ton	5.35	2.77	8.12
1680	10,000 S.F., 41.66 ton	4.84	2.78	7.62
1760	Bars and taverns, 500 S.F., 5.54 ton	10.25	2.97	13.22
1800	1,000 S.F., 11.08 ton	12.80	5.30	18.10
1840	3,000 S.F., 33.25 ton	10.85	4.20	15.05
1880	5,000 S.F., 55.42 ton	10.35	4.38	14.73
1920	10,000 S.F., 110.00 ton	10.10	4.33	14.43
2000	Bowling alleys, 500 S.F., 2.83 ton	7.15	2.41	9.56
2040	1,000 S.F., 5.66 ton	7.65	4.54	12.19
2080	3,000 S.F., 17.00 ton	7.25	3.79	11.04
2120	5,000 S.F., 28.33 ton	6.75	3.66	10.41
2160	10,000 S.F., 56.66 ton	6.45	3.75	10.20
2200	Department stores, 500 S.F., 1.46 ton	3.69	1.23	4.92
2240	1,000 S.F., 2.92 ton	3.96	2.33	6.29
2280	3,000 S.F., 8.75 ton	4	2.17	6.17
2320	5,000 S.F., 14.58 ton	3.74	1.95	5.69
2360	10,000 S.F., 29.17 ton	3.48	1.88	5.36
2440	Drug stores, 500 S.F., 3.33 ton	8.45	2.83	11.28
2480	1,000 S.F., 6.66 ton	9	5.35	14.35
2520	3,000 S.F., 20.00 ton	8.55	4.45	13
2560	5,000 S.F., 33.33 ton	8.85	4.39	13.24
2600	10,000 S.F., 66.66 ton	7.60	4.39	11.99
2680	Factories, 500 S.F., 1.66 ton	4.20	1.41	5.61
2720	1,000 S.F. 3.37 ton	4.56	2.70	7.26
2760	3,000 S.F., 10.00 ton	4.56	2.49	7.05
2800	5,000 S.F., 16.66 ton	4.27	2.23	6.50
2840	10,000 S.F., 33.33 ton	3.98	2.15	6.13
2920	Food supermarkets, 500 S.F., 1.42 ton	3.57	1.20	4.77
2960	1,000 S.F., 2.83 ton	4.76	2.70	7.46
3000	3,000 S.F., 8.50 ton	3.83	2.26	6.09
3040	5,000 S.F., 14.17 ton	3.87	2.11	5.98

MECHANICAL — A8.4 Cooling

8.4-230 Self-contained, Water Cooled Unit Systems

		COST PER S.F.		
		MAT.	INST.	TOTAL
3080	10,000 S.F., 28.33 ton	3.38	1.82	5.20
3160	Medical centers, 500 S.F., 1.17 ton	2.94	.99	3.93
3200	1,000 S.F., 2.33 ton	3.92	2.23	6.15
3240	3,000 S.F., 7.00 ton	3.15	1.87	5.02
3280	5,000 S.F., 11.66 ton	3.19	1.74	4.93
3320	10,000 S.F., 23.33 ton	2.98	1.56	4.54
3400	Offices, 500 S.F., 1.58 ton	4.01	1.35	5.36
3440	1,000 S.F., 3.17 ton	5.35	3.03	8.38
3480	3,000 S.F., 9.50 ton	4.34	2.37	6.71
3520	5,000 S.F., 15.83 ton	4.06	2.12	6.18
3560	10,000 S.F., 31.67 ton	3.79	2.05	5.84
3640	Restaurants, 500 S.F., 2.50 ton	6.35	2.12	8.47
3680	1,000 S.F., 5.00 ton	6.75	4.01	10.76
3720	3,000 S.F., 15.00 ton	6.85	3.73	10.58
3760	5,000 S.F., 25.00 ton	6.40	3.35	9.75
3800	10,000 S.F., 50.00 ton	4.67	3.08	7.75
3880	Schools and colleges, 500 S.F., 1.92 ton	4.84	1.62	6.46
3920	1,000 S.F., 3.83 ton	5.20	3.07	8.27
3960	3,000 S.F., 11.50 ton	5.25	2.85	8.10
4000	5,000 S.F., 19.17 ton	4.91	2.56	7.47
4040	10,000 S.F., 38.33 ton	4.45	2.56	7.01

For expanded coverage of these items see *Means Mechanical or Plumbing Cost Data 1997*

MECHANICAL — A8.4 Cooling

Self-Contained Air Cooled System

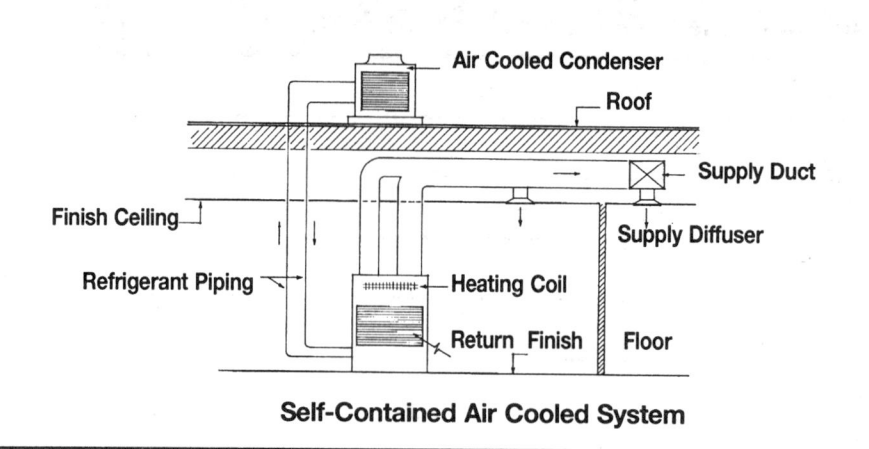

8.4-240 Self-contained, Air Cooled Unit Systems

Line	Description	MAT.	INST.	TOTAL
1300	Self-contained, air cooled unit			
1320	Apartment corridors, 500 S.F., .92 ton	4.56	2.23	6.79
1360	1,000 S.F., 1.83 ton	4.50	2.21	6.71
1400	3,000 S.F., 5.50 ton	3.82	2.06	5.88
1440	5,000 S.F., 9.17 ton	3.38	1.97	5.35
1480	10,000 S.F., 18.33 ton	2.93	1.81	4.74
1560	Banks or libraries, 500 S.F., 2.08 ton	8.45	2.82	11.27
1600	1,000 S.F., 4.17 ton	8.65	4.71	13.36
1640	3,000 S.F., 12.50 ton	7.70	4.49	12.19
1680	5,000 S.F., 20.80 ton	6.65	4.11	10.76
1720	10,000 S.F., 41.66 ton	6.30	4.11	10.41
1800	Bars and taverns, 500 S.F., 5.54 ton	15.60	6.15	21.75
1840	1,000 S.F., 11.08 ton	18.05	9	27.05
1880	3,000 S.F., 33.25 ton	14.65	7.85	22.50
1920	5,000 S.F., 55.42 ton	14.15	8.05	22.20
1960	10,000 S.F., 110.00 ton	14.30	7.95	22.25
2040	Bowling alleys, 500 S.F., 2.83 ton	11.55	3.86	15.41
2080	1,000 S.F., 5.66 ton	11.80	6.40	18.20
2120	3,000 S.F., 17.00 ton	9.10	5.60	14.70
2160	5,000 S.F., 28.33 ton	8.65	5.50	14.15
2200	10,000 S.F., 56.66 ton	8.45	5.60	14.05
2240	Department stores, 500 S.F., 1.46 ton	5.95	1.98	7.93
2280	1,000 S.F., 2.92 ton	6.10	3.30	9.40
2320	3,000 S.F., 8.75 ton	5.35	3.14	8.49
2360	5,000 S.F., 14.58 ton	5.35	3.14	8.49
2400	10,000 S.F., 29.17 ton	4.47	2.84	7.31
2480	Drug stores, 500 S.F., 3.33 ton	13.60	4.55	18.15
2520	1,000 S.F., 6.66 ton	13.85	7.55	21.40
2560	3,000 S.F., 20.00 ton	10.70	6.60	17.30
2600	5,000 S.F., 33.33 ton	10.20	6.50	16.70
2640	10,000 S.F., 66.66 ton	10	6.60	16.60
2720	Factories, 500 S.F., 1.66 ton	6.90	2.29	9.19
2760	1,000 S.F., 3.33 ton	7	3.78	10.78
2800	3,000 S.F., 10.00 ton	6.15	3.58	9.73
2840	5,000 S.F., 16.66 ton	5.30	3.30	8.60
2880	10,000 S.F., 33.33 ton	5.10	3.24	8.34
2960	Food supermarkets, 500 S.F., 1.42 ton	5.80	1.93	7.73
3000	1,000 S.F., 2.83 ton	6.95	3.43	10.38
3040	3,000 S.F., 8.50 ton	5.90	3.20	9.10
3080	5,000 S.F., 14.17 ton	5.20	3.04	8.24
3120	10,000 S.F., 28.33 ton	4.34	2.76	7.10
3200	Medical centers, 500 S.F., 1.17 ton	4.78	1.59	6.37
3240	1,000 S.F., 2.33 ton	5.75	2.83	8.58
3280	3,000 S.F., 7.00 ton	4.86	2.64	7.50

Important: See the Reference Section for critical supporting data – Location Factors & Historical Cost Indexes

MECHANICAL — A8.4 Cooling

8.4-240 Self-contained, Air Cooled Unit Systems

		COST PER S.F.		
		MAT.	INST.	TOTAL
3320	5,000 S.F., 16.66 ton	4.30	2.51	6.81
3360	10,000 S.F., 23.33 ton	3.73	2.31	6.04
3440	Offices, 500 S.F., 1.58 ton	6.45	2.16	8.61
3480	1,000 S.F., 3.16 ton	7.85	3.85	11.70
3520	3,000 S.F., 9.50 ton	5.85	3.41	9.26
3560	5,000 S.F., 15.83 ton	5.10	3.15	8.25
3600	10,000 S.F., 31.66 ton	4.86	3.09	7.95
3680	Restaurants, 500 S.F., 2.50 ton	10.15	3.40	13.55
3720	1,000 S.F., 5.00 ton	10.40	5.65	16.05
3760	3,000 S.F., 15.00 ton	9.20	5.35	14.55
3800	5,000 S.F., 25.00 ton	7.65	4.87	12.52
3840	10,000 S.F., 50.00 ton	7.65	4.92	12.57
3920	Schools and colleges, 500 S.F., 1.92 ton	7.80	2.61	10.41
3960	1,000 S.F., 3.83 ton	8	4.34	12.34
4000	3,000 S.F., 11.50 ton	7.05	4.12	11.17
4040	5,000 S.F., 19.17 ton	6.15	3.79	9.94
4080	10,000 S.F., 38.33 ton	5.80	3.76	9.56

For expanded coverage of these items see *Means Mechanical or Plumbing Cost Data 1997*

MECHANICAL — A8.4 Cooling

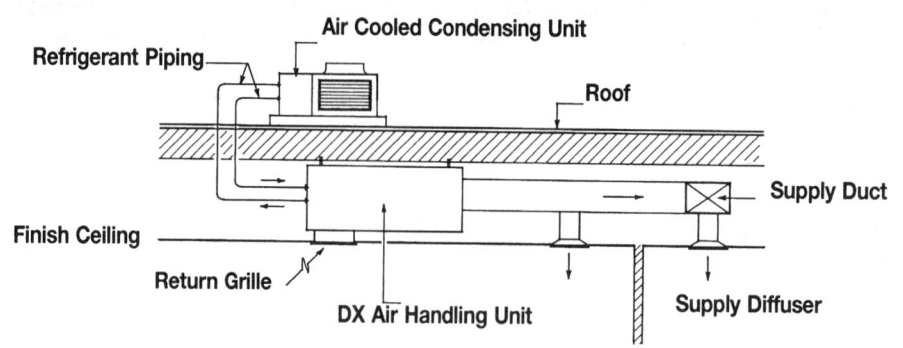

*Cooling requirements would lead to more than one system.

8.4-250 Split Systems With Air Cooled Condensing Units

		COST PER S.F.		
		MAT.	INST.	TOTAL
1260	Split system, air cooled condensing unit			
1280	Apartment corridors, 1,000 S.F., 1.83 ton	1.91	1.39	3.30
1320	2,000 S.F., 3.66 ton	1.71	1.41	3.12
1360	5,000 S.F., 9.17 ton	2.16	1.77	3.93
1400	10,000 S.F., 18.33 ton	1.99	1.93	3.92
1440	20,000 S.F., 36.66 ton	2.24	1.96	4.20
1520	Banks and libraries, 1,000 S.F., 4.17 ton	3.89	3.20	7.09
1560	2,000 S.F., 8.33 ton	4.91	4	8.91
1600	5,000 S.F., 20.80 ton	4.52	4.37	8.89
1640	10,000 S.F., 41.66 ton	5.10	4.44	9.54
1680	20,000 S.F., 83.32 ton	5.80	4.63	10.43
1760	Bars and taverns, 1,000 S.F., 11.08 ton	10.35	6.30	16.65
1800	2,000 S.F., 22.16 ton	10.75	7.55	18.30
1840	5,000 S.F., 55.42 ton	10.45	7.05	17.50
1880	10,000 S.F., 110.84 ton	13.25	7.40	20.65
1920	20,000 S.F., 220 ton*			
2000	Bowling alleys, 1,000 S.F., 5.66 ton	6.40	6	12.40
2040	2,000 S.F., 11.33 ton	6.65	5.45	12.10
2080	5,000 S.F., 28.33 ton	6.15	5.95	12.10
2120	10,000 S.F., 56.66 ton	6.90	6.05	12.95
2160	20,000 S.F., 113.32 ton	8.70	6.55	15.25
2320	Department stores, 1,000 S.F., 2.92 ton	2.75	2.20	4.95
2360	2,000 S.F., 5.83 ton	3.29	3.08	6.37
2400	5,000 S.F., 14.58 ton	3.44	2.80	6.24
2440	10,000 S.F., 29.17 ton	3.17	3.06	6.23
2480	20,000 S.F., 58.33 ton	3.56	3.11	6.67
2560	Drug stores, 1,000 S.F., 6.66 ton	7.50	7.05	14.55
2600	2,000 S.F., 13.32 ton	7.85	6.40	14.25
2640	5,000 S.F., 33.33 ton	8.10	7.10	15.20
2680	10,000 S.F., 66.66 ton	9.40	7.45	16.85
2720	20,000 S.F., 133.32 ton*			
2800	Factories, 1,000 S.F., 3.33 ton	3.14	2.51	5.65
2840	2,000 S.F., 6.66 ton	3.76	3.52	7.28
2880	5,000 S.F., 16.66 ton	3.62	3.50	7.12
2920	10,000 S.F., 33.33 ton	4.07	3.55	7.62
2960	20,000 S.F., 66.66 ton	4.71	3.72	8.43
3040	Food supermarkets, 1,000 S.F., 2.83 ton	2.67	2.14	4.81
3080	2,000 S.F., 5.66 ton	3.19	3	6.19
3120	5,000 S.F., 14.66 ton	3.34	2.72	6.06
3160	10,000 S.F., 28.33 ton	3.08	2.97	6.05
3200	20,000 S.F., 56.66 ton	3.45	3.02	6.47
3280	Medical centers, 1,000 S.F., 2.33 ton	2.22	1.73	3.95
3320	2,000 S.F., 4.66 ton	2.63	2.47	5.10
3360	5,000 S.F., 11.66 ton	2.75	2.25	5

MECHANICAL — A8.4 Cooling

8.4-250 Split Systems With Air Cooled Condensing Units

		MAT.	INST.	TOTAL
3400	10,000 S.F., 23.33 ton	2.53	2.46	4.99
3440	20,000 S.F., 46.66 ton	2.84	2.49	5.33
3520	Offices, 1,000 S.F., 3.17 ton	2.98	2.38	5.36
3560	2,000 S.F., 6.33 ton	3.57	3.34	6.91
3600	5,000 S.F., 15.83 ton	3.74	3.05	6.79
3640	10,000 S.F., 31.66 ton	3.44	3.34	6.78
3680	20,000 S.F., 63.32 ton	4.47	3.55	8.02
3760	Restaurants, 1,000 S.F., 5.00 ton	5.65	5.30	10.95
3800	2,000 S.F., 10.00 ton	5.90	4.82	10.72
3840	5,000 S.F., 25.00 ton	5.45	5.25	10.70
3880	10,000 S.F., 50.00 ton	6.10	5.35	11.45
3920	20,000 S.F., 100.00 ton	7.70	5.80	13.50
4000	Schools and colleges, 1,000 S.F., 3.83 ton	3.60	2.94	6.54
4040	2,000 S.F., 7.66 ton	4.51	3.70	8.21
4080	5,000 S.F., 19.17 ton	4.17	4.03	8.20
4120	10,000 S.F., 38.33 ton	4.67	4.10	8.77

For expanded coverage of these items see *Means Mechanical or Plumbing Cost Data 1997*

MECHANICAL — A8.5 Special Systems

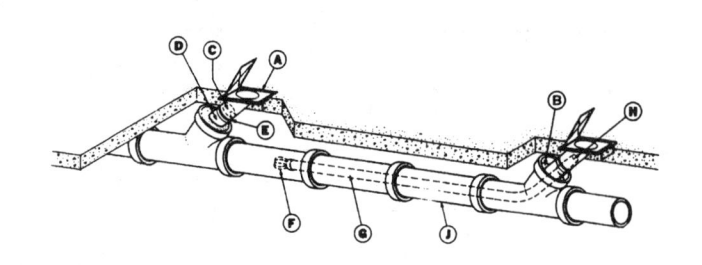

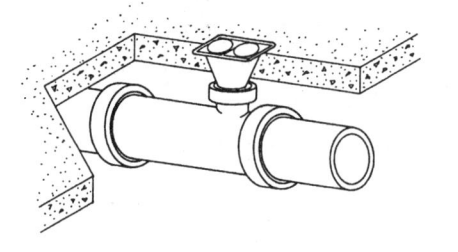

Vitrified Clay Garage Exhaust System

Dual Exhaust System

8.5-110	Garage Exhaust Systems	MAT.	INST.	TOTAL
1040	Garage, single exhaust, 3" outlet, cars & light trucks, one bay	1,650	745	2,395
1060	Additional bays up to seven bays	281	110	391
1500	4" outlet, trucks, one bay	1,675	745	2,420
1520	Additional bays up to six bays	289	110	399
1600	5" outlet, diesel trucks, one bay	1,900	745	2,645
1650	Additional single bays up to six	535	126	661
1700	Two adjoining bays	1,900	745	2,645
2000	Dual exhaust, 3" outlets, pair of adjoining bays	1,950	830	2,780
2100	Additional pairs of adjoining bays	525	126	651

COST PER BAY

For information about Means Estimating Seminars, see yellow pages 11 and 12 in back of book

Division 9
Electrical

ELECTRICAL A9.1 Service Distribution

Feeder Installation

Electric Service

Switchgear

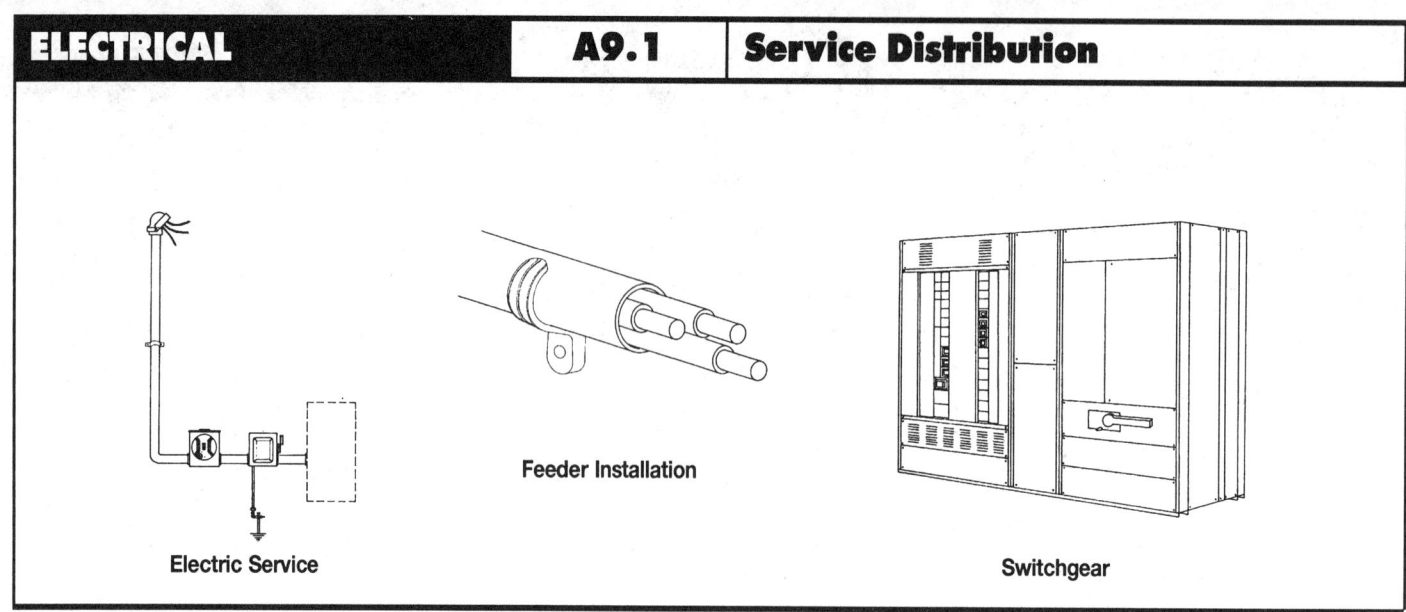

9.1-210	Electric Service, 3 Phase - 4 Wire	COST EACH		
		MAT.	INST.	TOTAL
0200	Service installation, includes breakers, metering, 20' conduit & wire			
0220	3 phase, 4 wire, 120/208 volts, 60 amp	490	555	1,045
0240	100 amps	600	670	1,270
0280	200 amps	895	1,025	1,920
0320	400 amps	2,075	1,900	3,975
0360	600 amps	3,925	2,550	6,475
0400	800 amps	5,300	3,075	8,375
0440	1000 amps	6,675	3,550	10,225
0480	1200 amps	8,075	3,625	11,700
0520	1600 amps	15,400	5,200	20,600
0560	2000 amps	16,900	5,925	22,825
0570	Add 25% for 277/480 volt			

9.1-310	Feeder Installation	COST PER L.F.		
		MAT.	INST.	TOTAL
0200	Feeder installation 600 volt, including conduit and wire, 60 amperes	3.02	6.75	9.77
0240	100 amperes	5.25	8.90	14.15
0280	200 amperes	11.45	13.75	25.20
0320	400 amperes	23	27.50	50.50
0360	600 amperes	47.50	45	92.50
0400	800 amperes	64.50	53.50	118
0440	1000 amperes	79	69	148
0480	1200 amperes	92.50	70.50	163
0520	1600 amperes	129	107	236
0560	2000 amperes	158	138	296

9.1-410	Switchgear	COST EACH		
		MAT.	INST.	TOTAL
0200	Switchgear inst., incl. swbd., panels & circ bkr, 400 amps, 120/208volt	3,375	2,275	5,650
0240	600 amperes	9,100	3,075	12,175
0280	800 amperes	11,200	4,350	15,550
0320	1200 amperes	13,900	6,650	20,550
0360	1600 amperes	19,100	9,325	28,425
0400	2000 amperes	23,800	11,900	35,700
0410	Add 20% for 277/480 volt			

Important: See the Reference Section for critical supporting data - Location Factors & Historical Cost Indexes

ELECTRICAL — A9.2 Lighting & Power

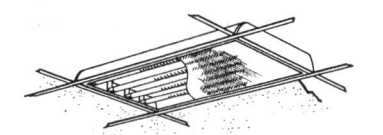

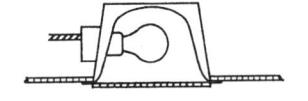

Fluorescent Fixture

Incandescent Fixture

9.2-213 Fluorescent Fixtures (by Wattage)

Line	Description	MAT.	INST.	TOTAL
0190	Fluorescent fixtures recess mounted in ceiling			
0200	1 watt per S.F., 20 FC, 5 fixtures per 1000 S.F.	.58	1.06	1.64
0240	2 watts per S.F., 40 FC, 10 fixtures per 1000 S.F.	1.15	2.09	3.24
0280	3 watts per S.F., 60 FC, 15 fixtures per 1000 S.F	1.72	3.15	4.87
0320	4 watts per S.F., 80 FC, 20 fixtures per 1000 S.F.	2.28	4.17	6.45
0400	5 watts per S.F., 100 FC, 25 fixtures per 1000 S.F.	2.86	5.25	8.11

9.2-223 Incandescent Fixture (by Wattage)

Line	Description	MAT.	INST.	TOTAL
0190	Incandescent fixture recess mounted, type A			
0200	1 watt per S.F., 8 FC, 6 fixtures per 1000 S.F.	.49	.86	1.35
0240	2 watt per S.F., 16 FC, 12 fixtures per 1000 S.F.	1	1.71	2.71
0280	3 watt per S.F., 24 FC, 18 fixtures, per 1000 S.F.	1.47	2.53	4
0320	4 watt per S.F., 32 FC, 24 fixtures per 1000 S.F.	1.97	3.38	5.35
0400	5 watt per S.F., 40 FC, 30 fixtures per 1000 S.F.	2.46	4.24	6.70

9.2-235 H.I.D. Fixture, High Bay (by Wattage)

Line	Description	MAT.	INST.	TOTAL
0190	High intensity discharge fixture, 16′ above work plane			
0200	1 watt/S.F., type D, 23 FC, 1 fixture/1000 S.F.	.69	.85	1.54
0240	Type E, 42 FC, 1 fixture/1000 S.F.	.81	.89	1.70
0280	Type G, 52 FC, 1 fixture/1000 S.F.	.81	.89	1.70
0320	Type C, 54 FC, 2 fixture/1000 S.F.	.98	1	1.98
0400	2 watt/S.F., type D, 45 FC, 2 fixture/1000 S.F.	1.38	1.71	3.09
0440	Type E, 84 FC, 2 fixture/1000 S.F.	1.64	1.82	3.46
0480	Type G, 105 FC, 2 fixture/1000 S.F.	1.64	1.82	3.46
0520	Type C, 108 FC, 4 fixture/1000 S.F.	1.96	1.99	3.95
0600	3 watt/S.F., type D, 68 FC, 3 fixture/1000 S.F.	2.07	2.52	4.59
0640	Type E, 126 FC, 3 fixture/1000 S.F.	2.46	2.69	5.15
0680	Type G, 157 FC, 3 fixture/1000 S.F.	2.46	2.69	5.15
0720	Type C, 162 FC, 6 fixture/1000 S.F.	2.94	2.98	5.92
0800	4 watt/ S.F., type D, 91 FC, 4 fixture/1000 S.F.	2.96	3.99	6.95
0840	Type E, 168 FC, 4 fixture/1000 S.F.	3.29	3.62	6.91
0880	Type G, 210 FC, 4 fixture/1000 S.F.	3.29	3.62	6.91
0920	Type C, 243 FC, 9 fixture/1000 S.F.	4.30	4.15	8.45
1000	5 watt/S.F., type D, 113 FC, 5 fixture/1000 S.F.	3.46	4.22	7.68
1040	Type E, 210 FC, 5 fixture/1000 S.F.	4.10	4.51	8.61
1080	Type G, 262 FC, 5 fixture/1000 S.F.	4.10	4.51	8.61
1120	Type C, 297 FC, 11 fixture/1000 S.F.	5.30	5.15	10.45

For expanded coverage of these items see Means Electrical Cost Data 1997

ELECTRICAL — A9.2 Lighting & Power

9.2-239 H.I.D. Fixture, High Bay (by Wattage)

		MAT.	INST.	TOTAL
0190	High intensity discharge fixture, 30' above work plane			
0200	1 watt/S.F., type D, 23 FC, 1 fixture/1000 S.F.	.76	1.09	1.85
0240	Type E, 37 FC, 1 fixture/1000 S.F.	.89	1.13	2.02
0280	Type G, 45 FC., 1 fixture/1000 S.F.	.89	1.13	2.02
0320	Type F, 50 FC, 1 fixture/1000 S.F.	.77	.87	1.64
0400	2 watt/S.F., type D, 40 FC, 2 fixtures/1000 S.F.	1.53	2.15	3.68
0440	Type E, 74 FC, 2 fixtures/1000 S.F.	1.78	2.26	4.04
0480	Type G, 92 FC, 2 fixtures/1000 S.F.	1.78	2.26	4.04
0520	Type F, 100 FC, 2 fixtures/1000 S.F.	1.55	1.80	3.35
0600	3 watt/S.F., type D, 60 FC, 3 fixtures/1000 S.F.	2.29	3.22	5.51
0640	Type E, 110 FC, 3 fixtures/1000 S.F.	2.68	3.40	6.08
0680	Type G, 138FC, 3 fixtures/1000 S.F.	2.68	3.40	6.08
0720	Type F, 150 FC, 3 fixtures/1000 S.F.	2.32	2.67	4.99
0800	4 watt/ S.F., type D, 80 FC, 4 fixtures/1000 S.F.	3.05	4.29	7.34
0840	Type E, 148 FC, 4 fixtures/1000 S.F.	3.57	4.54	8.11
0880	Type G, 185 FC, 4 fixtures/1000 S.F.	3.57	4.54	8.11
0920	Type F, 200 FC, 4 fixtures/1000 S.F.	3.09	3.60	6.69
1000	5 watt/ S.F., type D, 100 FC 5 fixtures/1000 S.F.	3.82	5.40	9.22
1040	Type E, 185 FC, 5 fixtures/1000 S.F.	4.46	5.65	10.11
1080	Type G, 230 FC, 5 fixtures/1000 S.F.	4.46	5.65	10.11
1120	Type F, 250 FC, 5 fixtures/1000 S.F.	3.86	4.47	8.33

COST PER S.F.

ELECTRICAL — A9.2 Lighting & Power

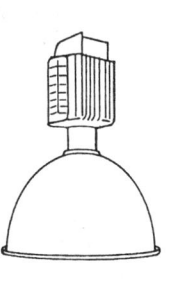

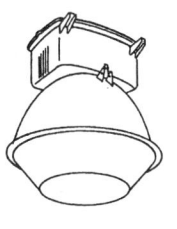

High Bay Fixture

Low Bay Fixture

9.2-242 H.I.D. Fixture, Low Bay (by Wattage)

		COST PER S.F.		
		MAT.	INST.	TOTAL
0190	High intensity discharge fixture, 8'-10' above work plane			
0200	1 watt/S.F., type H, 19 FC, 4 fixtures/1000 S.F.	1.80	1.82	3.62
0240	Type J, 30 FC, 4 fixtures/1000 S.F.	2.02	1.84	3.86
0280	Type K, 29 FC, 5 fixtures/1000 S.F.	1.95	1.66	3.61
0360	2 watt/S.F. type H, 33 FC, 7 fixtures/1000 S.F.	3.25	3.51	6.76
0400	Type J, 52 FC, 7 fixtures/1000 S.F.	3.60	3.38	6.98
0440	Type K, 63 FC, 11 fixtures/1000 S.F.	4.25	3.48	7.73
0520	3 watt/S.F., type H, 51 FC, 11 fixtures/1000 S.F.	5.05	5.35	10.40
0560	Type J, 81 FC, 11 fixtures/1000 S.F.	5.60	5.10	10.70
0600	Type K, 92 FC, 16 fixtures/1000 S.F.	6.20	5.10	11.30
0680	4 watt/S.F., type H, 65 FC, 14 fixtures/1000 S.F.	6.50	7.05	13.55
0720	Type J, 103 FC, 14 fixtures/1000 S.F.	7.20	6.70	13.90
0760	Type K, 127 FC, 22 fixtures/1000 S.F.	8.50	6.95	15.45
0840	5 watt/S.F., type H, 84 FC, 18 fixtures/1000 S.F.	8.30	8.85	17.15
0880	Type J, 133 FC, 18 fixtures/1000 S.F.	9.15	8.45	17.60
0920	Type K, 155 FC, 27 fixtures/1000 S.F.	10.45	8.60	19.05

For expanded coverage of these items see *Means Electrical Cost Data 1997*

ELECTRICAL — A9.2 Lighting & Power

9.2-244 H.I.D. Fixture, Low Bay (by Wattage)

Line	Description	MAT.	INST.	TOTAL
0190	High intensity discharge fixture, mounted 16' above work plane			
0200	1 watt/S.F., type H, 19 FC, 4 fixtures/1000 S.F.	1.88	2.07	3.95
0240	Type J, 28 FC, 4 fixt./1000 S.F.	2.11	2.08	4.19
0280	Type K, 27 FC, 5 fixt./1000 S.F.	2.17	2.34	4.51
0360	2 watts/S.F., type H, 30 FC, 7 fixt/1000 S.F.	3.41	4.01	7.42
0400	Type J, 48 FC, 7 fixt/1000 S.F.	3.80	4.04	7.84
0440	Type K, 58 FC, 11 fixt/1000 S.F.	4.66	4.78	9.44
0520	3 watts/S.F., type H, 47 FC, 11 fixt/1000 S.F.	5.30	6.10	11.40
0560	Type J, 75 FC, 11 fixt/1000 S.F.	5.90	6.10	12
0600	Type K, 85 FC, 16 fixt/1000 S.F.	6.80	7.10	13.90
0680	4 watts/S.F., type H, 60 FC, 14 fixt/1000 S.F.	6.80	8	14.80
0720	Type J, 95 FC, 14 fixt/1000 S.F.	7.60	8.05	15.65
0760	Type K, 117 FC, 22 fixt/1000 S.F.	9.30	9.55	18.85
0840	5 watts/S.F., type H, 77 FC, 18 fixt/1000 S.F.	8.75	10.35	19.10
0880	Type J, 122 FC, 18 fixt/1000 S.F.	9.70	10.15	19.85
0920	Type K, 143 FC, 27 fixt/1000 S.F.	11.45	11.90	23.35

Important: See the Reference Section for critical supporting data - Location Factors & Historical Cost Indexes

ELECTRICAL — A9.2 Lighting & Power

Light Pole

9.2-252 Light Pole (Installed)

		COST EACH		
		MAT.	INST.	TOTAL
0200	Light pole, aluminum, 20' high, 1 arm bracket	785	655	1,440
0240	2 arm brackets	870	655	1,525
0280	3 arm brackets	950	680	1,630
0320	4 arm brackets	1,025	680	1,705
0360	30' high, 1 arm bracket	1,375	830	2,205
0400	2 arm brackets	1,450	830	2,280
0440	3 arm brackets	1,525	850	2,375
0480	4 arm brackets	1,625	850	2,475
0680	40' high, 1 arm bracket	1,875	1,125	3,000
0720	2 arm brackets	1,975	1,125	3,100
0760	3 arm brackets	2,050	1,150	3,200
0800	4 arm brackets	2,125	1,150	3,275
0840	Steel, 20' high, 1 arm bracket	1,000	695	1,695
0880	2 arm brackets	1,100	695	1,795
0920	3 arm brackets	1,175	720	1,895
0960	4 arm brackets	1,300	720	2,020
1000	30' high, 1 arm bracket	1,200	880	2,080
1040	2 arm brackets	1,300	880	2,180
1080	3 arm brackets	1,375	900	2,275
1120	4 arm brackets	1,500	900	2,400
1320	40' high, 1 arm bracket	1,475	1,200	2,675
1360	2 arm brackets	1,575	1,200	2,775
1400	3 arm brackets	1,650	1,225	2,875
1440	4 arm brackets	1,775	1,225	3,000

For expanded coverage of these items see *Means Electrical Cost Data 1997*

ELECTRICAL — A9.2 Lighting & Power

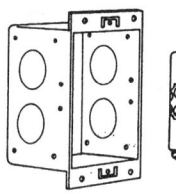

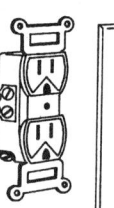

Duplex Wall Receptacle

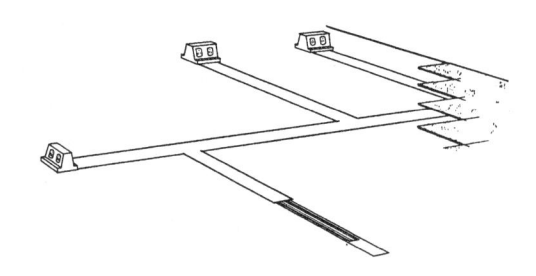

Undercarpet Receptacle System

9.2-522 Receptacle (by Wattage)

		COST PER S.F.		
		MAT.	INST.	TOTAL
0190	Receptacles include plate, box, conduit, wire & transformer when required			
0200	2.5 per 1000 S.F., .3 watts per S.F.	.22	.80	1.02
0240	With transformer	.25	.84	1.09
0280	4 per 1000 S.F., .5 watts per S.F.	.25	.94	1.19
0320	With transformer	.30	1	1.30
0360	5 per 1000 S.F., .6 watts per S.F.	.31	1.11	1.42
0400	With transformer	.37	1.19	1.56
0440	8 per 1000 S.F., .9 watts per S.F.	.33	1.23	1.56
0480	With transformer	.41	1.34	1.75
0520	10 per 1000 S.F., 1.2 watts per S.F.	.35	1.34	1.69
0560	With transformer	.49	1.52	2.01
0600	16.5 per 1000 S.F., 2.0 watts per S.F.	.41	1.67	2.08
0640	With transformer	.65	1.98	2.63
0680	20 per 1000 S.F., 2.4 watts per S.F.	.44	1.81	2.25
0720	With transformer	.72	2.17	2.89

9.2-524 Receptacles

		COST PER S.F.		
		MAT.	INST.	TOTAL
0200	Receptacle systems, underfloor duct, 5' on center, low density	2.26	1.77	4.03
0240	High density	2.49	2.28	4.77
0280	7' on center, low density	1.82	1.52	3.34
0320	High density	2.05	2.03	4.08
0400	Poke thru fittings, low density	.77	.77	1.54
0440	High density	1.55	1.50	3.05
0520	Telepoles, using Romex, low density	.64	.53	1.17
0560	High density	1.28	1.07	2.35
0600	Using FMT, low density	.66	.70	1.36
0640	High density	1.33	1.42	2.75
0720	Conduit system with floor boxes, low density	.67	.61	1.28
0760	High density	1.35	1.21	2.56
0840	Undercarpet power system, 3 conductor with 5 conductor feeder, low density	.96	.24	1.20
0880	High density	1.90	.45	2.35

Important: See the Reference Section for critical supporting data - Location Factors & Historical Cost Indexes

ELECTRICAL — A9.2 Lighting & Power

Description: Table 9.2-542 includes the cost for switch, plate, box, conduit in slab or EMT exposed and copper wire. Add 20% for exposed conduit.

No power required for switches.

Federal energy guidelines recommend the maximum lighting area controlled per switch shall not exceed 1000 S.F. and that areas over 500 S.F. shall be so controlled that total illumination can be reduced by at least 50%.

9.2-542	Wall Switch by Sq. Ft.	MAT.	INST.	TOTAL
0200	Wall switches, 1.0 per 1000 S.F.	.04	.13	.17
0240	1.2 per 1000 S.F.	.04	.14	.18
0280	2.0 per 1000 S.F.	.05	.21	.26
0320	2.5 per 1000 S.F.	.06	.26	.32
0360	5.0 per 1000 S.F.	.15	.57	.72
0400	10.0 per 1000 S.F.	.31	1.14	1.45

COST PER S.F.

9.2-582	Miscellaneous Power	MAT.	INST.	TOTAL
0200	Miscellaneous power, to .5 watts	.02	.06	.08
0240	.8 watts	.02	.10	.12
0280	1 watt	.03	.11	.14
0320	1.2 watts	.04	.14	.18
0360	1.5 watts	.05	.16	.21
0400	1.8 watts	.06	.20	.26
0440	2 watts	.07	.24	.31
0480	2.5 watts	.09	.28	.37
0520	3 watts	.10	.34	.44

COST PER S.F.

9.2-610	Central A. C. Power (by Wattage)	MAT.	INST.	TOTAL
0200	Central air conditioning power, 1 watt	.04	.14	.18
0220	2 watts	.05	.16	.21
0240	3 watts	.06	.19	.25
0280	4 watts	.09	.25	.34
0320	6 watts	.17	.34	.51
0360	8 watts	.20	.35	.55
0400	10 watts	.27	.40	.67

For expanded coverage of these items see *Means Electrical Cost Data 1997*

ELECTRICAL — A9.2 Lighting & Power

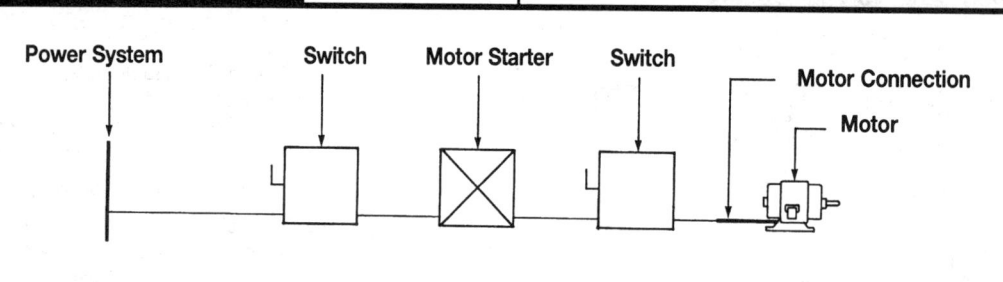

Motor Installation

9.2-710	Motor Installation	MAT.	INST.	TOTAL
0200	Motor installation, single phase, 115V, to and including 1/3 HP motor size	375	550	925
0240	To and incl. 1 HP motor size	390	550	940
0280	To and incl. 2 HP motor size	425	585	1,010
0320	To and incl. 3 HP motor size	485	595	1,080
0360	230V, to and including 1 HP motor size	375	555	930
0400	To and incl. 2 HP motor size	395	555	950
0440	To and incl. 3 HP motor size	440	600	1,040
0520	Three phase, 200V, to and including 1-1/2 HP motor size	435	610	1,045
0560	To and incl. 3 HP motor size	470	665	1,135
0600	To and incl. 5 HP motor size	520	740	1,260
0640	To and incl. 7-1/2 HP motor size	540	755	1,295
0680	To and incl. 10 HP motor size	865	950	1,815
0720	To and incl. 15 HP motor size	1,175	1,050	2,225
0760	To and incl. 20 HP motor size	1,450	1,200	2,650
0800	To and incl. 25 HP motor size	1,475	1,225	2,700
0840	To and incl. 30 HP motor size	2,400	1,425	3,825
0880	To and incl. 40 HP motor size	2,825	1,700	4,525
0920	To and incl. 50 HP motor size	5,175	1,975	7,150
0960	To and incl. 60 HP motor size	5,275	2,100	7,375
1000	To and incl. 75 HP motor size	6,650	2,375	9,025
1040	To and incl. 100 HP motor size	13,500	2,800	16,300
1080	To and incl. 125 HP motor size	13,800	3,100	16,900
1120	To and incl. 150 HP motor size	16,300	3,625	19,925
1160	To and incl. 200 HP motor size	19,500	4,425	23,925
1240	230V, to and including 1-1/2 HP motor size	420	605	1,025
1280	To and incl. 3 HP motor size	455	660	1,115
1320	To and incl. 5 HP motor size	505	735	1,240
1360	To and incl. 7-1/2 HP motor size	505	735	1,240
1400	To and incl. 10 HP motor size	810	900	1,710
1440	To and incl. 15 HP motor size	885	985	1,870
1480	To and incl. 20 HP motor size	1,375	1,175	2,550
1520	To and incl. 25 HP motor size	1,450	1,200	2,650
1560	To and incl. 30 HP motor size	1,475	1,225	2,700
1600	To and incl. 40 HP motor size	2,775	1,650	4,425
1640	To and incl. 50 HP motor size	2,875	1,750	4,625
1680	To and incl. 60 HP motor size	5,175	1,975	7,150
1720	To and incl. 75 HP motor size	6,150	2,250	8,400
1760	To and incl. 100 HP motor size	6,800	2,500	9,300
1800	To and incl. 125 HP motor size	13,700	2,900	16,600
1840	To and incl. 150 HP motor size	14,700	3,300	18,000
1880	To and incl. 200 HP motor size	15,700	3,650	19,350
1960	460V, to and including 2 HP motor size	505	610	1,115
2000	To and incl. 5 HP motor size	540	665	1,205
2040	To and incl. 10 HP motor size	580	735	1,315
2080	To and incl. 15 HP motor size	800	845	1,645
2120	To and incl. 20 HP motor size	825	905	1,730
2160	To and incl. 25 HP motor size	875	945	1,820
2200	To and incl. 30 HP motor size	1,150	1,025	2,175

Important: See the Reference Section for critical supporting data - Location Factors & Historical Cost Indexes

ELECTRICAL — A9.2 Lighting & Power

9.2-710 Motor Installation

		MAT. (Cost Each)	INST.	TOTAL
2240	To and incl. 40 HP motor size	1,450	1,100	2,550
2280	To and incl. 50 HP motor size	1,575	1,225	2,800
2320	To and incl. 60 HP motor size	2,475	1,425	3,900
2360	To and incl. 75 HP motor size	2,825	1,575	4,400
2400	To and incl. 100 HP motor size	3,050	1,750	4,800
2440	To and incl. 125 HP motor size	5,325	2,000	7,325
2480	To and incl. 150 HP motor size	6,375	2,225	8,600
2520	To and incl. 200 HP motor size	7,175	2,500	9,675
2600	575V, to and including 2 HP motor size	505	610	1,115
2640	To and incl. 5 HP motor size	540	665	1,205
2680	To and incl. 10 HP motor size	580	735	1,315
2720	To and incl. 20 HP motor size	800	845	1,645
2760	To and incl. 25 HP motor size	825	905	1,730
2800	To and incl. 30 HP motor size	1,150	1,025	2,175
2840	To and incl. 50 HP motor size	1,225	1,050	2,275
2880	To and incl. 60 HP motor size	2,450	1,425	3,875
2920	To and incl. 75 HP motor size	2,475	1,425	3,900
2960	To and incl. 100 HP motor size	2,825	1,575	4,400
3000	To and incl. 125 HP motor size	5,250	1,950	7,200
3040	To and incl. 150 HP motor size	5,325	2,000	7,325
3080	To and incl. 200 HP motor size	6,500	2,250	8,750

9.2-720 Motor Feeder

		MAT. (Cost per L.F.)	INST.	TOTAL
0200	Motor feeder systems, single phase, feed up to 115V 1HP or 230V 2 HP	1.28	4.27	5.55
0240	115V 2HP, 230V 3HP	1.36	4.33	5.69
0280	115V 3HP	1.56	4.51	6.07
0360	Three phase, feed to 200V 3HP, 230V 5HP, 460V 10HP, 575V 10HP	1.35	4.60	5.95
0440	200V 5HP, 230V 7.5HP, 460V 15HP, 575V 20HP	1.47	4.69	6.16
0520	200V 10HP, 230V 10HP, 460V 30HP, 575V 30HP	1.77	4.96	6.73
0600	200V 15HP, 230V 15HP, 460V 40HP, 575V 50HP	2.24	5.70	7.94
0680	200V 20HP, 230V 25HP, 460V 50HP, 575V 60HP	3.18	7.20	10.38
0760	200V 25HP, 230V 30HP, 460V 60HP, 575V 75HP	3.58	7.30	10.88
0840	200V 30HP	4.12	7.55	11.67
0920	230V 40HP, 460V 75HP, 575V 100HP	5.25	8.25	13.50
1000	200V 40HP	6	8.80	14.80
1080	230V 50HP, 460V 100HP, 575V 125HP	7	9.70	16.70
1160	200V 50HP, 230V 60HP, 460V 125HP, 575V 150HP	8.50	10.30	18.80
1240	200V 60HP, 460V 150HP	10.05	12.10	22.15
1320	230V 75HP, 575V 200HP	11.95	12.60	24.55
1400	200V 75HP	13.60	12.90	26.50
1480	230V 100HP, 460V 200HP	16.45	15	31.45
1560	200V 100HP	22.50	18.85	41.35
1640	230V 125HP	27.50	20.50	48
1720	200V 125HP, 230V 150HP	28.50	23.50	52
1800	200V 150HP	31.50	25.50	57
1880	200V 200HP	45	37.50	82.50
1960	230V 200HP	41.50	28	69.50

For expanded coverage of these items see *Means Electrical Cost Data 1997*

ELECTRICAL | A9.4 Special Electrical

9.4-100 Communication & Alarm Systems

		COST EACH		
		MAT.	INST.	TOTAL
0200	Communication & alarm systems, includes outlets, boxes, conduit & wire			
0210	Sound system, 6 outlets	4,075	4,550	8,625
0220	12 outlets	5,600	7,300	12,900
0240	30 outlets	9,650	13,800	23,450
0280	100 outlets	28,900	46,100	75,000
0320	Fire detection systems, 12 detectors	2,125	3,800	5,925
0360	25 detectors	3,725	6,400	10,125
0400	50 detectors	7,125	12,600	19,725
0440	100 detectors	13,000	22,600	35,600
0480	Intercom systems, 6 stations	2,325	3,125	5,450
0560	25 stations	7,925	12,000	19,925
0640	100 stations	30,300	43,900	74,200
0680	Master clock systems, 6 rooms	3,625	5,100	8,725
0720	12 rooms	5,350	8,700	14,050
0760	20 rooms	7,175	12,300	19,475
0800	30 rooms	11,300	22,700	34,000
0840	50 rooms	18,000	38,300	56,300
0920	Master TV antenna systems, 6 outlets	2,000	3,225	5,225
0960	12 outlets	3,750	6,000	9,750
1000	30 outlets	7,350	13,900	21,250
1040	100 outlets	24,300	45,300	69,600

ELECTRICAL — A9.4 Special Electrical

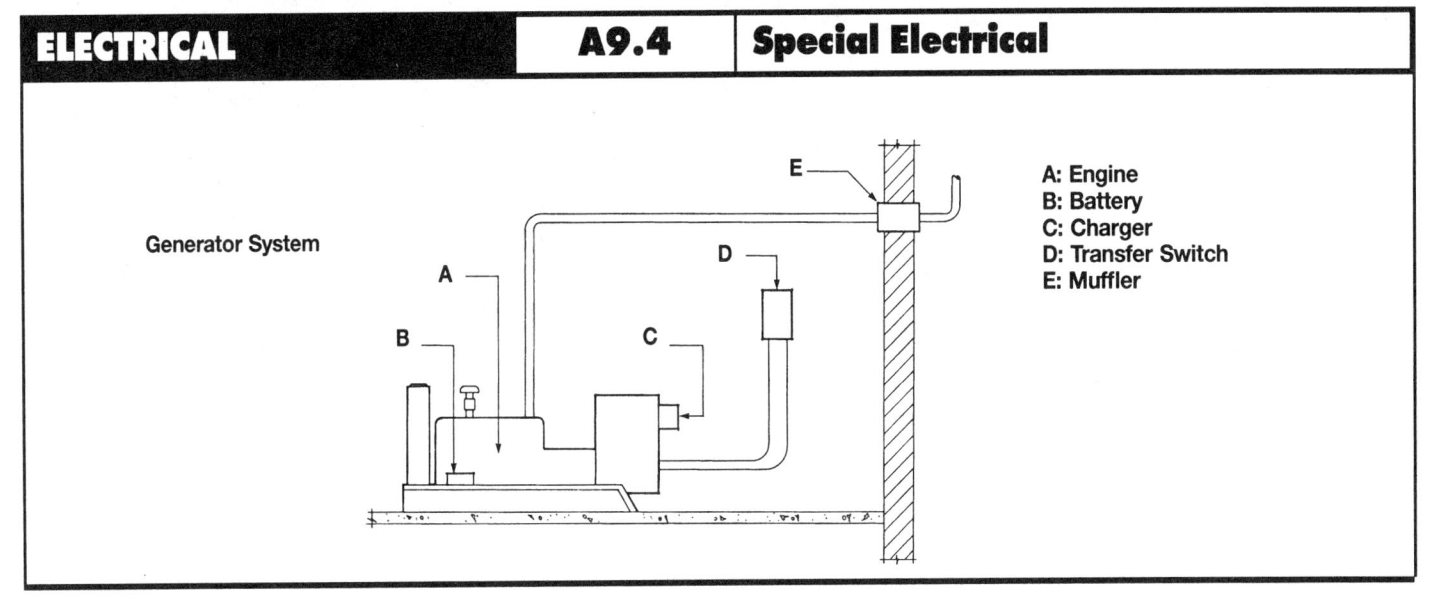

Generator System

A: Engine
B: Battery
C: Charger
D: Transfer Switch
E: Muffler

9.4-310	Generators (by KW)	MAT.	INST.	TOTAL
0190	Generator sets, include battery, charger, muffler & transfer switch			
0200	Gas/gasoline operated, 3 phase, 4 wire, 277/480 volt, 7.5 KW	845	162	1,007
0240	10 KW	865	145	1,010
0280	15 KW	680	108	788
0320	30 KW	485	61.50	546.50
0360	70 KW	345	36.50	381.50
0400	85 KW	335	36	371
0440	115 KW	450	31.50	481.50
0480	170 KW	525	24	549
0560	Diesel engine with fuel tank, 30 KW	585	61.50	646.50
0600	50 KW	430	48.50	478.50
0640	75 KW	375	39	414
0680	100 KW	315	33	348
0760	150 KW	257	26	283
0840	200 KW	209	21.50	230.50
0880	250 KW	183	17.70	200.70
0960	350 KW	169	14.55	183.55
1040	500 KW	167	11.35	178.35

For information about Means Estimating Seminars, see yellow pages 11 and 12 in back of book

For expanded coverage of these items see *Means Electrical Cost Data 1997*

Division 11
Special Construction

SPECIAL CONSTR. — A11.1 Specialties

11.1-100 Architectural Specialties/Each

		COST EACH		
		MAT.	INST.	TOTAL
1000	Specialties, bathroom accessories, st. steel, curtain rod, 5' long, 1" diam	28.50	25.50	54
1100	Dispenser, towel, surface mounted	42	20.50	62.50
1200	Grab bar, 1-1/4" diam., 12" long	36.50	13.70	50.20
1300	Mirror, framed with shelf, 18" x 24"	99	16.45	115.45
1340	72" x 24"	640	55	695
1400	Toilet tissue dispenser, surface mounted, single roll	10.55	10.95	21.50
1500	Towel bar, 18" long	30	14.30	44.30
1600	Canopies, wall hung, prefinished aluminum, 8' x 10'	1,250	1,025	2,275
1700	Chutes, linen or refuse incl. sprinklers, galv. steel, 18" diam., per floor	745	212	957
1740	Aluminized steel, 36" diam., per floor	1,225	265	1,490
1800	Mail, 8-3/4" x 3-1/2", aluminum & glass, per floor	550	148	698
1840	Bronze or stainless, per floor	860	165	1,025
1900	Control boards, magnetic, porcelain finish, framed, 24" x 18"	106	82	188
1940	96" x 48"	480	132	612
2000	Directory boards, outdoor, black plastic, 36" x 24"	590	330	920
2040	36" x 36"	680	440	1,120
2100	Indoor, economy, open faced, 18" x 24"	84.50	94	178.50
2300	Disappearing stairways, folding, pine, 8'-6" ceiling	88	82	170
2400	Fire escape, galvanized steel, 8'-0" to 10'-6" ceiling	1,175	660	1,835
2500	Automatic electric, wood, 8' to 9' ceiling	5,825	660	6,485
2600	Display cases, freestanding, glass and aluminum, 3'-6" x 3' x 1'-0" deep	490	82	572
2700	Wall mounted, glass and aluminum, 3' x 4' x 1'-4" deep	860	132	992
2900	Fireplace prefabricated, freestanding or wall hung, economy	970	262	1,232
2940	Deluxe	2,800	375	3,175
3000	Woodburning stoves, cast iron, economy	715	520	1,235
3040	Deluxe	2,575	850	3,425
3100	Flagpoles, on grade, aluminum, tapered, 20' high	660	390	1,050
3140	70' high	5,700	975	6,675
3240	59' high	5,725	860	6,585
3300	Concrete, internal halyard, 20' high	1,250	310	1,560
3340	100' high	12,400	775	13,175
3400	Lockers, steel, single tier, 5' to 6' high, per opening, min.	107	26.50	133.50
3440	Maximum	182	31	213
3700	Mail boxes, horizontal, rear loaded, aluminum, 5" x 6" x 15" deep	36	9.65	45.65
3740	Front loaded, aluminum, 10" x 12" x 15" deep	136	16.45	152.45
3800	Vertical, front loaded, aluminum, 15" x 5" x 6" deep	27.50	9.65	37.15
3840	Bronze, duranodic finish	52	9.65	61.65
3900	Letter slot, post office	202	41	243
4000	Mail counter, window, post office, with grille	545	164	709
4100	Medicine cabinets, sliding mirror doors, 20" x 16" x 4-3/4", unlighted	88.50	47	135.50
4140	24" x 19" x 8-1/2", lighted	149	66	215
4400	Partitions, shower stall, single wall, painted steel, 2'-8" x 2'-8"	610	148	758
4440	Fiberglass, 2'-8" x 2'-8"	500	165	665
4500	Double wall, enameled steel, 2'-8" x 2'-8"	625	148	773
4540	Stainless steel, 2'-8" x 2'-8"	1,125	148	1,273
4700	Tub enclosure, sliding panels, tempered glass, aluminum frame	269	185	454
4740	Chrome/brass frame-deluxe	770	247	1,017
5400	Projection screens, wall hung, manual operation, 50 S.F., economy	278	66	344
5440	Electric operation, 100 S.F., deluxe	1,550	330	1,880
5500	Scales, dial type, built in floor, 5 ton capacity, 8' x 6' platform	4,875	1,975	6,850
5540	10 ton capacity, 9' x 7' platform	9,250	2,825	12,075
5600	Truck (including weigh bridge), 20 ton capacity, 24' x 10'	9,300	3,300	12,600
5900	Security gates-scissors type, painted steel, single, 6' high, 5-1/2' wide	122	214	336
5940	Double gate, 7-1/2' high, 14' wide	390	430	820
6000	Shelving, metal industrial, braced, 3' wide, 1' deep	19.65	7.35	27
6040	2' deep	28	7.80	35.80
6200	Signs, interior electric exit sign, wall mounted, 6"	40.50	45	85.50
6240	Street, reflective alum., dbl. face, 4 way, w/bracket	77.50	22	99.50

Important: See the Reference Section for critical supporting data - Location Factors & Historical Cost Indexes

SPECIAL CONSTR. — A11.1 Specialties

11.1-100 Architectural Specialties/Each

		MAT.	INST.	TOTAL
6300	Letters, cast aluminum, 1/2" deep, 4" high	14.65	10.25	24.90
6340	1" deep, 10" high	44.50	18.25	62.75
6400	Plaques, cast aluminum, 20" x 30"	870	164	1,034
6440	Cast bronze, 36" x 48"	3,400	330	3,730
6700	Turnstiles, one way, 4' arm, 46" diam., manual	263	132	395
6740	Electric	925	550	1,475
6800	3 arm, 5'-5' diam. & 7' high, manual	2,425	660	3,085
6840	Electric	3,000	1,100	4,100

11.1-100 Architectural Specialties/S.F.

		MAT.	INST.	TOTAL
7100	Bulletin board, cork sheets, no frame, 1/4" thick	2.58	2.27	4.85
7200	Aluminum frame, 1/4" thick, 3' x 5'	6.85	2.75	9.60
7500	Chalkboards, wall hung, alum, frame & chalktrough	9.30	1.46	10.76
7540	Wood frame & chalktrough	8.60	1.57	10.17
7600	Sliding board, one board with back panel	30	1.40	31.40
7640	Two boards with back panel	46	1.40	47.40
7700	Liquid chalk type, alum. frame & chalktrough	9.90	1.46	11.36
7740	Wood frame & chalktrough	17.05	1.46	18.51

11.1-100 Architectural Specialties/L.F.

		MAT.	INST.	TOTAL
8600	Partitions, hospital curtain, ceiling hung, polyester oxford cloth	11.70	3.21	14.91
8640	Designer oxford cloth	22.50	4.06	26.56

11.1-200 Architectural Equipment/Each

		MAT.	INST.	TOTAL
0200	Arch. equip., appliances, range, cook top, 4 burner, economy	242	60	302
0240	Built in, single oven 30" wide, economy	495	164	659
0300	Standing, single oven-21" wide, economy	340	51.50	391.50
0340	Double oven-30" wide, deluxe	1,650	129	1,779
0400	Compactor, residential, economy	330	66	396
0440	Deluxe	550	110	660
0500	Dish washer, built-in, 2 cycles, economy	220	184	404
0540	4 or more cycles, deluxe	770	365	1,135
0600	Garbage disposer, sink type, economy	43	73.50	116.50
0640	Deluxe	226	73.50	299.50
0700	Refrigerator, no frost, 10 to 12 C.F., economy	390	51.50	441.50
0740	21 to 29 C.F., deluxe	3,025	171	3,196
0800	Automotive equipment, compressors, electric, 1-1/2 H.P., std. controls	2,700	615	3,315
0840	5 H.P., dual controls	3,800	925	4,725
0900	Hoists, single post, 4 ton capacity, swivel arms	3,850	2,300	6,150
0940	Dual post, 12 ton capacity, adjustable frame	6,600	6,150	12,750
1000	Lube equipment, 3 reel type, with pumps	6,600	1,850	8,450
1040	Product dispenser, 6 nozzles, w/vapor recovery, not incl. piping, installed	16,500		16,500
1100	Bank equipment, drive up window, drawer & mike, no glazing, economy	5,100	855	5,955
1140	Deluxe	7,400	1,700	9,100
1200	Night depository, economy	6,600	855	7,455
1240	Deluxe	10,500	1,700	12,200
1300	Pneumatic tube systems, 2 station, standard	20,800	2,775	23,575
1340	Teller, automated, 24 hour, single unit	38,100	2,775	40,875
1400	Teller window, bullet proof glazing, 44" x 60"	4,625	515	5,140
1440	Pass through, painted steel, 72" x 40"	2,750	1,075	3,825
1500	Barber equipment, chair, hydraulic, economy	440	13.70	453.70
1540	Deluxe	2,975	20.50	2,995.50
1700	Church equipment, altar, wood, custom, plain	1,375	235	1,610
1740	Granite, custom, deluxe	11,400	3,250	14,650
1800	Baptistry, fiberglass, economy	1,550	845	2,395
1840	Bells & carillons, keyboard operation	8,250	5,225	13,475
1900	Confessional, wood, single, economy	3,275	550	3,825
1940	Double, deluxe	16,400	1,650	18,050

For expanded coverage of these items see *Means Assemblies Cost Data 1997*

SPECIAL CONSTR. — A11.1 Specialties

11.1-200 Architectural Equipment/Each

		COST EACH		
		MAT.	INST.	TOTAL
2000	Steeples, translucent fiberglas, 30" square, 15' high	1,875	1,075	2,950
2100	Checkout counter, single belt	2,125	51.50	2,176.50
2140	Double belt, power take-away	3,625	57	3,682
2600	Dental equipment, central suction system, economy	2,250	310	2,560
2640	Compressor-air, deluxe	6,475	675	7,150
2700	Chair, hydraulic, economy	3,300	675	3,975
2740	Deluxe	7,600	1,350	8,950
2800	Drill console with accessories, economy	4,125	211	4,336
2840	Deluxe	10,100	211	10,311
2900	X-ray unit, portable	4,650	84.50	4,734.50
2940	Panoramic unit	11,800	565	12,365
3000	Detention equipment, cell front rolling door, 7/8" bars, 5' x 7' high	4,400	915	5,315
3040	Cells, prefab., including front, 5' x 7' x 7' deep	6,875	1,200	8,075
3100	Doors and frames, 3' x 7', single plate	3,275	455	3,730
3140	Double plate	4,050	455	4,505
3200	Toilet apparatus, incl wash basin	2,150	565	2,715
3240	Visitor cubicle, vision panel, no intercom	2,475	915	3,390
3300	Dock bumpers, rubber blocks, 4-1/2" thick, 10" high, 14" long	46	12.65	58.65
3340	6" thick, 20" high, 11" long	97	25.50	122.50
3400	Dock boards, H.D., 5' x 5', aluminum, 5000 lb. capacity	1,075		1,075
3440	16,000 lb. capacity	1,475		1,475
3500	Dock levelers, hydraulic, 7' x 8', 10 ton capacity	5,825	905	6,730
3540	Dock lifters, platform, 6' x 6', portable, 3000 lb. capacity	5,725		5,725
3600	Dock shelters, truck, scissor arms, economy	975	330	1,305
3640	Deluxe	1,375	660	2,035
3700	Kitchen equipment, bake oven, single deck	3,350	84	3,434
3740	Broiler, without oven	3,225	84	3,309
3800	Commercial dish washer, semiautomatic, 50 racks/hr.	4,800	520	5,320
3840	Automatic, 275 racks/hr.	19,300	2,100	21,400
3900	Cooler, beverage, reach-in, 6 ft. long	2,500	112	2,612
3940	Food warmer, counter, 1.65 kw	715		715
4000	Fryers, with submerger, single	3,375	96	3,471
4040	Double	4,350	135	4,485
4100	Kettles, steam jacketed, 20 gallons	4,400	156	4,556
4140	Range, restaurant type, burners, 2 ovens and 24" griddle	4,000	112	4,112
4200	Range hood, incl. carbon dioxide system, economy	2,575	224	2,799
4240	Deluxe	18,700	675	19,375
4300	Laboratory equipment, glassware washer, distilled water, economy	5,100	410	5,510
4340	Deluxe	18,800	735	19,535
4440	Radio isotope	12,500		12,500
4600	Laundry equipment, dryers, gas fired, residential, 16 lb. capacity	540	125	665
4640	Commercial, 30 lb. capacity, single	2,550	125	2,675
4700	Dry cleaners, electric, 20 lb. capacity	29,700	3,675	33,375
4740	30 lb. capacity	40,700	4,900	45,600
4800	Ironers, commercial, 120" with canopy, 8 roll	143,000	10,500	153,500
4840	Institutional, 110", single roll	25,300	1,800	27,100
4900	Washers, residential, 4 cycle	615	125	740
4940	Commercial, coin operated, deluxe	2,575	125	2,700
5000	Library equipment, carrels, metal, economy	195	66	261
5040	Hardwood, deluxe	770	82	852
5100	Medical equipment, autopsy table, standard	6,050	375	6,425
5140	Deluxe	8,250	625	8,875
5200	Incubators, economy	2,650		2,650
5240	Deluxe	13,800		13,800
5300	Station, scrub-surgical, single, economy	3,525		3,525
5340	Dietary, medium, with ice	11,200		11,200
5400	Sterilizers, general purpose, single door, 20" x 20" x 28"	9,625		9,625
5440	Floor loading, double door, 28" x 67" x 52"	154,000		154,000
5500	Surgery tables, standard	9,350	610	9,960
5540	Deluxe	22,000	855	22,855

Important: See the Reference Section for critical supporting data - Location Factors & Historical Cost Indexes

SPECIAL CONSTR. — A11.1 Specialties

11.1-200 Architectural Equipment/Each

		COST EACH		
		MAT.	INST.	TOTAL
5600	Tables, standard, with base cabinets, economy	985	219	1,204
5640	Deluxe	2,925	330	3,255
5700	X-ray, mobile, economy	8,800		8,800
5740	Stationary, deluxe	143,000		143,000
5800	Movie equipment, changeover, economy	380		380
5840	Film transport, incl. platters and autowind, economy	7,675		7,675
5900	Lamphouses, incl. rectifiers, xenon, 1000W	6,325	180	6,505
5940	4000W	10,400	241	10,641
6000	Projector mechanisms, 35 mm, economy	10,700		10,700
6040	Deluxe	14,200		14,200
6100	Sound systems, incl. amplifier, single, economy	1,975	400	2,375
6140	Dual, Dolby/super sound	13,600	900	14,500
6200	Seating, painted steel, upholstered, economy	99	18.80	117.80
6240	Deluxe	255	23.50	278.50
6300	Parking equipment, automatic gates, 8 ft. arm, one way	2,575	655	3,230
6340	Traffic detectors, single treadle	2,050	300	2,350
6400	Booth for attendant, economy	4,400		4,400
6440	Deluxe	22,000		22,000
6500	Ticket printer/dispenser, rate computing	6,425	515	6,940
6540	Key station on pedestal	360	176	536
6700	Frozen food, chest type, 12 ft. long	4,575	205	4,780
7000	Safe, office type, 1 hr. rating, 34" x 20" x 20"	1,975		1,975
7040	4 hr. rating, 62" x 33" x 20"	7,575		7,575
7100	Data storage, 4 hr. rating, 63" x 44" x 16"	12,200		12,200
7140	Jewelers, 63" x 44" x 16"	20,900		20,900
7200	Money, "B" label, 9" x 14" x 14"	480		480
7240	Tool and torch resistive, 24" x 24" x 20"	8,150		8,150
7300	Sauna, prefabricated, incl. heater and controls, 7' high, 6' x 4'	3,225	500	3,725
7340	10' x 12'	6,925	1,100	8,025
7400	Heaters, wall mounted, to 200 C.F.	485		485
7440	Floor standing, to 1000 C.F., 12500 W	1,450	120	1,570
7500	School equipment, basketball backstops, wall mounted, wood, fixed	540	575	1,115
7540	Suspended type, electrically operated	2,650	1,100	3,750
7600	Bleachers-telescoping, manual operation, 15 tier, economy (per seat)	44	21	65
7640	Power operation, 30 tier, deluxe (per seat)	86	37	123
7700	Weight lifting gym, universal, economy	3,675	515	4,190
7740	Deluxe	13,200	1,025	14,225
7800	Scoreboards, basketball, 1 side, economy	2,525	495	3,020
7840	4 sides, deluxe	27,500	6,775	34,275
8300	Vocational shop equipment, benches, metal	248	132	380
8340	Wood	425	132	557
8400	Dust collector, not incl. ductwork, 6' diam.	2,300	335	2,635
8440	Planer, 13" x 6"	1,550	164	1,714
8500	Waste handling, compactors, single bag, 250 lbs./hr., hand fed	6,250	385	6,635
8540	Heavy duty industrial, 5 C.Y. capacity	23,800	1,850	25,650
8600	Incinerator, electric, 100 lbs./hr., economy	14,400	1,850	16,250
8640	Gas, 2000 lbs./hr., deluxe	171,000	14,300	185,300
8700	Shredder, no baling, 35 tons/hr.	211,000		211,000
8740	Incl. baling, 50 tons/day	422,000		422,000

For expanded coverage of these items see *Means Assemblies Cost Data 1997*

SPECIAL CONSTR. — A11.1 Specialties

11.1-200 Architectural Equipment/S.F.

		MAT.	INST.	TOTAL
8910	Arch. equip., lab equip., counter tops, acid proof, economy	15.40	8	23.40
8940	Stainless steel	65.50	8	73.50
9000	Movie equipment, projection screens, rigid in wall, acrylic, 1/4" thick	34	3.18	37.18
9040	1/2" thick	39.50	4.77	44.27
9100	School equipment, gym mats, naugahyde cover, 2" thick	4.19		4.19
9140	Wrestling, 1" thick, heavy duty	4.76		4.76
9200	Stage equipment, curtains, velour, medium weight	12.10	1.10	13.20
9240	Silica based yarn, fireproof	27.50	13.15	40.65
9300	Stages, portable with steps, folding legs, 8" high	7.65		7.65
9340	Telescoping platforms, aluminum, deluxe	33	17.10	50.10

11.1-200 Architectural Equipment/L.F.

		MAT.	INST.	TOTAL
9410	Arch. equip., church equip. pews, bench type, hardwood, economy	53	16.45	69.45
9440	Deluxe	86	22	108
9500	Laboratory equipment, cabinets, wall, open	70.50	33	103.50
9540	Base, drawer units	230	36.50	266.50
9600	Fume hoods, not incl. HVAC, economy	705	122	827
9640	Deluxe incl. fixtures	1,100	274	1,374
9700	Library equipment, book shelf, metal, single face, 90" high x 10" shelf	68	27.50	95.50
9740	Double face, 90" high x 10" shelf	194	59.50	253.50
9800	Charging desk, built-in, with counter, plastic laminate	415	47	462
9900	Stage equipment, curtain track, heavy duty	24	36.50	60.50
9940	Lights, border, quartz, colored	132	18.05	150.05

11.1-500 Furnishings/Each

		MAT.	INST.	TOTAL
1000	Furnishings, blinds, exterior, aluminum, louvered, 1'-4" wide x 3'-0" long	32	33	65
1040	1'-4" wide x 6'-8" long	51.50	36.50	88
1100	Hemlock, solid raised, 1'-4" wide x 3'-0" long	53	33	86
1140	1'-4" wide x 6'-9" long	99	36.50	135.50
1200	Polystyrene, louvered, 1'-3" wide x 3'-3" long	40.50	33	73.50
1240	1'-3" wide x 6'-8" long	88	36.50	124.50
1300	Interior, wood folding panels, louvered, 7" x 20" (per pair)	24	19.35	43.35
1340	18" x 40" (per pair)	75.50	19.35	94.85
1800	Hospital furniture, beds, manual, economy	825		825
1840	Deluxe	1,450		1,450
1900	All electric, economy	1,375		1,375
1940	Deluxe	3,425		3,425
2000	Patient wall systems, no utilities, economy, per room	825		825
2040	Deluxe, per room	1,525		1,525
2200	Hotel furnishings, standard room set, economy, per room	1,500		1,500
2240	Deluxe, per room	7,850		7,850
2400	Office furniture, standard employee set, economy, per person	460		460
2440	Deluxe, per person	4,075		4,075
2800	Posts, portable, pedestrian traffic control, economy	82.50		82.50
2840	Deluxe	315		315
3000	Restaurant furniture, booth, molded plastic, stub wall and 2 seats, economy	400	164	564
3040	Deluxe	860	219	1,079
3100	Upholstered seats, foursome, single-economy	575	66	641
3140	Foursome, double-deluxe	3,100	110	3,210
3300	Seating, lecture hall, pedestal type, economy	108	18.80	126.80
3340	Deluxe	335	33	368
3400	Auditorium chair, veneer construction	108	18.80	126.80
3440	Fully upholstered, spring seat	159	18.80	177.80

11.1-500 Furnishings/S.F.

		MAT.	INST.	TOTAL
4010	Furnishings, blinds-interior, venetian-aluminum, stock, 2" slats, economy	1.29	.56	1.85
4040	Custom, 1" slats, deluxe	7.60	.75	8.35

SPECIAL CONSTR. — A11.1 Specialties

11.1-500 Furnishings/S.F.

		COST PER S.F.		
		MAT.	INST.	TOTAL
4100	Vertical, PVC or cloth, T&B track, economy	7.15	.71	7.86
4140	Deluxe	11.95	.82	12.77
4300	Draperies, unlined, economy	2.06		2.06
4440	Lightproof, deluxe	6.20		6.20
4700	Floor mats, recessed, inlaid black rubber, 3/8" thick, solid	21	1.66	22.66
4740	Colors, 1/2" thick, perforated	33	1.66	34.66
4800	Link-including nosings, steel-galvanized, 3/8" thick	5.70	1.66	7.36
4840	Vinyl, in colors	17.80	1.66	19.46
5000	Shades, mylar, wood roller, single layer, non-reflective	5.60	.48	6.08
5040	Metal roller, triple layer, heat reflective	9.85	.48	10.33
5100	Vinyl, light weight, 4 ga.	.40	.48	.88
5140	Heavyweight, 6 ga.	1.24	.48	1.72
5200	Vinyl coated cotton, lightproof decorator shades	1.24	.48	1.72
5300	Woven aluminum, 3/8" thick, light and fireproof	3.97	.94	4.91

11.1-500 Furnishings/L.F.

		COST PER L.F.		
		MAT.	INST.	TOTAL
5510	Furnishings, cabinets, hospital, base, laminated plastic	178	66	244
5540	Stainless steel	223	66	289
5600	Counter top, laminated plastic, no backsplash	30.50	16.45	46.95
5640	Stainless steel	85	16.45	101.45
5700	Nurses station, door type, laminated plastic	206	66	272
5740	Stainless steel	238	66	304
5900	Household, base, hardwood, one top drawer & one door below x 12" wide	110	26.50	136.50
5940	Four drawer x 24" wide	179	29.50	208.50
6000	Wall, hardwood, 30" high with one door x 12" wide	99	30	129
6040	Two doors x 48" wide	193	35.50	228.50
6100	Counter top-laminated plastic, stock, economy	5.20	10.95	16.15
6140	Custom-square edge, 7/8" thick	11	24.50	35.50
6300	School, counter, wood, 32" high	141	33	174
6340	Metal, 84" high	218	44	262
6500	Dormitory furniture, desk top (built-in), laminated plastc, 24"deep, economy	24	13.15	37.15
6540	30" deep, deluxe	105	16.45	121.45
6600	Dressing unit, built-in, economy	177	55	232
6640	Deluxe	525	82	607
7000	Restaurant furniture, bars, built-in, back bar	154	66	220
7040	Front bar	212	66	278
7200	Wardrobes & coatracks, standing, steel, single pedestal, 30" x 18" x 63"	75		75
7300	Double face rack, 39" x 26" x 70"	128		128
7340	Wall mounted rack, steel frame & shelves, 12" x 15" x 26"	53.50	4.66	58.16
7400	12" x 15" x 50"	42.50	2.44	44.94

For expanded coverage of these items see *Means Assemblies Cost Data 1997*

SPECIAL CONSTR. | A11.1 | Specialties

11.1-700 Special Construction/Each

		MAT.	INST.	TOTAL
1000	Special construction, bowling alley incl. pinsetter, scorer etc., economy	35,500	6,575	42,075
1040	Deluxe	40,400	7,300	47,700
1100	For automatic scorer, economy, add	6,600		6,600
1140	Deluxe, add	7,475		7,475
1200	Chimney, metal, high temp. steel jacket, factory lining, 24" diameter	5,350	3,025	8,375
1240	60" diameter	40,000	13,100	53,100
1300	Poured concrete, brick lining, 10' diam., 200' high			1,095,000
1340	20' diam., 500' high			4,887,500
1400	Radial brick, 3'-6" I.D., 75' high			184,000
1440	7' I.D., 175' high			555,500
1500	Control tower, modular, 12' x 10', incl. instrumentation, economy			280,000
1540	Deluxe			680,000
1600	Garage costs, residential, prefab, wood, single car economy	2,675	685	3,360
1640	Two car deluxe	7,950	1,375	9,325
1700	Grandstands, permanent, closed deck, steel, economy (per seat)			20
1740	Deluxe (per seat)			85
1800	Composite design, economy (per seat)			30
1840	Deluxe (per seat)			90
1900	Hangars, prefab, galv. steel, bottom rolling doors, economy (per plane)	8,950	3,125	12,075
1940	Electrical bi-folding doors, deluxe (per plane)	12,500	4,300	16,800
2000	Ice skating rink, 85' x 200', 55° system, 5 mos., 100 ton			340,000
2040	90° system, 12 mos., 135 ton			363,000
2100	Dash boards, acrylic screens, polyethylene coated plywood	88,000	22,700	110,700
2140	Fiberglass and aluminum construction	99,000	22,700	121,700
2200	Integrated ceilings, radiant electric, 2' x 4' panel, manila finish	44.50	14.45	58.95
2240	ABS plastic finish	81.50	28	109.50
2300	Kiosks, round, 5' diam., 1/4" fiberglass wall, 8' high	6,050		6,050
2340	Rectangular, 5' x 9', 1" insulated dbl. wall fiberglass, 7'-6" high	10,500		10,500
2400	Portable booth, acoustical, 27 db 1000 hz., 15 S.F. floor	3,175		3,175
2440	55 S.F. flr.	6,500		6,500
2500	Radio towers, guyed, 40 lb. section, 50' high, 70 MPH basic wind speed	1,575	855	2,430
2540	90 lb. section, 400' high, wind load 70 MPH basic wind speed	20,400	9,525	29,925
2600	Self supporting, 60' high, 70 MPH basic wind speed	3,275	1,675	4,950
2640	190' high, wind load 90 MPH basic wind speed	19,500	6,675	26,175
2700	Shelters, aluminum frame, acrylic glazing, 8' high, 3' x 9'	2,800	750	3,550
2740	9' x 12'	4,000	1,175	5,175
2800	Shielding, lead x-ray protection, radiography room, 1/16" lead, economy	4,500	2,450	6,950
2840	Deluxe	5,500	4,075	9,575
2900	Deep therapy x-ray, 1/4" lead, economy	9,350	7,650	17,000
2940	Deluxe	14,300	10,200	24,500
3000	Shooting range incl. bullet traps, controls, separators, ceilings, economy	11,600	4,325	15,925
3040	Deluxe	22,100	7,300	29,400
3100	Silos, concrete stave, industrial, 12' diam., 35' high	13,400	13,600	27,000
3140	25' diam., 75' high	50,500	29,900	80,400
3200	Steel prefab, 30,000 gal., painted, economy	11,200	3,525	14,725
3240	Epoxy-lined, deluxe	22,600	7,050	29,650
3300	Sport court, squash, regulation, in existing building, economy			14,500
3340	Deluxe			27,100
3400	Racketball, regulation, in existing building, economy	1,925	6,400	8,325
3440	Deluxe	25,800	12,700	38,500
3500	Swimming pool equipment, diving stand, stainless steel, 1 meter	4,475	244	4,719
3540	3 meter	6,300	1,650	7,950
3600	Diving boards, 16 ft. long, aluminum	1,275	244	1,519
3640	Fiberglass	640	244	884
3700	Filter system, sand, incl. pump, 6000 gal./hr.	940	415	1,355
3800	Lights, underwater, 12 volt with transformer, 300W	275	900	1,175
3900	Slides, fiberglass with aluminum handrails & ladder, 6' high, straight	835	410	1,245
3940	12' high, straight with platform	835	550	1,385
4000	Tanks, steel, ground level, 100,000 gal.			88,500
4040	10,000,000 gal.			2,100,000

SPECIAL CONSTR. — A11.1 Specialties

11.1-700 Special Construction/Each

		MAT.	INST.	TOTAL
4100	Elevated water, 50,000 gal.			179,000
4140	1,000,000 gal.			877,000
4200	Cypress wood, ground level, 3,000 gal.	5,500	6,725	12,225
4240	Redwood, ground level, 45,000 gal.	27,500	18,300	45,800

11.1-700 Special Construction/S.F.

		MAT.	INST.	TOTAL
4500	Special construction, acoustical, enclosure, 4" thick, 8 psf panels	26.50	13.70	40.20
4540	Reverb chamber, 4" thick, parallel walls	32	16.45	48.45
4600	Sound absorbing panels, 2'-6" x 8', painted metal	9.55	4.59	14.14
4640	Vinyl faced	7.75	4.11	11.86
4700	Flexible transparent curtain, clear	6	5.15	11.15
4740	With absorbing foam, 75% coverage	8.35	5.15	13.50
4800	Strip entrance, 2/3 overlap	4.17	8.25	12.42
4840	Full overlap	5.20	9.65	14.85
4900	Audio masking system, plenum mounted, over 10,000 S.F.	.57	.16	.73
4940	Ceiling mounted, under 5,000 S.F.	1.02	.30	1.32
5000	Air supported struc., polyester vinyl fabric, 24oz., warehouse, 5000 S.F.	13.20	.21	13.41
5040	50,000 S.F.	5.85	.16	6.01
5100	Tennis, 7,200 S.F.	14.30	.17	14.47
5140	24,000 S.F.	9.35	.17	9.52
5200	Woven polyethylene, 6 oz., shelter, 3,000 S.F.	6.50	.34	6.84
5240	24,000 S.F.	2.96	.17	3.13
5300	Teflon coated fiberglass, stadium cover, economy	33	.09	33.09
5340	Deluxe	38.50	.12	38.62
5400	Air supported storage tank covers, reinf. vinyl fabric, 12 oz., 400 S.F.	17.60	.29	17.89
5440	18,000 S.F.	3.91	.26	4.17
5500	Anechoic chambers, 7' high, 100 cps cutoff, 25 S.F.			1,925
5540	200 cps cutoff, 100 S.F.			1,000
5600	Audiometric rooms, under 500 S.F.	42	13.40	55.40
5640	Over 500 S.F.	39.50	10.95	50.45
5700	Comfort stations, prefab, mobile on steel frame, economy	33		33
5740	Permanent on concrete slab, deluxe	220	30	250
5800	Darkrooms, shell, not including door, 240 S.F., 8' high	35	5.50	40.50
5840	64 S.F., 12' high	82.50	10.25	92.75
5900	Domes, bulk storage, wood framing, wood decking, 50' diam.	27.50	1.08	28.58
5940	116' diam.	16.50	1.25	17.75
6000	Steel framing, metal decking, 150' diam.	29.50	7.10	36.60
6040	400' diam.	24	5.45	29.45
6100	Geodesic, wood framing, wood panels, 30' diam.	15.55	1.22	16.77
6140	60' diam.	12.90	.76	13.66
6200	Aluminum framing, acrylic panels, 40' diam.			75
6240	Aluminum panels, 400' diam.			30
6300	Garden house, prefab, wood, shell only, 48 S.F.	14.80	3.40	18.20
6340	200 S.F.	27	14.15	41.15
6400	Greenhouse, shell-stock, residential, lean-to, 8'-6" long x 3'-10" wide	34	19.35	53.35
6440	Freestanding, 8'-6" long x 13'-6" wide	26.50	6.10	32.60
6500	Commercial-truss frame, under 2000 S.F., deluxe			38
6540	Over 5,000 S.F., economy			23
6600	Institutional-rigid frame, under 500 S.F., deluxe			99
6640	Over 2,000 S.F., economy			39
6700	Hangar, prefab, galv. roof and walls, bottom rolling doors, economy	8.55	2.82	11.37
6740	Electric bifolding doors, deluxe	10.10	3.68	13.78
6800	Integrated ceilings, Luminaire, suspended, 5' x 5' modules, 50% lighted	3.88	7.70	11.58
6840	100% lighted	5.80	13.90	19.70
6900	Dimensionaire, 2' x 4' module tile system, no air bar	2.35	1.32	3.67
6940	With air bar, deluxe	3.40	1.32	4.72
7000	Music practice room, modular, perforated steel, under 500 S.F.	23	9.40	32.40
7040	Over 500 S.F.	17.45	8.20	25.65
7100	Pedestal access floors, stl pnls, no stringers, vinyl covering, >6000 S.F.	11.10	1.46	12.56
7140	With stringers, high pressure laminate covering, under 6000 S.F.	10.80	1.64	12.44

For expanded coverage of these items see *Means Assemblies Cost Data 1997*

SPECIAL CONSTR. | A11.1 | Specialties

11.1-700 Special Construction/S.F.

		COST PER S.F.		
		MAT.	INST.	TOTAL
7200	Carpet covering, under 6000 S.F.	12.45	1.64	14.09
7240	Aluminum panels, no stringers, no covering	22	1.32	23.32
7300	Refrigerators, prefab, walk-in, 7'-6" high, 6' x 6'	105	12	117
7340	12' x 20'	66	6	72
7400	Shielding, lead, gypsum board, 5/8" thick, 1/16" lead	5.45	3.83	9.28
7440	1/8" lead	10.90	4.38	15.28
7500	Lath, 1/16" thick	5.20	4.54	9.74
7540	1/8" thick	10.45	5.10	15.55
7600	Radio frequency, copper, prefab screen type, economy	26.50	3.65	30.15
7640	Deluxe	34	4.54	38.54
7700	Swimming pool enclosure, translucent, freestanding, economy	11	3.29	14.29
7740	Deluxe	275	9.40	284.40
7800	Swimming pools, residential, vinyl liner, metal sides	9.25	4.55	13.80
7840	Concrete sides	11.05	8.55	19.60
7900	Gunite shell, plaster finish, 350 S.F.	18.30	17.70	36
7940	800 S.F.	14.75	10.25	25
8000	Motel, gunite shell, plaster finish	22.50	22.50	45
8040	Municipal, gunite shell, tile finish, formed gutters	94	38.50	132.50
8100	Tension structures, steel frame, polyester vinyl fabric, 12,000 S.F.	7.35	1.52	8.87
8140	20,800 S.F.	9.50	1.37	10.87

11.1-700 Special Construction/L.F.

		COST PER L.F.		
		MAT.	INST.	TOTAL
8500	Spec. const., air curtains, shipping & receiving, 8'high x 5'wide, economy	171	206	377
8540	20' high x 8' wide, heated, deluxe	1,275	285	1,560
8600	Customer entrance, 10' high x 5' wide, economy	207	206	413
8640	12' high x 4' wide, heated, deluxe	375	265	640

For information about Means Estimating Seminars, see yellow pages 11 and 12 in back of book

Division 12
Site Work

SITE WORK — A12.3 Utilities

12.3-110 Trenching

		\multicolumn{3}{c	}{COST PER L.F.}
		MAT.	INST.	TOTAL
1310	Trenching, backhoe, 0 to 1 slope, 2' wide, 2' deep, 3/8 C.Y. bucket		2.13	2.13
1330	4' deep, 3/8 C.Y. bucket		3.84	3.84
1360	10' deep, 1 C.Y. bucket		8.13	8.13
1400	4' wide, 2' deep, 3/8 C.Y. bucket		4.32	4.32
1420	4' deep, 1/2 C.Y. bucket		6.63	6.63
1450	10' deep, 1 C.Y. bucket		16.35	16.35
1480	18' deep, 2-1/2 C.Y. bucket		25.50	25.50
1520	6' wide, 6' deep, 5/8 C.Y. bucket		14.75	14.75
1540	10' deep, 1 C.Y. bucket		21.90	21.90
1570	20' deep, 3-1/2 C.Y. bucket		35.35	35.35
1640	8' wide, 12' deep, 1-1/4 C.Y. bucket		26	26
1680	24' deep, 3-1/2 C.Y. bucket		78.50	78.50
1730	10' wide, 20' deep, 3-1/2 C.Y. bucket		61	61
1740	24' deep, 3-1/2 C.Y. bucket		99	99
3500	1 to 1 slope, 2' wide, 2' deep, 3/8 C.Y. bucket		3.23	3.23
3540	4' deep, 3/8 C.Y. bucket		7.92	7.92
3600	10' deep, 1 C.Y. bucket		35.85	35.85
3800	4' wide, 2' deep, 3/8 C.Y. bucket		5.48	5.48
3840	4' deep, 1/2 C.Y. bucket		9.52	9.52
3900	10' deep, 1 C.Y. bucket		37.60	37.60
4030	6' wide, 6' deep, 5/8 C.Y. bucket		20.45	20.45
4050	10' deep, 1 C.Y. bucket		44	44
4080	20' deep, 3-1/2 C.Y. bucket		116	116
4500	8' wide, 12' deep, 1-1/4 C.Y. bucket		49.50	49.50
4650	24' deep, 3-1/2 C.Y. bucket		241	241
4800	10' wide, 20' deep, 3-1/2 C.Y. bucket		139	139
4850	24' deep, 3-1/2 C.Y. bucket		258	258

12.3-310 Pipe Bedding

		COST PER L.F.		
		MAT.	INST.	TOTAL
1440	Pipe bedding, side slope 0 to 1, 1' wide, pipe size 6" diameter	.29	.40	.69
1460	2' wide, pipe size 8" diameter	.64	.88	1.52
1500	Pipe size 12" diameter	.67	.92	1.59
1600	4' wide, pipe size 20" diameter	1.67	2.30	3.97
1660	Pipe size 30" diameter	1.77	2.44	4.21
1680	6' wide, pipe size 32" diameter	3.09	4.26	7.35
1740	8' wide, pipe size 60" diameter	5.15	7.10	12.25
1780	12' wide, pipe size 84" diameter	10.10	13.95	24.05

12.3-710 Manholes & Catch Basins

		COST PER EACH		
		MAT.	INST.	TOTAL
1920	Manhole/catch basin, brick, 4' I.D. riser, 4' deep	695	1,025	1,720
1980	10' deep	1,425	2,425	3,850
3200	Block, 4' I.D. riser, 4' deep	710	835	1,545
3260	10' deep	1,350	2,000	3,350
4620	Concrete, cast-in-place, 4' I.D. riser, 4' deep	825	980	1,805
4680	10' deep	1,575	2,300	3,875
5820	Concrete, precast, 4' I.D. riser, 4' deep	830	705	1,535
5880	10' deep	1,600	1,625	3,225
6200	6' I.D. riser, 4' deep	1,300	1,125	2,425
6260	10' deep	2,500	2,650	5,150

SITE WORK — A12.5 Roads & Parking

12.5-110 Roadway Pavement

		\multicolumn{3}{c}{COST PER L.F.}
		MAT.	INST.	TOTAL
1500	Roadways, bituminous conc. paving, 2-1/2" thick, 20' wide	33.50	36	69.50
1580	30' wide	39.50	41.50	81
1800	3" thick, 20' wide	35.50	35	70.50
1880	30' wide	42	41.50	83.50
2100	4" thick, 20' wide	38.50	35	73.50
2180	30' wide	46.50	41.50	88
2400	5" thick, 20' wide	45	35.50	80.50
2480	30' wide	56.50	42.50	99
3000	6" thick, 20' wide	60.50	36.50	97
3080	30' wide	80	44	124
3300	8" thick 20' wide	73	36.50	109.50
3380	30' wide	98.50	44.50	143
3600	12" thick 20' wide	97.50	39	136.50
3700	32' wide	143	49.50	192.50

12.5-510 Parking Lots

		\multicolumn{3}{c}{COST PER CAR}
		MAT.	INST.	TOTAL
1500	Parking lot, 90° angle parking, 3" bituminous paving, 6" gravel base	236	199	435
1540	10" gravel base	255	233	488
1560	4" bituminous paving, 6" gravel base	281	197	478
1600	10" gravel base	300	231	531
1620	6" bituminous paving, 6" gravel base	380	206	586
1660	10" gravel base	400	239	639
1800	60° angle parking, 3" bituminous paving, 6" gravel base	272	218	490
1840	10" gravel base	293	257	550
1860	4" bituminous paving, 6" gravel base	325	216	541
1900	10" gravel base	345	256	601
1920	6" bituminous paving, 6" gravel base	440	226	666
1960	10" gravel base	465	265	730
2200	45° angle parking, 3" bituminous paving, 6" gravel base	310	237	547
2240	10" gravel base	330	281	611
2260	4" bituminous paving, 6" gravel base	370	235	605
2300	10" gravel base	390	280	670
2320	6" bituminous paving, 6" gravel base	500	247	747
2360	10" gravel base	525	291	816

For information about Means Estimating Seminars, see yellow pages 11 and 12 in back of book

For expanded coverage of these items see *Means Site Work & Landscape Cost Data 1997*

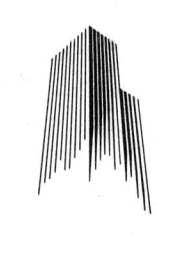

Reference Section

All the reference information is in one section making it easy to find what you need to know... and easy to use the book on a daily basis. This section is visually identified by a vertical gray bar on the edge of pages.

In this Reference Section, we've included General Conditions; Historical Cost Indexes for cost comparisons over time; Location Factors; a Glossary; and an explanation of all Abbreviations used in the book.

Table of Contents

General Conditions	428
Location Factors	429
Historical Cost Indexes	434
Glossary	442
Abbreviations	446
Forms	449

REFERENCE

Introduction: General Conditions

General Conditions (Overhead & Profit)

Total building costs in the Commercial/Industrial/Institutional section include a 15% allowance for general conditions. This allowance provides for the general contractor's overhead & profit and contingencies.

General contractor overhead includes indirect costs such as permits, Workers' Compensation, insurances, supervision and bonding fees. Overhead will vary with the size of project, the contractor's operating procedures and location. Profits will vary with economic activity and local conditions.

Contingencies provide for unforeseen construction difficulties which include material shortages and weather. In all situations, the appraiser should give consideration to possible adjustment of the 15% factor used in developing the Commercial/Industrial/Institutional models.

Architectural Fees

Tabulated below are typical percentage fees by project size, for good professional architectural service. Fees may vary from those listed depending upon degree of design difficulty and economic conditions in any particular area.

Rates can be interpolated horizontally and vertically. Various portions of the same project requiring different rates should be adjusted proportionately. For alterations, add 50% to the fee for the first $500,000 of project cost and add 25% to the fee for project cost over $500,000.

Architectural fees tabulated below include Engineering fees.

Insurance Exclusions

Many insurance companies exclude from coverage such items as architect's fees, excavation, foundations below grade, underground piping and site preparation. Since exclusions vary among insurance companies, it is recommended that for greatest accuracy each exclusion be priced separately using the unit-in-place section.

As a rule of thumb, exclusions can be calculated at 9% of total building cost plus the appropriate allowance for architect's fees.

Building Types	Total Project Size in Thousands of Dollars						
	100	250	500	1,000	5,000	10,000	50,000
Factories, garages, warehouses, repetitive housing	9.0%	8.0%	7.0%	6.2%	5.3%	4.9%	4.5%
Apartments, banks, schools, libraries, offices, municipal buildings	11.7	10.8	8.5	7.3	6.4	6.0	5.6
Churches, hospitals, homes, laboratories, museums, research	14.0	12.8	11.9	10.9	8.5	7.8	7.2
Memorials, monumental work, decorative furnishings	—	16.0	14.5	13.1	10.0	9.0	8.3

Location Factors

Costs shown in *Means cost data publications* are based on National Averages for materials and installation. To adjust these costs to a specific location, simply multiply the base cost by the factor for that city. The data is arranged alphabetically by state and postal zip code numbers. For a city not listed, use the factor for a nearby city with similar economic characteristics.

STATE/ZIP	CITY	Residential	Commercial
ALABAMA			
350-352	Birmingham	.82	.83
354	Tuscaloosa	.81	.79
355	Jasper	.76	.77
356	Decatur	.82	.83
357-358	Huntsville	.82	.83
359	Gadsden	.80	.81
360-361	Montgomery	.82	.80
362	Anniston	.74	.75
363	Dothan	.81	.79
364	Evergreen	.82	.80
365-366	Mobile	.83	.84
367	Selma	.81	.79
368	Phenix City	.81	.79
369	Butler	.81	.79
ALASKA			
995-996	Anchorage	1.28	1.27
997	Fairbanks	1.28	1.27
998	Juneau	1.27	1.26
999	Ketchikan	1.32	1.31
ARIZONA			
850,853	Phoenix	.94	.91
852	Mesa/Tempe	.90	.87
855	Globe	.93	.90
856-857	Tucson	.93	.90
859	Show Low	.94	.90
860	Flagstaff	.97	.93
863	Prescott	.94	.90
864	Kingman	.94	.90
865	Chambers	.94	.90
ARKANSAS			
716	Pine Bluff	.80	.80
717	Camden	.72	.72
718	Texarkana	.77	.76
719	Hot Springs	.72	.72
720-722	Little Rock	.80	.80
723	West Memphis	.82	.82
724	Jonesboro	.82	.82
725	Batesville	.78	.78
726	Harrison	.79	.79
727	Fayetteville	.71	.68
728	Russellville	.80	.77
729	Fort Smith	.83	.80
749	Poteau	.86	.82
CALIFORNIA			
900-902	Los Angeles	1.12	1.12
903-905	Inglewood	1.10	1.10
906-908	Long Beach	1.11	1.11
910-912	Pasadena	1.10	1.10
913-916	Van Nuys	1.12	1.12
917-918	Alhambra	1.11	1.11
919-921	San Diego	1.13	1.09
922	Palm Springs	1.13	1.09
923-924	San Bernardino	1.12	1.08
925	Riverside	1.15	1.11
926-927	Santa Ana	1.13	1.10
928	Anaheim	1.14	1.12
930	Oxnard	1.17	1.11
931	Santa Barbara	1.13	1.10
932-933	Bakersfield	1.13	1.07
934	San Luis Obispo	1.24	1.11
935	Mojave	1.12	1.08
936-938	Fresno	1.15	1.10
939	Salinas	1.14	1.14
940-941	San Francisco	1.23	1.26
942,956-958	Sacramento	1.13	1.12
943	Palo Alto	1.15	1.18
944	San Mateo	1.16	1.19
945	Vallejo	1.15	1.18
946	Oakland	1.17	1.20
947	Berkeley	1.30	1.33
948	Richmond	1.16	1.19
949	San Rafael	1.26	1.20
950	Santa Cruz	1.18	1.16
951	San Jose	1.23	1.21

STATE/ZIP	CITY	Residential	Commercial
CALIFORNIA (CONT'D)			
952	Stockton	1.15	1.11
953	Modesto	1.16	1.12
954	Santa Rosa	1.17	1.20
955	Eureka	1.13	1.12
959	Marysville	1.13	1.12
960	Redding	1.12	1.11
961	Susanville	1.12	1.10
COLORADO			
800-802	Denver	.97	.93
803	Boulder	.90	.86
804	Golden	.95	.91
805	Fort Collins	.98	.92
806	Greeley	.92	.86
807	Fort Morgan	.97	.91
808-809	Colorado Springs	.92	.90
810	Pueblo	.93	.91
811	Alamosa	.91	.89
812	Salida	.91	.89
813	Durango	.89	.87
814	Montrose	.87	.85
815	Grand Junction	.91	.87
816	Glenwood Springs	.97	.92
CONNECTICUT			
060	New Britain	1.07	1.08
061	Hartford	1.07	1.08
062	Willimantic	1.07	1.08
063	New London	1.07	1.06
064	Meriden	1.06	1.08
065	New Haven	1.07	1.08
066	Bridgeport	1.04	1.07
067	Waterbury	1.08	1.08
068	Norwalk	1.03	1.07
069	Stamford	1.05	1.09
D.C.			
200-205	Washington	.93	.95
DELAWARE			
197	Newark	.99	1.00
198	Wilmington	.98	.99
199	Dover	.99	1.00
FLORIDA			
320,322	Jacksonville	.86	.85
321	Daytona Beach	.90	.89
323	Tallahassee	.79	.81
324	Panama City	.73	.75
325	Pensacola	.88	.86
326	Gainesville	.87	.84
327-328,347	Orlando	.89	.87
329	Melbourne	.91	.90
330-332,340	Miami	.86	.88
333	Fort Lauderdale	.86	.88
334,349	West Palm Beach	.88	.85
335-336,346	Tampa	.84	.86
337	St. Petersburg	.85	.87
338	Lakeland	.83	.85
339	Fort Myers	.83	.83
342	Sarasota	.82	.84
GEORGIA			
300-303,399	Atlanta	.83	.88
304	Statesboro	.67	.69
305	Gainesville	.65	.69
306	Athens	.76	.81
307	Dalton	.68	.67
308-309	Augusta	.79	.81
310-312	Macon	.82	.82
313-314	Savannah	.82	.83
315	Waycross	.76	.76
316	Valdosta	.78	.78
317	Albany	.79	.81
318-319	Columbus	.78	.78

Location Factors

STATE/ZIP	CITY	Residential	Commercial
HAWAII			
967	Hilo	1.27	1.23
968	Honolulu	1.27	1.23
STATES & POSS.			
969	Guam	.86	.84
IDAHO			
832	Pocatello	.94	.93
833	Twin Falls	.82	.81
834	Idaho Falls	.85	.85
835	Lewiston	1.10	1.02
836-837	Boise	.94	.93
838	Coeur d'Alene	.99	.92
ILLINOIS			
600-603	North Suburban	1.07	1.06
604	Joliet	1.06	1.05
605	South Suburban	1.07	1.06
606	Chicago	1.09	1.08
609	Kankakee	.99	.99
610-611	Rockford	1.02	1.01
612	Rock Island	1.03	.95
613	La Salle	1.06	.99
614	Galesburg	1.04	.96
615-616	Peoria	1.05	.99
617	Bloomington	1.02	.98
618-619	Champaign	1.02	.99
620-622	East St. Louis	.98	.98
623	Quincy	.94	.92
624	Effingham	.98	.95
625	Decatur	1.00	.97
626-627	Springfield	1.00	.96
628	Centralia	.97	.97
629	Carbondale	.95	.95
INDIANA			
424	Henderson	.94	.92
460	Anderson	.94	.92
461-462	Indianapolis	.96	.94
463-464	Gary	1.01	.99
465-466	South Bend	.93	.91
467-468	Fort Wayne	.91	.92
469	Kokomo	.91	.90
470	Lawrenceburg	.91	.88
471	New Albany	.92	.88
472	Columbus	.92	.90
473	Muncie	.92	.91
474	Bloomington	.94	.91
475	Washington	.92	.92
476-477	Evansville	.94	.94
478	Terre Haute	.95	.94
479	Lafayette	.91	.91
IOWA			
500-503,509	Des Moines	.95	.91
504	Mason City	.88	.82
505	Fort Dodge	.85	.80
506-507	Waterloo	.89	.84
508	Creston	.92	.87
510-511	Sioux City	.90	.84
512	Sibley	.82	.80
513	Spencer	.83	.81
514	Carroll	.88	.83
515	Council Bluffs	.94	.87
516	Shenandoah	.83	.78
520	Dubuque	.97	.87
521	Decorah	.92	.83
522-524	Cedar Rapids	1.00	.91
525	Ottumwa	.95	.86
526	Burlington	.85	.80
527-528	Davenport	.96	.94
KANSAS			
660-662	Kansas City	.96	.94
664-666	Topeka	.87	.86
667	Fort Scott	.89	.87
668	Emporia	.84	.83
669	Belleville	.91	.85
670-672	Wichita	.89	.86
673	Independence	.82	.79
674	Salina	.87	.83
675	Hutchinson	.82	.78
676	Hays	.88	.84
677	Colby	.89	.85

STATE/ZIP	CITY	Residential	Commercial
KANSAS (CONT'D)			
678	Dodge City	.88	.85
679	Liberal	.81	.78
KENTUCKY			
400-402	Louisville	.93	.90
403-405	Lexington	.90	.87
406	Frankfort	.96	.90
407-409	Corbin	.81	.75
410	Covington	.96	.93
411-412	Ashland	.95	.96
413-414	Campton	.79	.76
415-416	Pikeville	.84	.85
417-418	Hazard	.79	.75
420	Paducah	.95	.91
421-422	Bowling Green	.94	.89
423	Owensboro	.92	.90
425-426	Somerset	.78	.74
427	Elizabethtown	.92	.88
LOUISIANA			
700-701	New Orleans	.88	.87
703	Thibodaux	.86	.86
704	Hammond	.85	.84
705	Lafayette	.87	.84
706	Lake Charles	.87	.87
707-708	Baton Rouge	.85	.84
710-711	Shreveport	.81	.81
712	Monroe	.80	.80
713-714	Alexandria	.79	.79
MAINE			
039	Kittery	.80	.82
040-041	Portland	.89	.91
042	Lewiston	.90	.91
043	Augusta	.81	.81
044	Bangor	.93	.93
045	Bath	.81	.81
046	Machias	.86	.86
047	Houlton	.83	.83
048	Rockland	.85	.85
049	Waterville	.82	.81
MARYLAND			
206	Waldorf	.89	.89
207-208	College Park	.91	.91
209	Silver Spring	.90	.90
210-212	Baltimore	.91	.91
214	Annapolis	.89	.90
215	Cumberland	.87	.88
216	Easton	.70	.71
217	Hagerstown	.90	.88
218	Salisbury	.77	.78
219	Elkton	.85	.86
MASSACHUSETTS			
010-011	Springfield	1.07	1.05
012	Pittsfield	1.03	1.03
013	Greenfield	1.05	1.03
014	Fitchburg	1.12	1.08
015-016	Worcester	1.14	1.10
017	Framingham	1.10	1.11
018	Lowell	1.12	1.12
019	Lawrence	1.12	1.12
020-022	Boston	1.18	1.19
023-024	Brockton	1.09	1.12
025	Buzzards Bay	1.07	1.09
026	Hyannis	1.09	1.10
027	New Bedford	1.11	1.12
MICHIGAN			
480,483	Royal Oak	1.01	1.00
481	Ann Arbor	1.03	1.02
482	Detroit	1.06	1.05
484-485	Flint	.98	.99
486	Saginaw	.96	.97
487	Bay City	.96	.97
488-489	Lansing	1.00	.97
490	Battle Creek	1.00	.94
491	Kalamazoo	1.00	.94
492	Jackson	.98	.95
493,495	Grand Rapids	.88	.85
494	Muskegon	.96	.93
496	Traverse City	.93	.89
497	Gaylord	.93	.94

Location Factors

STATE/ZIP	CITY	Residential	Commercial
MICHIGAN (CONT'D)			
498-499	Iron Mountain	.98	.95
MINNESOTA			
540	New Richmond	1.00	.93
550-551	Saint Paul	1.10	1.08
553-554	Minneapolis	1.13	1.10
556-558	Duluth	1.03	1.04
559	Rochester	1.06	1.02
560	Mankato	.98	.97
561	Windom	.89	.88
562	Willmar	.89	.88
563	St. Cloud	1.08	1.00
564	Brainerd	1.06	.98
565	Detroit Lakes	.89	.95
566	Bemidji	.91	.98
567	Thief River Falls	.88	.94
MISSISSIPPI			
386	Clarksdale	.72	.68
387	Greenville	.83	.79
388	Tupelo	.73	.74
389	Greenwood	.74	.70
390-392	Jackson	.83	.79
393	Meridian	.78	.77
394	Laurel	.74	.71
395	Biloxi	.85	.81
396	Mccomb	.70	.68
397	Columbus	.72	.73
MISSOURI			
630-631	St. Louis	.98	1.01
633	Bowling Green	.93	.96
634	Hannibal	1.02	.95
635	Kirksville	.86	.90
636	Flat River	.96	.99
637	Cape Girardeau	.94	.97
638	Sikeston	.90	.92
639	Poplar Bluff	.90	.92
640-641	Kansas City	.98	.95
644-645	St. Joseph	.88	.92
646	Chillicothe	.81	.84
647	Harrisonville	.94	.92
648	Joplin	.85	.87
650-651	Jefferson City	.97	.91
652	Columbia	.95	.89
653	Sedalia	.95	.88
654-655	Rolla	.89	.83
656-658	Springfield	.85	.87
MONTANA			
590-591	Billings	.99	.96
592	Wolf Point	.97	.94
593	Miles City	.98	.96
594	Great Falls	.97	.96
595	Havre	.95	.94
596	Helena	.97	.95
597	Butte	.95	.94
598	Missoula	.95	.94
599	Kalispell	.95	.93
NEBRASKA			
680-681	Omaha	.90	.89
683-685	Lincoln	.89	.83
686	Columbus	.76	.75
687	Norfolk	.86	.85
688	Grand Island	.88	.84
689	Hastings	.85	.81
690	Mccook	.76	.72
691	North Platte	.85	.81
692	Valentine	.80	.76
693	Alliance	.77	.73
NEVADA			
889-891	Las Vegas	1.03	1.02
893	Ely	.95	.96
894-895	Reno	.93	.98
897	Carson City	.95	.98
898	Elko	.93	.95
NEW HAMPSHIRE			
030	Nashua	.96	.97
031	Manchester	.96	.97
032-033	Concord	.95	.96

STATE/ZIP	CITY	Residential	Commercial
NEW HAMPSHIRE (CONT'D)			
034	Keene	.83	.84
035	Littleton	.85	.86
036	Charleston	.81	.82
037	Claremont	.80	.81
038	Portsmouth	.95	.94
NEW JERSEY			
070-071	Newark	1.14	1.12
072	Elizabeth	1.11	1.09
073	Jersey City	1.13	1.12
074-075	Paterson	1.13	1.13
076	Hackensack	1.11	1.11
077	Long Branch	1.11	1.09
078	Dover	1.13	1.11
079	Summit	1.11	1.09
080,083	Vineland	1.12	1.08
081	Camden	1.12	1.09
082,084	Atlantic City	1.12	1.09
085-086	Trenton	1.13	1.11
087	Point Pleasant	1.12	1.10
088-089	New Brunswick	1.13	1.11
NEW MEXICO			
870-872	Albuquerque	.88	.90
873	Gallup	.89	.91
874	Farmington	.89	.91
875	Santa Fe	.88	.90
877	Las Vegas	.88	.90
878	Socorro	.88	.90
879	Truth/Consequences	.88	.88
880	Las Cruces	.85	.85
881	Clovis	.90	.90
882	Roswell	.91	.91
883	Carrizozo	.92	.92
884	Tucumcari	.91	.91
NEW YORK			
100-102	New York	1.35	1.35
103	Staten Island	1.31	1.31
104	Bronx	1.29	1.29
105	Mount Vernon	1.23	1.23
106	White Plains	1.22	1.22
107	Yonkers	1.25	1.25
108	New Rochelle	1.24	1.24
109	Suffern	1.16	1.16
110	Queens	1.30	1.30
111	Long Island City	1.31	1.31
112	Brooklyn	1.30	1.30
113	Flushing	1.31	1.31
114	Jamaica	1.30	1.30
115,117,118	Hicksville	1.28	1.28
116	Far Rockaway	1.31	1.31
119	Riverhead	1.29	1.29
120-122	Albany	.99	.99
123	Schenectady	1.00	1.00
124	Kingston	1.14	1.12
125-126	Poughkeepsie	1.17	1.15
127	Monticello	1.13	1.11
128	Glens Falls	.97	.95
130-132	Syracuse	1.02	.99
133-135	Utica	.93	.96
136	Watertown	.95	.98
137-139	Binghamton	.97	.97
140-142	Buffalo	1.08	1.04
143	Niagara Falls	1.09	1.05
144-146	Rochester	1.03	1.04
147	Jamestown	.97	.94
148-149	Elmira	.98	.96
NORTH CAROLINA			
270,272-274	Greensboro	.78	.79
271	Winston-Salem	.78	.79
275-276	Raleigh	.79	.79
277	Durham	.78	.79
278	Rocky Mount	.67	.67
279	Elizabeth City	.67	.67
280	Gastonia	.77	.78
281-282	Charlotte	.77	.78
283	Fayetteville	.78	.78
284	Wilmington	.76	.78
285	Kinston	.67	.67
286	Hickory	.64	.65
287-288	Asheville	.76	.78

Location Factors

STATE/ZIP	CITY	Residential	Commercial
NORTH CAROLINA (CONT'D)			
289	Murphy	.66	.67
NORTH DAKOTA			
580-581	Fargo	.79	.84
582	Grand Forks	.79	.84
583	Devils Lake	.79	.84
584	Jamestown	.79	.84
585	Bismarck	.81	.85
586	Dickinson	.80	.84
587	Minot	.81	.85
588	Williston	.80	.84
OHIO			
430-432	Columbus	.96	.94
433	Marion	.90	.91
434-436	Toledo	.98	.97
437-438	Zanesville	.91	.90
439	Steubenville	.96	.96
440	Lorain	1.02	.96
441	Cleveland	1.08	1.01
442-443	Akron	1.00	.99
444-445	Youngstown	.99	.96
446-447	Canton	.96	.95
448-449	Mansfield	.95	.93
450	Hamilton	.98	.92
451-452	Cincinnati	.98	.92
453-454	Dayton	.92	.91
455	Springfield	.93	.91
456	Chillicothe	1.00	.94
457	Athens	.90	.89
458	Lima	.93	.92
OKLAHOMA			
730-731	Oklahoma City	.82	.84
734	Ardmore	.83	.82
735	Lawton	.84	.83
736	Clinton	.80	.82
737	Enid	.83	.82
738	Woodward	.82	.81
739	Guymon	.71	.70
740-741	Tulsa	.86	.84
743	Miami	.86	.83
744	Muskogee	.78	.75
745	Mcalester	.76	.78
746	Ponca City	.82	.81
747	Durant	.80	.81
748	Shawnee	.79	.81
OREGON			
970-972	Portland	1.09	1.07
973	Salem	1.07	1.06
974	Eugene	1.06	1.05
975	Medford	1.06	1.05
976	Klamath Falls	1.07	1.06
977	Bend	1.07	1.06
978	Pendleton	1.05	1.02
979	Vale	1.00	.98
PENNSYLVANIA			
150-152	Pittsburgh	1.05	1.03
153	Washington	1.03	1.01
154	Uniontown	1.01	.99
155	Bedford	1.04	.97
156	Greensburg	1.02	1.00
157	Indiana	1.05	.98
158	Dubois	1.04	.98
159	Johnstown	1.05	.98
160	Butler	1.01	.98
161	New Castle	1.01	.98
162	Kittanning	1.03	1.00
163	Oil City	.90	.95
164-165	Erie	.99	.98
166	Altoona	1.06	.97
167	Bradford	1.00	.98
168	State College	.98	.98
169	Wellsboro	.95	.96
170-171	Harrisburg	.98	.97
172	Chambersburg	.97	.96
173-174	York	.99	.97
175-176	Lancaster	.98	.96
177	Williamsport	.94	.93
178	Sunbury	.96	.95
179	Pottsville	.97	.96

STATE/ZIP	CITY	Residential	Commercial
PENNSYLVANIA (CONT'D)			
180	Lehigh Valley	1.04	1.03
181	Allentown	1.04	1.03
182	Hazleton	.98	.97
183	Stroudsburg	.98	.97
184-185	Scranton	.98	1.01
186-187	Wilkes-Barre	.94	.97
188	Montrose	.95	.98
189	Doylestown	.95	1.06
190-191	Philadelphia	1.13	1.11
193	Westchester	1.06	1.04
194	Norristown	1.08	1.06
195-196	Reading	.98	.99
RHODE ISLAND			
028	Newport	1.05	1.07
029	Providence	1.05	1.07
SOUTH CAROLINA			
290-292	Columbia	.74	.77
293	Spartanburg	.74	.77
294	Charleston	.76	.78
295	Florence	.73	.75
296	Greenville	.74	.77
297	Rock Hill	.66	.69
298	Aiken	.67	.70
299	Beaufort	.70	.72
SOUTH DAKOTA			
570-571	Sioux Falls	.89	.83
572	Watertown	.87	.81
573	Mitchell	.86	.80
574	Aberdeen	.88	.82
575	Pierre	.87	.81
576	Mobridge	.88	.81
577	Rapid City	.86	.80
TENNESSEE			
370-372	Nashville	.85	.85
373-374	Chattanooga	.85	.84
375, 380-381	Memphis	.86	.86
376	Johnson City	.83	.82
377-379	Knoxville	.82	.82
382	Mckenzie	.71	.71
383	Jackson	.70	.77
384	Columbia	.78	.78
385	Cookeville	.70	.70
TEXAS			
750	Mckinney	.91	.84
751	Waxahackie	.85	.85
752-753	Dallas	.90	.86
754	Greenville	.81	.76
755	Texarkana	.91	.80
756	Longview	.86	.76
757	Tyler	.92	.81
758	Palestine	.76	.76
759	Lufkin	.80	.80
760-761	Fort Worth	.85	.84
762	Denton	.90	.81
763	Wichita Falls	.82	.82
764	Eastland	.77	.76
765	Temple	.80	.79
766-767	Waco	.83	.82
768	Brownwood	.75	.74
769	San Angelo	.82	.78
770-772	Houston	.89	.90
773	Huntsville	.83	.83
774	Wharton	.79	.80
775	Galveston	.87	.88
776-777	Beaumont	.85	.87
778	Bryan	.83	.84
779	Victoria	.83	.83
780	Laredo	.80	.81
781-782	San Antonio	.84	.85
783-784	Corpus Christi	.82	.81
785	Mc Allen	.81	.79
786-787	Austin	.81	.83
788	Del Rio	.70	.70
789	Giddings	.76	.75
790-791	Amarillo	.82	.82
792	Childress	.78	.81
793-794	Lubbock	.81	.83
795-796	Abilene	.80	.80

Location Factors

STATE/ZIP	CITY	Residential	Commercial
TEXAS (CONT'D)			
797	Midland	.81	.82
798-799,885	El Paso	.81	.80
UTAH			
840-841	Salt Lake City	.88	.87
842,844	Ogden	.88	.86
843	Logan	.90	.87
845	Price	.83	.82
846-847	Provo	.89	.88
VERMONT			
050	White River Jct.	.76	.75
051	Bellows Falls	.77	.76
052	Bennington	.73	.72
053	Brattleboro	.77	.76
054	Burlington	.86	.87
056	Montpelier	.85	.86
057	Rutland	.88	.87
058	St. Johnsbury	.77	.78
059	Guildhall	.76	.77
129	Plattsburgh	.97	.95
VIRGINIA			
220-221	Fairfax	.89	.90
222	Arlington	.90	.91
223	Alexandria	.90	.91
224-225	Fredericksburg	.86	.87
226	Winchester	.81	.82
227	Culpeper	.79	.80
228	Harrisonburg	.76	.77
229	Charlottesville	.84	.82
230-232	Richmond	.86	.84
233-235	Norfolk	.83	.83
236	Newport News	.83	.82
237	Portsmouth	.81	.81
238	Petersburg	.85	.83
239	Farmville	.77	.75
240-241	Roanoke	.79	.78
242	Bristol	.81	.76
243	Pulaski	.72	.71
244	Staunton	.74	.73
245	Lynchburg	.82	.78
246	Grundy	.71	.71
WASHINGTON			
980-981,987	Seattle	1.00	1.05
982	Everett	.98	1.04
983-984	Tacoma	1.07	1.05
985	Olympia	1.07	1.04
986	Vancouver	1.10	1.04
988	Wenatchee	.97	1.01
989	Yakima	1.04	1.02
990-992	Spokane	1.01	1.00
993	Richland	1.02	1.01
994	Clarkston	1.01	1.00
WEST VIRGINIA			
247-248	Bluefield	.87	.87
249	Lewisburg	.91	.91
250-253	Charleston	.95	.95
254	Martinsburg	.78	.79
255-257	Huntington	.94	.96
258-259	Beckley	.92	.92
260	Wheeling	.93	.95
261	Parkersburg	.92	.94
262	Buckhannon	.96	.94
263-264	Clarksburg	.96	.94
265	Morgantown	.97	.94
266	Gassaway	.94	.94
267	Romney	.92	.92
268	Petersburg	.95	.92
WISCONSIN			
530,532	Milwaukee	.98	.97
531	Kenosha	.99	.98
534	Racine	1.03	.98
535	Beloit	.99	.97
537	Madison	.97	.95
538	Lancaster	.92	.90
539	Portage	.96	.94
541-543	Green Bay	.98	.95
544	Wausau	.95	.91
545	Rhinelander	.97	.93
546	La Crosse	.95	.92

STATE/ZIP	CITY	Residential	Commercial
WISCONSIN (CONT'D)			
547	Eau Claire	1.01	.94
548	Superior	1.01	.95
549	Oshkosh	.95	.92
WYOMING			
820	Cheyenne	.88	.83
821	Yellowstone Nat. Pk.	.84	.81
822	Wheatland	.85	.80
823	Rawlins	.85	.80
824	Worland	.81	.79
825	Riverton	.85	.81
826	Casper	.88	.84
827	Newcastle	.82	.78
828	Sheridan	.86	.84
829-831	Rock Springs	.85	.81
CANADIAN FACTORS (reflect Canadian currency)			
ALBERTA			
	Calgary	1.03	1.00
	Edmonton	1.03	1.00
BRITISH COLUMBIA			
	Vancouver	1.08	1.09
	Victoria	1.05	1.06
MANITOBA			
	Winnipeg	1.01	1.00
NEW BRUNSWICK			
	Moncton	.96	.94
	Saint John	.99	.97
NEWFOUNDLAND			
	St. John's	.98	.97
NOVA SCOTIA			
	Halifax	.99	.98
ONTARIO			
	Barrie	1.13	1.11
	Brantford	1.15	1.13
	Cornwall	1.13	1.11
	Hamilton	1.16	1.12
	Kingston	1.12	1.10
	Kitchener	1.08	1.06
	London	1.12	1.10
	North Bay	1.11	1.09
	Oshawa	1.13	1.11
	Ottawa	1.13	1.11
	Owen Sound	1.12	1.10
	Peterborough	1.12	1.10
	St. Catharines	1.08	1.06
	Sarnia	1.13	1.11
	Sudbury	1.07	1.05
	Thunder Bay	1.09	1.07
	Toronto	1.14	1.13
	Windsor	1.09	1.07
PRINCE EDWARD ISLAND			
	Charlottetown	.95	.93
QUEBEC			
	Chicoutimi	1.04	1.03
	Montreal	1.10	1.03
	Quebec	1.12	1.04
SASKATCHEWAN			
	Regina	.92	.92
	Saskatoon	.92	.92

Historical Cost Indexes

The following tables are the revised Historical Cost Indexes based on a 30-city national average with a base of 100 on January 1, 1993.

The indexes may be used to:

1. Estimate and compare construction costs for different years in the same city.
2. Estimate and compare construction costs in different cities for the same year.
3. Estimate and compare construction costs in different cities for different years.
4. Compare construction trends in any city with the national average.

EXAMPLES

1. Estimate and compare construction costs for different years in the same city.

    A. To estimate the construction cost of a building in Lexington, KY in 1970, knowing that it cost $900,000 in 1997.

    Index Lexington, KY in 1970 = 26.9
    Index Lexington, KY in 1997 = 96.4

    $$\frac{\text{Index 1970}}{\text{Index 1997}} \times \text{Cost 1996} = \text{Cost 1970}$$

    $$\frac{26.9}{96.4} \times \$900,000 = \$251,100$$

    Construction Cost in Lexington in 1970 = $251,100

    B. To estimate the current construction cost of a building in Boston, MA that was built in 1960 for $300,000.

    Index Boston, MA in 1960 = 20.5
    Index Boston, MA in 1997 = 131.7

    $$\frac{\text{Index 1997}}{\text{Index 1960}} \times \text{Cost 1960} = \text{Cost 1997}$$

    $$\frac{131.7}{20.5} \times \$300,000 = \$1,950,000$$

    Construction Cost in Boston in 1997 = $1,950,000

2. Estimate and compare construction costs in different cities for the same year.

    To compare the construction cost of a building in Topeka, KS in 1990 with the known cost of $600,000 in Baltimore, MD in 1990.

    Index Topeka, KS in 1990 = 83.2
    Index Baltimore, MD in 1990 = 85.6

    $$\frac{\text{Index Topeka}}{\text{Index Baltimore}} \times \text{Cost Baltimore} = \text{Cost Topeka}$$

    $$\frac{83.2}{85.6} \times \$600,000 = \$583,200$$

    Construction Cost in Topeka in 1990 = $583,200

3. Estimate and compare construction costs in different cities for different years.

    To compare the construction cost of a building in Detroit, MI in 1978 with the known construction cost of $4,000,000 for the same building in San Francisco, CA in 1973.

    Index Detroit, MI in 1978 = 52.6
    Index San Francisco, CA in 1973 = 40.7

    $$\frac{\text{Index Detroit 1978}}{\text{Index San Francisco 1973}} \times \text{Cost San Francisco 1973} = \text{Cost Detroit 1978}$$

    $$\frac{52.6}{40.7} \times \$4,000,000 = \$5,169,500$$

    Construction Cost in Detroit in 1978 = $5,169,500

4. Compare construction trends in any city with the national average.

    To compare the construction cost in Reno, NV from 1965 to 1979 with the increase in the National Average during the same time period.

    Index Reno, NV for 1965 = 21.6   For 1979 = 56.7
    Index 30 City Average for 1965 = 21.5   For 1979 = 54.9

    A. National Average Increase From 1965 to 1979
    $$= \frac{\text{Index 30 City 1979}}{\text{Index 30 City 1965}} = \frac{54.9}{21.5}$$

    National Average Increase From 1965 to 1979 = 2.55 or 255%

    B. Increase for Reno, NV From 1965 to 1979
    $$= \frac{\text{Index Reno, NV 1979}}{\text{Index Reno, NV 1965}} = \frac{56.7}{21.6}$$

    Reno Increase 1965–1979 = 2.63 or 263%

    Conclusion: Construction costs in Reno are higher than National Average and increased at a greater rate from 1965 to 1979 than the National Average.

Historical Cost Indexes

Year	National 30 City Average	Alabama Birmingham	Alabama Huntsville	Alabama Mobile	Alabama Montgomery	Alabama Tuscaloosa	Alaska Anchorage	Arizona Phoenix	Arizona Tuscon	Arkansas Fort Smith	Arkansas Little Rock	California Anaheim	California Bakersfield	California Fresno	California Los Angeles	California Oxnard
Jan 1997	111.7E	93.0E	93.4E	93.8E	89.1E	88.1E	141.0E	101.3E	100.2E	88.5E	89.0E	124.2E	119.3E	122.7E	125.1E	123.8E
1996	108.9	90.9	91.4	92.3	87.5	86.5	140.4	98.3	97.4	86.5	86.7	121.7	117.2	120.2	122.4	121.5
1995	105.6	87.8	88.0	88.8	84.5	83.5	138.0	96.1	95.5	85.0	85.6	120.1	115.7	117.5	120.9	120.0
1994	103.0	85.9	86.1	86.8	82.6	81.7	134.0	93.7	93.1	82.4	83.9	118.1	113.6	114.5	119.0	117.5
1993	100.0	82.8	82.7	86.1	82.1	78.7	132.0	90.9	90.9	80.9	82.4	115.0	111.3	112.9	115.6	115.4
1992	97.9	81.7	81.5	84.9	80.9	77.6	128.6	88.8	89.4	79.7	81.2	113.5	108.1	110.3	113.7	113.7
1991	95.7	80.5	78.9	83.9	79.8	76.6	127.4	88.1	88.5	78.3	80.0	111.0	105.8	108.2	110.9	111.4
1990	93.2	79.4	77.6	82.7	78.4	75.2	125.8	86.4	87.0	77.1	77.9	107.7	102.7	103.3	107.5	107.4
1989	91.0	77.5	76.1	81.1	76.8	73.7	123.4	85.3	85.6	75.6	76.4	105.6	101.0	101.7	105.3	105.4
1988	88.5	75.7	74.7	79.7	75.4	72.3	121.4	84.4	84.3	74.2	75.0	102.7	97.6	99.2	102.6	102.3
1987	85.7	74.0	73.8	77.9	74.5	71.7	119.3	79.7	80.6	72.2	72.8	100.9	96.1	98.0	100.0	101.1
1986	83.7	72.8	71.7	76.7	72.3	69.4	117.3	78.9	78.8	70.7	72.3	99.2	94.2	93.8	97.8	100.4
1985	81.8	71.1	70.5	72.7	70.9	67.9	116.0	78.1	77.7	69.6	71.0	95.4	92.2	92.6	94.6	96.6
1984	80.6	69.7	69.0	75.2	69.4	66.3	113.8	79.4	79.2	69.1	70.3	93.4	89.8	90.6	92.1	94.0
1983	78.2	67.4	68.8	72.9	69.4	65.5	104.5	77.5	77.9	66.8	68.7	90.9	88.8	88.3	89.7	91.8
1982	72.1	63.5	63.1	67.3	64.4	61.8	96.1	72.3	71.7	62.2	64.6	83.1	83.0	82.5	80.9	83.4
1981	66.1	60.0	58.1	62.2	61.3	58.1	91.5	69.0	67.5	57.7	60.2	75.7	76.5	75.2	74.8	77.1
1980	60.7	55.2	54.1	56.8	56.8	54.0	91.4	63.7	62.4	53.1	55.8	68.7	69.4	68.7	67.4	69.9
1979	54.9	49.9	48.9	52.5	50.8	48.7	82.5	55.9	56.5	47.8	49.5	62.9	61.9	62.5	61.7	62.9
1978	51.3	46.7	45.5	48.8	46.9	45.0	75.3	51.9	52.3	44.8	45.4	57.4	57.1	57.7	57.0	58.6
1977	47.9	43.1	43.2	45.1	42.4	40.5	70.6	48.7	48.5	41.6	42.2	53.2	52.8	53.2	53.7	53.4
1976	45.3	40.8	40.7	42.3	40.3	38.6	64.0	46.5	46.4	39.2	39.4	49.6	49.3	50.4	50.2	48.2
1975	43.7	40.0	40.9	41.8	40.1	37.8	57.3	44.5	45.2	39.0	38.7	47.6	46.6	47.7	48.3	47.0
1974	39.4	34.8	35.6	36.6	36.0	34.4	55.9	40.1	40.5	34.6	32.4	44.1	43.7	44.2	42.2	44.1
1973	36.3	31.2	32.7	33.2	33.1	31.5	52.4	36.5	37.1	31.7	28.9	40.4	40.1	40.5	38.2	39.4
1970	27.8	24.1	25.1	25.8	25.4	24.2	43.0	27.2	28.5	24.3	22.3	31.0	30.8	31.1	29.0	31.0
1965	21.5	19.6	19.3	19.6	19.5	18.7	34.9	21.8	22.0	18.7	18.5	23.9	23.7	24.0	22.7	23.9
1960	19.5	17.9	17.5	17.8	17.7	16.9	31.7	19.9	20.0	17.0	16.8	21.7	21.5	21.8	20.6	21.7
1955	16.3	14.8	14.7	14.9	14.9	14.2	26.6	16.7	16.7	14.3	14.5	18.2	18.1	18.3	17.3	18.2
1950	13.5	12.2	12.1	12.3	12.3	11.7	21.9	13.8	13.8	11.8	11.6	15.1	15.0	15.1	14.3	15.1
1945	8.6	7.8	7.7	7.8	7.8	7.5	14.0	8.8	8.8	7.5	7.4	9.6	9.5	9.6	9.1	9.6
1940	6.6	6.0	6.0	6.1	6.0	5.8	10.8	6.8	6.8	5.8	5.7	7.4	7.4	7.4	7.0	7.4

Year	National 30 City Average	California Riverside	California Sacramento	California San Diego	California San Francisco	California Santa Barbara	California Stockton	California Vallejo	Colorado Colorado Springs	Colorado Denver	Colorado Pueblo	Connecticut Bridge-Port	Connecticut Bristol	Connecticut Hartford	Connecticut New Britain	Connecticut New Haven
Jan 1997	111.7E	123.0E	124.7E	120.7E	139.5E	122.8E	123.7E	131.0E	100.5E	103.4E	101.6E	119.5E	119.7E	120.0E	119.9E	120.2E
1996	108.9	120.6	122.4	118.4	136.8	120.5	121.4	128.1	98.5	101.4	99.7	117.4	117.6	117.9	117.8	118.1
1995	105.6	119.2	119.5	115.4	133.8	119.0	119.0	122.5	96.1	98.9	96.8	116.0	116.5	116.9	116.3	116.5
1994	103.0	116.6	114.8	113.4	131.4	116.4	115.5	119.7	94.2	95.9	94.6	114.3	115.0	115.3	114.7	114.8
1993	100.0	114.3	112.5	111.3	129.6	114.3	115.2	117.5	92.1	93.8	92.6	108.7	107.8	108.4	106.3	106.1
1992	97.9	111.8	110.8	109.5	127.9	112.2	112.7	115.2	90.6	91.5	90.9	106.7	106.0	106.6	104.5	104.4
1991	95.7	109.4	108.5	107.7	125.8	110.0	111.0	113.4	88.9	90.4	89.4	97.6	97.3	98.0	97.3	97.9
1990	93.2	107.0	104.9	105.6	121.8	106.4	105.4	111.4	86.9	88.8	88.8	96.3	95.9	96.6	95.9	96.5
1989	91.0	105.2	103.1	103.6	119.2	104.6	103.7	109.8	85.6	88.7	87.4	94.5	94.2	94.9	94.2	94.6
1988	88.5	102.6	100.5	101.0	115.2	101.4	101.2	107.2	83.8	87.0	85.6	92.8	92.6	93.3	92.6	93.0
1987	85.7	100.3	98.7	99.0	111.0	99.3	99.6	103.2	81.6	84.8	83.8	92.4	93.0	93.3	93.0	92.6
1986	83.7	98.6	94.4	96.9	108.0	97.0	96.1	99.1	81.9	83.1	82.7	88.5	88.7	89.4	88.6	88.8
1985	81.8	94.8	92.2	94.2	106.2	93.2	93.7	97.0	79.6	81.0	80.4	86.1	86.2	87.2	86.1	86.3
1984	80.6	91.8	93.8	92.0	102.8	90.6	91.9	95.6	79.8	84.1	80.5	83.5	83.4	84.4	83.0	83.2
1983	78.2	89.4	92.0	90.3	101.0	88.8	91.0	93.0	77.3	80.5	77.4	79.6	79.9	81.1	79.5	79.3
1982	72.1	82.1	84.9	83.8	91.9	81.8	84.9	85.7	71.2	73.4	75.0	71.9	71.2	73.2	71.4	72.0
1981	66.1	75.5	77.7	74.5	82.8	75.9	77.4	78.5	65.5	66.3	65.4	67.0	66.3	67.3	66.3	67.0
1980	60.7	68.7	71.3	68.1	75.2	71.1	71.2	71.9	60.7	60.9	59.5	61.5	60.7	61.9	60.7	61.4
1979	54.9	62.2	64.0	61.6	67.5	65.0	64.2	65.2	54.9	56.0	54.0	56.0	55.3	56.4	55.4	55.8
1978	51.3	57.7	60.0	58.3	62.6	59.2	59.5	59.9	51.1	52.0	50.6	52.5	50.9	51.9	51.8	52.0
1977	47.9	53.3	55.7	53.4	58.1	53.0	54.8	54.2	47.2	48.3	47.0	49.3	48.6	48.6	48.3	49.0
1976	45.3	49.7	51.5	50.1	53.9	49.0	51.1	50.6	44.5	45.6	44.4	47.1	46.6	46.8	46.6	47.8
1975	43.7	47.3	49.1	47.7	49.8	46.1	47.6	46.5	43.1	42.7	42.5	45.2	45.5	46.0	45.2	46.0
1974	39.4	44.0	45.7	43.1	44.7	44.5	45.2	45.1	39.2	37.3	38.8	41.7	40.1	41.8	40.1	41.8
1973	36.3	40.4	41.9	39.3	40.7	40.8	41.5	41.4	35.9	33.4	35.6	38.3	36.7	38.5	36.8	38.4
1970	27.8	31.0	32.1	30.7	31.6	31.3	31.8	31.8	27.6	26.1	27.3	29.2	28.2	29.6	28.2	29.3
1965	21.5	23.9	24.8	24.0	23.7	24.1	24.5	24.5	21.3	20.9	21.0	22.4	21.7	22.6	21.7	23.2
1960	19.5	21.7	22.5	21.7	21.5	21.9	22.3	22.2	19.3	19.0	19.1	20.0	19.8	20.0	19.8	20.0
1955	16.3	18.2	18.9	18.2	18.0	18.4	18.7	18.6	16.2	15.9	16.0	16.8	16.6	16.8	16.6	16.8
1950	13.5	15.0	15.6	15.0	14.9	15.2	15.4	15.4	13.4	13.2	13.2	13.9	13.7	13.8	13.7	13.9
1945	8.6	9.6	10.0	9.6	9.5	9.7	9.9	9.8	8.5	8.4	8.4	8.8	8.7	8.8	8.7	8.9
1940	6.6	7.4	7.7	7.4	7.3	7.5	7.6	7.6	6.6	6.5	6.5	6.8	6.8	6.8	6.8	6.8

Historical Cost Indexes

Year	National 30 City Average	Connecticut Norwalk	Connecticut Stamford	Connecticut Waterbury	Delaware Wilmington	D.C. Washington	Florida Fort Lauderdale	Florida Jacksonville	Florida Miami	Florida Orlando	Florida Tallahassee	Florida Tampa	Georgia Albany	Georgia Atlanta	Georgia Columbus	Georgia Macon
Jan 1997	111.7E	119.1E	121.1E	120.5E	110.5E	106.3E	97.8E	94.4E	97.8E	97.1E	89.7E	95.1E	89.6E	98.1E	86.0E	91.5E
1996	108.9	117.0	119.0	118.4	108.6	105.4	95.9	92.6	95.9	95.1	88.0	93.6	86.6	94.3	83.9	88.3
1995	105.6	115.5	117.9	117.3	106.1	102.3	94.0	90.9	93.7	93.5	86.5	92.2	85.0	92.0	82.6	86.8
1994	103.0	113.9	116.4	115.8	105.0	99.6	92.2	88.9	91.8	91.5	84.5	90.2	82.3	89.6	80.6	83.7
1993	100.0	108.8	110.6	104.8	101.5	96.3	87.4	86.1	87.1	88.5	82.1	87.7	79.5	85.7	77.8	80.9
1992	97.9	107.2	109.0	103.1	100.3	94.7	85.7	84.0	85.3	87.1	80.8	86.2	78.2	84.3	76.5	79.6
1991	95.7	100.6	103.2	96.5	94.5	92.9	85.1	82.8	85.2	85.5	79.7	86.3	76.3	82.6	75.4	78.4
1990	93.2	96.3	98.9	95.1	92.5	90.4	83.9	81.1	84.0	82.9	78.4	85.0	75.0	80.4	74.0	76.9
1989	91.0	94.4	97.0	93.4	89.5	87.5	82.4	79.7	82.5	81.6	76.9	83.5	73.4	78.6	72.4	75.3
1988	88.5	92.3	94.9	91.8	87.7	85.0	80.9	78.0	81.0	80.1	75.4	81.9	71.8	76.9	70.8	73.6
1987	85.7	92.0	92.7	92.5	85.1	82.1	78.8	76.0	77.9	76.6	74.4	79.4	72.2	73.6	70.3	72.3
1986	83.7	88.2	89.7	87.8	83.8	80.8	78.9	75.0	79.9	76.2	72.3	79.1	68.5	72.0	67.5	69.8
1985	81.8	85.3	86.8	85.6	81.1	78.8	76.7	73.4	78.3	73.9	70.6	77.3	66.9	70.3	66.1	68.1
1984	80.6	82.6	83.7	82.5	79.7	79.1	73.8	72.6	75.6	73.0	69.6	76.5	65.9	68.6	65.2	67.5
1983	78.2	78.5	79.8	78.7	76.3	76.0	71.5	70.4	72.9	71.1	67.6	73.6	65.6	68.6	64.7	65.3
1982	72.1	71.7	72.0	71.8	69.1	69.4	65.4	65.2	65.7	66.0	62.8	67.7	60.1	61.9	59.3	60.0
1981	66.1	66.2	66.2	67.3	63.4	64.9	60.3	61.0	60.7	62.0	58.6	62.1	56.2	58.3	55.7	56.1
1980	60.7	60.7	60.9	62.3	58.5	59.6	55.3	55.8	56.5	56.7	53.5	57.2	51.7	54.0	51.1	51.3
1979	54.9	55.2	55.6	57.0	53.6	53.9	50.0	51.7	50.8	51.8	48.9	51.8	46.5	48.3	46.1	46.4
1978	51.3	51.5	51.6	52.9	50.0	51.5	47.0	47.7	47.6	48.0	45.4	48.4	42.7	45.0	42.5	42.4
1977	47.9	48.1	48.3	49.0	47.1	48.4	43.7	43.3	45.7	44.9	42.0	45.9	41.6	42.1	37.7	40.8
1976	45.3	46.3	45.9	47.4	43.3	45.7	42.2	42.1	43.9	42.8	40.1	43.0	39.9	39.6	35.8	37.9
1975	43.7	44.7	45.0	46.3	42.9	43.7	42.1	40.3	43.2	41.5	38.1	41.3	37.5	38.4	36.2	36.5
1974	39.4	39.9	40.0	41.0	38.4	38.5	36.5	34.2	39.6	37.2	34.9	35.8	33.8	34.6	32.2	33.2
1973	36.3	36.6	36.7	37.6	35.3	34.9	33.5	30.8	36.7	34.1	32.0	32.4	31.0	31.9	29.7	30.5
1970	27.8	28.1	28.2	28.9	27.0	26.3	25.7	22.8	27.0	26.2	24.5	24.2	23.8	25.2	22.8	23.4
1965	21.5	21.6	21.7	22.2	20.9	21.8	19.8	17.4	19.3	20.2	18.9	18.6	18.3	19.8	17.6	18.0
1960	19.5	19.7	19.7	20.2	18.9	19.4	18.0	15.8	17.6	18.3	17.2	16.9	16.7	17.1	16.0	16.4
1955	16.3	16.5	16.5	17.0	15.9	16.3	15.1	13.2	14.7	15.4	14.4	14.1	14.0	14.4	13.4	13.7
1950	13.5	13.6	13.7	14.0	13.1	13.4	12.5	11.0	12.2	12.7	11.9	11.7	11.5	11.9	11.0	11.3
1945	8.6	8.7	8.7	8.9	8.4	8.6	7.9	7.0	7.8	8.1	7.6	7.5	7.4	7.6	7.0	7.2
1940	6.6	6.7	6.7	6.9	6.4	6.6	6.1	5.4	6.0	6.2	5.8	5.7	5.7	5.8	5.5	5.6

Year	National 30 City Average	Georgia Savannah	Hawaii Honolulu	Idaho Boise	Idaho Pocatello	Illinois Chicago	Illinois Decatur	Illinois Joliet	Illinois Peoria	Illinois Rockford	Illinois Springfield	Indiana Anderson	Indiana Evansville	Indiana Fort Wayne	Indiana Gary	Indiana Indianapolis
Jan 1997	111.7E	91.9E	136.6E	103.7E	103.5E	120.1E	107.3E	116.8E	110.1E	112.0E	107.2E	102.1E	104.3E	101.8E	110.3E	104.5E
1996	108.9	88.6	134.5	102.2	102.1	118.8	106.6	116.2	109.3	111.5	106.5	100.0	102.1	99.9	107.5	102.7
1995	105.6	87.4	130.3	99.5	98.2	114.2	98.5	110.5	102.3	103.6	98.1	96.4	97.2	95.0	100.7	100.1
1994	103.0	85.3	124.0	94.8	95.0	111.3	97.3	108.9	100.9	102.2	97.0	93.6	95.8	93.6	99.1	97.1
1993	100.0	82.0	122.0	92.2	92.1	107.6	95.6	106.8	98.9	99.6	95.2	91.2	94.3	91.5	96.7	93.9
1992	97.9	80.8	120.0	91.0	91.0	104.3	94.4	104.2	97.3	98.2	94.0	89.5	92.9	89.9	95.0	91.5
1991	95.7	79.5	106.1	89.5	89.4	100.9	92.3	100.0	95.9	95.8	91.5	87.8	91.4	88.3	93.3	89.1
1990	93.2	77.9	104.7	88.2	88.1	98.4	90.9	98.4	93.7	94.0	90.1	84.6	89.3	83.4	88.4	87.1
1989	91.0	76.0	102.8	86.6	86.5	93.7	89.4	92.8	91.6	92.1	88.6	82.3	87.8	81.7	86.5	85.1
1988	88.5	74.5	101.1	83.9	83.7	90.6	87.5	89.8	88.5	87.9	86.8	80.8	85.5	80.0	84.6	83.1
1987	85.7	72.0	99.1	81.4	81.5	86.6	84.4	86.7	86.3	85.2	85.0	78.8	82.4	78.1	81.7	80.4
1986	83.7	70.8	97.5	80.6	80.3	84.4	83.5	85.3	85.1	84.5	83.4	77.0	80.9	76.5	79.6	78.6
1985	81.8	68.9	94.7	78.0	78.0	82.4	81.9	83.4	83.7	83.0	81.5	75.2	79.8	75.0	77.8	77.1
1984	80.6	67.9	90.7	76.6	76.8	80.2	80.4	82.2	83.6	80.6	80.1	73.5	77.4	73.5	77.3	75.9
1983	78.2	66.3	87.2	76.0	75.8	79.0	78.8	80.6	81.5	78.9	78.8	70.9	75.0	71.2	75.2	74.0
1982	72.1	61.1	79.3	71.0	70.1	75.0	72.7	74.4	75.3	72.6	72.7	66.4	69.8	67.1	70.5	68.2
1981	66.1	57.1	73.4	65.4	64.8	68.6	67.2	68.8	70.3	67.3	67.0	61.5	64.3	61.5	65.3	62.4
1980	60.7	52.2	68.9	60.3	59.5	62.8	62.3	63.4	64.5	61.6	61.1	56.5	59.0	56.7	59.8	57.9
1979	54.9	47.2	63.0	54.4	53.7	56.5	56.0	57.0	58.4	54.8	55.6	50.8	53.3	51.2	54.0	51.9
1978	51.3	43.6	58.9	50.0	49.9	52.9	51.9	52.7	53.8	50.4	51.5	46.7	49.1	46.9	49.4	48.7
1977	47.9	40.2	52.0	45.1	45.2	49.0	47.7	49.1	49.4	46.8	47.5	43.2	45.7	43.0	46.7	45.4
1976	45.3	38.3	45.7	41.9	41.9	47.3	45.0	46.0	46.5	44.1	44.1	40.6	42.8	41.4	43.9	43.1
1975	43.7	36.9	44.6	40.8	40.5	45.7	43.1	44.5	44.7	42.7	42.4	39.5	41.7	39.9	41.9	40.6
1974	39.4	31.0	40.6	37.9	37.7	41.7	39.7	40.6	41.2	39.0	39.2	36.0	36.5	36.3	38.4	36.6
1973	36.3	27.5	37.7	34.8	34.6	38.2	36.5	37.8	37.8	35.8	36.0	33.1	33.3	33.3	35.3	33.6
1970	27.8	21.0	30.4	26.7	26.6	29.1	28.0	28.6	29.0	27.5	27.6	25.4	26.4	25.5	27.1	26.2
1965	21.5	16.4	21.8	20.6	20.5	22.7	21.5	22.1	22.4	21.2	21.3	19.5	20.4	19.7	20.8	20.7
1960	19.5	14.9	19.8	18.7	18.6	20.2	19.6	20.0	20.3	19.2	19.3	17.7	18.7	17.9	18.9	18.4
1955	16.3	12.5	16.6	15.7	15.6	16.9	16.4	16.8	17.0	16.1	16.2	14.9	15.7	15.0	15.9	15.5
1950	13.5	10.3	13.7	13.0	12.9	14.0	13.6	13.9	14.0	13.3	13.4	12.3	12.9	12.4	13.1	12.8
1945	8.6	6.6	8.8	8.3	8.2	8.9	8.6	8.9	9.0	8.5	8.6	7.8	8.3	7.9	8.4	8.1
1940	6.6	5.1	6.8	6.4	6.3	6.9	6.7	6.8	6.9	6.5	6.6	6.0	6.4	6.1	6.5	6.3

Historical Cost Indexes

Year	National 30 City Average	Indiana Muncie	Indiana South Bend	Indiana Terre Haute	Iowa Cedar Rapids	Iowa Davenport	Iowa Des Moines	Iowa Sioux City	Iowa Waterloo	Kansas Topeka	Kansas Wichita	Kentucky Lexington	Kentucky Louisville	Louisiana Baton Rouge	Louisiana Lake Charles	Louisiana New Orleans
Jan 1997	111.7E	101.2E	101.3E	104.2E	101.3E	104.4E	101.2E	93.1E	92.8E	95.4E	95.8E	96.4E	100.2E	93.2E	96.8E	96.0E
1996	108.9	99.4	99.1	102.1	99.1	102.0	99.1	90.9	90.7	93.6	93.8	94.4	98.3	91.5	94.9	94.3
1995	105.6	95.6	94.6	97.2	95.4	96.6	94.4	88.2	88.9	91.1	91.0	91.6	94.6	89.6	92.7	91.6
1994	103.0	93.2	93.0	95.8	93.0	92.3	92.4	85.9	86.8	89.4	88.1	89.6	92.2	87.7	89.9	89.5
1993	100.0	91.0	90.7	94.4	91.1	90.5	90.7	84.1	85.2	87.4	86.2	87.3	89.4	86.4	88.5	87.8
1992	97.9	89.4	89.3	93.1	89.8	89.3	89.4	82.9	83.8	86.2	85.0	85.8	88.0	85.2	87.3	86.6
1991	95.7	87.7	87.5	91.7	88.5	88.0	88.4	81.8	81.6	84.4	83.9	84.1	84.4	83.2	85.3	85.8
1990	93.2	83.9	85.1	89.1	87.1	86.5	86.9	80.4	80.3	83.2	82.6	82.8	82.6	82.0	84.1	84.5
1989	91.0	81.9	83.5	86.7	85.3	84.9	85.3	79.0	78.8	81.6	81.2	81.3	80.1	80.6	82.7	83.4
1988	88.5	80.4	82.1	84.9	83.4	83.1	83.6	76.8	77.2	80.1	79.5	79.4	78.2	79.0	81.1	82.0
1987	85.7	78.5	79.6	83.0	82.4	81.1	81.2	75.4	75.6	78.9	78.4	77.3	76.3	77.1	78.0	81.3
1986	83.7	76.9	77.5	81.1	79.5	80.0	79.8	73.8	73.7	76.9	76.2	76.5	75.3	76.1	78.1	79.2
1985	81.8	75.2	76.0	79.2	75.9	77.6	77.1	72.1	72.3	75.0	74.7	75.5	74.5	74.9	78.5	78.2
1984	80.6	73.6	75.1	78.0	80.3	77.8	76.6	74.4	73.7	75.1	74.6	75.6	73.7	77.3	78.3	77.5
1983	78.2	71.2	74.5	76.1	79.6	77.0	75.5	74.6	73.0	73.2	72.4	74.9	74.1	75.1	76.6	74.7
1982	72.1	66.0	68.3	69.9	73.5	71.7	71.1	70.1	67.7	68.4	66.1	69.4	69.2	69.3	70.5	69.2
1981	66.1	61.0	63.1	65.1	68.6	66.7	67.1	65.5	63.5	63.0	62.2	64.7	65.1	64.0	64.8	62.7
1980	60.7	56.1	58.3	59.7	62.7	59.6	61.8	59.3	57.7	58.9	58.0	59.3	59.8	59.1	60.0	57.2
1979	54.9	50.7	52.6	54.0	56.6	54.1	55.6	53.6	52.0	53.5	52.9	53.2	54.5	53.5	54.5	52.6
1978	51.3	46.8	48.2	49.9	52.3	50.0	51.1	49.2	48.1	50.0	48.9	49.2	50.0	49.2	50.2	48.5
1977	47.9	43.4	44.7	46.4	48.1	45.7	47.6	45.3	44.2	46.2	45.8	45.8	46.1	45.2	45.2	46.0
1976	45.3	40.9	42.2	43.9	45.8	43.9	45.1	42.3	42.0	43.7	43.3	43.5	43.7	41.7	42.3	43.5
1975	43.7	39.2	40.3	41.9	43.1	41.0	43.1	41.0	40.0	42.0	43.1	42.7	42.5	40.4	40.6	41.5
1974	39.4	36.0	37.2	38.5	40.1	38.2	38.1	37.8	36.9	38.1	37.4	38.1	37.9	35.3	37.9	37.4
1973	36.3	33.0	34.1	35.3	36.7	35.1	34.9	34.6	33.8	35.1	34.0	35.0	34.6	32.2	34.8	34.3
1970	27.8	25.3	26.2	27.0	28.2	26.9	27.6	26.6	25.9	27.0	25.5	26.9	25.9	25.0	26.7	27.2
1965	21.5	19.5	20.2	20.9	21.7	20.8	21.7	20.5	20.0	20.8	19.6	20.7	20.3	19.4	20.6	20.4
1960	19.5	17.8	18.3	18.9	19.7	18.8	19.5	18.6	18.2	18.9	17.8	18.8	18.4	17.6	18.7	18.4
1955	16.3	14.9	15.4	15.9	16.6	15.8	16.3	15.6	15.2	15.8	15.0	15.8	15.4	14.8	15.7	15.6
1950	13.5	12.3	12.7	13.1	13.7	13.1	13.5	12.9	12.6	13.1	12.4	13.0	12.8	12.2	13.0	12.8
1945	8.6	7.8	8.1	8.4	8.7	8.3	8.6	8.2	8.0	8.3	7.9	8.3	8.1	7.8	8.3	8.2
1940	6.6	6.0	6.3	6.5	6.7	6.4	6.6	6.3	6.2	6.4	6.1	6.4	6.3	6.0	6.4	6.3

Year	National 30 City Average	Louisiana Shreveport	Maine Lewiston	Maine Portland	Maryland Baltimore	Massachusetts Boston	Massachusetts Brockton	Massachusetts Fall River	Massachusetts Lawrence	Massachusetts Lowell	Massachusetts New Bedford	Massachusetts Pittsfield	Massachusetts Springfield	Massachusetts Worcester	Michigan Ann Arbor	Michigan Dearborn
Jan 1997	111.7E	90.2E	101.3E	101.1E	101.5E	131.7E	124.0E	124.3E	124.2E	124.1E	124.2E	114.6E	116.9E	122.3E	113.4E	117.2E
1996	108.9	88.5	99.5	99.2	99.6	128.7	121.6	122.2	121.7	122.1	122.2	112.7	115.0	120.2	112.9	116.2
1995	105.6	86.7	96.8	96.5	96.1	128.6	119.6	117.7	120.5	119.6	117.2	110.7	112.7	114.8	106.1	110.5
1994	103.0	85.0	95.2	94.9	94.4	124.9	114.2	113.7	116.8	116.9	111.6	109.2	111.1	112.9	105.0	109.0
1993	100.0	83.6	93.0	93.0	93.1	121.1	111.7	111.5	114.9	114.3	109.4	107.1	108.9	110.3	102.8	105.8
1992	97.9	82.3	91.7	91.7	90.9	118.0	110.0	109.8	110.9	110.2	107.7	105.7	107.2	108.6	101.5	103.2
1991	95.7	81.6	89.8	89.9	89.1	115.2	105.9	104.7	107.3	105.5	105.1	100.6	103.2	105.7	95.1	98.3
1990	93.2	80.3	88.5	88.6	85.6	110.9	103.7	102.9	105.2	102.8	102.6	98.7	101.4	103.2	93.2	96.5
1989	91.0	78.8	86.7	86.7	83.5	107.2	100.4	99.7	102.9	99.7	99.8	94.7	96.4	99.1	89.6	94.6
1988	88.5	77.0	84.0	84.1	81.2	102.5	96.7	96.4	97.7	96.6	96.4	91.9	92.7	94.4	85.8	91.4
1987	85.7	74.5	81.3	81.4	78.6	97.3	94.5	94.8	94.9	94.8	95.0	89.4	90.5	92.2	87.4	88.7
1986	83.7	74.1	79.0	79.0	75.7	95.1	91.2	90.8	92.5	90.8	90.7	87.1	87.8	89.8	81.7	85.3
1985	81.8	73.6	76.7	77.0	72.7	92.8	88.8	88.8	89.4	88.2	88.6	85.0	85.6	86.7	80.1	82.7
1984	80.6	73.3	75.0	75.1	72.4	88.1	85.5	85.8	86.8	84.1	84.8	82.8	83.5	83.6	78.9	81.1
1983	78.2	72.2	73.1	72.9	70.6	84.0	82.5	82.2	82.7	80.9	81.3	79.5	80.6	81.3	77.7	79.4
1982	72.1	67.0	67.4	67.4	64.7	76.9	75.7	75.0	74.5	73.6	74.2	72.3	73.7	73.1	73.7	75.8
1981	66.1	62.5	62.2	63.1	59.0	67.8	68.7	69.2	68.4	67.1	68.7	66.3	67.0	66.7	68.6	70.1
1980	60.7	58.7	57.3	58.5	53.6	64.0	63.7	64.1	63.4	62.7	63.1	61.8	62.0	62.3	62.9	64.0
1979	54.9	53.1	52.3	52.6	48.2	57.9	58.0	57.5	57.3	57.0	57.4	56.2	55.9	56.0	57.0	57.8
1978	51.3	49.2	48.6	48.8	45.9	54.0	53.4	53.3	53.5	53.1	53.2	52.6	52.3	52.2	52.5	54.1
1977	47.9	44.8	46.2	45.2	44.4	49.3	50.4	50.1	50.5	49.5	49.9	49.2	50.0	49.4	48.6	48.6
1976	45.3	42.0	43.2	42.5	41.5	47.7	46.9	47.8	47.2	46.9	47.3	46.3	47.6	47.1	46.5	46.4
1975	43.7	40.5	41.9	42.1	39.8	46.6	45.8	45.7	46.2	45.7	46.1	45.5	45.8	46.0	44.1	44.9
1974	39.4	37.3	37.5	35.6	37.4	41.7	41.3	41.3	41.3	40.9	41.1	40.4	41.0	40.6	40.4	40.9
1973	36.3	34.3	34.5	32.1	34.4	38.1	38.0	38.0	38.0	37.6	37.8	37.2	37.2	37.4	37.2	37.6
1970	27.8	26.4	26.5	25.8	25.1	29.2	29.1	29.2	29.1	28.9	29.0	28.6	28.5	28.7	28.5	28.9
1965	21.5	20.3	20.4	19.4	20.2	23.0	22.5	22.5	22.5	22.2	22.4	22.0	22.4	22.1	22.0	22.3
1960	19.5	18.5	18.6	17.6	17.5	20.5	20.4	20.4	20.4	20.2	20.3	20.0	20.1	20.1	20.0	20.2
1955	16.3	15.5	15.6	14.7	14.7	17.2	17.1	17.1	17.1	16.9	17.1	16.8	16.9	16.8	16.7	16.9
1950	13.5	12.8	12.9	12.2	12.1	14.2	14.1	14.1	14.1	14.0	14.1	13.8	14.0	13.9	13.8	14.0
1945	8.6	8.1	8.2	7.8	7.7	9.1	9.0	9.0	9.0	8.9	9.0	8.9	8.9	8.9	8.8	8.9
1940	6.6	6.3	6.3	6.0	6.0	7.0	6.9	7.0	7.0	6.9	6.9	6.8	6.9	6.8	6.8	6.9

HISTORICAL COST INDEXES

Historical Cost Indexes

Year	National 30 City Average	Michigan Detroit	Flint	Grand Rapids	Kala-mazoo	Lansing	Sagi-naw	Minnesota Duluth	Minne-apolis	Roches-ter	Mississippi Biloxi	Jackson	Missouri Kansas City	St. Joseph	St. Louis	Spring-field
Jan 1997	111.7E	117.0E	110.3E	94.7E	104.2E	108.3E	107.6E	115.9E	122.4E	113.9E	89.9E	88.0E	105.6E	102.1E	112.8E	96.7E
1996	108.9	116.0	109.7	93.9	103.5	107.8	106.9	115.0	120.4	113.5	87.1	85.4	103.2	99.9	110.1	94.6
1995	105.6	110.1	104.1	91.1	96.4	98.3	102.3	100.3	111.9	102.5	84.2	83.5	99.7	96.1	106.3	89.5
1994	103.0	108.5	102.9	89.7	95.0	97.1	101.1	99.1	109.3	101.1	82.4	81.7	97.3	92.8	103.2	88.2
1993	100.0	105.4	100.7	87.7	92.7	95.3	98.8	99.8	106.7	99.9	80.0	79.4	94.5	90.9	99.9	86.2
1992	97.9	102.9	99.4	86.4	91.4	94.0	97.4	98.5	105.3	98.6	78.8	78.2	92.9	89.5	98.6	84.9
1991	95.7	97.6	93.6	85.3	90.0	92.1	90.3	96.0	102.9	97.3	77.6	76.9	90.9	88.1	96.4	83.9
1990	93.2	96.0	91.7	84.0	87.4	90.2	88.9	94.3	100.5	95.6	76.4	75.6	89.5	86.8	94.0	82.6
1989	91.0	94.0	90.2	82.4	85.8	88.6	87.4	92.5	97.4	93.7	75.0	74.1	87.3	85.1	91.8	80.9
1988	88.5	90.8	87.8	80.7	84.1	86.3	85.6	90.5	94.6	91.9	73.5	72.9	84.9	82.9	89.0	79.2
1987	85.7	87.8	85.4	79.7	82.6	85.2	84.5	88.5	92.0	90.6	72.4	71.3	81.7	82.7	85.5	78.2
1986	83.7	84.3	82.8	77.5	80.9	82.7	81.5	87.2	89.7	88.4	70.6	69.9	79.9	79.4	83.5	75.6
1985	81.8	81.6	80.7	75.8	79.4	80.0	80.5	85.2	87.9	86.5	69.4	68.5	78.2	77.3	80.8	73.2
1984	80.6	79.4	79.5	74.3	78.7	79.0	79.9	84.9	86.9	86.0	68.6	67.5	77.7	76.7	77.7	72.7
1983	78.2	77.0	77.4	74.1	76.2	77.1	78.0	81.6	81.4	82.6	68.0	67.3	76.2	76.8	76.2	71.6
1982	72.1	74.8	73.5	69.9	72.3	73.2	73.7	74.6	75.2	75.6	63.1	62.6	70.2	70.3	69.2	66.5
1981	66.1	68.6	68.2	64.6	67.1	67.5	68.5	68.8	69.6	69.8	58.7	59.1	63.5	65.7	64.5	61.4
1980	60.7	62.5	63.1	59.8	61.7	60.3	63.3	63.4	64.1	64.8	54.3	54.3	59.1	60.4	59.7	56.7
1979	54.9	56.2	56.4	53.7	55.5	55.8	56.8	59.2	58.5	58.7	49.3	49.3	54.4	54.6	54.9	51.3
1978	51.3	52.6	52.0	49.1	51.7	51.4	53.2	53.1	54.1	53.3	45.0	45.8	50.1	51.4	51.1	47.9
1977	47.9	48.7	48.4	44.9	48.4	47.4	49.1	49.5	50.0	48.9	41.2	42.3	46.8	48.8	48.2	44.5
1976	45.3	47.0	46.4	42.7	46.2	45.0	46.5	45.8	47.0	45.1	39.2	39.3	43.5	47.5	44.9	41.7
1975	43.7	45.8	44.9	41.8	43.7	44.1	44.8	45.2	46.0	45.3	37.9	37.7	42.7	45.4	44.8	41.1
1974	39.4	41.8	40.3	37.8	39.8	39.4	40.6	41.0	42.4	40.9	34.5	30.8	37.7	40.0	39.7	36.7
1973	36.3	38.9	37.1	34.8	36.7	36.3	37.4	37.7	39.2	37.6	31.8	27.3	34.2	36.8	36.5	33.8
1970	27.8	29.7	28.5	26.7	28.1	27.8	28.7	28.9	29.5	28.9	24.4	21.3	25.3	28.3	28.4	26.0
1965	21.5	22.1	21.9	20.6	21.7	21.5	22.2	22.3	23.4	22.3	18.8	16.4	20.6	21.8	21.8	20.0
1960	19.5	20.1	20.0	18.7	19.7	19.5	20.1	20.3	20.5	20.2	17.1	14.9	19.0	19.8	19.5	18.2
1955	16.3	16.8	16.7	15.7	16.5	16.3	16.9	17.0	17.2	17.0	14.3	12.5	16.0	16.6	16.3	15.2
1950	13.5	13.9	13.8	13.0	13.6	13.5	13.9	14.0	14.2	14.0	11.8	10.3	13.2	13.7	13.5	12.6
1945	8.6	8.9	8.8	8.3	8.7	8.6	8.9	8.9	9.0	8.9	7.6	6.6	8.4	8.8	8.6	8.0
1940	6.6	6.8	6.8	6.4	6.7	6.6	6.9	6.9	7.0	6.9	5.8	5.1	6.5	6.7	6.7	6.2

Year	National 30 City Average	Montana Billings	Great Falls	Nebraska Lincoln	Omaha	Nevada Las Vegas	Reno	New Hampshire Man-chester	Nashua	New Jersey Camden	Jersey City	Newark	Pater-son	Trenton	NM Albu-querque	NY Albany
Jan 1997	111.7E	108.1E	107.6E	92.7E	98.6E	113.1E	109.2E	108.2E	108.1E	121.2E	124.5E	124.6E	125.8E	123.8E	100.5E	110.2E
1996	108.9	108.0	107.5	91.4	97.4	111.8	108.9	106.5	106.4	119.1	122.5	122.5	123.8	121.8	98.6	108.3
1995	105.6	104.7	104.9	85.6	93.4	108.5	105.0	100.9	100.8	107.0	112.2	111.9	112.1	111.2	96.3	103.6
1994	103.0	100.2	100.7	84.3	91.0	105.3	102.2	97.8	97.7	105.4	110.6	110.1	110.6	108.2	93.4	102.4
1993	100.0	97.9	97.8	82.3	88.7	102.8	99.9	95.4	95.4	103.7	109.0	108.2	109.0	106.4	90.1	99.8
1992	97.9	95.5	96.4	81.1	87.4	99.4	98.4	90.4	90.4	102.0	107.5	107.0	107.8	103.9	87.5	98.5
1991	95.7	94.2	95.1	80.1	86.4	97.5	95.8	87.8	87.8	95.0	98.5	95.6	100.0	96.6	86.2	96.1
1990	93.2	92.9	93.8	78.8	85.0	96.3	94.5	86.3	86.3	93.0	93.5	93.8	97.2	94.5	84.9	93.2
1989	91.0	91.3	92.2	77.3	83.6	94.8	92.3	84.8	84.8	90.4	91.4	91.8	95.5	91.9	83.3	88.4
1988	88.5	89.5	90.4	75.8	82.0	93.0	90.5	83.2	83.2	87.7	89.5	89.7	92.5	89.5	81.7	86.8
1987	85.7	85.4	87.3	75.3	79.8	91.5	88.3	82.4	82.5	86.1	88.3	88.7	90.3	87.9	78.1	84.5
1986	83.7	86.2	87.3	72.6	78.8	90.1	87.2	79.7	79.7	84.6	86.1	86.7	87.8	86.1	78.9	81.7
1985	81.8	83.9	84.3	71.5	77.5	87.6	85.0	78.1	78.1	81.3	83.3	83.5	84.5	82.8	76.7	79.5
1984	80.6	83.2	82.9	70.9	77.6	85.9	83.3	76.4	75.5	78.2	80.3	80.1	81.3	78.5	75.9	77.3
1983	78.2	80.3	80.0	72.1	77.3	83.0	81.9	73.0	72.2	74.6	76.5	76.7	76.5	74.2	73.1	74.2
1982	72.1	74.3	74.1	67.8	72.5	76.9	75.8	67.0	66.7	67.9	69.8	70.4	70.1	69.1	67.1	69.2
1981	66.1	69.5	70.3	64.1	68.2	70.2	68.5	61.5	61.1	62.5	65.0	65.1	64.8	63.4	63.3	64.4
1980	60.7	63.9	64.6	58.5	63.5	64.6	63.0	56.6	56.0	58.6	60.6	60.1	60.0	58.9	59.0	59.5
1979	54.9	57.5	58.9	52.9	56.2	58.4	56.7	51.2	50.8	53.5	55.6	55.3	54.7	54.0	54.3	54.2
1978	51.3	53.4	54.0	49.3	52.2	53.9	51.9	47.5	47.1	50.1	51.7	51.4	51.7	50.8	49.8	50.6
1977	47.9	48.9	49.1	44.6	48.0	49.7	48.1	43.9	43.6	47.2	47.4	47.2	47.7	47.0	45.1	47.7
1976	45.3	45.4	46.2	42.1	45.2	46.3	45.1	40.8	41.1	44.6	44.8	45.4	45.3	45.2	41.9	45.1
1975	43.7	43.1	43.8	40.9	43.2	42.8	41.9	41.3	40.8	42.3	43.4	44.2	43.9	43.8	40.3	43.9
1974	39.4	40.4	41.0	37.3	37.7	40.0	39.7	36.3	36.3	38.4	39.3	40.5	39.4	39.4	35.5	40.4
1973	36.3	37.2	37.7	34.4	34.1	37.2	36.5	33.2	33.3	35.4	36.2	37.5	36.2	36.1	32.6	37.2
1970	27.8	28.5	28.9	26.4	26.8	29.4	28.0	26.2	25.6	27.2	27.8	29.0	27.8	27.4	26.4	28.3
1965	21.5	22.0	22.3	20.3	20.6	22.4	21.6	20.6	19.7	20.9	21.4	23.8	21.4	21.3	20.6	22.3
1960	19.5	20.0	20.3	18.5	18.7	20.2	19.6	18.0	17.9	19.0	19.4	19.4	19.4	19.2	18.5	19.3
1955	16.3	16.7	17.0	15.5	15.7	16.9	16.4	15.1	15.0	16.0	16.3	16.3	16.3	16.1	15.6	16.2
1950	13.5	13.9	14.0	12.8	13.0	14.0	13.6	12.5	12.4	13.2	13.5	13.5	13.5	13.3	12.9	13.4
1945	8.6	8.8	9.0	8.1	8.3	8.9	8.7	8.0	7.9	8.4	8.6	8.6	8.6	8.5	8.2	8.5
1940	6.6	6.8	6.9	6.3	6.4	6.9	6.7	6.2	6.1	6.5	6.6	6.6	6.6	6.6	6.3	6.6

Historical Cost Indexes

Year	National 30 City Average	New York - Binghamton	New York - Buffalo	New York - New York	New York - Rochester	New York - Schenectady	New York - Syracuse	New York - Utica	New York - Yonkers	North Carolina - Charlotte	North Carolina - Durham	North Carolina - Greensboro	North Carolina - Raleigh	North Carolina - Winston-Salem	North Dakota - Fargo	Ohio - Akron
Jan 1997	111.7E	107.5E	115.9E	150.3E	115.2E	111.4E	110.6E	107.2E	139.4E	86.8E	87.8E	87.9E	88.0E	87.7E	93.3E	110.2E
1996	108.9	105.5	114.0	148.0	113.3	109.5	108.2	105.2	137.2	85.0	85.9	86.0	86.1	85.8	91.7	107.9
1995	105.6	99.3	110.1	140.7	106.6	104.6	104.0	97.7	129.4	81.8	82.6	82.6	82.7	82.6	88.4	103.4
1994	103.0	98.1	107.2	137.1	105.2	103.5	102.3	96.5	128.1	80.3	81.1	81.0	81.2	81.1	87.0	102.2
1993	100.0	95.7	102.2	133.3	102.0	100.8	99.6	94.1	126.3	78.2	78.9	78.9	78.9	78.8	85.8	100.3
1992	97.9	93.7	100.1	128.2	99.3	99.3	98.1	90.5	123.9	77.1	77.7	77.8	77.8	77.6	83.1	98.0
1991	95.7	89.5	96.8	124.4	96.0	96.7	95.5	88.7	121.5	76.0	76.6	76.8	76.7	76.6	82.6	96.8
1990	93.2	87.0	94.1	118.1	94.6	93.9	91.0	85.6	111.4	74.8	75.4	75.5	75.5	75.3	81.3	94.6
1989	91.0	85.4	91.8	114.8	92.6	89.6	88.4	84.1	102.1	73.3	74.0	74.0	74.0	73.8	79.7	92.9
1988	88.5	83.6	89.4	106.5	87.9	87.7	86.5	82.4	99.9	71.5	72.2	72.2	72.2	72.0	78.2	91.1
1987	85.7	82.4	85.8	102.6	85.8	85.3	85.1	82.1	98.9	70.6	71.2	71.2	71.1	71.0	77.4	90.1
1986	83.7	80.2	84.4	99.8	83.7	82.4	83.3	79.2	96.6	68.4	69.1	69.1	69.1	68.9	75.4	87.7
1985	81.8	77.5	83.2	94.9	81.6	80.1	81.0	77.0	92.8	66.9	67.6	67.7	67.6	67.5	73.4	86.8
1984	80.6	75.4	81.3	91.1	80.6	78.5	79.0	76.1	89.5	66.3	66.6	66.8	66.2	66.0	72.2	82.8
1983	78.2	73.2	78.2	86.5	77.1	75.3	76.5	74.5	85.4	64.6	65.1	65.7	64.9	64.6	70.5	79.4
1982	72.1	67.5	70.3	78.3	71.3	69.8	70.2	68.4	77.8	58.7	60.0	60.5	59.5	59.4	66.1	73.2
1981	66.1	62.6	65.2	71.6	65.6	64.7	64.6	63.5	71.0	55.3	56.6	56.7	55.9	56.3	61.7	67.7
1980	60.7	58.0	60.6	66.0	60.5	60.3	61.6	58.5	65.9	51.1	52.2	52.5	51.7	50.8	57.4	62.3
1979	54.9	52.8	56.0	60.2	55.0	54.9	56.2	53.2	60.0	45.9	46.9	47.1	46.4	45.9	52.2	56.3
1978	51.3	50.0	52.6	56.4	51.7	50.8	52.2	49.1	54.7	41.9	43.0	42.9	42.4	41.9	47.8	52.1
1977	47.9	46.5	49.0	52.9	48.3	48.5	49.1	46.0	51.8	40.2	41.5	40.7	40.0	39.6	44.2	48.5
1976	45.3	44.1	46.2	50.6	45.6	45.3	46.6	43.8	48.5	37.5	37.9	38.1	37.6	37.0	41.3	45.8
1975	43.7	42.5	45.1	49.5	44.8	43.7	44.8	42.7	47.1	36.1	37.0	37.0	37.3	36.1	39.3	44.7
1974	39.4	38.2	41.8	45.4	41.5	39.3	40.3	38.1	42.5	30.3	33.5	33.5	33.1	32.6	36.5	40.1
1973	36.3	35.1	38.7	42.0	38.6	36.2	37.1	35.0	39.1	27.3	30.8	30.8	30.5	30.0	33.6	36.9
1970	27.8	27.0	28.9	32.8	29.2	27.8	28.5	26.9	30.0	20.9	23.7	23.6	23.4	23.0	25.8	28.3
1965	21.5	20.8	22.2	25.5	22.8	21.4	21.9	20.7	23.1	16.0	18.2	18.2	18.0	17.7	19.9	21.8
1960	19.5	18.9	19.9	21.5	19.7	19.5	19.9	18.8	21.0	14.4	16.6	16.6	16.4	16.1	18.1	19.9
1955	16.3	15.8	16.7	18.1	16.5	16.3	16.7	15.8	17.6	12.1	13.9	13.9	13.7	13.5	15.2	16.6
1950	13.5	13.1	13.8	14.9	13.6	13.5	13.8	13.0	14.5	10.0	11.5	11.5	11.4	11.2	12.5	13.7
1945	8.6	8.3	8.8	9.5	8.7	8.6	8.8	8.3	9.3	6.4	7.3	7.3	7.2	7.1	8.0	8.8
1940	6.6	6.4	6.8	7.4	6.7	6.7	6.8	6.4	7.2	4.9	5.6	5.6	5.6	5.5	6.2	6.8

Year	National 30 City Average	Ohio - Canton	Ohio - Cincinnati	Ohio - Cleveland	Ohio - Columbus	Ohio - Dayton	Ohio - Lorain	Ohio - Springfield	Ohio - Toledo	Ohio - Youngstown	Oklahoma - Lawton	Oklahoma - Oklahoma City	Oklahoma - Tulsa	Oregon - Eugene	Oregon - Portland	PA - Allentown
Jan 1997	111.7E	105.2E	102.8E	112.6E	104.2E	101.4E	106.3E	101.2E	108.1E	106.6E	92.6E	93.1E	93.0E	117.1E	118.9E	114.2E
1996	108.9	103.2	100.2	110.1	101.1	98.8	103.8	98.4	105.1	104.1	90.9	91.3	91.2	114.4	116.1	112.0
1995	105.6	98.8	97.1	106.4	99.1	94.6	97.1	92.0	100.6	100.1	85.3	88.0	89.0	112.2	114.3	108.3
1994	103.0	97.7	95.0	104.8	95.4	93.0	96.0	90.1	99.3	98.9	84.0	86.5	87.6	107.6	109.2	106.4
1993	100.0	95.9	92.3	101.9	93.9	90.6	93.9	87.7	98.3	97.2	81.3	83.9	84.8	107.2	108.8	103.6
1992	97.9	94.5	90.6	98.7	92.6	89.2	92.2	86.4	97.1	96.0	80.1	82.7	83.4	101.1	102.6	101.6
1991	95.7	93.4	88.6	97.2	90.6	87.7	91.1	84.9	94.2	92.7	78.3	80.7	81.4	99.5	101.1	98.9
1990	93.2	92.1	86.7	95.1	88.1	85.9	89.4	83.4	92.9	91.8	77.1	79.9	80.0	98.0	99.4	95.0
1989	91.0	90.3	84.6	93.6	85.8	82.6	87.9	81.0	91.4	88.4	75.8	78.5	78.4	95.2	97.0	92.4
1988	88.5	89.1	82.9	92.1	83.9	81.0	85.2	79.5	89.6	86.9	74.3	77.0	76.8	93.5	95.3	89.8
1987	85.7	87.7	80.9	89.6	81.5	79.5	86.9	78.0	86.1	85.4	72.6	74.4	74.7	91.4	91.9	86.9
1986	83.7	84.9	78.8	87.3	79.7	78.0	81.6	76.1	84.6	83.8	71.6	74.1	73.9	90.0	91.7	84.7
1985	81.8	83.7	78.2	86.2	77.7	76.3	80.7	74.3	84.7	82.7	71.2	73.9	73.8	88.7	90.1	81.9
1984	80.6	81.3	77.4	82.5	76.1	75.9	79.5	73.9	83.5	80.3	70.6	72.9	72.8	89.5	89.3	78.4
1983	78.2	76.9	74.9	77.9	74.5	71.6	77.2	71.9	80.4	77.3	69.1	71.5	72.2	89.0	89.2	75.0
1982	72.1	70.8	70.1	71.6	68.6	65.2	70.8	66.5	74.6	71.7	63.5	65.2	66.2	82.3	83.6	68.2
1981	66.1	65.8	65.3	66.0	63.9	60.6	65.7	61.3	69.4	67.2	57.0	60.3	61.5	74.2	74.5	62.7
1980	60.7	60.6	59.8	61.0	58.6	56.6	60.8	56.2	64.0	61.5	52.2	55.5	57.2	68.9	68.4	58.7
1979	54.9	55.2	54.3	54.8	52.7	51.2	55.2	51.0	57.9	56.0	47.7	49.8	51.4	61.9	60.6	54.1
1978	51.3	51.1	50.4	51.3	49.3	47.2	50.3	46.9	54.0	52.2	44.5	46.3	48.0	57.2	56.0	50.4
1977	47.9	47.7	46.9	48.2	45.3	44.3	46.7	42.9	50.2	48.7	40.9	43.0	44.8	51.5	51.7	46.7
1976	45.3	45.4	44.8	45.9	43.7	42.3	44.3	40.7	47.4	45.8	39.0	40.6	41.7	48.7	48.2	45.1
1975	43.7	44.0	43.7	44.6	42.2	40.5	42.7	40.0	44.9	44.5	38.6	38.6	39.2	44.6	44.9	42.3
1974	39.4	39.4	39.9	41.0	38.4	37.6	38.9	35.9	41.9	40.3	34.1	33.3	34.1	43.0	40.8	38.6
1973	36.3	36.3	37.1	38.6	35.2	35.1	35.7	33.1	38.7	37.2	31.4	29.9	30.6	39.5	37.6	35.5
1970	27.8	27.8	28.2	29.6	26.7	27.0	27.5	25.4	29.1	29.6	24.1	23.4	24.1	30.4	30.0	27.3
1965	21.5	21.5	20.7	21.1	20.2	20.0	21.1	19.6	21.3	21.0	18.5	17.8	20.6	23.4	22.6	21.0
1960	19.5	19.5	19.3	19.6	18.7	18.1	19.2	17.8	19.4	19.1	16.9	16.1	18.1	21.2	21.1	19.1
1955	16.3	16.3	16.2	16.5	15.7	15.2	16.1	14.9	16.2	16.0	14.1	13.5	15.1	17.8	17.7	16.0
1950	13.5	13.5	13.3	13.6	13.0	12.5	13.3	12.3	13.4	13.2	11.7	11.2	12.5	14.7	14.6	13.2
1945	8.6	8.6	8.5	8.7	8.3	8.0	8.5	7.8	8.6	8.4	7.4	7.1	8.0	9.4	9.3	8.4
1940	6.6	6.7	6.5	6.7	6.4	6.2	6.5	6.1	6.6	6.5	5.7	5.5	6.1	7.2	7.2	6.5

Historical Cost Indexes

Year	National 30 City Average	Pennsylvania Erie	Pennsylvania Harrisburg	Pennsylvania Philadelphia	Pennsylvania Pittsburgh	Pennsylvania Reading	Pennsylvania Scranton	RI Providence	South Carolina Charleston	South Carolina Columbia	South Dakota Rapid City	South Dakota Sioux Falls	Tennessee Chattanooga	Tennessee Knoxville	Tennessee Memphis	Tennessee Nashville
Jan 1997	111.7E	108.8E	108.1E	123.2E	114.0E	110.6E	112.3E	119.2E	86.9E	85.9E	89.2E	92.2E	93.6E	91.2E	96.0E	94.6E
1996	108.9	106.8	105.7	120.3	110.8	108.7	109.9	117.2	84.9	84.1	87.2	90.0	91.6	88.8	94.0	91.9
1995	105.6	99.8	100.6	117.1	106.3	103.6	103.8	111.1	82.7	82.2	84.0	84.7	89.2	86.0	91.2	87.6
1994	103.0	98.2	99.4	115.2	103.7	102.2	102.6	109.6	81.2	80.6	82.7	83.1	87.4	84.1	89.0	84.8
1993	100.0	94.9	96.6	107.4	99.0	98.6	99.8	108.2	78.4	77.9	81.2	81.9	85.1	82.3	86.8	81.9
1992	97.9	92.1	94.8	105.3	96.5	96.7	97.8	106.9	77.1	76.8	80.0	80.7	83.6	77.7	85.4	80.7
1991	95.7	90.5	92.5	101.7	93.2	94.1	94.8	96.1	75.0	75.8	78.7	79.7	81.9	76.6	83.0	79.1
1990	93.2	88.5	89.5	98.5	91.2	90.8	91.3	94.1	73.7	74.5	77.2	78.4	79.9	75.1	81.3	77.1
1989	91.0	86.6	86.5	94.2	89.4	87.9	88.8	92.4	72.1	72.8	75.7	77.0	78.5	73.7	81.2	74.8
1988	88.5	84.9	84.0	89.8	87.7	85.3	86.4	89.1	70.5	71.3	74.2	75.5	77.0	72.1	79.7	73.1
1987	85.7	82.6	81.8	86.8	85.6	82.9	83.3	86.3	70.2	70.2	73.6	74.9	74.4	70.2	77.5	71.0
1986	83.7	81.5	79.6	84.9	83.6	79.7	81.3	85.2	67.4	68.1	71.0	72.3	74.2	69.2	75.3	68.1
1985	81.8	79.6	77.2	82.2	81.5	77.7	80.0	83.0	65.9	66.9	69.7	71.2	72.5	67.7	74.3	66.7
1984	80.6	78.3	75.4	79.0	78.9	75.9	78.8	80.3	64.3	65.5	68.7	70.4	71.4	66.4	73.3	66.3
1983	78.2	76.3	72.2	74.4	75.5	74.0	75.0	76.3	65.3	65.2	67.7	69.7	69.3	64.1	72.8	65.1
1982	72.1	71.0	66.3	69.0	70.9	67.7	68.4	70.0	59.9	60.0	64.4	67.6	64.5	59.8	66.5	62.2
1981	66.1	64.6	61.7	63.4	66.3	62.5	62.3	64.3	56.3	55.9	59.7	63.6	59.9	55.9	60.3	57.3
1980	60.7	59.4	56.9	58.7	61.3	57.7	58.4	59.2	50.6	51.1	54.8	57.1	55.0	51.6	55.9	53.1
1979	54.9	54.0	52.6	54.2	55.6	53.8	53.5	53.3	45.7	44.4	49.7	51.5	49.8	46.4	51.0	47.0
1978	51.3	50.2	49.6	51.3	51.4	49.7	50.0	50.2	42.1	40.7	46.0	47.4	45.9	42.8	47.4	43.9
1977	47.9	46.6	45.2	48.1	48.3	46.5	46.9	47.0	39.2	39.4	42.5	44.3	42.5	40.2	45.0	40.7
1976	45.3	44.7	43.7	45.8	45.6	44.2	44.4	44.2	36.6	37.9	40.1	42.0	39.7	38.3	42.0	38.3
1975	43.7	43.4	42.2	44.5	44.5	43.6	41.9	42.7	35.0	36.4	37.7	39.6	39.1	37.3	40.7	37.0
1974	39.4	39.2	36.7	39.6	39.9	38.4	38.2	39.7	32.3	32.4	35.1	36.5	35.2	31.5	33.9	32.1
1973	36.3	36.2	32.9	36.0	36.6	35.3	35.2	36.7	29.7	29.8	32.3	33.6	32.4	28.4	30.1	29.1
1970	27.8	27.5	25.8	27.5	28.7	27.1	27.0	27.3	22.8	22.9	24.8	25.8	24.9	22.0	23.0	22.8
1965	21.5	21.1	20.1	21.7	22.4	20.9	20.8	21.9	17.6	17.6	19.1	19.9	19.2	17.1	18.3	17.4
1960	19.5	19.1	18.6	19.4	19.7	19.0	18.9	19.1	16.0	16.0	17.3	18.1	17.4	15.5	16.6	15.8
1955	16.3	16.1	15.6	16.3	16.5	15.9	15.9	16.0	13.4	13.4	14.5	15.1	14.6	13.0	13.9	13.3
1950	13.5	13.3	12.9	13.5	13.6	13.2	13.1	13.2	11.1	11.1	12.0	12.5	12.1	10.7	11.5	10.9
1945	8.6	8.5	8.2	8.6	8.7	8.4	8.3	8.4	7.1	7.1	7.7	8.0	7.7	6.8	7.3	7.0
1940	6.6	6.5	6.3	6.6	6.7	6.4	6.5	6.5	5.4	5.5	5.9	6.2	6.0	5.3	5.6	5.4

Year	National 30 City Average	Texas Abilene	Texas Amarillo	Texas Austin	Texas Beaumont	Texas Corpus Christi	Texas Dallas	Texas El Paso	Texas Fort Worth	Texas Houston	Texas Lubbock	Texas Odessa	Texas San Antonio	Texas Waco	Texas Wichita Falls	Utah Ogden
Jan 1997	111.7E	88.4E	91.3E	93.0E	96.9E	90.1E	95.9E	88.5E	93.2E	99.9E	91.8E	88.6E	94.4E	91.5E	91.6E	96.0E
1996	108.9	86.8	89.6	90.9	95.3	88.5	94.1	86.7	91.5	97.9	90.2	87.1	92.3	89.7	89.9	94.1
1995	105.6	85.2	87.4	89.3	93.7	87.4	91.4	85.2	89.5	95.9	88.4	85.6	88.9	86.4	86.8	92.2
1994	103.0	83.3	85.5	87.0	91.8	84.6	89.6	82.2	87.4	93.4	87.0	83.9	87.0	84.9	85.4	89.4
1993	100.0	81.4	83.4	84.9	90.0	82.8	87.8	80.2	85.5	91.1	84.8	81.9	85.0	83.0	83.5	87.1
1992	97.9	80.3	82.3	83.8	88.9	81.6	86.2	79.0	84.1	89.8	83.6	80.7	83.9	81.8	82.4	85.1
1991	95.7	79.2	81.1	82.8	87.8	80.6	85.9	78.0	83.3	87.9	82.7	79.8	83.3	80.6	81.5	84.1
1990	93.2	78.0	80.1	81.3	86.5	79.3	84.5	76.7	82.1	85.4	81.5	78.6	80.7	79.6	80.3	83.4
1989	91.0	76.6	78.7	79.7	85.1	77.8	82.5	75.2	80.6	84.0	80.0	77.0	78.7	78.1	78.7	81.9
1988	88.5	75.0	77.2	78.3	83.9	76.3	81.1	73.6	79.8	82.5	78.4	75.5	77.1	76.5	77.4	80.4
1987	85.7	75.9	76.4	76.0	81.3	75.6	79.9	72.2	77.7	81.3	77.3	74.6	75.7	75.2	77.4	79.1
1986	83.7	72.5	74.1	75.3	80.5	73.5	78.9	70.7	76.7	80.3	75.4	72.4	74.3	72.8	74.5	77.3
1985	81.8	71.1	72.5	74.5	79.3	72.3	77.6	69.4	75.1	79.6	74.0	71.2	73.9	71.7	73.3	75.2
1984	80.6	70.5	71.6	72.8	78.7	71.6	79.8	68.3	77.1	80.7	72.7	69.8	73.1	71.3	72.4	74.9
1983	78.2	68.3	70.6	71.0	75.0	70.2	77.8	66.3	74.9	79.8	71.0	69.0	71.8	69.4	70.1	76.1
1982	72.1	62.7	65.5	64.9	65.7	64.7	70.7	62.6	68.2	72.1	65.6	62.5	66.0	64.3	64.8	69.7
1981	66.1	57.9	60.1	59.3	62.4	59.6	63.1	58.6	61.0	65.1	61.0	58.3	60.5	59.2	60.0	65.3
1980	60.7	53.4	55.2	54.5	57.6	54.5	57.9	53.1	57.0	59.4	55.6	57.2	55.0	54.9	55.4	62.2
1979	54.9	48.6	49.8	49.0	52.3	48.7	51.5	48.5	51.4	53.3	50.3	51.1	49.2	49.7	49.4	53.3
1978	51.3	44.7	45.8	46.2	48.0	44.9	48.2	45.3	47.5	50.2	46.5	43.9	46.0	46.4	47.1	49.6
1977	47.9	42.9	42.5	43.1	43.9	42.2	44.9	41.7	44.7	47.6	42.9	41.5	42.5	42.1	41.1	45.4
1976	45.3	40.4	39.7	40.3	40.7	39.6	41.2	39.0	43.8	40.3	38.6	40.8	39.7	37.7	—	43.2
1975	43.7	37.6	39.0	39.0	39.6	38.1	40.7	38.0	40.4	41.2	38.9	37.9	39.0	38.6	38.0	40.0
1974	39.4	34.7	35.2	35.3	36.5	34.7	35.8	32.3	36.6	36.1	35.6	34.9	33.8	35.2	34.7	37.9
1973	36.3	31.9	32.4	32.5	33.6	31.9	32.3	29.4	33.7	32.9	32.7	32.1	30.5	32.4	32.0	34.9
1970	27.8	24.5	24.9	24.9	25.7	24.5	25.5	23.7	25.9	25.4	25.1	24.6	23.3	24.8	24.5	26.8
1965	21.5	18.9	19.2	19.2	19.9	18.9	19.9	19.0	19.9	20.0	19.4	19.0	18.5	19.2	18.9	20.6
1960	19.5	17.1	17.4	17.4	18.1	17.1	18.2	17.0	18.1	18.2	17.6	17.3	16.8	17.4	17.2	18.8
1955	16.3	14.4	14.6	14.6	15.1	14.4	15.3	14.3	15.2	15.2	14.8	14.5	14.1	14.6	14.4	15.7
1950	13.5	11.9	12.1	12.1	12.5	11.9	12.6	11.8	12.5	12.6	12.2	12.0	11.6	12.1	11.9	13.0
1945	8.6	7.6	7.7	7.7	8.0	7.6	8.0	7.5	8.0	8.0	7.8	7.6	7.4	7.7	7.6	8.3
1940	6.6	5.9	5.9	5.9	6.1	5.8	6.2	5.8	6.2	6.2	6.0	5.9	5.7	5.9	5.8	6.4

Historical Cost Indexes

Year	National 30 City Average	Utah Salt Lake City	Vermont Burlington	Vermont Rutland	Virginia Alexandria	Virginia Newport News	Virginia Norfolk	Virginia Richmond	Virginia Roanoke	Washington Seattle	Washington Spokane	Washington Tacoma	West Virginia Charleston	West Virginia Huntington	Wisconsin Green Bay	Wisconsin Kenosha
Jan 1997	111.7E	96.8E	97.2E	96.9E	101.0E	91.5E	91.7E	93.0E	86.9E	117.2E	111.1E	116.2E	105.4E	106.0E	105.1E	108.8E
1996	108.9	94.9	95.1	94.8	99.7	90.2	90.4	91.6	85.5	115.2	109.2	114.3	103.1	104.8	103.8	106.4
1995	105.6	93.1	91.1	90.8	96.3	86.0	86.4	87.8	82.8	113.7	107.4	112.8	95.8	97.2	97.6	97.9
1994	103.0	90.2	89.5	89.3	93.9	84.6	84.8	86.3	81.4	109.9	104.0	108.3	94.3	95.3	96.3	96.2
1993	100.0	87.9	87.6	87.6	91.6	82.9	83.0	84.3	79.5	107.3	103.9	106.7	92.6	93.5	94.0	94.3
1992	97.9	86.0	86.1	86.1	90.1	81.0	81.6	82.0	78.3	105.1	101.4	103.7	91.4	92.3	92.0	92.1
1991	95.7	84.9	84.2	84.2	88.2	77.6	77.9	79.8	77.3	102.2	100.0	102.2	89.7	88.6	88.6	89.8
1990	93.2	84.3	83.0	82.9	86.1	76.3	76.7	77.6	76.1	100.1	98.5	100.5	86.1	86.8	86.7	87.8
1989	91.0	82.8	81.4	81.3	83.4	74.9	75.2	76.0	74.2	96.1	96.8	98.4	84.5	84.8	84.3	85.6
1988	88.5	81.3	79.7	79.7	81.0	73.4	73.7	74.1	72.3	94.2	95.0	96.6	82.8	83.1	82.0	83.0
1987	85.7	79.8	79.0	79.0	77.9	71.0	71.7	72.7	70.2	91.9	92.4	92.9	81.1	81.3	80.2	81.7
1986	83.7	78.1	76.4	76.4	76.8	70.3	70.5	71.0	67.8	90.5	91.9	92.8	79.9	80.2	77.8	79.3
1985	81.8	75.9	74.8	74.9	75.1	68.7	68.8	69.5	67.2	88.3	89.0	91.2	77.7	77.7	76.7	77.4
1984	80.6	75.2	73.7	73.7	75.0	67.9	68.3	68.8	66.2	89.2	88.1	90.0	75.4	75.8	75.0	75.1
1983	78.2	74.5	70.9	71.5	72.9	65.9	66.5	68.9	66.0	87.6	86.9	88.8	73.8	74.4	73.4	73.2
1982	72.1	68.1	64.9	65.4	67.0	60.6	60.9	63.7	61.0	80.7	80.6	81.7	67.6	68.4	67.5	69.1
1981	66.1	61.9	59.3	59.7	61.6	56.0	56.8	58.2	55.6	75.7	72.9	73.8	62.6	63.3	63.6	63.5
1980	60.7	57.0	55.3	58.3	57.3	52.5	52.4	54.3	51.3	67.9	66.3	66.7	57.7	58.3	58.6	58.3
1979	54.9	52.5	49.8	51.2	51.7	47.6	47.6	49.1	46.8	61.1	60.1	60.0	51.8	52.5	52.7	53.2
1978	51.3	49.0	46.1	47.2	48.9	44.0	44.9	46.0	43.6	56.7	55.4	55.6	48.2	48.5	48.7	49.5
1977	47.9	45.0	44.5	46.3	44.6	41.2	41.5	42.5	41.2	51.8	51.1	50.7	44.7	43.2	44.8	45.5
1976	45.3	42.8	42.3	45.0	42.5	38.4	38.7	39.9	38.6	48.5	47.5	47.1	42.3	40.3	42.6	42.9
1975	43.7	40.1	41.8	43.9	41.7	37.2	36.9	37.1	37.1	44.9	44.4	44.5	41.0	40.0	40.9	40.5
1974	39.4	36.1	35.7	38.0	37.2	33.8	31.4	31.1	33.6	39.8	40.1	42.0	37.0	36.5	37.3	37.5
1973	36.3	33.2	32.1	34.9	34.2	31.1	28.0	29.6	30.9	36.0	36.8	38.6	34.1	33.6	34.3	34.6
1970	27.8	26.1	25.4	26.8	26.2	23.9	21.5	22.0	23.7	28.8	29.3	29.6	26.1	25.8	26.4	26.5
1965	21.5	20.0	19.8	20.6	20.2	18.4	17.1	17.2	18.3	22.4	22.5	22.8	20.1	19.9	20.3	20.4
1960	19.5	18.4	18.0	18.8	18.4	16.7	15.4	15.6	16.6	20.4	20.8	20.8	18.3	18.1	18.4	18.6
1955	16.3	15.4	15.1	15.7	15.4	14.0	12.9	13.1	13.9	17.1	17.4	17.4	15.4	15.2	15.5	15.6
1950	13.5	12.7	12.4	13.0	12.7	11.6	10.7	10.8	11.5	14.1	14.4	14.4	12.7	12.5	12.8	12.9
1945	8.6	8.1	7.9	8.3	8.1	7.4	6.8	6.9	7.3	9.0	9.2	9.2	8.1	8.0	8.1	8.2
1940	6.6	6.3	6.1	6.4	6.2	5.7	5.3	5.3	5.7	7.0	7.1	7.1	6.2	6.2	6.3	6.3

Year	National 30 City Average	Wisconsin Madison	Wisconsin Milwaukee	Wisconsin Racine	Wyoming Cheyenne	Canada Calgary	Canada Edmonton	Canada Hamilton	Canada London	Canada Montreal	Canada Ottawa	Canada Quebec	Canada Toronto	Canada Vancouver	Canada Winnipeg
Jan 1997	111.7E	105.9E	108.3E	108.9E	92.7E	110.9E	110.8E	124.6E	122.2E	114.9E	123.0E	115.6E	125.9E	121.0E	111.5E
1996	108.9	104.4	107.1	106.5	91.1	109.1	109.0	122.6	120.2	112.9	121.0	113.6	123.9	119.0	109.7
1995	105.6	96.5	103.9	97.8	87.6	107.4	107.4	119.9	117.5	110.8	118.2	111.5	121.6	116.2	107.6
1994	103.0	94.5	100.6	96.1	85.4	106.7	106.6	116.5	114.2	109.5	115.0	110.2	117.8	115.1	105.5
1993	100.0	91.3	96.7	93.8	82.9	104.7	104.5	113.7	111.9	106.9	112.0	107.0	114.9	109.2	102.8
1992	97.9	89.2	93.9	91.7	81.7	103.4	103.2	112.4	110.7	104.2	110.8	103.6	113.7	108.0	101.6
1991	95.7	86.2	91.6	89.3	80.3	102.1	102.0	108.2	106.7	101.8	106.9	100.4	109.0	106.8	98.6
1990	93.2	84.3	88.9	87.3	79.1	98.0	97.1	103.8	101.4	99.0	102.7	96.8	104.6	103.2	95.1
1989	91.0	81.8	86.4	85.0	77.8	95.6	94.7	98.2	96.5	94.9	98.5	92.5	98.5	97.6	92.9
1988	88.5	79.9	84.1	82.4	76.3	93.9	93.0	95.4	93.2	90.6	94.0	88.3	94.8	95.9	90.0
1987	85.7	78.1	81.1	81.1	76.7	90.2	89.3	89.7	90.1	87.2	89.6	85.0	90.4	94.2	87.1
1986	83.7	76.1	78.9	78.6	73.4	91.2	90.1	88.9	87.9	84.8	87.7	82.4	89.3	93.2	85.3
1985	81.8	74.3	77.4	77.0	72.3	90.2	89.1	85.8	84.9	82.4	83.9	79.9	86.5	89.8	83.4
1984	80.6	72.5	76.3	74.7	73.8	89.6	88.1	85.1	84.1	81.6	82.9	80.0	85.2	89.2	82.5
1983	78.2	71.9	74.0	73.0	73.8	84.6	82.8	80.1	79.1	76.4	77.6	75.8	79.9	83.3	77.9
1982	72.1	65.9	70.5	68.6	68.3	76.2	74.1	73.7	71.8	70.2	72.3	69.6	71.8	76.4	70.7
1981	66.1	61.4	64.9	63.3	62.7	70.6	68.1	68.8	67.1	64.6	65.8	64.3	67.5	70.3	65.4
1980	60.7	56.8	58.8	58.1	56.9	64.9	63.3	63.7	61.9	59.2	60.9	59.3	60.9	65.0	61.7
1979	54.9	51.4	53.0	52.8	51.2	58.9	58.5	58.1	56.6	54.4	55.5	53.9	56.0	59.2	56.1
1978	51.3	47.3	49.9	49.2	47.5	54.9	54.4	54.0	52.6	50.4	51.7	49.7	51.9	55.1	51.7
1977	47.9	44.0	46.9	45.7	44.8	49.9	49.9	48.8	48.2	43.7	48.0	43.0	48.2	50.2	47.8
1976	45.3	41.9	44.2	43.5	42.2	45.6	45.4	44.5	43.3	41.7	43.8	40.6	44.1	45.9	42.1
1975	43.7	40.7	43.3	40.7	40.6	42.2	41.6	42.9	41.6	39.7	41.5	39.0	42.2	42.4	39.2
1974	39.4	37.6	40.9	37.6	36.7	40.9	40.5	40.3	39.4	34.7	39.0	36.8	36.4	37.1	33.9
1973	36.3	34.9	38.1	34.6	33.8	37.6	37.2	37.1	36.2	31.6	35.9	33.9	32.8	33.6	30.5
1970	27.8	26.5	29.4	26.5	26.0	28.9	28.6	28.5	27.8	25.6	27.6	26.0	25.6	26.0	23.1
1965	21.5	20.6	21.8	20.4	20.0	22.3	22.0	22.0	21.4	18.7	21.2	20.1	19.4	20.5	17.5
1960	19.5	18.1	19.0	18.6	18.2	20.2	20.0	20.0	19.5	17.0	19.3	18.2	17.6	18.6	15.8
1955	16.3	15.2	15.9	15.6	15.2	17.0	16.8	16.7	16.3	14.3	16.2	15.3	14.8	15.5	13.3
1950	13.5	12.5	13.2	12.9	12.6	14.0	13.9	13.8	13.5	11.8	13.4	12.6	12.2	12.8	10.9
1945	8.6	8.0	8.4	8.2	8.0	9.0	8.8	8.8	8.6	7.5	8.5	8.0	7.8	8.2	7.0
1940	6.6	6.2	6.5	6.3	6.2	6.9	6.8	6.8	6.6	5.8	6.6	6.2	6.0	6.3	5.4

Glossary

Accent lighting–Fixtures or directional beams of light arranged to bring attention to an object or area.

Acoustical material–A material fabricated for the sole purpose of absorbing sound.

Addition–An expansion to an existing structure generally in the form of a room, floor(s) or wing(s). An increase in the floor area or volume of a structure.

Aggregate–Materials such as sand, gravel, stone, vermiculite, perlite and fly ash (slag) which are essential components in the production of concrete, mortar or plaster.

Air conditioning system–An air treatment to control the temperature, humidity and cleanliness of air and to provide for its distribution throughout the structure.

Air curtain–A stream of air directed downward to prevent the loss of hot or cool air from the structure and inhibit the entrance of dust and insects. Generally installed at loading platforms.

Anodized aluminum–Aluminum treated by an electrolytic process to produce an oxide film that is corrosion resistant.

Arcade–A covered passageway between buildings often with shops and offices on one or both sides.

Ashlar–A square-cut building stone or a wall constructed of cut stone.

Asphalt paper–A sheet paper material either coated or saturated with asphalt to increase its strength and resistance to water.

Asphalt shingles–Shingles manufactured from saturated roofing felts, coated with asphalt and topped with mineral granules to prevent weathering.

Assessment ratio–The ratio between the market value and assessed valuation of a property.

Back-up–That portion of a masonry wall behind the exterior facing-usually load bearing.

Balcony–A platform projecting from a building either supported from below or cantilevered. Usually protected by railing.

Bay–Structural component consisting of beams and columns occurring consistently throughout the structure.

Beam–A horizontal structural framing member that transfers loads to vertical members (columns) or bearing walls.

Bid–An offer to perform work described in a contract at a specified price.

Board foot–A unit or measure equal in volume to a board one foot long, one foot wide and one inch thick.

Booster pump–A supplemental pump installed to increase or maintain adequate pressure in the system.

Bowstring roof–A roof supported by trusses fabricated in the shape of a bow and tied together by a straight member.

Brick veneer–A facing of brick laid against, but not bonded to, a wall.

Bridging–A method of bracing joists for stiffness, stability and load distribution.

Broom finish–A method of finishing concrete by lightly dragging a broom over freshly placed concrete.

Built-up roofing–Installed on flat or extremely low-pitched roofs, a roof covering composed of plies or laminations of saturated roofing felts alternated with layers of coal tar pitch or asphalt and surfaced with a layer of gravel or slag in a thick coat of asphalt and finished with a capping sheet.

Caisson–A watertight chamber to expedite work on foundations or structure below water level.

Cantilever–A structural member supported only at one end.

Carport–An automobile shelter having one or more sides open to the weather.

Casement window–A vertical opening window having one side fixed.

Caulking–A resilient compound generally having a silicone or rubber base used to prevent infiltration of water or outside air.

Cavity wall–An exterior wall usually of masonry having an inner and outer wall separated by a continuous air space for thermal insulation.

Cellular concrete–A lightweight concrete consisting of cement mixed with gas-producing materials which, in turn, forms bubbles resulting in good insulating qualities.

Chattel mortgage–A secured interest in a property as collateral for payment of a note.

Clapboard–A wood siding used as exterior covering in frame construction. It is applied horizontally with grain running lengthwise. The thickest section of the board is on the bottom.

Clean room–An environmentally controlled room usually found in medical facilities or precision manufacturing spaces where it is essential to eliminate dust, lint or pathogens.

Close studding–A method of construction whereby studs are spaced close together and the intervening spaces are plastered.

Cluster housing–A closely grouped series of houses resulting in a high density land use.

Cofferdam–A watertight enclosure used for foundation construction in waterfront areas. Water is pumped out of the cofferdam allowing free access to the work area.

Collar joint–Joint between the collar beam and roof rafters.

Column–A vertical structural member supporting horizontal members (beams) along the direction of its longitudinal axis.

Combination door (windows)–Door (windows) having interchangeable screens and glass for seasonal use.

Common area–Spaces either inside or outside the building designated for use by the occupant of the building but not the general public.

Common wall–A wall used jointly by two dwelling units.

Compound wall–A wall constructed of more than one material.

Concrete–A composite material consisting of sand, coarse aggregate (gravel, stone or slag), cement and water that when mixed and allowed to harden forms a hard stone-like material.

Concrete floor hardener–A mixture of chemicals applied to the surface of concrete to produce a dense, wear-resistant bearing surface.

Conduit–A tube or pipe used to protect electric wiring.

Coping–The top cover or capping on a wall.

Craneway–Steel or concrete column and beam supports and rails on which a crane travels.

Curtain wall–A non-bearing exterior wall not supported by beams or girders of the steel frame.

Dampproofing–Coating of a surface to prevent the passage of water or moisture.

Dead load–Total weight of all structural components plus permanently attached fixtures and equipment.

Decibel–Unit of acoustical measurement.

Decorative block–A concrete masonry unit having a special treatment on its face.

Deed restriction–A restriction on the use of a property as set forth in the deed.

Depreciation–A loss in property value caused by physical aging, functional or economic obsolescence.

Direct heating–Heating of spaces by means of exposed heated surfaces (stove, fire, radiators, etc.).

Distributed load–A load that acts evenly over a structural member.

Dock bumper–A resilient material attached to a loading dock to absorb the impact of trucks backing in.

Dome–A curved roof shape spanning an area.

Dormer–A structure projecting from a sloping roof.

Double-hung window–A window having two vertical sliding sashes-one covering the upper section and one covering the lower.

Downspout–A vertical pipe for carrying rain water from the roof to the ground.

Glossary

Drain tile–A hollow tile used to drain water-soaked soil.

Dressed size–In lumber about 3/8" - 1/2" less than nominal size after planing and sawing.

Drop panel–The depressed surface on the bottom side of a flat concrete slab which surrounds a column.

Dry pipe sprinkler system–A sprinkler system that is activated with water only in case of fire. Used in areas susceptible to freezing or to avoid the hazards of leaking pipes.

Drywall–An interior wall constructed of gypsum board, plywood or wood paneling. No water is required for application.

Duct–Pipe for transmitting warm or cold air. A pipe containing electrical cable or wires.

Dwelling–A structure designed as living quarters for one or more families.

Easement–A right of way or free access over land owned by another.

Eave–The lower edge of a sloping roof that overhangs the side wall of the structure.

Economic rent–That rent on a property sufficient to pay all operating costs exclusive of services and utilities.

Economic life–The term during which a structure is expected to be profitable. Generally shorter than the physical life of the structure.

Economic obsolescence–Loss in value due to unfavorable economic influences occurring from outside the structure itself.

Effective age–The age a structure appears to be based on observed physical condition determined by degree of maintenance and repair.

Elevator–A mechanism for vertical transport of personnel or freight equipped with car or platform.

Ell–A secondary wing or addition to a structure at right angles to the main structure.

Eminent domain–The right of the state to take private property for public use.

Envelope–The shape of a building indicating volume.

Equity–Value of an owner's interest calculated by subtracting outstanding mortgages and expenses from the value of the property.

Escheat–The assumption of ownership by the state, of property whose owner cannot be determined.

Facade–The exterior face of a building sometimes decorated with elaborate detail.

Face brick–Brick manufactured to present an attractive appearance.

Facing block–A concrete masonry unit having a decorative exterior finish.

Feasibility study–A detailed evaluation of a proposed project to determine its financial potential.

Felt paper–Paper sheathing used on walls as an insulator and to prevent infiltration of moisture.

Fenestration–Pertaining to the density and arrangement of windows in a structure.

Fiberboard–A building material composed of wood fiber compressed with a binding agent, produced in sheet form.

Fiberglass–Fine spun filaments of glass processed into various densities to produce thermal and acoustical insulation.

Field house–A long structure used for indoor athletic events.

Finish floor–The top or wearing surface of the floor system.

Fireproofing–The use of fire-resistant materials for the protection of structural members to ensure structural integrity in the event of fire.

Fire stop–A material or member used to seal an opening to prevent the spread of fire.

Flashing–A thin, impervious material such as copper or sheet metal used to prevent air or water penetration.

Float finish–A concrete finish accomplished by using a flat tool with a handle on the back.

Floor drain–An opening installed in a floor for removing excess water into a plumbing system.

Floor load–The live load the floor system has been designed for and which may be applied safely.

Flue–A heat-resistant enclosed passage in a chimney to remove gaseous products of combustion from a fireplace or boiler.

Footing–The lowest portion of the foundation wall that transmits the load directly to the soil.

Foundation–Below grade wall system that supports the structure.

Foyer–A portion of the structure that serves as a transitional space between the interior and exterior.

Frontage–That portion of a lot that is placed facing a street, public way or body of water.

Frost action–The effects of freezing and thawing on materials and the resultant structural damage.

Functional obsolescence–An inadequacy caused by outmoded design, dated construction materials or over or undersized areas, all of which cause excessive operating costs.

Furring strips–Wood or metal channels used for attaching gypsum or metal lath to masonry walls as a finish or for leveling purposes.

Gambrel roof (mansard)–A roof system having two pitches on each side.

Garden apartment–Two or three story apartment building with common outside areas.

General contractor–The prime contractor responsible for work on the construction site.

Girder–A principal horizontal supporting member.

Girt–A horizontal framing member for adding rigid support to columns which also acts as a support for sheathing or siding.

Grade beam–That portion of the foundation system that directly supports the exterior walls of the building.

Gross area–The total enclosed floor area of a building.

Ground area–The area computed by the exterior dimensions of the structure.

Ground floor–The floor of a building in closest proximity to the ground.

Grout–A mortar containing a high water content used to fill joints and cavities in masonry work.

Gutter–A wood or sheet metal conduit set along the building eaves used to channel rainwater to leaders or downspouts.

Gunite–A concrete mix placed pneumatically.

Gypsum–Hydrated calcium sulphate used as both a retarder in Portland cement and as a major ingredient in plaster of Paris. Used in sheets as a substitute for plaster.

Hall–A passageway providing access to various parts of a building-a large room for entertainment or assembly.

Hangar–A structure for the storage or repair of aircraft.

Head room–The distance between the top of the finished floor to the bottom of the finished ceiling.

Hearth–The floor of a fireplace and the adjacent area of fireproof material.

Heat pump–A mechanical device for providing either heating or air conditioning.

Hip roof–A roof whose four sides meet at a common point with no gabled ends.

I-beam–A structural member having a cross-section resembling the letter "I".

Insulation–Material used to reduce the effects of heat, cold or sound.

Jack rafter–An unusually short rafter generally found in hip roofs.

Jalousie–Adjustable glass louvers which pivot simultaneously in a common frame.

Jamb–The vertical member on either side of a door frame or window frame.

Glossary

Joist–Parallel beams of timber, concrete or steel used to support floor and ceiling.

Junction box–A box that protects splices or joints in electrical wiring.

Kalamein door–Solid core wood doors clad with galvanized sheet metal.

Kip–A unit of weight equal to 1,000 pounds.

Lally column–A concrete-filled steel pipe used as a vertical support.

Laminated beam–A beam built up by gluing together several pieces of timber.

Landing–A platform between flights of stairs.

Lath–Strips of wood or metal used as a base for plaster.

Lead-lined door, sheetrock–Doors or sheetrock internally lined with sheet lead to provide protection from X-ray radiation.

Lean-to–A small shed or building addition with a single pitched roof attached to the exterior wall of the main building.

Lintel–A horizontal framing member used to carry a load over a wall opening.

Live load–The moving or movable load on a structure composed of furnishings, equipment or personnel weight.

Load bearing partition–A partition that supports a load in addition to its own weight.

Loading dock leveler–An adjustable platform for handling off-on loading of trucks.

Loft building–A commercial/industrial type building containing large, open unpartitioned floor areas.

Louver window–A window composed of a series of sloping, overlapping blades or slats that may be adjusted to admit varying degrees of air or light.

Main beam–The principal load bearing beam used to transmit loads directly to columns.

Mansard roof–A roof having a double pitch on all four sides, the lower level having the steeper pitch.

Mansion–An extremely large and imposing residence.

Masonry–The utilization of brick, stone, concrete or block for walls and other building components.

Mastic–A sealant or adhesive compound generally used as either a binding agent or floor finish.

Membrane fireproofing–A coating of metal lath and plaster to provide resistance to fire and heat.

Mesh reinforcement–An arrangement of wire tied or welded at their intersection to provide strength and resistance to cracking.

Metal lath–A diamond-shaped metallic base for plaster.

Mezzanine–A low story situated between two main floors.

Mill construction–A heavy timber construction that achieves fire resistance by using large wood structural members, noncombustible bearing and non-bearing walls and omitting concealed spaces under floors and roof.

Mixed occupancy–Two or more classes of occupancy in a single structure.

Molded brick–A specially shaped brick used for decorative purposes.

Monolithic concrete–Concrete poured in a continuous process so there are no joints.

Movable partition–A non-load bearing demountable partition that can be relocated and can be either ceiling height or partial height.

Moving walk–A continually moving horizontal passenger carrying device.

Multi-zone system–An air conditioning system that is capable of handling several individual zones simultaneously.

Net floor area–The usable, occupied area of a structure excluding stairwells, elevator shafts and wall thicknesses.

Non-combustible construction–Construction in which the walls, partitions and the structural members are of non-combustible materials.

Nurses call system–An electrically operated system for use by patients or personnel for summoning a nurse.

Occupancy rate–The number of persons per room, per dwelling unit, etc.

One-way joist connection–A framing system for floors and roofs in a concrete building consisting of a series of parallel joists supported by girders between columns.

Open web joist–A lightweight prefabricated metal chord truss.

Parapet–A low guarding wall at the point of a drop. That portion of the exterior wall that extends above the roof line.

Parging–A thin coat of mortar applied to a masonry wall used primarily for waterproofing purposes.

Parquet–Inlaid wood flooring set in a simple design.

Peaked roof–A roof of two or more slopes that rises to a peak.

Penthouse–A structure occupying usually half the roof area and used to house elevator, HVAC equipment, or other mechanical or electrical systems.

Pier–A column designed to support a concentrated load.

Pilaster–A column usually formed of the same material and integral with, but projecting from, a wall.

Pile–A concrete, wood or steel column usually less than 2' in diameter which is driven or otherwise introduced into the soil to carry a vertical load or provide vertical support.

Pitched roof–A roof having one or more surfaces with a pitch greater than 10°.

Plenum–The space between the suspended ceiling and the floor above.

Plywood–Structural wood made of three or more layers of veneer and bonded together with glue.

Porch–A structure attached to a building to provide shelter for an entranceway or to serve as a semi-enclosed space.

Post and beam framing–A structural framing system where beams rest on posts rather than bearing walls.

Post-tensioned concrete–Concrete that has the reinforcing tendons tensioned after the concrete has set.

Pre-cast concrete–Concrete structural components fabricated at a location other than in-place.

Pre-stressed concrete–Concrete that has the reinforcing tendons tensioned prior to the concrete setting.

Purlins–A horizontal structural member supporting the roof deck and resting on the trusses, girders, beams or rafters.

Rafters–Structural members supporting the roof deck and covering.

Resilient flooring–A manufactured interior floor covering material in either sheet or tile form, that is resilient.

Ridge–The horizontal line at the junction of two sloping roof edges.

Rigid frame–A structural framing system in which all columns and beams are rigidly connected; there are no hinged joints.

Roof drain–A drain designed to accept rainwater and discharge it into a leader or downspout.

Rough floor–A layer of boards or plywood nailed to the floor joists that serves as a base for the finished floor.

Sanitary sewer–A sewer line designed to carry only liquid or waterborne waste from the structure to a central treatment plant.

Sawtooth roof–Roof shape found primarily in industrial roofs that creates the appearance of the teeth of a saw.

Semi-rigid frame–A structural system wherein the columns and beams are so attached that there is some flexibility at the joints.

Septic tank–A covered tank in which waste matter is decomposed by natural bacterial action.

Service elevator–A combination passenger and freight elevator.

Sheathing–The first covering of exterior studs by boards, plywood or particle board.

Glossary

Shoring–Temporary bracing for structural components during construction.

Span–The horizontal distance between supports.

Specification–A detailed list of materials and requirements for construction of a building.

Storm sewer–A system of pipes used to carry rainwater or surface waters.

Stucco–A cement plaster used to cover exterior wall surfaces usually applied over a wood or metal lath base.

Stud–A vertical wooden structural component.

Subfloor–A system of boards, plywood or particle board laid over the floor joists to form a base.

Superstructure–That portion of the structure above the foundation or ground level.

Terrazzo–A durable floor finish made of small chips of colored stone or marble, embedded in concrete, then polished to a high sheen.

Tile–A thin piece of fired clay, stone or concrete used for floor, roof or wall finishes.

Unit cost–The cost per unit of measurement.

Vault–A room especially designed for storage.

Veneer–A thin surface layer covering a base of common material.

Vent–An opening serving as an outlet for air.

Wainscot–The lower portion of an interior wall whose surface differs from that of the upper wall.

Wall bearing construction–A structural system where the weight of the floors and roof are carried directly by the masonry walls rather than the structural framing system.

Waterproofing–Any of a number of materials applied to various surfaces to prevent the infiltration of water.

Weatherstrip–A thin strip of metal, wood or felt used to cover the joint between the door or window sash and the jamb, casing or sill to keep out air, dust, rain, etc.

Wing–A building section or addition projecting from the main structure.

X-ray protection–Lead encased in sheetrock or plaster to prevent the escape of radiation.

Abbreviations

A	Area Square Feet; Ampere	BTU	British Thermal Unit	Db.	Decibel
ABS	Acrylonitrile Butadiene Stryrene; Asbestos Bonded Steel	BTUH	BTU per Hour	Dbl., dbl.	Double
		BX	Interlocked Armored Cable	DC	Direct Current
A.C.	Alternating Current; Air-Conditioning; Asbestos Cement; Plywood Grade A & C	c	Conductivity	Demob.	Demobilization
		C	Hundred; Centigrade	d.f.u.	Drainage Fixture Units
		C/C	Center to Center	D.H.	Double Hung
		Cab.	Cabinet	DHW	Domestic Hot Water
A.C.I.	American Concrete Institute	Cair.	Air Tool Laborer	Diag.	Diagonal
A.C.T.	Acoustical Ceiling Tile	Calc	Calculated	Diam., dia.	Diameter
AD	Plywood, Grade A & D	Cap.	Capacity	Distrib.	Distribution
ADA	Americans with Disabilities Act	Carp.	Carpenter	Dk.	Deck
ADAAG	Americans with Disabilities Act Accessibility Guidelines	C.B.	Circuit Breaker	D.L.	Dead Load; Diesel
		C.C.A.	Chromate Copper Arsenate	Do.	Ditto
Addit., Add.	Additional	C.C.F.	Hundred Cubic Feet	Dp.	Depth
Adj.	Adjustable	cd	Candela	D.P.S.T.	Double Pole, Single Throw
af	Audio-frequency	cd/sf	Candela per Square Foot	Dr.	Driver
a.f.f.	Above Finished Floor	CD	Grade of Plywood Face & Back	Drink.	Drinking
A.G.A.	American Gas Association	CDX	Plywood, Grade C & D, exterior glue	D.S.	Double Strength
Agg.	Aggregate			D.S.A.	Double Strength A Grade
A.H.	Ampere Hours	Cefi.	Cement Finisher	D.S.B.	Double Strength B Grade
A hr.	Ampere-hour	Cem.	Cement	Dty.	Duty
A.H.U.	Air Handling Unit	CF	Hundred Feet	DWV	Drain Waste Vent
A.I.A.	American Institute of Architects	C.F.	Cubic Feet	DX	Deluxe White, Direct Expansion
AIC	Ampere Interrupting Capacity	CFM	Cubic Feet per Minute	dyn	Dyne
Allow.	Allowance	c.g.	Center of Gravity	e	Eccentricity
alt.	Altitude	CHW	Chilled Water; Commercial Hot Water	E	Equipment Only; East
Alum.	Aluminum			Ea.	Each
a.m.	Ante Meridiem	C.I.	Cast Iron	E.B.	Encased Burial
Amp.	Ampere, Amplifier	C.I.P.	Cast in Place	Econ.	Economy
Anod.	Anodized	Circ.	Circuit	EDP	Electronic Data Processing
Approx.	Approximate	C.L.	Carload Lot	EIFS	Exterior Insulation Finish System
Apt.	Apartment	Clab.	Common Laborer	E.D.R.	Equiv. Direct Radiation
Asb.	Asbestos	C.L.F.	Hundred Linear Feet	Eq.	Equation
A.S.B.C.	American Standard Building Code	CLF	Current Limiting Fuse	Elec.	Electrician; Electrical
Asbe.	Asbestos Worker	CLP	Cross Linked Polyethylene	Elev.	Elevator; Elevating
A.S.H.R.A.E.	American Society of Heating, Refrig. & AC Engineers	cm	Centimeter	EMT	Electrical Metallic Conduit; Thin Wall Conduit
		CMP	Corr. Metal Pipe		
A.S.M.E.	American Society of Mechanical Engineers	C.M.U.	Concrete Masonry Unit	Eng.	Engine
		Col.	Column	EPDM	Ethylene Propylene Diene Monomer
A.S.T.M.	American Society for Testing and Materials	C.O.	Clear Opening		
		CO_2	Carbon Dioxide	EPS	Expanded Polystyrene
Attchmt.	Attachment	Comb.	Combination	Eqhv.	Equip. Oper., Heavy
Avg.	Average	Compr.	Compressor	Eqlt.	Equip. Oper., Light
A.W.G.	American Wire Gauge	Conc.	Concrete	Eqmd.	Equip. Oper., Medium
A.W.W.A.	American Water Works Association	Cont.	Continuous; Continued	Eqmm.	Equip. Oper., Master Mechanic
Bbl.	Barrel	Corr.	Corrugated	Eqol.	Equip. Oper., Oilers
B. & B.	Grade B and Better; Balled & Burlapped	Cos	Cosine	Equip.	Equipment
		Cot	Cotangent	ERW	Electric Resistance Welded
B. & S.	Bell and Spigot	Cov.	Cover	Est.	Estimated
B. & W.	Black and White	CPA	Control Point Adjustment	esu	Electrostatic Units
b.c.c.	Body-centered Cubic	Cplg.	Coupling	E.W.	Each Way
B.C.Y.	Bank Cubic Yards	C.P.M.	Critical Path Method	EWT	Entering Water Temperature
BE	Bevel End	CPVC	Chlorinated Polyvinyl Chloride	Excav.	Excavation
B.F.	Board Feet	C.Pr.	Hundred Pair	Exp.	Expansion, Exposure
Bg. cem.	Bag of Cement	CRC	Cold Rolled Channel	Ext.	Exterior
BHP	Boiler Horsepower; Brake Horsepower	Creos.	Creosote	Extru.	Extrusion
		Crpt.	Carpet & Linoleum Layer	f.	Fiber stress
B.I.	Black Iron	CRT	Cathode-ray Tube	F	Fahrenheit; Female; Fill
Bit.; Bitum.	Bituminous	CS	Carbon Steel	Fab.	Fabricated
Bk.	Backed	Csc	Cosecant	FBGS	Fiberglass
Bkrs.	Breakers	C.S.F.	Hundred Square Feet	F.C.	Footcandles
Bldg.	Building	CSI	Construction Specifications Institute	f.c.c.	Face-centered Cubic
Blk.	Block			f'c.	Compressive Stress in Concrete; Extreme Compressive Stress
Bm.	Beam	C.T.	Current Transformer		
Boil.	Boilermaker	CTS	Copper Tube Size	F.E.	Front End
B.P.M.	Blows per Minute	Cu	Cubic	FEP	Fluorinated Ethylene Propylene (Teflon)
BR	Bedroom	Cu. Ft.	Cubic Foot		
Brg.	Bearing	cw	Continuous Wave	F.F.E.	Furniture, Fixtures, & Equipment
Brhe.	Bricklayer Helper	C.W.	Cool White; Cold Water	F.G.	Flat Grain
Bric.	Bricklayer	Cwt.	100 Pounds	F.H.A.	Federal Housing Administration
Brk.	Brick	C.W.X.	Cool White Deluxe	Fig.	Figure
Brng.	Bearing	C.Y.	Cubic Yard (27 cubic feet)	Fin.	Finished
Brs.	Brass	C.Y./Hr.	Cubic Yard per Hour	Fixt.	Fixture
Brz.	Bronze	Cyl.	Cylinder	Fl. Oz.	Fluid Ounces
Bsn.	Basin	d	Penny (nail size)	Flr.	Floor
Btr.	Better	D	Deep; Depth; Discharge		
		Dis.; Disch.	Discharge		

446

Abbreviations

F.M.	Frequency Modulation; Factory Mutual	Incl.	Included; Including	Maint.	Maintenance
Fmg.	Framing	Int.	Interior	Marb.	Marble Setter
Fndtn.	Foundation	Inst.	Installation	Mat; Mat'l.	Material
Fori.	Foreman, Inside	Insul.	Insulation	Max.	Maximum
Foro.	Foreman, Outside	I.P.	Iron Pipe	MBF	Thousand Board Feet
Fount.	Fountain	I.P.S.	Iron Pipe Size	MBH	Thousand BTU's per hr.
FPM, fpm	Feet per Minute	I.P.T.	Iron Pipe Threaded	MC	Metal Clad Cable
FPT	Female Pipe Thread	I.W.	Indirect Waste	M.C.F.	Thousand Cubic Feet
Fr.	Frame	J	Joule	M.C.F.M.	Thousand Cubic Feet per Minute
F.R.	Fire Rating	J.I.C.	Joint Industrial Council	M.C.M.	Thousand Circular Mils
FRK	Foil Reinforced Kraft	K	Thousand; Thousand Pounds; Heavy Wall Copper Tubing	M.C.P.	Motor Circuit Protector
FRP	Fiberglass Reinforced Plastic	K.A.H.	Thousand Amp. Hours	MD	Medium Duty
FS	Forged Steel	KCMIL	Thousand Circular Mils	M.D.O.	Medium Density Overlaid
FSC	Cast Body; Cast Switch Box	KD	Knock Down	Med.	Medium
Ft.	Foot; Feet	K.D.A.T.	Kiln Dried After Treatment	met.	Metric
Ftng.	Fitting	kg	Kilogram	MF	Thousand Feet
Ftg.	Footing	kG	Kilogauss	M.F.B.M.	Thousand Feet Board Measure
Ft. Lb.	Foot Pound	kgf	Kilogram Force	Mfg.	Manufacturing
Furn.	Furniture	kHz	Kilohertz	Mfrs.	Manufacturers
FVNR	Full Voltage Non-Reversing	Kip.	1000 Pounds	mg	Milligram
FXM	Female by Male	KJ	Kiljoule	MGD	Million Gallons per Day
Fy.	Minimum Yield Stress of Steel	K.L.	Effective Length Factor	MGPH	Thousand Gallons per Hour
g	Gram	K.L.F.	Kips per Linear Foot	MH, M.H.	Manhole; Metal Halide; Man-Hour
G	Gauss	Km	Kilometer	MHz	Megahertz
Ga.	Gauge	K.S.F.	Kips per Square Foot	Mi.	Mile
Gal.	Gallon	K.S.I.	Kips per Square Inch	MI	Malleable Iron; Mineral Insulated
Gal./Min.	Gallon per Minute	K.V.	Kilovolt	mm	Millimeter
Galv.	Galvanized	K.V.A.	Kilovolt Ampere	Mill.	Millwright
Gen.	General	K.V.A.R.	Kilovar (Reactance)	Min., min.	Minimum, minute
G.F.I.	Ground Fault Interrupter	KW	Kilowatt	Misc.	Miscellaneous
Glaz.	Glazier	KWh	Kilowatt-hour	ml	Milliliter
GPD	Gallons per Day	L	Labor Only; Length; Long; Medium Wall Copper Tubing; Liters	M.L.F.	Thousand Linear Feet
GPH	Gallons per Hour			Mo.	Month
GPM	Gallons per Minute			Mobil.	Mobilization
GR	Grade			Mog.	Mogul Base
Gran.	Granular	Lab.	Labor	MPH	Miles per Hour
Grnd.	Ground	lat	Latitude	MPT	Male Pipe Thread
Gyp.	Gypsum	Lath.	Lather	MRT	Mile Round Trip
H	High; High Strength Bar Joist; Henry	Lav.	Lavatory	ms	Millisecond
		lb.; #; lbs.	Pound, pounds	M.S.F.	Thousand Square Feet
H.C.	High Capacity	L.B.	Load Bearing; L Conduit Body	Mstz.	Mosaic & Terrazzo Worker
H.D.	Heavy Duty; High Density	L. & E.	Labor & Equipment	M.S.Y.	Thousand Square Yards
H.D.O.	High Density Overlaid	lb./hr.	Pounds per Hour	Mtd.	Mounted
Hdr.	Header	lb./L.F.	Pounds per Linear Foot	Mthe.	Mosaic & Terrazzo Helper
Hdwe.	Hardware	lbf/sq.in.	Pound-force per Square Inch	Mtng.	Mounting
Help.	Helpers Average	L.C.L.	Less than Carload Lot	Mult.	Multi; Multiply
HEPA	High Efficiency Particulate Air Filter	Ld.	Load	M.V.A.	Million Volt Amperes
		LE	Lead Equivalent	M.V.A.R.	Million Volt Amperes Reactance
Hg	Mercury	L.F.	Linear Foot	MV	Megavolt
HIC	High Interrupting Capacity	Lg.	Long; Length; Large	MW	Megawatt
H.O.	High Output	L & H	Light and Heat	MXM	Male by Male
Horiz.	Horizontal	L.H.	Long Span High Strength Bar Joist	MYD	Thousand Yards
HP	Handicapped Person	L.J.	Long Span Standard Strength Bar Joist	N	Natural; North
H.P.	Horsepower; High Pressure			nA	Nanoampere
H.P.F.	High Power Factor	L.L.	Live Load	NA	Not Available; Not Applicable
Hr.	Hour	L.L.D.	Lamp Lumen Depreciation	N.B.C.	National Building Code
Hrs./Day	Hours per Day	lm	Lumen	NC	Normally Closed
HSC	High Short Circuit	lm/sf	Lumen per Square Foot	N.E.M.A.	National Electrical Manufacturers Assoc.
Ht.	Height	lm/W	Lumen per Watt		
Htg.	Heating	L.O.A.	Length Over All	NEHB	Bolted Circuit Breaker to 600V.
Htrs.	Heaters	log	Logarithm	N.L.B.	Non-Load-Bearing
HVAC	Heating, Ventilation & Air-Conditioning	L.P.	Liquefied Petroleum; Low Pressure	MN	Non-Metallic Cable
		L.P.F.	Low Power Factor	nm	Nanometer
Hvy.	Heavy	LR	Long Radius	No.	Number
HW	Hot Water	L.S.	Lump Sum	NO	Normally Open
Hyd.;Hydr.	Hydraulic	Lt.	Light	N.O.C.	Not Otherwise Classified
Hz.	Hertz (cycles)	Lt. Ga.	Light Gauge	Nose.	Nosing
I.	Moment of Inertia	L.T.L.	Less than Truckload Lot	N.P.T.	National Pipe Thread
I.C.	Interrupting Capacity	Lt. Wt.	Lightweight	NQOD	Combination Plug-on/Bolt on Circuit Breaker to 240V.
ID	Inside Diameter	L.V.	Low Voltage		
I.D.	Inside Dimension; Identification	m	Meters	N.R.C.	Noise Reduction Coefficient
I.F.	Inside Frosted	M	Thousand; Material; Male; Light Wall Copper Tubing	N.R.S.	Non Rising Stem
I.M.C.	Intermediate Metal Conduit			ns	Nanosecond
In.	Inch	m/hr; M.H.	Man-hour	nW	Nanowatt
Incan.	Incandescent	mA	Milliampere	OB	Opposing Blade
		Mach.	Machine	OC; OC	On Center
		Mag. Str.	Magnetic Starter		

447

Abbreviations

OD	Outside Diameter		Residential Hot Water	T.E.M.	Transmission Electron Microscopy
O.D.	Outside Dimension	rms	Root Mean Square	TFE	Tetrafluoroethylene (Teflon)
ODS	Overhead Distribution System	Rnd.	Round	T. & G.	Tongue & Groove; Tar & Gravel
O & P	Overhead and Profit	Rodm.	Rodman		
Oper.	Operator	Rofc.	Roofer, Composition	Th.; Thk.	Thick
Opng.	Opening	Rofp.	Roofer, Precast	Thn.	Thin
Orna.	Ornamental	Rohe.	Roofer Helpers (Composition)	Thrded	Threaded
OSB	Oriented Strand Board	Rots.	Roofer, Tile & Slate	Tilf.	Tile Layer, Floor
O. S. & Y.	Outside Screw and Yoke	R.O.W.	Right of Way	Tilh.	Tile Layer, Helper
Ovhd.	Overhead	RPM	Revolutions per Minute	THW.	Insulated Strand Wire
OWG	Oil, Water or Gas	R.R.	Direct Burial Feeder Conduit	THWN; THHN	Nylon Jacketed Wire
Oz.	Ounce	R.S.	Rapid Start	T.L.	Truckload
P.	Pole; Applied Load; Projection	Rsr	Riser	Tot.	Total
p.	Page	RT	Round Trip	T.S.	Trigger Start
Pape.	Paperhanger	S.	Suction; Single Entrance; South	Tr.	Trade
P.A.P.R.	Powered Air Purifying Respirator	Scaf.	Scaffold	Transf.	Transformer
PAR	Weatherproof Reflector	Sch.; Sched.	Schedule	Trhv.	Truck Driver, Heavy
Part.	Partition	S.C.R.	Modular Brick	Trlr	Trailer
Pc.	Piece	S.D.	Sound Deadening	Trlt.	Truck Driver, Light
P.C.	Portland Cement; Power Connector	S.D.R.	Standard Dimension Ratio	TV	Television
P.C.F.	Pounds per Cubic Foot	S.E.	Surfaced Edge	T.W.	Thermoplastic Water Resistant Wire
P.C.M.	Phase Contract Microscopy	Sel.	Select		
P.E.	Professional Engineer; Porcelain Enamel; Polyethylene; Plain End	S.E.R., S.E.U.	Service Entrance Cable	Typ.	Typical
		S.F.	Square Foot	UCI	Uniform Construction Index
		S.F.C.A.	Square Foot Contact Area	UF	Underground Feeder
Perf.	Perforated	S.F.G.	Square Foot of Ground	U.H.F.	Ultra High Frequency
Ph.	Phase	S.F. Hor.	Square Foot Horizontal	U.L.	Underwriters Laboratory
P.I.	Pressure Injected	S.F.R.	Square Feet of Radiation	Unfin.	Unfinished
Pile.	Pile Driver	S.F. Shlf.	Square Foot of Shelf	URD	Underground Residential Distribution
Pkg.	Package	S4S	Surface 4 Sides		
Pl.	Plate	Shee.	Sheet Metal Worker	V	Volt
Plah.	Plasterer Helper	Sin.	Sine	V.A.	Volt Amperes
Plas.	Plasterer	Skwk.	Skilled Worker	V.C.T.	Vinyl Composition Tile
Pluh.	Plumbers Helper	SL	Saran Lined	VAV	Variable Air Volume
Plum.	Plumber	S.L.	Slimline	VC	Veneer Core
Ply.	Plywood	Sldr.	Solder	Vent.	Ventilation
p.m.	Post Meridiem	S.N.	Solid Neutral	Vert.	Vertical
Pord.	Painter, Ordinary	S.P.	Static Pressure; Single Pole; Self-Propelled	V.F.	Vinyl Faced
pp	Pages			V.G.	Vertical Grain
PP; PPL	Polypropylene	Spri.	Sprinkler Installer	V.H.F.	Very High Frequency
P.P.M.	Parts per Million	Sq.	Square; 100 Square Feet	VHO	Very High Output
Pr.	Pair	S.P.D.T.	Single Pole, Double Throw	Vib.	Vibrating
Prefab.	Prefabricated	SPF	Spruce Pine Fir	V.L.F.	Vertical Linear Foot
Prefin.	Prefinished	S.P.S.T.	Single Pole, Single Throw	Vol.	Volume
Prop.	Propelled	SPT	Standard Pipe Thread	W	Wire; Watt; Wide; West
PSF; psf	Pounds per Square Foot	Sq. Hd.	Square Head	w/	With
PSI; psi	Pounds per Square Inch	Sq. In.	Square Inch	W.C.	Water Column; Water Closet
PSIG	Pounds per Square Inch Gauge	S.S.	Single Strength; Stainless Steel	W.F.	Wide Flange
PSP	Plastic Sewer Pipe	S.S.B.	Single Strength B Grade	W.G.	Water Gauge
Pspr.	Painter, Spray	Sswk.	Structural Steel Worker	Wldg.	Welding
Psst.	Painter, Structural Steel	Sswl.	Structural Steel Welder	W. Mile	Wire Mile
P.T.	Potential Transformer	St.; Stl.	Steel	W.R.	Water Resistant
P. & T.	Pressure & Temperature	S.T.C.	Sound Transmission Coefficient	Wrck.	Wrecker
Ptd.	Painted	Std.	Standard	W.S.P.	Water, Steam, Petroleum
Ptns.	Partitions	STP	Standard Temperature & Pressure	WT., Wt.	Weight
Pu	Ultimate Load	Stpi.	Steamfitter, Pipefitter	WWF	Welded Wire Fabric
PVC	Polyvinyl Chloride	Str.	Strength; Starter; Straight	XFMR	Transformer
Pvmt.	Pavement	Strd	Stranded	XHD	Extra Heavy Duty
Pwr.	Power	Struct.	Structural	XHHW; XLPE	Cross-Linked Polyethylene Wire Insulation
Q	Quantity Heat Flow	Sty.	Story		
Quan.; Qty.	Quantity	Subj.	Subject	Y	Wye
Q.C.	Quick Coupling	Subs.	Subcontractors	yd	Yard
r	Radius of Gyration	Surf.	Surface	yr	Year
R	Resistance	Sw.	Switch	Δ	Delta
R.C.P.	Reinforced Concrete Pipe	Swbd.	Switchboard	%	Percent
Reconn.	Reconnect	S.Y.	Square Yard	~	Approximately
Rect.	Rectangle	Syn.	Synthetic	Ø	Phase
Reg.	Regular	S.Y.P.	Southern Yellow Pine	@	At
Reinf.	Reinforced	Sys.	System	#	Pound; Number
Rem.	Remove	t.	Thickness	<	Less Than
Repl.	Replace	T	Temperature; Ton	>	Greater Than
Req'd.	Required	Tan	Tangent		
Res.	Resistant	T.C.	Terra Cotta		
Resi.	Residential	T & C	Threaded and Coupled		
Rgh.	Rough	T.D.	Temperature Difference		
R.H.W.	Rubber, Heat & Water Resistant;	TDD, (or TTY)	Telecommunications Display Device		

FORMS

Means Forms
RESIDENTIAL COST ESTIMATE

OWNER'S NAME: _____ APPRAISER: _____

RESIDENCE ADDRESS: _____ PROJECT: _____

CITY, STATE, ZIP CODE: _____ DATE: _____

CLASS OF CONSTRUCTION	RESIDENCE TYPE	CONFIGURATION	EXTERIOR WALL SYSTEM
☐ ECONOMY	☐ 1 STORY	☐ DETACHED	☐ WOOD SIDING - WOOD FRAME
☐ AVERAGE	☐ 1-1/2 STORY	☐ TOWN/ROW HOUSE	☐ BRICK VENEER - WOOD FRAME
☐ CUSTOM	☐ 2 STORY	☐ SEMI-DETACHED	☐ STUCCO ON WOOD FRAME
☐ LUXURY	☐ 2-1/2 STORY		☐ PAINTED CONCRETE BLOCK
	☐ 3 STORY	OCCUPANCY	☐ SOLID MASONRY (AVERAGE & CUSTOM)
	☐ BI-LEVEL	☐ ONE FAMILY	☐ STONE VENEER - WOOD FRAME
	☐ TRI-LEVEL	☐ TWO FAMILY	☐ SOLID BRICK (LUXURY)
		☐ THREE FAMILY	☐ SOLID STONE (LUXURY)
		☐ OTHER _____	

* LIVING AREA (Main Building)
First Level	_____ S.F.
Second level	_____ S.F.
Third Level	_____ S.F.
Total	_____ S.F.

* LIVING AREA (Wing or Ell) ( )
First Level	_____ S.F.
Second level	_____ S.F.
Third Level	_____ S.F.
Total	_____ S.F.

* LIVING AREA (WING or ELL) ( )
First Level	_____ S.F.
Second level	_____ S.F.
Third Level	_____ S.F.
Total	_____ S.F.

* Basement Area is not part of living area.

MAIN BUILDING	COSTS PER S.F. LIVING AREA
Cost per Square Foot of Living Area, from Page _____	$
Basement Addition: _____ % Finished, _____ % Unfinished	+
Roof Cover Adjustment: _____ Type, Page _____ (Add or Deduct)	( )
Central Air Conditioning: ☐ Separate Ducts ☐ Heating Ducts, Page _____	+
Heating System Adjustment: _____ Type, Page _____ (Add or Deduct)	( )
Main Building: Adjusted Cost per S.F. of Living Area	$

MAIN BUILDING TOTAL COST $ _____ /S.F. x _____ S.F. x _____ = $ _____
Cost per S.F. Living Area   Living Area   Town/Row House Multiplier   TOTAL COST
(Use 1 for Detached)

WING OR ELL ( ) _____ STORY	COSTS PER S.F. LIVING AREA
Cost per Square Foot of Living Area, from Page _____	$
Basement Addition: _____ % Finished, _____ % Unfinished	+
Roof Cover Adjustment: _____ Type, Page _____ (Add or Deduct)	( )
Central Air Conditioning: ☐ Separate Ducts ☐ Heating Ducts, Page _____	+
Heating System Adjustment: _____ Type, Page _____ (Add or Deduct)	( )
Wing or Ell ( ): Adjusted Cost per S.F. of Living Area	$

WING OR ELL ( ) TOTAL COST $ _____ /S.F. x _____ S.F. x _____ = $ _____
Cost per S.F. Living Area   Living Area   TOTAL COST

WING OR ELL ( ) _____ STORY	COSTS PER S.F. LIVING AREA
Cost per Square Foot of Living Area, from Page _____	$
Basement Addition: _____ % Finished, _____ % Unfinished	+
Roof Cover Adjustment: _____ Type, Page _____ (Add or Deduct)	( )
Central Air Conditioning: ☐ Separate Ducts ☐ Heating Ducts, Page _____	+
Heating System Adjustment: _____ Type, Page _____ (Add or Deduct)	( )
Wing or Ell ( ): Adjusted Cost per S.F. of Living Area	$

WING OR ELL ( ) TOTAL COST $ _____ /S.F. x _____ S.F. x _____ = $ _____
Cost per S.F. Living Area   Living Area   TOTAL COST

TOTAL THIS PAGE _____

Page 1 of 2

449

FORMS

Means Forms
RESIDENTIAL COST ESTIMATE

		QUANTITY	UNIT COST	$
Total Page 1				
Additional Bathrooms: _____ Full _____ Half				
Finished Attic: _____ Ft. x _____ Ft.		S.F.		+
Breezeway: ☐ Open ☐ Enclosed _____ Ft. x _____ Ft.		S.F.		+
Covered Porch: ☐ Open ☐ Enclosed _____ Ft. x _____ Ft.		S.F.		+
Fireplace: ☐ Interior Chimney ☐ Exterior Chimney				
☐ No. of Flues ☐ Additional Fireplaces				+
Appliances:				+
Kitchen Cabinets Adjustments: (±)				
☐ Garage ☐ Carport: _____ Car(s) Description _____ (±)				
Miscellaneous:				+

ADJUSTED TOTAL BUILDING COST $ _____

REPLACEMENT COST

ADJUSTED TOTAL BUILDING COST	$ _____
Site Improvements	
(A) Paving & Sidewalks	$ _____
(B) Landscaping	$ _____
(C) Fences	$ _____
(D) Swimming Pools	$ _____
(E) Miscellaneous	$ _____
TOTAL	$ _____
Location Factor	x _____
Location Replacement Cost	$ _____
Depreciation	- $ _____
LOCAL DEPRECIATED COST	$ _____

INSURANCE COST

ADJUSTED TOTAL BUILDING COST	$ _____
Insurance Exclusions	
(A) Footings, Site work, Underground Piping	- $ _____
(B) Architects Fees	- $ _____
Total Building Cost Less Exclusion	$ _____
Location Factor	x _____
LOCAL INSURABLE REPLACEMENT COST	$ _____

SKETCH AND ADDITIONAL CALCULATIONS

FORMS

CII APPRAISAL

5. DATE: _____

1. SUBJECT PROPERTY: _____   6. APPRAISER: _____
2. BUILDING: _____
3. ADDRESS: _____
4. BUILDING USE: _____   7. YEAR BUILT: _____

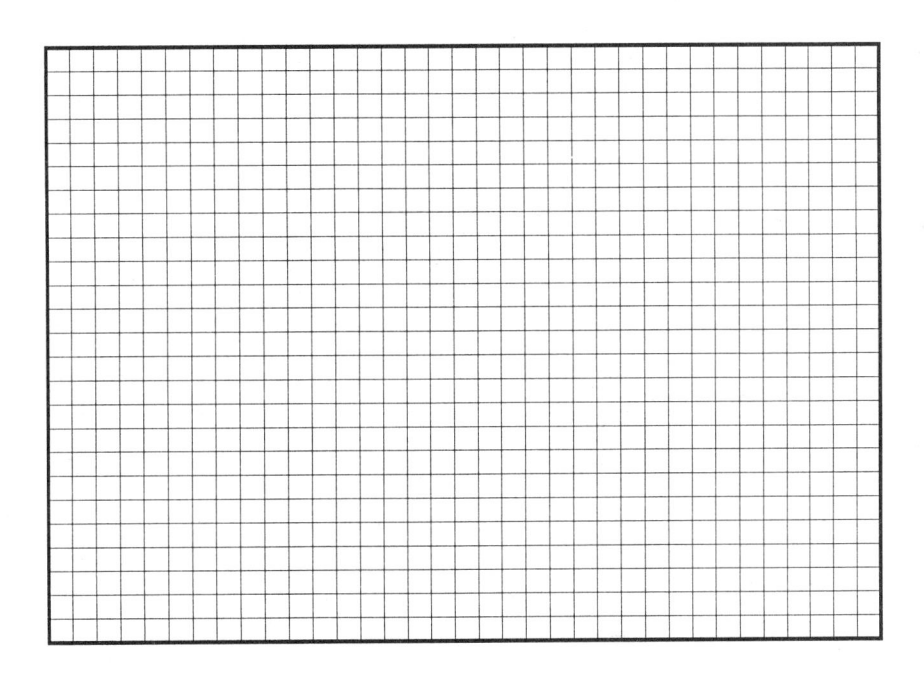

8. EXTERIOR WALL CONSTRUCTION: _____
9. FRAME: _____
10. GROUND FLOOR AREA: _____ S.F. 11. GROSS FLOOR AREA (EXCL. BASEMENT): _____ S.F.
12. NUMBER OF STORIES: _____   13. STORY HEIGHT: _____
14. PERIMETER: _____ L.F.   15. BASEMENT AREA: _____ S.F.
16. GENERAL COMMENTS: _____
_____
_____
_____
_____
_____
_____
_____

FORMS

| SQUARE FOOT AREA *(EXCLUDING BASEMENT)* FROM ITEM 11 _____ S.F. _____ |
| PERIMETER FROM ITEM 14 _____ L.F. _____ |
| ITEM 17 - MODEL SQUARE FOOT COSTS *(FROM MODEL SUB-TOTAL)* $ _____ |

Field Description & Calculation Section

NO.	SYSTEM/COMPONENT	DESCRIPTION	UNIT	UNIT COST	NEW S.F. COST	MODEL S.F. COST	+/- CHANGE
1.0	**FOUNDATION**						
.1	Bldg. Excavation	Depth					
		Area	S.F.				
	Trench Excavation	Depth Width Length	L.F.				
.2	Footings: Strip Width	L.F. Length	L.F.				
	Footings: Spread	Bay Size	S.F. Gnd.				
		Other					
.3	Foundation Wall	Material					
		Height Thickness	L.F. Wall				
.4	Slab on Grade	Material Thickness	S.F. Gnd.				
3.0	**SUPERSTRUCTURE**						
.1	Elevated Floors	Bay Size Material					
		Member Size	S.F. Floor				
.2	Roof Structure	Material					
		Bay Size					
		Member Size	S.F. Roof				
.3	Column/Bearing Wall						
	Column	Size	L.F. Col				
		Bay Size					
	Bearing Wall	Material					
		Thickness	L.F. Wall				
.4	Fireproofing	Material	S.F. Surf.				
4.0	**EXTERIOR WALLS**	Building Ht. _____ Ft.					
.1	Outer Wall	Material					
		Thickness	S.F. Wall				
.2	Interior Finish		S.F. Wall				
.3	Windows	Type					
		% of Wall	S.F. Wind.				
.4	Doors	Type					
		Density	S.F. Door				
5.0	**ROOFING**						
.1	Roof Covering	Material	S.F. Roof				

FORMS

NO.	SYSTEM/COMPONENT		DESCRIPTION	UNIT	UNIT COST	NEW S.F. COST	MODEL S.F. COST	+/- CHANGE
6.0	INTERIOR CONSTRUCTION							
.1	Partitions	Material		S.F. Part.				
		Density						
.2	Doors	Type		S.F. Door				
		Density						
.3	Floor Finish	Material		S.F. Floor				
		% of Floor						
.4	Ceiling Finish	Material		S.F. Ceil.				
		% of Ceiling						
.5	Partition Finish	Material		S.F. Surf.				
7.0	CONVEYING SYSTEMS							
.1	Vertical	Type		Each				
		Capacity						
		Stops						
8.0	MECHANICAL SYSTEMS							
.1	Plumbing			Each				
.2	Fire Protection			S.F.				
.3	Heating	Type		S.F.				
.4	Air Conditioning			S.F.				
9.0	ELECTRICAL							
.1	Lighting			S.F.				
10.0	MISCELLANEOUS							
11.0	SPECIALTIES							

Total Change ____

ITEM
18 Adjusted S.F. cost $ _____ item 17 +/- changes
19 Building area - from item 11 _____ S.F. x adjusted S.F. cost .... $ ____
20 Basement area - from item 15 _____ S.F. x S.F. cost $ .... $ ____
21 Base building sub-total - item 20 + item 19 .... $ ____
22 Miscellaneous addition (quality, etc.) .... $ ____
23 Sub-total - item 22 + item 21 .... $ ____
24 General conditions - 15 % of item 23 .... $ ____
25 Sub-total - item 24 + item 23 .... $ ____
26 Architect's fees _____ % of item 25 .... $ ____
27 Sub-total - item 26 + item 25 .... $ ____
28 Location modifier .... x ____
29 Local replacement cost - item 28 x item 27 .... $ ____
30 Depreciation _____ % of item 29 .... $ ____
31 Depreciated local replacement cost - item 29 less item 30 .... $ ____
32 Exclusions .... $ ____
33 Net depreciated replacement cost - item 31 less item 32 .... $ ____

FORMS

Base Cost per square foot floor area: *(from square foot table)*

Specify source: Page: _____, Model # _____, Area _____ S.F.,
Exterior wall _____, Frame _____

Adjustments for exterior wall variation:

Size Adjustment:
(Interpolate)

Height Adjustment: _____ + _____ = _____

Adjusted Base Cost per square foot:

Building Cost $ _____ x _____ = _____
                 Adjusted Base Cost   *Floor Area*
                 per square foot

Basement Cost $ _____ x _____ = _____
                 Basement Cost   *Basement Area*

Lump Sum Additions

TOTAL BUILDING COST *(Sum of above costs)*   _____

Modifications: *(complexity, workmanship, size)*   +/- _____ %   _____

Location Modifier: City _____ Date _____   x _____

Local cost of replacement   _____

Less depreciation: _____   - _____

Local cost of replacement less depreciation $ _____

Notes

Notes

Notes

Means Project Cost Report

By filling out and returning the Project Description, you can receive a discount of $20.00 off any one of the Means products advertised in the following pages. The cost information required includes all items marked (✔) except those where no costs occurred. The sum of all major items should equal the Total Project Cost.

$20.00 Discount per product for each report you submit.

DISCOUNT PRODUCTS AVAILABLE—FOR U.S. CUSTOMERS ONLY—STRICTLY CONFIDENTIAL

Project Description (No remodeling projects, please.)

✔ Type Building _____
✔ Location _____
   Capacity _____
✔ Frame _____
✔ Exterior _____
✔ Basement: full ☐ partial ☐ none ☐ crawl ☐
✔ Height in Stories _____
✔ Total Floor Area _____
   Ground Floor Area _____
✔ Volume in C.F. _____
   % Air Conditioned _____ Tons _____
   Comments _____

   Owner _____
   Architect _____
   General Contractor _____
✔ Bid Date _____
   Typical Bay Size _____
✔ Labor Force: _____ % Union _____ % Non-Union
✔ Project Description (Circle one number in each line)
    1. Economy   2. Average   3. Custom   4. Luxury
    1. Square   2. Rectangular   3. Irregular   4. Very Irregular

	✔ **Total Project Cost**			$
A	✔ **General Conditions**			$
B	✔ **Site Work**			$
BS	Site Clearing & Improvement			
BE	Excavation	(	C.Y.)	
BF	Caissons & Piling	(	L.F.)	
BU	Site Utilities			
BP	Roads & Walks Exterior Paving	(	S.Y.)	
C	✔ **Concrete**			$
C	Cast in Place	(	C.Y.)	
CP	Precast	(	S.F.)	
D	✔ **Masonry**			$
DB	Brick	(	M)	
DC	Block	(	M)	
DT	Tile	(	S.F.)	
DS	Stone	(	S.F.)	
E	✔ **Metals**			$
ES	Structural Steel	(	Tons)	
EM	Misc. & Ornamental Metals			
F	✔ **Wood & Plastics**			$
FR	Rough Carpentry	(	MBF)	
FF	Finish Carpentry			
FM	Architectural Millwork			
G	✔ **Thermal & Moisture Protection**			$
GW	Waterproofing-Dampproofing	(	S.F.)	
GN	Insulation	(	S.F.)	
GR	Roofing & Flashing	(	S.F.)	
GM	Metal Siding/Curtain Wall	(	S.F.)	
H	✔ **Doors and Windows**			$
HD	Doors	(	Ea.)	
HW	Windows	(	S.F.)	
HH	Finish Hardware			
HG	Glass & Glazing	(	S.F.)	
HS	Storefronts	(	S.F.)	

J	✔ **Finishes**			$
JL	Lath & Plaster	(	S.Y.)	
JD	Drywall	(	S.F.)	
JM	Tile & Marble	(	S.F.)	
JT	Terrazzo	(	S.F.)	
JA	Acoustical Treatment	(	S.F.)	
JC	Carpet	(	S.Y.)	
JF	Hard Surface Flooring	(	S.F.)	
JP	Painting & Wall Covering	(	S.F.)	
K	✔ **Specialties**			$
KB	Bathroom Partitions & Access.	(	S.F.)	
KF	Other Partitions	(	S.F.)	
KL	Lockers	(	Ea.)	
L	✔ **Equipment**			$
LK	Kitchen			
LS	School			
LO	Other			
M	✔ **Furnishings**			$
MW	Window Treatment			
MS	Seating	(	Ea.)	
N	✔ **Special Construction**			$
NA	Acoustical	(	S.F.)	
NB	Prefab. Bldgs.	(	S.F.)	
NO	Other			
P	✔ **Conveying Systems**			$
PE	Elevators	(	Ea.)	
PS	Escalators	(	Ea.)	
PM	Material Handling			
Q	✔ **Mechanical**			$
QP	Plumbing (No. of fixtures)			
QS	Fire Protection (Sprinklers)			
QF	Fire Protection (Hose Standpipes)			
QB	Heating, Ventilating & A.C.			
QH	Heating & Ventilating (BTU Output)			
QA	Air Conditioning	(	Tons)	
R	✔ **Electrical**			$
RL	Lighting	(	S.F.)	
RP	Power Service			
RD	Power Distribution			
RA	Alarms			
RG	Special Systems			
S	✔ **Mech./Elec. Combined**			$

Product Name _____
Product Number _____
Your Name _____
Title _____
Company _____
   ☐ Company
   ☐ Home   Street Address _____
City, State, Zip _____

☐ Please send _____ forms.

Please specify the Means product you wish to receive. Complete the address information as requested and return this form with your check (product cost less $20.00) to address below.

R.S. Means Company, Inc.,
Square Foot Costs Department
100 Construction Plaza, P.O. Box 800
Kingston, MA 02364-9988

R.S. Means Company, Inc... a tradition of excellence in Construction Cost Information and Services since 1942.

Table of Contents
Annual Cost Guides, Page 2
Reference Books, Page 6
Seminars, Page 11
Consulting Services, Page 13
Electronic Data, Page 14
Order Form, Page 16

Book Selection Guide

The following table provides definitive information on the content of each cost data publication. The number of lines of data provided in each unit price or assemblies division, as well as the number of reference tables and crews is listed for each book. The presence of other elements such as an historical cost index, city cost indexes, square foot models or cross-referenced index is also indicated. You can use the table to help select the Means' book that has the quantity and type of information you most need in your work.

Unit Cost Divisions	Building Construction Costs	Mechanical	Electrical	Repair & Remodel.	Square Foot	Site Work Landsc.	Assemblies	Interior	Concrete Masonry	Open Shop	Heavy Construc.	Residential	Light Commercial	Facil. Construc.	Plumbing	Western Construction Costs
1	985	515	515	525		935		405	970	985	1045	340	525	1445	585	985
2	3245	1320	445	2085		5155		1035	1550	3210	5055	1080	1145	4825	1560	3230
3	1480	85	65	740		1275		190	1865	1455	1400	260	205	1315	45	1455
4	850	30		655		785		630	1205	845	690	315	400	1115		840
5	1160	160	155	450		690		425	570	1125	995	260	285	1220	85	1140
6	1355	95	85	1240		470		1325		325	1335	590	1410	1360	60	1715
7	1360	60	10	1255		485		520	445	1355	355	785	1000	1350	70	1355
8	1815	60		1850		320		1680	700	1805	10	1090	1140	1930		1810
9	1645			1490		190		1740	325	1645	115	1305	1380	1820		1635
10	980	55	30	550		200		810	195	980		240	450	975	255	980
11	985	255	200	565		45		815	30	985	20	110	230	980	220	980
12	320			85		225		1610	30	320		95	90	1610		320
13	850	380	200	140		450		235	105	850	325	115	125	925	210	905
14	320	45		235		35		310		320	40	10	20	320	15	320
15	2715	13590	700	2300		1975		1415	80	2645	2285	985	1610	13040	10300	2600
16	1535	800	10520	1115		1045		1240	60	1555	990	705	1270	10100	705	1485
17	460	370	370							460				460	370	460
Totals	22060	17820	13295	15280		14280		14385	8455	21875	13915	9195	11285	44790	14480	22215

Assembly Divisions	Building Construction Costs	Mechanical	Electrical	Repair & Remodel.	Square Foot	Site Work Landsc.	Assemblies	Interior	Concrete Masonry	Open Shop	Heavy Construc.	Residential	Light Commercial	Facil. Construc.	Plumbing	Western Construction Costs
1				205	130	740	775		690		755	845	130	90		
2				85	35	35	50		50			590	30			
3				415	1115		3065	165	1250			1550	805	175		
4				715	1355		3175	255	1275			2045	1170	25		
5				290	230		465					1080	225	30		
6				1010	855		1310	1515	150			1085	750	335		
7				40	90		180	160				725	35	70		
8		2275	155	550	1700		2745	995				1695	1200	1115	2100	
9			1345	305	355		1285	305				240	360	320		
10													465	185		
11				205	465		730	290					85	120	855	
12		500	160	540	85	2400	700	105	705		540					
Totals		2775	1660	4360	6415	3175	14480	3790	4120		1295	9855	5255	2465	2955	

Reference Section	Building Construction Costs	Mechanical	Electrical	Repair & Remodel.	Square Foot	Site Work Landsc.	Assemblies	Interior	Concrete Masonry	Open Shop	Heavy Construc.	Residential	Light Commercial	Facil. Construc.	Plumbing	Western Construction Costs
Tables	156	46	85	70	4	87	223	61	84	156	56	51	74	88	49	158
Models					102							32	43			
Crews	341	341	341	341		341		341	341	341	341	341	341	341	341	341
City Cost Indexes	yes	yes	yes	yes	yes	yes	yes	yes	yes	yes	yes	yes	yes	yes	yes	yes
Historical Cost Indexes	yes	yes	yes	yes	yes	yes	yes	yes	yes	yes	yes	no	yes	yes	yes	yes
Index	yes	yes	yes	yes	yes	no	yes	yes	yes	yes	yes	yes	yes	yes	yes	yes

Annual Cost Guides

Means Building Construction Cost Data 1997

Available in Both Softbound and Looseleaf Editions

The "Bible" of the industry now comes in the standard softcover edition or the looseleaf edition.

Many customers enjoy the convenience and flexibility of the looseleaf binder, which increases the usefulness of *Means Building Construction Cost Data 1997* by making it easy to add and remove pages. You can insert your own cost information pages, so everything is in one place. Copying pages for faxing is easier also. Whichever edition you prefer, softbound or the convenient looseleaf edition, you'll get *The DI Quarterly/Change Notice* newsletter at no extra cost.

$79.95 per copy, Softbound
Catalog No. 60017

$109.95 per copy, Looseleaf
Catalog No. 61017

Means Building Construction Cost Data 1997

Offers you unchallenged unit price reliability in an easy-to-use arrangement. Whether used for complete, finished estimates or for periodic checks, it supplies more cost facts better and faster than any comparable source. Over 21,000 unit prices for 1997. The City Cost Indexes have been expanded to coverage of over 930 areas, for indexing to any project location in North America. Order and get *The DI Quarterly/Change Notice* newsletter sent to you FREE.

$79.95 per copy
Over 630 pages, illustrated, available Oct. 1996
Catalog No. 60017

Means Building Construction Cost Data 1997

Metric Version

The Federal Government has stated that all federal construction projects must now use metric documentation. The *Metric Version* of *Means Building Construction Cost Data 1997*, is presented in metric measurements covering all construction areas. Don't miss out on these billion dollar opportunities. Make the switch to metric today.

$99.95 per copy
Over 630 pages, illustrated, available Nov. 1996
Catalog No. 63017

Annual Cost Guides

Means Mechanical Cost Data 1997
• HVAC • Controls

Total unit and systems price guidance for mechanical construction... materials, parts, fittings, and complete labor cost information. Includes price for piping, heating, air conditioning, ventilation, and all related construction. With unit cost and system assembly costs, productivity data for labor cost comparisons, diagrams, notes, and explanations. Local cost indexing, easy-to-use subject look-up index, and invaluable references enhance cost information.

$83.95 per copy
Over 560 pages, available October 1996
Catalog No. 60027

Means Plumbing Cost Data 1997

Comprehensive unit prices and assemblies for plumbing, irrigation systems, commercial and residential fire protection, point-of-use water heaters, and the latest approved materials. This publication and its companion, *Means Mechanical Cost Data*, provide full-range cost estimating coverage for all the mechanical trades.

$83.95 per copy
Over 480 pages, illustrated, available October 1996
Catalog No. 60217

Means Electrical Cost Data 1997

Pricing information for every part of electrical cost planning: unit and systems costs with design tables; engineering guides and illustrated estimating procedures; complete labor-hour, materials, and labor costs for better scheduling and procurement. With the latest products and construction methods used in electrical work. More than 17,000 unit and systems costs, clear specifications and drawings.

$83.95 per copy
Over 450 pages, illustrated, available October 1996
Catalog No. 60037

Means Electrical Change Order Cost Data 1997

You are provided with electrical unit prices exclusively for pricing change orders—based on the recent, direct experience of contractors and suppliers. Analyze and check your own change order estimates against the experience others have had doing the same work. It also covers productivity analysis and change order cost justifications. With useful information for calculating the effects of change orders and dealing with their administration.

$89.95 per copy
Over 440 pages, available October 1996
Catalog No. 60237

Means Facilities Maintenance & Repair Cost Data 1997

Published in a looseleaf format, *Means Facilities Maintenance & Repair Cost Data* gives you a complete system to manage and plan your facility repair and maintenance cost and budget efficiently. Gives you guidelines for auditing a facility and developing an annual maintenance plan. Budgeting is also included along with reference tables on cost and management. You'll have information on frequency and productivity of maintenance operation.

$199.95 per copy
Over 600 pages, illustrated available December 1996
Catalog No. 50300

Means Square Foot Costs 1997

Provides you with construction costs for hundreds of commercial, residential, industrial, and institutional buildings. The costs are coordinated with drawings to speed identification of the desired type of structure. For detailed square foot estimates, the manual provides you with unit cost groupings of installed components covering many types of construction approaches.

New Enhancements for 1997

$95.95 per copy
Over 460 pages, illustrated, available November 1996
Catalog No. 60057

Annual Cost Guides

Means Repair & Remodeling Cost Data 1997
Commercial/Residential

You can use this valuable tool to estimate commercial and residential renovation and remodeling. By using the specialized costs in this manual, you'll find it's not necessary to force fit prices for new construction into remodeling cost planning. Provides comprehensive unit costs, building systems costs, extensive labor data and estimating assistance for every kind of building improvement.

$79.95 per copy
Over 600 pages, illustrated, available November 1996
Catalog No. 60047

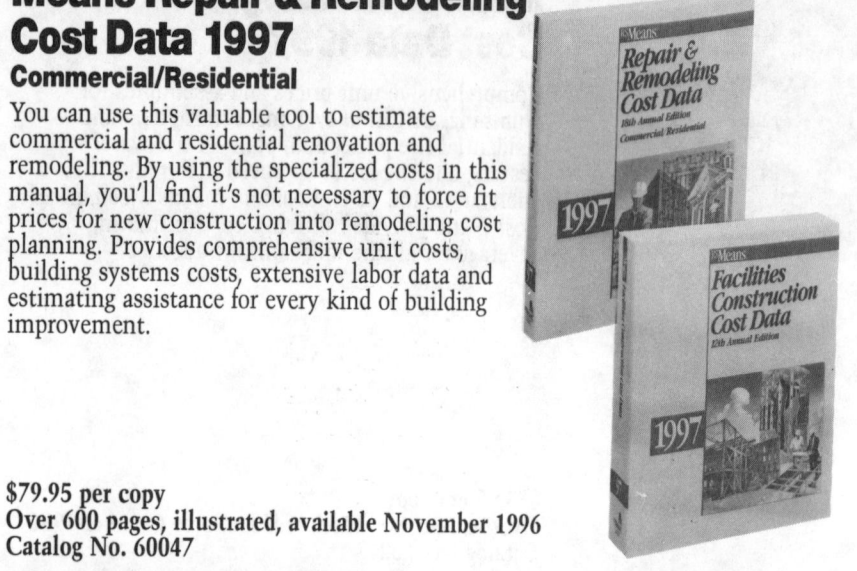

Means Facilities Construction Cost Data 1997

For the maintenance and construction of commercial, industrial, municipal, and institutional properties. Costs are shown for new and remodeling construction and are broken down into materials, labor, equipment, overhead, and profit. Special emphasis is given to sections on mechanical, electrical, furnishings, site work, building maintenance, finish work, and demolition. More than 40,000 unit costs plus assemblies and reference sections are included.

$199.95 per copy
Over 1080 pages, illustrated, available Nov. 1996
Catalog No. 60207

Means Residential Cost Data 1997

Speeds you through residential construction pricing with more than 100 illustrated complete house square-foot costs. Alternate assemblies cost selections are located on adjoining pages, so that you can develop tailor-made estimates in minutes. Complete data for detailed unit cost estimates is also provided.

$74.95 per copy
Over 550 pages, illustrated, available December 1996
Catalog No. 60177

Means Light Commercial Cost Data 1997

Specifically addresses the light commercial market, which is an increasingly specialized niche in the industry. Aids you, the owner/designer/contractor, in preparing all types of estimates, from budgets to detailed bids. Includes new advances in methods and materials. Assemblies section allows you to evaluate alternatives in early stages of design/planning.

$76.95 per copy
Over 600 pages, illustrated, available November 1996
Catalog No. 60187

Means Assemblies Cost Data 1997

With thousands of building assemblies costs and hundreds of exacting illustrations, this cost tool can help you work out estimates with pre-grouped components costs. It's ideal for developing cost comparisons, trade-offs, and design ideas. Features costs for virtually every conceivable building system, data on energy use and conservation, design ideas and costs, and space-planning data.

$134.95 per copy
Over 570 pages, illustrated, available October 1996
Catalog No. 60067

Means Site Work & Landscape Cost Data 1997

Hard-to-find costs are presented in an easy-to-use format for every type of site work and landscape construction. Costs are organized, described, and laid out for earthwork, utilities, roads and bridges, as well as grading, planting, lawns, trees, irrigation systems, and site improvements.

$86.95 per copy
Over 570 pages, illustrated, available November 1996
Catalog No. 60287

Annual Cost Guides

Means Open Shop Building Construction Cost Data 1997

The open-shop version of the *Means Building Construction Cost Data*. More than 21,700 reliable unit cost entries based on open shop trade labor rates. Eliminates time-consuming searches for these prices. The first book with open shop labor rates and crews. Labor information is itemized by labor-hours, crew, hourly/daily output, equipment, overhead and profit. For contractors, owners and facility managers.

$82.95 per copy
Over 600 pages, illustrated, available November 1996
Catalog No. 60157

Means Building Construction Cost Data 1997
Western Edition

This regional edition provides more precise cost information for western North America. Labor rates are based on union rates from 13 western states and western Canada. Included are western practices and materials not found in our national edition: tilt-up concrete walls, glu-lam structural systems, specialized timber construction, seismic restraints, landscape and irrigation systems.

$89.95 per copy
Over 600 pages, illustrated, available November 1996
Catalog No. 60227

Means Construction Cost Indexes 1997

Who knows what 1997 holds? What materials and labor costs will change unexpectedly? By how much?
• Breakdowns for 305 major cities.
• National averages for 30 key cities.
• Expanded five major city indexes.
• Historical construction cost indexes.

$198.00 per year/$49.50 individual quarters
Catalog No. 60147

Means Concrete & Masonry Cost Data 1997

Provides you with cost facts for virtually all concrete/masonry estimating needs, from complicated form work to various sizes and face finishes of brick and block, all in great detail. The comprehensive unit cost section contains more than 15,000 selected entries. The assemblies cost section is illustrated with isometric drawings. A detailed reference section supplements the cost data.

$77.95 per copy
Over 520 pages, illustrated, available December 1996
Catalog No. 60117

Means Heavy Construction Cost Data 1997

A comprehensive guide to heavy construction costs. Includes costs for highly specialized projects such as tunnels, dams, highways, airports, and waterways. Information on different labor rates, equipment, and material costs is included. Has unit price costs, systems costs, and numerous reference tables for costs and design. Valuable not only to contractors and civil engineers, but also to government agencies and city/town engineers.

$86.95 per copy
Over 440 pages, illustrated, available December 1996
Catalog No. 60167

Means Heavy Construction Cost Data 1997
Metric Verson

Make sure you have the Means industry standard metric costs for the federal, state, municipal and private marketplace. With thousands of up-to-date metric unit prices in tables by CSI standard divisions. Supplies you with assemblies costs using the metric standard for reliable cost projections in the design stage of your project. Helps you determine sizes, material amounts, and has tips for handling metric estimates.

$99.95 per copy
Over 450 pages, illustrated, available December 1996
Catalog No. 63167

Means Interior Cost Data 1997

Provides you with prices and guidance needed to make accurate interior work estimates. Contains costs on materials, equipment, hardware, custom installations, furnishings, labor costs . . . every cost factor for new and remodeled commercial and industrial interior construction, plus more than 50 reference tables. For contractors, facility managers, owners.

$83.95 per copy
Over 530 pages, illustrated, available November 1996
Catalog No. 60097

Means Labor Rates for the Construction Industry 1997

Complete information for estimating labor costs, making comparisons and negotiating wage rates by trade for over 300 cities (United States and Canada), 46 construction trades listed by local union number in each city, historical wage rates included for comparison. Market: contractors, government agencies, estimators, owners, libraries. No similar book is available through the trade.

$184.95 per copy
Over 330 pages, available December 1996
Catalog No. 60127

Reference Books

The ADA in Practice
(Revised, expanded edition of The New ADA: Compliance & Costs)

By Deborah S. Kearney, PhD

A unique publication to help you meet and budget for the requirements of the Americans with Disabilities Act. Shows you how to do the job right, right from the start, by understanding what the law actually requires and allows. Includes an objective "authoritative buyers guide" for 70 ADA-compliant products with specs and purchasing information. With sample evaluation forms and illustrations of ADA-compliant products.

$69.95 per copy
Over 600 pages, illustrated, Softcover
Catalog No. 67147A

Means ADA Compliance Pricing Guide

Gives you detailed cost estimates for budgeting modification projects, including estimates for each of 260 alternates. You get the 75 most commonly needed modifications for ADA compliance, each an assembly estimate with detailed cost breakdown including materials, labor hours and contractor's overhead. With 3,000 additional ADA compliance-related unit cost line items and costs easily adjusted to 900 cities and towns.

$69.95 per copy
Over 350 pages, illustrated, Softcover
Catalog No. 67310

Means Square Foot Estimating

By Billy J. Cox and F. William Horsley

Proven techniques for conceptual and design-stage cost planning

Means Square Foot Estimating helps you accomplish more in less time. It steps you through the entire square foot cost process, pointing out faster, better ways to relate the design to the budget. Now updated to the latest version of UniFormat.

NEW REVISED EDITION

$67.95 per copy
over 300 pages, illustrated, Hardcover
Catalog No. 67145A

How to Estimate with Metric Units

A great companion to *Means Building Construction Cost Data, Metric Version*, as well as a stand-alone reference on how to do quantity takeoffs and estimating in metric units for construction. It presents metric terminology, conversion methods and factors, and special considerations with applications to various types of construction and materials, organized by MasterFormat division. Timed to coincide with the federal government's requirement to metricate new federal construction projects.

$49.95 per copy,
200 pages, illustrated, Softcover
Catalog No. 67304

HVAC: Design Criteria, Options, Selection
Expanded 2nd Edition

By William H. Rowe III, AIA, PE

Now including Indoor Air Quality, CFC Removal, Energy Efficient Systems and Special Systems by Building Type.

The book that helps you solve a wide range of HVAC system design and selection problems effectively and economically. Gives you explanations of the latest ASHRAE standards.

$84.95 per copy
Over 600 pages, illustrated, Hardcover
Catalog No. 67306

Basics For Builders: Plan Reading & Material Takeoff

By Wayne J. DelPico

For Residential and Light Commercial Construction

A valuable tool for training those who are new to plan reading and quantity takeoff procedures, as well as for professionals in related fields, such as facility managers, interior designers, or anyone interested in a current, clear and complete reference on plan reading and takeoff.

- Step-by-step instructions.
- Takeoff procedures based on a full set of working drawings.

$35.95 per copy
Over 420 pages, Softcover
Catalog No. 67307

Reference Books

Cost Planning & Estimating for Facilities Maintenance

In this unique new book, a team of facilities management authorities shares their expertise at:
- Evaluating and budgeting operations
- Maintaining & repairing key building components
- Applying *Means Facilities Maintenance & Repair Cost Data*
- With the special maintenance requirements of 10 different building types

NEW TITLE

$79.95 per copy,
Over 450 pages, Hardcover
Catalog No. 67314

Facilities Planning & Relocation

By David D. Owen

An A–Z working guide—complete with the checklists, schematic diagrams and questionnaires to ensure the success of every office relocation. Complete with a step-by-step manual, 100-page technical reference section, over 50 reproducible forms in hard copy and on computer diskettes.

$109.95 per copy, Textbook-384 pages,
Forms Binder-137 pages, illustrated, Hardcover
Catalog No. 67301

Means Facilities Maintenance Standards

Unique features of this one-of-a-kind working guide for facilities maintenance

A working encyclopedia that points the way to solutions to every kind of maintenance and repair dilemma. With a labor-hours section to provide productivity figures for over 180 maintenance tasks. Included are ready-to-use forms, checklists, worksheets and comparisons, as well as analysis of materials systems and remedies for deterioration and wear.

$159.95 per copy, 600 pages, 205 tables,
checklists and diagrams, Hardcover
Catalog No. 67246

Maintenance Management Audit

At a time when many companies and institutions are reorganizing or downsizing, this annual audit program is essential for every organization in need of a proper assessment of its maintenance operation. The forms presented in this easy-to-use workbook allow managers to identify and correct problems, enhance productivity, and impact the bottom line.

$64.95 per copy,
125 pages, spiral bound, illustrated, Hardcover
Catalog No. 67299

The Facilities Manager's Reference

By Harvey H. Kaiser, PhD,
Senior Vice President for Facilities Administration, Syracuse University

The tasks and tools the facility manager needs to accomplish the organization's objectives, and develop individual and staff skills. Includes Facilities and Property Management, Administrative Control, Planning and Operations, Support Services, and a complete building audit with forms and instructions, widely used by facilities managers nationwide.

$84.95 per copy, over 250 pages, with prototype forms and line art graphics, Hardcover
Catalog No. 67264

Facilities Maintenance Management

By Gregory H. Magee, PE

Now you can get successful management methods and techniques for all aspects of facilities maintenance. This comprehensive reference explains and demonstrates successful management techniques for all aspects of maintenance, repair and improvements for buildings, machinery, equipment and grounds. Plus, guidance for outsourcing and managing internal staffs.

$84.95 per copy,
Over 280 pages with illustrations, Hardcover
Catalog No. 67429

Understanding Building Automation Systems

- Direct Digital Control
- Security/Access Control
- Energy Management
- Life Safety
- Lighting

By Reinhold A. Carlson, PE & Robert Di Giandomenico

The authors, leading authorities on the design and installation of these systems, describe the major building systems in both an overview with estimating and selection criteria, and in system configuration-level detail.

$74.95 per copy,
Over 200 pages, illustrated, Hardcover
Catalog No. 67284

Planning and Managing Interior Projects

By Carol E. Farren

Expert, up-to-date guidance for managing interior installation projects. Includes: project phases; winning client support; space planning and design; budgeting, bidding and purchasing, and more.
For interior designers, architects, facilities professionals.

$74.95 per copy,
Over 330 pages, illustrated, Hardcover
Catalog No. 67245

Reference Books

Quantity Takeoff for Contractors:
How to Get Accurate Material Counts
By Paul J. Cook

Contractors who are new to material takeoffs or want to be sure they are using the best techniques will find helpful information in this book, organized by CSI MasterFormat division. Covers commercial construction with detailed calculations, and sample takeoffs.

$35.95 per copy,
over 250 pages, illustrated, Softcover
Catalog No. 67262

Superintending for Contractors:
How to Bring Jobs in On-time, On-Budget
By Paul J. Cook

This book examines the complex role of the superintendent/field project manager, and provides guidelines for the efficient organization of this job. Includes administration of contracts, change orders, purchase orders, and more.

$35.95 per copy
over 220 pages, illustrated, Softcover
Catalog No. 67233

Bidding for Contractors:
How to Make Bids that Make Money
By Paul J. Cook

The author shares the benefit of his more than 30 years of experience in construction project management, providing contractors with the tools they need to develop competitive bids.

$35.95 per copy
over 225 pages with graphics, Softcover
Catalog No. 67180

Estimating for Contractors:
How to Make Estimates that Win Jobs
By Paul J. Cook

Estimating for Contractors is a reference that will be used over and over, whether to check a specific estimating procedure, or to take a complete course in estimating.

$35.95 per copy
over 225 pages, illustrated, Softcover
Catalog No. 67160

Roofing:
Design Criteria, Options, Selection
By R.D. Herbert, III

This book is required reading for those who specify, install or have to maintain roofing systems. It covers all types of roofing technology and systems. You'll get the facts needed to intelligently evaluate and select both traditional and new roofing systems.

$62.95 per copy,
Over 225 pages with illustrations, Hardcover
Catalog No. 67253

Means Forms for Contractors
For general and specialty contractors.

Means' editors have created and collected the most needed forms as requested by contractors of various-sized firms and specialties. This book covers all project phases. Includes sample correspondence and personnel administration.
Full-size forms on durable paper for photocopying or reprinting.

$79.95 per copy
over 400 pages, three-ring binder, more than 80 forms
Catalog No. 67288

Means Forms for Building Construction Professionals

Don't waste time trying to compose forms, we've done the job for you!
Forms for all primary construction activities—estimating, designing, project administration, scheduling, appraising. Forms compatible with Means annual cost books and other typical user systems.

$94.95 per copy
over 400 pages, three-ring binder
Catalog No. 67231

Means Mechanical Estimating
Second Edition

This guide assists you in making a review of plans, specs and bid packages with suggestions for takeoff procedures, listings, substitutions and pre-bid scheduling. Includes suggestions for budgeting labor and equipment usage. Compares materials and construction methods to allow you to select the best options for your job.

$62.95 per copy
over 350 pages, illustrated, Hardcover
Catalog No. 67294

HVAC Systems Evaluation
By Harold R. Colen, PE

You get direct comparisons of how each type of system works, with the relative costs of installation, operation, maintenance and applications by type of building. With requirements for hooking up electrical power to HVAC components. Contains experienced advice for repairing operational problems in existing HVAC systems, ductwork, fans, cooling coils, and much more!

$72.95 per copy
over 500 pages, illustrated, Hardcover
Catalog No. 67281

Means Construction Dictionary
Unabridged Edition

Written in contractor's language, the information is adaptable for report writing, specifications or just intelligent discussion. Contains over 17,000 construction terms, words, phrases, acronyms and abbreviations, slang, regional terminology, and hundreds of illustrations.

$99.95 per copy
over 700 pages, Hardcover
Catalog No. 67292

Reference Books

Means Estimating Handbook

This comprehensive reference covers a full spectrum of technical data for estimating, with information on sizing, productivity, equipment requirements, codes, design standards and engineering factors.
Means Estimating Handbook will help you: evaluate architectural plans and specifications, prepare accurate quantity takeoffs, prepare estimates from conceptual to detailed, and evaluate change orders.

$99.95 per copy
over 900 pages, Hardcover
Catalog No. 67276

Means Repair and Remodeling Estimating
By Edward B. Wetherill

Focuses on the unique problems of estimating renovations of existing structures. It helps you determine the true costs of remodeling through careful evaluation of architectural details and a site visit.

$69.95 per copy
over 450 pages, illustrated, Hardcover
Catalog No. 67265

Successful Estimating Methods:
From Concept to Bid
By John D. Bledsoe, PhD, PE

A highly practical, all-in-one guide to the tips and practices of today's successful estimator. Presents techniques for all types of estimates, *and* advanced topics such as life cycle cost analysis, value engineering, and automated estimating.
Disk available with ready-to-use estimate spreadsheets.

$64.95 per copy
over 300 pages, illustrated, Hardcover
Catalog No. 67287

Means Productivity Standards for Construction
New Expanded Edition (Formerly Man-Hour Standards)

Here is the working encyclopedia of labor productivity information for construction professionals, with labor requirements for thousands of construction functions in CSI MasterFormat.
Completely updated, with over 3,000 new work items.

$159.95 per copy
over 800 pages, Hardcover
Catalog No. 67236A

Means Electrical Estimating
Second Edition

Expanded new version includes sample estimates and cost information in keeping with the latest version of the CSI MasterFormat. Contains new coverage of Fiber Optic and Uninterruptible Power Supply electrical systems, broken down by components and explained in detail. A practical companion to *Means Electrical Cost Data*.

$59.95 per copy
over 325 pages, Hardcover
Catalog No. 67230A

Successful Interior Projects Through Effective Contract Documents
By Joel Downey & Patricia K. Gilbert

A proven system for enhancing project team relationships, administrating the project and fulfilling contract requirements.

$69.95 per copy,
496 pages, Hardcover
Catalog No. 67313

Means Scheduling Manual
Third Edition
By F. William Horsley

Fast, convenient expertise for keeping your scheduling skills right in step with today's cost-conscious times. Covers bar charts, PERT, precedence and CPM scheduling methods. Now updated to include computer applications.

$62.95 per copy
over 200 pages, Spiral-bound, Softcover
Catalog No. 67291

Means Graphic Construction Standards

Means Graphic Construction Standards bridges the gap between design and actual construction methods. With illustrations of unit assemblies, systems and components, you can see quickly which construction methods work best to meet design, budget and time objectives.

$124.95 per copy
over 540 pages, illustrated, Hardcover
Catalog No. 67210

Means Heavy Construction Handbook

Informed guidance for planning, estimating and performing today's heavy construction projects. Provides expert advice on every aspect of heavy construction work including hazardous waste remediation and estimating. To assist planning, estimating, performing, or overseeing work.

$72.95 per copy,
over 430 pages, illustrated, Hardcover
Catalog No. 67148

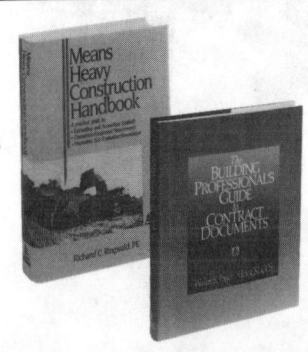

The Building Professional's Guide to Contract Documents
By Waller S. Poage, AIA, CSI, CCS

Includes latest changes to AIA prototype contracts. Directions for preparing specifications and technical requirements... description writing, proprietary and performance specs, standards. Guidance on owner, designer and contractor liability.

$62.95 per copy
over 400 pages, illustrated, Hardcover
Catalog No. 67261

Reference Books

Unit Price Estimating
Updated 2nd Edition
$59.95 per copy
Catalog No. 67303

Legal Reference for Design & Construction
By Charles R. Heuer, Esq., AIA
$109.95 per copy
Catalog No. 67266

Plumbing Estimating
By Joseph J. Galeno & Shelton T. Green
$59.95 per copy
Catalog No. 67283

Structural Steel Estimating
By S. Paul Bunea, PhD
$79.95 per copy
Catalog No. 67241

Landscape Estimating
2nd Edition
By Sylvia H. Fee
$62.95 per copy
Catalog No. 67295

Business Management for Contractors
How to Make Profits in Today's Market
By Paul J. Cook
$35.95 per copy
Catalog No. 67250

Understanding Legal Aspects of Design/Build
By Timothy R. Twomey, Esq., AIA
$74.95 per copy
Catalog No. 67259

Construction Paperwork
An Efficient Management System
By J. Edward Grimes
$49.95 per copy
Catalog No. 67268

Fire Protection:
Design Criteria, Options, Selection
By J. Walter Coon, PE
$74.95 per copy
Catalog No. 67286

Contractor's Business Handbook
By Michael S. Milliner
$42.95 per copy
Catalog No. 67255

Basics for Builders: How to Survive and Prosper in Construction
By Thomas N. Frisby
$34.95 per copy
Catalog No. 67273

Project Planning & Control for Construction
By David R. Pierce, Jr.
$62.95 per copy
Catalog No. 67247

Illustrated Construction Dictionary, Condensed
$59.95 per copy
Catalog No. 67282

Managing Construction Purchasing
By John G. McConville, CCC, CPE
$62.95 per copy
Catalog No. 67302

Hazardous Material & Hazardous Waste
By Francis J. Hopcroft, PE, David L. Vitale, M. Ed., & Donald L. Anglehart, Esq.
$89.95 per copy
Catalog No. 67258

Fundamentals of the Construction Process
By Kweku K. Bentil, AIC
$69.95 per copy
Catalog No. 67260

Construction Delays
By Theodore J. Trauner, Jr., PE, PP
$67.95 per copy
Catalog No. 67278

Basics for Builders: Framing & Rough Carpentry
By Scot Simpson
$24.95 per copy
Catalog No. 67298

Interior Home Improvement Costs
New 5th Edition
$19.95 per copy
Catalog No. 67308A

Exterior Home Improvement Costs
New 5th Edition
$19.95 per copy
Catalog No. 67309A

Concrete Repair and Maintenance Illustrated
By Peter H. Emmons
$64.95 per copy
Catalog No. 67146

Seminars

How to Develop Facility Assessment Programs

This two-day program concentrates on the management process required for planning, conducting, and documenting the physical condition and functional adequacy of buildings and other facilities. Regular facilities condition inspections are one of the facility department's most important duties. However, gathering reliable data hinges on the design of the overall program used to identify and gauge deferred maintenance requirements. Knowing where to look . . . and reporting results effectively are the keys. This seminar is designed to give the facility executive essential steps for conducting facilities inspection programs.

Inspection Program Requirements • Where is the deficiency? • What is the nature of the problem? • How can it be remedied? • How much will it cost in labor, equipment and materials? • When should it be accomplished? • Who is best suited to do the work?

Note: Because of its management focus, this course will not address trade practices and procedures.

Repair and Remodeling Estimating

Repair and remodeling work is becoming increasingly competitive as more professionals enter the market. Recycling existing buildings can pose difficult estimating problems. Labor costs, energy use concerns, building codes, and the limitations of working with an existing structure place enormous importance on the development of accurate estimates. Using the exclusive techniques associated with Means' widely acclaimed **Repair & Remodeling Cost Data**, this seminar sorts out and discusses solutions to the problems of building alteration estimating. Attendees will receive two intensive days of eye-opening methods for handling virtually every kind of repair and remodeling situation . . . from demolition and removal to final restoration.

Mechanical and Electrical Estimating

This seminar is tailored to fit the needs of those seeking to develop or improve their skills and to have a better understanding of how mechanical and electrical estimates are prepared during the conceptual, planning, budgeting and bidding stages. Learn how to avoid costly omissions and overlaps between these two interrelated specialties by preparing complete and thorough cost estimates for both trades. Featured are order of magnitude, assemblies, and unit price estimating. In combination with the use of **Means Mechanical Cost Data**, **Means Plumbing Cost Data** and **Means Electrical Cost Data**, this seminar will ensure more accurate and complete Mechanical/Electrical estimates for both unit price and preliminary estimating procedures.

Unit Price Estimating

This seminar shows how today's advanced estimating techniques and cost information sources can be used to develop more reliable unit price estimates for projects of any size. It demonstrates how to organize data, use plans efficiently, and avoid embarrassing errors by using better methods of checking.

You'll get down-to-earth help and easy-to-apply guidance for:
- making maximum use of construction cost information sources
- organizing estimating procedures in order to save time and reduce mistakes
- sorting out and identifying unusual job requirements to improve estimating accuracy.

Square Foot Cost Estimating

Learn how to make better preliminary estimates with a limited amount of budget and design information. You will benefit from examples of a wide range of systems estimates with specifications limited to building use requirements, budget, building codes, and type of building. And yet, with minimal information, you will obtain a remarkable degree of accuracy.

Workshop sessions will provide you with model square foot estimating problems and other skill-building exercises. The exclusive Means building assemblies square foot cost approach shows how to make very reliable estimates using "bare bones" budget and design information.

Scheduling and Project Management

This seminar helps you successfully establish project priorities, develop realistic schedules, and apply today's advanced management techniques to your construction projects. Hands-on exercises familiarize participants with network approaches such as the Critical Path Method. Special emphasis is placed on cost control, including use of computer-based systems. Through this seminar you'll perfect your scheduling and management skills, ensuring completion of your projects *on time* and *within budget*. Includes hands-on application of **Means Scheduling Manual** and **Means Building Construction Cost Data**.

Facilities Maintenance and Repair Estimating

With our new Facilities Maintenance and Repair Estimating seminar, you'll learn how to plan, budget, and estimate the cost of ongoing and preventive maintenance and repair for all your buildings and grounds. Based on R.S. Means' new groundbreaking cost estimating book, this two-day seminar will show you how to decide to either contract out or retain maintenance and repair work in-house. In addition, you'll learn how to prepare budgets and schedules that help cut down on unplanned and costly emergency repair projects. Facilities Maintenance and Repair Estimating crystallizes what facilities professionals have learned over the years, but never had time to organize or document. This program covers a variety of maintenance and repair projects, from underground storage tank removal, roof repair and maintenance, exterior wall renovations, and energy source conversions, to service upgrades and estimating energy-saving alternatives.

Managing Facilities Construction and Maintenance

In you're involved in new facility construction, renovation or maintenance projects and are concerned about getting quality work done on time and on or below budget, in Means' new seminar **Managing Facilities Construction and Maintenance** you'll learn management techniques needed to effectively plan, organize, control and get the most out of your limited facilities resources.

Learn how to develop budgets, reduce expenditures and check productivity. With the knowledge gained in this course, you'll be better prepared to successfully sell accurate project budgets, timing and manpower needs to senior management . . . plus understand how to evaluate the impact of today's facility professionals perspective.

Call 1-800-448-8182 for more information

Seminars

1997 Means Seminar Schedule

Location	Dates
Las Vegas, NV	March 24-27
Washington, DC	April 14-17
Denver, CO	April 28-May 1
Los Angeles, CA	May 19-22
New Orleans, LA	June 2-5
San Francisco, CA	June 16-19
Washington, DC	September 15-18
Orlando, FL	November 17-20

Registration Information

How to Register Register by phone today! Means toll-free number for making reservations is: **1-800-448-8182**

Individual Seminar Registration Fee $845 To register by mail, complete the registration form and return with your full fee (or minimum deposit of $300/person for one 2-day seminar, or minimum deposit of $475/person for two 2-day seminars) to: Seminar Division, R.S. Means Company, Inc., 100 Construction Plaza, P.O. Box 800, Kingston, MA 02364-0800.

Federal Government Pricing All federal government employees save 25% off regular seminar price. Other promotional discounts cannot be combined with Federal Government discount.

Team Discount Program Two to four seminar registrations: $745 per person—Five or more seminar registrations: $695 per person—Ten or more seminar registrations: Call for pricing.

Consecutive Seminar Offer One individual signing up for two separate courses at the same location during the designated time period pays only $1,345. You get the second course for only $500 (**a 40% discount**). Payment must be received at least ten days prior to seminar dates to confirm attendance.

Refunds Cancellations will be accepted up to ten days prior to the seminar start. There are no refunds for cancellations postmarked later than ten working days prior to the first day of the seminar. A $150 processing fee will be charged for all cancellations. Written notice or telegram is required for all cancellations. Substitutions can be made at any time before the session starts. No-shows are subject to the full seminar fee.

AACE Approved Courses The R.S. Means Construction Estimating and Management Seminars described and offered to you here have each been approved for 14 hours (1.4 CEU) of credit by the AACE International Certification Board toward meeting the continuing education requirements for re-certification as a Certified Cost Engineer/Certified Cost Consultant.

AIA Continuing Education R.S. Means is registered with the AIA Continuing Education System (AIA/CES) and is committed to developing quality learning activities in accordance with the CES criteria. R.S. Means seminars meet the AIA/CES criteria for Quality Level 2. AIA members will receive (28) learning units (LUs) for each two day R.S. Means Course.

Daily Course Schedule The first day of each seminar session begins at 8:30 A.M. and ends at 4:30 P.M. The second day is 8:00 A.M.-4:00 P.M. Participants are urged to bring a hand-held calculator since many actual problems will be worked out in each session.

Continental Breakfast Your registration includes the cost of a continental breakfast, a morning coffee break, and an afternoon cola break. These informal segments will allow you to discuss topics of mutual interest with other members of the seminar. (You are free to make your own lunch and dinner arrangements.)

Hotel/Transportation Arrangements R.S. Means has arranged to hold a block of rooms at each hotel hosting a seminar. To take advantage of special group rates when making your reservation be sure to mention that you are attending the Means Seminar. You are of course free to stay at the lodging place of your choice. (**Hotel reservations and transportation arrangements should be made directly by seminar attendees.**)

Important Class sizes are limited, so please register as soon as possible.

Registration Form

Call 1-800-448-8182, ext. 701 to register or FAX 1-617-585-7466

Please register the following people for the Means Construction Seminars as shown here. Full payment or deposit is enclosed, and we understand that we must make our own hotel reservations if overnight stays are necessary.

☐ Full payment of $ _____ enclosed.

☐ Deposit of $ _____ enclosed.

Balance Due is $ _____
U.S. Funds

Name of Registrant(s)
(To appear on certificate of completion)

Firm Name _____

Address _____

City/State/Zip _____

Telephone No. _____ Fax No. _____

E-Mail Address _____

Charge our registration(s) to: ☐ MasterCard ☐ VISA ☐ American Express ☐ Discover

Account No. _____ Exp. Date _____

Cardholder's Signature _____

Seminar Name   City   Dates

Please mail check to: R.S. MEANS COMPANY, INC., 100 Construction Plaza, P.O. Box 800, Kingston, MA 02364-0800 USA

Consulting Services Group

Proven Solutions for Managing the Costs of Construction and Facility Operations

Business and government leaders go to great lengths today to control costs and increase return on their construction-related activities. But they seldom have an opportunity to achieve optimum success. No matter what the activity . . . planning and design, new construction, facilities management, or introducing new building technologies . . . it comes down to the same problem: To control costs one must accurately predict them. R.S. Means Consulting Services Group has a proven record of success helping businesses meet this challenge while greatly increasing opportunity to maximize return on construction investments. With its extensive, highly specialized experience understanding both construction costs and relational database technology, Means Consulting Services Group offers an unparalleled opportunity for businesses to realize dramatic improvement in their construction/facilities cost control and valuation programs.

Database Management Solutions

- **Custom Database Development**—Means expertise in construction cost engineering and database management can be put to work creating customized cost databases and applications.
- **Data Licensing & Integration**—To enhance applications dealing with construction, any segment of Means vast database can be licensed for use, and harnessed in a format compatible with a previously developed proprietary system.
- **Cost Modeling**—Pre-built custom cost models provide organizations that expend countless hours estimating repetitive work with a systematic time-saving estimating solution.
- **Database Auditing & Maintenance**—For clients with in-house data, Means can help organize it, and by linking it with Means database, fill in any gaps that exist and provide necessary updates to maintain current and relevant proprietary cost data.

Cost Planning Solutions

- **Estimating Service**—Means expertise is available to perform construction cost estimates, as well as to develop baseline schedules and establish management plans for projects of all sizes and types. Conceptual, budget and detailed estimates are available.
- **Benchmarking**—Means can run baseline estimates on existing project estimates. Gauging estimating accuracy and identifying inefficiencies improves the success ratio, precision and productivity of estimates.
- **Project Feasibility Studies**—The Consulting Services Group can assist in the review and clarification of the most sound and practical construction approach in terms of time, cost and use.
- **Professional Review/Expert Witness**—Means is available to provide opinions of value and to supply expert interpretations or testimony in the resolution of construction cost claims, litigation and mediation.

Training and Development Solutions

- **Core Curriculum**—Means educational programs, delivered on-site, are designed to sharpen professional skills and to maximize effective use of cost estimating and management tools. On-site training cuts down on travel expenses and time away from the office. The broad curriculum covers such topics as repair, new construction, and conceptual estimating; scheduling and project management; facilities management; metrication; and delivery order contracting (DOC) methods.
- **Custom Curriculum**—Means can custom-tailor courses to meet the specific needs and requirements of clients. The goal is to simultaneously boost skills and broaden cost estimating and management knowledge while focusing on applications that bring immediate benefits to unique operations, challenges, or markets.
- **Development Programs**—In addition to custom curricula, Means can work with a client's Human Resources Department or with individual operating units to create programs consistent with long-term employee development objectives.

For more information and a copy of our capabilities brochure, please call 1-800-448-8182 and ask for the Consulting Services Group, or reach us at http://www.rsmeans.com

Means Data™
CONSTRUCTION COSTS FOR SOFTWARE APPLICATIONS

Your construction estimating software is only as good as your cost data. You can access and utilize MeansData™ directly through your spreadsheet, industry software or your own application.

MeansData™ for Spreadsheets

Turn your standard spreadsheet program into a powerful estimating tool and directly integrate the data from your R.S. Means Cost Books. Corporate site licenses available.

**TRY IT FOR 45 DAYS FOR JUST $9.95.
CALL 1-800-448-8182 TO ORDER!**

Integrates with:

Lotus®   Microsoft EXCEL
The Most Popular Spreadsheet for Windows™

DemoSource™

One-stop shopping for the latest cost estimating software for just $19.95

DemoSource™ — The evaluation tool for estimating software. Includes product literature and demo diskettes for ten or more estimating systems, all of which link to MeansData™.

CALL 1-800-448-8182 TO ORDER!

Visit our site on-line at *http://www.rsmeans.com/demo/* for product information and free demos.

Corporate Applications

If you have developed your own in-house software application and want to link with MeansData™ call Ed Damphousse at 1-800-448-8182, Ext. 888 for more information.

MeansData is a registered trademark of R.S. Means Co., Inc., A Southam Company. Other brand and product names are trademarks or registered trademarks of their respective holders.

FOR MORE INFORMATION ON ELECTRONIC PRODUCTS CALL 1-800-448-8182 OR FAX 1-617-585-7466

FOR MORE INFORMATION

Software Integration

A proven construction cost database is a mandatory part of any estimating package. We have linked MeansData™ directly into the industry's leading software applications. The following list of software providers can offer you MeansData™ with their estimating systems. Visit them on-line at *http://www.rsmeans.com/demo/* for more information and free demos. Or call their numbers listed below.

AEC DATA SYSTEMS
1-800-659-9001

BLUEGRASS TECHNOLOGIES
1-800-537-3785

BSD
Building Systems Design, Inc.
1-800-875-0047

CDCI
Construction Data Controls, Inc.
1-800-285-3929

CMS
Computerized Micro Solutions
1-800-255-7407

CONAC GROUP
1-604-273-3463

CONSTRUCTIVE COMPUTING, Inc.
1-800-456-2113

DATAQUIRE
1-401-253-8969

EAGLE POINT
1-800-678-6565

ESTIMATING SYSTEMS, Inc.
1-800-967-8572

G2, Inc.
1-800-657-6312

GEAC SOFTWARE SHOP SYSTEMS, Inc.
1-800-554-9865

GRANTLUN CORPORATION
1-602-897-7750

HANSEN INFORMATION TECHNOLOGIES, Inc.
1-800-821-9316

IQ BENECO
1-800-565-1122

MC^2
Management Computer Controls
1-800-225-5622

PDA Construction
Handheld Cost Estimating
1-909-653-5878

SANDERS SOFTWARE, Inc.
1-404-934-8423

STN, Inc.
Workline Maintenance Systems
1-800-321-1969

TIMBERLINE SOFTWARE CORP.
1-800-628-6583

TMA SYSTEMS, Inc.
1-918-494-2890

US Cost, Inc.
1-800-955-1385

VERTIGRAPH, Inc.
1-800-989-4243

WINESTIMATOR, Inc.
1-800-950-2374

1997 Order Form

ORDER TOLL FREE 1-800-334-3509
OR FAX 1-800-632-6732.

Qty.	Book No.	COST ESTIMATING BOOKS	Unit Price	Total
	60067	Assemblies Cost Data 1997	$134.95	
	60017	Building Construction Cost Data 1997	79.95	
	61017	Building Const. Cost Data–Looseleaf Ed. 1997	109.95	
	63017	Building Const. Cost Data–Metric Ed. 1997	99.95	
	60227	Building Const. Cost Data–Western Ed. 1997	89.95	
	60117	Concrete & Masonry Cost Data 1997	77.95	
	50140	Construction Cost Indexes 1997	198.00	
	60147A	Construction Cost Index–January 1997	49.50	
	60147B	Construction Cost Index–April 1997	49.50	
	60147C	Construction Cost Index–July 1997	49.50	
	60147D	Construction Cost Index–October 1997	49.50	
	60317	Contractor's Pricing Guide: Framing & Carpentry	34.95	
	60337	Contractor's Pricing Guide: Resid. Detailed Costs	36.95	
	60327	Contractor's Pricing Guide: Resid. S.F. Costs	39.95	
	COSTOS	COSTOS–Bldg. Costs in Mexico (subscrip.)	450.00	
	64027	ECHOS Assemblies Cost Book	250.00	
	64017	ECHOS Unit Cost Book	150.00	
	54000	ECHOS (combo set of both books)	350.00	
	58400	ECHOS (both books with softbooks)	450.00	
	60237	Electrical Change Order Cost Data 1997	89.95	
	60037	Electrical Cost Data 1997	83.95	
	60207	Facilities Construction Cost Data 1997	199.95	
	50300	Facilities Maintenance & Repair Cost Data 1997	199.95	
	60167	Heavy Construction Cost Data 1997	86.95	
	63167	Heavy Const. Cost Data–Metric Version	99.95	
	60097	Interior Cost Data 1997	83.95	
	60127	Labor Rates for the Const. Industry 1997	184.95	
	60187	Light Commercial Cost Data 1997	76.95	
	60027	Mechanical Cost Data 1997	83.95	
	60157	Open Shop Building Const. Cost Data 1997	82.95	
	60217	Plumbing Cost Data 1997	83.95	
	60047	Repair and Remodeling Cost Data 1997	79.95	
	60177	Residential Cost Data 1997	74.95	
	60287	Site Work & Landscape Cost Data 1997	86.95	
	60057	Square Foot Costs 1997	95.95	
		REFERENCE BOOKS		
	67147A	ADA in Practice	69.95	
	67310	ADA Pricing Guide	69.95	
	67305	Asia Pacific Constr. Costs Handbook	139.95	
	67298	Basics for Builders: Framing & Rough Carpentry	24.95	
	67273	Basics for Builders: How to Survive and Prosper	34.95	
	67267A	Basics for Builders: Insurance Repair	36.95	
	67307	Basics for Builders: Plan Reading & Takeoff	35.95	
	67180	Bidding for Contractors	35.95	
	67311CD	Builders Cost & Plan Kit	59.95	
	67261	Building Profess. Guide to Contract Documents	62.95	
	67312	Building Spec Homes Profitably	29.95	
	67250	Business Management for Contractors	35.95	
	67146	Concrete Repair & Maintenance Illustrated	64.95	
	V1000	Constr. & Develop. in Vietnam	85.00	
	V2000	Constr. Invest. Guide for Vietnam	140.00	
	67314	Cost Planning & Est. for Facil. Maint.	79.95	
	62136	DesignIntelligence (subscription)	199.00	
	67230A	Electrical Estimating Methods–2nd Ed.	59.95	

Qty.	Book No.	REFERENCE BOOKS (Cont'd)	Unit Price	Total
	67160	Estimating for Contractors	$ 35.95	
	67276	Estimating Handbook	99.95	
	67300	European Construction Costs Handbook	139.95	
	67249	Facilities Maintenance Management	84.95	
	67246	Facilities Maintenance Standards	159.95	
	67264	Facilities Manager's Reference	84.95	
	67301	Facilities Planning & Relocation	109.95	
	67286	Fire Protection: Design Criteria, Options, Select.	74.95	
	67231	Forms for Building Construction Professionals	94.95	
	67288	Forms for Contractors	79.95	
	67210	Graphic Construction Standards	124.95	
	67258	Hazardous Material & Hazardous Waste	89.95	
	67148	Heavy Construction Handbook	72.95	
	67308A	Home Improvement Costs–Interior Projects	19.95	
	67309A	Home Improvement Costs–Exterior Projects	19.95	
	67304	How to Estimate with Metric Units	49.95	
	67306	HVAC: Design Criteria, Options, Select.–2nd Ed.	84.95	
	67281	HVAC Systems Evaluation	72.95	
	67282	Illustrated Construction Dictionary, Condensed	59.95	
	67292	Illustrated Construction Dictionary, Unabridged	99.95	
	67237	Interior Estimating	62.95	
	67295	Landscape Estimating–2nd Ed.	62.95	
	67299	Maintenance Management Audit	64.95	
	67294	Mechanical Estimating–2nd Ed.	62.95	
	67245	Planning and Managing Interior Projects	74.95	
	67283	Plumbing Estimating	59.95	
	67236A	Productivity Standards for Constr.–3rd Ed.	159.95	
	67262	Quantity Takeoff for Contractors	35.95	
	67265	Repair & Remodeling Estimating	69.95	
	67253	Roofing: Design Criteria, Options, Selection	62.95	
	R1000	Russian Newsletter (subscription)	195.00	
	67291	Scheduling Manual–3rd Ed.	62.95	
	67145A	Square Foot Estimating–2nd Ed.	67.95	
	67241	Structural Steel Estimating	79.95	
	67287	Successful Estimating Methods	64.95	
	67313	Successful Interior Projects	69.95	
	67233	Superintending for Contractors	35.95	
	67284	Understanding Building Automation Systems	74.95	
	67303	Unit Price Estimating Methods–2nd Ed.	59.95	

MA residents add 5% state sales tax
Shipping & Handling**
Total (U.S. Funds)*

Prices are subject to change and are for U.S. delivery only. *Canadian customers may call for current prices. **Shipping & handling charges: Add 6.5% of total order for check and credit card payments. Add 9% of total order for invoiced orders.

Send Order To: ADDV-1000

Name (Please Print) _____

Company _____

☐ Company
☐ Home   Address _____

City/State/Zip _____

Phone # _____ P.O. # _____
(Must accompany all orders being billed)

Mail To: **R.S. Means Company, Inc.,** P.O. Box 800, Kingston, MA 02364-0800